Rosenfeld.

ENERGY
An Introduction to
PHYSICS

Pacific Gas and Electric Company

ENERGY
An Introduction to
PHYSICS

Robert H. Romer AMHERST COLLEGE

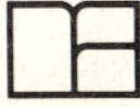

W. H. Freeman and Company San Francisco

Library of Congress Cataloging in Publication Data

Romer, Robert H
Energy.

Includes index.
1. Physics. 2. Force and energy. 3. Power (Mechanics) 4. Power sources. I. Title.
QC28.R64 530 75-35591
ISBN 0-7167-0357-2

Printed in the United States of America

9 8 7 6 5 4 3

Contents

11
The Nature of Light and Other Types of Radiation, and the Earth's Energy Balance

12
Atoms: Their Structure and Energy

13
Radioactivity and Nuclear Physics

14
Nuclear Fission and Nuclear Fission Power Plants

Preface

Physics and energy, energy as a scientific concept and as the center of the energy problem—these are the subjects of this book. The book is first of all an introductory text in physics, but it differs from most physics texts in that, from the very beginning, the focus is on energy and the laws that govern its behavior. This book is also about the energy problem, but it differs from most energy books in its emphases on the underlying science, a quantitative approach to the energy problem, and the necessity of preparing individuals to deal with the energy problem on their own, so that they will be liberated from a dependence on the "experts."

There are many approaches to the subject of physics. The one employed here, focusing on the interest everyone has in the energy problem, is especially effective now and will almost certainly continue to be so in the future. The energy problem, too, can be discussed from various points of view. I am convinced that the most meaningful way of learning about the energy problem is to study the relevant fundamental science, to become familiar with the various units in which energy is measured and with the amounts of energy we use, and to learn about such important matters as the meaning of a "degree-day" and the difference between a kilowatt and a kilowatt-hour.

Most of those who write about energy hold strong opinions on the subject, and I am no exception. I am convinced, for instance, that solar energy is the most promising large-scale source of energy in the long run, and for that reason the concluding chapter of the book is reserved for this topic. Rather than forcing my opinions on the readers, however, I have tried to prepare them to arrive at their own informed judgments, so that they will be able to confront the energy problem of the next decade as well as that of the 1970s.

Since the dimensions of the energy problem—the statistics of energy production and consumption—change from year to year, some of the specific numbers cited in this book will soon be out of date. But the *patterns* change much more gradually; for instance, the fraction of national energy consumption devoted to residential heating will probably not change markedly in the course of the next ten years. Although no significant decrease in the rate of growth of overall energy consumption has yet become evident, data for 1974 do show a slight decrease in United States energy consumption from the 1973 value. It remains to be seen whether this decrease is a temporary aberration or an indication that we have indeed recognized some of the facts we should have learned during the early 1970s.

The subject matter of this book can be treated in either a half-year or a full-year course; a full year permits more thorough discussion and provides more opportunities for independent projects or papers on the themes developed here. Although such a course can be given in conjunction with conventional laboratory work, and many appropriate exercises are available, the subject does lend itself to many unconventional investigations. If one is fortunate enough to have a river-cooled power plant nearby, for instance, a canoe and a thermometer are all the equipment needed for a study of waste heat and thermal pollution.

No mathematics beyond basic algebra is needed. What *is* required of the reader is a willingness to *do* arithmetic, to handle large and small numbers in powers-of-10 notation, to carry out the multiplications and divisions necessary to convert from one unit to another or to estimate the number of deuterons in the oceans. The ready availability of pocket calculators simplifies such calculations; rarely, though, is a precise result called for, and the reader is encouraged to become adept at doing "approximate arithmetic," to obtain useful numerical results with or without the aid of a calculator. It is my thesis that someone who cannot deal with numerical values, who cannot confidently convert quantities of energy from BTU's to joules, will be forever at the mercy of self-proclaimed experts and is destined to be a passive spectator in society's attempts to resolve the energy problem.

Many people have contributed, directly and indirectly, to the writing of this book. I am indebted to many students and colleagues who have influenced my views about physics and energy, in particular to Richard

Fink and Henry Yost with whom I taught an energy seminar for several years, and especially to the following persons who patiently read some or all of the manuscript and whose criticisms were immensely helpful in improving it: Francis Dart, Richard Duffy, John Fowler, Allan Franklin, Joel Gordon, Mario Iona, Gene Rochlin, David Romer, Diana Romer, Evan Romer, Theodore Romer, Edwin Taylor, and Mark Troll. Those who share my name also helped me in so many other ways that I can scarcely begin to acknowledge my indebtedness; without their help in collating innumerable pages of manuscript, checking calculations, and reading galley proofs, and without their constant quiet encouragement, this book might never have been completed.

January, 1976 Robert H. Romer

Notes to the Reader

Fundamental Premises of the Book

Energy is a scientific concept of central importance, not only in physics but in all the sciences. Energy is also important in modern society; the world does face an energy "problem," a problem that will undoubtedly become more severe in the future. The energy problem is not our only problem, but it is certainly a key one that must be solved along with the others. It is not simply a scientific and technical problem, but science sets constraints and provides a framework within which solutions (not only to the energy problem but to others as well) must be found. Without scientific and technical knowledge, we do not know even what our options are.

Science and technology—some as yet undiscovered or undeveloped—*may* relieve the immediacy of the energy problem and of other problems as well and make the very important contribution of giving us additional *time*—another century, perhaps—to come to grips with the much more complex problems of population growth and inequities in the distribution of wealth. This is my own position, one of "cautious technological optimism." Scientific developments may lead to more effective uses of energy and even to new sources of energy; thus science may provide temporary solutions to the energy problem, and temporary solutions may be almost as valuable as "permanent" ones if they give us enough time to learn how to resolve other critical issues.

In confronting the energy problem, there is no substitute for *rational quantitative* inquiry, though social, political, legal, and ethical approaches are also essential. Few if any "answers" to the energy problem are to be found in this book, but the more people there are who are familiar with the problem, who know some of the numbers, who are willing and able to check someone else's arithmetic, in short the more who are prepared in some way to participate in solving the energy problem rather than to act merely as spectators, the more probable it is that we will arrive at reasonable and humane solutions to the world's pressing problems.

The Approach to Physics

Physics is presented in this book not merely as a collection of unrelated facts and laws to be accepted on faith, but as an intellectual discipline, a way of thinking about natural phenomena, which has been highly successful in helping us to comprehend the universe we live in. How can we recognize the essential aspects of some physical phenomenon, and what are we gaining or losing by focusing on these and ignoring other aspects of the same phenomenon? What are we doing when we make idealizations, for example when we think of a baseball or a car or a planet as a small particle rather than as the complicated object it really is? Which scientific statements are convenient definitions, and which are descriptions of how things actually behave? What kinds of observations and analysis are used in arriving at physical "laws"? What is the nature of proof in science? What can be done to prove or disprove a physical law? Is it even possible to arrive at a definitive proof, or must all scientific laws be regarded as only *tentatively* correct?

Most of the physics discussed in this book is central to the concept of energy and to a discussion of the energy problem. Emphasis has been given to the fundamental laws governing the behavior of energy—the law of conservation of energy (or the first law of thermodynamics) and the second law of thermodynamics—laws that have been extremely valuable in all areas of science and that give every indication of permanence. In order that discussion of the energy problem can begin early in the book, sometimes an idea to be discussed carefully in a later chapter will be presented more loosely in an early chapter and immediately applied to the energy problem. We will often rely on the reader's prior knowledge of certain facts and ideas and will not attempt to follow all the reasoning which led to these concepts.

The Importance of the Questions

But be ye doers of the word,
and not hearers only,
deceiving your own selves.

—James 1:22

Numerous questions are included in this book, some within the text and others at the ends of the sections and chapters. The questions are highly varied. Some call for simple calculations, whereas others are much more open-ended and might require extensive library research, and some have no real "answers." Many of the questions constitute an important part of the argument. Most of those requiring a numerical calculation are not just "exercises in arithmetic"; instead the number obtained from the calculation is one to be studied and interpreted. Does it confirm or disprove an argument just given? Does it suggest that windmills could supply a significant amount of energy? "Plugging numbers into formulas" is worthwhile if the numbers are real numbers and if the answers are important. In some questions, a reference may be made to a specific appendix, where needed information may be found; for other questions, readers are expected to extract the information by themselves—from the appendixes, other reference sources, or their own prior knowledge.

Conversion of units is required in many of the questions, and the conversion tables given in Appendix A should be used freely. With few exceptions, approximate answers are all that are needed. You may be asked, for example, to show that a certain quantity, P, has the value 1200 watts. Do not waste time doing accurate longhand multiplication or division. If by the methods of approximate arithmetic you find that $P = 1500$ watts, that is generally quite adequate, but do *not* be satisfied with results such as $P = 150$ watts or $P = 15{,}000$ watts. If you have access to a slide rule, or better yet to a desk or pocket calculator, these tools can be very helpful, but they are by no means essential.

The Appendixes

The appendixes may constitute the most important part of the book. They include not only conversion tables and values of physical constants but also a wide range of information about the world in which we live and the ways in which energy is used by humans. Frequent reference to this information is made throughout the book. It is hoped that these appendixes will be useful not only as supplements to the text but also as a convenient "energy handbook" for future reference. Every effort has been made to provide reliable and accurate information, although for some of the data it is impossible to provide figures that are accurate to within a few percent. Although some of the statistics will soon be out of date, it is extremely unlikely that the scientific applications of the concept of energy will be significantly altered in the foreseeable future, and the energy *problem* is not going to disappear. That of the 1980s will differ from that of the 1970s in some details, but it will for the most part almost certainly be simply a larger version of the same problem.

1

The Concept of Energy and the Role of Energy in Society

T. W. Tenney

1 The Concept of Energy and the Role of Energy in Society

Energy Crisis Ahead; The Energy Joyride is Over; Energy Shortage Worsens; Are We Running Out?; The Oil Gap; Cold Homes?; Scrounging for Fuel; Rationing Plans Studied; Nearing the Limits.

Coal, the Stopgap Fuel; Nuclear Future Looms; Fusion Breakthrough?; Windmill Revival; Solar Energy to the Rescue.

The energy crisis is a fiction.

—M. A. Adelman, 1973

The energy of the world is constant.

—Clausius, 1865

§1.1 INTRODUCTION

Energy is crucial to our modern way of life: yet the world today faces an energy crisis, a crisis with which we shall be living for at least the remainder of this century. Energy is also a scientific concept of the greatest importance, a concept that is central to every field of science. In this book we shall concentrate on both aspects of this vital subject—(1) energy as the focus of an important social problem and (2) energy as a unifying scientific concept. To begin with, science as an intellectual enterprise is an exciting one, and of the various sciences physics is the most fundamental and the one in which the concept of energy was developed in the most basic way. Of all the ideas that come from physics, none has been more fruitful than the concept of energy and the laws that govern its behavior. But science does have applications, as evidenced not only by such technological wonders as the light bulb and the air conditioner, but also by the increasingly familiar environmental problems that accompany the large-scale use of energy.

As the United States and the world seek solutions to the energy problem, everyone should consider whether he or she wishes to be a passive spectator, leaving all the decisions to the "experts," or one of the participants, one who will join in the search and help in making the decisions. In order to become a participant—perhaps even one of the experts—one must begin by learning the meaning of the scientific concept of energy and the fundamental laws which govern its behavior, the constraints imposed by these laws, and the possible options that science may provide. One must also become familiar with the dimensions of the energy problem: the amounts of energy available from various sources and the amounts of energy used or needed for various purposes.

Besides the energy problem, the world is confronted by many others. Among the more obvious are the threat of nuclear war, the population explosion, food shortages, and inequities in the distribution of wealth. All of these are inextricably linked to one another, and, although we shall concentrate on the energy problem, it is not possible to discuss it completely in a single book, and it is impossible to discuss it at all without at least recognizing the existence of other, more perplexing problems and touching upon some of the related economic and social issues. In spite of the emphasis on these areas, this book is not just an "energy book"; it is also a physics book. It does not purport to cover physics completely, but energy is such a central concept that anyone who aspires to an understanding of the energy problem will inevitably learn a great deal of fundamental physics.

Energy comes in many forms: therefore, it is a particularly important and useful concept, but it is virtually impossible to give an initial accurate definition. Instead, the meaning of the term must be defined bit by bit, and the initial definition amplified and refined as we extend the concept from one type of situation to another. Not even at the end of this book will it be possible to give a simple concise definition. But knowledge of the origin of the concept, and of the ways in which its meaning was revised and extended, and familiarity with the numerical values of various kinds of energy and the ways in which they change in various situations will give us more information about what energy "is" than any dictionary possibly could.

In Chapter 2 we will begin to define energy and the units in which it is measured. But in this first chapter, we will provide a less rigorous introduction, by describing some of the different forms of energy, the role of energy in science, and the ways in which energy is important to society. Energy is not a "thing" or a "substance," but a concept that has been developed to describe in specific terms how fast something is moving, where it is, how hot it is, and so on. We have learned to identify various forms of energy. *Kinetic* energy is what a moving car has, simply because it is in motion. A rock sitting at the top of a hill, or water at the top of a dam, has *gravitational potential* energy. *Electrical* energy is what we buy from the power company; we convert it into *thermal* energy in an electric heater or stove or into the kinetic energy of a spinning motor. *Chemical* energy is what we obtain when we buy gasoline, and it is converted into kinetic energy of the car and into thermal energy of the tires and the parts of the car and the surrounding air. Atomic nuclei have *nuclear* energy, some of which can be converted into kinetic energy, thermal energy, or electrical energy. Sunlight brings us *light* energy (or energy in the form of "electromagnetic radiation"), which is converted into thermal energy when it strikes the ground or into chemical energy in a growing plant.

It is implicit in this description that various forms of energy can be *converted* into one another, and that the increase of one form of energy occurs only at the expense of a corresponding decrease of some other form. In technical language, energy is *conserved*. Total energy, the sum of all the various different kinds, has a numerical value that does not change even though conversions from one form to another may be taking place. "The energy of the world is constant," provided (and this is an important qualification) that the world is a "closed system" (that no energy is added or removed), or that for any energy added (by sunlight), an equal amount is reradiated into space. This is the "law of conservation of energy," also known as the "first law of thermodynamics." The energy of a *closed* system is constant; the energy of a system that is not closed increases or decreases in accordance with the net amount of energy added or subtracted.

It is much easier simply to proclaim that energy is conserved—to announce a law of conservation of energy—than it is to justify such an assertion. Much of this book is devoted to examining how it is possible to arrive at such a law, by defining first one form of energy, and then a succession of others. Energy is a concept that has been developed by human intelligence, but it is always experimental *facts* that provide the guide for formulating the most useful definitions and concepts, and it is *facts* on which we must rely to see whether what we have done is correct.

"Conservation" is a peculiar word, especially as used in the phrase "conservation of energy," a phrase that has two rather different meanings. In scientific terms, conservation of energy refers to the fact that the energy of a closed system *is* constant—that there is nothing we can do to change its total value, even though we can change energy from one form to another. But in everyday conversation, when we speak of conserving energy, we mean decreasing the rate at which we convert energy from useful forms into less useful forms, or saving particular *kinds* of energy for later use.

For the world as a whole, the amount of energy we receive from the sun is almost exactly equal to the amount we send off into space; the energy of the world remains nearly constant from one year to the next. What we are doing when we "use" energy is not to use it up, not to destroy it, but to convert it from one form to another. Some forms of energy are more useful than other forms, and there are definite limitations on the types of energy conversions that are possible. Some of these limitations are embodied in the *second* law of thermodynamics. An important part of this law is that there is something different about *thermal* energy, and that there is a universal tendency for all forms of energy to be converted into thermal energy, and thereby to be "degraded." The production of thermal energy by conversion of other forms of energy is often very easy. The reverse process—the conversion of thermal energy into kinetic energy, electrical energy, chemical energy, and other forms—is something that can be done only with difficulty and never completely. When a car is being driven, the chemical energy of gasoline is converted into other forms, partly into the kinetic energy of the car but mostly into thermal energy. Although no change in the total amount of energy has occurred, an important and irreversible process has taken place. There is no way in which this process can be completely reversed. Thermal energy is not as useful as other forms of energy, and "useful energy" has been irrevocably used up.

During the building of the pyramids, some of the chemical energy of the food eaten by the workers was converted into the gravitational potential energy of the elevated pieces of rock. Until quite recent times, most of the conversions of energy for which humans were responsible were carried out in this simple way. Today we have available numerous ways of *multiplying* our effectiveness as energy converters, and this multiplication has been crucial in modern history. By using an electric hoist or a bull-dozer fueled by gasoline, for example, a single person can do as much work in a day as many hundreds of unaided workers could perform. At present, energy conversions for which humans are responsible are being carried out, in the United States, at a rate more than 100 times greater than that at which energy could be converted by the unaided manual labor of the whole population. It is as if each one of us had 100 full-time slaves working for us. Although very few people could afford to have 100 full-time servants, most Americans can and do command the equivalent amount of energy.

There are three inescapable facts about the use of energy that are becoming obvious, but have been too often neglected in the past. First, any energy conversion has side effects that may be undesirable and that may not have been anticipated. In the current jargon, every energy conversion has an "environmental impact": air pollution in Los Angeles, coal smoke from a power plant, oil spills off the California coast, deaths of coal miners from mine explosions and from black lung disease, deaths of uranium miners from inhalation of radioactive gas, the strip mining of Kentucky and of Indian reservations in the Southwest, the alteration of river temperatures by the unwanted thermal energy produced by power plants, the possible disruption of caribou migration by a pipeline in Alaska, and so on. Different kinds of energy conversions have differing environmental impacts, and there is a serious lack of quantitative information about many of these effects. Even if the available information were as complete and accurate as possible, the balancing of one environmental effect against another, or deciding between the possible environmental effects of energy production and the social disruptions that might ensue from a cutback in energy consumption, would nevertheless present us with almost insuperably difficult choices. In sum, such decisions cannot be based on scientific facts alone, but are necessarily influenced by political and ethical considerations as well.

The second fact about energy use that we must all recognize is that the energy we use has to come from somewhere. The source of most of our energy is currently the chemical energy of the "fossil fuels": coal, oil, and natural gas. Not only do undesirable environmental effects result from the use of fossil fuels, but also these resources are nonrenewable. Their chemical energy is that of plants which grew hundreds of millions of years ago, energy which has been locked up, unconverted, until very recent times. Since the late 1800s, we have been using up the fossil fuels at a constantly increasing rate, and although large quantities still remain, they can be used up, and long before they are exhausted they will become scarce and more expensive. Energy in useful forms is not free, no matter how much it may have seemed so in the past. The effectiveness of these resources for multiplying human ability to convert energy will decrease, as it becomes necessary to employ more time and effort to find and recover usable quantities of fossil fuels.

Third, the energy we use has to go somewhere after we have used it. With almost no exceptions, the energy that we use sooner or later is converted into thermal energy. This is one of the many environmental impacts of energy use, but it is such an important and universal one that it is worthy of special mention. From a local perspective, thermal energy heats the air around a city; from a global one, this worldwide "thermal pollution" heats the earth as a whole. To date, the effect on the earth's overall climate is very slight, but it could become a serious problem if the rate at which we use energy were to become many times greater than it is at present.

§1.2 THE USE OF NUMBERS. THE ENERGY PROBLEM AS A QUANTITATIVE PROBLEM

It is impossible to confront the concept of energy without using numbers. Indeed, it is the relationships among the numbers we use to describe speed, height, mass, and so on that lead us to the concept of energy. Moreover, one cannot understand the energy crisis unless one has access to some of the numbers: for example, how *many* gallons of oil? how *many* kilowatt-hours of electrical energy? A number of "good ideas" for alleviating the energy problem have been proposed that have turned out to be useless when subjected to a quantitative examination. Because it is absolutely necessary to approach the subject of energy quantitatively, let us introduce some of the ways of dealing with numbers that are important in this book.

§1.2.A Powers-of-10 Notation and the Use of Decimal Multiples

We must often use numbers which are very large or very small: for example, the amount of oil used in the world in 1973 (about 900,000,000,000 gallons) or the mass of an oxygen atom (about 0.00000000000000000000000003 kilograms). It is almost essential to have efficient ways of dealing with such numbers. These numbers are much less awkward to handle if we write them as the product of a more reasonable number (say between 1 and 10) times some *power* of 10. Thus the mass of an oxygen atom is about 3×10^{-26} kilograms, and the amount of oil used in 1973 was about 9×10^{11} gallons. Arithmetic with numbers written in this way is used throughout this book, and facility at doing calculations with such numbers is essential.* This kind of notation is sometimes referred to as "scientific notation," a term which is unfortunate because it tends to obscure the fact that this method of dealing with very large and very small numbers is invaluable in any area in which numbers are used.

As an example of why it is so helpful to write numbers in powers-of-10 notation, suppose that for some reason we wanted to know the product of these two numbers. Longhand multiplication would be awkward at best. However, if we use powers-of-10 notation, the product is easy to calculate:

$$3 \times 10^{-26} \times 9 \times 10^{11} = 27 \times 10^{-15} = 2.7 \times 10^{-14}.$$

Similarly, it is easy to calculate the quotient of these two numbers:

$$\frac{3 \times 10^{-26}}{9 \times 10^{11}} = 0.33 \times 10^{-37} = 3.3 \times 10^{-38}.$$

---

*10^3 means 10 multiplied by itself three times:

$$10^3 = 10 \times 10 \times 10 = 1000.$$

10^{-2} means 1 divided by 10^2:

$$10^{-2} = \frac{1}{10^2} = \frac{1}{10 \times 10} = \frac{1}{100} = 0.01.$$

10^0 simply means the number 1. With these definitions, multiplication is carried out by adding exponents with due regard to sign:

$$10^3 \times 10^2 = 10^5;\ 10^4 \times 10^{-1} = 10^3;\ 10^4 \times 10^0 = 10^4.$$

To divide one such number by another, subtract the exponents:

$$\frac{10^5}{10^2} = 10^{5-2} = 10^3;$$

$$\frac{10^6}{10^{-2}} = 10^{6-(-2)} = 10^{6+2} = 10^8;$$

$$\frac{10^4}{10^4} = 10^{4-4} = 10^0 = 1.$$

For a further discussion of these fundamental rules, see §1.4.A.

Closely related to the use of powers-of-10 notation is the use of such terms as thousand, million, and billion, and the use of prefixes to denote decimal multiples and submultiples. 1 milligram (1 mg) is one thousandth of a gram (10^{-3} g), 1 kilowatt (1 kW) = one thousand watts = 10^3 W, 1 megawatt (1 MW) = one million watts = 10^6 W. Less familiar is the prefix "giga" for 10^9: 1 gigawatt (1 GW) = 10^9 W. A word of caution is in order here. The terms thousand (10^3) and million (10^6) are universally understood, but the terms for larger quantities may mean different things to different people. For example, in the United States, one billion means one thousand million (10^9), but in some countries it means one *million* million (10^{12}). The possibilities for confusion are serious, and it would be safer not to use the terms billion, trillion, and so on at all. To use the American definition of billion, the population of the world is about 4 billion; why not simply say 4×10^9? The fact remains, however, that these terms are employed in many discussions of the energy problem, and they will appear in a few places in this book. Here, as in any American work, 1 billion = 10^9, 1 trillion = 10^{12}, 1 quadrillion = 10^{15} and 1 quintillion = 10^{18}. Fortunately, however, the last two terms are rarely used.

§1.2.B Approximate Arithmetic, Order of Magnitude Estimates, and Significant Figures

Let us now discuss some very important methods of using numbers in two very common situations: (1) when it is more important to arrive at an *approximate* result *quickly* than it is to achieve the highest possible precision; (2) when the numbers one is using are uncertain enough that any result derived from them must also be regarded as imprecise. Such skills are all too rarely taught, partly because by the very nature of the subject it is impossible to establish rigid rules and also because the best way to learn these skills is to make approximate calculations in a variety of situations in which one has an interest in the result.

An informed student of the energy problem needs to be able to make quick estimates: "Will the amount of oil delivered by the Alaska pipeline be large enough to make the use of nuclear power plants unnecessary?"; "How large a windmill do I need in order to obtain about half of my electrical energy?" Even though physics is often described as an "exact science," physicists, too, spend much of their time doing rough calculations: "Is the predicted gravitational force between two lead spheres large enough so that it might be measured with this equipment?"; "If the temperature in the laboratory increases by 10°F, will the resulting expansion of this part of my apparatus be large enough so that I must correct for it?"

One of the most important mathematical symbols in this book is this one: $\simeq$ (sometimes written $\cong$, or just $\sim$). It means "approximately equal," an elusive phrase that can mean various things in various contexts. Is the following statement correct: $2.5 \simeq 3$? For many purposes, yes. We shall, in many places throughout this book, be concerned with making rough estimates of various quantities. If you estimate the amount of coal needed to run a power plant for a year and come up with a figure of 2.5×10^6 tons, whereas someone else gets 3×10^6 tons, your results do not disagree. If the calculation is not expected to be more than a very rough estimate, then answers of 2×10^6 tons, or 4×10^6 tons, or even 10×10^6 tons can also be considered as approximately equal to 3×10^6 tons, but a result of 3×10^8 tons would be wrong. In the appropriate context, then, the statement $3 \simeq 10$ can be a valid one.

The phrase "order of magnitude estimate" is often used to describe a calculation that is expected to be within a factor of about 5 or 10 of the true result, but not necessarily very much closer. Two numbers which differ by no more than a factor of 5 or so are said to be "of the same order of magnitude." This is obviously a rather loosely defined term, but little is to be gained by trying to define it precisely.

When we make a statement such as $x \simeq y$, its meaning may vary with the context. It may mean that x and y differ by no more than 10% or so, or it may mean that x and y are of the same order of magnitude. In contexts in which the danger of being misunderstood is slight, an equal sign will sometimes be used even though the *approximately* equal sign would be more appropriate. If, in estimating the area of the United States, you consider it as a rectangle 3000 miles wide by 1000 miles high, and state the result as $A = 3 \times 10^6$ square miles, this statement is obviously meant as an approximation, and no one is likely to be deceived.

What is the area, A, of a circle whose radius is $r = 2.5$ ft? Since the formula for the area of a circle is $A = \pi r^2$, the area is

$$A = \pi \times 2.5^2 = 3.14159265\ \ldots \times 6.25 = 19.63495406\ \text{ft}^2.$$

However, unless the radius of the circle has been measured with extremely high accuracy, most of the digits in the result are ridiculous. The usual convention is that if we write $r = 2.5$ ft, this means that r is probably closer to 2.5 than to 2.4 or 2.6. (We cannot always be this fussy. We might write $r = 2.5$ ft, when we know only that r is probably somewhere between 2.2 ft and 2.8 ft.) If we want to state that r is definitely no larger than 2.5001 and no less than 2.4999, we would not write $r = 2.5$ ft, but rather $r = 2.5000$ ft. In this example, if r is given as 2.5 ft (two "significant figures"), then all but the first two or three digits in the result are meaningless. $A = 20$ ft^2 is a proper answer, as is $A = 19.6$ ft^2, or *perhaps* $A = 19.63$ ft^2, but any more digits would give a false impression of the precision of the result. The extra digits do not do any real harm, *if* one remembers that they should not be taken seriously.

We can always make calculations with 8 or 10 digits and then round off the result. This is not a bad procedure if a calculating machine is available, but it is definitely not recommended if long division or multiplication must be done by hand. Life is much too short to be wasted on long division. Everyone needs to develop methods for doing *approximate arithmetic*, flexible methods that can be varied as circumstances require.

As one way to begin, consider two specific examples. First, if the radius of the earth's orbit around the sun is 93 million miles, what is the earth's speed in miles per hour? The circumference of its orbit, the distance that it travels in one year, is 2π times the radius, or (setting $\pi \simeq 3.14$): $6.28 \times 93{,}000{,}000$ miles. The number of hours in one year is equal to 24×365, so the speed in miles per hour is:

$$v = \frac{6.28 \times 93{,}000{,}000}{24 \times 365}.$$

The first step is to express every number in powers-of-10 notation and do the arithmetic with the powers of 10 first:

$$v = \frac{6.28 \times 9.3 \times 10^7}{2.4 \times 10^1 \times 3.65 \times 10^2} = \frac{6.28 \times 9.3}{2.4 \times 3.65} \times 10^4.$$

Now comes the *approximate* part of the arithmetic. 6.28 divided by 2.4 is about 2, and 9.3 divided by 3.65 is about 3:

$$v = \frac{\overset{2}{\cancel{6.28}} \times \overset{3}{\cancel{9.3}}}{\cancel{2.4} \times \cancel{3.65}} \times 10^4 \simeq 6 \times 10^4 \text{ miles/hr}.$$

It is an inherent consequence of doing approximate arithmetic that the same calculation could have been done differently and might have given a different result, for example:

$$v = \frac{\overset{2}{\cancel{6.28}} \times \overset{4}{\cancel{9.3}}}{\cancel{2.4} \times \cancel{3.65}} \times 10^4 \simeq 8 \times 10^4 \text{ miles/hr}.$$

In this context, both statements, $v \simeq 6 \times 10^4$ miles/hr and $v \simeq 8 \times 10^4$ miles/hr, are correct. (The "right" answer is $v = 6.67 \times 10^4$ miles/hr.)

As a second example, suppose that, for some reason, one needs to know the approximate value of a quantity Q given by the following expression:

$$Q = \frac{86000 \times (420)^2 \times 640}{0.0027 \times 500 \times 30{,}000}.$$

Again, the first thing to do is to collect the powers of 10:

$$Q = \frac{8.6 \times 10^4 \times (4.2 \times 10^2)^2 \times 6.4 \times 10^2}{2.7 \times 10^{-3} \times 5 \times 10^2 \times 3 \times 10^4}$$

$$= \frac{8.6 \times (4.2)^2 \times 6.4 \times 10^4 \times 10^4 \times 10^2}{2.7 \times 5 \times 3 \times 10^{-3} \times 10^2 \times 10^4}$$

$$= \frac{8.6 \times (4.2)^2 \times 6.4}{2.7 \times 5 \times 3} \times 10^7.$$

Next, apply approximate arithmetic to the factor in front: $8.6/2.7 \simeq 3$; $(4.2)^2 \simeq 16$ and so $(4.2)^2/5 \simeq 16/5 \simeq 3$; $6.4/3 \simeq 2$. Thus,

$$Q \simeq 3 \times 3 \times 2 \times 10^7 = 18 \times 10^7 = 1.8 \times 10^8.$$

For comparison, the correct result is $Q = 2.40 \times 10^8$ (to three significant figures). In other circumstances, the same sort of approximate arithmetic might yield a result either more accurate or less so than this, but the important objective is to develop sufficient skill in doing approximate calculations so that you have confidence that the result is correct to within a factor of two or so without having to check each calculation on a calculator. Many variations are possible, and the best way to develop this skill is to practice. A number of examples are given in the questions at the end of this chapter, together with the "correct" answers, and a great many more such calculations are required throughout the book.

§1.2.C Dimensions, Units, and Conversions of Units

We must of necessity work with numbers not merely as abstractions but as representations of physical quantities: lengths, masses, forces, areas, energies, times, and so on. Such physical quantities are said to have *dimensions*. This term originates from the description of lines, surfaces and volumes as having one, two, and three spatial dimensions respectively. Any distance has the physical dimension of length, an area has dimensions of length $\times$ length, or length^2, and a volume has dimensions of length^3. This usage has been extended to other types of physical quantities: speed is defined by dividing a distance by a time and has the dimensions of length/time, or length $\times$ time^{-1}. In contrast, a number such as π, the ratio of two similar quantities (the circumference and diameter of a circle), is said to be "dimensionless."

If a numerical value that is given to any physical quantity with dimensions is to make sense, we must be careful to specify the *units* in which it is measured. For the scientist as well as the nonscientist, nothing is more frustrating than the fact that for any one sort of quantity, many different units exist and are in common use. Distances may be measured in inches, feet, miles, centimeters, angstrom units, meters, kilometers, and so on; energies in joules, British thermal units, calories, kilocalories, kilowatt-hours, electron-volts, and so on. Many physics books shield the reader from these unpleasant realities by using only a very restricted number of units. This may make it easier to learn physics, but one may as a consequence be unprepared for reading about the energy crisis or even for understanding other physics books. It sometimes seems as if those who write and speak about physics and energy and environmental problems use a multitude of units in order to confuse their audience and restrict their message to those few who *can* deal with a variety of units. Whether deliberate or not, this is often the result, and unless usage is restricted by legislation to just one set of units, there is only one possible response: join the select circle! Become one of those who can correctly and easily convert from one unit to another. For this reason, various units are used in this book, and many conversions are required in the text and questions. An extensive collection of conversion factors is provided in Appendix A.

The fundamental system of units used in this book is the MKS system (the Meter-Kilogram-Second system), a particular version of the metric system. The basic units of length, mass and time are the meter (m),* slightly more than 3 ft, the kilogram (kg), approximately 2.2 lb,† and the second (sec). In addition to these, it is necessary to define units for measuring temperature and for electrical charges or currents. Units for other quantities (area, velocity, acceleration, force, energy, power, and so on) can be defined in terms of the basic set of units. Some of these are so important that they are given their own special names. The MKS units of energy, power and force, for instance, are the joule, the watt, and the newton. Units that are not part of the MKS system can be defined with reference to MKS units. Many of these units are familiar ones, for instance the centimeter, mile, and foot, for measuring distances, the hour or the year for measuring times. Others that are not as familiar will be defined as the need arises.

Many of the numerical values used in this book are given in units other than MKS units. It is very often possible to make a calculation without ever using MKS units at all, for example in converting the numerical value of an area from square feet to square miles. In case of doubt, it is always possible to convert the given information into MKS units, make the appropriate calculation, and then convert the result into whatever units are desired.

Practical procedures for converting values of physical quantities from one unit to another will be illustrated here with one example. For further discussion, see §1.4.B and some of the questions at the end of the chapter. The distance, *d*, from New York to San Francisco is about 3000 miles. What is *d* in feet? Since 1 mile is equal to 5280 ft, the value of *d* in feet is:

---

*Be careful with the abbreviation m; it stands for *meter* and should not be used for mile, although it is so used in the abbreviation mph for miles-per-hour. The letter "m" is one of those which is used for a variety of purposes: as the abbreviation for meter, as an abbreviation for "milli" (as in 1 mg = 1 milligram = 10^{-3} g), as a symbol for mass, often as a symbol for the slope of a straight line, and (as a capital M) as an abbreviation for "mega" (as in 1 MW = 1 megawatt = 10^6 W).

†Mass and weight are two terms that are regarded by many as synonyms. But to the physicist, mass is a measure of *inertia,* the resistance that an object has to forces that act on it, whereas the "weight" of an object is a particular *force,* the gravitational force that the earth exerts on it. Mass and weight are thus two logically distinct concepts, but there is a precise proportionality between the masses and weights of various objects. If the mass of one object is six times the mass of another, its weight is also six times greater. Thus a unit such as the pound can be used for either mass or weight. Some physicists object to the statement that 1 kg $\simeq$ 2.2 lb, pointing out that the pound was originally defined as a unit of force or weight, not of mass, but in this book we will regard the pound as a unit of either mass or weight, and therefore we will consider the statement 1 kg $\simeq$ 2.2 lb as a correct one. More is said about the mass-weight problem in §3.3.

$$d = 3000 \times 5280 = 3 \times 10^3 \times 5.28 \times 10^3 \simeq 16 \times 10^6 \text{ ft} = 1.6 \times 10^7 \text{ ft}.$$

In order to be confident of not making errors, especially if the needed conversion is more complicated, it is extremely valuable to have a method that can be applied readily and easily. The following is such a procedure.

The key to the method is this. Multiplication of both numerator and denominator by equal quantities produces no change in the original quantity, and we can perform this multiplication in various ways. For example, since 1 mile = 5280 ft,

$$\begin{aligned} d = 3000 \text{ miles} &= 3000 \cancel{\text{miles}} \times \frac{5280 \text{ ft}}{1 \cancel{\text{mile}}} \\ &= 3000 \times 5280 \text{ ft} \simeq 1.6 \times 10^7 \text{ ft}, \end{aligned}$$

as before. The rules of arithmetic allow us to multiply numerator and denominator by equal quantities; because of the particular way in which it has been done above, the units of miles "cancel" and we get the desired result.

What if we want d in centimeters rather than feet? From Table A.6 (Appendix A), we see that 1 mile = 1.609×10^5 cm, and therefore

$$\begin{aligned} d = 3000 \text{ miles} &= 3000 \cancel{\text{miles}} \times \frac{1.609 \times 10^5 \text{ cm}}{1 \cancel{\text{mile}}} \\ &= 3 \times 10^3 \times 1.609 \times 10^5 \text{ cm} \simeq 4.8 \times 10^8 \text{ cm}. \end{aligned}$$

In more complex examples, the trick of multiplying numerator and denominator by equal quantities may have to be performed several times, but this method is guaranteed to produce correct results. More examples are discussed in §1.4.B.

The existence of a multitude of units can be confusing and can require one to spend a good deal of time in the task of converting from one kind of unit to another. Although no one would deny that we have too many units, (that the quart, say, could be replaced by the liter, which is nearly equal in size), rational explanations for the existence of a variety of units and good reasons for their continued use do exist. Energy itself is an excellent example. It was not at first realized that different forms of energy, such as kinetic energy and thermal energy, were quantities that *could* be measured in the same units. That kinetic energy and thermal energy could be "converted into one another" was recognized only after the two subjects, mechanics and heat, had each been quantitatively investigated for many years, but meanwhile separate units had been devised and used. Some of our present energy units, such as the calorie and the British thermal unit are relics of this earlier time; they could now be abandoned, but there is no indication that this will happen in the near future.

One excellent reason for the continued use of various units is that it often makes very large and very small numbers easier to remember and to comprehend. The mass of a uranium atom, for instance, is about 3.95×10^{-25} kg, but it can also be expressed as about 238 "atomic mass units" (amu). If we also know that the lightest atom, the hydrogen atom, has a mass of about 1 amu, this immediately tells us that the mass of a uranium atom is about 238 times as great as that of a hydrogen atom, a piece of information that may convey more meaning than the quantity 3.95×10^{-25} kg. Similarly, in working with very large quantities of energy, we will find it convenient to use the units Q and milli-Q (mQ), units that will be defined later. The total amount of energy used in the United States during 1973 was about 73 mQ, an easier number to remember than the equally correct values, 7.3×10^{16} British thermal units (BTU) or 7.7×10^{19} joules (J).

Similarly, a number of "informal" units may be useful. The electrical generating capacity of the United States in 1973 was about 4.6×10^{11} W. It can be helpful to recognize that a large modern power plant has a capacity of about 1000 MW (10^9 W or 1 GW), to define 1 "power plant" as equal to 10^9 W, and then to state the electrical generating capacity of the United States as about 460 power plants. Likewise a "ton of coal" or a "barrel of oil" may be used as a unit of energy, equal of course to the amount of energy released when they are burned.

§1.2.D The Uses of Graphs and Equations, and the Hazards of Extrapolation

We are often interested in the changes in one quantity produced by changes in another. How has annual energy consumption in the United States varied with *time?* How does a car's gasoline mileage depend on its *speed?* How does one's monthly fuel bill vary with the average monthly *temperature?* There are three different ways in which we can portray the relationship between two related quantities. First, we can make a *table* of numerical values—the most fundamental way of presenting the data—but any regularities may be difficult to discern. Second, we can make a *graph*. Just as a picture is often worth a thousand words, a graph derived from a table of a thousand numerical entries may help us to see patterns that are not at all evident from an examination of the numbers themselves. The ability to read and interpret graphs, either those we make ourselves or others that have been made for us, is a highly valuable skill. As we will see, the same set of data can be graphed in various ways. Although this flexibility is very useful, it also poses a danger, because graphs can be used to mislead as well as to inform. Third, we may be able to obtain an *equation* relating the two quantities. It is not always possible to obtain one, but when we do, it can be very useful because we can then proceed to use algebra for further analysis.

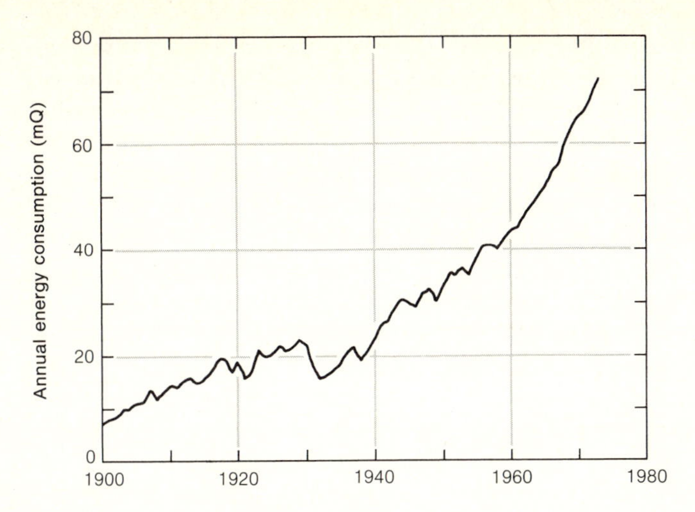

FIGURE 1.1
Annual consumption of energy in the United States (mQ/yr).

The possibility of using an equation to represent the relationship between two quantities will be particularly important as we begin the study of the concept of energy in Chapter 2.

Consider some examples of graphs of various types. Figure 1.1 shows annual energy consumption in the United States. We would describe this graph either as one showing energy consumption as a *function* of time, or as a graph of energy consumption *versus* time. It is clear that energy consumption has increased steadily throughout this period, with the exception of a few minor fluctuations. The largest fluctuation on this graph might be interpreted as the result of the depression of the 1930s. (Or was it perhaps the reduction in energy consumption that caused the depression, or was there perhaps no causal connection at all? We should be careful not to leap to conclusions. All that we can be certain of is that there was a drop in energy consumption in the 1930s, and we also know that there was an economic depression at the same time.)

Figures 1.2a and 1.2b show the average number of miles per gallon obtained by American passenger cars from 1940 to 1970. These two graphs are based on the same data, but Figure 1.2b looks more alarming. This variation shows one of the obvious ways in which the choices made in plotting a graph can change the impression conveyed.

Figures 1.3a and 1.3b show the total distance traveled by a particular car since it was purchased (its odometer reading) at the end of successive 6-month periods. Notice again that these two graphs are based on the same data; the apparent difference results simply from the different choices of scales on the two axes. These two graphs are especially interesting because both are *straight* lines. Another example of a straight-

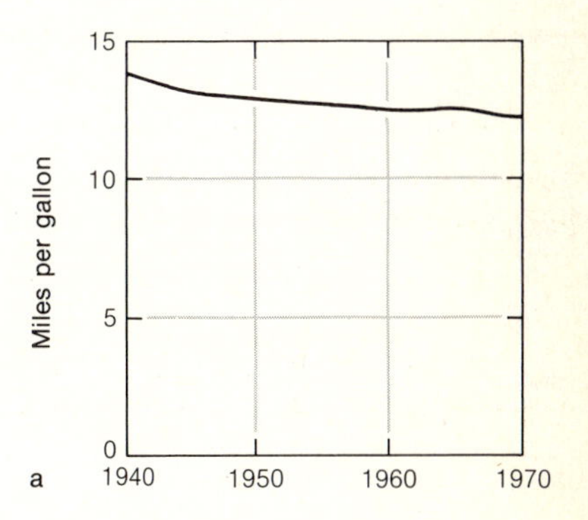

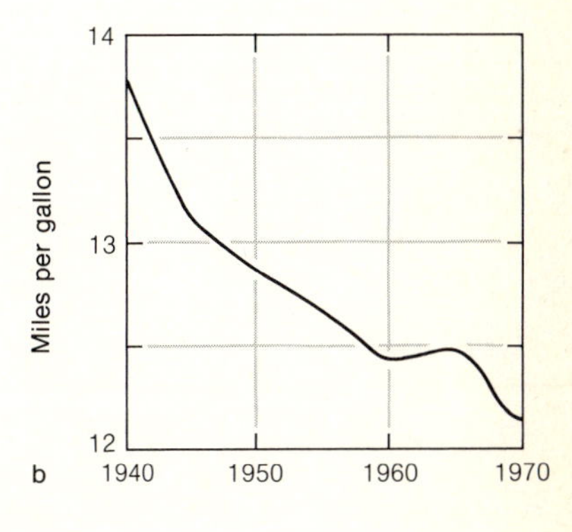

FIGURE 1.2
Average gasoline mileage of American cars.

line graph is shown in Figure 1.4, which shows the monthly cost, C, of a telephone as a function of the number, N, of calls made during the month, when there is a basic monthly charge of \$3 plus a charge of 10¢ for each call made.

Any straight-line graph, such as Figures 1.3a, 1.3b, or 1.4, is said to show a *linear* relationship between the two variables. If the graph is not only straight, but also passes through the origin (as in Figures 1.3a and 1.3b), the two quantities are said to be *proportional* to one another. We use a special symbol $\propto$ for proportionality: thus, in Figures 1.3a and b, $d \propto t$. The quantities C and N in Figure 1.4 are *linearly* related, but C is *not* proportional to N.

If a graph shows that two quantities are linearly related, the simplicity of this relationship suggests two fruitful ideas. First, it is easy to *extrapolate* such a relationship beyond the range of the original data, and, second, we can easily find a simple *equation* to express the relationship between the two quantities. Figures 1.5a and 1.5b show extrapolations of the linear relationship shown in Figure 1.3. (Notice that it is not at all obvious how to extrapolate graphs that are curved, such as Figures 1.1 and 1.2.) Extrapolation is useful but very dangerous. Will the odometer reading on the car be 20,000 miles at the end of 4 years, as shown in Figure 1.5a? We cannot be sure, but this is the most reasonable guess we can make, and it will probably be approximately correct unless the owner's life style changes in the course of the next year. But will the odometer reading be 200,000 miles at the end of 40 years, as suggested by the extrapolation in Figure 1.5b? The extrapolation is of the same type as in Figure 1.5a, but we can be fairly certain that this long-range extrapolation will turn out to be incorrect.

Extrapolation is useful, and we often have to make extrapolations, for instance to estimate future needs for electrical power plants, but whenever we make a simple extrapolation, we are implicitly assuming that whatever factors have been operating in the past will continue to operate in the future without change. Therefore, the greater the range of the extrapolation, the less justification there is for assuming that the extrapolation is correct. The clearest expression of the danger of long-range extrapolations was given by Mark Twain, in *Life on the Mississippi*. After pointing out that, from time to time, a bend in the river is cut off and that as a consequence the length of the river between New Orleans and Cairo is gradually decreasing, he made the following extrapolation.

> In the space of one hundred and seventy-six years, the Lower Mississippi has shortened itself two hundred and forty-two miles. That is an average of a trifle over one mile and a third per year. Therefore, any calm person, who is not blind or idiotic, can see that in the Old Oölitic Silurian Period, just a million years ago next November, the Lower Mississippi River was upward of one million three hundred thousand miles long, and stuck out over the Gulf

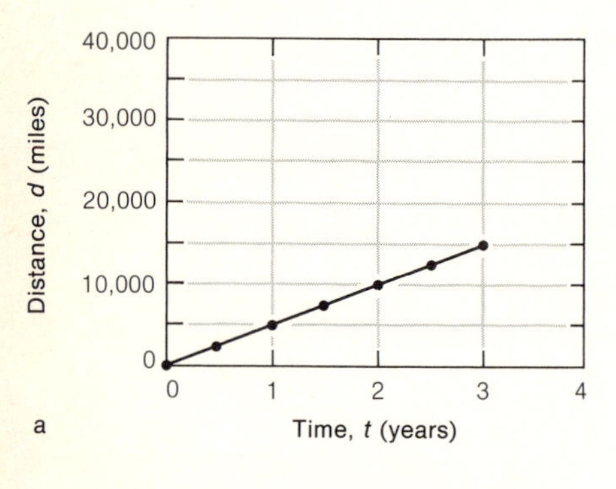

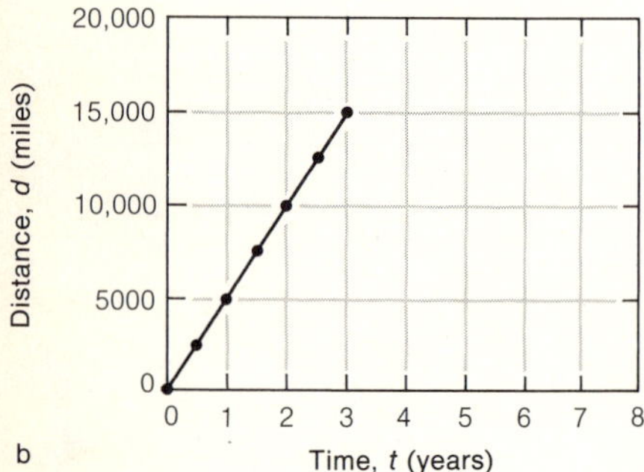

FIGURE 1.3
Total distance, d, traveled by a car as a function of time, t, since the car was purchased.

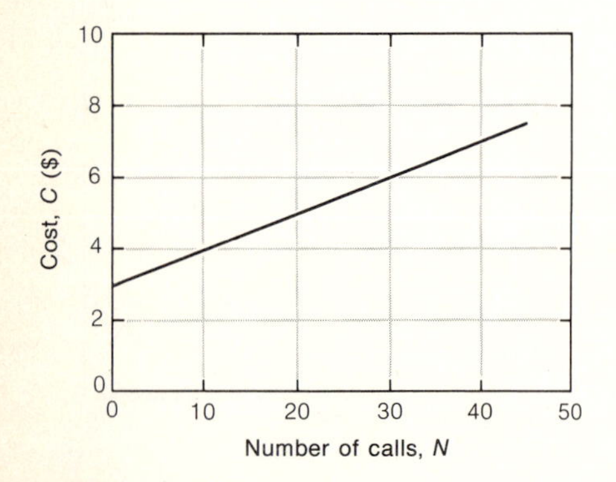

FIGURE 1.4
Monthly cost, C, of a telephone versus number, N, of calls made.

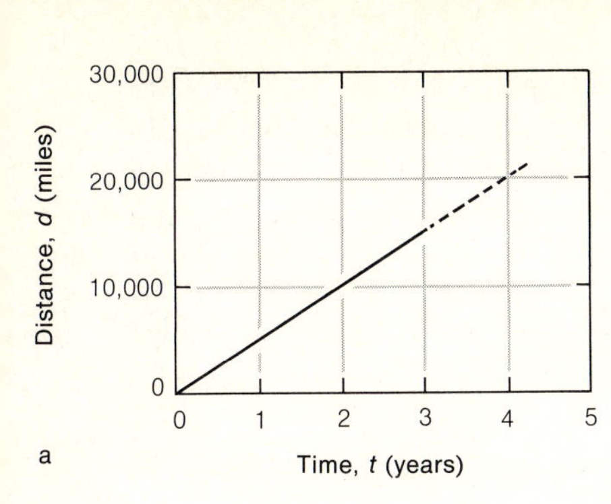

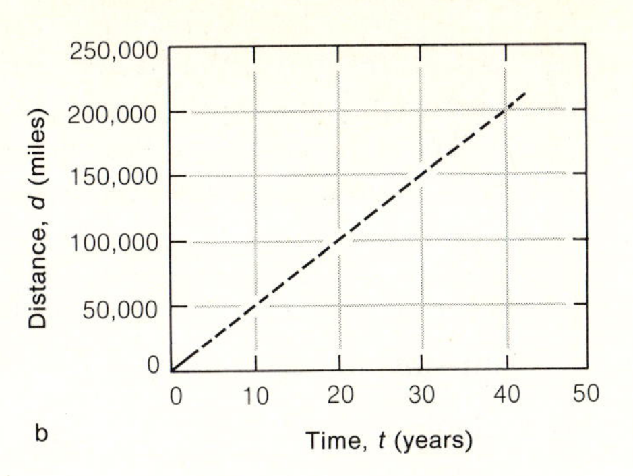

FIGURE 1.5
Extrapolations of the relationship shown in Figure 1.3.

> of Mexico like a fishing-rod. And by the same token any person can see that seven hundred and forty-two years from now the Lower Mississippi will be only a mile and three-quarters long, and Cairo and New Orleans will have joined their streets together, and be plodding comfortably along under a single mayor and a mutual board of aldermen. There is something fascinating about science. One gets such wholesale returns of conjecture out of such a trifling investment of fact.

If two quantities are linearly related, an examination of the straight-line graph will allow us to find a simple *equation* relating them; techniques for doing so are reviewed in §1.4.C.

QUESTION

1.1 Referring to §1.4.C if necessary, show that the relationships shown by Figures 1.3 and 1.4 can be described by the following equations:

$$d = 5000t; \qquad (1.1)$$

$$C = 0.1N + 3. \qquad (1.2)$$

Notice that when one quantity is *proportional* to another, the statement that d is proportional to t, as in Figure 1.3 ($d \propto t$), can always be rewritten as an equation simply by introducing a "constant of proportionality":

$$d = \text{constant} \times t.$$

In this example, the constant of proportionality is 5000. If the relationship is a linear one (but not a proportionality), the equation is similar, but there is an extra term (the number 3 in Equation 1.2), which must be added on.

When we are working, not with numbers in the abstract but with numbers that represent times, distances, dollars, or numbers of phone calls (as we are here), the numbers that appear in our equations also have *units*. The constant of proportionality in Equation 1.1, for instance, must be given the units of miles per year in order to make the units correct for the whole equation. Therefore we might better write Equation 1.1 as

$$d \text{ (miles)} = (5000 \text{ miles per year}) \times t \text{ (years)}, \qquad (1.3)$$

and Equation 1.2 as

$$C \text{ (dollars)} = [(0.1 \text{ dollars per call}) \times N \text{ (calls)}] + 3 \text{ dollars}. \qquad (1.4)$$

Just as extrapolations can be dangerous, equations too must be used with caution and for the same reason. An equation has a certain appearance of authority that may lead us, for example, to calculate the total distance traveled by the car after 1000 years simply by plugging $t = 1000$ years into Equation 1.3:

$$d = 5000 \times 1000 = 5 \times 10^6 \text{ miles},$$

which is clearly absurd. Every equation has a "hidden text" that sets out the range of validity of the equation. The hidden text for Equation 1.3 would be something like the following. "This is an equation that applies to one particular car, not to all. It appears to describe correctly the relationship between d and t during the first three years of the car's life but should be used cautiously if at all for values of t beyond this range." It is often easy to write the equation but more difficult to give the hidden text. As a result, it is all too often overlooked, but every equation does have some such set of qualifications and limitations that accompany it.

Even though straight-line graphs can be dangerously misleading, such graphs are so simple and it is so easy to use them to derive equations that it is often useful to search for ways in which we can create straight-line graphs even when the first "obvious" graph is not straight. The only restriction on the way in which graph paper is used is the elementary one of honesty; this is what turns an ordinary sheet of graph paper into one of the most potent scientific instruments. Consider an example. Measurements of the weight (W) of water contained in various spherical containers of radius r are shown in the first column of Table 1.1 and a graph of W versus r in Figure 1.6a. The relationship between W and r is obviously nonlinear, but there are graphs other than Figure 1.6a that we can make from the same data. Suppose, for instance, that we define a quantity x to be equal to r^2, and make a graph of W versus x. (We can *define* anything we like; whether the definition turns out to be useful or not is another question.) Thus the first and third columns of Table 1.1 give corresponding values of W and x, and the graph is shown in Figure 1.6b. This line is also curved, but less so than that in Figure 1.6a. Therefore let us define y as equal to r^3 and z as equal to r^4 and make graphs of W versus y and W versus z. These results are shown in

TABLE 1.1 Weight (W) of water contained in various spherical containers of radius, r.

W (lb)	r (ft)	$x = r^2$ (ft^2)	$y = r^3$ (ft^3)	$z = r^4$ (ft^4)
0	0	0	0	0
260	1.0	1.0	1.0	1.0
890	1.5	2.25	3.38	5.06
2100	2.0	4.0	8.0	16.0
4050	2.5	6.25	15.62	39.06
7000	3.0	9.0	27.0	81.0

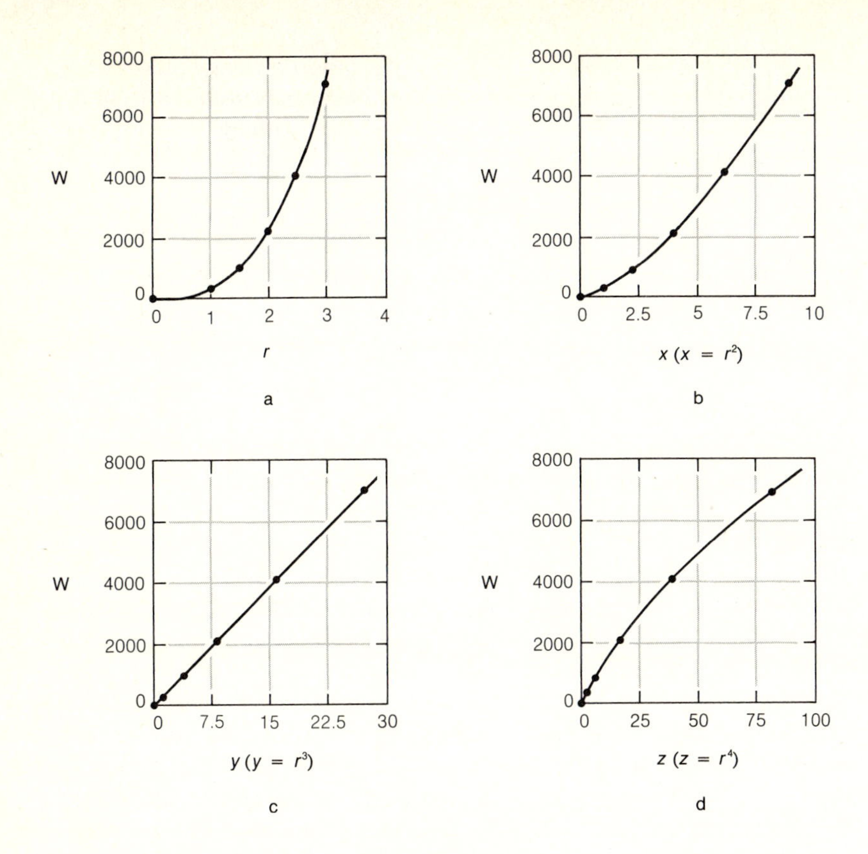

FIGURE 1.6
Various ways of presenting the relationship between the weight, W, of a spherical container of water and the radius, r.

Figures 1.6c and 1.6d. Figure 1.6c shows that the graph of W versus y is a straight line and therefore potentially useful. Remembering that y is just an abbreviation for r^3, we can regard Figure 1.6c as a graph of W versus r^3. Recall also that, since it is a straight line through the origin, we can immediately say that W is *proportional* to r^3, or, $W \propto r^3$. Furthermore, we can use the graph to determine the constant of proportionality and to write an equation relating W and r.

QUESTION

1.2 Show that the equation describing Figure 1.6c is $W \simeq 260\ r^3$. In what *units* should the number 260 be expressed?

In this simple example, we could have easily predicted that since the volume of a container is proportional to r^3 (volume $= \frac{4}{3}\pi r^3$), the weight of the water contained would also be proportional to r^3. Even if we had not known this fact to begin with, we could have discovered it by using the original data of Table 1.1. We will see important examples later in the book, demonstrating that the plotting of data in ways other than the most obvious can lead to the discovery of the mathematical relationship between two quantities.

§1.3 GROWTH, EXPONENTIAL GROWTH, AND THE DEPLETION OF WORLD OIL SUPPLIES

A country that runs on oil can't afford to run short.

Like it or not, we live in an age of growth: growth in population, in gross national product (GNP), in energy consumption (Figure 1.1), in numbers of automobiles, in air pollution, and so on. In recent years, increasing concern has been voiced about the direction in which the modern world is going. Is continued growth desirable? Is it physically possible for growth to continue indefinitely, or are there limits to growth, and if so, how close are we to these limits? Do we face an energy crisis now, or, if not, will continued growth produce one in the very near future?

Growth is not a new phenomenon, nor is the concern about its implications, although in the past those voicing concern have been greatly outnumbered by those extolling the merits of continued growth. Obviously, there are *some* limits to growth. Consumption of oil, for example, cannot continue to increase forever, because there is only a finite amount of oil in the ground. In thinking about growth, the *time scale* is of primary importance. A rate of growth that would bring us to a true limit in 10 years would require emergency measures, but a rate of growth that would bring us to the limit a million years from now would be little cause for concern. Perhaps the most difficult choices are presented by problems that appear to be manageable for another 50 to 100 years; should we leave these problems to our descendants, or should we begin now to use the available time to find solutions? Because of the importance of the time scale, it is important to assess as carefully as possible what the limits are and how rapidly we are approaching them, so that plans can be made if required and, equally important, so that drastic measures not be imposed unless needed.

For the most part we shall restrict ourselves to the relatively simple questions of the limits imposed on growth by natural laws. What is possible and what is impossible? We cannot avoid completely the much more difficult questions of what is *desirable*. What is the optimum population? Have we already exceeded it, and should we try to achieve a condition of zero-population-growth, or even a decreasing population? Should we attempt to slow down or stop economic growth? If so, can we achieve a more equitable distribution of material goods and services among the people of the world, or are we incapable of designing social systems better than those we now have? All of these questions are interrelated, but even in concentrating on questions that can be quantitatively studied ("How long before we run out of oil?" as opposed to "What is the highest possible quality of life?"), definitive answers are hard to come by.

Growth can take place in many different ways, continuously or in spurts, rapidly or slowly. One type of growth that is of particular interest—both because of its frequency of occurrence, and because of the way in which it can quickly turn an apparently modest rate of increase into an explosively rapid one—is "exponential growth." If a quantity is growing exponentially, it increases by the same *multiplicative factor* in equal intervals of time; this is the fundamental characteristic of exponential growth. Examples of this sort of growth are very common. As a simple example, consider a species of bacteria that reproduce by dividing in two, thus doubling the population with every division. If such a division occurs every hour, and if initially there is just one bacterium, then after one hour there are two; after two hours, four; after three hours, eight, and so on (see Table 1.2 and Figure 1.7). Notice that the population grows rather gradually at first and then dramatically increases. The same fundamental process of division is occurring all the time, once every hour, but a doubling of the population produces a much greater increase in the number of individuals when there are more individuals reproducing.

TABLE 1.2 Growth in number of bacteria, with doubling of population every hour.

Time, t (hr)	Number of bacteria, N
0	1
1	2
2	4
3	8
4	16
5	32
6	64
7	128
8	256
9	512
10	1,024
11	2,048
12	4,096
13	8,192
14	16,384
15	32,768
16	65,536
17	131,072
18	262,144
19	524,288
20	1,048,576

Exponential growth, whether of bacteria, people, or electrical power plants, cannot continue forever. Neither can any sort of growth, but something that is growing exponentially requires particular attention, because it can continue at a safe level for a long time and then quite abruptly encounter whatever the factors are that will limit the growth. In the example of the bacteria, suppose that a continuing supply of nutrients is available to support a population of 10^6 individuals. The population will reach this limit in about 20 hours. Just one hour before this limit is reached, the population is still "safely below" the limit. If we put ourselves in the position of a bacterium, we might be aware of the fact that our dish could support only 10^6 individuals and that some sort of birth control would have to be implemented if the population approached that limit. For a long time, there would seem to be nothing to worry about, for even after 19 hours the population would still be only half of the limiting value. But if at that point we suddenly began to understand the arithmetic of exponential growth, it might be too late to take corrective measures, and catastrophe would ensue.

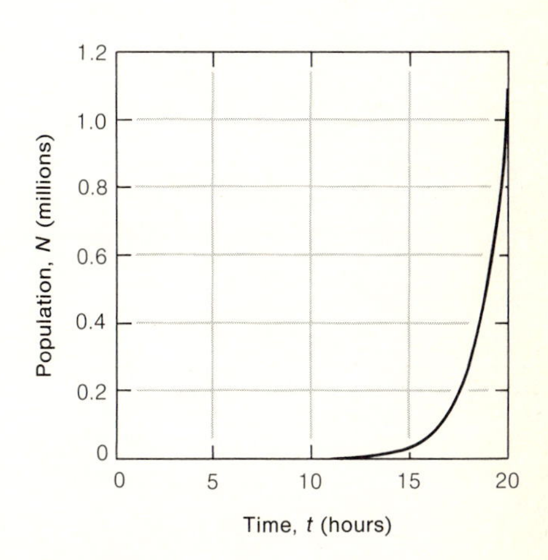

FIGURE 1.7
Growth of a population of bacteria. For the first 12 hours, the population is too small to be seen on this graph.

The origin of the term *exponential* growth can easily be seen from this example, by writing an equation relating N and t. The equation is simply this:

$$N = 2^t. \tag{1.5}$$

The time appears in the exponent. In other examples of exponential growth, the equation may be slightly more complicated, but time (t) always appears in this way—in the exponent of a number. In contrast, in the equation

$$N = t^2 \tag{1.6}$$

the t is not in the exponent; this equation does represent growth but not *exponential* growth.* Although one can write equations that describe exponential growth of all sorts, we will not need to do so. Graphical methods for discovering whether the growth is exponential will be discussed in the following paragraphs. Once the existence of exponential growth has been established, a combination of simple arithmetic and the use of graphs is all that one needs for studying exponential growth.

QUESTION

1.3 Make a single graph showing the behavior of these two equations for $t = 0, 1, 2 \ldots 10$.

The peculiarities of exponential growth have long been recognized in legends and riddles, though the obvious applications to human affairs have been made less frequently. One ancient story describes a Persian king who invited a trusted adviser to name his own reward for some particularly valuable service. The adviser brought out his chessboard and asked to be given one grain of wheat for the first square, two for the second, four for the next, eight for the next, and so on up to the 64th square. The king (not adept at arithmetic) smiled at the naiveté of his adviser and readily granted this modest request, only to discover that he was bankrupt long before the last square had been reached.

There are numerous other examples of exponential growth. A common one is the growth in the size of a savings account. Suppose that you put \$100 in a savings bank that pays an interest of 6% "compounded annually." At the end of one year, the bank credits you with an interest of \$6, giving you a new balance of \$106, 1.06 times larger than your original balance. At the end of the second year, you are credited with 6% interest on \$106, an interest of \$6.36, giving you a new balance of \$112.36; your balance has again increased by a factor of 1.06. As time passes, at the end of each year your balance is 1.06 times as large as at the

*N is a number (of bacteria) and t is a time, and so although Equation 1.5 correctly describes the relationship between the numbers in Table 1.2, it is dimensionally peculiar. We could rectify this by writing instead:

$$N = 2^{(t/T)},$$

where $T = 1$ hour. Similarly, Equation 1.6 ought more properly to be written:

$$N = (t/T)^2.$$

beginning of the year, and your savings grow as shown in Table 1.3. For the first few years, the balance grows slowly, but after five or six years, it is increasing at a rate substantially greater than \$6 per year. The balance doubles in size in about 12 years, whereas if it had grown at a steady rate of \$6 per year, it would have increased only to \$172 in 12 years. If it continues to increase exponentially at a rate of 6% per year, then after another 12 years it will have doubled again to \$400, in another 12 years to \$800, and so on. The exponential growth of this bank balance is shown in Figure 1.8, where it is compared with the *linear* growth that would result if interest were paid at the steady rate of \$6 per year.

Any growth that is exponential can be described by its "doubling time," the amount of time required for it to increase by a factor of two. The bacteria discussed earlier increase with a doubling time of 1 hour. Something growing exponentially at the rate of 6% per year has a doubling time of about 12 years. (In mathematical terms, this means that $1.06^{12} \simeq 2$; this is what is shown by the numbers in Table 1.3.) Likewise, a quantity that increases by 10% per year has a doubling time of about 7.3 years. The relationship between the percentage rate of increase per unit time and the corresponding doubling time is shown for various rates of exponential growth in Appendix O.

There are numerous important examples of exponential growth in the present-day world. The population of the world appears to be growing exponentially, with a doubling time of about 35 years. Energy consumption in the United States has been growing exponentially since about 1950, with a doubling time of approximately 20 years. In some examples of exponential growth, such as that of a human population, we can easily identify the mechanism that can lead to exponential growth. Thus if, on the average, every woman grows up and gives birth to more than one daughter, then the size of the population will increase exponentially. If the average number of daughters is two, the size of the population will double with each successive generation, that is, about every 25 years. If the average number of daughters is greater than one but less than two, the population will still grow exponentially but not as rapidly—that is, the doubling time will be longer. In this example, not only can we see how exponential growth might be expected, but also we can easily point to factors, such as food supply and living space, which will eventually limit the growth. In other examples, such as the growth of energy consumption, it is not as easy to see the mechanism responsible for exponential growth. Growth in population has caused some of the growth in energy consumption, but it is almost certain that other factors are important as well, because energy consumption in the United States is growing more rapidly (with a shorter doubling time) than is population. One of the causes of growth is that industries are encouraged to invest part of their profits in new capital equipment. Thus a successful automobile company builds new factories, produces more cars, makes larger profits, builds still more factories, and so on. This is an example

TABLE 1.3 The growth of a savings account from an initial deposit of \$100, at an annual interest rate of 6%.

Time (years)	Increase during one year (\$)	Balance (\$)
0		100.00
1	6.00	106.00
2	6.36	112.36
3	6.74	119.10
4	7.15	126.25
5	7.57	133.82
6	8.03	141.85
7	8.51	150.36
8	9.02	159.38
9	9.56	168.94
10	10.14	179.08
11	10.74	189.82
12	11.39	201.21

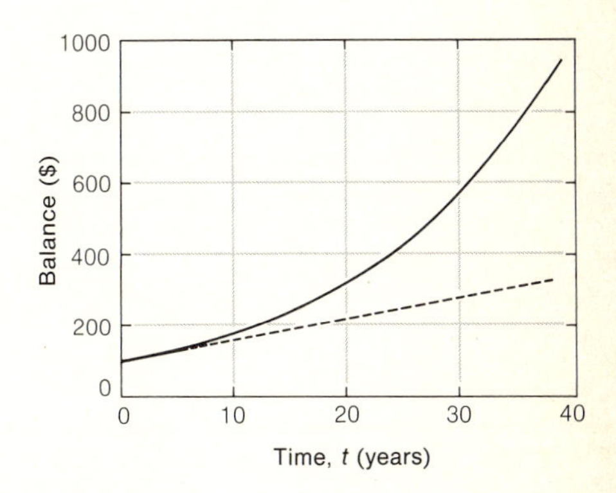

FIGURE 1.8
Growth of a savings account, with annual interest of 6%. The dotted line shows *linear* growth which would result from an annual interest payment of \$6 every year.

FIGURE 1.9
A Texas oil field in 1903. [Photograph by Fred A. Schell; American Petroleum Institute Historical Photo Library.]

of a "positive feedback loop": some of the output of the factory is "fed back" in such a way that the number of factories is increased. This sort of feedback loop is a basic part of a capitalistic economy, but in various ways it operates in all countries. Another contributing factor is that predictions of exponential growth have a way of becoming self-fulfilling prophecies. It takes so long to build an electrical power plant that utilities must make plans about ten years in advance. If consumption of electrical energy has doubled during the previous decade, utility executives feel obliged to play safe by building new plants on the assumption that this growth will continue. Once the plants are built, uses seem to be found for the electrical energy; if necessary the price will be lowered to stimulate demand. So exponential growth does indeed continue, and it seems that we are unable to do anything about it.

We shall be especially concerned in this book with growth in the use of energy: rates of growth, physical limits that may suddenly become important, new sources of energy that may make these limits more flexible, and ways in which our style of life may be altered so that we can slow down the exponential growth in the use of energy. Let us consider here just one of the many aspects of this problem, the world supply of oil, and try to arrive at a tentative answer to an important question: "When will we run out of oil?" In analyzing this question, we will introduce a graphical method that is extremely useful in studying exponential growth.

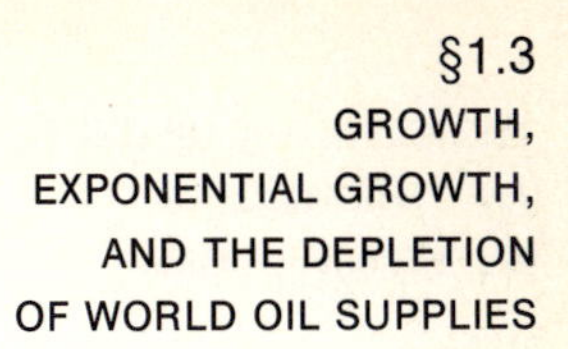

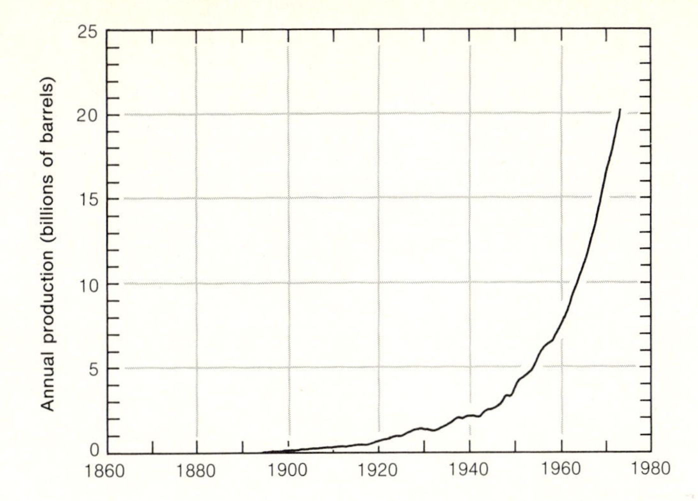

FIGURE 1.10
Annual world production of crude oil.

Annual world production* of crude oil is shown in Figure 1.10. Apart from some minor irregularities (such as the dip that occurred during the depression of the 1930s), the rate at which we have been using oil has been steadily increasing for the last century. Is this an example of *exponential* growth? The answer to this question cannot be obtained from a superficial examination of Figure 1.10. If oil production were growing *linearly* with time, the graph would be a straight line; since the graph is not straight, we can quickly see that the growth shown in Figure 1.10 is not linear. This curve looks rather similar to the growth curve for bacteria shown in Figure 1.7, but a more careful study is required to see whether Figure 1.10 represents exponential growth. Does this curve increase by the same *multiplicative factor* in equal intervals of time?

Although we can answer this question by examining Figure 1.10, fortunately there is a better way to make a sensitive graphical test for exponential growth. It is possible to construct a graph so that a *straight* line results if growth is exponential and not otherwise. The trick is to use a *distorted* scale on the vertical axis, a scale chosen so that equal vertical distances represent increases by the same multiplicative factor. If something is increasing exponentially, it must give a straight line when plotted in this way. Such a graph of world oil production is shown in Figure 1.11. Notice that the scale on the vertical axis does have the property described above. For example, the vertical distance between the points labeled 0.1 and 1 (a factor of 10) is the same as that between 1

*For the world as a whole, annual *consumption* of oil is virtually identical to annual production. Because of international trade, this may not be true for an individual nation. In recent years, oil consumption in the United States has been considerably greater than domestic oil production; the fact that we import a significant fraction of our oil is an important factor in our short-term energy problem.

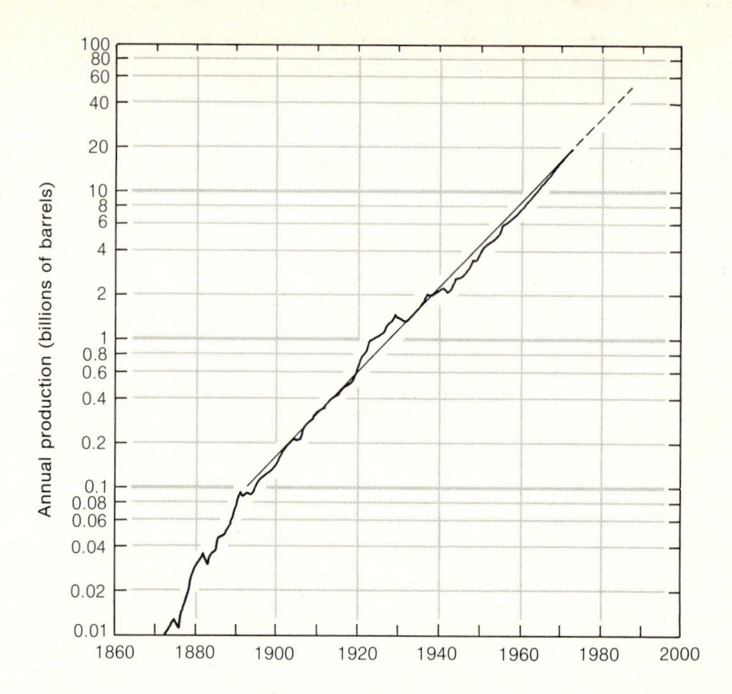

FIGURE 1.11
Annual world production of crude oil, semilogarithmic plot. A straight-line fit has been added.

and 10, between 2 and 20, and between any other pair of numbers differing by a factor of 10. Similarly, the vertical distance between 2 and 4 is equal to that between 4 and 8, and so on. Note that data for the earliest years, when annual production was less than 0.01 billion barrels, cannot be displayed on Figure 1.11. A similar graph showing the growth of the balance in a savings account (Figure 1.8) is shown in Figure 1.12.

Technically, graphs of this type are called "semilogarithmic" graphs. We need not take the space here to explore the properties of logarithms which are used in constructing this kind of graph, because "semilog" graph paper, like that shown in Figures 1.11 and 1.12, is widely available, and we can use it simply as a convenient tool to test for exponential growth.*

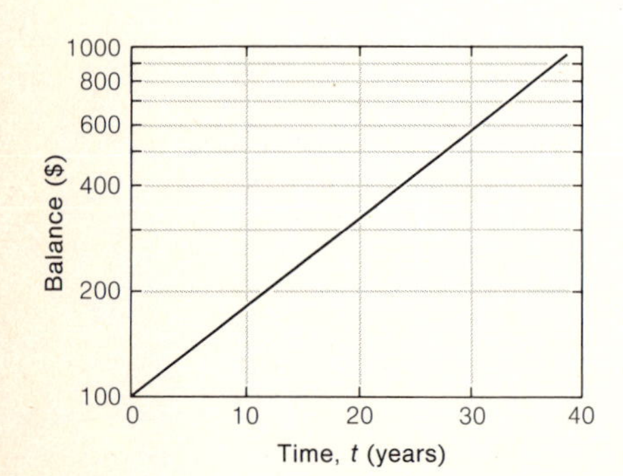

FIGURE 1.12
Data from Figure 1.8, replotted on a semilogarithmic graph.

If a semilog graph produces a straight line, then the quantity in question is increasing exponentially; if the semilog graph is not a straight line, then the growth is not exponential. What can we say about world oil production? If we look at the data from about 1890 to 1973, all of these results lie rather close to a single straight line, as indicated in Figure 1.11. Although the curve is not perfectly straight, it is sufficiently so that it is reasonable to say that Figure 1.11 demonstrates that world oil production has been growing exponentially for almost a century.

Once the existence of exponential growth has been established, we can use the straight line that has been added to Figure 1.11 to estimate

*In a *semi*logarithmic graph, only one of the two scales is distorted; in "log-log" graphs, which are occasionally used for other purposes, the scales on both the vertical and horizontal axes are similarly distorted.

the *doubling time*. This line crosses the 0.2 level at about 1903, the 0.4 level at about 1913, the 0.8 level at about 1923, and so on: thus the doubling time is about 10 years. Once we know the doubling time, we can make extrapolations either graphically or by simple arithmetic; the extrapolation is that the 1980 value will be twice the 1970 value, the 1990 value four times as great as the 1970 value, and so on.

In attempting to learn when we may begin to run out of oil, it is useful to present the data on oil production in terms of *cumulative* production, the total amount that has been produced up to each given date. This information is easy to obtain from the figures for production in individual years and is presented in Figure 1.13 and (as a semilog graph) in Figure 1.14. These figures show, for instance, a cumulative value of about 65 billion barrels for 1950; this is the total amount of oil produced in all years up to and including 1950.

Notice that although the four graphs shown in Figures 1.10, 1.11, 1.13, and 1.14 are all based on the same original data, they do not at all resemble one another. The irregularities that are so apparent in the graphs of annual production are virtually unnoticeable in the graphs of cumulative production. Another difference is that the semilog graphs (Figures 1.11 and 1.14) appear much less alarming than do the conventional graphs of Figures 1.10 and 1.13. Figure 1.13 is almost a self-explanatory announcement of imminent catastrophe, whereas Figure 1.14, presenting the very same data, seems to show a much more gradual rate of growth.

The semilog graph of cumulative production (Figure 1.14) is fairly close to a straight line for recent decades. That is, cumulative produc-

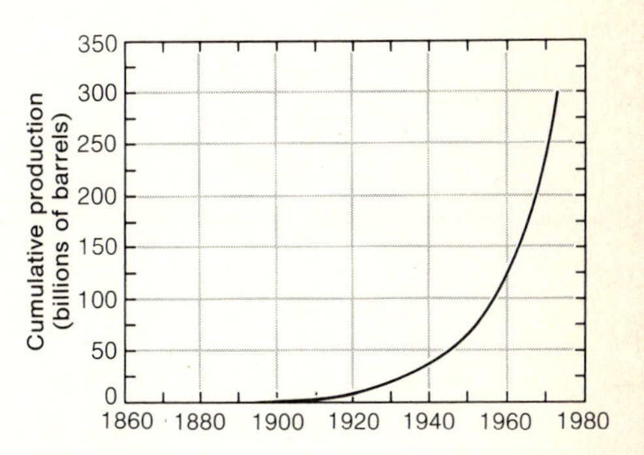

FIGURE 1.13
Cumulative world production of crude oil.

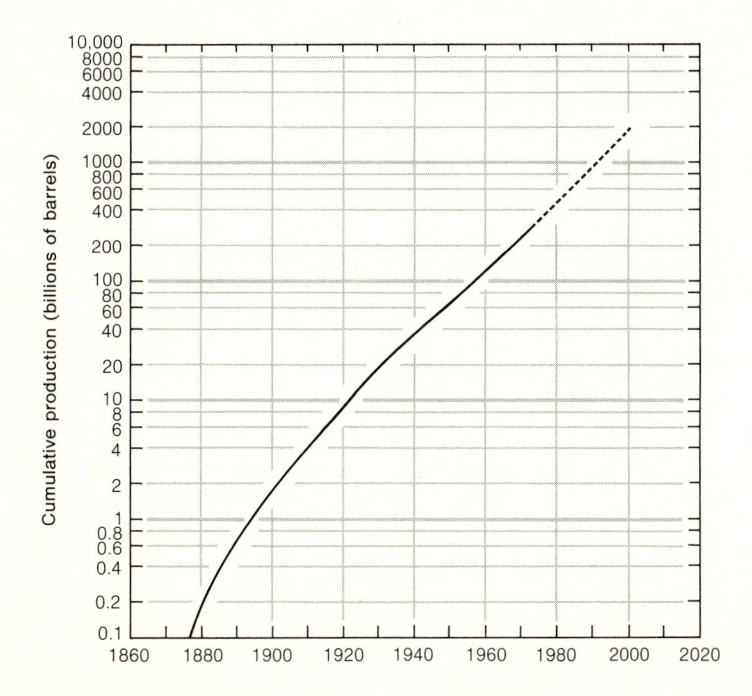

FIGURE 1.14
Cumulative world production of crude oil, semilogarithmic graph.

tion, like annual production, has been increasing exponentially, and such a straight line can readily be extrapolated into the future (see the dashed line in the figure). This straight line corresponds to a doubling time of about 10 years.*

In the years through 1956, the world had used up about 100 billion barrels of oil; by 1966, the cumulative production had reached about 200 billion barrels. In other words, we used as much oil in the single decade 1956–66 (100 billion barrels) as in all the previous years. If this trend continues (and for the sake of argument, let us tentatively assume that the extrapolation shown in Figure 1.14 *is* correct), then the amount of oil used in *any* one 10-year interval will be as great as all the oil used in all preceding years. This is the stark significance of exponential growth with a 10-year doubling time. If it continues unabated, the annual production rate and the cumulative production 10 years from now will be at twice their present values, in 20 years four times as large, in 30 years eight times as large, and so on. Similar increases will take place in levels of air pollution and numbers of oil spills *if* exponential growth continues with no other changes.

One other important piece of information is necessary in order to estimate when we will run out of oil: what is the total amount of oil available? This figure is much more difficult to obtain than that for the amount of oil that has been used in the past. Many books have been written in the attempt to formulate this question in more precise terms and then to provide tentative answers, for oil and for other important resources as well. The reasons for the difficulty in ascertaining the amount of oil left are obvious. Realistic estimates must include large quantities of oil not yet discovered, and estimates of the amount of oil to be produced must be based on geological knowledge and on past experience. The literature of this subject abounds with estimates that have subsequently been found to be drastically in error, as well as with words that reflect the uncertain nature of the subject, such as "probable," "inferred," "marginal," "potential," and so on. What has been estimated quite accurately is the amount of "proved recoverable reserves," meaning the oil contained in oil fields already under production or in fields that have been explored and mapped. Unknown is the total amount of oil in the world, a quantity that is large but finite. Indeed, the numerical value of this total is probably not important, since much of this oil is so hard to extract that more energy would be needed to recover it than would subsequently be obtained by burning it.

In spite of all these difficulties, it is better to have an uncertain number than none at all, and for the sake of argument we shall use a fairly careful analysis made in 1970 which led to the estimate of 2100

---

*It is not a coincidence that the doubling times for annual and cumulative production are nearly equal. It can be shown that, if the annual production rate grows in precisely exponential fashion, then cumulative production will also show exponential growth with the *same* doubling time.

billion barrels for the eventual cumulative world production of oil. This figure includes proved and probable reserves and estimates of future discoveries, in both offshore and land areas. One could easily argue for a higher or lower figure, but one of the consequences of exponential growth is that, even if this estimate is too high or too low by a factor of two or more, the conclusions will be affected in only a minor way.

The consequences of assuming a figure of 2100 billion barrels for the eventual cumulative world production, together with the assumption of the continuation of present exponential growth, are readily apparent from an examination of Figure 1.14. Although, in all the years before 1970, we used up only about 10% of the total, we will run out of oil in about the year 2000. As of 1970, an "ample" supply of oil remained—almost 90% of the oil originally formed some hundreds of millions of years ago—but the preceding calculation suggests that we may be less than 30 years away from the end. It is important to realize that this conclusion is not at all sensitive to the precise value assumed for eventual total production. If the figure is twice as large (4200 billion barrels), the end will be postponed only by one doubling time, to about 2010. If it is half as large, the supply will be exhausted 10 years sooner, in 1990.

QUESTIONS

1.4 This conclusion does depend critically on the assumption that exponential growth will continue with a 10-year doubling time. Show that if oil production had leveled off in 1970 and had then continued at the 1970 rate (about 17 billion barrels per year), we would not run out of oil until about the year 2080. How would *this* result be affected by assuming a value (a) twice as large or (b) half as large as 2100 billion barrels for eventual total production?

1.5 When would we run out of oil if growth in oil production had suddenly changed in 1970 from a 10-year doubling time to a 20-year doubling time?

The calculations discussed are in a sense just an academic exercise. We have been asking what would happen *if* the present exponential growth were to continue with no change. Of course, exponential growth will not simply continue unchecked until the day on which all the wells suddenly run dry. If it did, then the overall history of oil production would be as shown in Figure 1.15, a sudden catastrophe occurring in about the year 2000. Our calculation is an important one nonetheless, because in refusing to believe that exponential growth will continue in this way, we are asserting that changes of some sort will occur and will occur soon. If our calculation of the hypothetical date on which all the oil would be exhausted had given us as a result the year 3000 instead of 2000, the situation would be less critical. As it is, it is reasonably safe to predict that, political developments in the Middle East aside, oil will begin to become scarce within a decade or two, more wells will have to

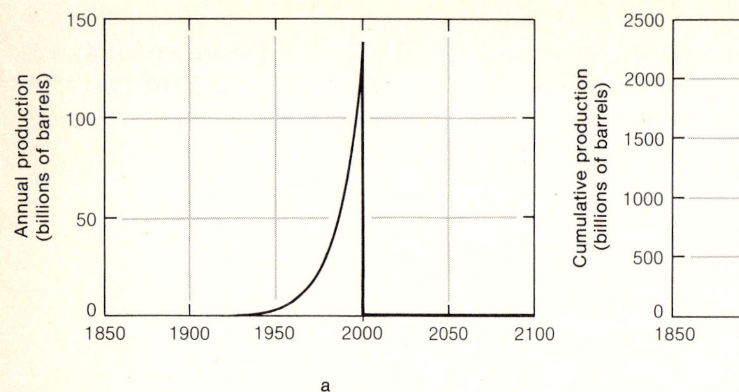

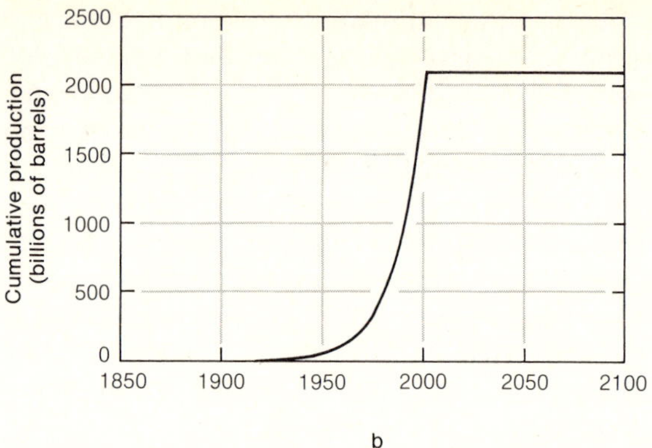

FIGURE 1.15
Projected history of world oil production *if* exponential growth with a 10-year doubling time were to continue: (a) annual production; (b) cumulative production.

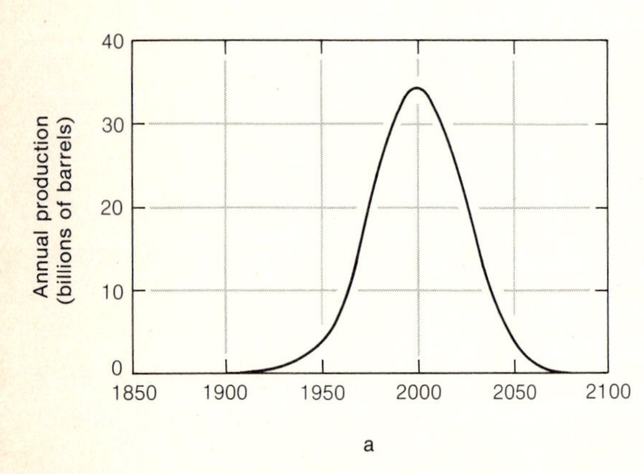

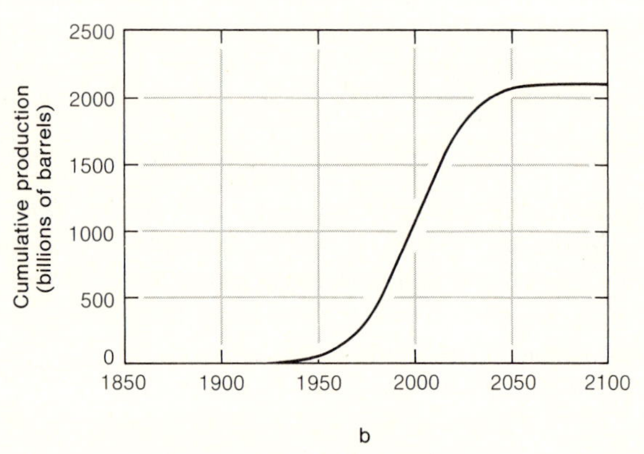

FIGURE 1.16
Projected history of world oil production with gradual depletion of oil supplies: (a) annual production; (b) cumulative production.

be drilled to recover the same amount of oil, prices will rise, and oil production will begin to level off. Future patterns will probably look more like those shown in Figure 1.16. The annual production rate will probably very soon begin to fall below the projected exponential growth rate; oil production will probably reach a peak by the end of this century and then begin to decline.

Oil shortages were experienced at least as early as 1973, when Americans learned that it was sometimes necessary to wait in line to buy gasoline. These were artificial short-term shortages, caused more by the policies of the oil companies and the nature of American foreign policy toward the Arab nations than from a lack of oil in the ground. M. A. Adelman—quoted at the beginning of this chapter—was referring to this sort of problem when he pronounced the energy crisis a "fiction." However, although the short-term crisis of 1973–74 may indeed have been a fiction, a genuine crisis may not be far away. If the growth in oil consumption continues, then even a doubling of the figure assumed for the total amount of available oil only postpones the crisis for a decade.

Oil is such a large source of energy and such a convenient one that a genuine oil shortage will require changes in our way of living. Half of the energy used in the world in 1973 was derived from oil, and in the United States, 30% of the oil used was for gasoline for private automobiles. One can safely predict, among other things, that the exponential growth of the number of automobiles on the road will not continue (because of air pollution and lack of space for highways and parking lots, as well as because of fuel shortages) and that the familiar gasoline-powered automobile will eventually become a relic of the past. Alternative sources of energy will be increasingly important, sources such as coal and nuclear fission, and—perhaps—nuclear fusion, solar energy, and windmills. We can also be confident that total consumption of energy will begin to level off, or at least increase at a slower rate, in the not very distant future.

§1.4 SUPPLEMENTARY TOPICS

§1.4.A Powers-of-10 Notation

The powers-of-10 notation, (writing 6.5×10^9 instead of 6,500,000,000 for instance), is used to simplify the writing of very large and very small numbers, and to simplify many numerical calculations. 10^2 means a product of two factors of 10 ($10 \times 10 = 100$), 10^3 means $10 \times 10 \times 10 = 1000$, and so on. 10^1 is just one factor of 10, that is, $10^1 = 10$.

Multiplication is easily performed by *adding* the exponents. For example:

$$10^2 \times 10^3 = 10^5.$$

That is,

$$100 \times 1000 = (10 \times 10) \times (10 \times 10 \times 10) = 10^5,$$

and the general rule is simply this:

$$10^a \times 10^b = 10^{a+b}.$$

Two such numbers can be divided by *subtracting* the exponents:

$$\frac{10^6}{10^2} = 10^{6-2} = 10^4.$$

That is,

$$\frac{10^6}{10^2} = \frac{10 \times 10 \times 10 \times 10 \times \cancel{10} \times \cancel{10}}{\cancel{10} \times \cancel{10}} = 10^4,$$

and the general rule is:

$$\frac{10^a}{10^b} = 10^{a-b}.$$

10^0 is defined as simply equal to 1. This definition is chosen so that the same rules for multiplying and dividing still apply. For example:

$$10^0 \times 10^3 = 10^{0+3} = 10^3;$$

that is, $1 \times 1000 = 1000$.

10^{-1} means $\frac{1}{10^1} = 0.1$, $10^{-2} = \frac{1}{10^2} = 0.01$, and so on. A negative exponent in the numerator is equivalent to a positive exponent in the denominator. Similarly a negative exponent in the denominator is equivalent to a positive exponent in the numerator: $\frac{1}{10^{-2}} = 10^2$. Thus any power of 10 can be moved from numerator to denominator or vice versa as desired, provided that the *sign* of the power is changed.

The various powers of 10 form an unending sequence that can be used to represent numbers of any size:

. . .	10^{-3}	10^{-2}	10^{-1}	10^0	10^1	10^2	10^3	. . .
	0.001	0.01	0.1	1	10	100	1000	

The basic rules for multiplication and division,

$$10^a \times 10^b = 10^{a+b} \qquad \text{and} \qquad \frac{10^a}{10^b} = 10^{a-b},$$

are universally applicable. Here are some examples.

$$10^3 \times 10^{-2} = 10^{3+(-2)} = 10^1; \text{ that is, } 1000 \times \frac{1}{100} = 10.$$

$$\frac{10^3}{10^5} = 10^{3-5} = 10^{-2}; \text{ that is, } \frac{1000}{100000} = \frac{1}{100} = 0.01.$$

$$10^{-3} \times 10^{-1} = 10^{-3+(-1)} = 10^{-4}; \text{ that is, } \frac{1}{1000} \times \frac{1}{10} = \frac{1}{10000} = 0.0001.$$

$$\frac{10^3}{10^{-2}} = 10^{3-(-2)} = 10^5; \text{ that is, } \frac{1000}{(1/100)} = 1000 \times 100 = 100000.$$

Most of the numbers we deal with are not exactly equal to a power of 10, but any number can be written as the product of a number of "reasonable" size and some power of 10:

$$6{,}270{,}000 = 6.27 \times 10^6$$

$$0.00075 = 7.5 \times 10^{-4}.$$

The same number can be written in many different ways: for example, $6{,}270{,}000 = 6.27 \times 10^6 = 0.627 \times 10^7 = 627 \times 10^4$. The choice is ours. It is frequently convenient to use certain powers of 10, such as 10^3, 10^6, or 10^9, since these particular numbers have familiar names. Thus the population of the United States in 1974 was 212 million; writing this as 212×10^6 may be preferable to 2.12×10^8, though both are equally correct. Or, we can write a power of 26,500 W as 26.5×10^3 W (26.5 kW).

The rules for changing the ways in which numbers are written are very simple. Suppose you wish to rewrite the number 8.65×10^9 as some number multiplied by 10^6. In order to compensate for reducing the power-of-10 part of the number by a factor of 1000, the number in front must be increased by the same factor:

$$8.65 \times 10^9 = (8.65 \times 10^3) \times 10^6 = 8650 \times 10^6.$$

In other words, just move the decimal point in the first number, three places to the right in this example. Similarly, $8.65 \times 10^9 = 0.865 \times 10^{10}$; the power-of-10 part of the number is increased by a factor of 10, and so the number in front must be reduced by the same factor. The same procedure is applied to numbers containing negative exponents:

$$5 \times 10^{-5} = 0.5 \times 10^{-4};$$

$$5 \times 10^{-5} = 50 \times 10^{-6}.$$

It is easy to multiply or divide numbers written in powers-of-10 notation, as demonstrated by the following examples.

$$(6 \times 10^5) \times (7 \times 10^8) = (6 \times 7) \times (10^5 \times 10^8) = 42 \times 10^{13}$$
$$= 4.2 \times 10^{14};$$

$$\frac{3 \times 10^{35}}{4 \times 10^9} = \frac{3}{4} \times \frac{10^{35}}{10^9} = 0.75 \times 10^{26} = 7.5 \times 10^{25};$$

$$\frac{5 \times 10^6}{6 \times 10^{22}} = \frac{5}{6} \times \frac{10^6}{10^{22}} = 0.83 \times 10^{-16} = 8.3 \times 10^{-17}.$$

Notice an important practical point. The powers of 10 can be multiplied or divided merely by adding or subtracting integers, and the remaining numbers in the calculation are neither very large nor very small. If you wish to make use of the speed and accuracy of an electronic calculator, you do not need one that can handle numbers up to 10^{35} or more; any calculator will do the job.

Multiplication or division of very large or very small numbers is most conveniently handled by first writing all numbers in powers-of-10 form:

$$320{,}000 \times 0.00000002 = (3.2 \times 10^5) \times (2 \times 10^{-8})$$
$$= (3.2 \times 2) \times 10^{-3} = 6.4 \times 10^{-3} = 0.0064;$$

$$\frac{320{,}000}{0.00000002} = \frac{3.2 \times 10^5}{2 \times 10^{-8}} = \frac{3.2}{2} \times 10^{13} = 1.6 \times 10^{13}.$$

Addition or subtraction requires more care. If the numbers are initially written with different powers of 10, one or the other must be rewritten:

$$(7 \times 10^5) + (8 \times 10^3) = (700 \times 10^3) + (8 \times 10^3) = 708 \times 10^3$$
$$= 7.08 \times 10^5;$$
$$(5 \times 10^3) - (3 \times 10^{-2}) = (500000 \times 10^{-2}) - (3 \times 10^{-2})$$
$$= 499997 \times 10^{-2} = 4.99997 \times 10^3.$$

What if we need to take a number written in powers-of-10 notation and square it or cube it?

$$(10^5)^3 = 10^5 \times 10^5 \times 10^5 = 10^{15}$$

or, in general,

$$(10^a)^b = 10^{ab}.$$

The number in front of the power of 10 can be raised to the appropriate power in a separate operation:

$$(3 \times 10^5)^3 = 3^3 \times (10^5)^3 = 27 \times 10^{15}.$$

The meaning of *fractional* exponents is defined so that all the previously given rules still apply. The figure $10^{\frac{1}{2}}$, for instance, denotes the square root of 10:

$$(10^{\frac{1}{2}})^2 = 10^{\frac{1}{2} \times 2} = 10^1 = 10.$$

Similarly, $10^{\frac{1}{3}}$ denotes the cube root of 10. Thus the square root, cube root and higher roots of any numbers can be written as fractional powers:

$$(10^6)^{\frac{1}{3}} = 10^{6 \times \frac{1}{3}} = 10^2 = 100.$$

(100 is the cube root of 10^6.) In many calculations of roots, the powers do not work out so neatly. What, for instance, is the square root of 10^{13}?

$$\sqrt{10^{13}} = (10^{13})^{\frac{1}{2}} = 10^{\frac{13}{2}}.$$

This is awkward (though correct), and it helps to write the original number in terms of an *even* power of 10:

$$10^{13} = 10 \times 10^{12};$$

$$(10^{13})^{\frac{1}{2}} = 10^{\frac{1}{2}} \times 10^6 \simeq 3.2 \times 10^6.$$

Similarly, to find the cube root of 10^{11}:

$$10^{11} = 100 \times 10^9;$$

$$(10^{11})^{\frac{1}{3}} = 100^{\frac{1}{3}} \times 10^3 \simeq 4.6 \times 10^3.$$

§1.4.B Dimensions, Units, and Conversions of Units

Physical quantities that are to be added or treated as equivalent to one another must be quantities of the same dimensions, and if numerical values are used, they must be expressed in the same units. Equations relating physical quantities must be dimensionally consistent. If you see an equation supposedly giving the relationship between the volume and radius of a sphere as $V = 4\pi r^2$, you know it is wrong because it is dimensionally inconsistent; the left-hand side has dimensions of length^3, whereas the right-hand side has the dimensions of length^2. The correct formula is $V = \frac{4}{3}\pi r^3$, which is dimensionally consistent. (An equation may be dimensionally consistent but still wrong, $V = 4\pi r^3$, for example.)

A speed is calculated by dividing a distance traveled by an elapsed time, and therefore its dimensions are length/time, or length-time^{-1} (note that in simple expressions such as this, a hyphen is often used to indicate multiplication of dimensions or units). The equation $d = vt$, relating distance traveled, d, speed or velocity, v, and time, t, is a correct equation. We can see that its dimensions are correct, for the dimensions of the left side are length and those of the right side (length/time) $\times$ time = length. Equations such as $d = vt^2$ or $d = v/t$ can be rejected at once as dimensionally inconsistent.

A speed has dimensions of length/time, and many important physical quantities are defined in such a way, with dimensions which contain various combinations of the fundamental dimensions of mass, length, and time. Momentum, for instance, is the product of mass and velocity. Its dimensions are mass $\times$ length/time, or mass-length-time^{-1}, and its MKS units are kg-m/sec or kg-m-sec^{-1}.

The units of measurement must be specified in giving the numerical value of any physical quantity with dimensions. Such statements as "my height is 173" and "my speed is 60" are incomplete. Instead, "my height is 173 centimeters (or 68 inches)," and "my speed is 60 miles per hour." A height can be correctly given in terms of any unit that has the dimensions of length—expressed, for example, in centimeters, feet, or yards. A speed can be expressed in any set of units with the correct dimensions, such as miles per hour, centimeters per second, or kilometers per week.

Acceleration is defined as the rate of change of speed. A car whose speed changes from 50 ft/sec to 65 ft/sec in 3 seconds has an acceleration a, given by

$$a = \frac{65 \text{ ft/sec} - 50 \text{ ft/sec}}{3 \text{ sec}} = \frac{15 \text{ ft/sec}}{3 \text{ sec}} = 5 \text{ ft/sec}^2 \text{ or } 5 \text{ ft-sec}^{-2}.^*$$

Any units with the dimensions length/time^2 can be used to describe acceleration; 4 miles/hr-sec is a perfectly correct way to state the

*One may see such expressions as "5 feet per second per second" or "5 ft/sec/sec." This usage is dangerous because it is ambiguous, just as the numerical expression 10/4/2 is ambiguous. It might mean 10/(4/2), which is equal to 5, or it might mean (10/4)/2, which is equal to 1.25.

acceleration of a car that can start at rest and reach a speed of 60 miles/hr in 15 sec. The acre-foot is a perfectly proper unit for measuring volumes. The acre has dimensions of area (length^2), so the acre-foot has dimensions of length^3. (This unit is often used in measuring amounts of water in rivers or reservoirs. It is the volume of water that will cover an area of 1 acre to a depth of 1 ft.) As long as the units have the correct dimensions, there are no other restrictions except those of common courtesy. (Why give the speed of a car in millimeters per week unless you wish to annoy your audience?)

Until it is agreed by everyone to use one and only one set of units, we must know how to convert quantities from one unit to another. The only essentials for this purpose are a table of conversion factors and a reliable method for using them. A large collection of conversion factors for various units is given in Appendix A. The key to a foolproof method for using such tables is this: we produce no change in a quantity if we simultaneously multiply and divide it by equal quantities, even though the quantities are expressed in different units. For example, if the height, h, of a waterfall is 200 ft, what is h in miles? Since 1 mile = 5280 ft, then it is legitimate to multiply the quantity 200 ft by 1 mile if we simultaneously divide it by 5280 ft:

$$h = 200\ \text{ft} = 200\ \cancel{\text{ft}} \times \frac{1\ \text{mile}}{5280\ \cancel{\text{ft}}} = \frac{200}{5280}\ \text{miles} = 3.788 \times 10^{-2}\ \text{miles}.$$

The units of feet "cancel" one another, and we get the desired result.* Table A.6 contains another entry which we could use: 1 ft = 1.894×10^{-4} miles. Thus $h = 200\ \text{ft} = 200\ \cancel{\text{ft}} \times \frac{1.894 \times 10^{-4}\ \text{miles}}{1\ \cancel{\text{ft}}} = 200 \times 1.894 \times 10^{-4}\ \text{miles} = 3.788 \times 10^{-2}$ miles, the same result as before. Each of the seven lists in Table A.6 is based on the same information; any one of the seven would be adequate, but the extra lists are provided to save time and effort.

Consider another example. It is stated that water flows over Niagara Falls at the rate of 1.2×10^8 gal/min. What is this flow rate, f, in cubic meters per second? Table A.4 contains conversion factors for fluid flow rate, but gallons per minute is not included. However, we do not really need such a table as long as separate tables for time and volume are available. From Tables A.12 and A.14:

$$f = 1.2 \times 10^8 \frac{\text{gal}}{\text{min}}$$

---

*Another way of describing this process is this. The quantity h has been multiplied by a ratio whose value is 1 $\left(\frac{1\ \text{mile}}{5280\ \text{ft}} = 1\right)$, and multiplication by the number 1 produces no change. This is the true meaning of the term "conversion *factor*," a quantity that is equal to 1 and that can be introduced as a multiplicative factor for the sole purpose of changing units.

$$= 1.2 \times 10^8 \frac{\text{gal}}{\text{min}} \times \frac{3.785 \times 10^{-3}\ \text{m}^3}{1\ \text{gal}} \times \frac{1\ \text{min}}{60\ \text{sec}}$$

$$= \frac{1.2 \times 10^8 \times 3.785 \times 10^{-3}}{60}\ \text{m}^3/\text{sec} \simeq 7600\ \text{m}^3/\text{sec}.$$

The possibility of error in this method is slight. About the worst that can happen is that the units do not cancel out as we want them to. This is usually caused by inadvertently interchanging the factors by which the quantity is multiplied and divided, as in the following example:

$$h = 200\ \text{ft} = 200\ \text{ft} \times \frac{5280\ \text{ft}}{1\ \text{mile}} = 200 \times 5280\ \text{ft}^2/\text{mile}\ (?)$$

This result is not really wrong, but a height in units of square feet per mile is not very useful. We can easily see in this case how to rewrite the two factors in order to obtain the desired result.

To recapitulate, you can find in a table of conversion factors a wide variety of pairs of quantities whose values are equal (such as 5280 ft = 1 mile). In converting a quantity from one unit of measure to another, remember that it is always legitimate to multiply and divide it simultaneously by equal quantities, and that this procedure can be repeated as many times as needed until the units you want to eliminate have been canceled, and the desired units remain.

In many calculations that are not really problems in converting units, the cancellation of units can be a great help. Suppose, for example, that it is known that the "energy content" of oil is 5.6×10^6 BTU/barrel. How many barrels of oil do we need to get an energy of 10^8 BTU? It is not hard to see that the required number of barrels is $\frac{10^8}{5.6 \times 10^6} \simeq 18$ barrels. If we put in the units, we can be more confident of having done it correctly:

$$\frac{10^8\ \text{BTU}}{5.6 \times 10^6\ \text{BTU/barrel}} = \frac{10^8}{5.6 \times 10^6} \frac{\cancel{\text{BTU}}\text{-barrel}}{\cancel{\text{BTU}}} \simeq 18\ \text{barrels}.$$

We might have made a mistake (a genuine error) of multiplying by 5.6×10^6 instead of dividing:

$$10^8\ \text{BTU} \times 5.6 \times 10^6\ \text{BTU/barrel} = 5.6 \times 10^{14}\ \text{BTU}^2/\text{barrel}\ (?)$$

The units are nonsensical, and this gives us a signal that a mistake has been made.

§1.4.C Straight-Line Graphs, Equations of Straight Lines, and Proportionalities

Suppose that the dependence of one quantity, y, on another quantity, x, is expressed by the equation:

$$y = 2x + 3. \tag{1.7}$$

We can describe this relationship by making a table of corresponding values of x and *y*:

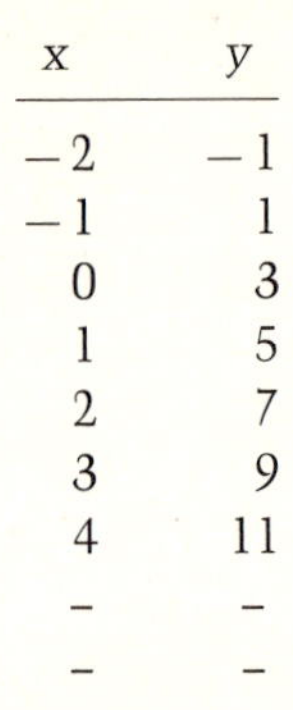

x	y
−2	−1
−1	1
0	3
1	5
2	7
3	9
4	11
–	–
–	–

We can also use these values to draw a graph showing "*y* as a function of x," a graph of "*y* versus x" (Figure 1.17). The equation relating x and *y* in this example is described as a linear one, because the graph is a straight line. One can see from the equation why the graph must be straight. Take any value of x and calculate the corresponding value of *y*. Now take a new value of x, 1 greater than the first value; the new value of *y* must be 2 greater than before. The graph must be one which rises at a steady rate, 2 units upward for every 1 unit to the right.

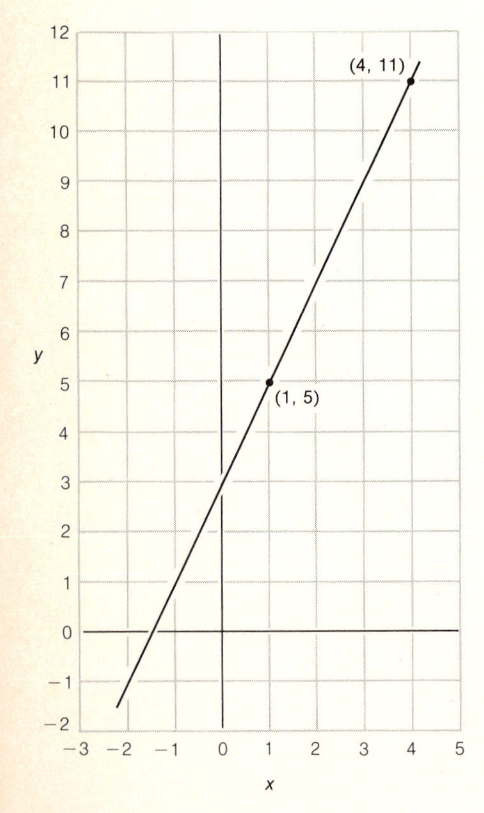

FIGURE 1.17
Graph of the equation $y = 2x + 3$.

We define the *slope* of a straight line as a measure of its steepness. The slope is the amount of "rise" per unit "run" (to the right); in this example the slope has the value 2. The slope tells us the rate of change of *y*: a change of 2 units for every change of 1 unit in x. The slope is formally defined by the following procedure. Choose any two points on the line and find their x and *y* coordinates, (x_1, y_1) and (x_2, y_2). The slope (often denoted by the symbol *m*) is then defined by

$$m = \text{slope} = \frac{y_2 - y_1}{x_2 - x_1}. \tag{1.8}$$

From Figure 1.17 we have in this example,

$$m = \text{slope} = \frac{11 - 5}{4 - 1} = 2.$$

If we had chosen any other pair of points on the line, we would have found the same value for the slope.

A straight line is completely determined if we know its slope and if we also know one point that is on the line. One particular point that is often especially convenient is the "*y*-intercept," the point for which $x = 0$, the point at which the line crosses the vertical axis. In Equation 1.7, when $x = 0$, $y = 3$; the *y*-intercept is 3. By inspecting Equation 1.7, we can see the significance of the numbers that appear in the equation: the multiplier of x (2) is the slope; the additive constant (3) is the *y*-intercept.

Any equation of this form has a graph that is a straight line. The graph of $y = 5x + 3$ has a slope of 5 and a y-intercept of 3; the graph of $y = 2x + 1$ has a slope of 2 (and is therefore parallel to the graph of Equation 1.7) and a y-intercept of 1. More generally, any equation of the form

$$y = mx + b, \tag{1.9}$$

where m and b are constants, has a graph which is a straight line, with a slope equal to m and a y-intercept equal to b. Some equations may have a different appearance but can be rewritten to be put in this form. The equation $3y - 7 = 8x$ can be rewritten as

$$y = \tfrac{8}{3}x + \tfrac{7}{3}.$$

Its graph has a slope of $\frac{8}{3}$ and a y-intercept of $\frac{7}{3}$. Any equation containing only the first powers of x and y (with no terms such as x^2, $1/x$ and so on) can be written in this way.

Even if m or b or both have negative values, these conclusions are still valid. If $y = -3x + 6$, then as x increases, y decreases, as shown in Figure 1.18. As before, the slope can be calculated from the coordinates of any two points on the line:

$$m = \frac{y_2 - y_1}{x_2 - x_1} = \frac{0 - 6}{2 - 0} = -3.$$

If the slope is negative, the line slopes downward to the right. The equation $y = -3x - 2$ can be rewritten as $y = -3x + (-2)$, that is, $b = -2$. Its slope is -3 and its y-intercept is -2, as shown in Figure 1.19. If $m = 0$, the equation is simply represented by a horizontal

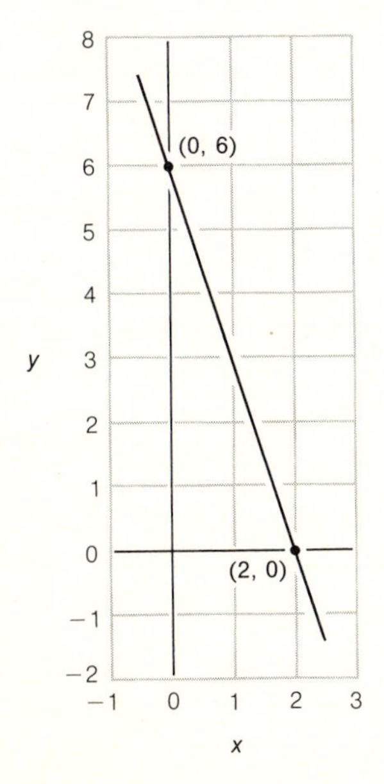

FIGURE 1.18
Graph of the equation $y = -3x + 6$.

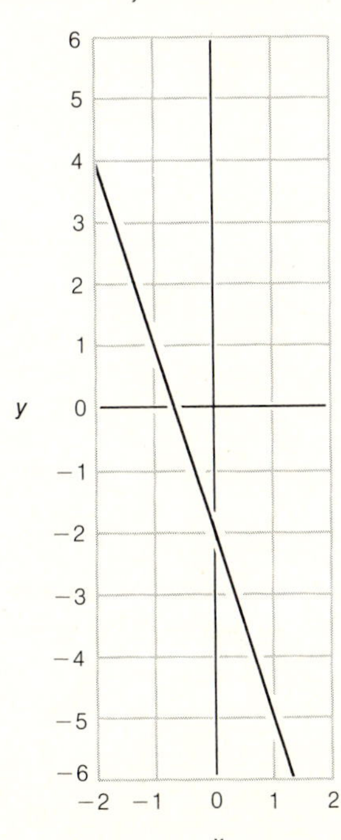

FIGURE 1.19
Graph of the equation $y = -3x - 2$.

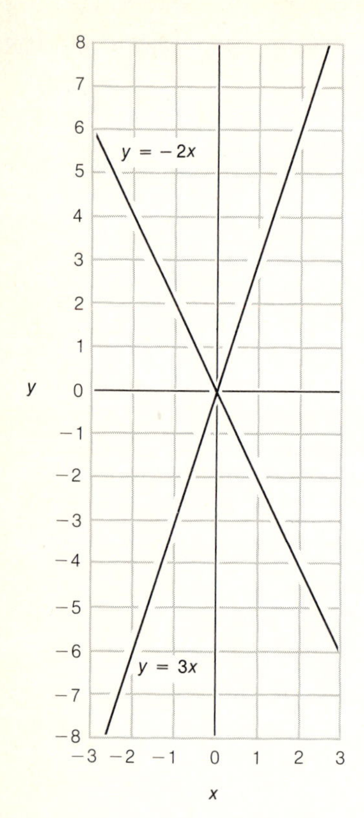

FIGURE 1.20
Graphs of the equations $y = 3x$ and $y = -2x$.

straight line. On the other hand if $b = 0$, we have simple equations such as $y = 3x$ or $y = -2x$, and any equation like this has a graph (Figure 1.20) that goes through the origin of the coordinate system.

To summarize this description of linear equations, if we know the equation relating x and y and if it is of the form $y = mx + b$ (or if it can be put into this form), then we can immediately say that it can be represented by a straight line whose slope is m and whose y-intercept is b. We can also do the reverse; if we have a straight-line graph, then we can quickly write an equation relating the two quantities. Consider Figure 1.21a, for instance. The y-intercept is 1, and the slope, calculated from the two points shown, is $\frac{2}{3}$. The equation of this line is therefore

$$y = \tfrac{2}{3}x + 1.$$

Another example is shown in Figure 1.21b. The y-intercept is 6 and the slope is -3. The equation is therefore

$$y = -3x + 6.$$

In pure mathematics, x and y usually represent abstract numbers, but we can also use the same ideas in discussing relationships between physical quantities: heights, times, speeds, masses, and so on. Suppose h represents height and t represents time. If the equation between h and t is linear ($h = 2t + 4$, for instance), we know that a graph showing h as a function of time will be a straight line and that its slope is 2 and its y-intercept (or its h-intercept in this case) is 4. Conversely, if we look at the graph and see that it is a straight line and then determine its slope and intercept, we can easily write an equation relating the two quantities. We must be careful, though, to keep track of the units, for if the two quantities being studied have units, the slope and the intercept will

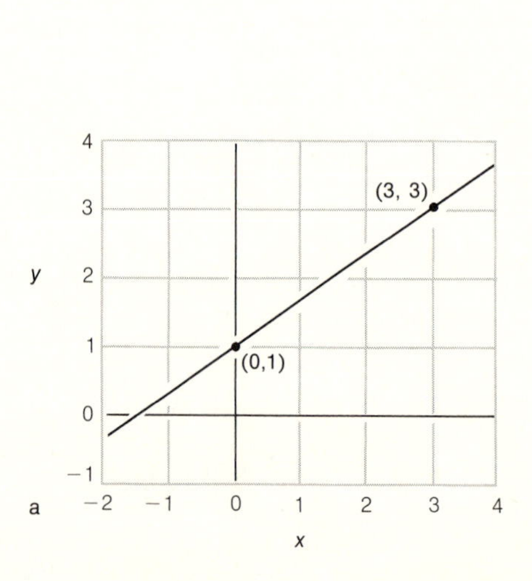

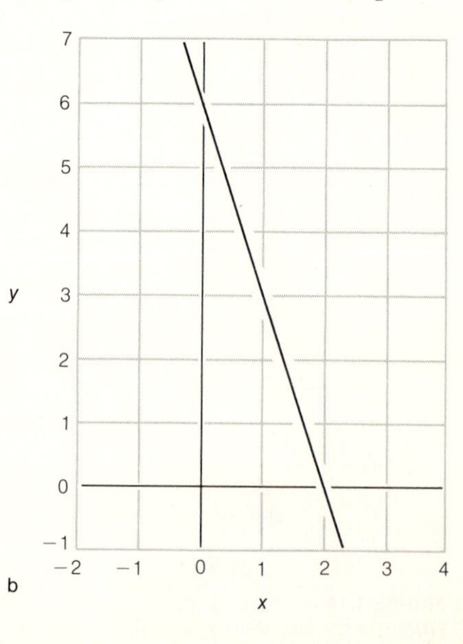

FIGURE 1.21
Two examples of straight-line graphs.

also have units. Consider Figure 1.22a. The h-intercept is +4 ft. We can calculate the slope from its definition (Equation 1.8), using the coordinates of any two points:

$$\text{slope} = \frac{h_2 - h_1}{t_2 - t_1} = \frac{8\text{ ft} - 4\text{ ft}}{2\text{ sec} - 0\text{ sec}} = \frac{4\text{ ft}}{2\text{ sec}} = 2\text{ ft/sec}.$$

Thus we should write the equation of this line as

$$h = [(2\text{ ft/sec}) \times t] + 4\text{ ft},$$

where it is understood that h and t are measured in feet and seconds. We may choose to write the equation simply as $h = 2t + 4$, but it must be remembered that the numbers 2 and 4 really have units attached to them, and it must be made clear, at least from the context, what these units are. In this example, the slope is the time rate of change of h; the slope is the *speed*. This motion is that of something whose height is 4 ft at t = 0 sec, that rises at a steady speed of 2 ft/sec.

If the two related quantities are simply pure numbers, then we can always make a graph in which the scales on the vertical and horizontal axes are identical. From the appearance of the graph, we can estimate the slope of the line. A line sloping upwards at an angle of 45°, for instance, has a slope of 1. However, if the two quantities are physical quantities of different sorts, we must be very careful. One inch on the vertical axis may represent 1 ft, but, in this example, one inch on the horizontal axis represents a *time*, something completely different. Figure 1.22b shows the same relationship between h and t as does Figure 1.22a. The equations of the two lines are identical, and so are the slopes (2 ft/sec). We always follow the definition of slope given in Equation 1.8; the same numbers can be used to calculate the slope of the line in Figure 1.22b as in 1.22a. The two figures contain exactly the same information, even though one of the lines is steeper than the other.

Whether the two related quantities are simply numbers or whether they are physical quantities, the important conclusions of this discussion are the following. If the equation is a linear one (for example, $y = 2x + 5$, $h = 3t + 7$, or $N = 85p + 117$), the corresponding graph must be a straight line, and we can determine the slope and intercept of the line by inspecting the equation. Conversely, if the graph (of y versus x, h versus t, or N versus p) is a straight line, we need only make some simple measurements on the graph to find the slope and the intercept on the vertical axis, and then we can write down the equation relating the two quantities.

There is one more very important point about the use of graphs. Even when the relationship between two variables is nonlinear, we can often (but not always) find a way of constructing a graph so that a straight line will be obtained. An example was given in §1.2.D. Consider another example. Suppose that y and x are related by the equation $y = \frac{1}{2}x^2 + 3$.

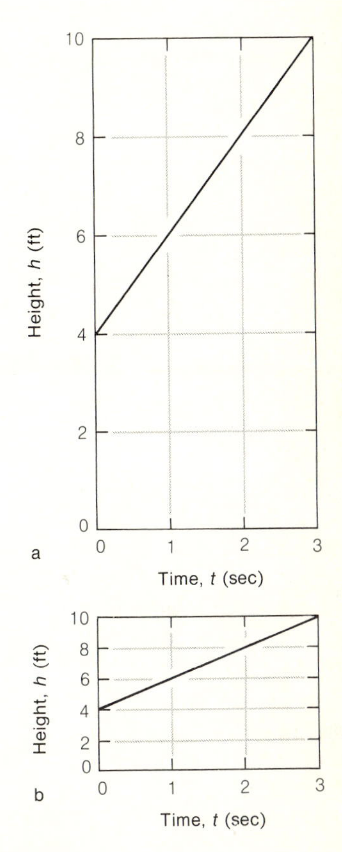

FIGURE 1.22
Two graphs of the equation $h = 2t + 4$.

A graph of y versus x (Figure 1.23a) is not straight. However, if we calculate corresponding values of y and x^2, as shown in the table, and

x	x^2	y
0	0	3
1	1	3.5
2	4	5
3	9	7.5

make a graph of y versus x^2, then we do get a straight line, as shown in Figure 1.23b, a line whose slope is $\frac{1}{2}$ and whose y-intercept is 3. This idea is most useful when used the other way around, for the purpose of discovering the equation relating the two variables. Suppose that we did not know the equation but had only a table of a few values of x and y. We could use the table to make a graph of y versus x, but we could also use it to make other graphs: y versus x^2, y^2 versus x, and so on. In this example, the graph of y versus x^2 turns out to be a straight line with a slope of $\frac{1}{2}$ and an intercept of 3. The relationship between y and x^2 is linear, and we can readily see that the equation is $y = \frac{1}{2}x^2 + 3$. This method of finding an equation relating two variables may seem to be just a "trick"; it is a trick, but an extremely useful one. The difficulty with using it in practice is that we have no sure way of knowing what kind of graph to make in order to get a straight line, and it may be that there is *no* simple way of getting a straight-line graph.

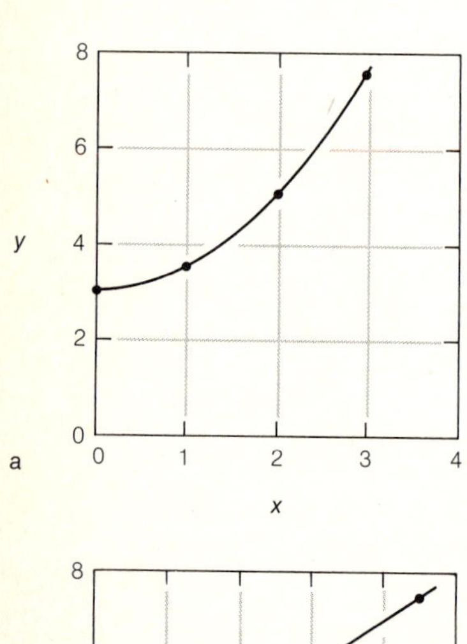

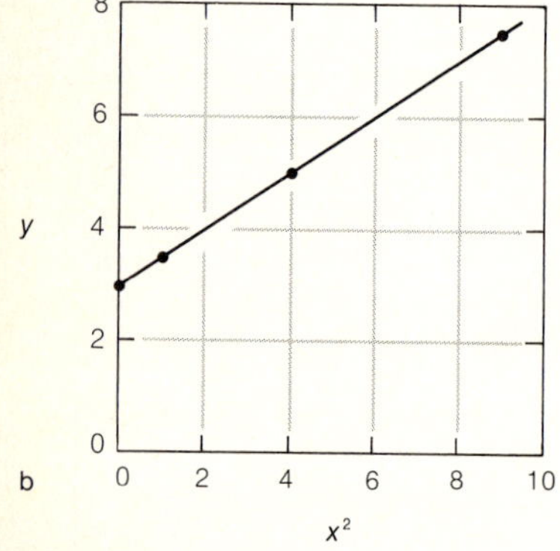

FIGURE 1.23
Two ways of showing the relationship $y = \frac{1}{2}x^2 + 3$: (a) y versus x; (b) y versus x^2.

Let us return to the special case of linear equations in which the y-intercept (b) is 0, equations whose graphs are straight lines passing through the origin. In such equations, y is described as being directly proportional to x ($y \propto x$): if x is doubled, so is y; if x is tripled, so is y, and so on. For instance, the equation $y = 6x$ is represented by a straight line through the origin, with a slope of 6. y is directly proportional to x ($y \propto x$) with a proportionality constant of 6, another way of saying that y and x are related by the equation $y = 6x$. Direct proportionality is a special type of linear relationship. If y is directly proportional to x, then the relationship between x and y is a linear one, but the converse is not true. (If $y = 2x + 3$, x and y are linearly related, but y is not proportional to x. If x increases from 1 to 2, the value of y is not doubled.)

If the equation is one such as $y = 3/x$, y is described as being inversely proportional to x ($y \propto 1/x$). If x is doubled, y is halved. This is a nonlinear relationship between x and y; the graph of y versus x is not a straight line. If the equation is $y = 3x^2$, y is proportional to the square of x ($y \propto x^2$). If x is doubled, y is increased by a factor of 4. Again, the relationship between x and y is nonlinear, and a graph of y versus x is not a straight line. (A graph of y versus x^2 *is* a straight line, however.)

It should be noted that any statement of proportionality can always be rewritten as an equation: for example, $y \propto x$ can be rewritten as $y = kx$, where k is some constant. We may or may not know the value of k, the proportionality constant, but we can always introduce a symbol for it. Similarly, the statement $y \propto 1/x^2$ can be rewritten as an equation, $y = k'/x^2$, where k' is some constant. One remark about terminology: if a statement is made about proportionality with no qualifications, it is always a *direct* proportionality which is meant.

FURTHER QUESTIONS

1.6 The number 8.65×10^9 can be written in various other ways, for instance as 86.5×10^8. Express the same number using each of the following powers of 10: (a) 10^6; (b) 10^{10}; (c) 10^{11}; (d) 10^{-2}.

1.7 Use a combination of powers-of-10 notation and approximate arithmetic to calculate the values of the following expressions. The results (with two significant figures) are given at the end of this section.

(a) $\dfrac{4000 \times 0.0005 \times 392}{75000 \times 22}$;

(b) $\dfrac{0.00027 \times 1200}{0.004 \times 85}$;

(c) $365 \times 86400 \times 0.0037$;

(d) $\dfrac{\sqrt{6500}}{1560}$;

(e) $\dfrac{260}{13.5 \times 0.00078}$.

1.8 Make an estimate of the total area occupied by buildings in your city. How many significant figures should be retained in the result?

1.9 The term "percent" means literally "per hundred." "47% of the students are males" means that of every 100 students, 47 are males; that is, the fraction of males is 0.47. Many fractions can be expressed either as ordinary fractions or as percentages. The efficiency of a certain motor (the ratio of useful energy output to energy input) is 0.35 or 35%; the two statements are equivalent.

(a) Express the following fractions as percentages: 0.5, 0.99, 0.01, 10^{-5}.

(b) Express the following percentages as fractions: 5%, 0.1%, 0.0028%, 10^{-6} %.

1.10 As applied to the number 2, the fundamental rules for dealing with exponents are:

$$2^a \times 2^b = 2^{a+b};$$

$$\frac{2^a}{2^b} = 2^{a-b};$$

$$(2^a)^b = 2^{ab}.$$

Construct some examples to illustrate these rules, using various combinations of positive and negative integers for a and b.

1.11 The 1500-meter race is often referred to as the "metric mile." Is this distance less than or greater than a mile? What is the percentage difference?

1.12 The speed of light is 3×10^8 m/sec. What is this speed in inches per century?

1.13 A "light-year" is a unit of distance, the distance light can travel in one year at a speed of 3×10^8 m/sec. Make a conversion table that gives the value of 1 light-year in centimeters, meters, miles, and kilometers.

1.14 The Alaska pipeline is 790 miles long and has a radius of 24 in. What is the volume of this pipe in cubic inches? in cubic miles?

1.15 Given that the mass of the earth is 6.6×10^{21} tons and its radius 4000 miles, estimate the average density of the earth in g/cm^3.

1.16 Two different sources each give estimates of the volume of water per unit time carried by the Gulf Stream. One is 10^9 ft^3/sec, and the other is 2200 km^3/day. Do the two sources agree with each other? (Do not forget that even if the two figures are in approximate agreement with one another, they might *both* be wrong. Erroneous results are all too frequently accepted, copied, converted into other units, and passed on to the unsuspecting reader.)

1.17 Sketch the graphs of the following equations:

(a) $y = -2x + 7$;

(b) $y = \frac{1}{2}x - 3$;

(c) $y = 2x - 5$;

(d) $y = x^2 - 5$.

1.18 Three different straight lines (a, b, and c) are shown in the following figure. Determine the slope and intercept of each one, and write equations relating y and x.

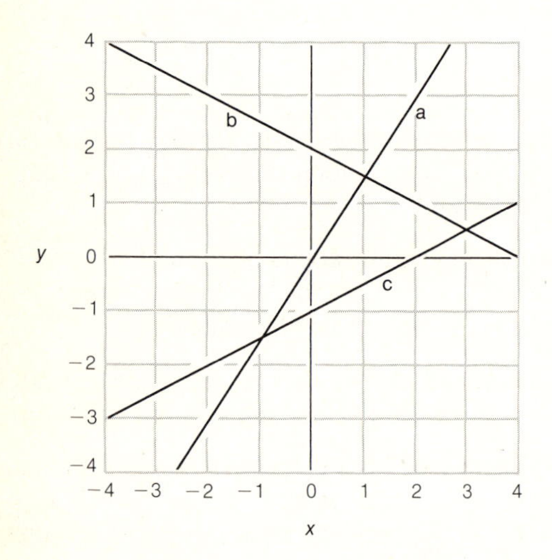

1.19 The following are various possible descriptions of the relationship between two variables:

(a) the relationship between x and y is linear;

(b) the relationship between x and y is nonlinear;

(c) $y \propto x$;

(d) $y \propto x^2$;

(e) $y \propto \frac{1}{x^2}$.

Which of the preceding descriptions can be properly applied to each of the following equations?

(1) $y = 16x$;

(2) $y = 3x^2$;

(3) $y = \frac{3}{x^2}$;

(4) $y = 2x^2 - 7$;

(5) $y = 2x - 7$.

1.20 A driver sets out on a trip at a time called $t = 0$ hours and drives at a steady speed of 50 miles/hr for 10 hours. At the beginning of the trip, the odometer read 1300 miles. Make a graph showing the odometer reading as a function of t. What is the slope of this straight-line graph? Write an equation relating the odometer reading and the time. What units should be given to the various terms in this equation?

1.21 The average retail price of gasoline for a number of years is given in Appendix P. Make two different graphs of these data, first a graph designed to dramatize recent increases in price, and second a graph designed to show that these increases have been relatively insignificant.

1.22 A TV rental service will rent you a set for a basic monthly charge of \$5 plus a charge of 10¢ per hour of use. Plot a graph showing the size of the monthly bill as a function of the number of hours used. Write an equation to describe this relationship. Specify the units of all terms in the equation.

1.23 The following data show a boy's height, h, as a function of his age, A.

A (years)	h (inches)
2	34.0
3	36.6
4	39.1
5	41.7

Make a graph of these data, and use the graph to estimate his height at the time he was born, at the age of 3.5 years, at the age of 6 years, and at the age of 20 years. Find an equation relating h and A, and use it to test the estimates made from the graph. Discuss the reliability of the various estimates. What would be the "hidden text" accompanying your equation?

1.24 The data below show the areas, A, of various circles of radius, r. A simple graph of A versus r is not a straight line; can you plot these data in a way that *does* give a straight line?

r (cm)	A (cm^2)
2	12.6
4	50.3
7	154
8	201

1.25 The number of miles of railroad track in operation in the United States for the period 1860–1890 is shown below.

Year	Miles of track
1860	30,626
1870	52,922
1880	93,262
1890	166,703

Was this growth exponential? (Make a semilog graph.) With what doubling time? By extrapolating this growth pattern, "predict" the date at which the entire area of the 48 states will be completely covered with railroad tracks. (Some specific value must be assumed for the width of the right of way needed for a track.)

1.26 The total number of cars in use in the United States is shown below. Make a semilog graph to determine whether this number has been growing exponentially, and if so, with what doubling time.

Year	Millions of cars
1940	27.5
1950	40.3
1960	61.7
1970	89.2

1.27 It seems "intuitively obvious" that a quantity increasing at the rate of 10% per year will double in size in approximately 10 years. This is not true, however; it will double in about seven years. Explain briefly why the "obvious" conclusion is false.

1.28 In signing on with a ship's staff for a two-year cruise, you are offered the choice of two salary plans:

(a) $1000 for the first month, $2000 for the second month, $3000 for the third, and so on;

(b) $1 for the first month, $2 for the second, $4 for the third, $8 for the fourth, and so on.

Which plan will produce the greater total pay?

1.29 According to the legend, Dutch explorers purchased Manhattan from the Indians in 1626 for about $24 worth of beads. If the Indians had put their money in a savings bank that paid an annual interest of 6%, what would their balance be now? (There is a hard way to do such a calculation: $24 \times 1.06 \times 1.06 \times 1.06 \ldots$. It is easier to find the approximate value of the doubling time and then find how many doubling times have elapsed.)

1.30 About how much wheat was the king's adviser asking for when he requested 1 grain for the first square of a chessboard, 2 grains for the second, 4 for the third, and so on?

(a) Show first that the number of grains requested for the 64th square is 2^{63}.

(b) Show that the number of grains requested for any one square is approximately equal to the total requested for all the preceding squares together, and that therefore the total number of grains requested is almost exactly twice the number requested for the last square, that is, $2 \times 2^{63} = 2^{64}$.

(c) A trick can be used to obtain an approximate value for this number in powers-of-10 notation. 2^{10} is very nearly an exact power of 10: $2^{10} = 1024 \simeq 10^3$. Using this approximation,

$$2^{64} = 2^4 \times 2^{60} = 2^4 \times (2^{10})^6 \simeq 2^4 \times (10^3)^6 \simeq 16 \times 10^{18}.$$

If this wheat were equally divided among all of the people now living in the world, how many grains of wheat would each person receive? Is this enough wheat to feed a person for a day? For a year? (Make an approximate guess of the weight of a grain of wheat, say a thousandth of an ounce.)

1.31 (a) In recent decades, the population of the world has been increasing exponentially with a doubling time of about 35 years. By what percentage does the population increase each year?

(b) If this growth were to continue, by what date would we be standing shoulder-to-shoulder (one person per square foot)? (The 1973 world population is given in Table K.1 and the land area of the world in Table E.1.)

(c) How much additional time would we gain by covering the oceans with floating platforms?

(d) How soon would the mass of the world's population be equal to the mass of the earth itself?

1.32 Make a semilog graph of the data in Table 1.2, showing the growth of a population of bacteria.

1.33 When we make a "semilog graph of y versus x," what we are in fact making is a graph of the *logarithm* of y as a function of x. Suppose that y and x are related as follows:

x	y
0	10
1	100
2	1,000
3	10,000
–	–
–	–

Since y increases by a factor of 10 every time x increases by 1, this is an example of exponential growth. Define z as the logarithm of y to the base 10. Then make an ordinary graph of z versus x; it should be a straight line.

Answers to Question 1.7

(a) 4.8×10^{-4}; (b) 0.95; (c) 1.2×10^{5}; (d) 5.2×10^{-2}; (e) 2.5×10^{4}.

2

Conservation of Mechanical Energy: Kinetic Energy and Gravitational Potential Energy

William Tenney

2

Conservation of Mechanical Energy: Kinetic Energy and Gravitational Potential Energy

There is a boy throwing a ball into the air. This event conveys great beauty to me—it elates me to watch that ball hovering in the air on its path of rhythmic precision, and the movement of its flight is as much alive to me as when I see the flight of a bird. I find the vivid energy of the boy's hand continuing its life in the speed of the ball, and my experience of it is as real to me as my breathing. What is this phenomenon and how can I let others see and live through it as I do in my experience?

—Naum Gabo, *Of Divers Arts* (1962)

§2.1 INTRODUCTION

The concept of energy was first introduced in the study of motion, of the ways in which things move and how their motions are influenced by other objects. Physics is concerned with motions of all kinds, motions spanning a vast range of times and distances and speeds. The motions of electrons and the nuclei of atoms, the motions of cars, baseballs, and rockets, the motions of the planets around the sun, the motion of our sun and of other stars within our galaxy—all these motions are of interest to physicists. The study of physics is not a cataloging of the kinds of motion that occur in nature, but rather it is a search for patterns and regularities, for physical "laws" to describe the features shared by the motions of electrons, apples, and planets. The law of conservation of energy is one of the most important unifying principles we have, probably *the* most important principle, for describing these phenomena. Energy is not something that can be defined in a single brief statement. We must begin by studying some examples of motion; we will gradually develop and broaden the concept as we proceed.

It is clear that we should investigate rather *simple* motions to begin with, but it is not easy to say exactly what is meant by "simple." Something that is simple from one point of view is highly complex from another. Motions such as those of the earth around the sun, of a baseball thrown through the air, or of a pendulum swinging back and forth, are simple enough for study, provided that we do not insist on looking at all the details right from the start. If we were to examine the details of how the air affects the motion of a baseball, or if we were to inquire into the motions of the electrons and atomic nuclei of which the ball is composed, the simple example of a moving ball would become very complex. We shall begin by ignoring the microscopic composition of the objects to be studied, and concentrate on the macroscopic objects themselves.* Furthermore, we will often describe a moving object by picking out one point in the object, such as the center of a ball, and describing the motion of this point while ignoring effects which may arise from the ball's spin about its center. In selecting various motions for study, we should choose those that are sufficiently *simple* that they can be measured and analyzed, and yet that are *complicated* enough to be interesting, so that analysis may yield principles of general interest, that can be applied to other motions.

§2.2 THE REPETITIVE BEHAVIOR OF A SWINGING PENDULUM

Consider first the motion of a *pendulum*. Any object suspended in such a way that it is free to swing back and forth is a pendulum. A child on a swing, a rock tied to a piece of string and suspended from a finger, the pendulum of a grandfather clock, a light fixture hung from the ceiling by a wire—each of these is an example of a pendulum. From one point of view, the back-and-forth motion of a rock on the end of a string is not very exciting; it is, after all, only a rock tied to a piece of string, but if we are willing to think about this motion, to make measurements and to look for patterns in the data, such a motion can be very interesting. The motion of a pendulum does satisfy the criterion of being simple enough for its regularities to be discovered, but complicated enough to be interesting and to suggest principles that can help us understand other types of motion.

A fundamental and obvious fact about the motion of a pendulum is that its speed varies as it swings back and forth, that there is a relationship between its speed and its height above the ground. Its speed is greatest when it is closest to the ground, and least when it is farthest from the ground; at the end of each swing, it stops momentarily as it

*The terms *microscopic* and *macroscopic* will be used to distinguish two points of view which we may adopt in studying various phenomena. Microscopic is used in the general sense of "very small" (rather than to refer specifically to what can be seen through a microscope); from a microscopic perspective, a baseball consists of a tremendous number of atoms, which are themselves composed of still smaller particles. From a macroscopic, large-scale, point of view, a baseball is merely an approximately spherical object that weighs about 5 oz and has a diameter of about 3 in.

FIGURE 2.1
Stroboscopic photograph of a swinging pendulum. [Education Development Center.]

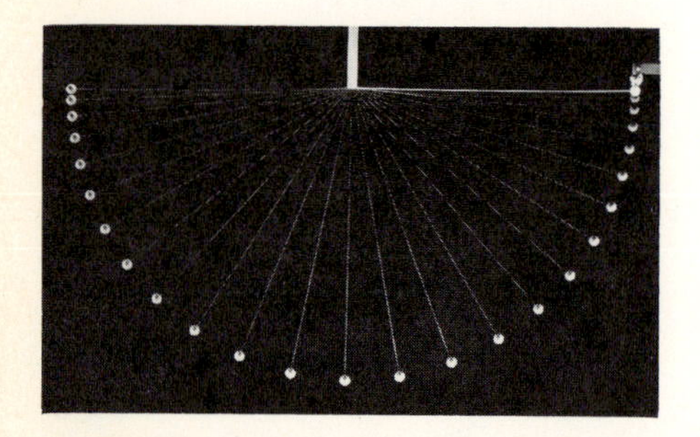

FIGURE 2.2
Stroboscopic photograph of a pendulum swinging through a large angle. Length of pendulum: 1.3 m. Time interval between flashes: 0.05 sec. [Education Development Center.]

reverses direction to begin a new swing. The relationship between speed and height is clearly shown in Figures 2.1 and 2.2, stroboscopic photographs of two pendulums, one of which is swinging through a rather small angle and the other through a large one. Each photograph was made by taking a time exposure of a single swing, with the pendulum illuminated by brief flashes of light, occurring at equally spaced instants of time. The farther apart the images, the greater the distance covered by the pendulum between flashes—and the greater the speed. At either end of the swing, where the pendulum's speed is the least, the images are closest together.

The qualitative nature of a pendulum's motion is illustrated in the figures. Notice that when it is at point A in Figure 2.1, it is at its maximum height. As it swings downward toward B, the speed increases. Then the pendulum rises toward C, losing speed as the height increases, until the speed momentarily becomes zero at the end of its swing. But speed has been lost only temporarily, for the pendulum reverses direction and swings downward again, losing height but regaining speed as it moves to the right. This pattern of events is repeated again and again as the pendulum swings back and forth. In some sense, in the course of the pendulum's motion, speed and height are converted back and forth: as the pendulum moves downward toward the center, height is converted into speed, and as it moves upward, speed is converted into height. The speed that it has at B is not permanently lost as it rises, but is temporarily "stored" in the form of height; the height that it has at A or C is not permanently lost as it swings downward, because the pendulum somehow retains the ability to rise to the same height again. It is possible that there is some quantity that is "conserved," some combination of "speed-plus-height" that retains the same value throughout the motion.

This description of the motion of the pendulum is intended to suggest a productive approach to a more careful analysis of this motion. It is only in a very general sense that speed and height are converted back and forth, that the combination "speed-plus-height" is constant. It does not make much sense simply to *add* speed and height, any more than it makes sense to add your weight in pounds and your age in years. Yet this description contains the essence of the law of conservation of energy. We will see in §2.3 that it is possible to define a quantity *related* to the speed, a quantity that we shall call kinetic energy, *KE*, and to define another quantity related to the height, a quantity called gravitational potential energy, *GPE*, such that the sum of the two (the "total mechanical energy" of the pendulum) does remain almost constant as the pendulum swings, so that, as it swings back and forth, these two "forms of energy" are converted back and forth from one form to the other.*

*In order to emphasize the relationships between different forms of energy and the fact that they can be converted into one another, special multiple-letter abbreviations such as *KE* and *GPE* have been introduced in this book. Such abbreviations must of course be read as a whole; *KE* does *not* mean "*K* times *E*."

However, it is in fact not quite true that every time the pendulum passes through position B it always has the same speed. If the pendulum is a "poor" one, it may slow down appreciably during a very few swings. If it is a "good" pendulum, it may continue to swing for a long time, but sooner or later it will slow down and eventually come to rest. It is impossible, then, for the sum of kinetic energy and gravitational potential energy to remain precisely constant during the entire motion of the pendulum. This sum, which will in some way represent "speed-plus-height," is large when the pendulum is swinging back and forth with a large amplitude; it is less when the pendulum is swinging with a smaller amplitude, and still less when it has finally come to rest. Furthermore, when we start the pendulum moving, it is our intervention that produces a change in the "speed-plus-height." We push it to give it some initial speed, or we pull it aside, increasing its height. Since there must be an increase in the "energy" of the pendulum when we start it, and a gradual decrease as it slows down, it cannot be true that its energy remains precisely constant.

What attitude shall we take toward these observations, simple facts that raise doubts about the idea of energy even before we have really begun? Let us see how much we can learn by ignoring these effects for the time being, concentrating instead on the fact that during any one swing, or for several successive swings, "speed-plus-height" is *almost* constant. But it is very important that we ignore these effects only temporarily. Ultimately, we must take them into consideration. One of the keys to success in science, and indeed in almost any intellectual endeavor, is to develop a sense of when to concentrate on certain central aspects of a problem, deliberately ignoring certain details, and when (if ever) to incorporate such details into the final solution. For some problems, those aspects that have been temporarily disregarded are so subtle that they need never be explained. For others, we find we can make initial progress by ignoring certain aspects, but when we re-examine them later, we see they provide us with deeper insight into the problem as a whole. This approach turns out to be the most fruitful one for developing the concept of energy.

§2.3 KINETIC ENERGY AND GRAVITATIONAL POTENTIAL ENERGY OF A VERTICALLY THROWN BALL

We shall return to the example of the pendulum in §2.10, after having developed the concepts of kinetic energy and gravitational potential energy in some simpler cases. Consider first the motion of a ball thrown vertically up into the air. We will analyze this example not because it is itself especially important, but rather because a careful study of even this very simple sort of motion reveals patterns and regularities that we will be able to apply in the study of many other motions. We will not make much progress by making qualitative observations; a detailed numerical study is essential and will lead to some general ideas about energy and the law of conservation of energy. The goal is not to learn "why" the ball turns around and falls back to the ground, but rather to find out *how* it moves. Although it is often said that the ball falls "because of gravity" or "because of the gravitational attraction of the earth," such statements should really be understood as alternative ways of saying simply that objects *do* fall toward the earth. In what particular way do they move? What regularities can we find in the motion of one ball that can be applied to the motion of other balls, and to other types of motions?

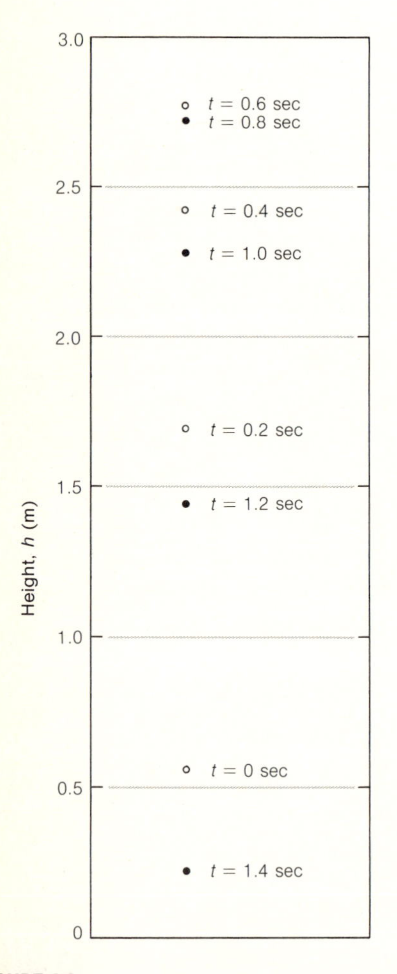

FIGURE 2.3
Representation of a stroboscopic photograph of a vertically thrown baseball ($m = 5$ oz $= 0.14$ kg). The horizontal line at the bottom shows the level from which heights (h) are measured. The time interval between flashes is 0.2 sec. Recorded positions of the ball as it moves upward are designated by the light circles; the dark circles show the positions of the ball as it moves downward. Times (t) are assigned to various positions, beginning with $t = 0$ sec for the first recorded position.

In some respects, the motion of a ball thrown vertically into the air is quite similar to that of a pendulum. The ball rises, gaining height and losing speed, until it reaches its maximum height, where its speed is momentarily zero. Then it falls back to the ground, regaining speed while losing height. This motion is not repetitive like that of a pendulum, but it is simpler to analyze because the up-and-down motion of the ball is confined to a single vertical straight line rather than the arc of a circle. The relationship between speed and height must be studied quantitatively, to see if a definite mathematical regularity can be found. In order to do this, we must obtain numerical values of heights and speeds for particular examples. In other words, it is time to develop a quantitative language for describing motion. Figure 2.3 is adapted from a stroboscopic photograph of a vertically thrown baseball. The height of the ball is recorded as a function of time. The slower speed of the ball near the top of its path is apparent from the diminished spacing between the symbols for the images, and from the figure we can obtain quantitative information about its speed and the corresponding values of height. If we arbitrarily label the time of the first flash $t = 0$ sec, then the times of succeeding flashes are $t = 0.2$ sec, $t = 0.4$ sec, and so on. (Note that $t = 0$ is the time of the first flash of light, not the time at which the ball was thrown.) Table 2.1 shows the height, h, as a function of time. From these data we can obtain information about the variation in speed.

Speed is a quantity that indicates how fast something is moving, and is measured, for example, in miles per hour or in meters per second. In common speech, the word "velocity" is often synonymous with "speed." For scientific purposes, it is useful to differentiate slightly between the two terms. In many problems we are interested not only in how *fast* something is moving but also in its *direction* (up or down, north or east, and so on). The term "velocity" is used when information about the direction of motion is included, whereas "speed" is understood as describing only how fast it is moving. Velocity is an example of a *vector* quantity, a quantity that has a direction as well as a magnitude. A specification of the velocity of an object tells us more about its motion (50 miles/hr toward the east, or 50 miles/hr upward) than does a specification of its speed (50 miles/hr). Speed and velocity are closely related terms and will be denoted by similar symbols. The symbol v will be used for speed, but that for velocity (indicating that information about the direction of motion is to be included as well) will be the same letter but in boldface type: **v**. This notation indicates that the quantity is to be considered as a *vector* quantity and that information about its direction is to be included.* Suppose that at one instant the ball is moving upward at a rate of 2 m/sec and at a later instant downward at 2 m/sec. At both instants, the speed is

$$v = 2 \text{ m/sec.}$$

The velocities, however, are different. At the first instant

$$\mathbf{v} = 2 \text{ m/sec} \quad \text{(upward)},$$

and at the second instant,

$$\mathbf{v} = 2 \text{ m/sec} \quad \text{(downward)}.$$

There is one additional technical term which is often useful for indicating the direction of a vector quantity. If a ball is moving upward at a speed of 2 m/sec, we say that the "*upward component* of its velocity" is +2 m/sec (v_{up} = +2 m/sec), and if it is moving down at the same speed, the upward component of its velocity is −2 m/sec (v_{up} = −2 m/sec). In the preceding example, there are only two directions of the velocity, up and down, and the direction of motion can be conveniently specified by the algebraic sign of the upward component of the velocity. If a motion is not confined to a single straight line, there are many possible directions of motion, and more care is required to specify the direction. Techniques for describing such situations will be developed in Chapter 3.†

If the ball in this example (Table 2.1) were moving at a steady speed, we could simply note its height in meters at the beginning of a time interval and the height at the end of the time interval, subtract the initial height from the final height, and divide by the length of the time interval to give the speed in meters per second. A serious problem arises in this example, however, because the speed is constantly changing. What do we mean by the speed at some particular instant when the speed is quite obviously not constant? The best we can do is to take

TABLE 2.1 The motion of a vertically thrown ball.*

t (sec)	h (m)
0.0	0.56
0.2	1.69
0.4	2.42
0.6	2.77
0.8	2.72
1.0	2.28
1.2	1.44
1.4	0.22

*Data are from Figure 2.3.

*Two other common ways of indicating that a quantity is to be considered as a vector are those of drawing an arrow above the symbol ($\vec{v}$) and drawing a wavy line underneath it ($\underset{\sim}{v}$).

†We frequently need to refer to various aspects of a particular vector quantity such as the velocity: (1) the vector itself, denoted by a symbol printed in boldface type, **v**; (2) the size or *magnitude* of the vector; the speed is the magnitude of the velocity vector and is denoted by v; (3) the component of the vector in a particular direction, such as the upward component of the velocity, v_{up}. Note that the magnitude of a vector is represented by a number (positive or possibly zero, but never negative), and that the component of a vector in any given direction is represented by a number which may be positive, negative or zero. The *magnitude* of a vector is represented by a symbol in ordinary lightfact type (e.g., v) because it is not itself a vector; the same is true of a *component* of a vector.

the change in height that occurs in a short time interval and make the calculation as if the speed were constant during this interval. For example, in Table 2.1, we see that between $t = 0.4$ sec and $t = 0.6$ sec, the height changes from $h = 2.42$ m to $h = 2.77$ m. Thus the calculated speed is

$$v = \frac{2.77\text{ m} - 2.42\text{ m}}{0.6\text{ sec} - 0.4\text{ sec}} = \frac{0.35\text{ m}}{0.2\text{ sec}} = 1.75\text{ m/sec}.$$

Obviously the ball is actually moving faster than 1.75 m/sec at $t = 0.4$ sec and more slowly at $t = 0.6$ sec. Let us assign this speed of 1.75 m/sec to a time halfway between these two times, that is to the instant $t = 0.5$ sec. Furthermore, since our plan is to study the relationship between speed and height, we must decide with what value of height this speed should be associated. Presumably the ball has an instantaneous speed of 1.75 m/sec when it is somewhere between $h = 2.42$ m and $h = 2.77$ m. We will not make a serious error if we assign this value of the speed to a value of height halfway in between, that is, to the height

$$h = \frac{2.42 + 2.77}{2} = 2.595\text{ m}.$$

This is not a completely rigorous procedure, but as long as the instants of time for which data are available are quite closely spaced, no serious error is incurred.*

Let us apply the same procedure to another time interval from Table 2.1, this time the interval from $t = 1.0$ sec to $t = 1.2$ sec. If again we subtract the height at the beginning of the interval from the height at the end of the interval, we obtain

$$\frac{1.44\text{ m} - 2.28\text{ m}}{0.2\text{ sec}} = \frac{-0.84\text{ m}}{0.2\text{ sec}} = -4.2\text{ m/sec}.$$

It is clear that the negative sign arises from the fact that the ball is now falling downward. In other words -4.2 m/sec is the upward component of the velocity, v_{up}. The *speed* is simply 4.2 m/sec. Our major interest is simply in the variation of speed with height, but for future reference it will be useful to retain the information about the direction of motion. Thus Table 2.2, which gives the completed set of calculations based on

TABLE 2.2 Corresponding values of speed and height, derived from Table 2.1.

t (sec)	h (m)	v (m/sec)	v_{up} (m/sec)
0.1	1.125	5.65	+5.65
0.3	2.055	3.65	+3.65
0.5	2.595	1.75	+1.75
0.7	2.745	0.25	−0.25
0.9	2.50	2.2	−2.2
1.1	1.86	4.2	−4.2
1.3	0.83	6.1	−6.1

*How large an error might we be making? How close together in time must our observations be in order that this procedure not lead us astray? A single snapshot is not adequate for determining the ball's speed; we must record its position at least twice, at two different times. In what sense, then, is it meaningful to speak of an *instantaneous* speed? These are some exceedingly delicate questions, questions that bothered Greek philosophers and that provided much of the motivation for the development of the calculus by Newton and his contemporaries. We shall not attempt to give a completely satisfactory resolution of these questions. A perfectly acceptable though not completely rigorous definition of the instantaneous speed of the ball is given by the procedure followed in the text. Take a *small* time interval which includes the instant in question, and take the instantaneous speed to be the distance covered in that time interval divided by the length of the time interval.

Figure 2.3, includes a column showing the upward component of the velocity vector, with the direction of motion indicated by the signs + and −. The second and third columns of Table 2.2 provide the necessary information on the relationship between speed and height. In a graph of these data (Figure 2.4) all the points appear to lie on a single smooth curve, and, as nearly as one can determine from the graph, the speed of the ball at any particular value of h is the same for the downward motion as for the upward motion.

What else can we learn from these data? It would be extremely useful if we could obtain an *equation* relating speed and height, because if that is possible, the equation may reveal regularities that are not apparent from an inspection of the tabulated data or from the graph shown in Figure 2.4. In order to obtain an equation, we draw upon what we know about graphs, including the fact that we can easily use a *straight-line* graph to find an equation. Several aspects of the use and interpretation of graphs were discussed in Chapter 1. For this discussion, the most important of these aspects are the following. (1) Even though it is the relationship between speed and height in which we are interested, we are by no means restricted to making the obvious graph of speed versus height shown in Figure 2.4. We can make a graph of the square of the speed versus the height, or of the speed versus the square of the height, or of any other pair of quantities we choose. Such graphs may or may not be useful, but nothing should inhibit us from trying any of the possibilities to see whether or not any interesting patterns emerge. (2) If we happen to discover a way of plotting the data that gives a straight line, then we can easily find an equation relating the two quantities.

Our initial graph of speed versus height (Figure 2.4) does not give a straight line. It is certainly not obvious which, if any, of the many other possible ways of presenting the same data will give a straight line, but it is easy, though perhaps tedious, to try various ideas. Figure 2.5 shows several other graphs, all based on the same speed and height data from Table 2.2. With the exception of Figure 2.5b (a graph of the square of the speed versus the height), all of these graphs are definitely curved; consequently graph 2.5b is the one we should consider more carefully,

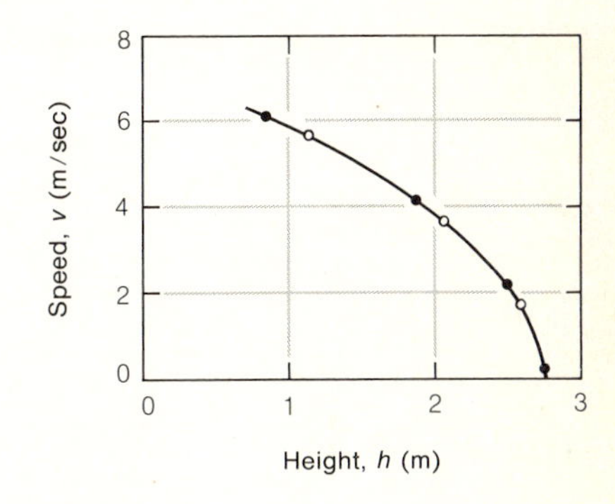

FIGURE 2.4
Graph showing the variation in the speed of a ball with height, based on data from Table 2.2.

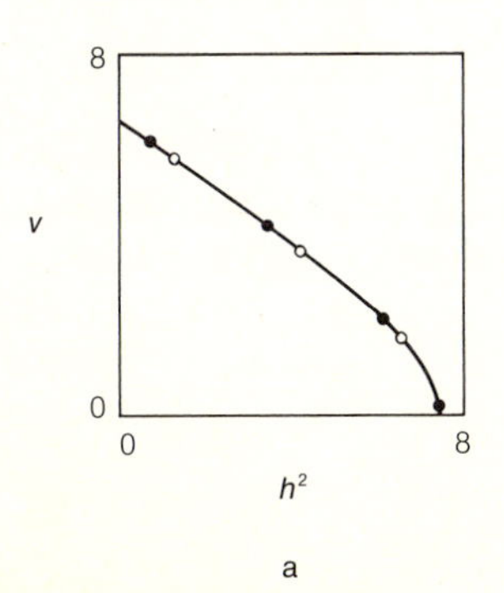

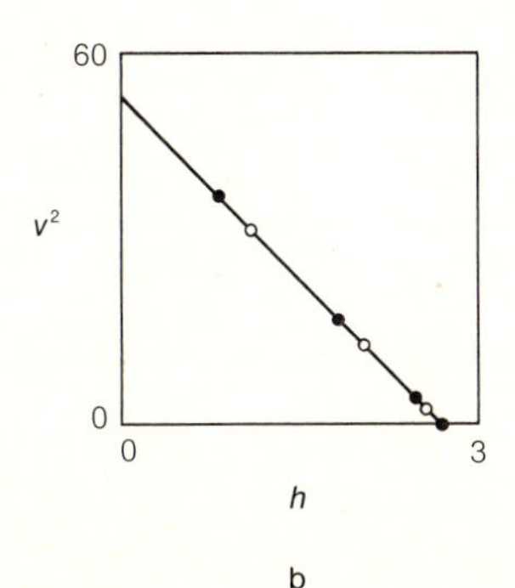

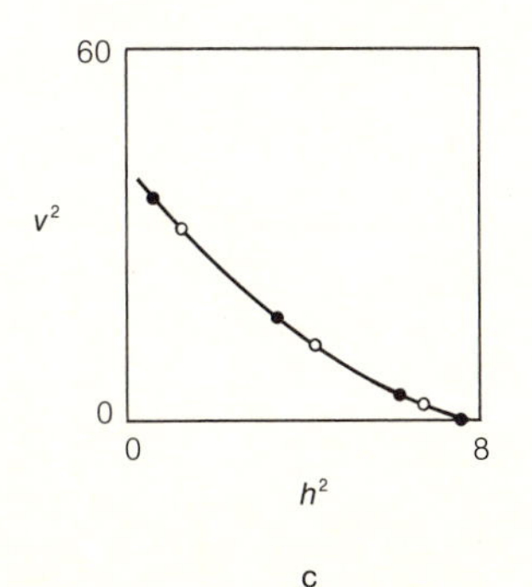

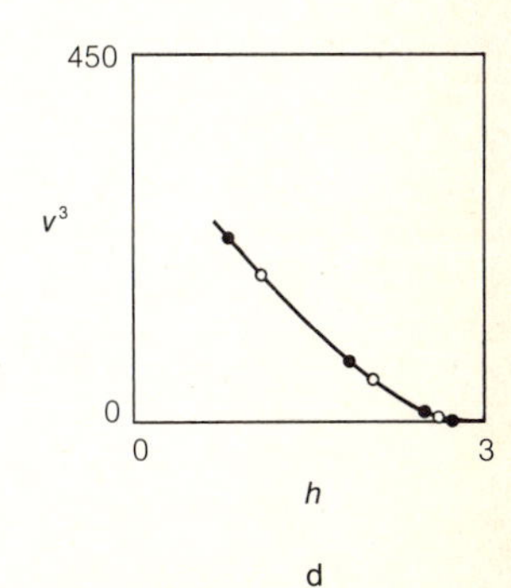

FIGURE 2.5
Various graphic representations of speed and height data from Table 2.2.

and it is shown in more detail in Figure 2.6. (Note that the coordinates correspond to the data in Table 2.2. For example, from the first row of the table, $h = 1.125$ m, and $v^2 = (5.65 \text{ m/sec})^2 = 31.9 \text{ m}^2/\text{sec}^2$, and one of the points on the graph has the coordinates 1.125 and 31.9.)

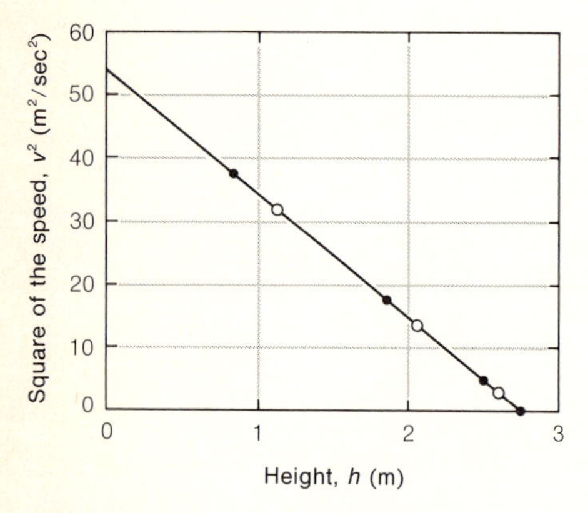

FIGURE 2.6
Graph of v^2 versus h from Table 2.2.

From one point of view, the graph of v^2 versus h (Figure 2.6) is less informative than the original graph of v versus h (Figure 2.4). The speed is a quantity closer to our direct perceptions than is the square of the speed, and if our only objective was to summarize the results of this one particular experiment, Figure 2.4 would be perfectly adequate. However, if our purpose is to obtain some clues from this experiment that are of more general interest, the graph of v^2 versus h is much more useful, because we can use it to derive an equation that relates v and h. The special advantage of having an equation is that, once we have it, we can employ it in various ways that may reveal regularities not previously apparent. For example, we might add equal quantities to both sides of the equation or multiply both sides of the equation by equal quantities and thus obtain new equations. We can look at other motions of the same kind, or at other kinds of motions, and try to find equations for these new examples, to see in what respects the various motions are similar to each other and in what respects they differ.

Using methods discussed in §1.4, we can easily use the slope and intercept of the straight line in Figure 2.6 to find the corresponding equation. Since this is a graph of v^2 versus h, the equation can be written in the form

$$v^2 = \text{slope} \times h + \text{intercept on the vertical axis}. \tag{2.1}$$

The intercept is approximately 54, and its units must be those of the quantity plotted vertically, that is, the same units as v^2, meter² per second². The slope of the line can readily be obtained from the points at which the line crosses the two axes:

$$\begin{aligned} \text{slope} &= \frac{0 \text{ m}^2/\text{sec}^2 - 54 \text{ m}^2/\text{sec}^2}{2.75 \text{ m} - 0 \text{ m}} \\ &\simeq -19.6 \, \frac{\text{m}^2/\text{sec}^2}{\text{m}} \\ &= -19.6 \text{ m/sec}^2. \end{aligned}$$

Just as the intercept has units and dimensions, so too does the slope, for the slope in this case is obtained by dividing a change in the square of a velocity by a change in a distance. The slope has a negative sign, as expected from the fact that the straight line slopes downward to the right.

Thus, if we put these results into Equation 2.1, we have an equation that describes the straight line:

$$v^2 = -19.6h + 54, \tag{2.2}$$

where the units attached to the numbers 19.6 and 54 have been omitted for the present. Such an equation gives a concise description of the relationship between speed and height in this particular experiment. We can use it to calculate the speed at a particular height or the height which the ball had at the time its speed had some specified value. A result of considerably more interest emerges when we take the apparently trivial step of adding $19.6h$ to both sides of Equation 2.2, so that the quantities that are varying are grouped together:

$$v^2 + 19.6h = 54. \qquad (2.3)$$

In this equation, we have the essentials of the idea of conservation of energy. As the ball moves, both its height, h, and its speed, v, vary, but always in precisely such a way that the rather peculiar looking combination, $v^2 + 19.6h$, keeps the same numerical value (a value of 54 in this case). As the ball goes up, the first term, v^2, decreases and the second one, $19.6h$, increases, the decrease in one term always being offset by the increase in the other. As the ball falls, the reverse occurs. This behavior of the numerical values can be seen in Table 2.3, which is based on the same data for v and h (Table 2.2) that we have been using all along. The entries in the last column of Table 2.3 are not all *precisely* the same, but they are sufficiently close to one another that it is reasonable to attribute the minor deviations to inaccuracies in measurement and to rounding errors in the arithmetic.

The result that the particular combination, $v^2 + 19.6h$, retains the same numerical value throughout the motion is much more specific than the qualitative observation that as the height increases, the speed decreases, and vice versa. We now see one possible meaning of the term "speed-plus-height" introduced in §2.2: the square of the speed plus the height multiplied by the factor 19.6. We could not have arrived at this specific conclusion without having made some fairly careful measurements, or—equally important—without having carried out a thoughtful analysis of the data.

Although we are most interested in the numerical regularities discovered in such data, it is very helpful to introduce some terminology for describing these regularities. The word "energy" has not yet been

TABLE 2.3 Calculations based on Table 2.2.

t	h	v	v^2	$19.6h$	$v^2 + 19.6h$
0.1	1.125	5.65	31.9	22.1	54.0
0.3	2.055	3.65	13.3	40.3	53.6
0.5	2.595	1.75	3.06	50.9	54.0
0.7	2.745	0.25	0.0625	53.8	53.9
0.9	2.50	2.2	4.84	49.0	53.8
1.1	1.86	4.2	17.6	36.5	54.1
1.3	0.83	6.1	37.2	16.3	53.5

used in referring to Equation 2.3, but it is from this equation, and from others like it, which arise in the analysis of other experiments, that the whole idea of energy emerges. We will say that this ball has two "kinds of energy" during its motion, one type related to its speed and the other to its height above the ground. It is clear that the energy that it has as a result of its motion, which we will call kinetic energy, *KE*, should be defined either as v^2 or as some quantity *proportional* to v^2. Defining kinetic energy merely as equal to v^2 would be perfectly adequate as long as we were concerned with only one particular object. However, we will surely want our definition of kinetic energy to apply to all objects, and it is certainly reasonable to suppose that if we have various objects, all moving with the same speed, we should ascribe more kinetic energy to the heavier objects than to the light objects. Consider several objects joined together; if kinetic energy is defined as proportional to mass, then the kinetic energy of the composite object is equal to the sum of the kinetic energies of the individual parts.

Therefore, we will define kinetic energy as proportional to the mass (m) of the object as well as to the square of the speed. We could then define kinetic energy simply as mv^2. Historically, this definition was the first to be used, but for reasons that will become clear shortly, it has become conventional to define kinetic energy as just one *half* of mv^2, that is,

$$KE = \tfrac{1}{2}mv^2. \tag{2.4}$$

The inclusion of the factor $\frac{1}{2}$ in the definition of kinetic energy may seem quite arbitrary and totally pointless. We will see later why this particular definition is the most useful. In principle, we have complete freedom in making definitions. Kinetic energy can be defined in any way we choose, but one particular definition may turn out to be more *useful* than another. "Right" and "wrong" are meaningful adjectives with which to describe definitions only in the important sense that people who wish to communicate with each other must agree on their definitions; once a definition has been adopted, no matter how arbitrary the reasons for choosing it may have been, any other definition is wrong unless all parties agree to the change.

The dimensions of kinetic energy are

$$\text{mass} \times \text{velocity}^2 = \text{mass} \times \frac{\text{length}^2}{\text{time}^2}.$$

In the MKS system of units, the units of kinetic energy are

$$\text{kg} \times \frac{\text{m}^2}{\text{sec}^2}.$$

Any form of energy must be a quantity with the same dimensions as kinetic energy, and if two or more quantities of energy are to be added or compared, all their numerical values must be expressed in the same units. Even though energies can be expressed in terms of the basic units of mass, length and time, energy is such an important quantity that a variety of names have been adopted for particular quantities of energy. The MKS unit of energy is the joule (abbreviated as J):

$$1 \text{ J} = 1 \text{ kg-m}^2/\text{sec}^2.$$

For example, at the moment when the ball we have been considering, whose mass is 0.14 kg, has a speed of 3.65 m/sec, its kinetic energy is the following:

$$KE = \tfrac{1}{2} \times 0.14 \times (3.65)^2 \simeq 0.93 \text{ J}.$$

Having now adopted a definition of kinetic energy, we can modify the equation describing the motion of the baseball to get an equation in which the first term is the kinetic energy of the ball. If we multiply both sides of Equation 2.3 by $\frac{1}{2}m$, it becomes

$$\tfrac{1}{2}mv^2 + \tfrac{1}{2}m \times 19.6 \times h = \tfrac{1}{2}m \times 54 = 0.07 \times 54 = 3.8$$

or

$$\tfrac{1}{2}mv^2 + m \times 9.8 \times h = 3.8 \text{ J}. \qquad (2.5)$$

The first term on the left side of this equation ($\frac{1}{2}mv^2$) is the kinetic energy. The second term ($m \times 9.8 \times h$) we also label as a "form" of energy. It is the energy that an object has more of when its height is greater, an energy that we attribute to the effects of the earth's gravity and that we call gravitational potential energy, *GPE.* (The word "potential" reflects the fact that kinetic energy is in some sense the most fundamental form of energy; an elevated object is *potentially* capable of acquiring more kinetic energy by falling to a lower level.)

We can then describe the motion of the ball as one in which kinetic energy is converted into gravitational potential energy as the ball rises, and in which gravitational potential energy is converted back into kinetic energy as the ball falls, the sum of these two forms of energy remaining the same throughout the motion. The number 9.8 (which is just half of the number 19.6 and must therefore be given the same units, m/sec^2) is derived from our numerical analysis of the motion of the ball, and it is one that appears again and again in descriptions of motion under the influence of the earth's gravity. When experiments such as this one are repeated—say, when the same ball is thrown upward at a different initial speed, or when another ball, heavier or lighter, is used—it is always found that the equation describing the motion can be put into a form exactly like that of Equation 2.5, with the same number 9.8 m/sec^2 but with differing values for the constant on the right-hand side of the equation. The number 9.8 m/sec^2 tells us how the gravitational potential energy of an object varies with height, and it is given the symbol g:

$$g = 9.8 \text{ m/sec}^2.$$

Thus gravitational potential energy is defined as

$$GPE = mgh. \qquad (2.6)$$

We can then write Equation 2.5 as

$$\tfrac{1}{2}mv^2 + mgh = 3.8 \text{ J} \quad \text{(a constant throughout the motion of the ball in this experiment),} \qquad (2.7)$$

or as

$$KE + GPE = E = \text{constant}. \qquad (2.8)$$

The symbol E has been introduced to represent the "total (mechanical) energy" of the ball. This definition of E as the sum of the kinetic energy and the gravitational potential energy will suffice for the present; it will be necessary to include other forms of energy later. For the example we have been analyzing, the variations in kinetic energy and gravitational potential energy and the constancy of E, their sum, are shown explicitly in Table 2.4 and Figure 2.7.

A somewhat different way of expressing the law of conservation of energy (Equation 2.8) comes from the realization that if E keeps the same value throughout the motion, then although the kinetic energy and the gravitational potential energy may be individually changing, the sum of the *changes* in these two quantities for any particular interval of time must be zero. It is convenient to introduce a special symbol, Δ(delta), to denote a *change*. Thus $\Delta(KE)$ denotes the change in the value of kinetic energy. (Δ is not a number, and $\Delta(KE)$ does not mean Δ "times" KE. The expression $\Delta(KE)$ must be read as a whole.) Similarly, if an object moves from $h = 2$ m to $h = 2.5$ m, we describe its change in height by writing $\Delta h = 0.5$ m, or if it moves from $h = 6$ m to $h = 5.5$

TABLE 2.4 Variation of kinetic energy and gravitational potential energy with time.*

t (sec)	h (m)	v (m/sec)	KE ($\frac{1}{2}mv^2$) (J)	GPE (mgh) (J)	$E = KE + GPE$ (J)
0.1	1.125	5.65	2.24	1.54	3.78
0.3	2.055	3.65	0.93	2.82	3.75
0.5	2.595	1.75	0.21	3.56	3.77
0.7	2.745	0.25	0.004	3.77	3.77
0.9	2.50	2.2	0.34	3.43	3.77
1.1	1.86	4.2	1.24	2.55	3.79
1.3	0.83	6.1	2.60	1.14	3.74

*Data are derived from Table 2.3, with $m = 0.14$ kg and $g = 9.8$ m/sec^2. The total mechanical energy ($E = KE + GPE$) is nearly constant.

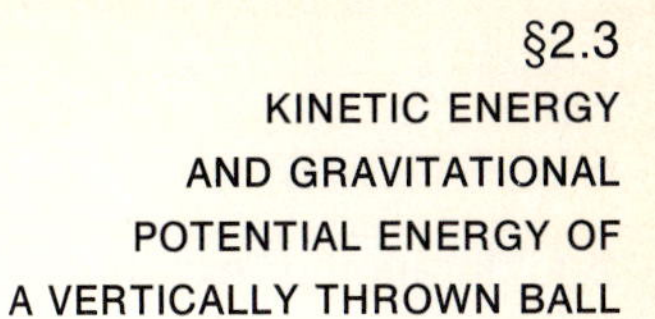

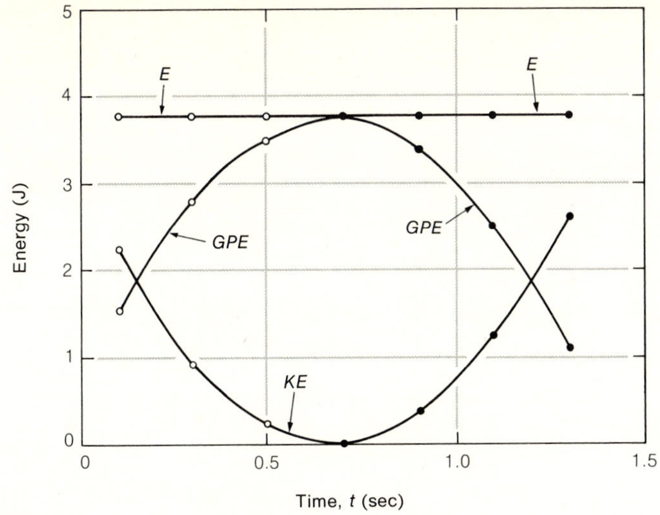

FIGURE 2.7
Variation of *KE* and *GPE*, and the constancy of their sum, *E*.

m, $\Delta h = -0.5$ m. If this notation is used, we can write

$$\Delta(KE) + \Delta(GPE) = \Delta E = 0, \tag{2.9}$$

since ΔE, the change in a constant, must be zero. It is to be understood in equations such as this that the change in kinetic energy, the change in gravitational potential energy, and the change in the total energy are all calculated for the *same* interval of time. We can also write, from Equation 2.9,

$$\Delta(KE) = -\Delta(GPE). \tag{2.10}$$

That is, in any time interval, whatever the change in kinetic energy may be, the gravitational potential energy experiences an equal and opposite change. Referring to Table 2.4, for instance, consider the time interval from $t = 0.1$ sec to $t = 0.9$ sec ($\Delta t = 0.8$ sec). For this interval

$$\Delta(KE) = 0.34 - 2.24 = -1.90 \text{ J}$$

and

$$\Delta(GPE) = 3.43 - 1.54 = +1.89 \text{ J}.$$

With allowance for slight errors, the changes in kinetic energy and gravitational potential energy are equal in size and opposite in sign; the sum of these changes is zero. Note that as long as the value of E is truly constant throughout the motion, the time interval in question need not be a *short* one.

QUESTIONS

2.1 In another experiment like that shown in Figure 2.3, the same ball was used, but the height and time data, shown below, were different.

Time, t (sec)	Height, h (m)
0.0	0.34
0.2	1.22
0.4	1.72
0.6	1.83
0.8	1.54
1.0	0.85

Analyze these data to see whether or not a graph of v^2 versus h is a straight line with a slope of approximately -19.6 m/sec^2. What was the total energy ($E = KE + GPE$) in this experiment? Was the ball initially thrown more rapidly or less rapidly in this experiment than in the one described by Table 2.1?

2.2 (a) Here are some hypothetical "data," supposedly of the same sort as those shown in Table 2.1. Analyze these data, and show that a graph of v^2 versus h does *not* give a straight line.

Time, t (sec)	Height, h (m)
0.0	0.10
0.3	1.42
0.6	2.20
0.9	2.14
1.2	1.26

(b) It is a more challenging problem to take the numbers for v and h and see if you *can* find a way of plotting them to give a straight line. If you can find such a graph, how would you define kinetic energy and gravitational potential energy in a world in which thrown balls behaved in this way?

§2.4 THE SIGNIFICANCE OF g

The most fundamental significance of the quantity g can be seen from its appearance in the definition of gravitational potential energy ($GPE = mgh$): the quantity g tells us how much the gravitational potential energy of a given object changes for a specified change in height. We note that g is not just a pure number (9.8) but a number with dimensions (length/time^2) and units (meter/second^2 in the MKS system). Furthermore, we might well suspect that g is not a truly universal "constant of nature." It appears in the equation that tells us how the speed of a thrown ball varies with height; if the experiment were to be repeated on the surface of the moon, this relationship would be different, and an analysis of a similar experiment would lead to a different result. We shall examine this point in more detail in Chapter 5; for now it will suffice simply to use the value $g = 9.8$ m/sec^2.

What "meaning" can we give to g, aside from its appearance in the definition of gravitational potential energy? There is another way of interpreting g, one that seems to be quite different but is in fact closely related. The units of g are meters per second^2. These are the same units we would use to measure *acceleration,* the rate of change of velocity with time. That is, the acceleration of an object is the change in velocity divided by the length of time during which this change occurred. If, for instance, the velocity changes from 25 m/sec to 28 m/sec during a time interval of 2 sec, the acceleration is

$$\frac{28 \text{ m/sec} - 25 \text{ m/sec}}{2 \text{ sec}} = \frac{3 \text{ m/sec}}{2 \text{ sec}} = 1.5 \frac{\text{m/sec}}{\text{sec}},$$

or, more concisely, 1.5 m/sec^2.

As we will see, we can describe the motion of a vertically thrown ball as one which has a uniform downward acceleration of 9.8 m/sec^2; thus the quantity g gains a new significance. Acceleration, like velocity, is something that has a direction as well as a magnitude. On the downward part of its journey, the ball is falling with increasing speed, and it is said to have an acceleration whose direction is downward. Similarly, on the upward part of its trip it is slowing down at the rate of 9.8 m/sec^2, and its acceleration is also described as being in the downward direction. Adopting the same procedure used in discussing the vector nature of the velocity, we define the "upward component of acceleration" as the rate of change of the upward component of velocity. To find this rate of change, we plot in Figure 2.8 the upward component of velocity as a function of time, using the data from columns 1 and 4 of Table 2.2. Indeed, the upward component of velocity decreases steadily, from +5.65 m/sec to 0, then to negative values. The rate of change can be calculated simply as the *slope* of this line. If we use the two points indicated on Figure 2.8, for instance, this slope is

$$\frac{-6.1 \text{ m/sec} - 5.65 \text{ m/sec}}{1.3 \text{ sec} - 0.1 \text{ sec}} = \frac{-11.75 \text{ m/sec}}{1.2 \text{ sec}} = -9.8 \text{ m/sec}^2.$$

That is, the upward component of acceleration is equal to -9.8 m/sec^2; in simpler terms, the acceleration is in the downward direction and has a magnitude of 9.8 m/sec^2. Thus g represents the magnitude of the downward acceleration due to gravity.

It is curious that if we use this procedure to calculate the acceleration, we find that the whole motion of the ball can be described as one of *constant* acceleration, because during the upward part of its motion the ball is slowing down whereas during its fall it is speeding up. We might say that during the upward part of its motion it loses speed at the rate of 9.8 m/sec^2 and that during its fall it gains speed at this rate, as shown in Figure 2.9. The description of the whole motion as one in which the acceleration is the same at all times, always downward directed and always of the same size, is a useful one, however. One needs simply to realize that a downward acceleration results in an increase in speed at times when the ball's velocity *is* downward, but a decrease in speed when it is moving upward.

If we look at other examples of the same type of motion, we find that no matter what ball we use and no matter what its initial speed—whether it is thrown upward with a large or small initial speed, whether it is simply dropped, or even if it is thrown downward—its motion can always

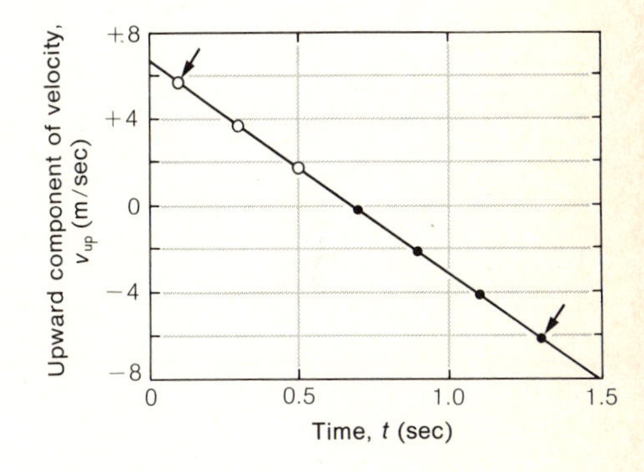

FIGURE 2.8
Variation of the upward component of velocity, v_{up}, with time, *t*, from Table 2.2. The two points marked by the arrows are the ones that have been used to calculate the slope of the line.

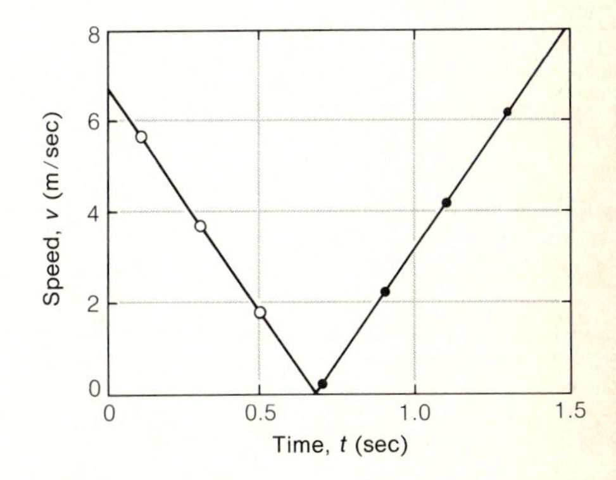

FIGURE 2.9
Variation of speed with time, from Table 2.2.

be described as one with a constant downward acceleration of magnitude 9.8 m/sec^2. Thus the single quantity, g, has two different (though related) meanings: first, as the factor to be included along with mass and height in calculating gravitational potential energy, and second, as the magnitude of the downward acceleration due to gravity. One can now see the advantage of including the factor $\frac{1}{2}$ in the definition of kinetic energy. With this particular choice, the factor g which appeared in the definition of gravitational potential energy turned out to have an alternative interpretation as the downward acceleration due to gravity. A different definition of kinetic energy would have been quite acceptable but would not have led to quite such a simple interpretation.

We have seen from an examination of Figure 2.8 that the quantity g (9.8 m/sec^2) can be interpreted as the downward acceleration due to gravity. There is another way of arriving at the same conclusion. The argument is somewhat subtle but worthy of attention. What we will do is to show that if mechanical energy is conserved, it follows as a consequence that objects *do* fall with such an acceleration. We mean by acceleration the change of velocity per unit time. To be more specific, velocity and acceleration are both *vector* quantities, and if we denote the upward component of the velocity at some time, t_1, by v_1 and at a somewhat later time, t_2, by v_2, then the upward component of acceleration is

$$a = \frac{v_2 - v_1}{t_2 - t_1}.$$

The law of conservation of energy as applied to this case,

$$\tfrac{1}{2}mv^2 + mgh = E = \text{constant}, \qquad (2.11)$$

can be used to tell us how the velocity changes with height. It does not directly tell us how the velocity changes with *time.* Let us see how to extract this information; can we use Equation 2.11 to derive another equation containing the acceleration, $(v_2 - v_1)/(t_2 - t_1)$? In order to calculate the acceleration we need information about the velocities at two closely spaced times, t_1 and t_2. Since $(\frac{1}{2}mv^2 + mgh)$ has the same numerical value throughout the whole motion, then it certainly has the same value at times t_1 and t_2. That is,

$$\tfrac{1}{2}mv_1^2 + mgh_1 = \tfrac{1}{2}mv_2^2 + mgh_2,$$

where h_1 and h_2 are the heights at these two instants. We can put this equation into a form more useful for our present purposes by canceling out the common factor, m, and by regrouping terms so that the two terms containing velocities appear together as do the terms involving heights:

$$\tfrac{1}{2}(v_2^2 - v_1^2) = g(h_1 - h_2). \qquad (2.12)$$

In order to calculate the acceleration we need to know $(v_2 - v_1)$, and this is one of the two factors of the left-hand side of Equation 2.12:

$$\tfrac{1}{2}(v_2^2 - v_1^2) = \tfrac{1}{2}(v_2 - v_1)(v_2 + v_1).$$

Thus we can rewrite Equation 2.12 as

$$\tfrac{1}{2}(v_2 - v_1)(v_2 + v_1) = g(h_1 - h_2).$$

If we now divide both sides of this equation by $(t_2 - t_1)$, we obtain an equation in which the desired quantity, the acceleration, appears:

$$\frac{1}{2}\frac{(v_2 - v_1)(v_2 + v_1)}{(t_2 - t_1)} = g\frac{(h_1 - h_2)}{(t_2 - t_1)}$$

or

$$\tfrac{1}{2}a(v_2 + v_1) = g\frac{(h_1 - h_2)}{(t_2 - t_1)}. \tag{2.13}$$

If the two times, t_1 and t_2, are very close together (as we have assumed from the beginning of the argument), then v_1 and v_2 are very nearly equal to each other, and so, with an error which can be made as small as we wish by taking t_1 and t_2 close enough together, we can say that

$$v_2 + v_1 \simeq 2v,$$

where v stands for v_1 or v_2 or for some velocity in between these two; we need not be more specific since v_1 and v_2 are nearly equal. Thus the left-hand side of Equation 2.13 can be written,

$$\tfrac{1}{2}a(v_2 + v_1) \simeq \tfrac{1}{2}a \times 2v = av.$$

The right-hand side of Equation 2.13 can also be simplified. Except for the algebraic sign, the combination $(h_1 - h_2)/(t_2 - t_1)$ which appears there is just the change in height divided by the length of the time interval, that is, the upward component of the velocity:

$$g\frac{(h_1 - h_2)}{(t_2 - t_1)} = -g\frac{(h_2 - h_1)}{(t_2 - t_1)} = -gv.$$

Again, if t_1 and t_2 are very close together, v is the velocity at either t_1 or t_2 or at some instant in between. Thus Equation 2.13 can be written

$$av = -gv,$$

and after dividing through by v:

$$a = -g.$$

Thus we have derived from the equation that expresses the conservation of mechanical energy ($\tfrac{1}{2}mv^2 + mgh = \text{constant}$) the result that at any instant during this motion, the upward component of the acceleration has the value $-g$, that is, that the acceleration is directed downward and has the magnitude g. It is not a coincidence that g has two meanings: it is one of the factors in the expression for gravitational potential energy

(mgh) and it is also the downward acceleration due to gravity.

The critical reader will notice that it was essential in this argument first to take t_1 and t_2 as being two *different* times, since otherwise $(v_2 - v_1)$ and $(t_2 - t_1)$ would both be zero, and the "acceleration" we wanted to calculate would have appeared as the meaningless fraction $\frac{0}{0}$. At another point in the argument, we assumed that t_1 and t_2 were so close together that the quantity $(v_2 + v_1)$ could be replaced by $2v$ and that $(h_2 - h_1)/(t_2 - t_1)$ could be replaced by v. There is an apparent inconsistency here, but one may argue that any possible error in the result would become smaller and smaller as t_1 and t_2 become closer and closer together. The problems that arise in this argument are closely related to those mentioned in §2.3 in connection with the problem of defining speed when the speed is changing—problems that require calculus for a truly satisfactory resolution.

QUESTIONS

2.3 If an object that is dropped has a uniform downward acceleration of g(9.8 m/sec^2), how long will it take to reach a speed of 60 miles/hr? If it were to continue to fall with the same acceleration, how long would it take to reach a speed equal to that of light (3×10^8 m/sec)? (Obtain a numerical result in seconds and then interpret it. Is it a day? a year? a century?)

2.4 A car accelerating from rest to a speed of 60 miles/hr in 20 sec has an acceleration of

$$a = \frac{60 \text{ miles/hr}}{20 \text{ sec}} = 3 \text{ miles/hr-sec.}$$

These units are perfectly acceptable though rather peculiar. What is this acceleration in miles per hour^2? in miles per second^2? in meters per second^2?

§2.5 LIMITATIONS ON THE LAW OF CONSERVATION OF MECHANICAL ENERGY

Our most important conclusion thus far is that for objects moving along a vertical path under the influence of the earth's gravity, the variations in speed and height are related in such a way that the quantity $(\frac{1}{2}mv^2 + mgh)$, the sum of kinetic energy and gravitational potential energy, retains the same numerical value throughout the motion. We also saw that the quantity g appearing in the definition of gravitational potential energy, a quantity that has the value 9.8 m/sec^2, can be interpreted as the magnitude of the uniform downward acceleration with which objects move when acted upon by gravity.

As pointed out in §1.2.D, all equations—indeed, all "physical laws"—have associated with them a "hidden text," which describes the range of their validity, and tells us what limitations there are to their use. What is the hidden text that goes with the conclusions developed in this chapter? The first limitation is that the experiments described were carried out quite close to the surface of the earth. Our conclusions would require modification if we wished to describe similar experiments carried out on the surface of the moon, or if we tried to describe the motion of a rocket fired upward at a speed great enough to reach an altitude of several thousand miles, an altitude comparable to the dimensions of the earth itself. This is an important limitation, but we shall see in Chapter 5 that our analysis can be extended to such situations by making suitable modifications in the definition of gravitational potential energy. In any event, the definition of gravitational potential energy as mgh is quite satisfactory for the motions of objects that do not get far from the earth's surface.

A second limitation is quite fundamental and considerably more troubling. Even for objects such as the baseball that only goes up and down by a few meters and travels at very modest speeds, it is *not quite true* that the sum of $\frac{1}{2}mv^2$ and mgh retains precisely the same value throughout the motion, and it is not quite true that all objects fall in precisely the same way, with a precisely uniform acceleration, one which has the same value for all objects.* Just as the motion of the pendulum described in §2.2 ceases in time, when measurements of the highest accuracy are made, it is found that the sum of $\frac{1}{2}mv^2$ and mgh is slightly less when the ball returns to earth than it was at the beginning. For a baseball which only goes a few meters, these discrepancies are difficult to discern, and they are not evident in the data that have been presented here, but the discrepancies are real. The lighter the object and the farther it travels, the more noticeable they are. The failure of the law of conservation of mechanical energy is very definitely noticeable if we try to analyze the same experiment carried out with a Ping-Pong ball or an experiment in which a penny is dropped from the top of a tall building.

When the ball comes back down, its total mechanical energy, $(\frac{1}{2}mv^2 + mgh)$, is less than when the ball was initially thrown upward. It appears that some of the energy has been lost. What has happened to the missing energy? To ask the question implies that it has an answer, that some of the energy has "gone somewhere," that there is indeed a law of conservation of energy that is strictly correct if properly used. (It could well be that the question is not appropriate, that the only law is an approximate one, useful in many cases but not universally applicable.) It is tempting to say merely that energy appears to have been lost "because of air resistance" or "because of air friction" and to say that the

*Galileo is reputed to have dropped two different balls from the leaning tower of Pisa to show that all objects fall with the same acceleration. If you try this with two markedly different objects (a baseball and a Ping-Pong ball), you will find that they do not strike the ground at the same time. Galileo guessed correctly that, if there were no air resistance, if the balls could be dropped in a vacuum, then they would fall together. He had the insight to see that the important fact was not that one ball or the other fell slightly faster but that even with objects of quite different mass, one perhaps twice that of the other, they fell at *almost* the same rate. As far as we know, he did not actually perform a public experiment, probably because he could not expect his audience to be sophisticated enough to recognize what was important about the result.

missing energy has been transferred to the molecules of the air through which the ball passes. Such a statement *can* be justified, but not easily. If we are not careful, if we just *say* that the ball's lost energy has actually been given to the surrounding molecules, we run the risk of turning the law of conservation of energy into an empty tautology, a law that is "correct" simply because we juggle the figures to *make* it correct. To use an analogy, if there is a "law of conservation of money," then your checkbook should balance at the end of each month. You can use the law, for example, to deduce the amount of one check which you remember having written but whose amount you neglected to record. Suppose, though, that the following procedure is used to make the checkbook balance. Whenever it seems to be out of balance, say by $2.17, invent a category called "unexplained" and give it the value of $2.17. I myself often do this, but then the fact that my checkbook balances in no way confirms the law of conservation of money.

Fortunately for the state of science and for the status of the law of conservation of energy, we can do better than this. In Chapter 6 we will discuss the crucial experiments that justify the generalization of this law. However, for our present purposes, suffice it to say that when the ball loses mechanical energy or when the pendulum's motion dies down, the air becomes slightly warmer or the support from which the pendulum swings gets warmer, and it *is* possible to define a quantity we will call *thermal* energy, so that when it is included along with kinetic and gravitational potential energies, the sum does remain constant. This is not the only additional form of energy that will have to be introduced. We shall, for example, describe the initial kinetic energy of the thrown ball as being delivered by the hand that propels it and as coming from the store of *chemical* energy of the thrower's body. Of all the various forms of energy, thermal energy is perhaps the most important if only because it is so often produced, at least in small amounts, by almost everything we do, and also because almost all the energy we use is ultimately converted into thermal energy. For example, when the falling ball hits the ground (if it does not bounce), whatever kinetic energy it has is converted into thermal energy.

These reservations about the law of conservation of energy can be stated in a slightly different way. In any particular application, the number of forms of energy that must be included depends on the circumstances and on the precision which is desired. The law of conservation of energy in its simplest form, $KE + GPE =$ constant, applies only as long as the gravitational force of the earth is the most important force acting and as long as we do not demand that deductions made with the law be absolutely accurate. This simple law of conservation of energy does apply almost perfectly to a large number of examples—in which "air resistance" and "friction" are small enough to be neglected—and we can and will use it as an important tool in analyzing such motions. This alone would be enough to make the law a useful one; fortunately inclusion of other forms of energy will enable us to transform it into a universally correct law. For the present we shall not be concerned explicitly with other forms of energy, but we need to remember that there are limitations on the applicability of the simple form of the law of conservation of energy.

§2.6 THE CHOICE OF A "ZERO-LEVEL" FOR GRAVITATIONAL POTENTIAL ENERGY

We have defined the gravitational potential energy (*GPE*) of an object as mgh. The symbol m denotes the mass of the object, and g stands for 9.8 m/sec^2, a quantity that, as we have seen, can be interpreted as the acceleration due to gravity. But what is h? In our development, h is the height of the object above the floor, (let us say the floor of a laboratory on the third story of a building). Therefore the gravitational potential energy of the object is positive if the object is above the floor, negative if it is somewhere below the floor, and zero if it is at the level of the floor (Figure 2.10a). Obviously there is nothing special about the floor of the laboratory, but since it is necessary to measure heights from somewhere, we must arbitrarily choose a certain level as $h = 0$. We might have made a different arbitrary choice: we might have designated the laboratory ceiling, or the floor of a room on the story below, as $h = 0$. Whatever height we decide to call $h = 0$, the gravitational potential energy of an object will be zero if it is at that level, positive if above, and negative if below (Figure 2.10b). Let us consider the effect of this freedom to choose the "zero-level" or "reference level" for gravitational potential energy.

Although this freedom may cause some complications, it will not upset our most important conclusions. The most important result is that as the ball moves up and down, the sum of its kinetic energy, *KE*, and its gravitational potential energy, *GPE*, remains constant:

$$\tfrac{1}{2}mv^2 + mgh = E = \text{constant},$$

Positive values of h

Positive *GPE*

Laboratory

$h = 0$

Zero-level, $GPE = 0$

Negative values of h

Negative *GPE*

a

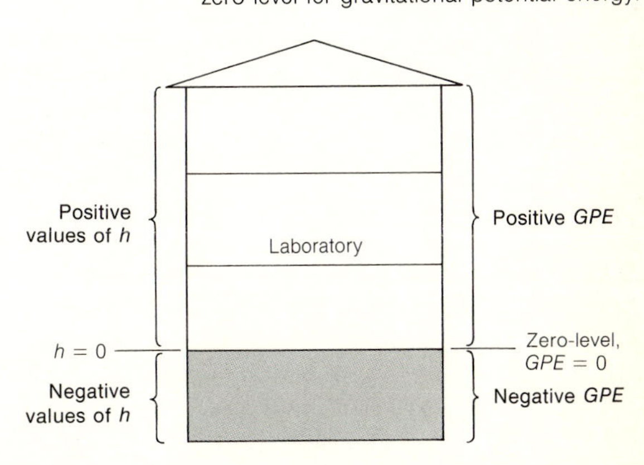

FIGURE 2.10
Two different possible choices of the zero-level for gravitational potential energy.

or that

$$\Delta(KE) + \Delta(GPE) = 0.$$

If we had decided initially to measure heights from some different zero-level, *all* of the h's would be changed, but they would all be changed by the *same* amount. All the numerical values of gravitational potential energy would also be changed, but each value of gravitational potential energy would undergo the same change. The numerical values of gravitational potential energy depend on the arbitrary choice of the zero-level and therefore so does the value of the total mechanical energy. But whatever zero-level we choose, the sum of the ball's kinetic and gravitational potential energies remains constant throughout its motion.

In other words, various forms of energy are of interest primarily when they *change*. To say that water at the top of a dam has a certain gravitational potential energy is of little interest if the water remains forever at this position. If the water can come down over the dam, it will lose gravitational potential energy and therefore gain kinetic energy. If we know how much its gravitational potential energy decreases as it falls, we can calculate how much its kinetic energy will increase and therefore how fast the water will be going when it gets to the bottom. The *change* in gravitational potential energy as it falls will not be affected in any way by the choice we may have made in choosing some particular height to be called $h = 0$. Since gravitational potential energy is defined as mgh, then

$$\Delta(GPE) = \Delta(mgh) = mg(\Delta h),$$

and Δh is the same no matter what our zero-level may be, and so the calculated change in gravitational potential energy is independent of the choice of zero-level.

There is nothing wrong with negative values of gravitational potential energy. It is changes in gravitational potential energy that are of interest, and if an object falls from $h = -6$ m to $h = -8$ m, it loses exactly the same amount of gravitational potential energy as it would if it fell from $h = +5$ m to $h = +3$ m. In a particular application, it may be *convenient* to choose the zero-level for gravitational potential energy in such a way that no negative values arise, but it is not *necessary* to do so. But in selecting the zero-level of kinetic energy, the choice is a natural and obvious one. Given our definition of kinetic energy as $\frac{1}{2}mv^2$, the kinetic energy is zero when the object is at rest, and it can never be negative; any other zero-level for kinetic energy would be awkward to deal with.*

*Some arbitrariness is involved even in the definition of kinetic energy. From our perspective, a rock on the surface of the earth is at rest, but to someone watching from the center of the solar system, the rock is moving and has a kinetic energy that is not zero. In most cases, however, we will assume that motions are measured with respect to the surface of the earth.

It is not important what level we choose as the zero-level for gravitational potential energy. But remember that if the gravitational potential energy of an object is to be calculated, the zero-level must be stated; for example, a statement that the gravitational potential energy of a rock is 3 J is meaningless if the zero-level is not specified. Our freedom to choose the zero-level can cause complications, but it is sometimes very convenient to have this flexibility, for we can make whatever choice is most convenient for our purposes. If we are doing experiments of this type in a laboratory, it may be most convenient to choose the zero-level at the floor. If we are dropping rocks out the window and want to know how fast they will be going when they reach the ground, it may be more convenient to choose the ground as the zero-level. But once we have chosen a zero-level, it is essential that we do not change it in the midst of an analysis.

QUESTION

2.5 (a) Table 2.4 refers to the motion shown in Figure 2.3, as described by an observer who chooses as the zero-level ($h = 0$) for gravitational potential energy the floor of the room in which the experiment is done. Suppose that you choose as a zero-level the floor of the room below, 4 m down, and construct a similar table to describe the same motion; in what ways will your table be different from Table 2.4 and in what ways similar? In the original description, the total energy of the ball ($E = KE + GPE$) was found to be approximately constant with a value of about 3.8 J. What value would you give for E, with your choice of zero-level?

(b) What would the answers to these questions be if you were to use as your zero-level the ceiling of the room in which the experiment is done, 3 m above the original zero-level?

---

§2.7 APPLICATIONS OF THE LAW OF CONSERVATION OF ENERGY

Now that we have developed a law of conservation of energy applicable to the motion of bodies moving along vertical paths, consider some of the ways in which it can be used to answer some specific questions about such motions. Consider the example already analyzed, the motion of a ball whose mass, m, is 0.14 kg, moving with a total energy, E, of 3.8 J. What kinds of information can we extract from the conservation of energy equation

$$\tfrac{1}{2}mv^2 + mgh = E(=3.8 \text{ J in this example})? \qquad (2.14)$$

First, suppose we wish to know the speed the ball had when its height was 1.5 m. Equation 2.14 can be solved for v:

$$v^2 = \frac{2}{m}(E - mgh) = \frac{2E}{m} - 2gh \qquad (2.15)$$

and we can then substitute for h the value 1.5 m:

$$v^2 = \frac{2 \times 3.8}{0.14} - 2 \times 9.8 \times 1.5$$

$$\simeq 54.3 - 29.4$$

$$= 24.9$$

and so

$$v = \pm \sqrt{24.9} \simeq \pm 5 \text{ m/sec}.$$

What is the significance of the *two* possible results, one positive and the other negative? We have defined speed to be a positive quantity (the magnitude of the velocity), so that $v = +5$ m/sec is the result we want. However, the equation is trying to tell us something by giving us two answers, namely, that the ball passed this point twice, once on the way up and once on the way down. If Equation 2.15 is considered as an equation for the *upward component* of the velocity, then the two roots can easily be understood in this way. We can see from Figure 2.4 that at $h = 1.5$ m, the speed was about 5 m/sec, on both the upward and downward parts of the ball's trip.

Now, what is the speed corresponding to $h = -5$ m? In this problem, we are pushing the equation beyond the limits of the original data, and we must be careful. The level $h = -5$ m would be 5 m below floor level, and, in our example, the ball never reached that point. We could extend the experiment, though, by allowing the ball to drop through a hole in the floor, and we can use our results to predict the speed it will then have. We simply substitute $h = -5$ m in Equation 2.15:

$$v^2 = \frac{2E}{m} - 2gh$$

$$= 54.3 - (2 \times 9.8 \times -5) = 54.3 + 98 = 152.3.$$

The speed is then the positive square root, $v \simeq 12.3$ m/sec. If the experiment is repeated with this modification, this predicted speed turns out to be approximately correct.

What is the speed corresponding to a height of $h = 3.0$ m? Again we follow the same procedure and substitute $h = 3.0$ m in Equation 2.15:

$$v^2 = \frac{2E}{m} - 2gh$$

$$= 54.3 - 2 \times 9.8 \times 3$$

$$= 54.3 - 58.8$$

$$= -4.5.$$

Now we have a problem. We are being asked to take the square root of a *negative* number, an impossible task.* Upon reflection, however, we can understand this strange result. The ball never got as high as $h = 3.0$ m in the course of the entire experiment, and there is no way it could do so unless it had more energy to begin with. We should be greatly relieved that our equation does not give us any answer at all when we ask for the speed at this height. Any result that it did yield—whether positive, negative, or zero—would be wrong. Our equation appears to give us right answers when we ask sensible questions, and it stubbornly refuses to give us any answer at all when we ask a question that has no answer!

These ideas can be used to answer another question. What was the maximum height reached by the ball? The larger the value of h, the smaller the value of v^2, and the smallest possible value of v^2 is zero; it cannot be negative. Thus if we set v^2 equal to zero in Equation 2.14, we can solve for h and this will give us h_{max}, the maximum value of h:

$$0 + mgh_{max} = 3.8$$

$$h_{max} = \frac{3.8}{mg} = \frac{3.8}{0.14 \times 9.8}$$

$$\simeq 2.8 \text{ m.}$$

If we inspect the data in Figure 2.4, we can see that this is at least approximately the correct value for the maximum height reached by the ball.

But now consider another question. What is the speed when $h = -400$ m? Following exactly the same procedure, we find that $v \simeq 89$ m/sec, but this is simply incorrect. If the ball were thrown upward and then allowed to fall below the starting point and to keep falling for another 400 m (the height of one of our tallest skyscrapers), its downward speed should be about 89 m/sec. This is the speed that the equation gives us, but if a ball is actually thrown from the top of a 400-m building, it hits the ground with a substantially slower speed.

Again, equations are excellent tools if used with caution, common sense, and an awareness of their possible limitations. It is interesting and often useful to see what an equation predicts in a problem such as this, but it is dangerous to *assume* that it can be safely extrapolated far beyond the range of the data on which it is based. The effects of air resistance—quite minor and safely ignored when a ball travels only a meter or so—become vitally important when the ball travels a much greater distance. When a ball is dropped from the top of a tall building, for a period of time it gains kinetic energy as it loses gravitational potential energy, but as time goes on, its speed no longer continues to increase. Even though the ball continues to lose gravitational potential energy as it falls, it does not pick up a corresponding amount of kinetic energy. The

*That is, we cannot calculate the square root of a negative number without using "imaginary" numbers, and it would be most peculiar to have to represent the speed by an imaginary number.

law of conservation of energy must be modified before it can be of much use in making quantitative predictions in cases in which air resistance and similar effects are so extremely important.

To examine yet another example, suppose that you want to throw a one-pound rock straight upward to a height at least as high as the top of a tree, 40 ft above your head. How fast must it be thrown? What initial kinetic energy must be given to the rock? First, it is safest to convert the data into MKS units before doing the calculations. The mass of the rock is

$$1 \text{ lb} = 0.4536 \text{ kg};$$

the height to be reached is

$$40 \text{ ft} = 40 \text{ ft} \times \frac{0.3048 \text{ m}}{1 \text{ ft}} = 12.2 \text{ m}.$$

Now we can reason in the following way. After the rock is released, its mechanical energy will be conserved. As it rises, it will gain gravitational potential energy as it loses kinetic energy. If it does get to the top of the tree, the increase in gravitational potential energy will be

$$\Delta(GPE) = mg(\Delta h) = 0.4536 \times 9.8 \times 12.2 = 54.2 \text{ J}.$$

(In this problem, we did not even need to choose a zero-level for gravitational potential energy; $\Delta h = 12.2$ m and $\Delta(GPE) = 54.2$ J, no matter what zero-level we use.) This increase in gravitational potential energy must be offset by the loss of an equal amount of kinetic energy. Since kinetic energies cannot be less than zero, the rock cannot lose more kinetic energy than it had to begin with, and therefore its initial kinetic energy must be at least 54.2 J. That is, if v_1 denotes the initial speed,

$$\tfrac{1}{2}mv_1^2 \geq 54.2 \text{ J}.$$

Thus

$$v_1^2 \geq \frac{2 \times 54.2}{m}$$

and, with $m = 0.4536$ kg:

$$v_1^2 \geq \frac{2 \times 54.2}{0.4536} = 239,$$

and so

$$v_1 \geq \sqrt{239} = 15.5 \text{ m/sec}.$$

The initial speed, v_1, must be at least 15.5 m/sec.

Consider one more illustration of the use of the conservation of energy equation. We shall look at this example first with numbers and then in terms of symbols, to examine some of the ways in which a symbolic algebraic solution can be especially informative. If you throw a 3-kg rock down from your window, 12 m above the ground, with an initial speed of 8 m/sec, what will its speed be when it hits the ground? Its initial kinetic energy is

$$\tfrac{1}{2}mv^2 = \tfrac{1}{2} \times 3 \times 8^2 = 96 \text{ J}.$$

As its height decreases by 12 m, it will lose an amount of gravitational potential energy given by

$$3 \times 9.8 \times 12 = 353 \text{ J}.$$

It must therefore gain an equal amount of kinetic energy, and so its kinetic energy when it hits the ground will be

$$96 + 353 = 449 \text{ J},$$

and thus, as it hits the ground,

$$\tfrac{1}{2}mv^2 = 449 \text{ J},$$

or

$$v = \sqrt{\frac{2 \times 449}{3}} = \sqrt{299} = 17.3 \text{ m/sec}.$$

It is instructive to work the same problem through without inserting any numerical values until the last possible moment. One reason for doing this is that, if we use symbols for the various quantities, we can work out the answer to *all* problems of this type; we will not have to start over again if we want to see what happens if the initial speed is 10 m/sec instead of 8 m/sec. In addition, and this is really much more interesting, if we can obtain a symbolic expression for the speed of the rock when it hits the ground, we can look at this expression, interpret it, and see if it seems to make sense.

Let us use the subscript "i" (initial) to designate the instant at which the rock is thrown, and the subscript "f" (final) for the instant at which it hits the ground. By the law of conservation of mechanical energy,

$$\tfrac{1}{2}mv_i^2 + mgh_i = \tfrac{1}{2}mv_f^2 + mgh_f.$$

Note that the mass, m, is a common factor in each term and therefore can be canceled:

$$\tfrac{1}{2}v_i^2 + gh_i = \tfrac{1}{2}v_f^2 + gh_f. \tag{2.16}$$

Let us arbitrarily choose the gound level as $h = 0$; then in this example, $h_f = 0$ m, and $h_i = 12$ m. If we set $h_f = 0$ and then solve Equation 2.16 for v_f, the desired result, we have

$$v_f = \sqrt{v_i^2 + 2gh_i}, \tag{2.17}$$

where we take the positive square root because we are trying to calculate the final speed, a positive quantity. We can easily substitute the numerical values appropriate to the particular example:

$$v_f = \sqrt{8^2 + 2 \times 9.8 \times 12} = \sqrt{64 + 235} = \sqrt{299} = 17.3 \text{ m/sec},$$

as we found earlier.

Thus we can calculate the specific numerical result for this particular problem, but what is considerably more interesting is the "reading," the interpretation, of Equation 2.17. There is much more to physics than just "plugging numbers into equations," and reading equations is one of the arts which makes physics enjoyable. We can look at Equation 2.17 and see how the result for v_f would change if the given conditions were varied; we can see if these variations "make sense." One good reason for solving problems symbolically, rather than always putting in numerical values right at the beginning, is an extremely practical one; if we have made a mistake, it is quite likely that our equation will *not* make sense, and we can thus be alerted to the error.

We notice first that the mass of the rock does not appear in Equation 2.17. This might seem upsetting at first, until we recall the remarkable similarity in the behavior of all objects moving under the influence of the earth's gravity. Since they all have the same downward acceleration, and they all fall in the same way, it is not surprising that the mass is missing from this equation. The speed with which it hits the ground will be the same whether it is a small rock or a large one. How does the result for v_f vary if we consider various possible initial speeds of the rock? If the rock is thrown downward with a higher speed, v_i^2 will be greater, and therefore, according to Equation 2.17, the rock will be going faster when it hits the ground. Certainly this is what we would expect.

(Suppose, by the way, that we had made an algebraic error and had come up with the incorrect formula

$$v_f = \sqrt{2gh_i - v_i^2} \quad \text{(wrong)}.$$

If we merely plugged numbers into this equation, we might never notice the mistake. But a reading of this erroneous equation would immediately indicate that something was wrong. It would tell us that, for larger initial speeds, v_f would be smaller—a ridiculous conclusion—and that if the initial speed were large enough, the quantity under the square root sign would be negative!)

Returning now to the correct equation (Equation 2.17), we can ask how v_f would vary if the height of the window above the ground were varied. According to Equation 2.17, if h_i is made larger, v_f will be larger. Again, this result makes sense for surely, other things being equal, a rock coming from a higher level will be going faster when it hits the ground. The equation tells us something else that is not so obvious. What if, instead of throwing the rock *down* with a speed of 8 m/sec, you were to throw it upward with the same speed and let it come down past the window and hit the ground? How would this affect v_f? Not at all, according to Equation 2.17, for v_i^2 has the same value (64 m^2/sec^2), whether the rock is thrown up or down. This result, though perhaps unanticipated, also makes sense, for if you throw a rock upward from the window, it will rise for a while and then fall, and at the instant it passes the window its kinetic energy will be the same as it was when you released it. (Its gravitational potential energy will be the same at that instant as it was initially, and so, because of conservation of energy, the value of its kinetic energy must also be the same as its initial value, even though the direction of its motion has been reversed.) From that moment on, its motion is just the same as if you had thrown it downward, and it will hit the ground with the same speed as it would if it had been thrown downward to begin with. What would happen if g were greater? This is a difficult question to answer experimentally, but our natural expectation is that if g, the acceleration due to gravity, were greater, the rock would be going faster when it hit the ground, and this is what Equation 2.17 tells us.

There is another useful test, which is easier to make with an equation in symbolic form than it would be if we merely had the numbers. Is Equation 2.17 dimensionally correct? On the left side, we have a speed, and thus the dimensions of the left side are length/time. Therefore the dimensions of the expression under the square root sign on the right side ought to be length^2/time^2. The first term, v_i^2, is obviously correct. The dimensions of the second term, $2gh_i$, are

$$\frac{\text{length}}{\text{time}^2} \times \text{length} = \frac{\text{length}^2}{\text{time}^2},$$

and thus Equation 2.17 is dimensionally correct. Once again, such a dimensional check can help us to catch some kinds of mistakes. If we had found the erroneous equation:

$$v_f = \sqrt{v_i^2 + 2mgh_i} \quad \text{(wrong)},$$

we would see that the dimensions of $2mgh_i$ are

$$\frac{\text{mass} \times \text{length}^2}{\text{time}^2}$$

and it would be clear that a mistake of some kind had been made.

Reading equations in this way can provide unanticipated insights, and it can protect us from making some kinds of mistakes, as we have seen. It is not a foolproof method for catching all mistakes, though, as we can see from the following illustration. Suppose that through a careless error we had derived the erroneous equation:

$$v_f = \sqrt{v_i^2 + \frac{gh_i}{2}} \quad \text{(wrong)}.$$

This equation for v_f varies in a perfectly reasonable way if we consider the effect of a change in v_i or a change in h_i or in g. It is also dimensionally correct. It is also wrong, but we would never be able to catch this particular error by "reading" equations in the way that we have been doing here.

QUESTIONS

2.6 A number of foolhardy people have gone over Niagara Falls in a barrel, a drop of 190 ft (and a few of them survived the attempt). Estimate the speed of a barrel when it gets to the bottom of the falls, assuming that the barrel is nearly at rest as it begins to fall and that the law of conservation of mechanical energy is applicable. Find the speed both in meters per second and miles per hour.

2.7 In the preceding discussion it was shown that a 1-lb rock must be given an initial upward speed of at least 15.5 m/sec (an initial kinetic energy of at least 54.2 J) if it is to reach the top of a 40-foot tree. How large must the initial speed and the initial kinetic energy be if a 2-lb rock is used instead? (First analyze the problem with symbols; that is, consider a rock of mass m and a tree of height H. Do the required values of speed and kinetic energy depend in a reasonable way on m, H, and g?)

2.8 The period, T, of a pendulum is the length of time it takes to go through one full cycle of its motion, from right to left and back again. Here are six theoretical equations purporting to give the period in terms of the length, l, of the pendulum and the acceleration due to gravity, g. Which of these can be rejected on grounds of dimensional inconsistency? Of those that pass this test, can you decide which is closest to the correct result by making a simple pendulum and measuring its period?

$$T = 2lg \qquad T = lg \qquad T = 2\sqrt{l/g}$$

$$T = 2\pi\sqrt{l/g} \qquad T = 4l/g \qquad T = \pi l/g$$

2.9 (a) Suppose you could fire a rock straight up with an initial speed of 2000 miles/sec. Show that if the law of conservation of mechanical energy discussed in this chapter is applicable, the rock will rise to a maximum altitude of about 3.3×10^8 miles.
(b) Would such a rock actually behave this way? On what grounds might you be suspicious of this result?

§2.8 PROJECTILE MOTION

We have studied the motion of an object moving under the influence of gravity, when the motion is simply up or down along a single straight vertical path, and we have seen that the law of conservation of mechanical energy,

$$\tfrac{1}{2}mv^2 + mgh = E,$$

provides a good description of such motions. We can use this law to describe many of the features of the motion of an object that is thrown upwards, thrown downwards, or simply dropped. We have been considering motions restricted to a vertical straight line for reasons of simplicity. We can now consider other situations in which some object moves under the influence of gravity, and we can see whether this law has wider application.

What happens if we throw something not straight up or straight down, but at some angle, or perhaps horizontally? We do not have to look far to find examples of such "projectile motions," as they are called: the motion of a bullet after it leaves a gun, the motion of a bomb dropped from a moving airplane, the motion of a baseball after having been hit or

FIGURE 2.11
Stroboscopic photograph of a projectile. [Education Development Center.]

thrown, and so on. Figure 2.11 is a stroboscopic photograph (made in the same way as Figures 2.1 and 2.2) of such a projectile motion. Our previous experience suggests that we should find corresponding values of v^2 and height, h (where h is the vertical distance above some arbitrarily chosen zero-level), and see whether or not the resulting graph is a straight line. The numerical details will be omitted; the result, though, is that a graph of v^2 versus h is indeed a straight line, and that its slope is again -19.6 m/sec². Thus our earlier conclusions apply to this example as well, and we can once again write the law of conservation of energy for this motion:

$$\tfrac{1}{2}mv^2 + mgh = E = \text{constant}.$$

The same law of conservation of energy is found to apply to other examples of projectile motion. Of course, in each example, the mass may be different, and the constant total energy will depend on the initial kinetic energy imparted to the object and also on the arbitrary choice of the zero-level for gravitational potential energy. For each, the relationship between v^2 and h is given by an equation of the form

$$\tfrac{1}{2}mv^2 + mgh = E = \text{constant},$$

with g having the same value, 9.8 m/sec².

There is an important difference between examples of projectile motion and that discussed earlier, in which a ball was thrown *vertically* into the air. If a projectile is fired at some angle with respect to the vertical, the direction of its velocity is not simply up or down. As a consequence, we cannot calculate the speed just by subtracting one value of h from another; we must measure the actual distance the object has moved. Furthermore, more care is required if we wish to specify the direction of the velocity. In Chapter 3 we will develop a method for describing the magnitude *and* the direction of the velocity for such examples.

Many of the calculations that we performed in the simpler case of the vertically thrown ball can be repeated here. If we know the speed of the object at any one height, we can use the law of conservation of energy to calculate its speed at any other height. For example, if a rock is thrown out of a window with a known speed, we can figure out how fast it will be going when it hits the ground. There is one significant difference, though, between the information we can obtain by means of the law of conservation of energy in the case of projectile motion and what we can learn in the case of an object thrown vertically upward. What is the maximum height the object reaches before it starts to come down? In the earlier example, we were able to determine it by recognizing that, at the top of its path, the object would momentarily have zero speed, and so we could set $v^2 = 0$ in the conservation of energy equation and solve for h. But a ball thrown at an angle does not stop at the top of its path; instead its horizontal motion continues.* The law of conservation of energy cannot by itself be used to find the maximum height of the ball.

There are other questions we might well ask about motions under the influence of gravity, questions to which the law of conservation of energy cannot by itself provide the answers. Neither in an example of projectile motion nor in one of a ball thrown vertically upward does the law of conservation of energy tell us *when* the object will arrive somewhere, *when* it will hit the ground. It does enable us to find out how *fast* it will be going when it gets there but not how long it takes. If "coverage" of physics were the goal, it would be appropriate to develop additional techniques for handling questions such as these. We shall not pursue this point, however, because it would divert us from the purpose at hand—to investigate the concept of energy as fully as possible. However, it is important to recognize that although the law of conservation of energy is a very powerful tool for answering many questions and describing many of the features of a wide variety of phenomena, it does have limitations and cannot by itself provide answers to *all* the questions we might ask.

§2.9 NEWTON'S FIRST LAW OF MOTION: THE PRINCIPLE OF INERTIA

Another kind of motion to which these results can be applied is that of something sliding on an inclined surface. Objects tend to slide downhill because of gravity, and we might well anticipate that the law of conservation of energy, in the same form used thus far, would apply to this situation as well. If we try to conduct an experiment of this sort—for

*As a matter of fact, if we focus our attention on the horizontal motion of the ball (left to right in Figure 2.11), we see that the horizontal motion is uniform. That is, between any one flash and the next, the ball moves the same distance to the right.

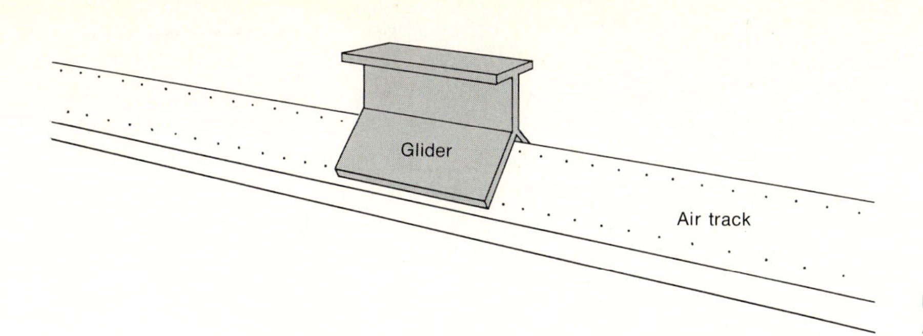

FIGURE 2.12
An air track and a glider.

example, by simply placing a book on a tilted board—the book may not slide at all, and if it does, the friction between the book and the board will play a very important role. Just as the law of conservation of energy in the form used thus far,

$$\tfrac{1}{2}mv^2 + mgh = E = \text{constant},$$

is not quite correct for vertically thrown balls and projectiles unless the effects of air resistance are small enough to be neglected, it does not accurately describe the motion of objects on tilted surfaces unless the effects of friction are extremely small.

Some experimental precautions are necessary if we are to study this type of motion without being bothered by friction. (We shall not be permanently satisfied with a "law" of conservation of energy that works only approximately, or that works only in idealized examples in which the effects of friction are small or zero. Our philosophy here, as throughout this chapter, is to postpone consideration of these effects while acknowledging their existence and their eventual importance.) We might hope to approximate the frictionless case by studying, perhaps, the motion of a skier on an icy slope, or by letting a smooth block of metal slide down a well oiled surface. One excellent way of reducing frictional effects to a minimum is to use an air-supported glider on an "air track" (Figure 2.12). An end view of such a glider on a track is shown in Figure 2.13. Air is continually blown into the inside of the track and comes out through a multitude of tiny holes in the surfaces of the track, thus supporting the glider on a thin cushion of air. With this method of support, frictional effects are extremely small. The glider does encounter some air resistance as it moves, just as a falling ball does, but this effect is also very small as long as the glider's speed is not too great.

The air track can be tilted at various angles, and we can then study the motion of such a glider and determine whether or not the law of conservation of energy still applies. Figure 2.14 is a stroboscopic photograph of the motion of a glider released at the top of a tilted track. Once again, we can determine speeds and corresponding heights (heights above an arbitrary zero-level). The measurements and calculations are

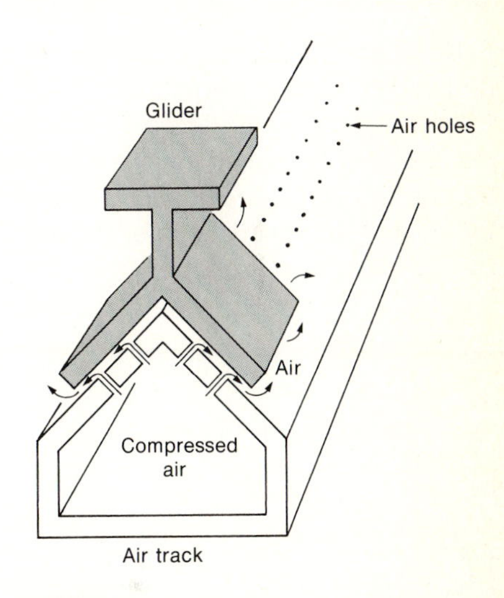

FIGURE 2.13
End view of a glider on an air track. Compressed air flowing out through many tiny holes in the surface of the track forms a thin film of air on which the glider rides.

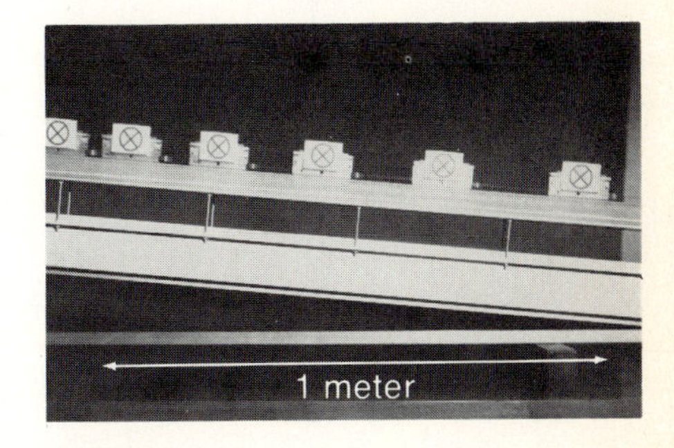

FIGURE 2.14
Stroboscopic photograph of the motion of a glider on a tilted air track. The time interval between flashes: 0.2 sec.

left as a question (Question 2.10); the result is that again a graph of v^2 versus h is a straight line with a slope of approximately -19.6 m/sec^2, just as before. Thus in this case too, the law of conservation of energy in the form

$$\tfrac{1}{2}mv^2 + mgh = E = \text{constant}$$

provides a correct description of the motion.

QUESTION

2.10 Using Figure 2.14, draw up a table of speeds and corresponding heights, make a graph of v^2 versus h, and determine the slope of the resulting line. Remember that the speed must be calculated by measuring the distance traveled along the track and dividing by the length of the time interval, and that heights are to be measured vertically upward from an arbitrarily chosen horizontal line. Each value of speed should be associated with a value of h midway between the two values of h at the beginning and end of the interval.

The motion of a glider on an air track is very much like the motion of a vertically thrown ball, in that the presence of the track constrains the motion to be along a straight line. The difference is that the track is not vertical, and the effects of gravity are correspondingly "diluted." The speed is still related to h (the vertical height above the zero-level) by the same equation, but the velocity is directed along the track, not up or down. If the glider's height is to change by some amount—say, 10 cm—it must travel much more than 10 cm along the sloping track. Thus the acceleration of the glider along the track is considerably less than it would be if it were falling freely.

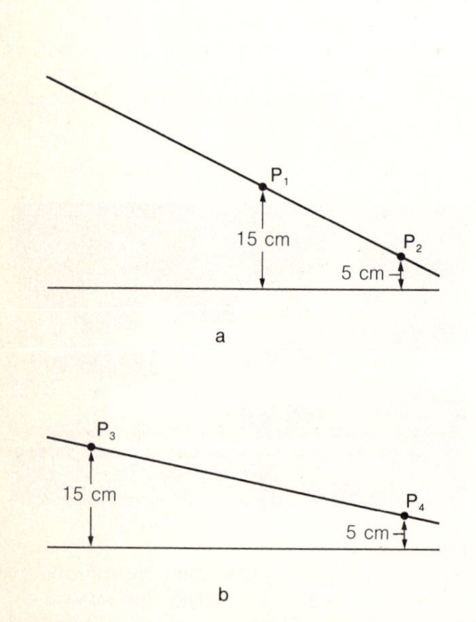

FIGURE 2.15
Two air tracks, tilted at different angles.

The more gentle the slope of the track, the greater the distance along the track the glider must go in order for its height to change by a specified amount. Compare, for instance, Figures 2.15a and b. As the glider travels from P_1 to P_2, in Figure 2.15a, its height changes by the same amount (10 cm) as does that of a glider traveling from P_3 to P_4 in Figure 2.15b; but in the second example, in which the track makes a small angle with the horizontal, the glider has to travel much farther along the track. The law of conservation of energy specifies that the amount by which the speed changes as the glider moves depends on how much the vertical height changes and only indirectly on the distance the glider travels. This suggests a very interesting question. If we look at the extreme case in which the track is perfectly horizontal, what information does the law of conservation of energy give us? If the track is perfectly horizontal, then no matter where we measure h from, h has the same value at all points along the track. Then in the energy equation,

$$\tfrac{1}{2}mv^2 + mgh = E = \text{constant},$$

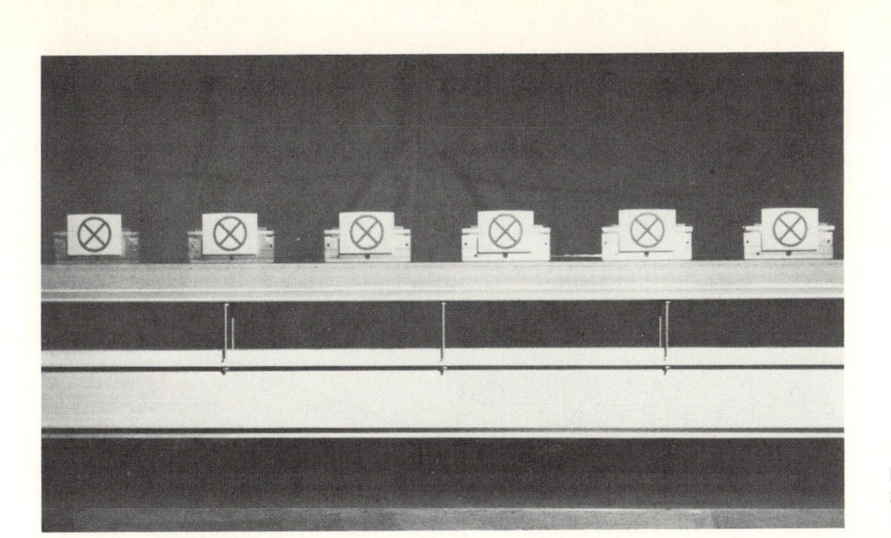

FIGURE 2.16
Stroboscopic photograph of motion on a horizontal air track.

the second term on the left (mgh) is a constant, and thus $\frac{1}{2}mv^2$ must remain constant as the glider moves. (To express it another way, if the sum of the kinetic energy and the gravitational potential energy remains constant, and if the gravitational potential energy remains constant because the height does not change, then the kinetic energy must remain constant as well.) Thus the speed of the glider will neither increase nor decrease as it moves! Of course, if you simply release a glider on a level track, it will not start to move; neither direction along the track is "down." But if you give it an initial velocity in either direction, the equation tells us that it will continue to move with unchanging speed. Real air tracks, such as those shown in Figure 2.12, are only 1 or 2 m long, but if we imagine a track 10 miles long, the argument suggests that, on such a track, the glider will continue to move the whole length of the track with constant speed.

Is this conclusion correct, or have we been misled by the equations? Certainly it is difficult to test it directly. Even if we had air tracks 10 miles long, the slightest bit of air drag would eventually cause the glider to slow down. When we contemplate an idealized experiment that we know we cannot actually carry out, we are conducting a "thought experiment," to try to understand what *would* happen if the idealized situation could be achieved. If a real horizontal track is used, a glider *is* observed to move with an almost constant velocity (Figure 2.16) until it strikes the obstacle at the end of the track. When we contemplate what would happen if we could make an extremely long level track, we are necessarily extrapolating from what we can actually observe. What our analysis suggests is that as we come closer and closer to eliminating frictional effects and air resistance, it will be more and more nearly true that an object, once set in motion, will continue to move with unchanging speed unless something acts to speed it up or to slow it down.

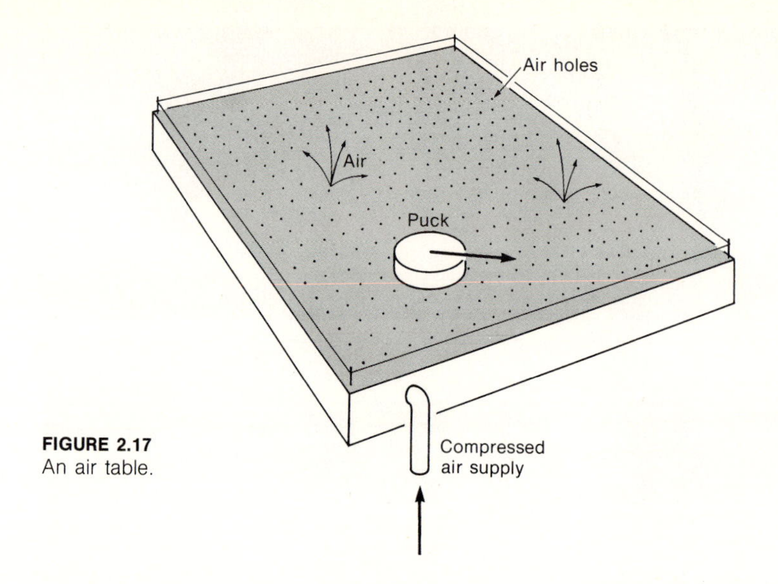

FIGURE 2.17
An air table.

We can make an even stronger statement by considering the motion of an air-supported puck on an "air table" (Figure 2.17), which is really a two-dimensional version of an air track. Air is blown upward through numerous tiny holes in the table to support the puck. Unlike the glider on the air track, which could not move except along the line of the track, the air-supported puck is free to move anywhere on the surface of the table. When we make such a table as level as possible and then watch a puck, we find that if the puck is at rest, it remains at rest, but once it is moving, it continues to move in the *same direction* and with constant speed until it hits the edge of the table.

With both of these devices (the level air track and the level air table), we have been considering the motion of objects in situations in which all the *forces* acting on the object have been reduced or canceled out as much as possible. These observations suggest a general principle, called the "principle of inertia" or "Newton's first law of motion": an object that is at rest remains at rest unless acted upon by a net* force; an object that is in motion remains in motion, moving in the same direction with constant speed unless acted upon by a net force.

This principle is called the principle of inertia because it describes the tendency of an object to maintain whatever velocity it may have as long as outside influences are absent. It was first formulated by Galileo, who did so much to establish the quantitative study of motion as a basic part of physics. Later it was included by Newton as one of his three laws of motion. The word "force" appears in the statement of the principle of inertia, and we have yet to define this term quantitatively. In Chapter 3, we will discuss Newton's second and third laws of motion,

*The term "*net* force" is used because the puck on the air table, for instance, is acted upon by the downward force of gravity, but this downward force is exactly balanced by the upward supporting force exerted by the stream of air; the *net* force is zero.

which deal with the meaning of force and the effects of forces. We may think of the principle of inertia as defining what we mean by *zero* force: *if* an object moves with constant speed and without any change in direction, then the net force acting on it is zero. Note that an object that is not moving at all represents one special example of such a motion, but the principle refers to *any* motion of constant speed without change in direction.

However, if we recognize that a force is some sort of push or pull, we see that the principle of inertia is more than simply a definition of "zero force." The principle states that if all pushes and pulls on an object can be eliminated, then its motion will be one of constant speed along a straight line. We can rephrase this by saying that when the net force on an object is zero, it moves at a constant velocity, that is, with a velocity that is constant both in magnitude and direction. The reverse is also true: if an object is observed to move with constant velocity, the net force acting on it is zero. If we want to change either the direction or the magnitude of the velocity, some sort of force is required. For example, if the puck on the air table is at rest, it has to be pushed to be set in motion, and then it continues to move with the same velocity until it is pushed again. Push in the direction in which it is moving and it speeds up; push in the opposite direction and it slows down; push sideways and the *direction* of its velocity changes. A careful sideways push may simply deflect the puck, changing the direction of the velocity without changing the speed, but even though the speed may not be changed, a push of some kind is needed to change the direction.

The principle of inertia seems an entirely reasonable one, in the light of experience with air tracks and air tables, where friction is extremely small. It would not seem such an obvious principle if we considered more familiar events, in which friction is often the dominant influence. If you push a book across a table, the book will stop almost immediately after you stop pushing; a steady push is needed just to keep it moving. It might seem more reasonable to say that a moving object comes to rest unless acted upon by a force to maintain its motion; this view, the one held by the followers of Aristotle, prevailed until Galileo's time. Galileo had the insight to propose that it would instead be more useful to consider the tendency of an object to maintain its velocity as the truly fundamental fact, and to regard the slowing down, which all objects experience to a greater or lesser extent, as being due to the unavoidable effects of a particular kind of force, a frictional force. On this basis, the motion of a book being pushed along a table at a constant velocity is described as a motion under the influence of zero *net* force; the force exerted by the hand that pushes it is of exactly the correct magnitude to cancel out the retarding frictional force. It is remarkable that Galileo, who had no such devices as air tracks* to work with, was able to get to

*Galileo avoided the problem of friction by studying round objects *rolling* down sloping surfaces; we have not discussed such experiments here because of the complication produced by the fact that a rolling object has an additional energy coming from its rotation.

the heart of the problem and to see that the basic laws could best be discovered by minimizing frictional effects or imagining what would happen if these effects were absent.

§2.10 ANOTHER LOOK AT THE PENDULUM, AND A SUMMARY

We began this chapter with a qualitative description of the motion of a pendulum. In its repetitive to-and-fro motion we saw the possibility of identifying some combination of speed and height that was "conserved" as the pendulum swung back and forth. In pursuing this idea, we first examined simpler cases such as the vertically thrown ball, and we found that indeed there is such a combination that is conserved. The basic law of conservation of energy as it has been developed in this chapter is

$$KE + GPE = E = \text{constant}, \tag{2.18}$$

where $KE = \frac{1}{2}mv^2$ and $GPE = mgh$. The motion of the pendulum, like the motions that led us to this law, is one that is controlled by the gravitational attraction of the earth. Does the law of conservation of mechanical energy also apply to the pendulum? The answer is "yes," and the evidence is found in Figure 2.2. We can again determine values of speed and corresponding values of height. The details will be omitted, but the result is that the graph of v^2 versus h (Figure 2.18) is again a straight line, again with a slope of approximately -19.6 m/sec^2. Thus it follows that, for the pendulum too, the law of conservation of energy in the form of Equation 2.18 provides a correct description of the motion. Just as anticipated in §2.2, kinetic energy and gravitational potential energy are converted back and forth as the pendulum swings.

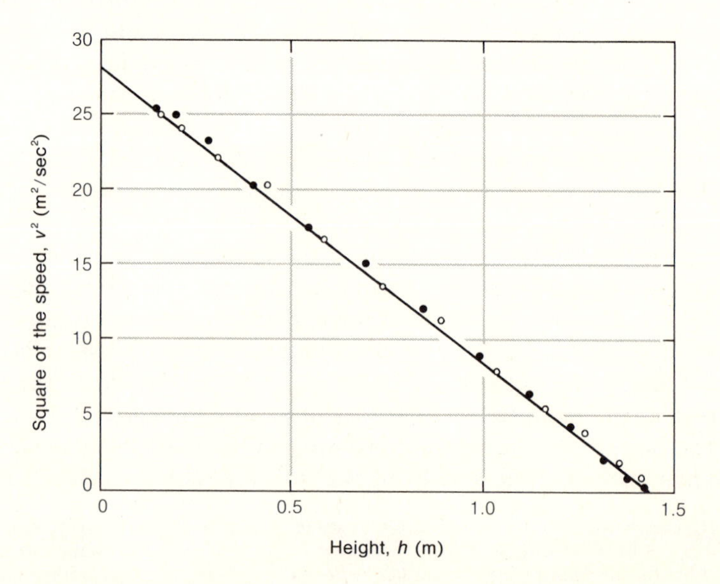

FIGURE 2.18
Graph of the square of the speed, v^2, versus the height, h, for the pendulum shown in Figure 2.2. The slope of the straight line is approximately -19.6 m/sec^2. The dark circles describe the downward swing (the right-hand half of the photograph); the light circles refer to the upward swing. Heights are measured from the bottom of the photograph.

QUESTION

2.11 Figure 2.18 was derived from measurements of the photograph of the swinging pendulum (Figure 2.2), but the details of the calculations were omitted. Make your own careful measurements on Figure 2.2, use them to construct a graph of v^2 versus h, and determine the slope of the resulting line.

Let us conclude this chapter by reviewing the energy concept as it has been developed thus far, and pointing out some of the additional work that needs to be done. For a large number of motions of a certain type, motions of objects under the influence of the earth's gravitational attraction, we have found that a certain quantity, $E = \frac{1}{2}mv^2 + mgh$, retains a constant value throughout the motion. *Why* does this quantity E behave in this simple way? That is a question that, in the final analysis, we can never hope to answer. The facts of the physical world as we find them expressed in the experimental data are such that E *does* behave in this way, and we can use this knowledge to give concise descriptions of many of the features of the motions to which this law of conservation of energy applies.

Throughout this chapter, we have been concerned with motions "caused" by the earth's gravity, and we have repeatedly used terms such as "gravity," "gravitational attraction," and so on. We should not delude ourselves into thinking that the use of the label "gravity" explains anything. We do not know why objects are accelerated downward; we only know that they *are* accelerated downward, that as they move the sum of $\frac{1}{2}mv^2$ and mgh *does* remain constant. The use of a term such as gravity is often extremely convenient because it enables us to give concise descriptions. When we say that various motions are all "caused by gravity," what we are really saying is that all these motions exhibit certain common patterns. Recognition of the fact that these motions have something in common with one another and the use of the word gravity to express this fact can be very fruitful in suggesting further questions. Does the motion of the moon or the motion of an earth satellite have something in common with these motions? Are they also caused by "gravity," whatever gravity may be? It was this line of questioning, this search for features common to apparently unrelated phenomena, that led Newton to the formulation of a "law of universal gravitation," which we shall discuss in detail in Chapter 5.

What *is* energy? Is it some sort of mysterious "vital fluid" that an object in motion possesses? As a matter of fact, one of the early names for kinetic energy was *vis viva*, literally "living force." Although such terminology might add a certain liveliness to our descriptions, it should be used, if at all, with great caution. It might suggest, if taken too literally, that energy is an actual substance, a liquid for example, which can be stored in a bottle and added to an object to make it move. A more cautious approach to the identification of energy is to recognize that energy is really an abstraction, a mental bookkeeping device that *we* have invented. Objects move in certain observed ways. By careful measurements, we find that if we *define* kinetic energy and gravitational potential energy in a certain way, then the statement that their sum remains constant provides a concise description of some aspects of motion. The speeding up and slowing down of a pendulum are correctly and concisely described by saying that every decrease in gravitational potential energy is compensated for by a corresponding increase in kinetic energy, and vice versa. The books balance, the sum of the two kinds of energy remains the same. The description of energy as a bookkeeping device does not diminish the importance of the concept. Rather, the fact that such a simple description of a variety of phenomena is possible is one of the marvels of nature, and that we have been able to formulate such a description is one of the marvels of the human mind.

Thus far we have concentrated on a very special type of motion: the motion of a single object under the influence of the earth's gravity. Alternatively, we might say that we have considered the gravitational interaction of two "objects," one of which is a ball, a rock, or an air-track glider, and the other the earth. Implicit in our discussion has been the understanding that the earth is large enough that, although we could think of it as influencing the motion of the smaller object, we need not worry about any effect the rock might have on the earth. However, it is clear that we will have to examine what happens when the motion of each of two objects is affected by the other. We will have to consider other ways in which objects can interact with one another, not just gravitationally but by colliding with each other, by exerting electrical forces on each other, and so on.

The law of conservation of energy in its present form is restricted to motions fairly close to the earth's surface—and to situations in which only a single force, gravity, is of significance. The process of imparting energy to a ball by throwing it, the loss of energy when it hits the ground, or the gradual loss of energy that the ball and the pendulum undergo as they move through the air are important features that cannot be explained by the present treatment. We will see in later chapters that it *is* possible to extend the concept of energy to take these various problems into account and to define other forms of energy so that the law of conservation of energy is, to the best of our knowledge, universally valid.

We may think of the general program for developing and extending the concept of energy and the idea of energy conservation in the following way. The most fundamental form of energy is kinetic energy, the energy an object has simply because it is in motion.* For an air glider on a frictionless horizontal track, kinetic energy alone is conserved.

*In the experiments discussed so far, the motions are those of objects traveling at fairly low speeds. At speeds approaching the speed of light, both experimental results and the theory of relativity require a revised definition of kinetic energy, a definition that gives results virtually indistinguishable from $\frac{1}{2}mv^2$ except for the very highest speeds.

If the track is tilted or if we consider a thrown ball, kinetic energy alone is *not* conserved. If we ask "where did the extra kinetic energy come from?" or "where did the missing kinetic energy go?" the definition of another form of energy, gravitational potential energy, provides an answer. Changes in kinetic energy can be accounted for by equal and opposite changes in gravitational potential energy; the sum $(KE + GPE)$ is conserved during the motion. Then when we consider the finer details of these motions or consider how they begin and end, we are led to define other forms of energy so that when the new forms are included as well, the books can be balanced once again, so that it will be possible to say that when all appropriate forms of energy are included, the total energy is conserved.

FURTHER QUESTIONS

2.12 Make rough estimates of the following energies, in joules:

(a) the kinetic energy of a car traveling at 60 miles/hr;

(b) the kinetic energy of a 747 jet in flight ($m = 250$ tons);

(c) the gravitational potential energy of a 747 jet at cruising altitude;

(d) the kinetic energy of a high-speed elevator ascending a tall building;

(e) the gravitational potential energy of an elevator at the top of a very tall building.

2.13 Estimate the kinetic energy in joules of a pole-vaulter just before he begins his jump. Estimate his gravitational potential energy as he clears the bar.

2.14 What is the value of g (9.8 m/sec^2) in feet per second^2? In miles per hour^2?

2.15 A 2-ton car runs out of gasoline while traveling at 50 miles/hr. It could coast to a gas station, but there is a 100-ft hill that it must get over. Assume (optimistically) that there is no friction of any sort. Does the car have enough kinetic energy to get over the hill? Would the result be any different for a lighter car?

2.16 (a) Show that if a rock is thrown upward with an initial speed $v_i = 12$ m/sec, it will rise to a maximum height above its starting point given by $h_{max} = 7.3$ m.

(b) Show that the total time, T, required for it to rise to the top of its path and then return to the starting point is $T = 2.4$ sec.

(c) Generalize these results. Show that for any initial upward speed,

$$h_{max} = v_i^2/2g$$

$$T = 2v_i/g.$$

(d) By eliminating v_i from these two equations, derive an equation relating h_{max} and T.

(e) The equation just derived can be easily tested, at least approximately, and no special equipment is required. Throw a rock upward and estimate the maximum height to which it rises and the total time that elapses until it returns. How well do the observed values of h_{max} and T fit the equation derived in (d)? Try this experiment with several rocks, and with a number of initial speeds.

2.17 A joule is a very small amount of energy in comparison with the amounts of energy used in this country. The average American uses about 10^4 J of energy every second. This is an average figure, computed from the amount used during all hours of night and day, and it includes not only the energy used in the home for heating, operating electrical appliances, etc., but also the chemical energy of the gasoline we burn, the energy used to run factories, the energy that is wasted (converted directly into thermal energy) in electrical generating plants, and so on. Some of this energy is obtained by converting the gravitational potential energy of water behind a dam or at the top of a waterfall into electrical energy in hydroelectric generating plants. To obtain a very general idea of the possible contribution of this source of energy, make the following calculations.

(a) The height of Niagara Falls is about 190 ft. How many kilograms of water would have to come down over the falls every second if all of your own energy (10^4 J every second) were to be derived from this source?

(b) How large a volume of water would this be? (The density of water—see Table A.2—is 1000 kg/m^3.)

(c) The river with the largest rate of flow in the United States is the Mississippi; the average rate of discharge of water at the mouth of the river is about 600,000 ft^3/sec. If the Mississippi could somehow be diverted to go over Niagara Falls, approximately how many Americans, each using 10^4 J/sec, could be supplied with energy from this source?

3
Momentum, Force and Work

Harold E. Edgerton, MIT, Cambridge, Mass.

3
Momentum, Force and Work

§3.1 INTRODUCTION

We discussed in Chapter 2 a number of situations in which a single object moves under the influence of the earth's gravitational force. The law of conservation of mechanical energy is a unifying principle that can be applied to all such motions, at least as long as effects such as friction and air resistance can be neglected. Now we must consider other forces besides the earth's gravitational force: for example, the force your hand must exert on a rock in order to throw it into the air, the force the surface of the earth exerts on a rock as the rock hits the ground, forces that two rocks exert on each other in a collision, the frictional force that acts to slow down a book sliding across a table, and so on. We will begin by examining collisions between two objects, in which for a brief period of time the interaction between the two objects is so large that any other effects can be ignored. In the majority of such collisions, mechanical energy is not conserved, and we must therefore digress from the discussion of energy and look for other sorts of regularities.

Although mechanical energy is often not conserved in collisions, we will define a different quantity that *is* conserved: *momentum*. When two bodies interact with one another, the total momentum is conserved, but the individual momenta may change; the *force* acting on any one object will be defined as the rate of change of the object's momentum.

This definition of force leads to Newton's famous "second law of motion," expressing the relationship between force and acceleration. Finally, we shall return to the concept of energy by defining the *work* done by a force; we will see that the work done on an object is equal to the change in energy. Work is a measure of energy *transfer,* and the work done by a force can be either positive or negative. Thus we will arrive at a more generally applicable statement about energy. While a thrown ball is moving through the air, under the influence of the earth's gravitational force, its mechanical energy is constant (if air resistance is small), and furthermore the change in its mechanical energy as it is thrown is equal to the work done on it by the hand that throws it; the subsequent change in energy when it hits the ground (a *negative* change) is equal to the work done on it by the ground (a *negative* amount of work).

§3.2 CONSERVATION OF MOMENTUM IN COLLISIONS

§3.2.A One-Dimensional Collisions

Consider first a collision between two gliders on an air track. This is an especially simple example because the motion of both objects is restricted to a single straight line. A further simplification is made by restricting attention to experiments in which the air track is horizontal; thus no changes in gravitational potential energy occur, and each glider moves with unchanging velocity except during a collision. Air gliders can be made to interact in a variety of ways during a collision. For example, the end of each glider can be equipped with a spring bumper, which is a springy loop of steel (Figure 3.1), so that the two gliders will bounce cleanly off one another. Or, lumps of putty might be placed on the ends of the gliders so that they stick together and move as a single larger object after the collision.

Consider first the second type of collision, a "sticking" collision. Results of one such experiment are shown in Figure 3.2. Glider B (whose mass, m_B, is 0.4 kg) is initially at rest and is struck by glider A (m_A = 0.2 kg) whose initial speed is 0.3 m/sec. After the collision, the composite object (with a mass of 0.6 kg) is observed to move with a speed of

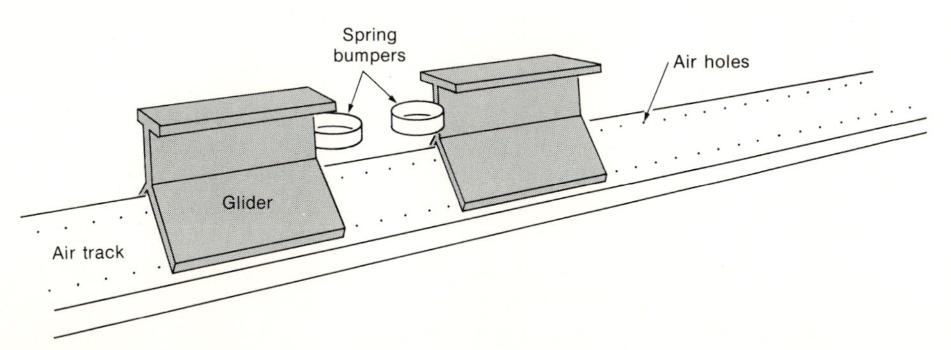

FIGURE 3.1
Air-track gliders with spring bumpers.

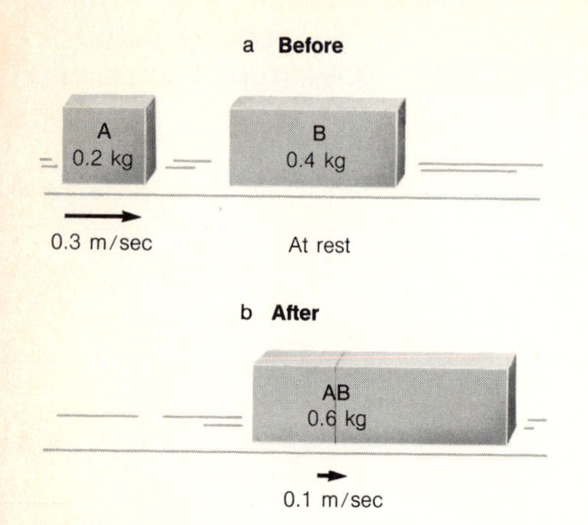

FIGURE 3.2
A sticking collision.

approximately 0.1 m/sec. Kinetic energy is definitely not conserved in this collision. Before the collision, glider B has no kinetic energy, whereas A has a kinetic energy equal to:

$$\tfrac{1}{2}mv^2 = \tfrac{1}{2} \times 0.2 \times (0.3)^2 = 0.009 \text{ J}.$$

After the collision, the kinetic energy is that of a single object with a mass of 0.6 kg moving with a speed of 0.1 m/sec:

$$\tfrac{1}{2}mv^2 = \tfrac{1}{2} \times 0.6 \times (0.1)^2 = 0.003 \text{ J}.$$

It is not even approximately true that kinetic energy is conserved in this interaction; two-thirds of the initial kinetic energy disappears.

Just as in the example of the pendulum, in which the sum of kinetic and gravitational potential energies gradually decreases, the law of conservation of energy can be salvaged by appealing to the observation that the temperatures of the gliders increase very slightly during this collision and by attributing the missing kinetic energy to the appearance of "thermal energy." However, this cannot properly be done without careful measurements of temperature and a careful definition of thermal energy. Kinetic energy is not conserved in a sticking collision, and the principle of conservation of mechanical energy does not provide a useful guide in analyzing this sort of collision. Kinetic energy decreases in such a collision, and that is about all we can say about the energy at this point.

One feature of these results, however, suggests a different conservation law. In this experiment, the initial speed of the incoming glider was 0.3 m/sec, three times as great as the speed of the composite object after the collision (0.1 m/sec). The mass of the composite object (0.6 kg) is larger than the mass of the incoming glider (0.2 kg) by exactly the same factor:

$$\frac{\text{initial speed}}{\text{final speed}} = \frac{\text{final mass}}{\text{initial mass of moving glider}}.$$

If we add up the products of mass times speed, we have the same values before and after the collision:

$$\begin{array}{llll} \text{Before:} & \text{A} & 0.2 \times 0.3 & = 0.06 \text{ kg-m/sec} \\ & \text{B} & 0.4 \times 0 & = 0 \\ & \text{Total:} & & 0.06 \text{ kg-m/sec} \\ \text{After:} & (\text{A} + \text{B}) & 0.6 \times 0.1 & = 0.06 \text{ kg-m/sec}. \end{array}$$

The product of mass times speed appears to be an interesting quantity. We tentatively give this the name momentum, and these results suggest that the total momentum (the sum of the individual momenta) is conserved in such a collision, that it has the same value before and after the collision.

This conclusion is incomplete, however, as we can see if we consider a sticking collision between two gliders of equal mass (0.2 kg), both moving with the same initial speed (0.3 m/sec), as shown in Figure 3.3.

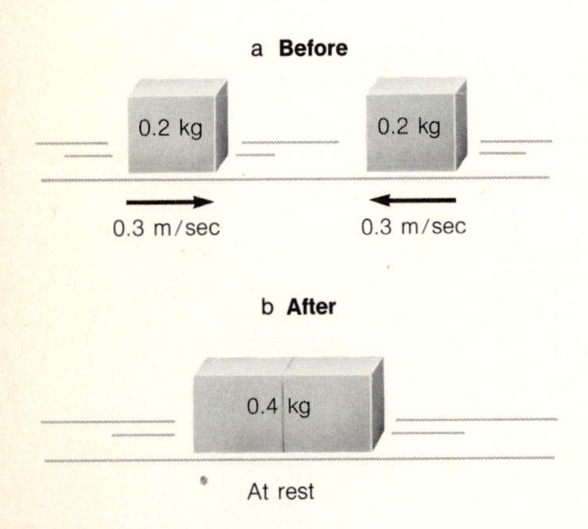

FIGURE 3.3
A symmetrical sticking collision.

The situation is perfectly symmetrical. After they collide and stick together, there is no reason why the new composite object should be moving one way rather than the other, and in fact they do not move at all after the collision. They simply stick together and stop. Thus it is definitely not true in this example that the sum of the products of mass times speed remains the same during the collision.

If we take the direction of motion into account and define the momentum to be a *vector* quantity—mass times *velocity* rather than mass times speed—then we *can* say that the total momentum is conserved in this collision. Before the collision, the two gliders have momentum vectors that are equal in magnitude but opposite in direction. If we define the "sum" of two such vectors to be zero, if we regard them as canceling each other out just as when we add the number -8 to the number $+8$, then the total momentum of A and B together is zero before the collision, just as it is afterwards.

We introduce the symbol $\mathbf{p}$ to denote momentum:

$$\mathbf{p} = m\mathbf{v}.$$

The units of momentum, in the MKS system, are kg-m/sec. If we reexamine the previous example (Figure 3.2) with this definition of momentum, we see that before the collision the total momentum is simply that of A, a vector of magnitude 0.06 kg-m/sec directed to the right, and that after the collision, the total momentum of A and B together is also of magnitude 0.06 kg-m/sec and also directed to the right.

§3.2.B Two-Dimensional Collisions and Vector Addition

Care is needed in extending these ideas to more complex examples, in which the colliding objects are not constrained to move in a straight line. Consider a collision in which two air pucks on an air table, both initially in motion as shown in Figure 3.4, collide and stick together. Notice first that, just as in the sticking collision of two gliders on an air track, kinetic energy is not conserved in this collision. Before the collision,

$$KE_{\mathrm{A}} = \tfrac{1}{2} \times 0.2 \times (0.4)^2 = 0.016 \text{ J}$$

$$KE_{\mathrm{B}} = \tfrac{1}{2} \times 0.3 \times (0.5)^2 = 0.0375 \text{ J}$$

$$\text{Total } KE = 0.0535 \text{ J}$$

and after the collision:

$$KE_{\mathrm{AB}} = \tfrac{1}{2} \times 0.5 \times (0.34)^2 = 0.0289 \text{ J}.$$

Momentum, however, *is* conserved in this collision. The sum of the separate momenta of A and B before the collision is equal to the momentum of A and B together after the collision, if we are careful to define

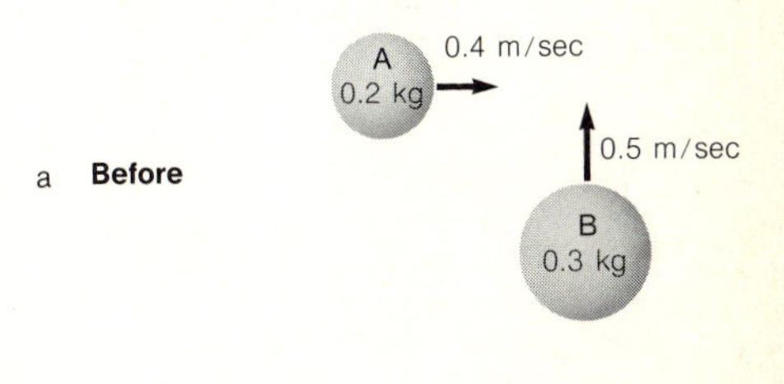

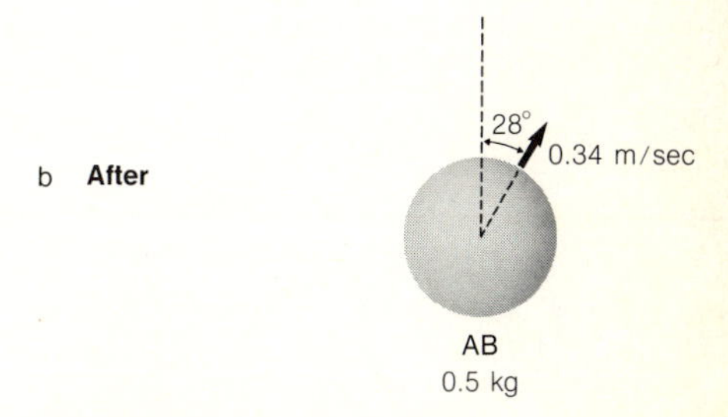

FIGURE 3.4
Representation of a sticking collision between two pucks on an air table.

what we mean by the "sum" of two momentum vectors. The momentum vector of puck A before the collision has a magnitude equal to

$$p = mv = 0.2 \times 0.4 = 0.08 \text{ kg-m/sec}$$

and a direction toward the right (Figure 3.4a):

$$\mathbf{p}_{\mathrm{A}} = 0.08 \text{ kg-m/sec (to the right).}$$

Likewise (as also shown in Figure 3.4a) the momentum of B before the collision is the vector

$$\mathbf{p}_{\mathrm{B}} = 0.15 \text{ kg-m/sec (upward).}$$

The total momentum of A and B together after the collision ($\mathbf{p}_{\mathrm{AB}}$) has a magnitude of

$$0.5 \times 0.34 = 0.17 \text{ kg-m/sec}$$

and the direction shown in Figure 3.4b. Is momentum conserved in this collision? The sum of the *magnitudes* of $\mathbf{p}_{\mathrm{A}}$ and $\mathbf{p}_{\mathrm{B}}$ is not equal to that of $\mathbf{p}_{\mathrm{AB}}$ ($0.08 + 0.15 \neq 0.17$), but we must take the directions of vectors into account in calculating their "sum."

We *define* the process of vector addition in the following way, as illustrated by the vector addition of $\mathbf{p}_{\mathrm{A}}$ and $\mathbf{p}_{\mathrm{B}}$ in this example. We represent a vector by an arrow, choosing the length of the arrow to show the magnitude of the vector (Figure 3.5). Vectors are completely determined by their magnitude and direction; we can move the arrows that represent them as much as we like, as long as we do not alter their magnitudes or directions. To add two vectors, place the second vector so that its "tail" coincides with the "head" of the first and draw a vector from the tail of the first to the head of the second. This new vector is defined to be the "vector sum" of the two individual vectors. Observe that the vector sum of the momenta of A and B before the collision is indeed equal in both magnitude and direction to $\mathbf{p}_{\mathrm{AB}}$, the momentum of A and B together after the collision.

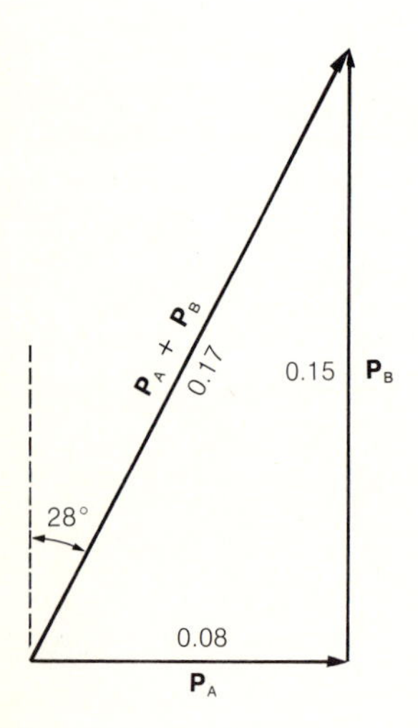

FIGURE 3.5
Vector addition of $\mathbf{p}_{\mathrm{A}}$ and $\mathbf{p}_{\mathrm{B}}$ (Figure 3.4).

The same procedure can be used when the two vectors to be added point in precisely the same direction or in precisely opposite directions. As shown in Figure 3.6a, the vector sum of **C**, a vector 3 units long pointing upward, and **D**, a vector 2 units long, also pointing upward, is a

FIGURE 3.6
Vector addition of two vectors in the same direction, and in opposite directions.

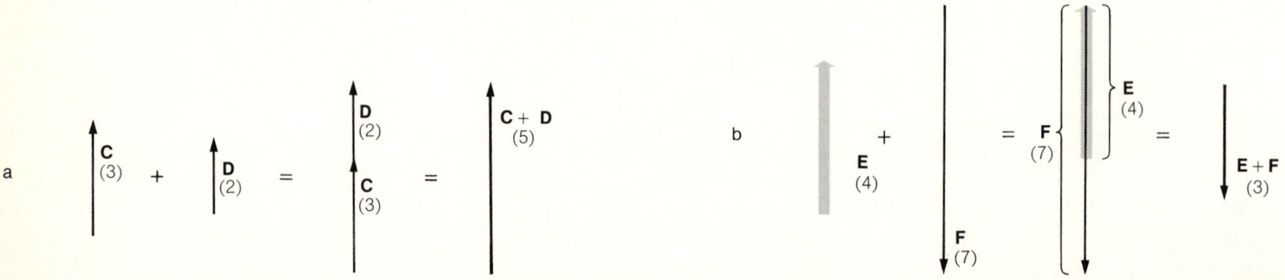

a C (3) + D (2) = D (2) C (3) = C + D (5)

b E (4) + F (7) = F (7) E (4) = E + F (3)

vector 5 units long, which points upward. Figure 3.6b shows that the sum of a vector **E**, 4 units long pointing upward, and **F**, a vector 7 units long that points downward, is a vector 3 units long in the downward direction. The definition of vector addition can easily be extended to the addition of three or more vectors, as shown in Figure 3.7.

Many other quantities besides momentum can be treated as vectors and are added in the same way. Walk two miles east and then three miles north; the overall displacement can be found by calculating the vector sum of the two separate displacements. Forces also combine in the same way; the motion of an object subject to two or more forces is the same as if a single force were applied, equal to the vector sum of the individual forces.

With this definition of vector addition and with the definition of momentum as a vector quantity, $\mathbf{p} = m\mathbf{v}$, we can state the law of conservation of momentum. Consider two objects that interact only with each other. Then the total momentum of these two objects, the vector sum of the separate momenta, is conserved; it is the same vector (the same magnitude and the same direction) at all times. The law of conservation of momentum can be generalized to a collection of any number of objects that constitute a *closed system,* that is, a set of objects that may interact with one another but not with any objects outside the system. The total momentum of the system, the vector sum of the individual momentum vectors, is conserved.

It is easy to see that the law must be restricted to a closed system. For example, consider a single glider on a frictionless air track. As long as it is alone on the track, this glider itself constitutes a closed system;* its momentum is conserved, because it moves at a steady velocity in accordance with the principle of inertia. When it hits the end of the track, though, its momentum is very definitely not conserved; if it bounces back, its momentum is completely reversed in direction. The law of conservation of momentum cannot be applied to the interaction of the glider with the obstacle at the end of the track unless we define the system to include not only the glider but also the track and the table on which the track sits. (And then we should include the building and eventually the whole earth.) We can define the boundaries of the system to be anywhere we wish, but with one choice we may have a closed system within which the law of conservation of momentum holds, and with another choice we may not have a closed system. Similarly, two gliders colliding on an air track constitute only an approximation to a closed system, even though a very good approximation. To the extent that the minute effects of air drag can be ignored, the predictions made by applying the law of conservation of momentum to a system consisting only of the two gliders will be correct.

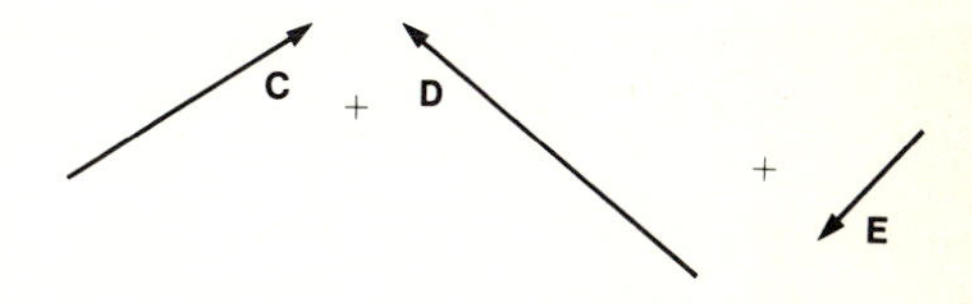

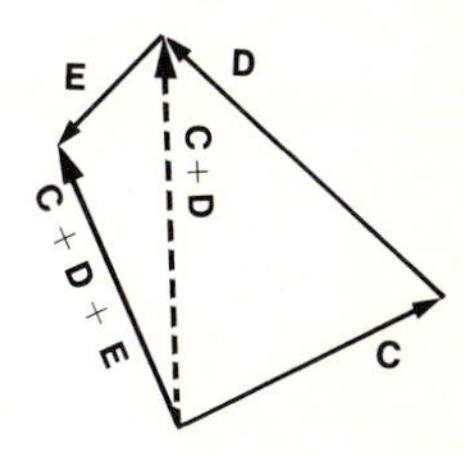

FIGURE 3.7
By extending the "head-to-tail" rule, any number of vectors can be added together. Here **C** and **D** are added to form **C** + **D**; then **E** is added to **C** + **D** to form the vector **C** + **D** + **E**.

*The glider does interact with the earth (which pulls it down) and the stream of air from the track (which supports it). However, these two interactions have canceling effects, and so they can be ignored.

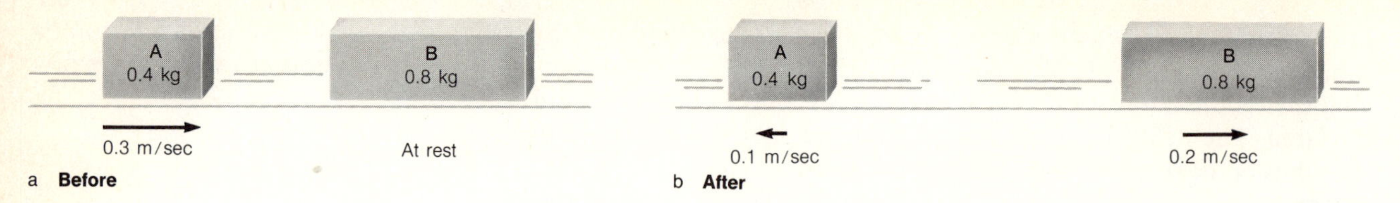

FIGURE 3.8
An elastic collision between two gliders.

The law of conservation of momentum has been suggested on the basis of results of a particular sort of interaction, the "sticking collision," a kind of collision in which kinetic energy is very definitely not conserved. The law applies, however, to all collisions and in fact to all kinds of interactions, even when there is no sharply defined collision, but rather a much more gradual sort of interaction.* There is an important class of collisions in which kinetic energy, as well as momentum, is conserved. Collisions in which kinetic energy is conserved (or nearly conserved) can be produced by equipping air gliders with spring bumpers, or by observing collisions between steel balls, which can bounce cleanly off one another and have no tendency to stick together. Collisions in which kinetic energy is precisely conserved are referred to as "perfectly elastic" collisions.

A perfectly elastic collision between two air gliders is shown in Figure 3.8. We can see that kinetic energy is conserved in this collision, for the kinetic energy before the collision is:

$$KE_A = \tfrac{1}{2} \times 0.4 \times (0.3)^2 = 0.018 \text{ J}$$
$$KE_B = 0$$
$$\text{Total } KE = 0.018 \text{ J};$$

and after the collision, the kinetic energy is:

$$KE_A = \tfrac{1}{2} \times 0.4 \times (0.1)^2 = 0.002 \text{ J}$$
$$KE_B = \tfrac{1}{2} \times 0.8 \times (0.2)^2 = 0.016 \text{ J}$$
$$\text{Total } KE = 0.018 \text{ J}.$$

Momentum is also conserved, for the momentum before the collision is just that of A:

$$\mathbf{p}_A = 0.4 \times 0.3 = 0.12 \text{ kg-m/sec} \quad \text{(to the right);}$$

and after the collision:

$$\mathbf{p}_A = 0.04 \text{ kg-m/sec} \quad \text{(to the left)}$$
$$\mathbf{p}_B = 0.16 \text{ kg-m/sec} \quad \text{(to the right);}$$

and the sum of these two momentum vectors after the collision is a vector of magnitude 0.12 kg-m/sec, directed to the right.

*The two gliders might carry electrical charges or magnets. If so, the gliders would begin to interact with each other even before coming into physical contact.

In a sticking collision, a large fraction of the original kinetic energy may be lost, or even all of it, as in the head-on collision shown in Figure 3.3. In a perfectly elastic collision, kinetic energy is precisely conserved. Many collisions are between these two extremes, and a partial loss of kinetic energy occurs. But in all collisions, whatever the nature of the interaction, it is found that the law of conservation of momentum is valid. The law of conservation of momentum is thus on quite a different footing from the "law" of conservation of kinetic energy. As long as we are careful to consider a *closed* system, the total momentum is conserved, but there is no real "law" of conservation of kinetic energy. If we want to know whether kinetic energy is conserved or not, either we must watch and find out, or else we have to know something about the nature of the interaction, for example whether the gliders are equipped with good springs or with pieces of putty. There *is* a law of conservation of *total* energy, but, except in special circumstances, the amount of any one form of energy may vary. The analogous problem does not arise with momentum; there are no "other forms of momentum."* The law of conservation of momentum is thus less ambiguous and easier to work with than the law of conservation of energy, but it is also more prosaic and less interesting.

Are there collisions in which the kinetic energy actually increases, collisions that we might call "superelastic"? Indeed there are, but examples are not as easy to find. Consider two gliders of equal mass, approaching one another on an air track with equal and opposite velocities. Suppose, as shown in Figure 3.9, that one of the gliders carries a firecracker and the other a lighted match. When the firecracker explodes, the two gliders may be blasted apart from each other with speeds greater than their initial speeds. The total kinetic energy of the system has been increased.

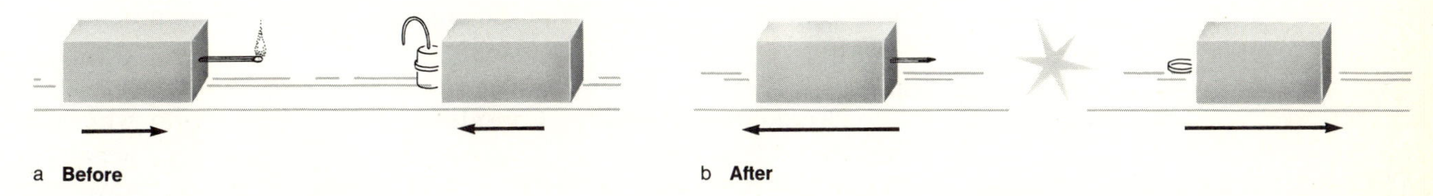

FIGURE 3.9
A superelastic collision between two gliders: the one on the left carries a lighted match; the one on the right, a firecracker.

There are numerous other examples of superelastic collisions. Suppose that you are standing on a perfectly smooth surface, perhaps in the center of a frozen pond. How can you get to shore? One approach is to take off a

*Momentum ($\mathbf{p} = m\mathbf{v}$) is often referred to as *linear* momentum. There is a physical quantity called *angular* momentum (not discussed in this book), but, in spite of its name, it is a quantity quite different from linear momentum. Whereas various forms of energy can be measured in the same units and can be added to find the total energy, angular momentum and linear momentum are quite different from one another; they even have different units and dimensions, and they can never be added together.

shoe and throw it horizontally as hard as you can. You and your shoe constitute a closed system. The total momentum of this system is zero before you throw the shoe, and therefore it remains zero after you throw the shoe. You will recoil in the opposite direction from the shoe, with a momentum equal in magnitude to that of the shoe but opposite in direction; if the ice is truly smooth, you will eventually reach the shore. One might object that it is somewhat artificial to call the interaction between you and your shoe a "collision." But whether we call it a collision or something else, it conforms to the same pattern as our other collisions: there are two objects, each with some definite velocity prior to their interaction; the two objects then interact briefly with each other, so that there is a clear separation of "before" and "after."

The firing of a cannon is another example in which kinetic energy is created. Initially both the cannon and the cannonball are at rest; when the cannon is fired, it recoils and the total momentum remains zero. In all such examples, an explanation of the increase in kinetic energy must ultimately be sought in the *decrease* of some other form of energy, a decrease in the chemical energy of the components of the firecracker, for example.

§3.2.C Analysis of Three Important Kinds of Collisions

Three kinds of collisions that are particularly interesting, and worth studying in more detail to see what happens to the kinetic energy as the masses of the interacting objects are varied, are sticking collisions, perfectly elastic collisions, and superelastic collisions. For the sake of simplicity in discussing sticking and elastic collisions, we shall consider only head-on collisions in which one of the two objects is initially at rest. Also, the discussion of superelastic collisions will be restricted to examples in which both objects are initially at rest. With these assumptions, the motion in each example will be along a single straight line; thus some of the complications arising from the vector nature of momentum will be removed.

STICKING COLLISIONS. Consider a sticking collision in which a moving body A (mass $= m_A$) strikes a body B (mass $= m_B$) which is initially at rest, as shown in Figure 3.10. In order to handle the vector aspect of momentum conservation, we shall now use symbols such as v_i and v_f to denote the *rightward component* of velocity. Positive values therefore indicate velocities directed to the right, and negative values, velocities directed to the left. Here the subscripts "i" and "f" will be used to denote the initial velocity of A and the final velocity of the composite object. An example of a sticking collision is a collision of a moving vehicle with a stationary one on an icy road, a collision in which the two vehicles lock bumpers.

Before such a collision, the total momentum of the system is just that of A ($m_A v_i$); after the collision the momentum is that of A and B to-

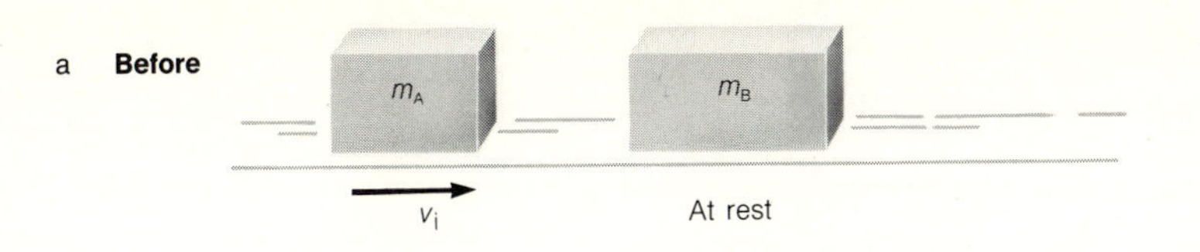

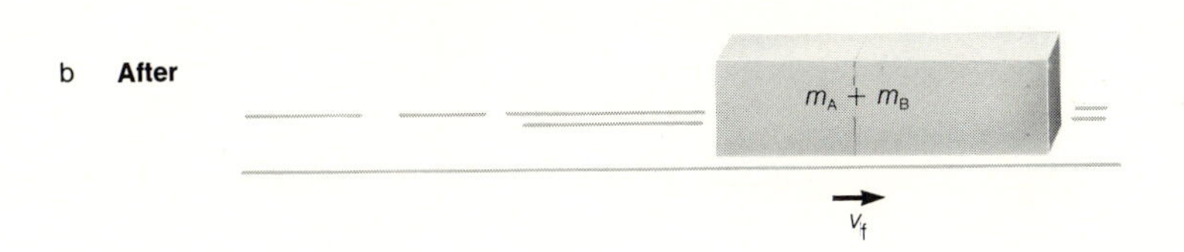

FIGURE 3.10
A sticking collision, in which a moving object (A) strikes a stationary object (B).

gether [$(m_A + m_B)v_f$]. By the law of conservation of momentum these two expressions can be equated, and we can find the final velocity:

$$m_A v_i = (m_A + m_B)v_f;$$

$$v_f = \left(\frac{m_A}{m_A + m_B}\right)v_i. \tag{3.1}$$

To interpret Equation 3.1, the ratio $\left(\frac{m_A}{m_A + m_B}\right)$ is always less than 1, so v_f, the final velocity, is always less than the initial velocity. If we let m_A be much greater than m_B ($m_A \gg m_B$), our results can be applied to the example of a 10-ton truck running into a stalled Volkswagen "beetle." The ratio $\left(\frac{m_A}{m_A + m_B}\right)$ is only very slightly less than 1, and so v_f is almost as large as v_i. To a first approximation, the truck is not affected at all by the collision; it simply gathers up the Volkswagen and continues on its way. But if it is the Volkswagen that hits the stationary truck, the situation is different. Now $m_A \ll m_B$, the ratio $\left(\frac{m_A}{m_A + m_B}\right)$ is much smaller than 1, so v_f is very much smaller than v_i. The Volkswagen is almost completely stopped by the collision. These equations tend to confirm what we intuitively already know: if there must be collisions between small cars and large trucks, one is better off as the driver of the truck than as the driver of the car.

PERFECTLY ELASTIC COLLISIONS. Now consider a perfectly elastic collision between a body of mass m_A, with an initial velocity v_i (directed to the right), and a stationary body of mass m_B. In this case, the two bodies do not stick together, and the final velocities of both objects (call them v_{fA} and v_{fB}, as shown in Figure 3.11) must be determined in terms of the initial quantities. In an elastic collision, both momentum and kinetic energy are conserved, and we can write two equations:

$$m_A v_i = m_A v_{fA} + m_B v_{fB}; \tag{3.2}$$

$$\tfrac{1}{2}m_A v_i^2 = \tfrac{1}{2}m_A v_{fA}^2 + \tfrac{1}{2}m_B v_{fB}^2. \tag{3.3}$$

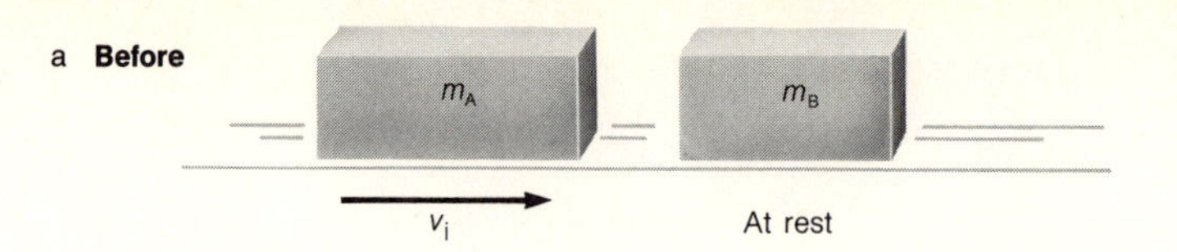

b After
m_A m_B
v_{fA} v_{fB}

FIGURE 3.11
An elastic collision: a moving object (A) strikes a stationary object (B).

Equations 3.2 and 3.3 are two equations in two unknowns (v_{fA} and v_{fB}), and the solution of this pair of equations can be written as follows:

$$v_{fA} = \left(\frac{m_A - m_B}{m_A + m_B}\right)v_i; \tag{3.4}$$

$$v_{fB} = \left(\frac{2m_A}{m_A + m_B}\right)v_i. \tag{3.5}$$

QUESTION

3.1 Show by substitution that these expressions for v_{fA} and v_{fB} do constitute a solution of Equations 3.2 and 3.3.

What can we make of Equations 3.4 and 3.5? Observe that if the two objects have equal masses ($m_A = m_B$), then according to these equations:

$$v_{fA} = 0;$$

$$v_{fB} = v_i.$$

The incoming body stops, and all of its kinetic energy and momentum are simply transferred to B.

What if A is much more massive than B? Then $m_A \gg m_B$, and the factor in parentheses in Equation 3.4 is almost equal to 1, whereas the factor in parentheses in Equation 3.5 is almost equal to 2:

$$v_{fA} \simeq v_i$$

$$v_{fB} \simeq 2v_i.$$

The incoming object is scarcely slowed down at all, although the object that is struck speeds off to the right with a speed nearly twice as great as that of A.

If B is more massive than A, the factor in parentheses in Equation 3.4 becomes negative, in fact nearly equal to -1 if $m_B \gg m_A$. This means that v_{fA}, the final velocity of A, is to the left. When a light object has an elastic collision with a heavy object, it bounces back. In the limiting case in which B is much more massive than A,

$$v_{fA} \simeq -v_i$$

whereas the factor in parentheses in Equation 3.5 is extremely small:

$$v_{fB} \simeq 0.$$

In this extreme case (like that of a tennis ball hitting a truck), the light incoming object is simply "reflected." Its momentum is drastically changed (completely reversed in direction), whereas it retains almost all of its kinetic energy. The heavy object is set in motion only very slightly, enough so that its momentum to the right is large enough to account for the change in momentum experienced by A.

In an elastic collision of this sort between a moving object and a stationary one, all the kinetic energy is transferred from one to the other if the masses are equal. If the masses are unequal, the moving object keeps some of its initial kinetic energy. Let us define a quantity α as a measure of the effectiveness of such elastic collisions at transferring kinetic energy from A to B. It is defined as the fraction of A's initial kinetic energy which is transferred to B:

$$\alpha = \frac{\text{final } KE \text{ of B}}{\text{initial } KE \text{ of A}} = \frac{\frac{1}{2}m_B v_{fB}^2}{\frac{1}{2}m_A v_i^2}. \tag{3.6}$$

This quantity α cannot be greater than 1 and, as seen above, it achieves this value when the two masses are equal. It is easy to discover the general way in which α varies by substituting v_{fB} as given by Equation 3.5 into Equation 3.6. The result is that

$$\alpha = \frac{4m_A m_B}{(m_A + m_B)^2}. \tag{3.7}$$

This equation can be rewritten so that it contains only the *ratio* of the two masses:

$$\alpha = \frac{4\dfrac{m_A}{m_B}}{\left(\dfrac{m_A}{m_B} + 1\right)^2}. \tag{3.8}$$

QUESTION

3.2 Verify Equations 3.7 and 3.8 by making the appropriate substitutions in Equation 3.6.

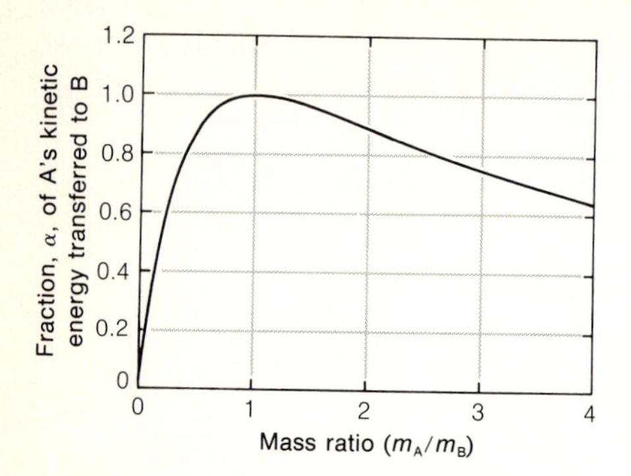

FIGURE 3.12
In an elastic collision, the fraction, α, of A's kinetic energy that is transferred to B depends on the mass ratio.

The way in which α depends on the mass ratio (m_A/m_B) is shown in Figure 3.12. We can see from this graph (or by substituting $m_A/m_B = 1$ into Equation 3.8) that if $m_A/m_B = 1$, then $\alpha = 1$; all of A's kinetic energy is transferred to B, as we saw earlier. The best way to transfer kinetic energy by means of elastic collisions is to use objects of equal mass, but a large fraction of the kinetic energy is still transferred even when the masses are quite different from each other. Even for a mass ratio as high as 3 or as low as $\frac{1}{3}$, 75% of the kinetic energy is transferred from one object to the other.

The toy found on many coffee tables and known as the "Swinging Wonder" (Figure 3.13) is an amusing illustration of elastic collisions. If ball 1 is pulled aside and released, it has an elastic collision with ball 2 and gives up all its kinetic energy. Ball 2 starts to move but almost instantly has an elastic collision with ball 3. As a result of a series of elastic collisions, the kinetic energy of ball 1 is eventually given to ball 5, which swings out and then returns to send the kinetic energy back through the line of balls from right to left. If both balls 1 and 2 are pulled aside, it is observed that balls 4 and 5 are knocked out from the other end of the line. This effect too can be analyzed as a series of elastic collisions. Ball 2 hits ball 3 and stops for an instant but then is struck by ball 1; ball 2 moves again, colliding with ball 3, which has just given up its kinetic energy to ball 4, and so on.

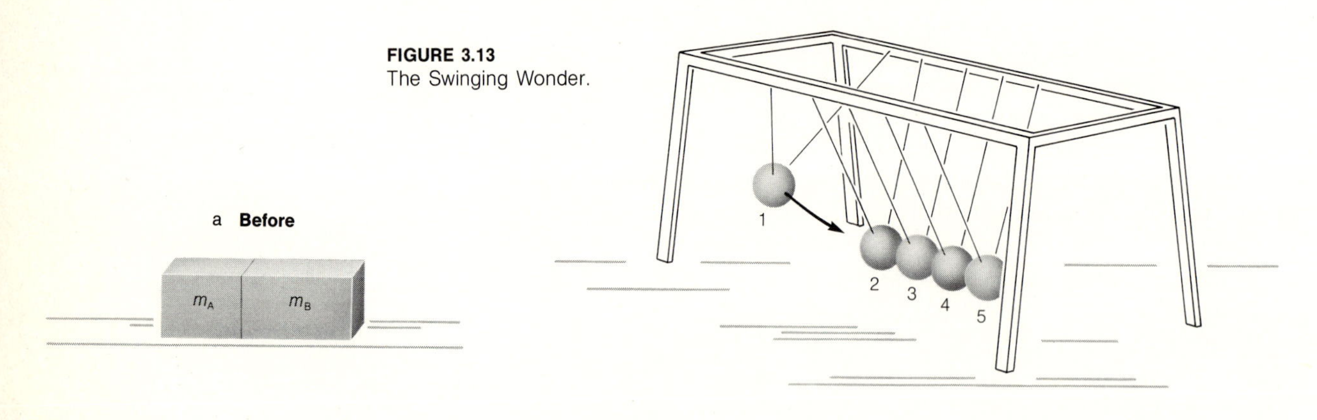

FIGURE 3.13
The Swinging Wonder.

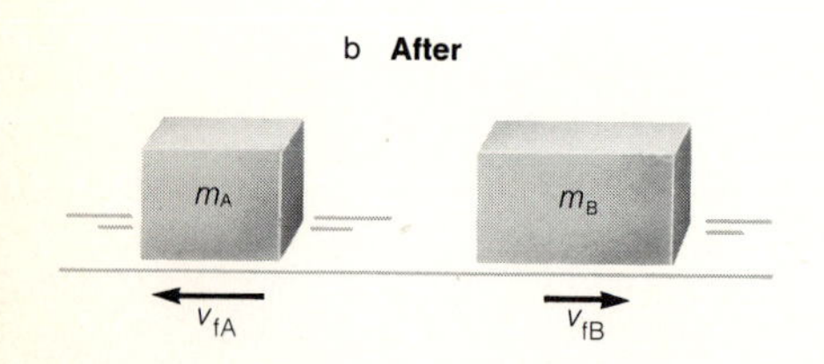

FIGURE 3.14
A superelastic collision, with both objects initially at rest.

SUPERELASTIC COLLISIONS. Now consider a "collision" in which both objects are initially at rest—a superelastic collision in which kinetic energy is created (Figure 3.14). The firing of a cannon or the throwing of a shoe by a person stranded on the ice are good examples. The total momentum is zero to begin with, and it is therefore zero afterwards; the individual momenta must be equal in magnitude and opposite in direction:

$$m_B v_{fB} = -m_A v_{fA}. \tag{3.9}$$

We cannot determine much more about the final velocities without knowing how hard the shoe was thrown or how much gunpowder was in the cannon (in short, how much kinetic energy was created), but we *can* find out how the kinetic energy is shared between the two objects. Let us calculate the ratio of the two kinetic energies, say the ratio of A's kinetic energy to that of B:

$$\frac{KE_A}{KE_B} = \frac{\frac{1}{2}m_A v_{fA}^2}{\frac{1}{2}m_B v_{fB}^2} = \frac{m_A}{m_B}\left(\frac{v_{fA}}{v_{fB}}\right)^2. \tag{3.10}$$

From Equation 3.9 we can find an expression for the ratio of the two final velocities,

$$\frac{v_{fA}}{v_{fB}} = -\frac{m_B}{m_A}$$

and when this is substituted in Equation 3.10, we obtain

$$\frac{KE_A}{KE_B} = \frac{m_A}{m_B} \times \frac{m_B^2}{m_A^2} = \frac{m_B}{m_A}. \tag{3.11}$$

The kinetic energy is distributed between A and B in inverse proportion to their masses. For example, when you throw your shoe in an attempt to get to shore, most of the kinetic energy is delivered to the shoe and only a very small fraction to you. Your momentum is the same size as that of the shoe, but your kinetic energy is much smaller.

When you throw your shoe, you recoil in the opposite direction; when a cannon is fired, it recoils. In each case, the total momentum of the system (shoe-plus-person or cannon-plus-cannonball) remains equal to zero. What happens when you throw a ball straight up in the air? This is another example of a superelastic collision. Do you, together with the earth, recoil in the opposite direction? In principle, yes, but for most purposes the effect can be ignored. Not only is the earth's recoil superimposed on its motion around the sun and on any other recoils (resulting, perhaps, from similar experiments being performed in Australia), but also the recoil velocity and the recoil energy are extremely small. We can apply Equation 3.9 to this example, letting B represent the ball and letting A represent you and the earth, as shown in Figure 3.15. The recoil *momentum* of the earth is equal in magnitude to the momentum of the ball, but the recoil velocity is

$$v_{fA} = -\frac{m_B}{m_A}v_{fB}$$

and m_B/m_A, the ratio of the mass of the ball to that of the earth, is extremely small. Furthermore, according to Equation 3.11, the recoil energy of the earth is also very small. Although the earth does recoil slightly when you throw a ball, it is an extremely good approximation to ignore this effect and to regard the earth simply as the "stage" on which we live.

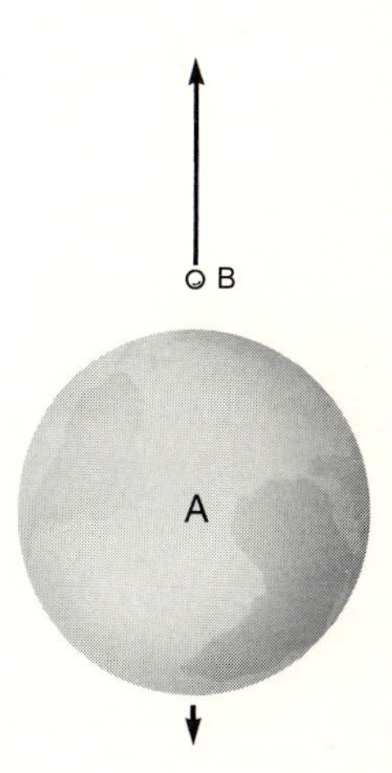

FIGURE 3.15
The earth (A) recoils when a ball (B) is thrown upward.

If the mass of the earth were much smaller, its recoil would have to be taken into account, and the simple act of throwing a ball in the air could have a significant effect on the motion of the earth. This is not an academic matter for occupants of a spaceship, since the mass of things they throw overboard is not totally negligible in comparison with the mass of their "earth." Astronauts deliberately make use of the principles discussed here to alter the motion of a spaceship, and if they want to get rid of some excess fuel, for instance, without affecting the spaceship's motion, they take care to do so by means of a "nonpropulsive vent." That is, the fuel is ejected through two nozzles pointing in opposite directions so that no recoil of the spaceship occurs.

QUESTION

3.3 In §2.3 we discussed the example of a baseball (m = 0.14 kg) thrown vertically into the air with an initial kinetic energy of 3.8 J. Calculate the approximate recoil velocity of the earth in that experiment.

---

§3.3 THE CONCEPT OF FORCE. NEWTON'S SECOND AND THIRD LAWS OF MOTION

§3.3.A The Definition of Force and the Formulation of Newton's Laws of Motion

According to the law of conservation of momentum, in any interaction the sum of all the momenta (the total momentum of the system of interacting objects) remains constant. The momenta of the individual bodies change as they interact with one another; it is only the sum, the vector sum, that does not change. It is valuable to have a way of describing and calculating how the momentum of any *one* of the objects changes, and it is for this purpose that the concept of *force* is used. We have already introduced the term *force,* but no careful definition has yet been given. Our definition of the concept here will be in accord with common usage, in which a force is some sort of push or pull.

Consider, for example, the collision of two gliders on an air track (Figure 3.11). Before the collision, glider A is moving to the right and glider B is at rest; after the collision, A is moving more slowly and B is moving to the right. During the collision, A exerts a force on B, thereby setting it in motion. We will define force to be a vector quantity; in this case the force on B is directed to the right. (We will concentrate our attention on a single object, B, but notice that the motion of A is also affected by the interaction; at the same time that A exerts a force on B that speeds it up, B exerts a force on A in the opposite direction, a force that slows A down.)

If the springs on the gliders are very stiff, the duration of the collision is very brief. B experiences a very strong force but only for a brief period of time, and its momentum changes very rapidly to its final value. But if the gliders have large easily deformable springs, B experiences a rather weak force but for a longer period of time, and its momentum changes more gradually. The overall change in B's momentum is the same; the change simply takes place more gradually if the force is weak. These observations suggest a formal definition of force: the net force acting on an object is a vector equal (in both magnitude and direction) to the *rate of change* of the object's momentum vector. Specifically, if the momentum of an object changes from $\mathbf{p}_1$ to $\mathbf{p}_2$ during a time interval of length Δt,

$$\mathbf{F}_{\text{net}} = \frac{\mathbf{p}_2 - \mathbf{p}_1}{\Delta t} = \frac{\Delta \mathbf{p}}{\Delta t}, \tag{3.12}$$

where $\Delta\mathbf{p}$ is an abbreviation for $(\mathbf{p}_2 - \mathbf{p}_1)$, the change in momentum during the time interval being considered.

Equation 3.12 requires several comments. First, the qualifying adjective "net" has been used because there may be several forces acting simultaneously on one object, and it is found that the rate of change of momentum is the same as if there were a single force, $\mathbf{F}_{\text{net}}$, equal to the *vector sum* of the separate forces. Second, as long as $\mathbf{F}_{\text{net}}$ is not changing, Equation 3.12 is perfectly satisfactory, but if the force is varying, it should be defined by applying Equation 3.12 to a very short time interval. (In the same way, in Chapter 2, §2.3, we calculated the velocity in a situation in which the velocity was varying by using short time intervals.) Alternatively, if $\mathbf{F}_{\text{net}}$ is varying, Equation 3.12 defines its *average* value during the time interval Δt.

Third, Equation 3.12 calls for the *subtraction* of two vectors ($\mathbf{p}_1$ and $\mathbf{p}_2$), and although we have defined earlier the meaning of vector addition, vector subtraction has not yet been defined. We define vector subtraction so that vector addition and subtraction are related to each other in exactly the same way as are addition and subtraction of ordinary numbers. The statement $7 - 5 = 2$ means that 2 is what must be added to 5 to make $7(5 + 2 = 7)$. Similarly, if **A**, **B** and **C** are three vectors, the statement that $\mathbf{B} - \mathbf{A} = \mathbf{C}$ means that **C** is the vector which must be added (vectorially) to **A** to make $\mathbf{B}(\mathbf{A} + \mathbf{C} = \mathbf{B})$. For example, in Figure 3.16a, notice that from the rules of vector addition given earlier, the vector sum of **A** and **C** is equal to **B**, so that this figure can be interpreted as showing either that $\mathbf{A} + \mathbf{C} = \mathbf{B}$ or that $\mathbf{B} - \mathbf{A} = \mathbf{C}$, two equivalent statements. Figure 3.16b demonstrates another important point: **E** and **D** have equal magnitudes, but their difference (**F**) is not zero. Only if two vectors are truly equal, in both magnitude and direction, is their difference equal to zero. Vector subtraction is easy if the vector being subtracted has a magnitude of zero. Glider B in Figure 3.11 was initially at rest ($\mathbf{p}_1 = 0$); as the momentum changes from zero to $\mathbf{p}_2$, the change in momentum is simply $\Delta\mathbf{p} = \mathbf{p}_2 - \mathbf{p}_1 = \mathbf{p}_2$.

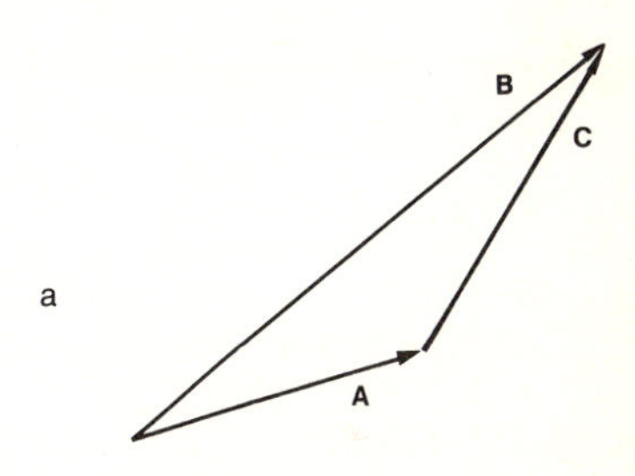

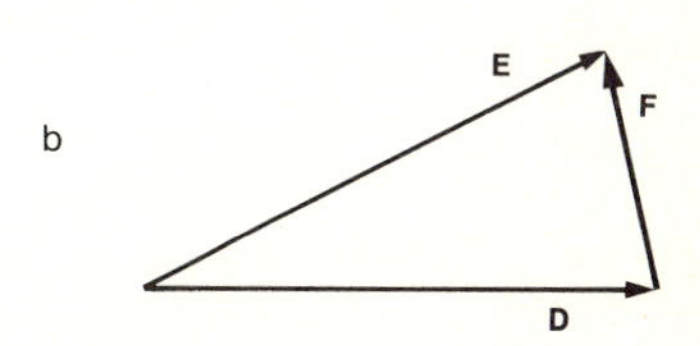

FIGURE 3.16
Two examples of vector subtraction: (a) $\mathbf{B} - \mathbf{A} = \mathbf{C}$; (b) $\mathbf{E} - \mathbf{D} = \mathbf{F}$.

Since $\mathbf{p} = m\mathbf{v}$, $\Delta\mathbf{p} = m\Delta\mathbf{v} = m(\mathbf{v}_2 - \mathbf{v}_1)$, and Equation 3.12 may also be written in the form:

$$\mathbf{F}_{\text{net}} = \frac{m\Delta\mathbf{v}}{\Delta t} = m\left(\frac{\Delta\mathbf{v}}{\Delta t}\right) = m\mathbf{a}, \tag{3.13}$$

where $\mathbf{a}$ is the acceleration vector, defined as the vector change in the velocity vector divided by the length of time during which the change takes place:

$$\mathbf{a} = \frac{\Delta \mathbf{v}}{\Delta t} = \frac{\mathbf{v}_2 - \mathbf{v}_1}{\Delta t}.$$

Equation 3.12 (or its equivalent, Equation 3.13) is known as Newton's second law of motion. It states that the acceleration of an object is in the same direction as the net force acting on it, and that the *magnitudes* of the net force and the acceleration are related by the equation

$$F_{\text{net}} = ma. \qquad (3.14)$$

For a specified value of F_{net}, the acceleration is inversely proportional to the mass of the object on which the force acts:

$$a = \frac{F_{\text{net}}}{m}. \qquad (3.15)$$

The dimensions of force are mass $\times$ velocity/time or mass $\times$ acceleration, that is, mass $\times$ length/time^2. The MKS units of force are therefore kilogram-meters per second^2, and because force is such an important quantity, this combination is given its own name, the *newton* (abbreviated N):

$$1 \text{ N} = 1 \text{ kg-m/sec}^2.$$

Notice in particular that according to Equation 3.13, the net force on an object is equal to zero if the *acceleration* is zero. An object that is simply at rest has zero acceleration, but so does any object moving at constant velocity, that is, at constant speed and with no change in direction. A force is required to set something in motion, but something that is once set in motion will continue to move at constant velocity when the force is removed. This is simply the principle of inertia, Newton's first law of motion, discussed in Chapter 2. In other words, Newton's second law of motion ($\mathbf{F}_{\text{net}} = m\mathbf{a}$) contains the first law; the first law is obtained from the second by applying the second law of motion to the case in which the net force is equal to zero. Why then retain the "first law of motion" as if it were a separate law, when it can logically be regarded simply as a special case of the second law? The explanation of this custom probably lies in the fact that the principle of inertia, the first law of motion, appears so contrary to common sense that it is useful to continue to draw attention to this principle by labeling it as one of the laws of motion.

A few examples will demonstrate how the concept of force, defined as the rate of change of the momentum vector, corresponds to one's intuitive idea of force as a push or a pull. Suppose a frictionless surface, or the nearly frictionless surface of an air table, is available. If a puck is at rest on the table, a push toward the right will accelerate it in that direction, give it a momentum directed toward the right. If the force is removed, the puck will then move at constant speed in a straight line

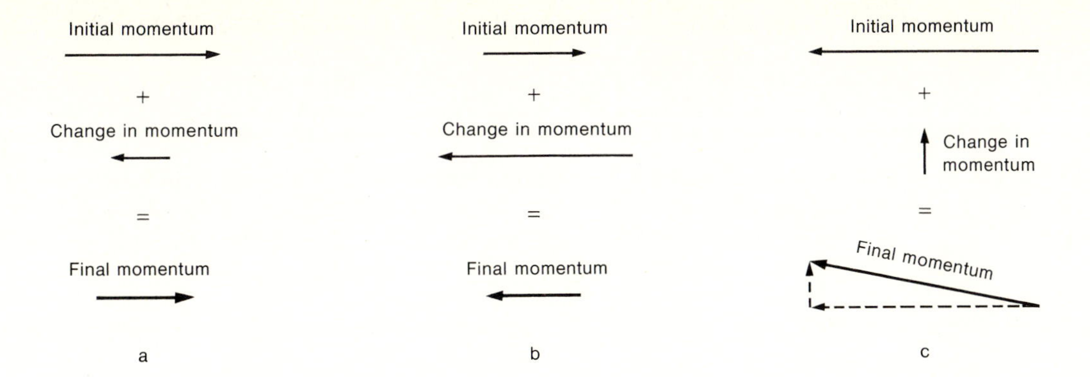

FIGURE 3.17
The final momentum depends on both the initial momentum and the change in momentum imparted by the applied force.

($\mathbf{F}_{net} = 0$, therefore $\mathbf{a} = 0$ or $\mathbf{v} =$ constant). If we then wish to slow it down, we must push to the left, so that its *change* in momentum will be toward the left, as shown in Figure 3.17a. If the puck is moving to the right and we wish to reverse its direction and leave it with the same speed it had before, then we must push harder. We must in fact give it a *change* in momentum directed to the left, whose magnitude is twice as large as its original momentum, as shown in Figure 3.17b. What if the puck is moving to the left and we push sideways? Consider Figure 3.17c. A sideways push changes the direction of motion with little change in speed.

In all these examples, the equations are consistent with what we observe; the direction in which we push is the same as the direction of the resulting change in momentum. Our experience also tells us that the more massive the object, the smaller the acceleration which is produced by a given force, and this too is in accord with the equations.

Since momentum is conserved in the interaction of two bodies, then the momentum change of either one is accompanied by an equal and opposite change in the momentum of the other. Since force is defined as the rate of change of momentum, it follows that if body A exerts a force on body B, changing B's momentum, the simultaneous change in A's momentum can be attributed to an equal and opposite force exerted by B, a force exerted on A. This principle, a corollary of the law of conservation of momentum, is Newton's third law of motion: whenever body A exerts a force on body B, body B at the same time exerts a force on A of equal magnitude but in the opposite direction. This law is applicable not only to the forces which arise during collisions but to all forces. When you sit in a chair, the chair pushes upward on you; at the same time, you are pushing downward on the chair. When a car going north collides with a telephone pole, the car exerts a northward force on the pole, and at the same time the telephone pole exerts a southward force on the car. When glider A hits glider B in Figure 3.11, A exerts a force to the right on B, and at the same time B exerts a leftward force on A. If you throw a shoe, you exert a force on the shoe, and the shoe simultaneously exerts a force on you in the opposite direction.

§3.3.B Gravitational Forces. The Distinction Between Mass and Weight

What can we say about the gravitational force acting on something falling under the influence of gravity? An object that is falling freely has a downward acceleration, and so the force (simply the gravitational force exerted by the earth) is directed downward. The magnitude of the downward acceleration is simply g (9.8 m/sec^2), and so, according to Equation 3.14, the magnitude of the gravitational force is

$$F_{\text{grav}} = mg. \tag{3.16}$$

If we consider various objects with various masses, the gravitational forces the earth exerts on them vary in proportion to their masses (according to Equation 3.16) since g is the same for all objects.

What is the gravitational force exerted by the earth on an object whose mass is 1 kg? If we put $m = 1$ kg and $g = 9.8$ m/sec^2 in Equation 3.16, we see that for this object

$$F_{\text{grav}} = 1 \times 9.8 \text{ kg-m/sec}^2 = 9.8 \text{ N}.$$

Similarly, the gravitational force on my body ($m \simeq 160$ lb $\simeq 73$ kg) is

$$F_{\text{grav}} = 73 \times 9.8 \simeq 715 \text{ N}.$$

It is clear that force and mass are quite different quantities, with different units (newtons for force, kilograms for mass). A force is some sort of push or pull that acts on an object; a force is a vector quantity. An object's mass is a measure of its "inertia," the tendency it has to maintain a constant velocity. If equal forces are applied to two different objects, the one with the larger mass will experience the smaller change in velocity.

It is appropriate to digress briefly to consider two difficult questions: the use of the term "weight," and the use of units other than the newton and the kilogram for measuring force and mass. "Weight" and "mass" are two terms commonly used almost interchangeably; but for some scientific purposes, an object's "weight" is defined as the magnitude of the *gravitational force* exerted on that object by the earth.* Since mass and force are different sorts of quantities, this can and does cause considerable confusion. The problem is not quite as serious as one might think, however, because the gravitational forces exerted by the earth on various objects are directly proportional to their masses (Equation 3.16). If object A has twice the mass of B, the gravitational force on A (its "weight") is also twice as large as the gravitational force on B. Since gravitational forces vary with position, the statement is correct only if both objects are at the same position. In the hopes of avoiding some confusion, "weight" is a term which will usually be avoided in this

*If this terminology is used, my mass is about 73 kg, but my weight is 715 N.

book; when we wish to refer to the gravitational force on an object, we will express it as F_{grav}.

The fact that the gravitational forces on various objects are directly proportional to their masses means that in practice we can determine the mass of one object relative to another by an indirect procedure, by comparing the gravitational forces which the earth exerts on them (by "weighing" them). This procedure is very convenient, but one consequence is that we find such units as the pound and the ton in use as units of both force and mass. This dual usage can be confusing, because force and mass are different concepts. A 1-lb object is one whose mass is about 0.45 kg; it might be described either as an object whose *mass* is 1 lb or as an object on which the *force* of gravity is 1 lb. Some of the confusion can be avoided by using two different units: the "pound (mass)" and the "pound (force)." ["I have a mass of 160 pounds (mass); the gravitational force on me is 160 pounds (force)."] Needless to say, most of us do not have the time to be this precise in our speech. Often the ambiguity presents no problem, for everyone knows quite well what I mean when I say that I weigh 160 pounds.

In case of doubt, one can always use MKS units and express masses in kilograms and forces in newtons; this is the safest thing to do when using equations such as $\mathbf{F} = m\mathbf{a}$. Even in the MKS system, though, one will occasionally find the kilogram used as if it were a unit of force, meaning the gravitational force exerted on something of mass 1 kg.

In converting tons of coal into pounds of coal or kilograms of coal, the ambiguity is quite tolerable. 1 ton is equal to 2000 lb, whether one is discussing force or mass, and the meaning of the statement that 1 ton = 907.2 kg is also completely clear from the context. The conversion tables in Appendix A contain entries for units such as the pound in both the force and mass tables. One pound (mass), for instance, is approximately equal to 0.45 kg, and one pound (force) is equal to the gravitational force exerted on such an object. In MKS units, this force is

$$1 \text{ lb (force)} = mg = 0.45 \text{ kg} \times 9.8 \text{ m/sec}^2 \simeq 4.4 \text{ N.*}$$

§3.3.C One-Dimensional Applications of Newton's Second Law of Motion

Let us now consider a few simple examples that illustrate Newton's second law.

(1) A 2-lb book sits at rest on a table. What forces act on it? We can deduce from Newton's second law (Equation 3.13) that the *net* force on the book must be zero, for the book's acceleration is zero, but this net

*The acceleration due to gravity varies somewhat from place to place, even on the surface of the earth. The conversion factors given in Appendix A are based on what has been accepted as the "standard value" of g (9.80665 m/sec²) and on the more precise conversion factor between pounds and kilograms (1 lb = 0.4536 kg).

force is composed of two parts: the downward gravitational force exerted by the earth ($\mathbf{F}_{grav}$) and an upward supporting force exerted by the table (let us call it $\mathbf{F}_{table}$). A diagram of the forces that act on the book is shown in Figure 3.18. The magnitude of the gravitational force can be deduced from Equation 3.16 if m is expressed in kilograms:

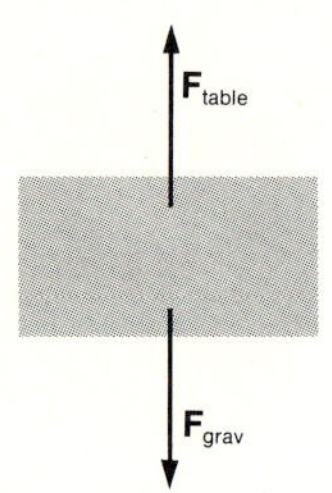

FIGURE 3.18
Force diagram of a book resting on a table.

$$m = 2 \text{ lb} = 2 \text{ lb} \times \frac{1 \text{ kg}}{2.205 \text{ lb}} \simeq 0.9 \text{ kg}.$$

Therefore $F_{grav} = 0.9 \text{ kg} \times 9.8 \text{ m/sec}^2 \simeq 8.8 \text{ N}$, and we can immediately conclude that the force exerted by the table has exactly the same size.

(2) A 2-lb glider moves at constant velocity along a horizontal air track. What forces act on the glider? As far as the *forces* are concerned, this example is identical to the one just discussed; the force diagram shown in Figure 3.18 applies to this example as well. The acceleration of the glider is zero, as is that of the book at rest on the table. Therefore the net force is again equal to zero, and the upward force of the supporting stream of air must have a magnitude of 8.8 N, equal to that of the downward gravitational force.

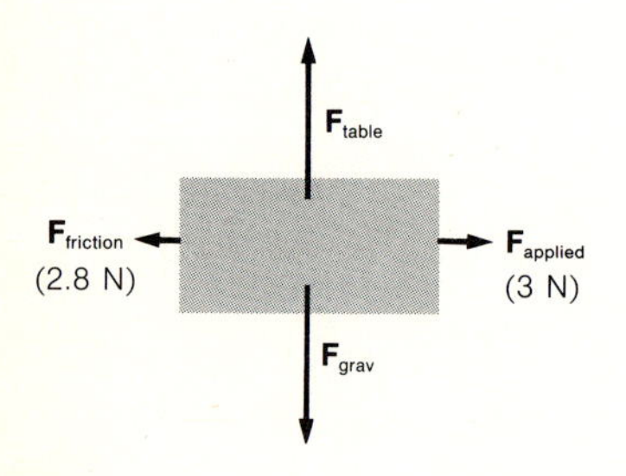

FIGURE 3.19
Force diagram of a book being pushed across a table.

(3) A 2-lb book ($m = 0.9$ kg) is being pushed across a table by a horizontal force of 3 N. Unlike the example of the glider on an air track, friction is extremely important in this case. Suppose that the retarding frictional force has a magnitude of 2.8 N (Figure 3.19). What can we say about the motion of the book? Again the upward force exerted by the table and the downward force of gravity cancel each other; the *net* force is a horizontal force of 0.2 N. The acceleration of the book can therefore be calculated from Equation 3.15:

$$a = \frac{F_{net}}{m} = \frac{0.2 \text{ N}}{0.9 \text{ kg}} = 0.22 \text{ m/sec}^2.$$

(4) Consider the forces acting on a 160-lb person ($m = 73$ kg) in an elevator that is *accelerating* upward at a rate of 3 m/sec². The *net* force acting on the person can immediately be calculated from Newton's second law; it is in the upward direction and it has a magnitude given by

$$F_{net} = ma = 73 \times 3 = 219 \text{ N}.$$

This net upward force of 219 N (Figure 3.20a) results from the combination of two forces: the *downward* force of gravity,

$$F_{grav} = mg = 73 \times 9.8 = 715 \text{ N},$$

and an upward force exerted by the floor of the elevator. In order that the *net* upward force be 219 N, the force exerted by the floor must exceed the force of gravity (715 N) by 219 N:

$$F_{floor} = 219 + 715 = 934 \text{ N},$$

as shown in Figure 3.20b.

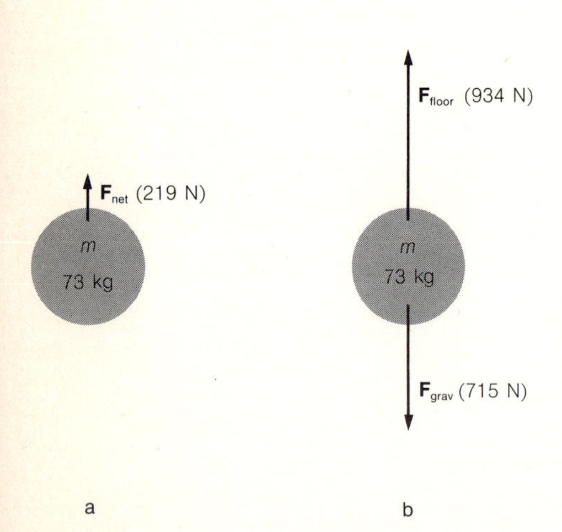

FIGURE 3.20
Forces on a person in an elevator that is accelerating upward.

If the elevator were standing still, or if it were moving at any *constant* velocity (up or down), the net force on the person would be zero, and the upward force exerted by the floor would therefore be precisely equal to F_{grav}, 715 N (a force of 160 lb, equal to the person's "weight"). The upward force exerted by the floor is larger than this while the elevator is accelerated upward. We can thus understand why one "feels heavier" for a few seconds as a rapidly accelerating elevator starts upward; it is not that the gravitational force is any larger but rather that the floor is pushing upward on our feet with an abnormally large force.

(5) What if the elevator's cable breaks, so that the car is falling freely? What will be your experience as a passenger during your last glorious moments? In this case, you and the elevator are both accelerating downward with an acceleration equal to g, 9.8 m/sec^2. The net force on you is therefore in the downward direction and, if your mass is 73 kg, the magnitude of the net force is

$$F_{net} = ma = 73 \times 9.8 = 715 \text{ N},$$

and this is precisely equal to the known force exerted by gravity:

$$F_{grav} = mg = 715 \text{ N}.$$

There can therefore be no other force acting on you. The floor of the elevator exerts no force on you at all: if it did, the net force would not be equal to 715 N. This is what is commonly referred to as a state of "weightlessness." The clues that you normally rely on to sense the presence of gravity (such as the floor pushing upward against your shoes) are absent; as far as you can tell, the elevator might be coasting through interstellar space. "Free fall" experiments of this sort have been carried out to study human reactions to the apparent absence of gravity. The absence of gravity is only apparent: in such an experiment the gravitational force is actually the *only* force. (These experiments are necessarily of very brief duration, and in order that the subjects survive, the falling elevator must be brought to a gradual stop at the bottom.) The same sensation of "weightlessness" is experienced by astronauts in earth satellites; in this case, too, the earth's gravitational force is the only force acting on them.

§3.3.D Force and Acceleration in Uniform Circular Motion

Let us consider one final important application of Newton's second law of motion. Suppose that an object is moving at *constant speed* in a circular orbit ("uniform circular motion"). Tie a rock to a piece of string and twirl it around your raised finger; the rock executes uniform circular motion. The moon executes nearly perfect uniform circular motion around the earth, as do many artificial earth satellites, and the planets in their motion around the sun.

How does Newton's second law of motion apply to the uniform circular motion of a rock attached to a piece of string? The important thing to recognize is that, even though the rock is moving at constant *speed,* the direction in which it is moving is constantly changing, and therefore the momentum (a *vector* quantity) is not constant. Another way of putting this is to say that the acceleration vector is not zero, even though the rock is neither speeding up nor slowing down, because the *direction* of the velocity keeps changing. The relationship between force and acceleration in uniform circular motion is correctly described by Newton's second law; this provides the best possible reason for insisting that an object whose direction of motion is changing has an acceleration, even though its speed may be constant.

The two most important facts about the acceleration vector of an object in uniform circular motion are, (1) that it is not zero, and (2) that it always points toward the center of the circle. Consider the two velocity vectors $\mathbf{v}_1$ and $\mathbf{v}_2$ shown in Figure 3.21a. The difference between them, $\Delta\mathbf{v}$, is found by subtracting the two vectors (Figure 3.21b). The vector $\Delta\mathbf{v}$ is certainly not zero, and if it is associated with the point P′ (midway between P_1 and P_2 in Figure 3.21c), it is clear that $\Delta\mathbf{v}$ points toward the center of the circle. If P_1 and P_2 are closer together (Figure 3.21d), $\Delta\mathbf{v}$ is smaller but still directed toward the center, and so the acceleration vector is always directed toward the center. For this reason, it is described as a "centripetal" (center-seeking) acceleration.

The *magnitude* of the acceleration vector can be calculated in terms of the object's speed (v) and the radius (r) of its orbit. The calculation is given at the end of this section; the result is that the centripetal acceleration in uniform circular motion is given by the simple equation

$$a = v^2/r. \tag{3.17}$$

We can understand why the acceleration depends on v and r in this particular way, even without going through each step of the calculation. Consider two objects moving in uniform circular motion with orbits of the same radius, but with one object moving at twice the speed of the other. The change in velocity, $\Delta\mathbf{v}$, which takes place during, say, a quarter of a revolution, is twice as large for the faster object. This explains one of the two factors of v in Equation 3.17. Furthermore, the faster object takes only half as long to execute this motion because of

FIGURE 3.21
Changes in the velocity of an object executing uniform circular motion.

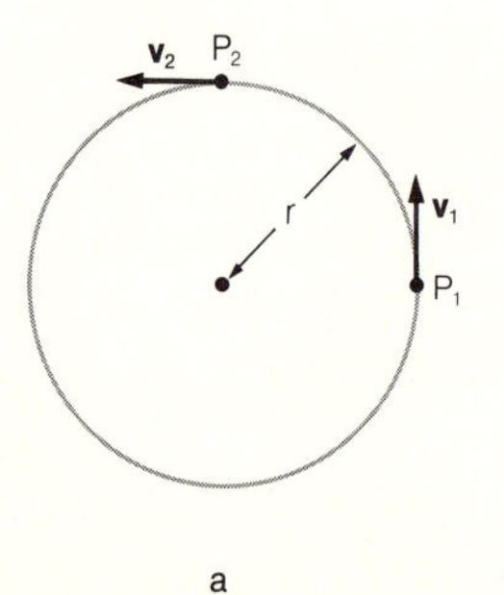

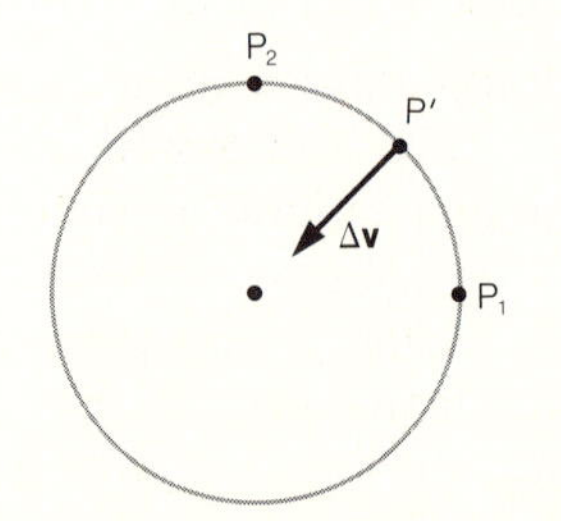

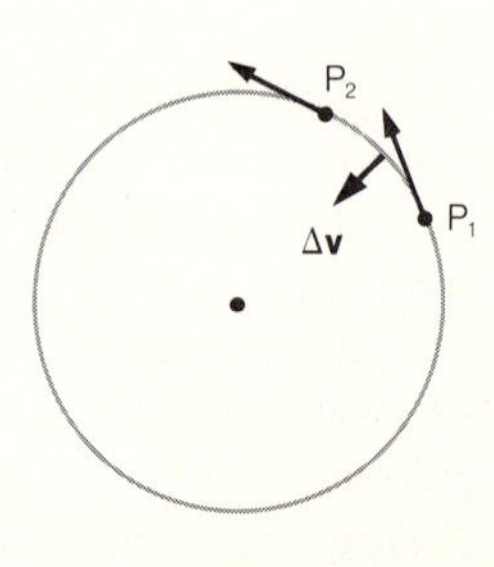

its greater speed. Together, these two observations explain why the acceleration is proportional to the *square* of the speed. Now think of two objects moving at equal speeds but in orbits of different sizes. The one with the smaller orbit takes less time to complete each orbit, and this explains why the acceleration is *inversely* proportional to r. These arguments show that the acceleration is *proportional* to the combination v^2/r. They do not prove that the constant of proportionality in Equation 3.17 is equal to 1; to do so, one must actually work through the detailed calculation.

QUESTION

3.4 Check the dimensions of Equation 3.17. Show that v^2/r does have the dimensions of acceleration.

Any object that is executing uniform circular motion has an acceleration even though its speed is not changing. If there is an acceleration toward the center, there must likewise be a net force toward the center, and the magnitude of this force is given by Newton's second law:

$$F = ma = \frac{mv^2}{r}. \tag{3.18}$$

Whatever its origin, such a force—one which produces a centripetal acceleration—is often referred to as a "centripetal force." For the rock on the end of a string, this force is exerted by the string. For satellites moving around the earth and the planets moving around the sun, this force is a gravitational force. An earth satellite is constantly "falling" toward the center of the earth, falling in the sense that it has an acceleration toward the center of the earth. If one were to take a satellite aloft and simply drop it, it would literally fall to the earth, but if it is instead given the proper sideways velocity, the downward force will simply maintain it in a circular orbit, its velocity constantly changing in direction but not in size, its distance from the earth never changing.

Although it may at first seem somewhat artificial to say that a rock traveling at constant speed has an "acceleration," anyone who has ever attempted to whirl a rock on the end of a string knows that an inward directed force must be applied. Newton's second law can be quantitatively checked as well, by measuring the force exerted by the string and comparing it with the value calculated from Equation 3.18. The best way to test these ideas is to study the motions of planets and satellites, and we shall examine these motions more carefully in Chapter 5.

What happens if the inward force is suddenly removed, if the string breaks? There is no longer a force acting, and the rock then moves in accord with the principle of inertia. Whatever velocity it has when the

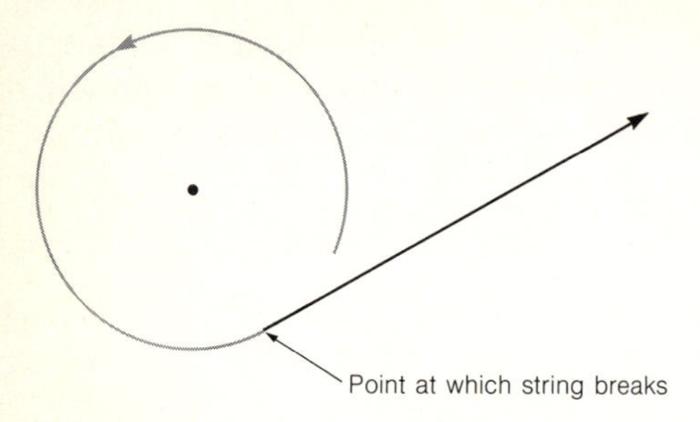

FIGURE 3.22
The moment the force is removed, the rock's motion becomes one of constant velocity—motion in a straight line at constant speed.

string breaks, it continues onward with the same velocity, as shown in Figure 3.22. (The rock does eventually fall to the ground, of course; in this discussion, we have been ignoring the ever-present gravitational force.) Although it is sometimes said that when the string breaks, the rock "flies outward" because of a "centrifugal" force (a "center-fleeing" force), there is no need to invoke any such imaginary force. When the string breaks, the rock moves as it does—at constant speed in a straight line—simply because there is no force present to cause it to do anything else.

It is worth noting that uniform circular motion is an interesting and important type of motion in which no energy changes occur. Even though an object in uniform circular motion has an acceleration, its speed and its kinetic energy do not change; in this respect, uniform circular motion is indistinguishable from motion at constant speed in a straight line. This is another reminder that there is more to physics than the study of energy and energy changes.

Let us now return to the derivation of Equation 3.17, which gives the acceleration of an object undergoing uniform circular motion. Consider the velocity vector at two closely spaced instants of time (Figure 3.23a). Let θ denote the angle between the two radii, $\overline{OP_1}$ and $\overline{OP_2}$. Since $\mathbf{v}_1$ and $\mathbf{v}_2$ are perpendicular to $\overline{OP_1}$ and $\overline{OP_2}$ respectively, the angle between them is also θ, and so the vector diagram for calculating $\Delta\mathbf{v} = \mathbf{v}_2 - \mathbf{v}_1$ is as shown in Figure 3.23b. If we take a very small time interval, the angle θ will become very small, $\mathbf{v}_1$ and $\mathbf{v}_2$ will be almost parallel to one another, and in the limit of very short time intervals, the vector $\Delta\mathbf{v}$ becomes perpendicular to both $\mathbf{v}_1$ and $\mathbf{v}_2$. Thus we can see that the instantaneous acceleration is perpendicular to the instantaneous velocity and is directed toward the center of the orbit. We can also calculate the magnitude of the instantaneous acceleration vector. The triangle shown

FIGURE 3.23
Geometrical constructions for calculating the acceleration of an object undergoing uniform circular motion.

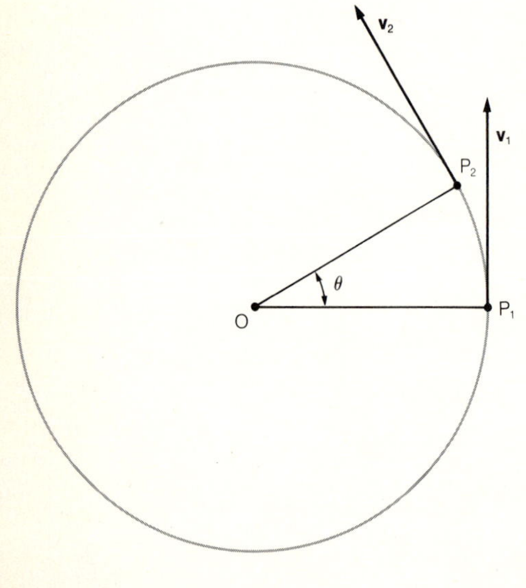

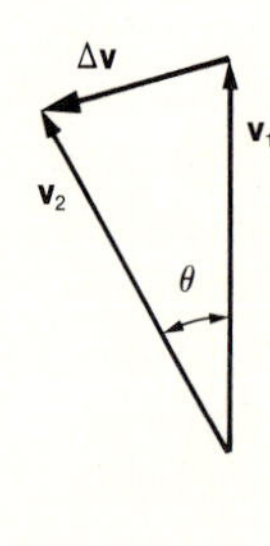

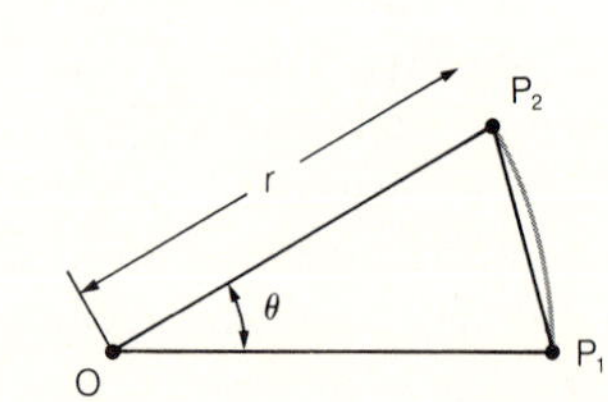

in Figure 3.23b is *similar* to the triangle OP_1P_2, which is shown separately in Figure 3.23c. From the properties of similar triangles,

$$\frac{\text{magnitude of } \Delta\mathbf{v}}{v} = \frac{\overline{P_1P_2}}{r}$$

or

$$\text{magnitude of } \Delta\mathbf{v} = \frac{v\overline{P_1P_2}}{r}, \tag{3.19}$$

where $\overline{P_1P_2}$ is the straight-line distance between P_1 and P_2. The object actually travels along the curved arc joining P_1 and P_2, and if Δt is the length of the time interval in question, the distance traveled by the object is $v\Delta t$. If P_1 and P_2 are very close together, the straight-line distance between P_1 and P_2 is nearly equal to the length of the curved arc, and thus in the limit, we can substitute $v\Delta t$ for $\overline{P_1P_2}$ in Equation 3.19:

$$\text{magnitude of } \Delta\mathbf{v} = \frac{v^2\Delta t}{r}. \tag{3.20}$$

But a, the magnitude of the acceleration vector, is just equal to the magnitude of $\Delta\mathbf{v}$ divided by the length of the time interval, so from Equation 3.20:

$$a = \frac{v^2}{r}.$$

The instantaneous acceleration vector has a magnitude equal to v^2/r and a direction toward the center of the orbit. The same result would apply, no matter where on the circular orbit we were, always the same magnitude and always directed toward the center, always perpendicular to the direction of the instantaneous velocity vector.

§3.4 THE CONCEPT OF WORK. WORK AS A MEASURE OF ENERGY TRANSFER

Our central concern is the description of physical processes in terms of energy, but energy has played a rather minor role so far in this chapter. Is it possible to find a relationship between the force acting on an object and the change in energy? Force is defined in terms of its effect in producing changes in momentum:

$$\mathbf{F}_{\text{net}} = \frac{\Delta\mathbf{p}}{\Delta t}.$$

If the kinetic energy of an object is changing, its momentum and velocity vectors are changing, and so there must be a non-zero net force acting. The converse of this statement, though, is not necessarily correct. There may be a net force acting without any change in kinetic energy, as seen in the case of uniform circular motion discussed in §3.3.D. Because of this complication, it is easiest to grasp the relationship between forces and energy changes by considering separately two cases: first the case in which an object moves in a straight line and is subjected to a force either in the direction of its motion (which tends to speed it up) or in the opposite direction (which tends to slow it down); second, cases like that of uniform circular motion in which the force is perpendicular to the velocity.

If an object moves along a straight line and experiences forces directed along this line, the problem can readily be handled without the full machinery of vector addition and subtraction. Consider an object moving along a line as indicated in Figure 3.24 and let v denote the component of its velocity along this line, positive if it is moving to the right and negative if to the left. Whether v is positive or negative, the kinetic energy is given by $\frac{1}{2}mv^2$. Suppose for simplicity that a steady force of magnitude F, directed toward the right, acts on this object for a time interval of length Δt. This force will cause a change in kinetic energy, and we want to relate the change in kinetic energy to the value of the force. We can write the change in kinetic energy as

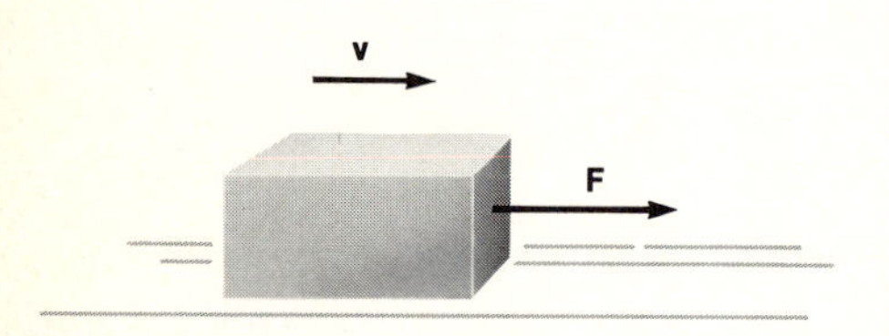

FIGURE 3.24
A force in the direction of motion results in an increase in kinetic energy.

$$\Delta(KE) = \Delta(\tfrac{1}{2}mv^2) = \tfrac{1}{2}mv_f^2 - \tfrac{1}{2}mv_i^2 = \tfrac{1}{2}m(v_f^2 - v_i^2), \qquad (3.21)$$

where v_i and v_f refer to the beginning and end of the particular time interval. Now we know that

$$F_{net} = m\frac{\Delta v}{\Delta t} = m\frac{(v_f - v_i)}{\Delta t}. \qquad (3.22)$$

A connection between Equations 3.21 and 3.22 can be found by factoring Equation 3.21:

$$\Delta(KE) = \tfrac{1}{2}m(v_f - v_i)(v_f + v_i). \qquad (3.23)$$

According to Equation 3.22,

$$(v_f - v_i) = \frac{F_{net}\Delta t}{m}$$

and we can substitute this result into Equation 3.23:

$$\Delta(KE) = \tfrac{1}{2}m\frac{F_{net}\Delta t}{m}(v_f + v_i)$$

$$= F_{net}\Delta t\left(\frac{v_f + v_i}{2}\right). \qquad (3.24)$$

Now $(v_f + v_i)/2$ is the arithmetic mean of the velocities at the beginning and end of the time interval; if the velocity changes steadily during this time interval, this is equal to the average velocity, and the distance traveled is equal to the average velocity multiplied by Δt, the length of the time interval:

$$\text{distance} = \text{average velocity} \times \Delta t$$

$$= \left[\frac{v_f + v_i}{2}\right]\Delta t.$$

Therefore from Equation 3.24,

$$\Delta(KE) = F_{\text{net}} \times \text{distance}. \tag{3.25}$$

The quantity on the right-hand side of Equation 3.25, $F_{\text{net}} \times$ distance, is called the *work* done by the net force, and in this equation we have the desired relationship between forces and changes in energy:

$$\text{Work done by } F_{\text{net}} = \Delta(KE). \tag{3.26}$$

Two important qualifications must immediately be added to the definition of work. First, it is easy to see that, if the force acts in the direction opposite to the direction of motion, the result is that the object slows down and the kinetic energy decreases. In this case, $\Delta(KE)$ is negative, but we can retain Equation 3.26 if we agree to count work as negative when the force acts in the direction opposite to the direction of motion. Second, if the force is always *perpendicular* to the direction of motion, no change in kinetic energy results, as we saw in the discussion of uniform circular motion in §3.3.D. If the force is at some other angle with respect to the direction of motion (Figure 3.25), we can treat the force as the vector sum of two forces, one along the line of motion and one perpendicular. Only the component of the force along the line of motion (a positive quantity in Figure 3.26a, a negative one in Figure 3.26b) is to be counted in determining the work done.

Thus the general definition of the work done by a force is

$$\text{Work} = (\text{the component of } \mathbf{F} \text{ in the direction of motion}) \times (\text{distance traveled by object on which force acts}). \tag{3.27}$$

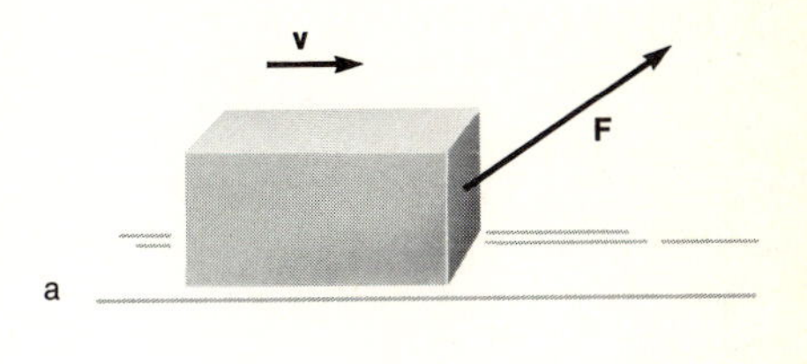

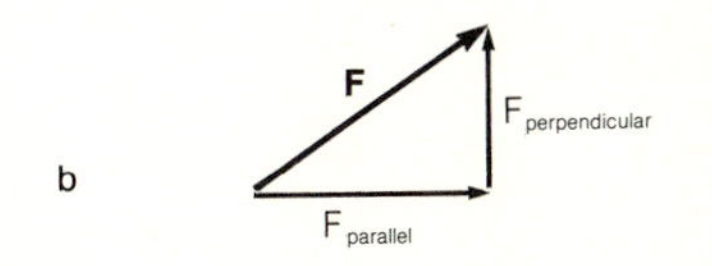

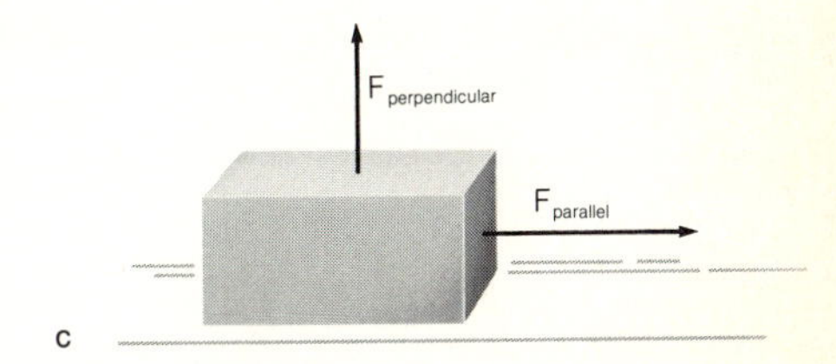

FIGURE 3.25
Any force can be treated as the vector sum of two forces, one parallel to the direction of motion and another perpendicular to this direction.

With this definition, Equation 3.26 can be applied in all cases. This equation is a valuable tool for describing how energy is given to an object. If you push a glider along an air track, you do work on it, and the result is an increase in the glider's kinetic energy. Energy has been transferred from your body's store of energy to the glider. Work represents a way of *transferring* energy.

Although work and energy are closely related concepts, they are not the same. They must have the same dimensions and must be measured in the same units, as we can see from Equation 3.26, in which work is set equal to a change in kinetic energy. Work and energy differ in the same way that a bank *deposit* and a bank *balance* differ. Deposits and balances are measured in the same units (dollars), but they are different concepts; in strict analogy with Equation 3.26, the *change* in the balance is equal to the amount of the deposit. For a positive deposit of \$10, the balance increases by \$10; for a "deposit" of minus \$10 (a withdrawal), the balance decreases by this amount.

Equation 3.26 relates the change in kinetic energy to the work done by the *net* force. If the net force is composed of several separate forces, a similar result holds: the change in kinetic energy is equal to the sum of the amounts of work done by the separate forces. In many ways, the use of the word "work" does not represent a happy choice, for although

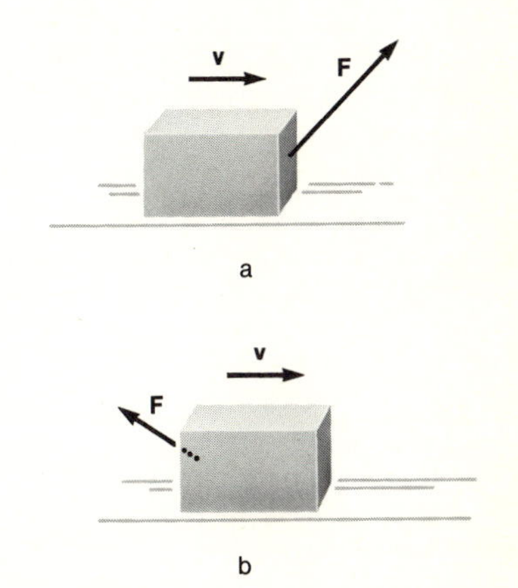

FIGURE 3.26
The component of the force in the direction of motion is positive in (a), negative in (b).

the concept as defined here is related to the ordinary meaning of the term, the correspondence is not perfect, as can be seen from a couple of examples. Suppose that you are pushing very hard on a stalled car, but not hard enough to move it; the opposing frictional forces are large enough to counteract the force you apply. Although the force you apply may be very large, in the technical sense you are not doing any work on the car, work being defined as the product of force and distance. As long as the car does not move, no work is done. In the ordinary sense of the term, you are certainly working very hard. The chemical processes going on in your muscles are much the same whether you succeed in moving the car or not, but as long as the car does not move, you are merely converting energy from one form to another inside your body; you are not delivering energy to the car.

To take a few more examples, when you lift a suitcase from the floor to a table, you *are* doing work on the suitcase, because you are exerting a force on the suitcase in the direction in which it is moving. However, if you just hold the suitcase, you are *not* doing any work on it, though it may be almost as tiring to hold it as to lift it. Furthermore, if you carry the suitcase from one place to another, then as long as you stay on the same level, technically you do no work on the suitcase. Although you exert a force and the suitcase *is* moving, your force is upward, perpendicular to the direction of motion, and therefore does no "work." When you take a suitcase from a table and lower it to the floor, you are doing a *negative* amount of work; the force you exert is upward, but the direction of motion is downward. When you twirl a rock on the end of a string, or in any other example of uniform circular motion, the force acting does no work, for it is always perpendicular to the direction of motion. In accord with Equation 3.26, this centripetal force causes no change in kinetic energy.

An especially interesting result emerges if Equation 3.26 is applied to the vertical motion of an object. Let us re-examine the example of a ball that is thrown upward, considering not only how it moves after you let it go but also the way in which it gets started. In order for the ball to be thrown, an upward force must be applied to it (by your hand). Let us call the work done by this external force W_{ext}. This is a positive quantity, since the force is applied in the direction in which the ball moves. The *net* force on the ball also includes the force of gravity, and if we call the work done by gravity W_{grav}, then from Equation 3.26 it follows that

$$W_{\text{ext}} + W_{\text{grav}} = \Delta(KE). \tag{3.28}$$

The quantity W_{grav} can easily be calculated. The magnitude of the gravitational force is equal to mg (Equation 3.16), and its direction is downward. Therefore the work done by the gravitational force as you throw the ball is *negative;* during the time that the ball's height changes by Δh, W_{grav} is given by force $\times$ distance but with a negative sign:

$$W_{\text{grav}} = -mg\Delta h.$$

Thus Equation 3.28 can be written

$$W_{\text{ext}} - mg\Delta h = \Delta(KE). \qquad (3.29)$$

If we now add $mg\Delta h$ to both sides of Equation 3.29, we have

$$W_{\text{ext}} = \Delta(KE) + mg\Delta h.$$

The term $mg\Delta h$, which we can equally well write as $\Delta(mgh)$, is just the change in gravitational potential energy:

$$W_{\text{ext}} = \Delta(KE) + \Delta(GPE) = \Delta(KE + GPE). \qquad (3.30)$$

In other words, by concentrating our attention on the work done by the external force, we see that the work done by the external force shows up as a change in the sum of the kinetic and gravitational potential energies of the ball. The work that you do with your hand serves to transfer energy from you to the ball, to increase both its kinetic energy and its gravitational potential energy. Observe that the change in gravitational potential energy, $\Delta(mgh)$, appears here as the negative of "the work done by the gravitational *force*." Motions such as those of objects moving under the influence of gravity can be described in either of two ways, in terms of changes in *energy* or in terms of *forces*. Both descriptions are valid, though one or the other may be more convenient in particular circumstances.

If Equation 3.30 is applied to some time interval *after* you let go, W_{ext} is equal to zero, and then we have simply

$$0 = \Delta(KE + GPE)$$

or

$$KE + GPE = \text{constant}.$$

Once again we have the principle of conservation of mechanical energy for a ball moving under the influence of gravity. We have a more powerful result, though, because in Equation 3.30 we have a way of describing how energy is transferred to the ball in order to start it moving.

We can even apply Equation 3.30 to the final part of the ball's motion, when it returns to the ground. If you stop the ball by catching it, you must exert an upward force to slow it down. Since this force is in the direction opposite to that in which the ball is moving, the work it does is negative. If W_{ext} in Equation 3.30 is negative, then so is the change in the total mechanical energy of the ball. That is, the ball's energy decreases as it is stopped. If the ball is stopped by letting it hit the ground, then it is the surface of the earth that does negative work on it.

Thus the whole history of the ball is described by Equation 3.30. As the ball is thrown, a positive amount of work is done on it by the external force, increasing its total mechanical energy. After the ball is released, its total mechanical energy remains constant: kinetic energy is converted into gravitational potential energy on the way up, and gravitational potential energy is converted back to kinetic energy as the ball falls. Finally, as the ball is stopped, negative work is done on it, decreasing its total mechanical energy.

Equation 3.30 describes the essence of the law of conservation of energy and of the way in which energy transfers are described. As long as $W_{ext} = 0$, the ball is a closed system as far as energy is concerned; its total mechanical energy remains constant. When W_{ext} is *not* equal to zero, the ball is not a closed system, energy is transferred to or from the ball, and the change in the ball's energy is simply equal to the work done on it (positive or negative) by external forces. The full energy story emerges when Equation 3.30 is generalized to include other forms of energy besides kinetic energy and gravitational potential energy (chemical energy, electrical energy, thermal energy, etc.) and by recognizing that energy can be transferred by "heat" as well as by work.

FURTHER QUESTIONS

3.5 A 1-ton sedan going north at 60 miles/hr smashes head-on into a 10-ton truck headed south at the same speed. Predict the direction and magnitude of the velocity of the wreckage immediately after the collision.

3.6 A 1-ton sedan is headed east through an intersection at 30 miles/hr when it is rammed in the side by a 10-ton truck headed north at 60 miles/hr. The two vehicles stick together. Predict the direction and magnitude of their velocity immediately after the collision.

3.7 (a) Father and daughter are coasting on their bicycles along a level road at 10 ft/sec, when the father gives his daughter a forward push. Immediately afterwards, the daughter's speed is 20 ft/sec. The masses of father and daughter (bicycles included) are 200 lb and 100 lb, respectively. Assuming that there are no frictional effects to worry about, what is the father's new speed?

(b) The kinetic energy of the daughter has increased and that of the father has decreased. Has the total kinetic energy increased, decreased, or remained the same?

(c) If the father had pushed hard enough so that just afterwards he himself had come to a stop, what would have been the daughter's new speed? What would be the change in the total kinetic energy in this situation?

3.8 In a nuclear power plant, fast neutrons are released in the fission of uranium nuclei. These neutrons can cause fission of other uranium nuclei and thus produce a continuing chain reaction, but slow neutrons are more effective in causing fission than fast ones. The best way of slowing a neutron down is to let it have an elastic collision with something of equal mass (another neutron), but a proton (the nucleus of an ordinary hydrogen atom) can also be used, for its mass is almost equal to that of a neutron. Although protons are often employed for this purpose, they have one drawback: it occasionally happens that the neutron is absorbed by the proton instead of undergoing an elastic collision. (In the language of this chapter, the neutron sometimes "sticks" to the proton.) Because we want only to slow the neutrons down and do not wish to have them absorbed, let us consider other things with which the neutrons might collide, and see how effective they might be at reducing the kinetic energy of the neutrons.

(a) One possible choice is the deuteron, the nucleus of "heavy hydrogen," whose mass is approximately twice as large as that of the neutron. How good are deuterons at slowing down neutrons? That is, in a head-on elastic collision between a neutron and a deuteron, what fraction of the neutron's kinetic energy is lost?

(b) Another substance sometimes used to slow down neutrons is graphite, a form of carbon. A normal carbon nucleus is approximately twelve times as massive as a neutron. What fraction of a neutron's kinetic energy is lost in an elastic collision with a carbon nucleus?

3.9 Show that Equation 3.8 gives the same result for α ($\alpha = 0.64$) if m_A/m_B has the value 4 or $\frac{1}{4}$. Is this true in general? That is, does α have the same value for mass ratios 5 and $\frac{1}{5}$, for 10 and $\frac{1}{10}$, etc.?

3.10 In principle, if you are stranded in the middle of a frozen pond, 100 ft from shore, you can escape by throwing a shoe very hard. If there were no friction, about how long would it take you to reach the shore?

3.11 A 30,000-ton ocean liner is 10 ft from the dock when it is abandoned by its tugboats. The captain calls the 1500 passengers on deck and asks them to throw their 3000 shoes as hard as possible in the direction away from the dock. Assuming that the law of conservation of momentum applies in this situation and that once the ship has started, it moves without any decrease in speed, estimate the speed with which the ship recoils and the length of time it takes to reach the dock.

3.12 (a) When you alone jump up in the air, the recoil velocity of the earth is extremely small. But suppose that all the people of the world were to congregate in Pennsylvania and jump up at the same time. Estimate the recoil velocity of the earth in such an experiment.

(b) Is there enough room in the state of Pennsylvania for all the people of the world?

3.13 Kinetic energy is produced when a polonium nucleus disintegrates into a nucleus of lead (whose mass is 206 "atomic mass units") and an α-particle whose mass is 4 atomic mass units. What fraction of the kinetic energy is given to the α-particle? To make this calculation, is it necessary to know the conversion factor between kilograms and atomic mass units?

3.14 The apparatus shown to the right above, a "ballistic pendulum," was once used to measure the

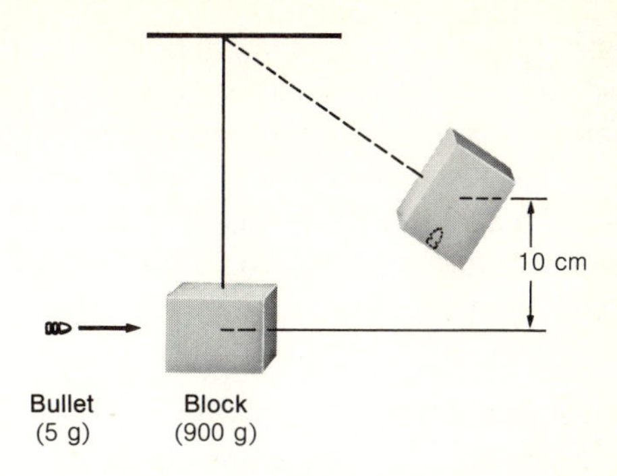

speeds of bullets by an indirect method. A rifle bullet slams into a wooden block supported as a pendulum. This is a sticking collision; momentum is conserved, but kinetic energy is not. Then the block and the imbedded bullet swing upward; during the swing, mechanical energy is conserved. We can use the conservation laws to work backwards from the observed maximum height to find the initial speed of the bullet.

(a) The results of one such experiment are shown in the figure. By applying the law of conservation of energy to the swing of the pendulum, find the speed of the block (together with the bullet) immediately after the collision.

(b) Use this result together with the law of conservation of momentum to find the initial speed of the bullet.

(c) What fraction of the initial kinetic energy was lost during the sticking collision between the bullet and the block?

3.15 A particular kind of string is just strong enough to support a 10-kg mass. If you use such a string to swing a 2-kg rock in a horizontal circle of radius 1.5 m, what is the maximum speed the rock can have without breaking the string? At this speed, how long does it take the rock to complete one orbit?

3.16 As a car rounds a curve at a steady speed, traveling for a few seconds along the arc of a circle, the car and its passengers are briefly traveling with uniform circular motion. There must therefore be a force of some kind acting on the car and on each passenger, directed toward the center of the circular curve. Estimate the magnitude of this force on a 60-

kg passenger in a car traveling at 60 miles/hr around a curve whose radius is 300 ft. How does this force compare in magnitude to the gravitational force exerted on the passenger? Think of your sensations as you round a sharp curve in a car; what is it in the car that exerts a force on you, a force directed toward the *center* of the curve?

3.17 How many joules of work do you do in lifting a 20-kg suitcase onto a table 0.8 m high? How much work do you do when you lower the suitcase back to the floor?

3.18 Work, according to its scientific definition, is a quantity that can be positive, zero, or even negative. Which of the three terms describes the work you do in the following situations?

(a) The work you do on a baseball in throwing it; in catching it; in holding it.

(b) The work you do on a soft cushion as you sit down; while sitting on it; while getting up.

3.19 A ball thrown upward slows down as it rises. Is this because it is gaining gravitational potential energy and therefore must be losing kinetic energy? Or is it because there is a downward gravitational force acting on it? Are both of these explanations correct? If so, are there two separate processes going on simultaneously, or are these merely two different ways of describing the same effect?

3.20 If you are sitting in the seat of an airplane, the only forces acting on you are the earth's gravitational force and the forces exerted by the back of your seat, the seat belt, and the cushion underneath you. For each of the following situations, sketch diagrams showing the direction and relative importance of each of these four forces. In each case, state whether the work being done on you by each of the forces is positive, negative, or zero.

(a) The plane is accelerating down the runway to take off.

(b) The plane is ascending at uniform velocity.

(c) The plane is cruising at a fixed altitude with a uniform velocity.

(d) The plane is descending at uniform velocity.

(e) The plane is braking to a stop after landing.

in describing a *closed* system, one for which there is no transfer of energy in or out. Then $W = 0$, $\Delta E = 0$, or

$$E = KE + GPE = \text{constant}. \tag{4.2}$$

The total energy of the system is conserved. The sum of the kinetic and gravitational potential energies is constant; any increase in one form of energy is compensated by an equal decrease in the other.

The difficulty with Equation 4.2 is that it is often incorrect. Two gliders interacting with each other on an air track constitute a closed system, but the total energy of this system, if the total energy includes only kinetic energy and gravitational potential energy, decreases markedly when they have an inelastic collision. When such cases are studied closely, it is found that the gliders are slightly warmer after the collision than before, and the law of conservation of energy can be rescued by defining another form of energy, *thermal* energy (TE).* Once it is recognized that thermal energy should be included as a form of energy, we realize that not only is thermal energy a form of energy that may be created in this way but also that energy can be transferred from a hot object to a cooler one just by placing them next to each other, by a flow of *heat* (symbolized by H).

Thus we recognize that there are other forms of energy besides kinetic and gravitational potential energy and other ways of transferring energy than by the performance of work, and this suggests that Equation 4.1 be generalized as follows:

$$W + H = \Delta E = \Delta(KE + GPE + TE). \tag{4.3}$$

Still other forms of energy must be included as well. When two gliders are blown apart by a firecracker, we attribute the resulting increase in kinetic energy to a decrease in the *chemical* energy of the firecracker, which is also the source of the bang and the flash of light. We will also need to define *electrical* energy and other forms of energy as well, and thus we generalize the basic energy equation as follows:

$$W + H = \Delta E = \Delta(KE + GPE + TE + \text{Chemical Energy} + \text{Electrical Energy} + \cdots). \tag{4.4}$$

Here H denotes the flow of heat, whereas W represents energy transfers of all other types. Heat, like work, can be either positive or negative, depending on whether heat flows into or out of the system. If a number of different processes are occurring at the same time, W denotes the net amount of work done on the system and H the net amount of heat flowing into the system. The dots in Equation 4.4 are significant. They mean that other forms of energy may have to be included as other kinds of phenomena are investigated.

*_Internal_ energy is a term nearly synonymous with thermal energy. The adjective "internal" reflects the fact that an object's thermal energy is really a manifestation of the rapid random motions of the atoms and molecules of which it is composed.

As applied to a closed system (where there are no energy transfers into or out of the system), Equation 4.4 takes a simple form:

$$\Delta E = 0, \; E = \text{constant}.$$

The total energy of a closed system is conserved. Any increase in, say, thermal energy, must be attributed to a decrease of equal size in the sum of all the other forms of energy. Any or all of the various forms of energy may be changing, but—in a closed system—always in such a way that the sum of all forms has a constant value.

The distinction between the terms "heat" and "thermal energy" is an important one. A flow of heat, like the performance of work, represents a *transfer* of energy; thermal energy, on the other hand, is one of the kinds of energy which an object may have, and changes in thermal energy are usually revealed by changes in temperature. The word "heat" is often used indiscriminately to refer to either concept. It is important to maintain the distinction between the two, because thermal energy can be increased in ways other than by a flow of heat: a sticking collision between two gliders results in an increase in thermal energy; hammer on a metal plate, do *work* on it, and its thermal energy increases just as it would if you put it near a fire. In short, what is implied by Equation 4.4 is that work and heat represent two different ways of transferring energy.

The most important evidence for the validity of this generalization derives from a long and careful study of heat and thermal energy, and the demonstration that under certain circumstances work and heat may be equivalent, that doing work on a system and allowing heat to flow into it can produce identical effects, producing what we would now describe as an increase in the total energy of the system. We will examine these crucial experiments in Chapter 6.

In any particular application of the energy equation, it may take a certain amount of skill and experience to see which of the various forms of energy are of importance, to see, in other words, which forms of energy are *changing*. (If in a particular process, one form of energy is constant, we can disregard it altogether, as we ignored gravitational potential energy when considering collisions between gliders on a horizontal air track.) The best way to get a feeling for the usefulness of the law of conservation of energy is to see how it is used, and so we shall now briefly consider a number of simple applications.

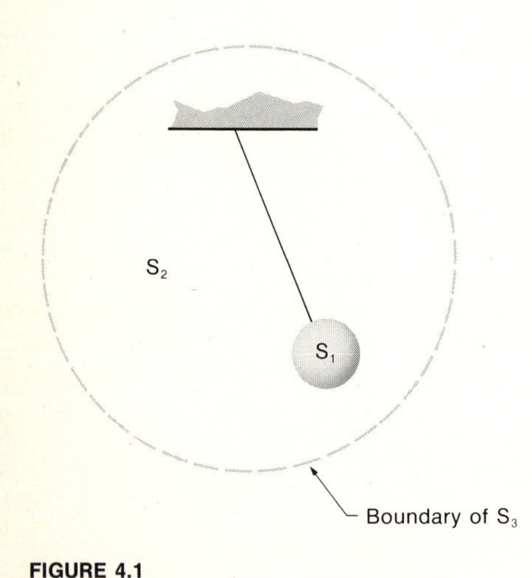

FIGURE 4.1
Three possible systems for describing the energy conversions that occur during the motion of a pendulum. The first system S_1 is the pendulum alone, S_2 is the air surrounding the pendulum, and S_3 is a composite of the other two systems, both air and pendulum.

Consider a pendulum swinging back and forth, the sum of its kinetic energy and gravitational potential energy gradually decreasing. As the motion dies out, the air around the pendulum gets slightly warmer. We could choose the pendulum itself as the system of interest; call this system S_1, as shown in Figure 4.1. The pendulum pushes against the air as it moves, doing work on the air, whereas the work done by the air *on* the pendulum is negative. That is, for system S_1 (the pendulum), W is negative, and ΔE is also negative, as the pendulum loses energy. We could perfectly well study the system S_2, defined as including the air around the pendulum but not the pendulum itself. For this system, W is positive (the pendulum does work on the air), and the thermal energy of the air increases. As a third choice, we could choose a larger system (S_3), which includes both the pendulum and all the air in its vicinity. S_3 is a closed system, for it the total energy is conserved, and there are three contributions to this total energy:

$$\text{For } S_3,\ E = \text{constant};$$

$$KE + GPE + TE = \text{constant}.$$

No change in the total energy of S_3 takes place. But within the system conversions of energy are taking place, as kinetic energy and gravitational potential energy are converted into one another, while at the same time the sum of these two forms of energy slowly diminishes and the thermal energy increases.

What happens when you rub two sticks together? You are doing work, transferring energy from your body to the sticks. The store of chemical energy in your body decreases, and at the same time the thermal energy of the sticks increases. If the system includes your body together with the sticks and the surrounding air, the total energy is constant, but energy is being converted from chemical energy to thermal energy and being transferred from one part of the system to another, from your body to the sticks.

Consider a container of water with a partition in the middle, hot water on one side and cold water on the other (Figure 4.2). There is a flow of heat from region 1 to region 2. For the system S_1, H is a negative quantity and so is $\Delta(TE)$. For system S_2 the reverse is true. If we choose a system whose boundaries contain the whole apparatus (S_3), we again have a closed system. The total energy of S_3 is constant, but within the system, a transfer of energy is taking place; the thermal energy of region 1 is decreasing while the thermal energy of region 2 is increasing. In this case, as well as in others, the question "What should the boundaries of the system be?" has no right or wrong answer. We can make whatever choice is most convenient, much as we have the freedom to choose the zero-level for gravitational potential energy. Different choices for the boundaries of the system result in different descriptions; as long as we are careful, the choice is ours to make.

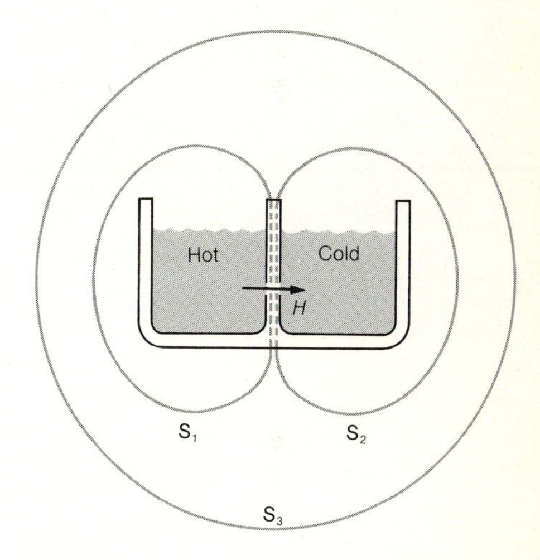

FIGURE 4.2
Three possible systems for describing the flow of heat from hot water to cold water.

Consider one more example. When a ball is dropped, it loses gravitational potential energy while gaining kinetic energy until it hits the floor. Then it comes to a momentary stop; all of its kinetic energy is gone, and its gravitational potential energy also has its minimum value. The ball is temporarily deformed at this instant, though, and it will regain its original shape a moment later. As it does, the kinetic energy reappears. During the impact, we describe the energy as being in the form of *elastic* energy, which is then converted back into kinetic energy. (Similarly, during the elastic collision of two gliders on an air track, some energy is temporarily stored in the elastic energy of their springy bumpers during the collision.) A "perfectly elastic" ball would rebound to its original height, and in a sequence of bounces we would have a conversion again and again between kinetic energy, gravitational potential energy, and elastic energy:

$$GPE \longrightarrow KE \longrightarrow \text{Elastic Energy} \longrightarrow KE \longrightarrow GPE \longrightarrow KE \longrightarrow \text{Elastic Energy} \longrightarrow KE \cdots.$$

Real balls (Figure 4.3) are not perfectly elastic. At each bounce, a small amount of kinetic energy is not recovered but is converted to thermal energy instead. If we choose our system in this case to include the ball, the air around it and the floor on which it bounces, then the total energy of this system ($KE + GPE +$ Elastic Energy $+ TE$), keeps the same value throughout the motion, from beginning to end.

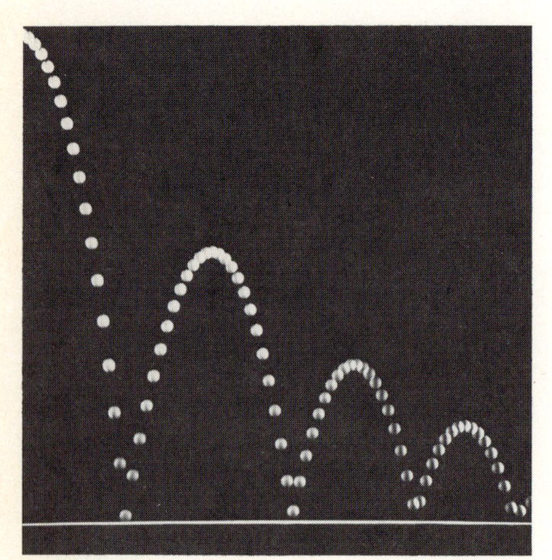

FIGURE 4.3
Energy conversions observed in successive bounces of a partially elastic ball. [*PSSC Physics* (D. C. Heath & Co., Lexington, Mass., 1965), and the Education Development Center.]

§4.1.B The Concept of Power

Energies are of importance when they are changing, or at least when the possibility of change exists. A gallon of oil has a large amount of chemical energy, a fact that is important because it can be converted into other forms of energy by burning, and the water behind a dam has gravitational potential energy, which can be converted into other forms by letting it fall. When energy conversions are taking place, when energy is being transferred into or out of a system, it is often the *rate* at which this occurs that is of interest, and the term *power* is used to describe the rate of any process of transferring energy or converting energy from one form to another.

The MKS unit of power is called the *watt* (W), which is defined as one joule per second. If you do work on something (pushing on it as it moves), performing, say, 5 J of work every second, you are delivering a power (P) equal to 5 W. Although the word "power" often conveys this idea of a rate of performance of mechanical work, the rate of *any* energy transfer or energy conversion can be expressed in watts. In Figure 4.4, heat is flowing from A to B at the rate of 150 J/sec = 150 W. The energy of A is decreasing at a rate of 150 J/sec, or 150 W; the energy of B is increasing at the same rate.

The fundamental energy equation (Equation 4.4) can be written in terms of power by referring to some specified interval of time, Δt:

$$P = \frac{W + H}{\Delta t} = \frac{\Delta E}{\Delta t}. \tag{4.5}$$

The power being delivered to the system, the net amount of energy transferred into the system per unit time by the combined effects of work and heat, is equal to the rate of change of the total energy of the system.

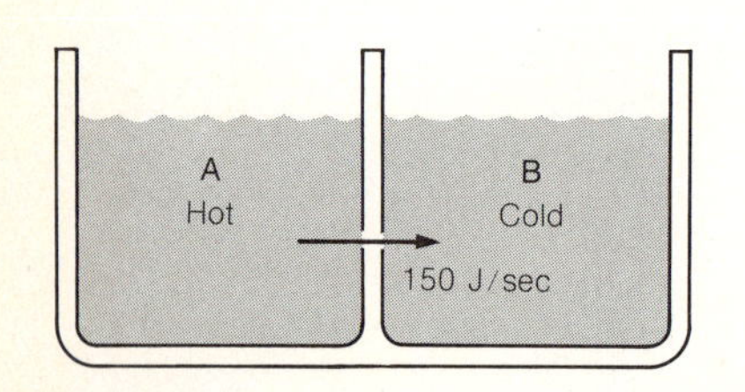

FIGURE 4.4
Heat is flowing from A to B at the rate of 150 W.

Power and energy are intimately related concepts, and as it suits our purposes, we can switch back and forth from one description to the other. Since power is a *rate* of transferring energy, the total amount of energy transferred by a power P in a time interval of length Δt is just $P\,\Delta t$:

$$\text{Total amount of energy transferred in time } \Delta t = P\,\Delta t.$$

This is consistent with Equation 4.4; simply multiply Equation 4.5 through by Δt and the result is

$$P\,\Delta t = W + H = \Delta E.$$

The *rate* of delivery of energy multiplied by the length of time is equal to the total amount of energy delivered, and this is equal to the overall change in energy of the system during this time.

Power and energy are different quantities, measured in different units. It is important to keep the distinction in mind, but it is not always easy to do so because any one process can often be correctly described with either term. Compare the two following statements. (1) During the past 10 sec, an electrical energy of 750 J has been delivered to this light bulb; (2) an electrical power of 75 W has been delivered to this light bulb for the past 10 sec. If the energy was delivered at a steady rate, the two statements are equivalent to one another.

Whenever one wants to relate a statement about energy to a statement about power, *some* interval of *time* must be used. Either the interval must be stated explicitly or it must be one whose value is clear from the context. If this simple fact is kept in mind, much of the common confusion between energy and power can be avoided.

§4.1.C Units of Energy and Power

The fundamental MKS units of energy and power are the joule and the watt. Life would be far simpler if only these units and their decimal multiples (such as the megajoule or the kilowatt) were permitted, but the fact is that a multitude of units are in common use for both energy and power. Undoubtedly, this is due to the importance of these concepts in so many fields. One has no choice but to become familiar with many of these units and to become adept at converting from one unit to another when necessary. A few of the more important units will be described here, and others are included in the conversion tables in Appendix A.

The erg (10^{-7} J) is the metric unit of energy that arises naturally in the CGS system, with distances and masses measured in centimeters and grams instead of meters and kilograms. The foot-pound (ft-lb), the work done by a force of 1 lb acting through a distance of 1 ft, arises in the British system of units; fortunately the ft-lb is rarely used either in scientific work or in discussions of the energy problem. The calorie and the British Thermal Unit (BTU) are two widely used energy units, both of which were first introduced specifically to measure heat and thermal energy. The calorie was originally defined as the amount of heat needed to raise the temperature of 1 g of water through 1°C, and the BTU as the amount of heat needed to raise the temperature of 1 lb of water through 1°F. After it was discovered that thermal energy and other forms of energy could be converted into one another, and that heat and work are two different ways of transferring energy, independent units for various forms of energy were no longer necessary. The calorie and the BTU ought logically to have been relegated to the history books; instead they have been retained, though they are now *defined* in terms of the joule. One calorie, for example, is by definition equal to 4.184 J. It still takes 1 cal to raise the temperature of 1 g of water 1°C, but this statement is no longer one which is correct by *definition*. The kilocalorie (10^3 cal) is a useful multiple of the calorie, so useful that the prefix *kilo-* is often omitted, a circumstance that naturally can lead to confusion. The "calories" with which nutritionists and dieters are familiar are actually kilocalories. Various other terms are occasionally used in referring to kilocalories: "large" calories (as opposed to the ordinary "small" calories), Calories (with a capital C, abbreviated Cal) and sometimes "food calories."

For discussing large quantities of energy, multiples of the BTU are often used:

$$1 \text{ Q} = 10^{18} \text{ BTU};$$

$$1 \text{ milli-Q} = 1 \text{ mQ} = 10^{-3} \text{ Q} = 10^{15} \text{ BTU};$$

$$1 \text{ MBTU} = 1 \text{ million BTU} = 10^6 \text{ BTU}.$$

The origin of the unit Q is obscure; it is possible that it was originally intended as an abbreviation for "quintillion" BTU, this term being used with its American meaning.* The Q and the milli-Q are useful in discussing extremely large amounts of energy. The United States uses about 70 mQ (0.07 Q) of energy each year; the worldwide figure is about 240 mQ (0.24 Q). The amount of energy available in unmined coal in the United States is estimated to be about 40 Q, enough to supply us with *all* our energy needs for about 600 years (at the *present* rate!).

There are also a number of "informal" units of energy such as 1 barrel of oil (approximately 5.6×10^6 BTU), the amount of chemical energy released when a barrel of oil is burned. It is sometimes informative to convert quantities of energy into an amount of food having the same energy value. The "jelly donut," for instance, is an informal energy unit designed for this purpose: by definition, 1 jelly donut $= 10^6$ J.

QUESTION

4.1 How many kilocalories ("food calories") are there in 1 "jelly donut"? Is this a reasonable value for a real jelly donut?

*The Q is a useful unit but only so long as everyone defines it in the same way. Some careless authors, perhaps thinking that Q was meant as an abbreviation for "quadrillion," have used this unit to mean 10^{15} BTU. Be careful!

The MKS unit of power is the watt (1 joule per second), and its decimal multiples are also very useful. (1 kilowatt = 1 kW = 10^3 W; 1 megawatt = 1 MW = 10^6 W; 1 gigawatt = 1 GW = 10^9 W.) Another unit of power is the horsepower (hp), originally defined as the rate at which a good horse could do work and now defined as equal to 746 W.

Since power is a rate of transferring energy or a rate of energy conversion, a unit of power can be formed from any unit of energy together with any unit of time. We can measure power in calories per second, BTU's per year, joules per week, and so on. The possibilities are endless, and since every group seems to have its own favorite, it is frequently necessary to be able to convert from one unit to another. Many conversion factors are given in Table A.9, but there is not enough room to include all the conceivable combinations.

Just as any unit of energy can be divided by a unit of time to form a unit of power, so too any unit of power can be multiplied by a unit of time to form an *energy* unit. One watt-second is the energy delivered by a power of 1 W during a time of 1 sec. This is obviously equal to 1 J:

$$1 \text{ W-sec} = 1\frac{\text{J}}{\text{sec}} \times \text{sec} = 1 \text{ J}.$$

Another such hybrid is one of our most familiar units of energy, the kilowatt-hour (kWh), the energy delivered by a power of 1 kW during a time of 1 hr.

QUESTION

4.2 Show that 1 kWh = 3.6×10^6 J.

There is almost no limit to the units for energy which can be created in this way, and they range from useful ones like the kilowatt-hour to abominations such as the horsepower-hour. There are also many informal units of power, such as the equivalent number of barrels of oil per day, a unit appearing frequently in newspapers whenever the energy crisis is on the public mind. One of the most bizarre of all units is used in annual reports of electric utilities: the kilowatt-hour per year. The kilowatt is a unit of power, the kilowatt-hour a unit of energy, and so the kilowatt-hour per year is another unit of power! Two different time units are involved; the kilowatt-hour per year differs from the kilowatt only by a factor equal to the number of hours in a year:

$$1 \text{ kWh/yr} = 1\frac{\text{kWh}}{\text{yr}} \times \frac{1 \text{ yr}}{8766 \text{ hr}} = 0.0001141 \text{ kW} = 0.1141 \text{ W}.$$

"During 1973, electrical energy was generated in the United States at an average rate of 1.95×10^{12} kWh/yr." Translation: Electrical energy was generated at an average rate of

$$1.95 \times 10^{12} \text{ kWh/yr} \times \frac{0.1141 \text{ W}}{1 \text{ kWh/yr}} = 2.22 \times 10^{11} \text{ W}.$$

It is easy to confuse energy and power; anyone who doubts this statement should consider the existence of units such as the kilowatt-hour per year. Because of the intimate relationship between energy and power, it is impossible to avoid altogether statements that are—at least technically—incorrect. Consider the following statement: "The energy needed to keep this house warm is 10^6 BTU/day." The meaning of this statement is completely clear. However, because the BTU is a unit of energy and the BTU per day a unit of power, if we want to be meticulously accurate, we should instead say, "The power required is 10^6 BTU/day," or "The amount of energy needed to keep this house warm for one day is 10^6 BTU." What *must* be avoided are ambiguous statements whose meaning is not clear from the context: "This house needs 10^6 BTU to keep it warm." 10^6 BTU per what? Per microsecond? Per hour? Per day? Per century? It does make a difference!

§4.1.D Energy and Power Use in the United States

The total amount of energy used in the United States during the year 1973 was approximately 73 mQ (73×10^{15} BTU, 7.7×10^{19} J, or whatever unit one prefers). An annual energy consumption of 73 mQ, an average power use of 73 mQ/yr, these are two equivalent ways of conveying the same information. During the past two decades, annual energy consumption has been increasing at a rate of about 3.5% per year, a rate of growth corresponding to a doubling time of approximately 20 years.

QUESTIONS

4.3 Show that 73 mQ/yr $\simeq 2.44 \times 10^{12}$ W.

4.4 The population of the United States is about 210 million. Show that the average per capita rate of energy consumption is approximately 11.5 kW, and that on the average, each person uses about 3.7×10^{11} J during a year.

Even though these numbers change from year to year, it is worth remembering their approximate values:

annual total consumption of energy:	approximately 70 mQ, almost 10^{20} J;
average nationwide power consumption:	approximately 70 mQ/yr, or 2×10^{12} W;
average per capita power consumption:	approximately 10 kW.

These figures refer to energy consumption of all sorts: for driving cars, operating electrical generating plants, mining coal, heating homes, and so on. The per capita figure of 10 kW, for instance, is not just the power use of which individuals are directly aware (the power used to drive cars or heat houses, and so on) but each person's share of *all* the power used in this country. These important numbers will almost certainly continue to increase, but how much they will grow from year to year is an open question. (We do not even know a great deal about *why* these figures will increase. Past experience strongly suggests that they *will* increase, and if we knew more about the reasons for the past increases, we might do more to moderate the rate of increase in the future.) In the remainder of this chapter, we will discuss three important energy topics: hydroelectricity (the generation of electrical energy from the energy of falling water); transportation; the heating of buildings ("space heating"). In each area, the amounts of energy involved will be related to the overall figures.

§4.2 HYDROELECTRICITY

§4.2.A Conventional Hydroelectric Power

One of the oldest methods of taking energy in one form and converting it to another form for our own purposes is to take advantage of the loss of gravitational potential energy experienced by water as it travels down a river to the sea. "Hydropower" has been used at least since the first century B.C., most often for grinding grain, a burdensome domestic chore made much easier with the invention of the water-driven grinding stone, and waterfalls supplied the energy to run the machinery of many 19th-century textile mills. A very important characteristic of this form of energy is that it is *renewable.* When we use coal, oil, natural gas, or uranium, we are consuming a resource that cannot be replaced. But the water that flows into the oceans is evaporated by the heat of the sun, clouds form, and some of the subsequent rainfall occurs over the land—the entire process is a cyclical one. In using the energy of falling water, we are indirectly using solar energy. We need not worry about depleting the supply; it will last as long as the sun continues to shine.

Today, virtually all hydropower is hydro*electric* power. The falling water strikes the blades of a turbine (Figure 4.5), which is connected to the shaft of an electrical generator. Thus the gravitational potential energy of the water is converted into kinetic energy, the water does work on the turbine and produces electrical energy. In Chapter 9 we will discuss the way in which the turbine's rotation generates electrical energy. For the present, if we accept the fact that gravitational potential energy can be converted into electrical energy, then knowledge of the amounts of electrical energy used by our society, together with information about the rate of water flow and the height through which the water falls will enable us to assess the importance of existing or planned hydroelectric plants. Gravitational potential energy can be converted into electrical energy with an efficiency of about 85%; that is, about 15% of

FIGURE 4.5
A hydroturbine, prior to installation at Wheeler Dam. [Tennessee Valley Authority.]

the energy is wasted and directly converted into thermal energy. In making rough estimates of the amounts of electrical energy that can be generated, this inefficiency is not terribly important; as an approximation, we will treat the conversion of gravitational potential energy into electrical energy as if the efficiency were 100%.

In this chapter, we will be concerned primarily with the generation of electrical energy in hydroelectric plants. These plants are important, but they supply only 14% of our electrical energy; the rest is generated in steam-electric plants, fueled by coal, oil, natural gas, or uranium. More will be said about steam-electric plants in subsequent chapters, but it should be pointed out that steam-electric power plants are much less efficient than hydroelectric plants. Only about one-third of the chemical energy released by burning fuels or of the nuclear energy released by nuclear fission can be converted into electrical energy.

QUESTION

4.5 During 1973, electrical energy was generated in the United States at an average rate of about 2.2×10^{11} W. Show that the average per capita consumption of electric power is approximately 1000 W. (This does not mean that the average home uses electrical energy at the rate of 1000 watts per occupant; 1000 W is every individual's share of all the electrical power used—not only for homes, but also for factories, stores, streetlights, and so on.)

The *gigawatt* is a convenient unit to use in discussing extremely large values of power. One gigawatt (1 GW) is equal to 10^9 W (1 billion watts in common American terminology), and thus the United States uses electrical energy at the rate of about 220 GW. The "installed generating capacity" (the power which all the electrical generating plants could produce if they were all operating simultaneously with maximum output) is considerably larger, about 460 GW, of which 62 GW is represented by hydroelectric plants. On the average, our electrical generating plants operate at about 50% of full capacity ($220/460 \simeq 0.5$). This does not mean that we have a surplus of generating capacity, because at any given time some plants are shut down for repairs (or for refueling, in the case of nuclear plants), and furthermore, consumption of electrical energy does not take place at a *steady* rate of 220 GW. This is only an average figure for a whole year, and demand varies widely from hour to hour during the day and from month to month during the year.

A useful perspective on these figures is gained by recognizing that a single rather large modern power plant has a generating capacity of about 1 GW (also described as 1000 MW or 1 million kW). Many plants have smaller capacities, a few are larger, but 1 GW is typical of the large power plants constructed in the early 1970s. We will define an informal unit, the "power plant," as equal to 1 GW, and with this unit, the installed generating capacity of the United States is about 460 power plants.

FIGURE 4.6
Hoover Dam. [U.S. Department of the Interior, Bureau of Reclamation.]

As a specific example of a large hydroelectric plant, consider Hoover Dam on the Colorado River, one of the largest generating plants in the United States. Behind Hoover Dam is Lake Mead, an artificially created lake stretching 115 miles up the river, with a surface area of 265 square miles and a volume of 8.5 cubic miles. Although the volume of water stored is important for some purposes, the two critical facts that must be known in order to estimate the power generating capability are (1) the height through which the water can fall, that is, the difference in water levels above and below the dam, a height referred to as the "head" of the power plant, and (2) the average rate of flow of water, a rate that can be specified either in volume per unit time (gallons per second, cubic kilometers per year, etc.), or mass per unit time (kilograms per second, tons per day). The head at Hoover Dam is about 570 ft (174 m). The average rate of flow of water in the Colorado River at this location, in the units in which river surveys are often reported, is approximately 17×10^6 acre-ft/yr. The acre-foot is a peculiar unit, the volume of water which will cover an area of one acre to a depth of one foot. Since 1 acre $= 4.356 \times 10^4$ ft^2, 1 acre-ft $= 4.356 \times 10^4$ ft^3.

QUESTIONS

4.6 Show that 17×10^6 acre-ft $\simeq 7.4 \times 10^{11}$ ft^3 $\simeq 2.1 \times 10^{10}$ m^3.

4.7 The density of water is 1000 kg/m^3. Show that the average rate of flow of water at this point on the Colorado River is about 6.6×10^5 kg/sec.

Now we can estimate the rate at which electrical energy can be generated by letting water flow through the turbines at this rate. In one second, a mass of 6.6×10^5 kg falls through a height of 174 m, and therefore the loss of gravitational potential energy every second is

$$mgh = 6.6 \times 10^5 \times 9.8 \times 174 \simeq 10^9 \text{ J}.$$

Thus the power generated is 10^9 W = 1 GW, about 1 "power plant." The actual installed capacity of the generators at Hoover Dam is somewhat larger than this, about 1.34 GW. Power can be generated at this larger rate for short periods of time, but only by allowing a larger than average flow of water, thus temporarily lowering the level of the lake.

QUESTIONS

4.8 Sometimes the rate of flow of water through the turbines is larger than the average flow rate of the river and at other times less. Some of the river's flow is not used at all for generating energy. In recent years, the average rate of generation of electrical energy at Hoover Dam has been about 4×10^9 kWh/yr. What is this rate in watts?

4.9 Make a rough estimate of the maximum conceivable rate at which hydroelectric energy could be generated in the United States, according to the following line of reasoning. If every drop of rain were to flow to the sea, it would lose an amount of gravitational potential energy equal to mgh, with m the mass of the water and h the height above sea level of the spot where it hits the ground. Suppose that it were possible to capture *all* of this energy and convert it into electrical energy. If one knows the average rainfall for the United States, it is possible to estimate the total mass of the rain which falls during the course of a year. As water flows to the sea, the water that starts from a point near the coast loses very little gravitational potential energy, and that which starts from the top of the Rocky Mountains loses a great deal; it is therefore necessary to guess the average elevation of the United States. On this basis, estimate the maximum number of joules of energy per year available from hydroelectric power.

4.10 It is by no means true that every drop of rainfall flows down the rivers to the sea. About two-thirds of the water returns to the atmosphere either by evaporation or by "transpiration," the process by which plants lose water to the atmosphere; the "runoff" is only about one-third as large as the rainfall. Revise the estimate made in the preceding question accordingly.

4.11 Convert the preceding result into an average rate of production of energy in watts. Compare this result with the actual installed hydroelectric capacity of the United States and with the overall rate at which energy of all forms is used in this country.

There are several good reasons for making calculations such as those suggested above. First, suppose that these calculations, which were meant to indicate the maximum rate at which hydroelectric energy can be obtained, yield a result notably smaller than the actual installed hydroelectric capacity of the United States. Then something is wrong, either with the values of some of the relevant quantities (for example, the area of the United States), or with the arithmetic, or with the understanding of the basic underlying concepts. Tracking down the error in a calculation can often be worthwhile and can lead to improved understanding. Second, if the calculation were to give a result enormously greater, say 1000 times greater, than the actual installed hydroelectric capacity, it would suggest that there *might* be opportunities for a significant expansion of our hydroelectric capacity. Of course, such a result would not by itself provide a compelling argument for building more dams; no one would want the energy of every mountain brook to be "harnessed," even if it were economically feasible.

A third reason for being interested in calculations of this sort is more general. Whenever any new way of obtaining energy is proposed, no matter how wild it may seem, or when an expansion of an established method is suggested, one of the first steps to be taken is to make an estimate (even from imperfect and approximate data, if necessary) of the *amount* of energy or power that *might* thus be obtained, to see how significant a contribution might be made. If it turns out that a substantial amount of energy might be available, there may still be good reasons for not going ahead. The proposal may be technologically infeasible, or it may be possible but prohibitively expensive, or the environmental side effects may be unacceptable. If, on the other hand, the maximum amount of energy obtainable is trivially small, then there is probably no point in going further no matter how appealing the idea seemed to begin with.

Generation of electrical energy from hydropower has been important for a long time. In 1920, 33% of our electrical energy came from this source. Although the actual amount of power generated in this way continues to increase, the relative importance of hydropower has steadily declined. Most of the best sites are already in use, and some of those which are not should never be developed. (The falls of the Yellowstone River in Yellowstone National Park would be an excellent site for a power plant.) We can all envy the man in Maine who had the foresight to purchase a waterfall and build his own small hydroelectric plant, but there are simply not enough waterfalls to go around. We could perhaps achieve a hydroelectric capacity twice as large, possibly three times as large, as we have at present, an increase which would be significant but which could not supply anything like a major share of the energy consumption predicted for the last decades of this century. The Tennessee Valley Authority presents a good example of this situation. In the late 1930s, during the early days of TVA, most of the electrical energy generated by TVA came from hydroelectric plants. The hydroelectric plants are still in use, but many new steam plants have been built, and now 80% of TVA's electrical energy is generated in steam-electric plants, largely fueled by strip-mined coal from Ohio and Kentucky. It might be noted that South America and Africa, two continents that are genuinely in need of more energy, have many undeveloped hydroelectric sites.

In one sense, hydroelectric energy is a "good" kind of energy. Its consumption does not deplete an exhaustible resource, and it produces no smoke or ashes. There are undesirable side effects, however: for example, many scenic canyons have been permanently lost. There can be other environmental impacts as well, and the Aswan Dam on the Nile is a classic example of a large dam with many unanticipated consequences. One of the effects of this dam has been that the annual spring floods have been largely eliminated, and the quantity of nutrients formerly carried to the lower Nile and to the Mediterranean Sea has been greatly reduced; as a result, the catch of fish in the eastern Mediterranean has declined considerably. Disease is another side effect of the Aswan Dam. The dam provides water for irrigation as well as for power generation, and the incidence of bilharziasis, a serious disease carried by snails that breed in irrigation canals, has increased dramatically.

In the past few years, it has become fashionable to subject various proposals to a "cost-benefit analysis." Too often it is the benefits (the kilowatt-hours) that can be accurately predicted, whereas the costs are underestimated, either because they are not anticipated at all, or because, like the destruction of a canyon, they are costs to which it is difficult to attach a price tag. Some of the benefits, too, are impossible to quantify. National pride may be one of the motives for building a large dam in an underdeveloped country: who is wise enough to weigh the incidence of bilharziasis against the increased sense of dignity of an Egyptian farmer?

§4.2.B Pumped-Storage Hydropower

An important feature of the energy problem, especially from the perspective of an electric power company, is that consumer demand for electricity varies from day to day and from hour to hour. It is not sufficient for a utility to have enough generating plants to meet the *average* power demand. It must supply energy at greater than the average rate on a hot summer afternoon, when air conditioners are on, and at dinnertime, when many electric stoves are in use. Figure 4.7 shows the week-to-week variations in the "load" of one particular power company for a number of different years. One can see in this figure both the steady increase in demand that has occurred over the years, and the summertime peaks, resulting from the increased use of air conditioning.

A more detailed representation of the hour-to-hour variation in load during one particular week is shown in Figure 4.8. During the small hours of the morning, only the *base* load needs to be supplied, the steady minimum level of power needed to operate street lights, a few factories, and so on. In order to meet the *peak* load during the afternoon, additional generating units are needed that must be shut down or operated at less than full capacity at other times of the day. Because of the varying load, a power company needs more generating *capacity* than it would to supply the same total amount of energy at a steady rate.

Supermarkets have the same problem of a fluctuating load. They sell groceries at a rapid rate from about 9 A.M. to 9 P.M. for six days a week and not at all the rest of the time. Their solution to this problem is easy: store the food on the shelves overnight. This solution is not available to the power company, for electrical energy cannot easily be stored. Energy storage is an important aspect of the energy problem. One of the

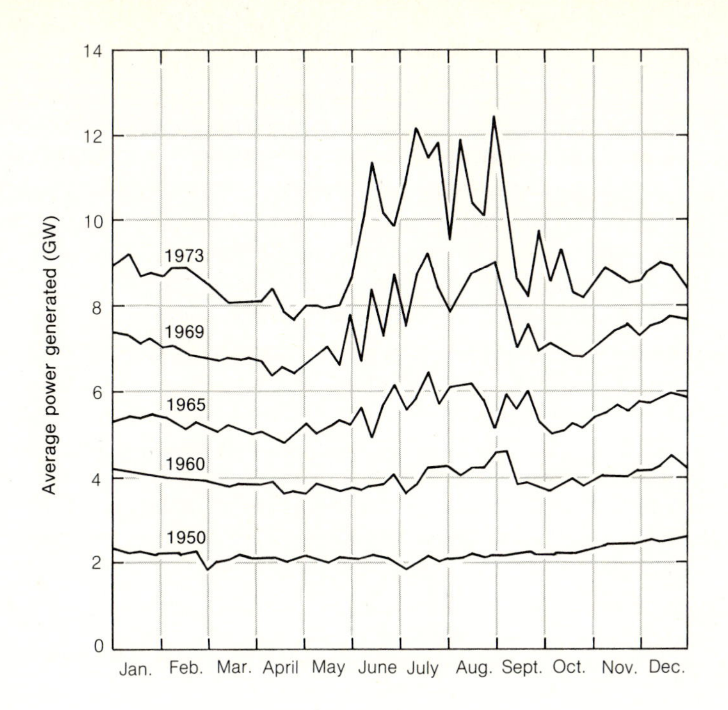

FIGURE 4.7
Week-to-week variations in the load of a power company for five selected years. [Commonwealth Edison Company, Chicago.]

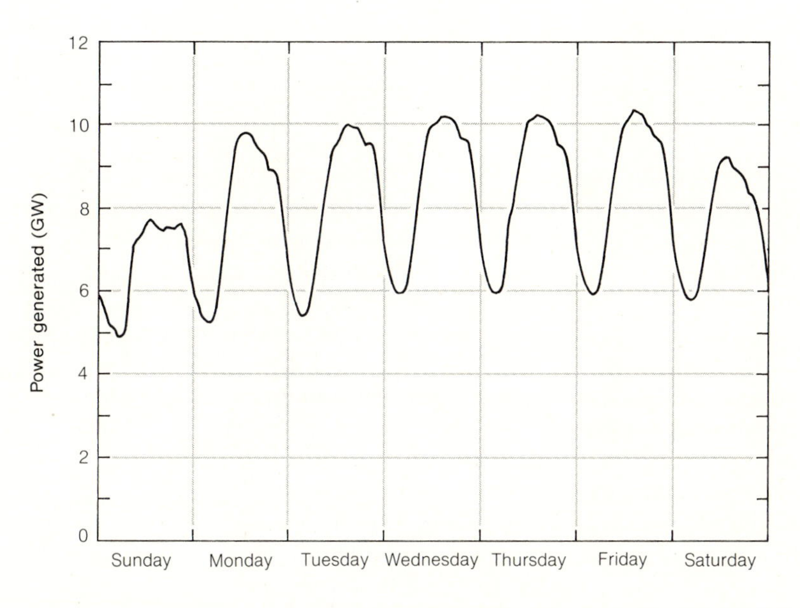

FIGURE 4.8
Hourly variations in load during one particular week.

obvious reasons for the importance of oil is that a tank of oil or a tank of gasoline represents stored chemical energy, energy that may lie idle for days but that is ready when we press on the starter or turn on the furnace. Power companies can and do store oil, coal, and uranium to fuel their steam-electric power plants, but no matter how much fuel may be on hand, the rate at which a generating plant can deliver energy is limited to a maximum value that is dependent on the size of the electrical generating equipment. Sometimes a power company may buy energy from another whose peak load comes at a different time. Since this requires the transporting of electrical energy across large distances, the problems of energy storage and energy transportation are related.

Most large power companies have a number of generating units, some of them steam plants fueled by coal, oil, natural gas, or nuclear fuel, and some hydroelectric plants. One response to the problem of fluctuating demand is to operate the steam plants almost continuously* to take care of the base load but to run some or all of the hydroelectric plants intermittently. During the night or on the weekends, a hydroelectric plant may be shut down, and energy is stored (as gravitational potential energy) by stopping or slowing the flow of the river and allowing the water level behind the dam to build up. Then, when large amounts of power are called for, the hydroelectric plants come into operation, letting through water at a greater than average rate. For a brief period of time the system as a whole can supply more power than it could if all the plants operated continuously.

QUESTION

4.12 Discuss some of the side effects that might result from artificially manipulating the rate of flow of a river.

An interesting variation on this procedure is the use of *pumped-storage* hydropower, a technique now used at various places in the United States. The idea is a simple one, one which is based on the fact that, just as it is easy to use the gravitational potential energy of water to produce electrical energy, it is also quite easy to reverse the process—to use electrical energy to run pumps, to pump water uphill and produce gravitational potential energy. The idea is to dig a reservoir on top of a mountain, a reservoir which will then be periodically filled and emptied as circumstances require. During the night, some of the power generated by the steam plants is used to drive pumps that raise the water to the top of the mountain; during periods of peak load, the reservoir is emptied, and the gravitational potential energy lost by the water as it comes downhill is used to generate electrical energy. This energy is supplied to consumers in *addition* to that supplied by the steam plants.

This seems like a great idea, but a moment's sober thought may raise doubts. This might be one of those "great ideas" that are found to be of

*Any power plant can be temporarily shut down, though it is not a simple matter to turn a large plant off and on. Such a measure also poses a financial problem, because a power plant requires a large capital investment, for which money must be borrowed and interest paid. Whether the plant is operating or not, it is a continuing drain on the company's funds, and it is therefore advantageous to have it operating as much of the time as possible.

no use at all when examined *quantitatively.* How do the *numbers* work out? This is the crucial question to which at least a rough answer must be obtained. Consider a specific example. A pumped-storage plant at Northfield Mountain in western Massachusetts went into operation in 1973. When water comes down from the reservoir at this site, it falls about 800 ft to the turbines, where its energy is converted into electrical energy. Can we estimate about how large the reservoir needs to be if this plant is to make a significant contribution to the power needs of western Massachusetts? Our calculation will require some arbitrary assumptions and a few guesses; the objective is not to obtain a precise figure but an order-of-magnitude result. If it should turn out that a reservoir the size of a bathtub would suffice, then pumped storage would obviously be of great interest. If a reservoir the size of the Atlantic Ocean is needed, we ought to forget the whole idea and simply wonder why any power companies have bothered with it.

The population of western Massachusetts is about 1.4×10^6 people. If Massachusetts residents are typical Americans, each person uses electrical energy at an average rate (averaged over night and day) of about 1 kW. Thus the daily use of electrical energy in western Massachusetts is about

$$24 \times 1.4 \times 10^6 = 33.6 \times 10^6 \text{ kWh},$$

or, expressed in joules,

$$33.6 \times 10^6 \text{ kWh} \times \frac{3.6 \times 10^6 \text{ J}}{1 \text{ kWh}} \simeq 1.2 \times 10^{14} \text{ J}.$$

We will arbitrarily define the pumped-storage plant as "significant" if a daily emptying can provide about 20% of this amount, that is, about 2.4×10^{13} J. What volume of water is needed in order that it lose this much gravitational potential energy in descending 800 ft?

QUESTIONS

4.13 Show that the volume of water required is about 10^7 m^3.

4.14 To make sense out of this number, show that a reservoir with an area of 1 mile^2 and a volume of 10^7 m^3 must have an average depth of about 4 m (about 13 ft).

This looks promising. A 1-mile^2 reservoir 4 m deep is large but not out of the question. The actual reservoir at Northfield Mountain has a smaller area (about 0.5 mile^2) but a greater depth and has a usable volume of about 1.3×10^7 m^3. This is 30% larger than the figure used in the preceding calculations, and thus one emptying of this reservoir can supply about 3.1×10^{13} J or 8.6×10^6 kWh.

On the basis of these results, we can sketch the design of a power system for western Massachusetts. We have estimated that the residents use about 33.6×10^6 kWh of electrical energy per day. Let us suppose that, of this energy, 75% (25.2×10^6 kWh) is delivered at a steady rate between 8 A.M. and 8 P.M. and the remainder (8.4×10^6 kWh) at a steady rate during the night and early morning. Thus the peak load (the daytime load) is $\dfrac{25.2 \times 10^6 \text{ kWh}}{12 \text{ hr}} = 2.1 \times 10^6 \text{ kW} = 2.1 \text{ GW}$ and the off-peak load one third as large, 0.7 GW. The system will be operated by emptying the pumped-storage reservoir during the day, delivering 8.6×10^6 kWh at a rate of

$$\frac{8.6 \times 10^6 \text{ kWh}}{12 \text{ hr}} \simeq 0.7 \times 10^6 \text{ kW} = 0.7 \text{ GW}.$$

To supply the remaining 1.4 GW of peak load, two steam-electric plants will be constructed, each with a capacity of 0.7 GW, and the system will operate as shown in Figure 4.9. During the day, the steam plants and the pumped-storage plant all deliver power to the consumers. At night, one of the steam plants supplies the off-peak load while the other is used to refill the pumped-storage reservoir. (Note, however, that we have made a number of obvious oversimplifications: for example, in reality power is often transmitted across state borders, and the shift from peak load to off-peak load is never so precisely defined.)

Contrast the pumped-storage system (Figure 4.9) with the conventional system of the same capacity but without the pumped-storage facility (Figure 4.10). Three steam plants would be needed to supply the daytime load, two of which would be shut down at night. With the use of a pumped-storage plant, construction of one relatively expensive steam plant can be avoided.

The pumped-storage idea is basically a good one and often a useful one. It is not true, however, that we are getting something for nothing, as some enthusiastic advertisements for the electric utilities would seem

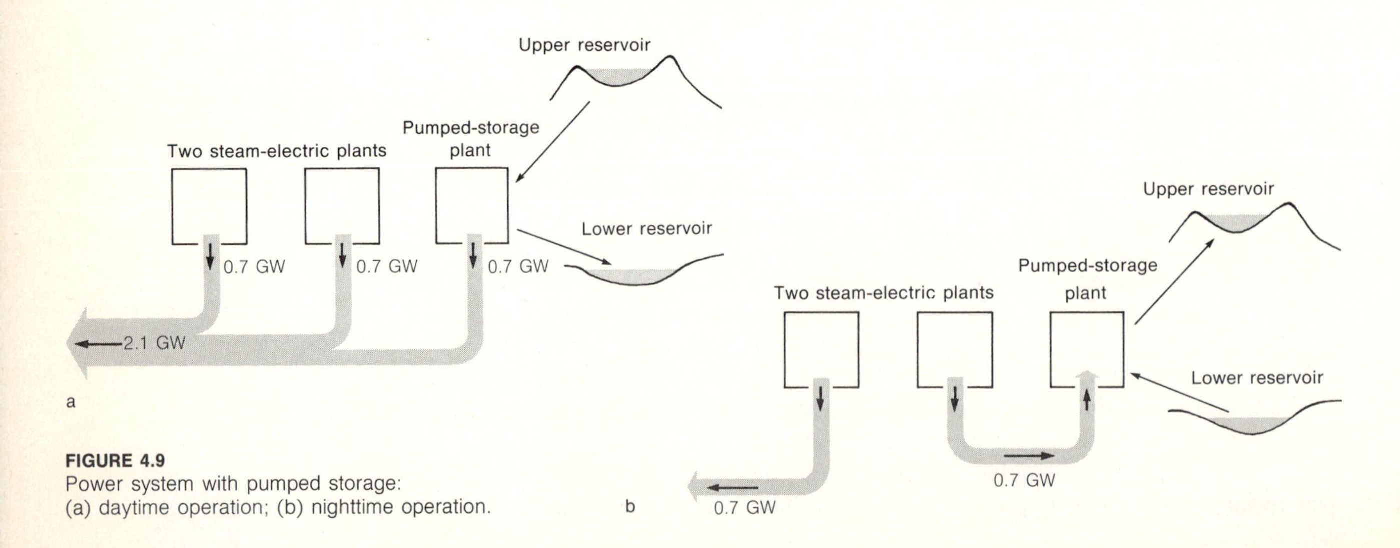

FIGURE 4.9
Power system with pumped storage:
(a) daytime operation; (b) nighttime operation.

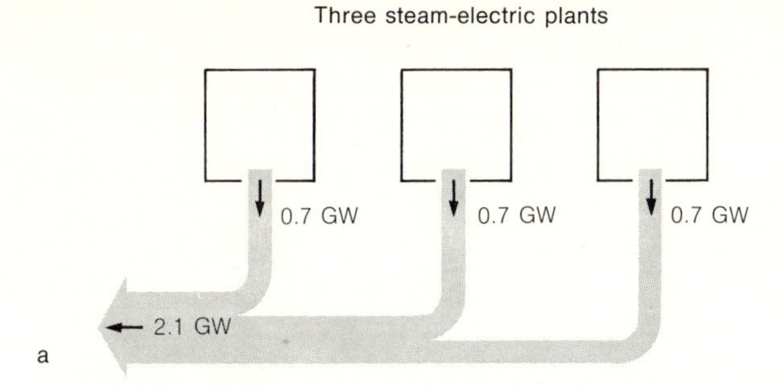

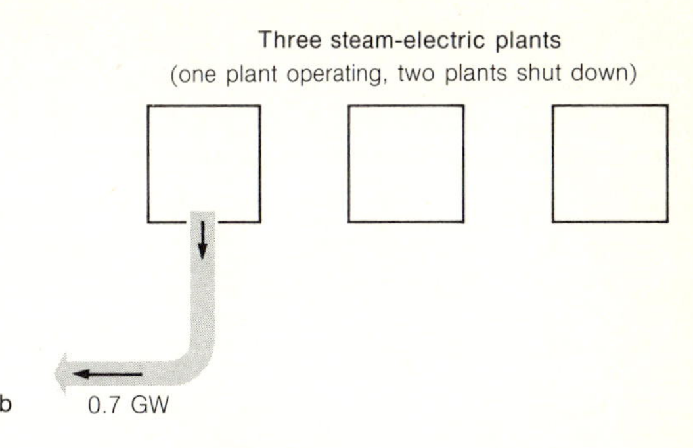

FIGURE 4.10
Conventional power system (without pumped storage): (a) daytime operation; (b) nighttime operation. Note that this system supplies the same load as that shown in Figure 4.9.

to suggest. The system shown in Figure 4.9 requires just as much fuel during a whole 24-hour day as does the system shown in Figure 4.10. Indeed, the system that includes the pumped-storage plant actually uses somewhat more fuel. This results from a fact that has been omitted from the preceding calculations and that is invariably omitted from the advertisements: somewhat more energy is required to pump water uphill than can be retrieved when we let it flow back down again. About 25% of the energy needed to run the pumps is wasted in overcoming friction, producing thermal energy in the water and in the machinery. Our proposed power system therefore requires some modification, it does not work out quite as neatly as suggested, and additional steam-electric generating capacity is required. This does not, however, alter the conclusion that pumped storage provides a good way of meeting the problem of a varying load if one of our concerns is that of minimizing the number of steam-electric plants that must be built; but pumped storage may not be a good idea at all if a shortage of *fuel* is the dominant problem.

§4.3 TRANSPORTATION

§4.3.A The Conventional Automobile

It has been recognized for a long time that numerous environmental and social problems have been produced by our present system of transportation. Transportation is also one of the major ways in which *energy* is used, a fact well known to everyone who has waited in line at a filling station. Of the total amount of energy used in the United States (about 73 mQ per year in 1973), about 25% is used for transportation. There are many components of the transportation system (cars, trucks, buses, airplanes, railroads, trains, pipelines, bicycles, and so on), but in the system as a whole, by far the largest user of energy is the ordinary automobile. Of all the energy used for transportation, more than half (14% of the total amount of energy consumed) is used for automobiles.

Let us therefore consider that familiar all-purpose device, the American car, a technological marvel that is not only a convenient means of transportation but also a status symbol, a prime source of air pollution, the essential ingredient of every traffic jam and highway accident, and a gluttonous consumer of petroleum, steel and other resources. We cannot give a complete treatment of the automobile problem without inquiring into the psyche of the American driver, the movement of the population away from the cities and into suburbia, and the factors that have left most of our cities with woefully inadequate public transit systems. In this discussion, however, we will be concerned only with the energy conversion processes that take place during the operation of an automobile.

The two forms of energy with which our discussion of the concept of energy began, kinetic energy and gravitational potential energy, are clearly relevant to the discussion. Kinetic energy is what we seek when we press on the accelerator pedal; increased gravitational potential energy is what we must have in order to get to the top of a hill.

One crucial piece of data that we will need, but that we cannot derive from fundamental principles, is the amount of chemical energy of a gallon of gasoline. According to the data in Appendix H, the energy of gasoline is about 5.2×10^6 BTU/barrel. This figure can be converted into joules per gallon:

$$5.2 \times 10^6 \frac{\text{BTU}}{\text{barrel}} = 5.2 \times 10^6 \frac{\text{BTU}}{\text{barrel}} \times \frac{1 \text{ barrel}}{42 \text{ gal}} \times \frac{1054 \text{ J}}{1 \text{ BTU}}$$

$$\simeq 1.3 \times 10^8 \text{ J/gal.}$$

This amount of energy, about 1.3×10^8 J, is released during the complete combustion of one gallon of gasoline. With this fact, together with an approximate knowledge of the number of miles to the gallon a car can obtain, we can make some quantitative statements about automotive energy consumption.

What is the kinetic energy of a 2-ton car traveling at 60 miles/hr? Using conversion factors from Appendix A, we obtain the following in MKS units:

$$KE = \tfrac{1}{2}mv^2 = \tfrac{1}{2} \times \left(2 \text{ tons} \times \frac{907.2 \text{ kg}}{1 \text{ ton}}\right)$$

$$\times \left(60 \text{ miles/hr} \times \frac{0.447 \text{ m/sec}}{1 \text{ mile/hr}}\right)^2$$

$$\simeq \tfrac{1}{2} \times 1814 \times 26.8^2 \simeq 6.5 \times 10^5 \text{ J.}$$

If the chemical energy of gasoline could be completely converted into the kinetic energy of the car, the amount of gasoline needed would be

$$\frac{6.5 \times 10^5 \text{ J}}{1.3 \times 10^8 \text{ J/gal}} \simeq 0.005 \text{ gal.}$$

QUESTIONS

4.15 Not very much gasoline, only 0.005 gal. How much is this? A cupful? A teaspoonful? A drop?

4.16 On the basis of the same assumption (that the chemical energy of gasoline can be completely converted to other forms of energy), show that the gravitational potential energy needed to raise a 2-ton car from sea level to the top of one of our highest mountain passes (altitude $\simeq$ 12,000 ft) could be obtained from about half a gallon of gasoline.

From these admittedly rough calculations, we can see that there is much more to the operation of a car than the simple conversion of chemical energy into kinetic and gravitational potential energy. It is true that, as a car goes up and down hills and as it slows down and speeds up again, continuing production of kinetic energy and gravitational potential energy is required, but the fact that the kinetic energy of a car traveling at 60 miles/hr is equivalent to the chemical energy of only 0.005 gal of gasoline strongly suggests the importance of other forms of energy as well. When a car goes down a hill or stops at a traffic light, its kinetic or gravitational potential energy is usually converted immediately into thermal energy. It would be nice if there were some way of storing energy so that the lost kinetic or gravitational potential energy could subsequently be retrieved without having to burn more gasoline, but present cars have no such storage devices. If the operation of a car required only the conversion of chemical energy into kinetic energy, we could use 0.005 gal to attain a speed of 60 miles/hr, and then, if the road were level, we could simply coast at a steady speed (obeying the principle of inertia).

Unfortunately the chemical energy of gasoline is not converted into mechanical energy at anything like 100% efficiency, and it also takes a large amount of gasoline simply to drive on the level at a steady speed, about $\frac{1}{15}$ or $\frac{1}{20}$ of a gallon for every mile. Under these conditions, essentially all of the energy that we use goes into heating up the air around us, the road, the tires, and the body of the car itself. That is, the chemical energy of the gasoline is steadily converted into thermal energy. Let us look at the flow of energy that takes place as the typical American automobile—say an intermediate-sized sedan or station wagon built in the early 1970s—is driven along a level road at a steady speed of 40 miles/hr. The numbers that will be cited are meant to be representative of actual cars and should not be regarded as applying precisely to any particular model. Let us assume to begin with that at a steady speed of 40 miles/hr, this car's "gasoline mileage" is 20 miles/gal.

QUESTION

4.17 We can immediately derive from this figure the rate in watts at which this car is using chemical energy. The rate at which gasoline is consumed is 2 gal/hr. Using the energy value of gasoline in joules per gallon, show that chemical energy is being used at the rate of about 72,000 J/sec = 72 kW.

Without making further calculations, we can immediately see that the family car represents a very significant part of the energy problem. According to the results discussed in §4.1.D, the *average* rate at which the typical American uses energy of all sorts amounts to about 10 kW. While driving a car, you are using energy, in gasoline alone, at about seven times this rate! In just $3\frac{1}{2}$ hours (140 miles) of driving, you will use up your daily share of energy, just in the chemical energy of the gasoline you burn. (No allowance has been made for the hidden additional amounts of energy consumed in driving—the energy used to make steel for the car and concrete for the highways, to operate the petroleum refinery, to transport the gasoline to the service station, and so on.)

QUESTION

4.18 Use these numbers to make a rough check on the statement made earlier that about 14% of the total national energy consumption is used for automobiles. Suppose every other person owns a car. Approximately how many miles per day for each car are needed to account for this energy? How many miles per year? Does this result seem to be of the correct order of magnitude? That is, is this a reasonable value for the number of miles per year that a car is normally driven?

The "typical car," while traveling at 40 miles/hr, uses chemical energy at the rate of 72 kW. Figure 4.11 shows approximately where this energy goes. About 1% is simply lost in the form of gasoline evaporating from the carburetor. (This energy remains in the form of chemical energy, but it is lost to us just as completely as if it were burned, and it also contributes to the air pollution problem.) The heart of the car is the internal-combustion engine, where the gasoline is burned and its chemical energy converted partly into thermal energy and partly into mechanical energy, as the hot gases expand, doing work on the pistons of the engine. This is where most of the energy is wasted. In a typical internal-combustion engine about 80% of the energy is lost, some of it in the form of unburned hydrocarbons which, like the gasoline evaporating from the carburetor, disperse their chemical energy uselessly in the atmosphere, but most of the lost energy is dissipated as thermal energy of the exhaust gases and of the air around the engine. Of the power fed into the engine, only about 20%, 14.2 kW, is obtained as mechanical output. Of this about 2.2 kW is used to operate various items such as the pump to circulate cooling water through the engine. Of the 12 kW left, about 25% is lost (converted into thermal energy) in the transmission that couples the engine to the wheels, and 75%, about 9 kW, is available to do what we want, to drive the wheels. (In the units often used in describing cars, this is approximately 12 horsepower.) This power of 9 kW (the "net propulsive power") is also eventually converted into thermal energy. About 4.4 kW goes into overcoming road and tire friction. As a tire goes around, each part of the tire is deformed as it comes in contact with the road and then relaxes as the wheel rolls

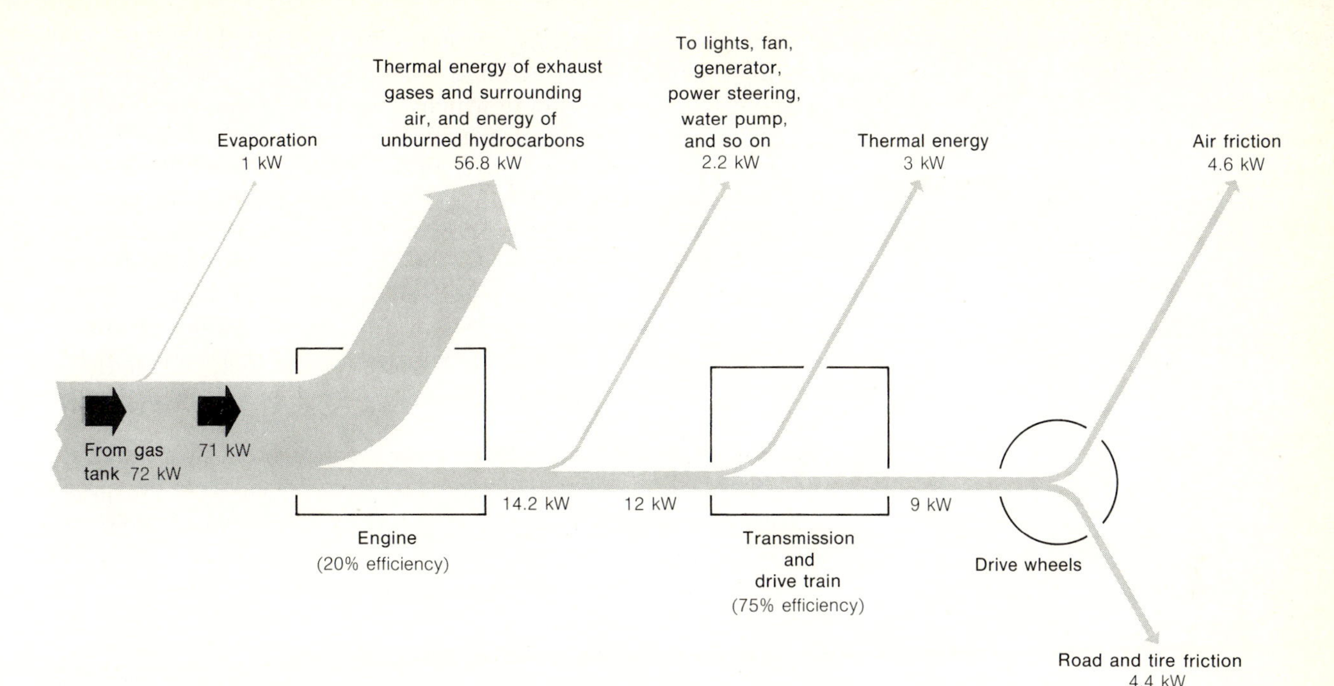

FIGURE 4.11
Energy flow in a 2-ton car traveling at 40 miles/hr.

on, and a large amount of heat is generated by this process. The rest of the power (4.6 kW) is used in overcoming air friction; that is, the car exerts a force on the air in front of it, doing work on the air, and this work is converted into thermal energy.

What is the "efficiency" of this car? Efficiency is a word that must be used with caution. The efficiency of the engine itself is about 20% (14.2/71 = 0.2). But if the efficiency is calculated by comparing the net propulsive power of the car (9 kW) with the rate at which chemical energy is delivered from the gas tank (72 kW), the efficiency is only 9/72 = 12.5%. From yet another point of view, the efficiency is zero, because *all* the energy is eventually converted into thermal energy if the car is traveling on a level road at a steady speed.

We need only a propulsive power of 9 kW at the rear wheels to drive the car, but to accomplish this, energy must be used at an overall rate of 72 kW. As Figure 4.11 shows, it is the low efficiency of the engine that is primarily responsible for this difference. Efficiencies of gasoline engines can be improved, but not to any great extent, and every improvement must be watched closely to see that it does not worsen the pollution problem. For example, other things being equal, higher temperatures in the combustion process lead to higher efficiencies but also tend to increase the production of nitrogen oxides. The trade-off between fuel consumption and pollution control is a delicate one, one which cannot be decided on technical grounds alone. Some (but by no means all) of the trend toward higher levels of fuel consumption can be traced to imposition of pollution controls in the early 1970s with less emphasis on engine efficiencies.

It is a familiar fact that cars use more gasoline per mile when driven at very high speeds. The preceding analysis can be extended in order to understand some of the ways in which "gas mileage" varies with speed. What happens if this car is driven at 80 miles/hr instead of 40? The power required to overcome road and tire friction will be approximately doubled, from 4.4 kW to 9 kW. (A certain amount of energy is needed to deform a tire; if the car is going twice as fast, the rate at which energy is used will be twice as large.) If the power at the rear wheels were required only for this purpose, then if the efficiency of the engine and the transmission were unchanged, the total rate of energy consumption for the whole car would approximately double; since it is going twice as fast, the "gas mileage," the number of miles per gallon, would be unchanged. However at higher speeds, air resistance gradually becomes more significant. The force the car must exert on the air to push it aside increases approximately in proportion to the *square* of the speed. The work done by this force is given by force $\times$ distance; the *rate* at which this force must do work is thus

$$\frac{\text{force} \times \text{distance}}{\text{time}} = \text{force} \times \text{speed}.$$

Since the force itself is proportional to the square of the speed, the power needed to overcome air resistance varies approximately as the cube of the speed, eight times as great at 80 miles/hr as at 40 miles/hr, 37 kW instead of 4.6 kW. The net propulsive power required is thus $9 + 37 = 46$ kW (62 hp). If we assume that the efficiencies of the transmission and the engine are unchanged at higher speeds and that the amounts of power used on the accessories are the same as before, we can work backwards from the net propulsive power of 46 kW to determine the rate at which energy must be delivered from the gas tank, which turns out to be 320 kW. The overall rate of energy consumption has increased from 72 kW to 320 kW, a factor of 4.4. This is the rate per unit time; since the car is going twice as fast, the amount of energy used per mile has increased by a factor half this large, a factor of 2.2. Thus instead of 20 miles/gal, the car driven at 80 miles/hr will only get $20/2.2 \simeq 9$ miles/gal.

By continuing in this way, we can estimate fuel consumption for various speeds. The decrease in gasoline mileage with increasing speed results from the combined effects of road friction, which tends to give a constant value of miles per gallon, and air friction, which causes a sharp decline at high speeds. At speeds lower than 30 miles/hr, the effects of air friction become smaller and smaller; however, the efficiency of the engine decreases at very low speeds, and gasoline mileage is therefore poor at both very low and very high speeds. Gasoline mileage figures vary markedly from one car to another, and it should not be assumed that the numbers given here can be applied directly to any particular car. Nevertheless, we can understand from these arguments the general nature of the variation in gasoline mileage with speed, as shown in Figure 4.12.

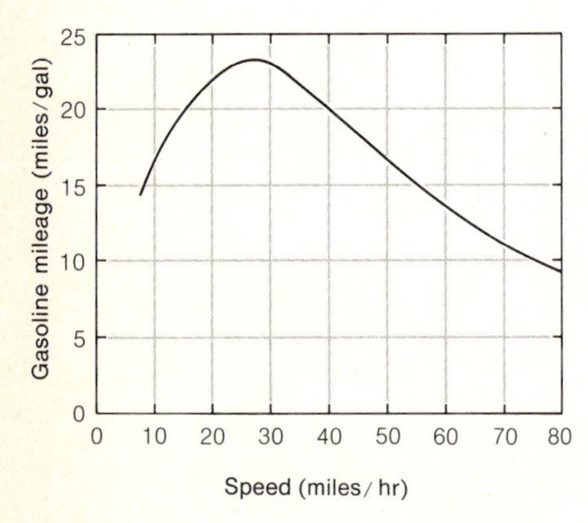

FIGURE 4.12
Gasoline mileage versus speed for a particular car at different constant speeds.

What can be done to lower gasoline consumption? As long as the conventional internal-combustion engine is used, no great improvement in efficiency can be expected. However, even without higher efficiency, a reduction in the propulsive power required would lead to a reduction in fuel consumption. At speeds up to about 40 miles/hr, road and tire friction are the most important factors. Some improvement can be achieved by keeping the tires well inflated, but reducing the weight of the car has a much more dramatic effect. The power necessary to overcome road friction in driving a 1-ton car is little more than half that for a 2-ton car driven at the same speed. Lighter cars are also smaller and thus the effects of air friction are reduced at the same time. Hence the lighter car may get almost twice as many miles to the gallon. This fact has long been known to automotive engineers. Until quite recently, though, gasoline has been so abundant and its cost so low (in comparison with other costs involved in operating cars) that fuel economy has been a minor consideration in the decisions made by most designers and buyers of cars in this country.

The calculations made at the beginning of this section suggested that production of kinetic energy itself is not a major factor in the operation of a car, and it is true that once a car has attained a certain speed, all the chemical energy delivered from the gas tank is being converted into thermal energy in one way or another. Nevertheless, energy is required for acceleration, and one can get an idea of the importance of this problem by considering how much additional propulsive power is necessary to accelerate from rest to a speed of 60 miles/hr in 15 sec. As we saw earlier, the kinetic energy of a 2-ton car traveling at 60 miles/hr is 6.5×10^5 J; to deliver this much energy in 15 sec requires a propulsive power of

$$\frac{6.5 \times 10^5 \text{ J}}{15 \text{ sec}} = 4.3 \times 10^4 \text{ W} = 43 \text{ kW},$$

over and above the power required to overcome road and air friction. This is the additional propulsive power required at the rear wheels, five times as large as the 9 kW of propulsive power needed to drive at a steady speed of 40 miles/hr. This result confirms what everyone knows, that if you are so foolish as to drive by accelerating, slamming on the brakes (generating thermal energy in the brake linings) and then accelerating again, your gasoline consumption will be greatly increased. No one drives in a fashion quite this absurd, but the message nevertheless is clear that frequent accelerations are to be avoided.

QUESTION

4.19 Additional fuel is also required to drive uphill. Estimate the additional propulsive power required to drive a 2-ton car up a moderately steep hill at 40 miles/hr. How does this figure compare with the propulsive power of 9 kW shown in Figure 4.11, the power needed to drive this car on a level road at this speed?

§4.3.B The Flywheel Car

The popularity enjoyed by the gasoline engine, in spite of its low efficiency, can be largely attributed to the fact that gasoline provides such a convenient way to store energy, to carry chemical energy about so that it is available when and where it is wanted. Nothing could be simpler than a gas tank; because of its very simplicity and the fact that it never needs a tune-up, it is perhaps the most important component of the modern car. In any discussion of alternatives to gasoline as the source of energy, the energy storage question is one of the first that must be faced. One suggested alternative is the electric car. Energy storage is achieved by charging storage batteries, which are then used to drive the wheels by means of an electric motor. We shall discuss the electric car in more detail in Chapter 8, but here let us consider another suggestion, the flywheel-operated car.

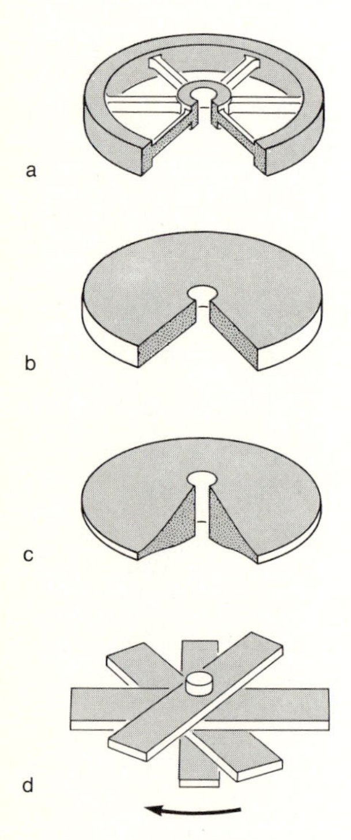

FIGURE 4.13
Flywheel designs: (a) rim-loaded flywheel; (b) solid disk; (c) tapered disk; (d) multi-spoke "super flywheel." [From R. F. Post and S. F. Post, "Flywheels." Copyright © 1973 by Scientific American, Inc. All rights reserved.]

A rotating flywheel can store kinetic energy, and it is possible to build a coupling to transfer energy from the flywheel to the rear axle of the car and also to transfer energy *from* the car *to* the flywheel when the car is slowing down or going downhill. In other words, the flywheel can be used not only as an accelerator but also as a brake—not the ordinary sort of brake in which kinetic or gravitational potential energy is uselessly dissipated into thermal energy, but one in which energy is stored for subsequent use. Coupling a flywheel to a car is easier said than done, and there are also gyroscopic problems associated with a large rotating wheel. Setting these problems aside, let us simply estimate how large a flywheel must be and how fast it must rotate in order to store a useful amount of energy. What is required if, say, we ask for enough stored energy to drive a 2-ton car for an hour on a level road at 40 miles/hr? The power required at the rear wheels is about 9 kW, 9000 J/sec, and therefore the energy needed for an hour's operation is

$$9000\frac{\text{J}}{\text{sec}} \times 3600 \text{ sec} \simeq 3.2 \times 10^7 \text{ J}.$$

We will make the optimistic assumption that there are no significant losses in the coupling of the flywheel to the car.

What is the kinetic energy of a spinning flywheel? Flywheels can be made in various shapes, and the simplest one to analyze is the rim-loaded flywheel (Figure 4.13a) in which most of the mass is concentrated at the rim. For a wheel of this shape, whatever the tangential speed (v) of a point on the rim, almost all the mass moves at this speed,* and the kinetic energy is simply $KE = \frac{1}{2}mv^2$. The speed, v, of the rim can be expressed in terms of the radius, R, of the wheel, and the number of revolutions per second (the frequency of rotation, denoted by ν).

*This is obviously not true for wheels of other shapes, for a wheel in the form of a solid disk for instance (Figure 4.13b), since the parts of the wheel near the axle do not move as fast.

QUESTION

4.20 Show that v is related to R and ν in the following way:

$$v = 2\pi R\nu.$$

Thus the kinetic energy of the flywheel can be expressed in terms of the frequency of rotation:

$$KE = \tfrac{1}{2}m(2\pi R\nu)^2 = 2\pi^2 mR^2\nu^2.$$

In order to store an energy of 3.2×10^7 J, we require that

$$2\pi^2 mR^2\nu^2 = 3.2 \times 10^7 \text{ J}.$$

Since we probably cannot afford to use more than half the mass of the vehicle in the flywheel, let us use a 1-ton steel flywheel (m = 1 ton = 907 kg). If the radius of the flywheel is 1 m, this is a diameter of 2 m (over 6 ft), and it is unreasonable to use a value that is any larger. With these values (m = 907 kg and R = 1 m), we can find the required value of ν, the number of revolutions per second:

$$\nu^2 = \frac{3.2 \times 10^7}{2\pi^2 \times 907 \times 1^2} \simeq 1800$$

$$\nu \simeq \sqrt{1800} \simeq 42 \text{ revolutions/sec.}$$

This figure does not seem to be ridiculously large. Admittedly we have made some rather generous estimates of the mass and the size of the wheel, but it appears possible that enough energy could be stored in this way to drive a 2-ton car for a distance of 40 miles.

QUESTION

4.21 With ν equal to 42 revolutions/sec, the speed of a point on the rim would be rather large. How large? How large in comparison with the speed of sound?

Such a flywheel does carry a good deal of energy (though less than the chemical energy of a gallon of gasoline!), and we could not afford to have the wheel disintegrate. A wheel *will* disintegrate if it is spun rapidly enough, for, when it spins, every particle of the wheel executes uniform circular motion, has an acceleration toward the center, and must therefore be subject to an inward force. If the material of which the wheel is made is not strong enough, it cannot supply this force and the wheel breaks apart. In order to know whether or not this flywheel will hold together, we need information about the strength of steel. The details will be omitted, but the conclusion is that such a wheel can hold together at this speed, though we could not spin it very much faster without risking disintegration.

It appears that the use of a flywheel is a possibility worthy of further study—one which may be helpful but not one which provides any easy solutions. However, a closer study of the physics of flywheels reveals that the rim-loaded flywheel is not the best one. In order to maximize the amount of stored kinetic energy, it seems reasonable to use a heavy material such as steel and to concentrate the mass near the outer edge of the wheel, but one can actually do better with lighter materials and with wheels of different shapes. A lighter wheel can be spun faster without disintegrating and can provide as much or more kinetic energy without adding so much to the mass of the car, and, as we have seen earlier, this is an important consideration. (The amount of kinetic energy stored is important, but so is the amount of kinetic energy stored *per unit mass* of the flywheel.) The rim-loaded wheel is not the best choice in terms of strength. For the wheel shown in Figure 4.13a, the inward centripetal force on the whole rim must all be supplied by the relatively thin spokes; more strength and therefore a higher safe speed can be achieved by using thicker spokes. If we make the "spokes" thick enough, the wheel eventually becomes a simple uniform disk (Figure 4.13b); even better is a disk tapered so as to be thinner at the outside edges (Figure 4.13c). Still better flywheels have been made with the design shown in Figure 4.13d, one which hardly looks like a wheel at all.

No matter how good the flywheel, the energy to start it going must come from somewhere, and the wheel must periodically be "recharged." Nevertheless, the use of flywheels could help to ease the energy problem in a number of ways. If the flywheel is recharged by connecting it to an electric motor using electrical energy from an oil-fueled power plant, this electrical energy can be obtained from the chemical energy of oil at an efficiency of about 33%, as compared to the 20% efficiency of an internal combustion engine. Less oil may be used, and even for equal amounts of oil, the resulting air pollution is likely to be smaller if this oil is used at a power plant where pollution can be more easily controlled. Furthermore this pollution is generated at the power plant, out in the countryside perhaps, rather than in the midst of an already polluted city. Another important point is that power plants need not use oil; they can be run equally well with coal or nuclear fuel. In this way the direct dependence of our transportation system on one particular resource, oil, can be reduced. If flywheels are adequate to power vehicles for a whole day, all the flywheels could be recharged at night. This would mean that we would not necessarily need to build new electric power plants for recharging flywheels, since, as we saw in §4.2.B, electric power companies often have generating capacity to spare during the night.

A number of experimental flywheel cars are currently being designed and tested, and for a number of years a flywheel bus was used in Switzerland. This bus had a flywheel with enough energy to drive the bus for half a mile, after which a 2- or 3-minute recharging was needed. An interesting variation on the flywheel car would be a hybrid, with both an internal-combustion engine and a flywheel. For intercity driving, the conventional engine would be used both to drive the car and to get the flywheel up to speed; upon entering a city one would switch over to the flywheel for as long as possible, thus easing the city's pollution problems (though increasing somewhat the level of pollution outside the city).

Flywheels might also be used for purposes other than transportation, namely as an alternative to pumped storage as a way of easing the power companies' problem of a varying load. Just as for pumped storage, the excess nighttime generating capacity would be used to deliver energy to a flywheel (or to a bank of flywheels); then during the day electrical energy would be generated by the flywheel and delivered to the consumers along with the energy generated by the conventional plants. An obvious advantage over pumped-storage hydropower is that flywheels can be located anywhere: it is not necessary to excavate a mountaintop. The design problem for such flywheels would be quite different from that of designing flywheels for cars or buses. Size and weight limitations would be much less severe, but an extremely large amount of energy must be stored to make this idea useful.

QUESTION

4.22 In the preceding discussion we considered a 1-ton flywheel with a radius of 1 m that could store an energy of 3.2×10^7 J. How many such flywheels would be needed to provide a storage capacity equal to that of the pumped-storage reservoir at Northfield Mountain, discussed in §4.2.B?

§4.3.C Other Modes of Transportation

When all is said and done, the modern family car *is* an impressive piece of technology (its inefficient and highly polluting gasoline-powered internal-combustion engine notwithstanding). The ease with which a car can be driven and the thousands of times that the pistons go back and forth and the wheels go around before requiring maintenance are factors that should be a source of pride to the engineering profession. One can only regret that comparable attention has not been paid by the automobile companies and by the public at large to the negative aspects of automobiles.

The major difficulty with cars is that there are so many of them (95 million in the United States in 1973) and that each one carries so few passengers, often just one and rarely more than four. It is absurd to drive a 2-ton vehicle with a 200-lb payload. We might nibble away at the energy and pollution problems associated with transportation by making minor improvements here and there in engine efficiency, by better streamlining, conceivably with flywheel cars and so on, but the overall transportation problem is only going to get worse until we outgrow the idea that we have the right to drive and park our cars whenever and wherever we wish. Our society must eventually turn to increased use of one form or another of public transit for most purposes, with private cars saved for a few special purposes. The motivation for this shift will only come in part from the energy problem; other factors will be the air pollution problem, frequent traffic jams, lack of parking space, and a revulsion at the paving of the country with expressways.

Such a shift will have a very significant impact on the energy problem, for there are methods of transportation that use far less energy, and it is useful to make some comparisons. For all sorts of driving taken together, American cars get about 12 miles/gal on the average. An intercity bus may only get 5 miles/gal, significantly worse than the car except for the fact that a bus can carry more passengers. The appropriate figure for comparison is the number of *passenger*-miles per gallon. If we suppose that a car has an average occupancy of two passengers and the intercity bus has 30, then the bus gets 150 passenger-miles/gal as opposed to 24 for the car.

QUESTIONS

4.23 Walking and bicycling are two useful methods of transportation, and the number of "passenger-miles per gallon" can be estimated with commonly available knowledge. One hiking group states that a hiker covering 30 miles/day on easy level ground requires about 2000 Cal of food energy, over and above what would be needed for sitting at a desk all day. (These are kilocalories, or "food-calories". One such Calorie is equal to 4184 J.) Show that the equivalent amount of gasoline, at the rate of 1.3×10^8 J/gal, is about 0.064 gal, and that the number of miles per "gallon" is approximately 470.

4.24 A bicyclist, with the same amount of supplementary food energy, can cover about 100 miles/day. Show that this is equivalent to approximately 1560 miles/gal.

Estimates of the number of passenger-miles per gallon of gasoline (or the equivalent) for various methods of passenger transportation are collected in Table N.1. Various types of transportation are often compared in terms of the number of BTU's needed per passenger-mile; except for the conversion of BTU's to the equivalent amount of gasoline, this figure is the reciprocal of the number of passenger-miles per gallon, and these values too are given in the table. Similar information for freight transportation is shown in Table N.2; here the appropriate measure is the number of *ton*-miles rather than the number of passenger miles.

The bicycle is far and away the best method of getting around, and where this is impractical, the advantages of trains and buses over automobiles are apparent from these figures. Similarly, oil pipelines (where they can be used), boats and trains are much better ways of moving freight than is the truck. In the years since World War II, an increasingly large fraction of our passenger traffic has been carried by cars, and trucks have steadily replaced trains for carrying freight. Tremendous numbers of new roads have been built, and many of our railroad lines abandoned. It is apparent from the data in Appendix N that a nation faced with an energy problem (in particular, with oil shortages) ought to take steps to save the railroads and to discourage the use of trucks and private cars.

§4.4 SPACE HEATING

§4.4.A Introduction

One of the most important uses of energy is for the heating of houses and other buildings during the winter. Space heating is a necessity, and it is also a very large item in the overall energy picture. Of all the energy consumed in the United States, about 20% is used for this mundane purpose. Most of this energy is derived from oil and from natural gas, though electrical space heating has been increasing in recent years. Although space heating does not require any exotic modern physics, it is an area in which there are real possibilities of alleviating the energy problem through changes in technology. It is also an area in which individual citizens, with their hands on the nation's thermostats and the choices they make in buying or building houses, have an opportunity to make a significant contribution.

Let us consider, therefore, the process of heating a building. Most houses contain a number of lights, a wide variety of electrical appliances, a stove, a few human bodies, perhaps a dog and a cat—all of which contribute to the energy balance—but for simplicity let us disregard these factors. Consider a house that has been empty for some time, with the furnace off, so that the temperature indoors is the same as that outdoors. What happens after the furnace is turned on?

We will assume that the house is heated by an oil furnace. When the furnace is turned on, the burning of oil begins, a chemical reaction in which chemical energy is converted into thermal energy of the air or water with which the burning oil is in contact. Whether by means of circulating air or hot water or steam, heat flows into the interior of the house. If we think of the house itself as the system and of the furnace as something attached to the house from which heat can flow into the house, we can represent this as shown in Figure 4.14. Energy is being added to the house by the flow of heat, H_1, which increases the thermal energy of the house, an increase reflected in an increase in temperature. The energy equation in this case is simply

$$H_1 = \Delta(TE). \tag{4.6}$$

In any period of time, the heat that flows into the house from the furnace is equal to the increase in thermal energy, $\Delta(TE)$, of the house.

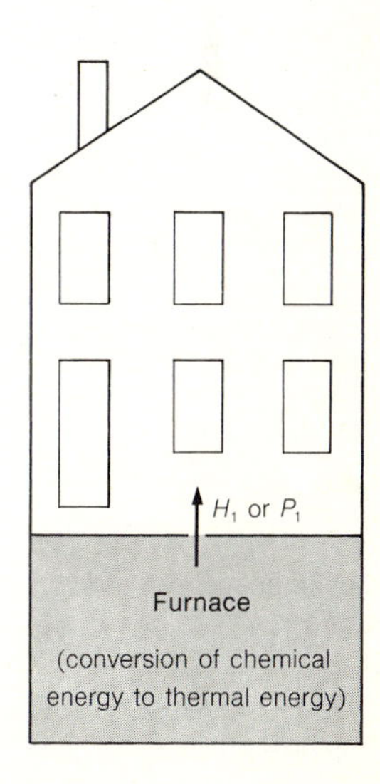

FIGURE 4.14
Initial heating of a house (in which the indoor temperature is still approximately equal to the outdoor temperature).

The same information can be conveyed by describing the *rate* at which energy is being added to the house. The rate at which heat is flowing into the house is an amount of energy per unit time, a *power*, that will be denoted by P_1. This power is equal to the *rate of change* of the thermal energy of the house:

$$P_1 = \text{Rate of change of } (TE).$$

If heat flows into the house at the rate of 10 BTU/sec, the thermal energy of the house increases at the same rate.

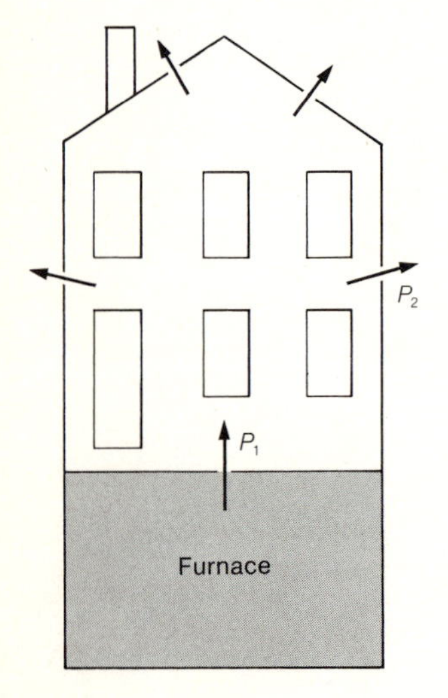

FIGURE 4.15
Heating a house: net rate of heating = $P_1 - P_2$.

As the house gets warmer, heat begins to flow out through the walls and windows to the cooler air outside. Now there are two heat flows to be considered (Figure 4.15). If P_2 represents the rate at which heat is flowing out of the house, then the net rate at which heat is flowing into the house is just the difference between P_1 and P_2:

$$P_1 - P_2 = \text{Rate of change of } (TE). \qquad (4.7)$$

P_2, the rate at which energy is lost to the outside, depends on the temperature of the house. As the house gradually gets warmer, P_2 steadily increases; the temperature of the house continues to increase but more and more slowly, for according to Equation 4.7, the rate at which the thermal energy of the house increases depends on the amount by which P_1 exceeds P_2. Eventually a steady state is reached, when the temperature of the house is just high enough so that P_2 cancels out P_1, and the temperature of the house levels off at this point, as shown in Figure 4.16. What the steady-state temperature inside the house will be depends on the value of P_1. If we want to raise this temperature, we can increase P_1, and then the temperature of the house will rise until P_2 has also increased to equal the new larger value of P_1.

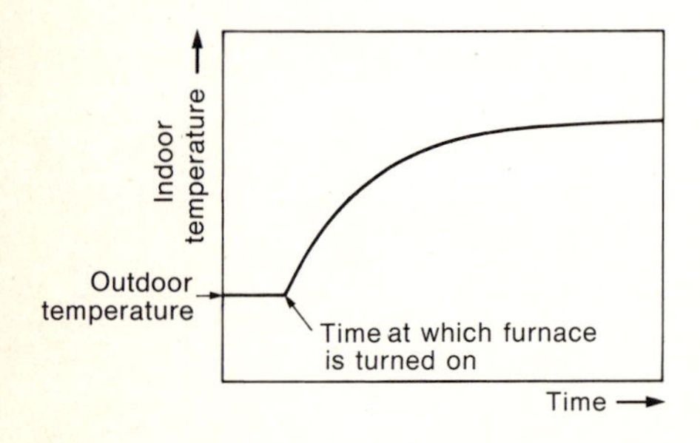

FIGURE 4.16
Indoor temperature as a function of time.

Heating a house is precisely analogous to pouring water into a leaky bucket, while the bucket is held partially submerged in a lake. As long as no water is being added, the water level inside the bucket is the same as that outside (Figure 4.17a), just like the temperatures inside and outside the house. As water pours into the bucket at a steady rate, the water first rises and then reaches a steady state in which water is lost through the hole as fast as it is added (Figure 4.17b). In order to raise the steady-state water level, one must increase the rate at which water is added. The

FIGURE 4.17
The leaky-bucket model of heating a house.

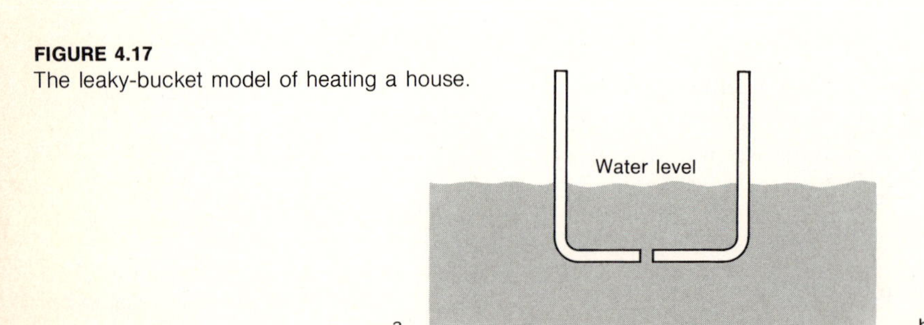

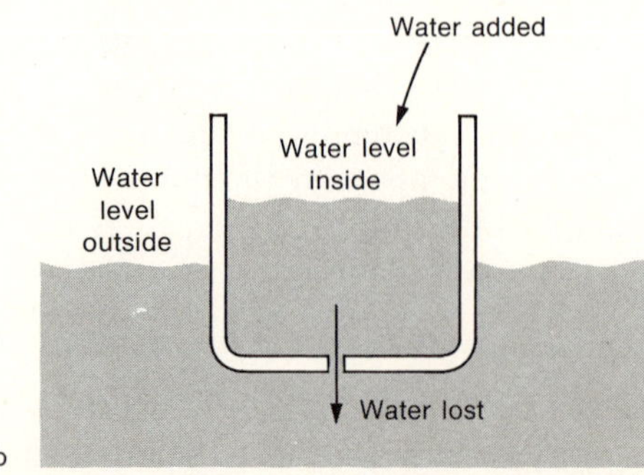

level then gradually rises, water flows more rapidly out the hole, and a new higher steady level is eventually reached.

Once a steady state has been reached, Equation 4.7 takes a very simple form:

$$P_1 - P_2 = 0, \qquad \text{or} \qquad P_1 = P_2. \tag{4.8}$$

Energy flows into the house from the furnace at a rate equal to the rate at which energy flows out through the walls and windows. This deceptively simple equation is the fundamental equation that describes how houses are kept warm in the winter. Actual houses vary widely in size and in the quality of their insulation, as well as in the severity of the climate to which they are exposed. An approximate value for P_2 (or P_1) for a medium-sized one-family house with an indoor temperature of 70°F and an outdoor temperature of 15°F is $P_2 = P_1 \simeq 10^6$ BTU/day. Since architects and heating engineers usually measure energies in BTU's and use the hour or the day or the year as the unit of time, the concerned citizen, too, must be familiar with these units. For the home owner who is not interested in such matters, the gas or oil company conveniently translates heat flow in BTU's per day into an equivalent in dollars per month.

QUESTION

4.25 Consider the value of P_2 cited above as an example, 10^6 BTU/day. What is this value of P_2 in watts? Is this a significant part of a family's overall average rate of energy consumption, about 10 kW per person?

It is useful to make some qualitative observations about Equation 4.8. The rate of flow of heat, P_1, from the furnace to the house, is what we must provide and what we must pay for. In the steady state, $P_1 = P_2$, and thus any reduction in P_2 reduces our rate of energy consumption and our fuel bill. The rate at which heat flows from the house to the outside, P_2, depends on the temperature difference between inside and outside. A reduction in P_2, and hence a reduction in P_1 and in our fuel bill, can be achieved by settling for a lower temperature inside the house, that is, by lowering the thermostat.

Another simple and obvious observation is this: the heat delivered to the house (P_1) is all "wasted". This energy flows into the house and then promptly flows to the outside.* If the insulation could be made thick

*Thus we might legitimately claim that the "efficiency" of any heating system is 0%. However, if the law of conservation of energy is valid, then no energy is ever really *lost*; it is only converted from one form to another or transported from one place to another. On this basis, we might claim that the "efficiency" of *any* process is 100%. Whenever the word "efficiency" is used, one must be careful to find out just what is meant by the term; it is a word that is too often used carelessly.

enough and if we did not need fresh air, then P_2 (and thus P_1) could be reduced to zero, even with a large difference in temperature between the inside and the outside. To put it another way, the "ecologically perfect" method of keeping one's house at the right temperature would be to build a house with enormously thick walls, wait for a day in September when the outside temperature is 70°F, and then close the doors. If the insulation were perfect and if there were no other sources of heat, the inside temperature would remain at 70°F, no matter how low the exterior temperature (or, for that matter, no matter how high the exterior temperature). No one would seriously suggest trying to do it in precisely this way (for one thing because human bodies generate heat), but the point is that if we are willing to pay enough for insulation, nearly any degree of reduction in P_2 and therefore in the amount of energy consumed is *possible.* How large a reduction is economically sensible is another matter, but there is no fundamental scientific principle that would prevent us from lowering our fuel consumption almost as much as we wish. In the real world, we cannot afford insulation thick enough to make P_2 always equal to zero, and some supply of heat will be needed.

§4.4.B The Heating Needs of a House. Thermal Conductance and the Degree-Day

Since P_1 is what we pay for and since P_1 must equal P_2, it is important to look at the factors on which P_2, the rate at which heat flows out of the house, depends. The most important of these is the difference in temperature between inside and outside.* To a good approximation, P_2 is just proportional to this temperature difference:

$$P_2 \propto \Delta T,$$

where ΔT is the difference in temperature. This proportionality can be rewritten as an equation,

$$P_2 = k\,\Delta T. \tag{4.9}$$

Hidden in the constant of proportionality, k, are the properties of the house itself. If the house is a large one with many windows and poor insulation, k will be large; that is, for a given value of ΔT, P_2 will be large. If k is small, heat is lost less rapidly. This constant of proportionality, k, is called the "thermal conductance" of the house; it is a measure of how easily heat is conducted through the walls.

Equation 4.9 refers to the rate of energy flow, in BTU's per day or in watts, for example. It can be rewritten to describe the total amount of heat H_2 flowing out through the walls in a time interval of length Δt by

*The heat loss also depends to a lesser extent on other factors such as wind speed, which we will not take into consideration.

multiplying both sides of the equation by Δt:

$$P_2 \, \Delta t = H_2 = k \, \Delta T \, \Delta t. \tag{4.10}$$

If the temperature difference, ΔT, is constant, then during a given time interval, the total amount of energy required is of course proportional to the length of the time interval.

Experience shows that quite accurate calculations can be made by taking ΔT in Equations 4.9 and 4.10 to be the difference between an indoor temperature of 65°F and the outdoor temperature, whatever the latter may be. For example, if the temperature outside is 55°F, we set $\Delta T = 10°\text{F}$; if the temperature is 45°F, $\Delta T = 20°\text{F}$. If the temperature outside is 65°F or above, we set $\Delta T = 0$, because little or no heating is required under these conditions. Since American thermostats have traditionally been set at about 70°F, it might seem more reasonable to calculate ΔT by using an indoor temperature of 70°F rather than 65°F. The reason that 65°F is more appropriate is that human activity in the house —operation of lights and stoves, the production of heat by human bodies— usually makes up the difference.

The rate at which heat must be supplied by the furnace varies as the outside temperature changes. A good estimate of the heat needed during a whole 24-hour period can be made by ignoring the hour-to-hour variations and taking the outside temperature to be the arithmetic mean of the maximum and minimum temperatures for that day. If, for example, the maximum and minimum temperatures on a given day are 45°F and 25°F, we take the "average" temperature for that day to be 35°F and set $\Delta T = 65° - 35° = 30°\text{F}$ in Equation 4.10. A very practical reason for this procedure is that the maximum and minimum temperatures for the day are the temperatures commonly available from weather records.

If the average daily temperature is the same for several successive days, Equation 4.10 can be applied to the whole period. If Δt is measured in *days* and ΔT in *degrees* Fahrenheit, the product of these two terms has the units degrees × days, or "degree-days" (abbreviated DD). A three-day period with a temperature of 35°F ($\Delta T = 30°\text{F}$) contributes 90 DD to the heating needs of the house. The same total amount of heat is required for any other period which has the same total number of degree-days, for example a two-day period with a temperature of 20°F ($\Delta T = 45°\text{F}$). The total amount of heat needed for a week, a month or whole winter heating season can be calculated by finding the number of degree-days for each day and adding them up:

$$H_2 = k \times \text{total number of degree-days}. \tag{4.11}$$

Suppose, for example, that the average temperatures during a week have the values shown in Table 4.1 on the next page; a total of 48 DD is accumulated during the week.

To calculate the total amount of heat needed for a whole heating season, it is the total number of degree-days accumulated during a year

TABLE 4.1 Example of daily numbers of degree-days and the weekly total

	Average outside temperature (°F)	ΔT (°F)	Number of degree-days
Sunday	55	10	10
Monday	58	7	7
Tuesday	68	0	0
Wednesday	73	0	0
Thursday	64	1	1
Friday	54	11	11
Saturday	46	19	19
		Total:	48

that is relevant. At any one spot, this total depends on the severity of the particular winter. The total number of degree-days per year for the "typical year" varies widely from place to place within the United States, from about 200 DD/yr in southern Florida, where the average daily temperature is rarely below 65°F, to over 10,000 DD/yr in northern Minnesota. Appendix G contains a map showing the average number of degree-days per year in the United States, together with a table of monthly values for selected cities.

What are the units of k, the "thermal conductance" of the house? If H_2 is measured in BTU's (the unit most commonly used in such calculations), then it can be seen from Equation 4.11 that k must be measured in BTU's per degree-day:

$$H_2 \text{ (BTU)} = k \text{ (BTU's per degree-day)} \times \text{total number of degree-days.} \quad (4.12)$$

In this equation, the total amount of heat needed is very conveniently expressed as the product of two factors. The first is characteristic of the house and would have the same value if the identical house were built somewhere else; the second does not depend on the nature and size of the house but only on the severity of the climate. The house used as an example (§4.4.A) required an amount of heat equal to 10^6 BTU on a day when the outside temperature was 15°F ($\Delta T = 50$°F). Since H_2 is 10^6 BTU for 50 DD, the value of k for this house is

$$k = \frac{10^6 \text{ BTU}}{50 \text{ DD}} = 20{,}000 \text{ BTU/DD}.$$

It is, for short, "a 20,000 BTU/DD house." This value is typical for one-family middle-income American homes, most of which probably have thermal conductances of between 15,000 and 30,000 BTU/DD.

If we know the thermal conductance of the house (k = 20,000 BTU/DD in this example) and the number of degree-days in the heating season—say, 6000 DD for places such as Boston, Cleveland, and Salt Lake

City—we can estimate the total amount of heat needed for one winter heating season: $H_1 = 20{,}000 \times 6{,}000 = 120 \times 10^6$ BTU, or, expressed in millions of BTU's (1 MBTU = 1 million BTU), $H_1 = 120$ MBTU.

§4.4.C Furnace Efficiencies, Individual Strategies for Reducing Energy Consumption, and the Value of a Storm Window

Knowing how many BTU's are available from the combustion of a gallon of oil or a cubic foot of natural gas, we can calculate the amount of fuel needed except for one significant factor that has thus far been omitted. Furnaces do not completely convert the chemical energy available in the fuel into thermal energy delivered to the house. Some of the oil or gas is simply not burned. Because of this incomplete combustion, a fraction of the fuel's chemical energy remains in the form of chemical energy but is lost as the unburned fuel goes up the chimney, just as chemical energy is lost in a car when unburned hydrocarbons are emitted in the exhaust or when gasoline evaporates from the carburetor. In addition, some of the thermal energy produced by combustion is not delivered to the house but goes up the chimney, where it merely heats the outdoors.* These factors are taken into consideration in Figure 4.18, which shows the energy flow in more detail.

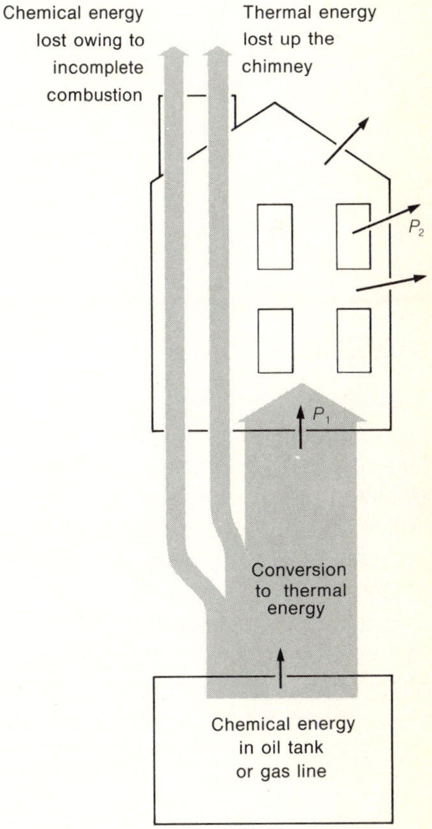

We will define the "efficiency" of the furnace as the ratio of the energy we want (H_1, the heat delivered to the house) to the total chemical energy of the oil used:

$$\text{efficiency} = \frac{H_1}{CE_{\text{total}}}. \tag{4.13}$$

Efficiencies of most modern well-adjusted gas or oil furnaces are between 0.7 and 0.8 (70% and 80%), but many home furnaces are quite old and are rarely kept in perfect adjustment. Furnaces also operate somewhat less efficiently for the first few minutes after they are turned on, and the average efficiency is somewhat lower when the thermostat operates to turn the furnace on for frequent brief intervals than it would be if the furnace could be left running continuously at a low level. For these various reasons, typical furnaces have efficiencies in the range of 55% to 65%.

If we know k (the thermal conductance of the house), the efficiency of the furnace and the number of degree-days per year, we can calculate the amount of fuel needed, and if we also know the current price of oil or natural gas, we can estimate the cost of keeping a house warm during the winter. The number of degree-days we can do nothing about, and as individual citizens we have equally little control over

FIGURE 4.18
The energy flow that takes place in the heating of a house. (The efficiency of the furnace is taken into consideration.)

*In this connection, Benjamin Franklin pointed out that, with wood as the fuel, a stove in the middle of the room is far superior to a fireplace at the edge of the room, directly below the chimney.

the price of fuel. We have some control over the efficiency of the furnace: if we are buying a new one, we can try to purchase a high-efficiency one, and if we already own one, we can have it tuned up.

Suppose the furnace in the house discussed earlier has an efficiency of 60%. Then to supply a total of 120 MBTU to the house, the total amount of chemical energy needed is (from Equation 4.13):

$$CE_{\text{total}} = \frac{120 \text{ MBTU}}{0.6} = 200 \text{ MBTU}.$$

This figure can be translated into gallons of oil. The chemical energy of oil is about 5.6 MBTU/barrel, or, since 1 barrel = 42 gal, $5.6/42 \simeq 0.13$ MBTU/gal. The amount of oil needed to keep this house warm for the winter is:

$$\frac{200 \text{ MBTU}}{0.13 \text{ MBTU/gal}} \simeq 1540 \text{ gal}.$$

If fuel oil sells for 30¢/gal, this will cost

$$0.3 \text{ \$/gal} \times 1540 \text{ gal} \simeq \$460.$$

QUESTION

4.26 It was remarked above that of the three factors that determine the total amount of energy required to keep a house warm during the winter (the efficiency of the furnace, the thermal conductance of the house, and the total number of degree-days), you can do nothing whatever about the total number of degree-days, which is simply a measure of the severity of the climate. Although it is true that you cannot change the outdoor temperature, you can in *effect* change the number of degree-days by manipulating the thermostat. If you are willing to lower the thermostat by 5°F from its present setting, this is equivalent to reducing the number of degree-days by 5 for each day on which heating is normally required. On about how many days during the winter is heat needed where you live? What fraction of your total heating needs could be eliminated in this way? How much (in gallons of oil, cubic feet of natural gas, kilowatt-hours of electricity, or dollars) would this save per year for the typical 20,000 BTU/DD house, or for the house in which you live?

The basic quantity that characterizes the house is the value of k, the thermal conductance of the house, the number of BTU's lost per degree-day. The larger the value of k, the more expensive it is to heat the house. To some extent, the value of k depends on the habits of the occupants; if you insist on leaving windows wide open in midwinter, your house will lose many more BTU's than it would otherwise. Aside from effects of this sort, the value of k is dependent on the amount and quality of the insulation—whether or not there are open cracks in the walls, the state of the weather stripping on the doors, the number of windows, whether storm doors and storm windows are used, and so on. Of course

k also depends on the size of the house; other things being equal, a house will lose heat to the outside at a rate proportional to its total outside surface area. In designing a house or trying to estimate the heating needs of an existing one, a heating engineer will consider the total surface area, the insulation, the number of windows, and other features. It is then possible to make quite an accurate calculation of the value of k, based on the known physical properties of glass, brick, and various kinds of insulation. An experienced engineer can also make a reasonably good estimate of that value by making a brief tour of the house and comparing it to other houses that have been studied more carefully.

QUESTION

4.27 Even without access to the information that the heating engineer has available, you can indirectly estimate the thermal conductance of a house if you know the efficiency of the furnace—or if you simply assume it to be 60%—and if you know about how much fuel is used during the winter and the number of degree-days. Try to obtain such information about a particular house, even if your numbers are quite rough, and estimate the value of its thermal conductance. Is it of the same order of magnitude as the figure we have been using as an illustration, 20,000 BTU/DD?

We will not go into the details of the calculations that must be made to calculate directly the thermal conductance of a house, but let us consider as an example the effect of putting storm windows on the house. One ordinary window contributes about 350 BTU/DD to the total thermal conductance of a house; putting on a storm window has a very significant effect and reduces this figure to about 150 BTU/DD. If the number of degree-days per year is 6000, this reduction of 200 BTU/DD saves

$$200 \times 6000 = 1.2 \times 10^6 \text{ BTU} = 1.2 \text{ MBTU}$$

in the amount of heat per year that must be supplied to the house, and this is the saving from a single storm window.

Is it a good idea to install the storm window? At first glance, obviously yes; you simply save this much energy. However, energy is also required to produce the aluminum and glass and to ship the window from the factory. It is extremely difficult to make an exact comparison of the amounts of energy consumed. (Do you count the energy used in shipping the fuel oil? A share of the energy for the lights in the office of the storm-window company?) When trying to make a comparison in which some of the factors are next to impossible to estimate, we can follow the lead of the economists and simply compare the cost of a storm window with the cost of the fuel which can be saved. It has become a truism in recent years to observe that prices do not necessarily reflect true costs, that the price of a gallon of oil should include—but does not include—the damage to health and property produced by burning the oil, by coastal oil-spills, and so on. However imperfect it may be, the marketplace does provide *one* way of comparing alternatives. Although some individual citizens may choose to conserve energy even at a financial sacrifice, most people are not likely to buy storm windows if it is significantly cheaper for them not to do so.

QUESTION

4.28 Show that if the furnace efficiency is 60%, a reduction of 1.2 MBTU in the amount of heat needed will save about 15 gal of oil, worth $4.50 at a price of 30¢/gal.

Suppose that a storm window costs $30 and that one such window will save $4.50 per year in fuel bills. The simplest way of making the comparison is to say that a storm window will pay for itself in six or seven years at a saving of $4.50 every year. If the window can be expected to last seven years or more, then it is obviously a good buy. Twenty years is a more reasonable lifetime, and on this basis the storm window costs $1.50 per year and saves $4.50 per year on fuel bills; the comparison comes out in favor of the storm window by a factor of three to one. There may be other factors which are more difficult to evaluate. If the prospective buyer has reason to believe that the price of oil will go up, purchase of the storm window will appear even more attractive; conversely, if there is a chance that the price of oil will go down, the storm window is not such a good buy.

There is a very basic objection to this method of making the comparison, however. A storm window represents a *capital investment* of $30. How is this to be compared with the alternative of a *continuing* operating cost of $4.50 *per year?* The point is that spending $30 *now* is not the same as spending $1.50 per year for the next 20 years. Suppose that you have $30 available. If instead of buying a storm window, you loan it to a bank which pays interest, you can withdraw $1.50 each year for 20 years and still have some money left. You can take out more than $1.50 each year, and if you make the interest calculations correctly, you can arrange it so that your bank balance is neatly reduced to zero at the end of 20 years. Alternatively, if you do not have the $30 but must borrow it, you will have to pay interest. If you agree to pay back the loan in 20 equal installments, you will find no bank that will agree to 20 payments of $1.50 each; the bank will insist on larger payments so that you will be paying interest on the outstanding balance.

One way of translating a capital cost into an equivalent operating cost is based on this last idea. If you must borrow the money and if you make annual payments to the bank (the same amount each year),* what sum must you pay per year? Although it may seem absurd to go to all this trouble to calculate the value of a storm window, the same principles apply if you are buying 20 storm windows instead of one, or if you are considering the purchase of a solar-heated home (with a high capital cost and a low operating cost). Utility executives face similar problems in trying to decide between construction of two different kinds of power plants, perhaps between a nuclear power plant and a coal-fueled plant. The capital investment for a nuclear plant is significantly larger, but the operating cost may be lower. We must have a rational way of comparing capital and operating costs.

We cannot convert a capital cost into an equivalent annual cost unless we know the annual interest rate and the number of years over which the loan is to be repaid. Once we have this information, the comparison can be made. For purposes of this discussion, there is no need to make the arithmetical calculations; tables to facilitate this comparison for various periods of time and various rates of interest are available, and such a table is given in Appendix Q. According to the table, if we assume a 20-year period and an annual interest rate of 10%, then, a loan of $1 can be repaid in 20 equal installments of 11.7¢ each, a total payment of 20 × 11.7¢ = $2.34. (Note that the total amount repaid is more than twice as large as the amount borrowed to begin with, for this interest rate and for this period of time.) A loan of $30 requires payments 30 times as large, annual payments of approximately 30 × 11.7¢ ≃ $3.50 (a total cost of about $70 in the course of 20 years). On the basis of these figures, a capital investment of $30 in a storm window is equivalent to an *annual* cost of $3.50. If this window saves $4.50 per year in fuel bills, then purchase of the window is still advantageous, but not as decisively so as our first estimate indicated. It should be noted that taking the cost of borrowing money into account changed the estimate of the annual cost by more than a factor of two, from $1.50 per year to $3.50 per year, by no means a negligible correction. The effect can easily be larger or smaller than this, depending on the time period and the interest rate. These two figures may not be known when one tries to make the estimate, but it is safer to assume some plausible values rather than to ignore the effects of interest charges altogether.

Even if the comparison had turned out the other way, one might still argue for the storm window. From the individual's point of view, the storm window provides partial security against continuing rises in oil prices or rationing of oil supplies. Also, since storm windows do reduce energy consumption, their use would obviously benefit society as a whole in the long run.

*During the first few years, a large fraction of each payment is for interest and only a small amount goes toward reducing the size of the balance; during the last few years, the reverse is true.

§4.4.D National Strategies for Reducing Energy Consumption in Space Heating

As has been pointed out earlier, *all* the energy used in heating buildings is really wasted. Because about 20% of the total amount of energy used in this country is used for space heating, very real savings could be made by getting used to lower indoor temperatures and by improving insulation. Adding storm windows is one way of improving insulation. Much more significant is the type of insulation installed when the building is constructed. Several careful studies have shown that it would be economically advantageous to use insulation sufficient to lower the thermal conductances of buildings to about 50% of what is typical at present. This must necessarily be regarded as a long-range program: buildings are made to last a long time, about half of the buildings that will be in use in the year 2000 will have been built before 1970, and it is much easier to put good insulation in a new building than to modify an old building.

Another long-range strategy is the use of solar energy to supply a significant fraction of our space heating, a possibility that will be discussed in more detail in Chapter 16. Here we have an extreme case of the comparison made in evaluating storm windows: the comparison of solar heating (with a large capital investment but with no fuel cost except for whatever is needed for heating during long periods of cloudy weather), and conventional heating that requires a large annual expense for fuel.

FURTHER QUESTIONS

4.29 Estimate the power which you can deliver by running upstairs, increasing your gravitational potential energy. How does your power compare with that of the "standard horse"? (1 hp = 746 W.)

4.30 From the basic facts that 1 W = 1 J/sec, 1 cal = 4.184 J, 1 BTU = 252 cal and 1 hr = 3600 sec, show that 1 kWh $\simeq$ 3413 BTU.

4.31 Just as power companies report the average rate at which they generate electrical energy in kilowatt-hours per year, customers often think of their average rate of consumption of electrical energy in another peculiar unit of power: kilowatt-hours per month. An electric bill shows that a particular family has used 450 kWh during a one-month period; to what average power in watts does this correspond?

4.32 Examine some current news articles on the energy problem. What units of energy and power are used? Are there any instances of confusion between the concepts of energy and power or the units in which they are measured? Give revised versions of any such statements.

4.33 Niagara Falls is another example of a large hydroelectric generating site. Here the head is 190 ft and the average rate of flow of water is 7600 m^3/sec. Estimate the average rate at which electrical energy could be generated at this site. How does Niagara Falls compare with Hoover Dam?

4.34 There is one difference between the generation of electrical energy from a natural waterfall and from a dam, such as Hoover Dam, a difference that might be important. The water used at Hoover Dam is essentially at rest when it starts to fall, whereas the water at Niagara Falls already has some kinetic energy, to which the kinetic energy it acquires as it falls is added. By making an estimate of the speed of the river above

Niagara Falls, even a rough guess, try to decide whether this initial kinetic energy is large enough to be of any importance or whether it is insignificant.

4.35 Is there a hydroelectric generating plant in your vicinity? (In all probability, it is a good deal smaller than the very large ones at Hoover Dam and Niagara Falls). Estimate the head and the rate of flow of the water and thus the possible generating capacity of the plant. What *is* the generating capacity, and how well did you do at estimating it? You can make a rough estimate of the rate of flow of water if you can estimate the width and depth of the river and the speed with which the river flows.

4.36 In an era when the complexity of modern life sometimes seems overwhelming, when our energy supply is more and more dependent on electrical power grids which tie together large parts of the United States and on negotiations between huge oil companies and Middle Eastern chiefs of state, it is very appealing to try to think of ways in which small communities, even individual families, might satisfy their energy needs on a local basis. Here is one idea that anyone who lives in a one or two-family house might try. When rain falls onto the roof of a building, it is usually allowed to flow down to the ground in drain pipes, wasting its gravitational potential energy, which has all been converted into thermal energy by the time it gets to the ground. Why not cover the roof with a collecting tank and let the water flow to the ground through a small hydroelectric generator? Without worrying about the feasibility of making family-size generators, estimate the average rate at which electrical energy could be generated, using a typical American house located at a typical place in the United States (or at your location). Is this an absolutely absurd idea, or is the reason it is not in use the result of some other factor, such as the expense of small generators or a conspiracy on the part of the electric utilities? If the initial calculation leads to a discouraging result, why not build a tower and catch the rain at a higher level?

4.37 In order to fill or empty the Northfield Mountain pumped-storage reservoir during a 12-hour period, the rate of flow of water must be quite large. How does the average rate of flow in this case compare with that of the Colorado River?

4.38 A furnace for a medium-sized house is advertised as having a "capacity of 65,000 BTU." What important variable has been left out of this description? Can you guess what the writer intended to say?

4.39 Space heating can also be accomplished with electrical energy, and electrical energy can be converted into thermal energy at an efficiency of 100%. Electrical energy is usually purchased by the kilowatt-hour: 1 kWh = 3413 BTU. Find out the prevailing residential prices of oil, natural gas and electricity in your area, and estimate the relative costs of heating with these three methods. (If current prices are unavailable, use data from Appendix P.)

4.40 Even home owners who use gas or oil inevitably obtain some of their heating electrically, from lights and other electrical appliances. They can reduce their gas or oil bills by leaving the lights on all the time. About how many 100-W light bulbs should be left on to eliminate the need for oil or gas in a 20,000 BTU/DD house on a day when the temperature outside is 15°F? Is this an economically sound procedure? Is the replacement cost of the light bulbs important, or is it only the cost of the electrical energy that needs to be considered?

4.41 For the average American family (or for your family), which consumes the most energy in the course of a year—the operation of an automobile or keeping the house warm? (Be sure to state whatever numerical values you use in making this comparison.)

4.42 Suppose that you live in an oil or gas-heated 20,000 BTU/DD house in an area in which there are 6000 DD/yr. How many dollars is an annual furnace tune-up worth if this results in improving the efficiency of the furnace from 55% to 65%?

4.43 Houses do not necessarily have to be heated by oil, gas or electricity. Buffalo chips were once a common fuel in the Great Plains; wood and waste-paper are possible alternatives now. (Fuel values are given in Appendix H.)

(a) About how many tons of wood would be needed to keep a house warm during a 6000-DD winter?

(b) A typical 1-acre wood lot produces about 2 tons of wood per year. How large a lot would be needed for this house?

(c) Estimate the amount of wastepaper generated by the average family (newspapers, junk mail, and so on), and consider the use of this fuel for heating the house.

4.44 Some people turn down the thermostat at night, arguing that it is wasteful of energy to keep the house warm while they are asleep. Others argue that it is best to leave the thermostat at a fixed temperature, because if the house is cold in the morning, a lot of energy must be used just to warm it up again. Discuss this controversy as carefully as possible. If you conclude that one of the arguments is erroneous, try to make a case for your own position that will be as convincing as possible to someone who holds the opposite view.

4.45 Compare the energy consumption of a 747 jet with the energy used in heating a house. Would the energy used for one transcontinental flight be enough to keep a typical house warm for an entire winter?

4.46 Repeat the storm window calculation (§4.4.C) for another geographical location. Are storm windows a good buy in your area? In Miami? In Barrow, Alaska?

4.47 Consider some possible governmental actions aimed at reducing energy consumption for space heating. Examples might include the following: building codes requiring storm windows and good insulation; condemnation of buildings with thermal conductances above some specified value; subsidized low-interest loans for buildings with good insulation or for solar-heated homes; lower property-tax rates on such homes; free annual furnace adjustments. Discuss these and any other possibilities which you can think of. Besides their effectiveness at saving energy, be sure to consider other aspects as well; for example, the ways in which each suggested plan would affect differently those who are relatively well-off and those who are not.

4.48 Your share of the national power consumption is about 10 kW, 10^4 J/sec. Since 1 day $= 8.64 \times 10^4$ sec, you are "entitled" to 8.64×10^8 J of energy every day. At this rate, if you fly from New York to Los Angeles on a full 747 jet, how many days' worth of your allotment is represented simply by your share of the fuel consumed?

5

Newton's Law of Universal Gravitation

Hale Observatories

5
Newton's Law of Universal Gravitation

Orbits are not difficult to comprehend. It is gravity which stirs the depths of insomnia.

—Norman Mailer,
Of a Fire on the Moon, 1969

§5.1 FORMULATION OF THE LAW OF UNIVERSAL GRAVITATION

Gravity is surely one of the most important of all the physical phenomena that govern our lives. Gravity is of great practical importance, but perhaps even more important is the part played by the study of gravitational phenomena in the development of modern science. In the late 17th century, Newton showed that the motions of the moon around the earth and of planets and comets around the sun, together with the motions of falling rocks and apples, could all be described by a single fundamental law. For the first time, there was convincing evidence that the fundamental laws governing the motions of heavenly bodies were the same as those applicable to everyday objects. Newton developed a law of *universal* gravitation, according to which there is an attractive gravitational force between *every* pair of objects in the universe. Moreover this force can be precisely calculated; it is directly proportional to the masses of both objects and inversely proportional to the square of the distance between them. Physics and astronomy were revolutionized by this development, but it was not only science that was affected. Newton's work had an impact on nearly all areas of thought, philosophy and religion in particular. More important even than the law itself was the style which Newton set for thought, and the confidence he inspired in humans, helping them realize that their minds were capable of discovering and comprehending regularities in the apparently chaotic world in which they lived. In two famous lines written in 1730, the British poet Alexander Pope described the significance of Newton's work:

> Nature, and Nature's Laws lay hid in Night.
> God said, *Let Newton be!* and All was *Light*.

The law of gravitation was the consequence of two great ideas. The first was the general idea that perhaps there is something in common between whatever it is that makes rocks and apples fall to the ground, the mysterious thing that we call "gravity," and whatever it is that governs the motion of the moon in its orbit around the earth and the planets in their orbits around the sun. Perhaps these various motions are all particular manifestations of the same basic phenomenon, perhaps they can all be explained as particular applications of one broadly applicable law of nature. The second idea follows from the first and is more sophisticated. If we accept the idea that all these phenomena have a common origin, what can we do to raise this idea from the level of a vague speculation to a quantitative hypothesis that can be tested? We can *say* that all these phenomena occur because of "gravity," but this does not help very much; anyone can invent labels. We shall tentatively accept the first of the two ideas just mentioned, the idea that it may be possible to understand a wide variety of phenomena as part of a single coherent structure, and we will try to implement the second idea, to test the implications of this assumption.

The objects with which we deal in everyday life, located at or near the surface of the earth, fall downward (toward the center of the earth) with an acceleration equal to g (9.8 m/sec^2) when acted upon solely by the gravitational force of the earth. Thus we can write an equation for the gravitational force exerted by the earth under these circumstances:

$$F_{grav} = mg. \qquad (5.1)$$

The gravitational force exerted by the earth is proportional to the mass of the object on which it acts. If this were not so, if for instance the gravitational force exerted on a large rock were equal to that exerted on a pebble, the large rock would fall more slowly because of its greater mass. As seen in Chapters 2 and 3, the effect of gravity on objects near the surface of the earth can be described either in terms of the gravitational force acting on them (Equation 5.1), or in terms of their gravitational potential energy,

$$GPE = mgh, \qquad (5.2)$$

where h is the height above some convenient zero-level. These equations describe the effect of gravity on objects near the earth's surface, but in order to develop a more general law of gravitation, we must consider objects that are not at the surface of the earth and find the appropriate modifications of Equations 5.1 and 5.2. The earth's gravitational force is presumably weaker on objects that are farther away; we need more data in order to find out how *much* weaker. According to Newton's hunch, it is "gravity" which acts to keep the moon in its orbit around the earth. Before turning our attention to the moon, let us first consider some objects that Newton could not study: artificial earth satellites. (Newton had no way of launching artificial satellites, but almost 300 years before Sputnik, he discussed the possibility of such satellites and the characteristics of their orbits.)

Consider the very first artificial satellite, Sputnik I, launched in 1957. Was it the gravitational force of the earth that kept Sputnik in orbit? Did Sputnik "fall" toward the center of the earth with an acceleration of 9.8 m/sec^2? Sputnik traveled in a nearly circular orbit, at an average altitude of about 350 miles. Its orbital period, the time required to complete one full orbit, was about 96 minutes. As it orbited the earth, it maintained a constant distance from the earth's center, but—as we saw in §3.3.D—an object in uniform circular motion does in a sense "fall" toward the center. Though its speed never changes, it has an acceleration—in the technical sense in which acceleration has been defined—whose direction is toward the center of the earth and whose magnitude can be calculated from Equation 3.17:

$$a = v^2/R$$

where v is its speed and R the radius of its orbit. We can easily calculate the numerical value of Sputnik's acceleration. The radius of the earth (R_E) is about 4000 miles, and the radius of Sputnik's orbit is 350 miles larger:

$$R \simeq 4350 \text{ miles} = 4350 \text{ miles} \times \frac{1609 \text{ m}}{1 \text{ mile}} \simeq 7 \times 10^6 \text{ m}.$$

In 96 minutes, it travels once around its orbit, a distance equal to $2\pi R$, and thus its speed is

$$v = \frac{2\pi \times 7 \times 10^6 \text{ m}}{96 \text{ min}} \times \frac{1 \text{ min}}{60 \text{ sec}} \simeq 7640 \text{ m/sec},$$

and so its acceleration, in m/sec^2, is

$$a = v^2/R = \frac{(7640)^2}{7 \times 10^6} \simeq 8.3 \text{ m/sec}^2.$$

It does *not* fall with an acceleration of 9.8 m/sec^2, but it does fall with an acceleration fairly close to this value. The discrepancy is even in the direction we should have anticipated. We would expect Sputnik to fall with an acceleration of 9.8 m/sec^2 if it were right at the surface of the earth (if, for instance, we were to *drop* it in the laboratory), but we expect the force of gravity to be somewhat weaker on objects farther away. To obtain more definitive evidence on the way in which the force of gravity varies with distance, consider an earth satellite that is much farther away, the moon.

QUESTION

5.1 The moon travels in a nearly circular orbit around the earth, with a radius of approximately 240,000 miles, and completes one orbit in about 27.3 days. Show that the acceleration of the moon toward the center of the earth is approximately 0.0027 m/sec^2.

The moon's acceleration toward the earth is far smaller than 9.8 m/sec^2. The ratio of these two accelerations is

$$\frac{9.8 \text{ m/sec}^2}{0.0027 \text{ m/sec}^2} \simeq 3600.$$

An object at the surface of the earth (4000 miles from the center) falls with an acceleration about 3600 times greater than does the moon, 60 times farther away (240,000 miles). The fact that the ratio of the accelerations (3600) is nearly equal to the square of the ratio of the distances is too striking to be overlooked. It appears that the gravitational force of the earth on an object is inversely proportional to the square of its distance from the earth's center. Let us tentatively assume that this conclusion is correct, and see what consequences follow, and what deductions we can make from this postulate in order to test whether or not it is correct. Let us first add a few important steps to the argument. Apparently the gravitational force exerted by the earth varies in direct proportion to the mass of the object on which the force is exerted and in inverse proportion to the square of the object's distance from the earth's center. What if the mass of the earth were different? Of course, we cannot change the mass of the earth, but it is reasonable to suppose that if the gravitational force is proportional to the mass of one of the two objects in question, it is also proportional to the mass of the other. Let us go even further, then, and suppose that the gravitational interaction of *any* two objects follows the same pattern.

This line of argument leads to the law of universal gravitation. Between any two objects there exists an attractive gravitational force whose magnitude is proportional to the mass of each object and inversely proportional to the square of the distance, r, between them:

$$F_{\text{grav}} \propto m_1 m_2 / r^2.$$

If we write this proportionality as an equation we have

$$F_{\text{grav}} = \frac{G m_1 m_2}{r^2}, \tag{5.3}$$

where G is a "universal gravitational constant."

What audacity! On the basis of an extremely limited amount of information about rocks, baseballs, Sputnik I, and the moon, we write down a law of *universal* gravitation, saying that between *any* two objects (the earth and the sun, one baseball and another, a rock here on earth and a distant galaxy, etc.) there exists an attractive gravitational force whose magnitude in every case is determined by one and the same equation. How legitimate is this sort of "reasoning," leaping to a very general conclusion on the basis of a very limited amount of evidence? Such an argument is often described by the elegant term "inductive reasoning," but the fact remains that what we have really been doing is intelligent guesswork. We have not *proved* that the general result is valid; we can only claim that this is one possible way of correlating a limited amount of data.

It would be irresponsible to accept the general law as if it were proven, but if we do as Newton did and accept it as a working hypothesis, we can proceed to use it, to deduce from it other results, which can then be compared with observations. The greater the number and variety of such deductions we can make that are in accord with experimental facts, the more reason we have to believe in the generalization, the more nearly we can consider it as "proven." However, no matter how many such tests may be successfully passed by the general law, we can never be certain that a new experimental result may not be found that is inconsistent with the law, nor can we be certain that a different law might not explain the observations just as well. In this sense, any scientific law—no matter how many times it has been tested—must in principle be regarded as tentative and subject to possible change. Newton's law of gravitation is itself an excellent example of these ideas. It does work amazingly well, but, as we will point out at the end of this chapter, it is now known that it is not the final theory of gravitation and that there are situations in which its predictions are not in agreement with the observed facts.

The importance of the proposed law of gravitation was quickly recognized, and Newton, and his contemporaries and his successors, began to test it in every possible way. In the course of the several hundred years that followed, the law of universal gravitation "passed" an enormous number of tests (some of which will be described in the course of this chapter). With the aid of this law it was possible to correlate and "understand" a great many previously unrelated facts, all on the basis of the one generalization expressed by Equation 5.3.

Newton himself recognized that in proposing the law of universal gravitation he was not "explaining" what gravity is, or *why* there should be an attractive force between any pair of objects. His proposal was simply this—see what consequences follow if we suppose that there *is* such an attractive force between any two objects, see how many different phenomena can be considered as manifestations of one single generalization. Call it "gravity" or call it something else, the name does not matter, and postpone until a later date, postpone indefinitely perhaps, an attempt to understand at a deeper level *why* such a force should exist.

§5.2 GRAVITATIONAL FORCES EXERTED BY SPHERICAL BODIES

Before we consider further applications of the law of universal gravitation, there is one difficult question, essentially a mathematical question, which must be faced. The law of universal gravitation states that between any two objects, there is an attractive force given by

$$F_{\text{grav}} = \frac{G\, m_1 m_2}{r^2}. \tag{5.3}$$

What, precisely, is the meaning of r, "the distance between the two objects"? If the two objects are separated by a distance that is large in comparison with the size of either one, it should be a good approximation to treat each one as if it were a mathematical point. If, however, one of the objects is the earth and the other a baseball a few feet above the earth's surface, the situation is more complex. In effect, we "induced" the law of universal gravitation by supposing that the force of the earth on a baseball can be calculated by thinking of the earth as a point, with all the earth's mass concentrated at the center, so that the relevant value of r for the earth and the baseball is approximately 4000 miles, the radius of the earth.

If we face honestly the implications of the law of universal gravitation, we must realize that a baseball just above the earth's surface is attracted by each and every bit of the earth—by a bit of rock nearby and by the rocks on the opposite side of the earth, 8000 miles away. Because of the inverse square nature of the gravitational force, there is a tremendous variation in the magnitudes of these forces. Furthermore, the various parts of the earth pull on the baseball in various directions. What must be done is to calculate the gravitational force on the baseball exerted by each little bit of the earth and then add these forces to find the total gravitational force the earth exerts on the ball. What we have is an extremely complicated problem in vector addition, as indicated schematically in Figure 5.1.

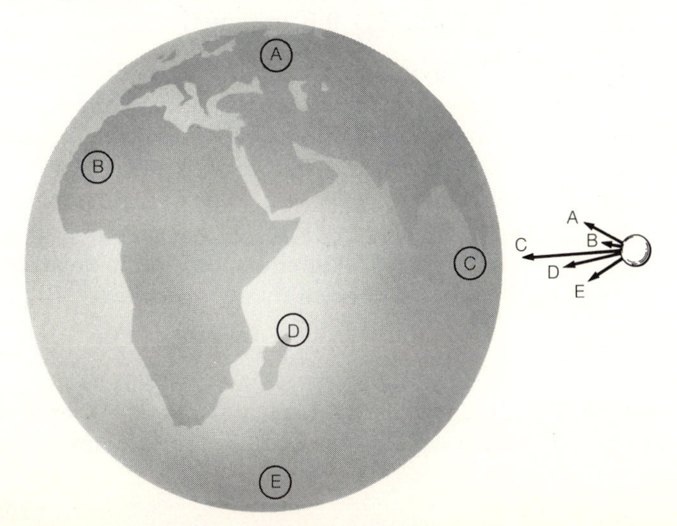

FIGURE 5.1
Every bit of the earth exerts a force on a baseball; this figure shows just a few of the many forces that must be added vectorially to determine the total gravitational force exerted by the earth (not drawn to scale).

There is no need to be embarrassed if a fruitful method of attack does not seem obvious, or if the "answer" is not immediately apparent. Newton himself wrestled with this problem for a decade and invented the branch of mathematics called "integral calculus" in order to solve it. The result, one of the most miraculously simple answers to a complicated question that could be imagined, is one we can easily state but cannot prove in this text. A large spherically symmetric object exerts a net gravitational force on an outside object—no matter how close to the surface it may be—precisely as if the spherical object were a mathematical point with all its mass concentrated at the center.

Perhaps it should be "obvious" that the result of the calculation must turn out this way? In the hopes of dispelling any illusions on this score, we will just remark that this conclusion—that we can treat the earth simply as if it were a geometrical point, with all its mass at the center—depends on the fact that the fundamental law of gravitational force is expressed in terms of the inverse *square* of the distance; any other dependence on distance would lead to a different and more complicated result. How fortunate we are that the gravitational force varies with the inverse square of the distance and not in some other way. In situations like this, it is hard for the tough-minded scientist to avoid a feeling such as that expressed by the 18th-century Ukrainian philosopher, Gregory Skovoroda: "We must be grateful to God that He created the world in such a way that everything simple is true and everything complicated is untrue," a sentiment that can be applied equally well to the whole of Newton's work.

In applications of the law of universal gravitation, we must exercise some caution. For small objects of any shape, it is a good approximation to treat them as point masses and to use Equation 5.3. For spherical objects, even large ones, Equation 5.3 is similarly applicable if we measure the distance from center to center.

§5.3 WHAT IS THE NUMERICAL VALUE OF *G*? WHY IS *g* EQUAL TO 9.8 m/sec²?

The universal gravitational constant, G, is a fundamental physical quantity of the highest importance. Can we deduce its numerical value from familiar information, for example from the fact that the acceleration of objects at the earth's surface, g, is equal to 9.8 m/sec^2? There is certainly a relationship between the two quantities (G and g), as we can see by trying to calculate g from the law of universal gravitation. According to this law, the gravitational force on some object (for example, a baseball of mass m) is given by Equation 5.3:

$$F_{\text{grav}} = \frac{GmM_{\text{E}}}{r^2},$$

where M_{E} is the mass of the earth, and r is the distance to the center of the earth, almost equal to the earth's radius (denoted by R_{E}) if the ball is at or near the surface. Thus if the ball is dropped (so that the gravitational force is the only one acting), its acceleration will be

$$a = \frac{F_{\text{grav}}}{m} = \frac{GM_{\text{E}}}{R_{\text{E}}^2}$$

and this must be equal to g. Thus we have a theoretical expression for g, the acceleration due to gravity of objects at the earth's surface, in terms of more fundamental quantities:

$$g = \frac{GM_{\text{E}}}{R_{\text{E}}^2}. \tag{5.4}$$

Knowing the values of g and of R_{E}, we could use this equation to deduce the value of G *if* we knew the mass of the earth ($G = gR_{\text{E}}^2/M_{\text{E}}$). The mass of the earth was not known in Newton's time; it is a sad fact that Newton died without knowing the numerical value of "his" gravitational constant, G.

QUESTION

5.2 Newton was not completely ignorant of the numerical value of G, because he could make an estimate of the mass of the earth. The size of the earth was known; if we know the average density, we can calculate the mass. The densities of ordinary solids and liquids found at the surface range from about 1 g/cm^3 for water to 10 or 20 g/cm^3 for heavy metals, such as silver and gold. In the absence of better information, it seems reasonable to guess that the average density of the earth lies somewhere within this range. On this basis, estimate a range of possible values for the mass of the earth, and thus—using the known value of g and Equation 5.4—the order of magnitude of the gravitational constant, G.

We cannot determine the value of G from the known value of g unless we also know M_{E}, but we could determine G if we could measure the gravitational force of attraction between two objects, *both* of whose masses are known. The difficulty is that the force of gravity that acts between two objects of ordinary size is extremely weak; only when at least one of the objects is something huge is the gravitational force strong enough to be readily measured. A century after Newton's time, in 1798, Cavendish devised an exceedingly delicate instrument for measuring small forces and was able to measure the gravitational force between two balls of lead, each only two inches in diameter, separated by distances of a few inches. Cavendish found that there is a force between two such objects, and that it varies as the inverse square of the distance between them. This in itself was a triumph for the law of universal gravitation; until that time, the existence of gravitational forces had been demonstrated only in situations in which at least one of the two objects was an "astronomical" one. Having measured the force, and knowing the two masses and their separation, Cavendish could then use Equation 5.3 to calculate the value of G, for which the best modern value is

$$G = 6.673 \times 10^{-11} \text{ m}^3/\text{kg-sec}^2.$$

QUESTIONS

5.3 Note that G has units and dimensions and that it is not the same sort of quantity as is g, whose dimensions are those of length²/time. By studying Equations 5.3 and 5.4, show that the units attributed to G are correct.

5.4 Using the numerical value of G given above, estimate the value of the gravitational force of attraction between two "ordinary" objects, perhaps two cars passing one another on the highway. Is it a small force? Small compared to what?

5.5 Cavendish's important experiment is sometimes given the grandiose description, "weighing the earth." This is a rather peculiar term, because in fact the earth played no role at all in his experiment except to support the apparatus. That was the whole point, to measure the gravitational force between two objects *neither* of which was the earth. The explanation for the phrase "weighing the earth" is that once G was known, it was possible for the first time to calculate the mass of the earth, by means of Equation 5.4. Use this equation to calculate the mass of the earth.

§5.4 APPLICATIONS OF THE LAW OF UNIVERSAL GRAVITATION

In this section we shall discuss a number of applications of the law of universal gravitation. Some of these we will treat quantitatively, others only qualitatively. It is important to notice that every successful application of the law is also a *test*, and with every test which the law passes, the more nearly it is "proven correct." The most impressive and precise tests of the law of gravitation came from "experiments" carried out in that marvelous physics laboratory of which we are a part, the solar system. It contains a small number of objects (the sun, the planets, the moons of the various planets, and an occasional comet). The only forces of any significance are the gravitational forces of attraction between all these objects. The effects of these forces can be accurately calculated, and the motions have been studied with great care for long periods of time, so that any failure of the law would be expected to show up.

Recall that Newton used observations of the earth's moon in arriving at the law of universal gravitation. He also had excellent data about the motions of the various planets; in fact, he used this information together with information about the moon in arriving at his results. To a first approximation, each planet travels in a circular orbit around the sun, executing uniform circular motion. Let us call the average radius of a planet's orbit r_p, and its centripetal acceleration toward the sun a_p. The force exerted by the sun (of mass M_S) on a planet (of mass M_p) is

$$F_{grav} = \frac{GM_S M_p}{r_p^2},$$

and if we attribute the planet's acceleration to the gravitational attraction of the sun, we can equate this force to the mass of the planet times its acceleration, a_p:

$$M_p a_p = \frac{GM_S M_p}{r_p^2};$$

and thus

$$a_p = \frac{GM_S}{r_p^2}. \tag{5.5}$$

The only quantity on the right-hand side of Equation 5.5 that changes from one planet to another is r_p; the centripetal accelerations of the various planets ought to vary in proportion to the inverse square of their distances from the sun.

Equation 5.5 can be rewritten by expressing a_p in terms of the planet's orbital radius and its orbital period, T, the length of time it takes to travel once around its orbit:

$$a_p = \frac{v^2}{r_p}$$

but

$$v = \frac{\text{circumference}}{T} = \frac{2\pi r_p}{T}$$

and so

$$a_p = \frac{4\pi^2 r_p^2}{T^2 r_p} = \frac{4\pi^2 r_p}{T^2}.$$

When this result is substituted into Equation 5.5, we have

$$\frac{4\pi^2 r_p}{T^2} = \frac{GM_S}{r_p^2}$$

or

$$T^2 = \left(\frac{4\pi^2}{GM_S}\right) r_p^3. \tag{5.6}$$

For the various planets, the squares of the periods should be proportional to the cubes of the radii. In other words, the combination (T^2/r_p^3) should have the same numerical value for all the planets.

QUESTIONS

5.6 The observed values of T and r_p are indeed related in this way; this regularity was familiar to Newton as "Kepler's third law." Check this relationship by looking up numerical values of T and r_p for several planets in Appendix D.

5.7 This equation can also be tested graphically. According to Equation 5.6, a graph of T^2 versus r_p^3 should be a straight line through the origin. Use data from Appendix D to test this prediction.

5.8 With the known value of G and with the numerical values of T and r_p for any one planet, we can calculate the mass of the sun from Equation 5.6. Make this calculation, and compare your result with the accepted value. (This is in fact the best way of determining the mass of the sun.)

Further details about planetary motions are contained in Kepler's first and second laws, which describe, for instance, the fact that planets actually travel in slightly elliptical orbits rather than in perfect circles. Although we shall not discuss these laws here, Newton was able to show that these features of planetary behavior could be explained by the law of universal gravitation. The motions of the moons of the various planets also provide useful information. Jupiter, for instance, has 13 moons. Their centripetal accelerations toward the center of Jupiter should be in proportion to the inverse squares of their distances from its center. (Equivalently, the periods and radii should be related by Kepler's third law, with the mass of Jupiter rather than the mass of the sun appearing in the proportionality constant in Equation 5.6.) This expectation is confirmed, and from the numerical values, we can determine the mass of Jupiter itself, in the same way that the mass of the sun was calculated in Question 5.8.

The paths of comets can also be understood on the basis of the law of universal gravitation. Comets have been seen since ancient times and usually regarded as signs from heaven, as warnings of wars or plagues. It was a remarkable discovery that comets, which look so different from planets (and which *are* different in composition), move in accord with the same law that describes the motions of planets and moons and terrestrial objects. Some comets could in fact be classified as minor planets, traveling on highly elongated orbits, spending most of their time in the outer reaches of the solar system. Once in each orbit, they approach the sun, whip around it, and then go back into space. Halley's comet, whose orbit is shown in Figure 5.2, is one such body, returning to our vicinity every 76 years. Its appearances have been recorded for over two thousand years, but it was not until Halley's observations in 1682 that it was realized that the same comet was returning again and again. Halley's comet appeared most recently in 1910; it will come again in 1986. Other comets visit the sun only once in their history. Such a comet may circulate for millennia in the outer reaches of the solar system, be perturbed in some way by one of the stars closest to our sun, and be set on a course toward the vicinity of the sun. We may see it for a few weeks as it approaches the sun; then it is deflected by the sun and moves away again, perhaps to depart from the solar system forever.

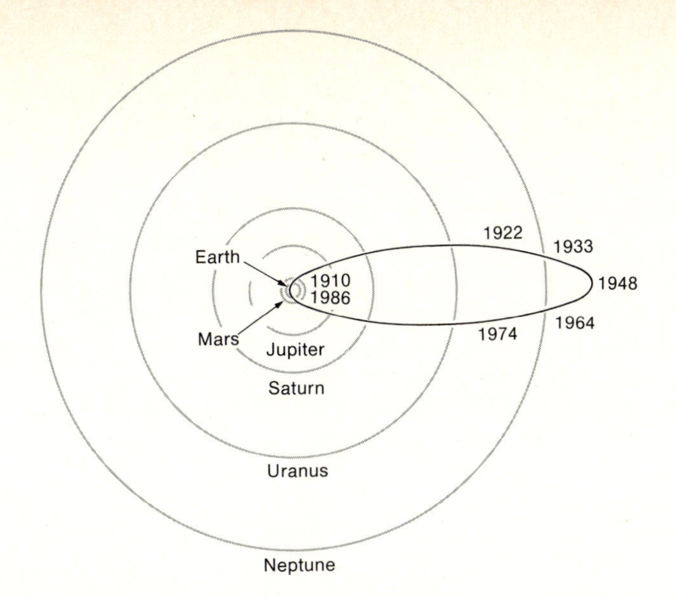

FIGURE 5.2
The orbit of Halley's comet. [From J. C. Brandt and S. P. Maran, *New Horizons in Astronomy*. W. H. Freeman and Company. Copyright © 1972.]

To a good approximation, the solar system can be described as one in which each planet travels in an orbit around the sun, under the influence of the sun's gravitational force. However, any one planet is subject to the gravitational attraction not only of the sun but also of all the other planets. The sun's force is by far the most important, because of the sun's great mass, but the other forces are also present. Planets do *not* move as they would if only the sun were present—only approximately so—and the deviation can be quantitatively understood, with the aid of the law of universal gravitation, as the result of the "perturbing" effects of the other planets. Taking these perturbations into account greatly complicates the computational problem, and we shall not take up this matter in detail. Suffice it to say that the successful explanation of most of the planetary perturbations was a triumph for the law of universal gravitation. There is a pattern to this development that is worth noting. The gravitational law was discovered (perhaps created by Newton's intellect would be a better way of putting it) by first treating each planet as moving independently under the influence of the sun, by temporarily ignoring the interactions of the planets with one another. Once this had been done, attention was turned to these secondary effects, and it was found that, rather than making it necessary to modify the postulated law, these complicating effects helped to confirm its correctness.

One apparent failure in this string of successes became apparent in the early 19th century. The planet Uranus, the most distant planet then known, was not following the plan. No matter how carefully the calculations were done, the perturbations of its orbit could not be accounted for. There are at least three possible responses to this situation. First, perhaps the law of universal gravitation is wrong, by a very small amount. After all, it does work extremely well; why should we be upset if we find that it is not quite correct when we consider distances ranging from an inch or so (in Cavendish's experiment) to 10^9 miles? Second, perhaps for some reason there is some kind of nongravitational force that acts on Uranus. Third, perhaps the law of universal gravitation *is* correct and the unexplained perturbations are caused by a planet as yet undiscovered.

No one can say *a priori* which response is right, but of the three, only the third lends itself to a direct test. If we have faith in the law of universal gravitation, then from the observed behavior of Uranus, we can determine where such a new planet would have to be in order to explain the perturbations. The calculation was not easy, but it was completed, in 1846 (almost simultaneously by two astronomers working independently of one another), telescopes were pointed in the appropriate direction, and thus the planet Neptune was discovered. After this demonstration, it was difficult to find anyone who would openly express serious doubts about Newton's law. In the 20th century, this story was repeated. In order to explain some remaining peculiarities of the motion of Uranus and of Neptune itself, the existence of a still more distant planet was suggested, and thus Pluto was discovered in 1930.*

In the space age, detailed observations of all the slight anomalies in the motions of artificial satellites have produced new information about the shape and composition of the earth and moon. We now know, for instance, not only that the earth bulges slightly at the equator but also that it is slightly pear-shaped, the southern hemisphere being a bit larger than the northern. Unexpected wobblings of spaceships circling the moon led to the discovery of "mascons," concentrations of mass at various spots near the surface of the moon. Another test of Newton's law is provided by the fact that since the force of gravity varies with the distance from the center of the earth, the measured value of g should vary with altitude; g should be smaller at the top of a mountain than at sea level. The decrease is slight but easily measurable, and the theoretical prediction has been confirmed.

The tides can also be understood with the aid of Newton's law of gravitation. Long before Newton's time, it was apparent that the tides were in some way caused by the moon. The earth exerts a gravitational force on the moon, and the moon exerts an equal and opposite force on the earth. The force exerted by the moon is not equally strong at all points on the earth's surface, because some places are closer to the moon than others. It is this slight variation in the strength of the moon's pull that is responsible for the tides. Newton showed how the general features of tidal motion could be understood, though the details are impossible to calculate. The forces exerted by both the sun and the moon must be considered (though the sun is less important than the moon in producing tides), but the most serious problem is that the time and height of the tide depend very much on the details of the local geography, as every sailor knows.

---

*Luck also played a role in the discovery of Pluto. The available observations of the orbits of Neptune and Uranus were not sufficiently precise to allow an accurate calculation of the undiscovered planet's position, but through good fortune Pluto was in the region of the sky that was most carefully searched.

QUESTIONS

5.9 Because of the increased distance from the center of the earth, the value of g ought to be smaller at the top of Mount Everest than at sea level. By about what percentage?

5.10 We saw earlier that Sputnik's downward acceleration was only about 8.3 m/sec² instead of 9.8 m/sec². Can the value of this acceleration be quantitatively understood on the basis of the inverse square law?

5.11 The moon is apparently made of material not very different from that of the earth. Suppose for the sake of argument that the average density of the moon is exactly equal to that of the earth. What would be the value of "g" on the surface of the moon, the downward acceleration of a rock dropped by a moon explorer?

5.12 The result of Question 5.11 is that "g" on the surface of the moon should be smaller than on the earth; this conclusion has been confirmed by the experiences of astronauts. However, one might argue that since gravitational forces vary as the inverse square of the distance and since the surface of the moon is much closer to its center than is the case on earth, the value ought to be considerably *larger* than 9.8 m/sec². What is wrong with this argument?

5.13 It has long been recognized that the tides can be used as a source of energy, and tidal power plants for turning mill wheels were used along Europe's Atlantic coast at least as early as the 11th century. The tides can also be used to generate hydroelectric power. With a dam across the mouth of a bay, at every high tide the bay can be closed off. About six hours later, the water level in the ocean outside will have dropped, and the water behind the dam can be released, generating electrical energy as it falls just as in a conventional hydroelectric plant. There are variations on this plan, but this is the simplest method. Newton's law of gravitation is needed to understand why we have tides at all, but we can calculate the amount of gravitational potential energy lost as a tidal pool is emptied by using the familiar results of Chapter 2 ($GPE = mgh$), just as for a conventional hydroelectric plant.

(a) The only large-scale tidal power plant now in operation is one completed in 1966 at the mouth of the Rance River in Brittany. Here the difference in height between high and low water is very large, about 35 ft, and the area of the pool behind the dam is about 9 square miles. Estimate the amount of gravitational potential energy lost during one emptying of this tidal pool. (Note that not all the water falls 35 ft as the pool is emptied; the average drop is only half this large.)

(b) How does this energy compare with that obtained from one emptying of the Northfield Mountain pumped-storage reservoir described in Chapter 4?

(c) There are about two full tidal cycles per day. (The tides are primarily caused by the moon, not the sun, and the average time between one high tide and the next is about 12½ hours.) At what average rate can electrical energy be generated by the tidal power plant described in (a)? How does this rate compare with that of the generating plant at Hoover Dam?

5.14 Tidal power could be generated everywhere along a coastline, but in most locations it is simply too expensive; the "fuel" is free, but the necessary dam costs too much for the amount of power which can be generated. What is needed is a fairly large body of water, which empties into the sea through a narrow outlet, and a fairly large difference in height between high and low tides. Estimate the average amount of power that could be generated by a tidal power plant at the Golden Gate, the outlet of San Francisco Bay, where the average difference between high and low water levels is approximately 3 ft.

FIGURE 5.3
La Rance, the French tidal power plant, under construction. [Société pour le développement des applications de l'électricité, Paris.]

§5.5 MODIFICATION OF THE EXPRESSION FOR GRAVITATIONAL POTENTIAL ENERGY

In our initial discussions of gravity (Chapter 2), we found that the effect of gravity on the motion of objects near the earth's surface could be described by attributing a gravitational potential energy to such an object, equal to mgh, where h was the height above some arbitrarily chosen reference level. Then in Chapter 3 we saw that we could equally well describe such motions by saying that an object experiences a downward gravitational force of magnitude, mg. We have the option of using either concept, force or energy, to provide two different ways of discussing the same phenomena, each of which has its own special advantages and disadvantages. All of the discussion of the general law of gravitation thus far in this chapter has been given in terms of the concept of force. We have learned that it is not correct to say that an object of mass, m, always experiences a downward force of the same size, given by its mass multiplied by g (9.8 m/sec^2). This is only an approximation, valid near the surface of the earth, and therefore the earlier expression for gravitational potential energy ($GPE = mgh$) must also be only an approximation to a more general expression.

If gravitational potential energy were given exactly by the expression mgh, a graph of gravitational potential energy versus h (Figure 5.4) would yield a straight line, A, rising steadily at the same rate no matter how large h might be. This would be correct if the force of gravity were always equal to mg, but since the earth's gravitational force does get gradually weaker with increasing h, the true graph must be something like B (Figure 5.4). The algebraic expression for gravitational potential energy can be modified in the following simple way to produce this behavior:

$$GPE = \frac{mgh}{1 + \frac{h}{R_E}} \text{ (plus some constant, if desired),} \qquad (5.7)$$

where R_E is the radius of the earth and h the height above the earth's surface. The parenthetical remark in Equation 5.7 is a reminder that since it is only changes in gravitational potential energy that are of significance, we can always add a constant term if we wish. This simply means that we are free to choose any convenient zero-level. As we shall see, it will be useful to take advantage of this freedom; for now, let us ignore the possible additive constant.

Equation 5.7 does behave in the desired way. As long as h is very small compared to R_E, the denominator is almost exactly equal to 1, and so in this case $GPE \simeq mgh$. However, as h becomes larger, the denominator becomes larger, and if h is very large compared to R_E, the denominator is nearly equal to h/R_E. In this limit,

$$GPE \simeq \frac{mgh}{h/R_E} = mgR_E \text{ (for large } h\text{).}$$

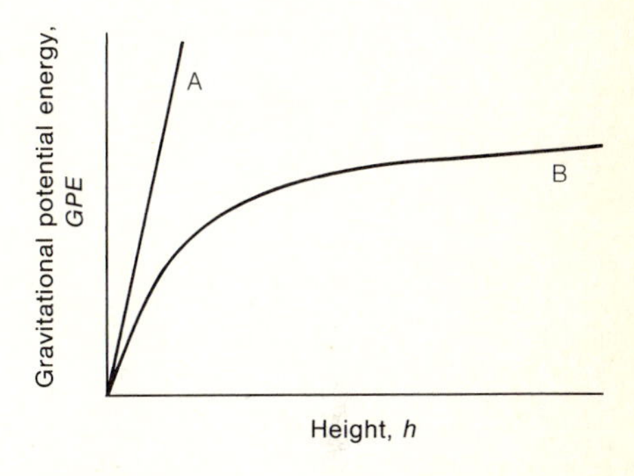

FIGURE 5.4
If the gravitational force were always equal to mg, a graph of gravitational potential energy versus h would be a straight line, A; in actuality, the graph must have the shape shown by curve B.

That is, as h gets extremely large, the expression for gravitational potential energy approaches a constant value, independent of h. This is the behavior suggested by Figure 5.4. Since the modification could have just as readily been made in a different way, however, Equation 5.7 must be tested further before we can be sure that it is correct.

Equation 5.7 can be rewritten by multiplying numerator and denominator by R_E:

$$GPE = \frac{mgR_E h}{R_E + h}. \qquad (5.8)$$

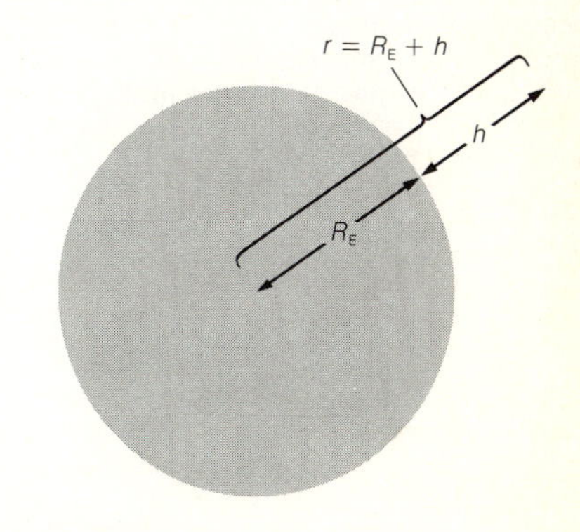

FIGURE 5.5
The relationship between r, R_E, and h.

Now $(R_E + h)$ is just equal to r, the distance from the center of the earth (Figure 5.5):

$$R_E + h = r;$$

$$h = r - R_E.$$

With these substitutions, Equation 5.8 becomes

$$GPE = \frac{mgR_E(r - R_E)}{r}$$

$$= mgR_E - \frac{mgR_E^2}{r}. \qquad (5.9)$$

The only quantity that varies with position is the r in the second term;

the first term (mgR_E) is simply a constant. Now we can simplify the expression by taking advantage of the option of adding a constant. Simply let this constant have the value ($-mgR_E$), and Equation 5.9 becomes

$$GPE = -\frac{mgR_E^2}{r}. \tag{5.10}$$

By adding a constant term, we have changed the zero-level for gravitational potential energy. This point should be kept in mind, but it has no effect on *changes* in gravitational potential energy. This expression for gravitational potential energy (Equation 5.10) is always negative. It approaches zero as r becomes larger and larger; in other words, the new zero-level is "at infinity" and is, strictly speaking, unattainable. Equation 5.10 is plotted in Figure 5.6. Even though with the revised expression, gravitational potential energies are always negative, energy changes are still given correctly by the new expression. A rock falling from A to B (in Figure 5.6) experiences a decrease in gravitational potential energy, from a negative value to a more negative value. In discussions of motion very near the earth's surface, it is convenient to choose the zero-level at or near the earth's surface, but in discussing the action of gravity over great distances, another choice for the zero-level (at infinity) is more convenient, and that is the choice which has been made here.

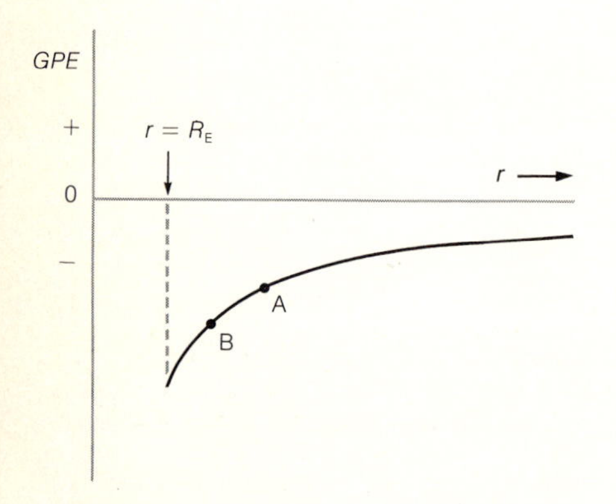

FIGURE 5.6
Variation of gravitational potential energy with distance, r, from the center of the earth; the zero-level is only attained for indefinitely large values of r. The gravitational potential energy of a rock decreases as it falls toward the earth (from A to B).

Equation 5.10 can be written in terms of the fundamental constant G by substituting the expression for g from Equation 5.4:

$$g = \frac{GM_E}{R_E^2}.$$

Then Equation 5.10 becomes

$$GPE = -\frac{m\frac{GM_E}{R_E^2}R_E^2}{r} = -G\frac{mM_E}{r}. \tag{5.11}$$

This is the expression for the gravitational potential energy of an object whose mass is m at a distance r from the center of the earth. In general, for any two objects (masses m_1 and m_2), the expression is

$$GPE = -G\frac{m_1 m_2}{r}, \tag{5.12}$$

with the usual qualification that the two objects must be either very far apart or spherical in shape, and also that we have chosen the zero-level such that it is reached only when the two objects are "infinitely" far apart.

Thus far we have shown only that, if the modified expression from which we began (Equation 5.7) is correct, then the results (Equations 5.11 and 5.12) follow. Equation 5.11 does give the right result for motions near the earth's surface, because in that context it simply says that $GPE \simeq mgh$ (plus a constant). To show that Equations 5.11 and 5.12 really are correct, we must consider motions that are not restricted to being very close to the surface. This can be done in two ways. One is to consider the motion of a rock dropped from a great height, to assume that the sum of kinetic energy and gravitational potential energy is constant (with gravitational potential energy given by Equation 5.11), and to show that the rock's downward acceleration is just that expected if it is acted upon by the downward gravitational force, $F = \frac{GmM_E}{r^2}$. This is not an easy calculation, and it is left as a question (Question 5.16).

A second way of testing the revised expression for gravitational potential energy is to consider the motion of an earth satellite traveling in a highly elliptical orbit (Figure 5.7). It can be seen from these data that the satellite travels relatively slowly at its greatest distance from the earth. As it approaches the earth, it picks up speed (gaining kinetic energy as it loses gravitational potential energy), and its maximum speed is achieved when its altitude is least. This behavior is similar to that of a thrown rock, with the obvious difference that the satellite in orbit does not strike the earth. It is deflected by the earth's gravitational attraction but manages to escape catastrophe and continue its orbital motion.

As the satellite moves, the sum of its kinetic energy and gravitational potential energy should always keep the same value:

$$(\tfrac{1}{2}mv^2) + \left(-\frac{GmM_E}{r}\right) = \text{constant},$$

$$\tfrac{1}{2}mv^2 - \frac{GmM_E}{r} = \text{constant}$$

or, equivalently,

$$\tfrac{1}{2}v^2 - \frac{GM_E}{r} = \frac{\text{constant}}{m} = \text{another constant}. \qquad (5.13)$$

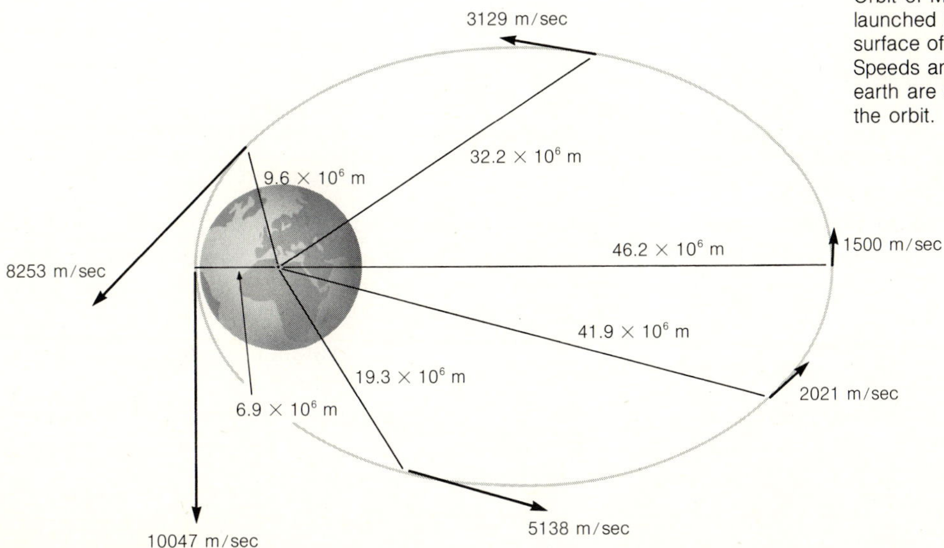

FIGURE 5.7
Orbit of Molniya 1E, a Russian satellite launched in 1968, shown in relationship to the surface of the earth. Orbital period: 11.8 hours. Speeds and distances from the center of the earth are shown for a number of points on the orbit.

Now GM_E has the known value of

$$6.673 \times 10^{-11} \frac{m^3}{\text{kg-sec}^2} \times 5.98 \times 10^{24}\ \text{kg} = 4 \times 10^{14} \frac{m^3}{\text{sec}^2},$$

and thus we can calculate the value of the combination on the left side of Equation 5.13 for each selected point on the orbit. Table 5.1 shows some of the results of this calculation. The prediction is borne out; the calculated values of $\left(\frac{1}{2}v^2 - \frac{GM_E}{r}\right)$ are almost exactly the same. The revised expression for gravitational potential energy (Equation 5.11) does work. That is, with this definition, the sum of kinetic and gravitational potential energies remains constant during the satellite's motion.

TABLE 5.1 Use of satellite data (Figure 5.7) to test the modified expression for gravitational potential energy. (Several of the entries have been omitted; see Question 5.15.)

r (m)	v (m/sec)	$\frac{1}{2}v^2$	$\frac{GM_E}{r}$	$\left(\frac{1}{2}v^2 - \frac{GM_E}{r}\right)$
6.9×10^6	10047	50.5×10^6	58.0×10^6	-7.5×10^6
9.6×10^6	8253	—	—	—
19.3×10^6	5138	13.2×10^6	20.7×10^6	-7.5×10^6
32.2×10^6	3129	—	—	—
41.9×10^6	2021	—	—	—
46.2×10^6	1500	1.1×10^6	8.7×10^6	-7.6×10^6

QUESTIONS

5.15 Complete the calculations to fill in the blank spaces in Table 5.1.

5.16 Another way of verifying the revised expression for gravitational potential energy. Suppose that a rock of mass, m, is released at a considerable height above the earth's surface. If we assume that the correct expression for gravitational potential energy is $(-GmM_E/r)$, where r is the distance from the center of the earth, we can calculate the resulting acceleration. Let t_1 and t_2 denote two closely spaced instants of time, and let r_1, v_1, r_2 and v_2 denote the rock's distance from the earth's center and its speeds at the two instants. From the law of conservation of energy:

$$KE_1 + GPE_1 = KE_2 + GPE_2;$$

$$\tfrac{1}{2}mv_1^2 - \frac{GmM_E}{r_1} = \tfrac{1}{2}mv_2^2 - \frac{GmM_E}{r_2}.$$

From this equation, using the argument given in §2.4 as a guide, show that the downward acceleration of the rock is equal to GM_E/r^2. This is just what is expected, since the mass of the rock is m, the force is GmM_E/r^2, and the acceleration is equal to the force divided by the mass. In other words, the revised expression for gravitational potential energy is consistent with the fundamental expression for the gravitational force given by Newton's law of gravitation.

Although the new expression for gravitational potential energy ($-GmM_E/r$) is more accurate than the old one (mgh), and though there are many circumstances in which the new expression must be used, the old expression can be used with virtually no error for motions close to the earth's surface. As applied to the motion of an object of mass, m, near the surface of the earth, the two expressions differ in that the zero-level for the old expression is some point on or near the surface of the earth, whereas for the new one it is at "infinity" (or at a very large distance). The zero-level should be kept in mind but is not important in calculating changes in gravitational potential energy.

The fact that these two expressions give virtually identical results for motions near the surface of the earth provides an important lesson for those who make use of physics. Consider the problem faced by engineers who need to know the force of gravity on the roadway of a bridge, or the amount of energy needed to lift an elevator. Which version of the laws of gravity should they use, the old one ($F_{grav} = mg$, $GPE = mgh$) or the new ($F_{grav} = Gm_1m_2/r^2$, $GPE = -GmM_E/r$)? The new version is more accurate, but the old version is much simpler to use. Even though it is in principle incorrect, the difference between the two is insignificant in these situations. Highway engineers would therefore be foolish to insist on using the "more accurate" version; if they did, they might take so much time designing the bridge that the job would never get done. On the other hand, astronautical engineers must use the new version if they expect their moon rocket to reach its destination.

§5.6 THE "ESCAPE VELOCITY"

Now that we have the modified expression for gravitational potential energy, we can apply it to a question that is interesting in its own right and that may have some practical importance. How fast must you throw a rock in order that it keep on going and never fall back to earth? To simplify the problem, we will consider only the effect of the earth; we will not try to predict what will happen if the rock eventually approaches another planet.

In order that the rock reach any particular height, its initial kinetic energy must be at least large enough so that it can gain the necessary amount of gravitational potential energy without using up its kinetic energy; otherwise, when it gets to the point at which its kinetic energy is zero, it stops momentarily and falls back to earth. We discussed a similar question in Chapter 2; what is the minimum speed that must be imparted to a rock in order for it to reach the top of a tree? If the exact expression for gravitational potential energy were mgh, nothing could ever escape from the earth. No matter how large its initial kinetic energy, there would always be some value of h at which its kinetic energy would

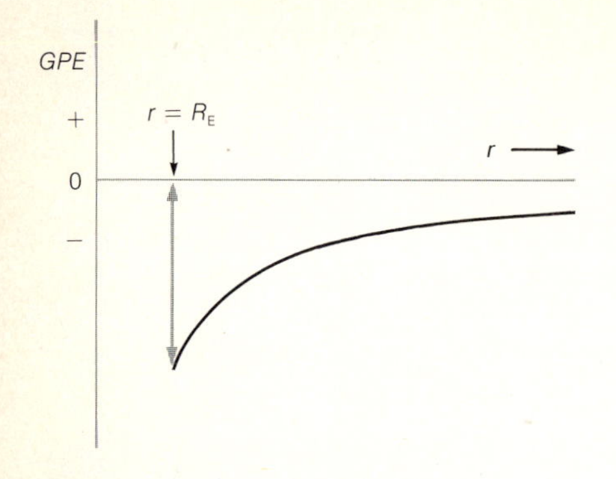

FIGURE 5.8
For an object to escape from the earth, its gravitational potential energy must increase by the amount shown by the two-headed arrow; its initial kinetic energy must therefore be at least this large.

become equal to zero, and when it reached this height, it would turn around and fall to earth.

The actual situation, however, is shown in the gravitational potential energy diagram of Figure 5.8.* If the object has enough kinetic energy, it can get indefinitely far away from the earth without stopping. All that is required is that its initial kinetic energy be at least as large as the increase in gravitational potential energy which it must experience in escaping, an amount of gravitational potential energy indicated by the arrow in Figure 5.8. If it does escape, its gravitational potential energy will then be zero; its initial gravitational potential energy is $-GmM_E/R_E$, and so the necessary increase in gravitational potential energy is simply

$$Gm\,M_E/R_E.$$

We need only require that its initial kinetic energy be at least this large, that is, that

$$\tfrac{1}{2}mv^2 \geq \frac{GmM_E}{R_E},$$

or that

$$v \geq \sqrt{\frac{2GM_E}{R_E}}. \tag{5.14}$$

QUESTION

5.17 Show from Equation 5.14 that the "escape velocity" is approximately 1.12×10^4 m/sec, or about 7 miles/sec.

The escape velocity is important in planning for space exploration. It has also been suggested—apparently in all seriousness—that our growing quantities of deadly radioactive wastes be disposed of once and for all by sending them off to some other part of the solar system. Quite aside from the horrifying consequences that could ensue if a rocket were to misfire and crash back onto the earth, this method of disposal requires energy, and knowledge of the escape velocity enables us to calculate the amount of energy—the absolute minimum amount—that would have to be used for this purpose. We shall return to this question in Chapter 14.

*Although we can use any zero-level we wish for gravitational potential energy, the problem discussed here indicates one advantage of choosing the zero-level to be at infinity. With respect to outer space, we do live in a deep "hole," and so it is perfectly reasonable to use an expression for gravitational potential energy that gives negative values for things near the earth's surface.

QUESTION

5.18 On a more prosaic level, could we get rid of our wastepaper, bottles and cans in this way? On the average, each American generates about five pounds of solid waste per day. What is the minimum amount of energy that would have to be used to dispose of solid waste in outer space? Would this constitute a significant, or a trivial, increment in the national use of energy? (It is obvious that the amount of energy required would actually be considerably greater, because not only the wastes but the rocket itself must be disposed of. Even if this plan should turn out to be technologically feasible, it is essential that we eventually learn to recycle our beer cans and not just throw them away, especially in such an irrevocable manner.)

§5.7 WHAT IS THE SOURCE OF THE SUN'S ENERGY? (AN INTERESTING IDEA THAT DOES NOT WORK)

Though we usually take the sun for granted, in the last analysis no other energy source is more important to us. Not only does the sun provide us with light and keep us warm, but also it is the sun's energy that plays the essential role in all plant life, and hence in the existence of all forms of life on earth. The energy "locked up" in coal, oil, and natural gas—the energy that we are now squandering at a fantastic rate—came from the sun hundreds of millions of years ago. If the sun's energy output were to be cut off, or even to increase or decrease by any appreciable amount, all other "energy crises" would be quickly forgotten. Needless to say, this is not a cause for immediate concern, but it is nonetheless of great interest to know where the sun's energy comes from.

The sun is pouring out energy at the enormous rate of 4×10^{26} W, and has been doing so for a long time. (The sun emits energy in all directions. A small fraction of this energy hits the earth, and by measuring this we can calculate the total amount emitted in all directions.) If the law of conservation of energy is valid, then either the sun's energy is constantly being replenished from some unknown source, or else the amount of energy remaining in the sun must be steadily decreasing. Is there some kind of energy to which we can point that is simultaneously *decreasing* at the rate of 4×10^{26} W? One suggestion, put forward in the middle of the 19th century, was that the energy emitted by the sun could be accounted for by a simultaneous decrease in its gravitational potential energy. To be specific, suppose that the sun were once much larger than it is now, that it was originally a widely dispersed cloud of atoms, and that it has been gradually shrinking all these years. As all the atoms of the sun approach one another, their gravitational potential energy decreases—just as does that of a rock as it approaches the earth; perhaps it is this decrease of gravitational potential energy that accounts for the energy emitted by the sun.

An interesting idea, but how can we test it? The first question that must be asked is a quantitative one. Could this mechanism provide *enough* energy to account for the energy emitted by the sun? If the answer to this question is affirmative, we must still subject the idea to closer scrutiny, but if the answer is negative, we need go no farther—the suggestion must be rejected. There are two quantities that we must calculate—at least approximately—and compare in order to answer this question. The first is relatively easy: the total amount of energy emitted by the sun during its known lifetime. The sun is now emitting energy at the rate of 4×10^{26} W, and we have reason to believe that it has been doing so at about the same rate for about 10 billion years (10×10^9 yr $\simeq 3 \times 10^{17}$ sec). Thus the total amount of energy that has been emitted by the sun during this time is approximately

$$4 \times 10^{26}\ \text{J/sec} \times 3 \times 10^{17}\ \text{sec} \simeq 10^{44}\ \text{J}.$$

The second quantity is a more difficult one to estimate: the amount of gravitational potential energy lost during the hypothesized contraction of the sun. This is a difficult calculation to make because the loss of gravitational potential energy results from the simultaneous gravitational interaction of countless numbers of atoms. Although we cannot make a rigorous calculation of this quantity, there are two arguments we can use to guess the approximate size of the result. First, we might well expect that the loss of gravitational potential energy as the sun contracted to its present size ought to depend on the total mass of the sun, M_S, and its present radius, R_S, and—presumably—also on the universal gravitational constant, G. We already know that an expression of the form

$$\frac{G \times \text{mass}^2}{\text{distance}}$$

has the dimensions and units of energy (see, for example, Equation 5.12), and thus we might anticipate that the quantity we are trying to calculate has roughly the value

$$GM_S^2/R_S.$$

This expression might well be in error by a numerical factor of, say, 2 or 4, but it would be surprising if it were wrong by a truly enormous factor.

Another argument leading to nearly the same result is this. Since it appears impossibly difficult to calculate *exactly* the gravitational potential energy lost during the contraction of such a cloud, let us estimate instead the amount of gravitational potential energy lost during a process which is itself an approximation of the supposed contraction. Suppose that the sun, instead of being initially completely dispersed, originally had the form of two widely separated parts, each of mass $M_S/2$, which came together to form the sun. Initially, their gravitational potential energy was zero or nearly so; when they are combined, we may think of them as being about a distance R_S apart (Figure 5.9), with a gravitational potential energy of

$$\frac{-G\frac{M_S}{2} \times \frac{M_S}{2}}{R_S} = \frac{-GM_S^2}{4R_S}.$$

That is, the amount of gravitational potential energy lost as they come together is equal to $GM_S^2/4R_S$. This is hardly a rigorous calculation, but to within a factor of 4, it gives the same result as suggested by our earlier argument. These arguments strongly suggest that the largest possible amount of gravitational potential energy that the sun could have lost during such a contraction* is approximately equal to GM_S^2/R_S, two or three times as large perhaps, but not much more.

QUESTIONS

5.19 Is this a plausible result? Read this expression. How would its magnitude change if M_S were larger or smaller? If R_S were different? If the value of G were different?

5.20 Show that this quantity of energy is about 4×10^{41} J.

This is a prodigious amount of energy, but smaller by a factor of 250 than our estimate of the total amount of energy given off by the sun. Our calculations have been approximate ones, but there is no way of modifying the result by such a large factor. Ingenious as this idea may be, it fails for the simple quantitative reason that not enough energy is available. Some other unsuccessful ideas will be examined elsewhere in the book, and nuclear fusion, the explanation that appears to be the correct one, will be discussed in Chapter 15.

It often happens (in politics, in science and in other areas as well) that the same evidence can be used to support very different conclusions. The calculation made here was once used for a different purpose. At one time during the 19th century, when the age of the solar system was not well known and when Darwin's theory of evolution was still controversial, these calculations were used to attack the evolutionists. The argument was the following. The sun is giving off energy at the rate of 4×10^{26} J/sec, and the total amount of energy available for this purpose is only about 4×10^{41} J. (This is the energy calculated above as the amount available from gravitational contraction. No larger source

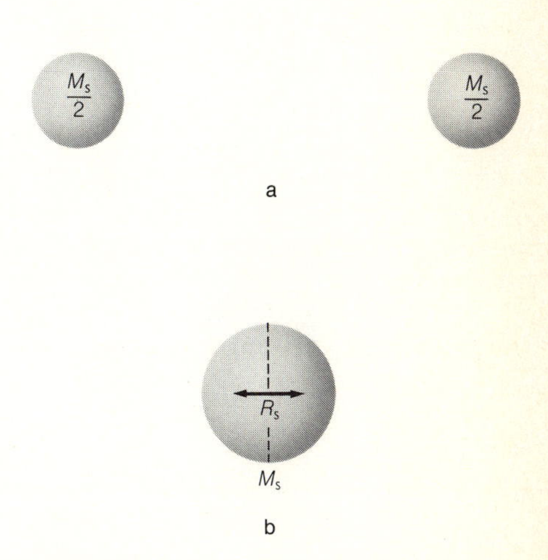

FIGURE 5.9
(a) The hypothetical initial state of the sun: two halves very far apart; (b) the present state of the sun: the two halves are separated by approximately R_S.

*A contraction to its *present* size, with a radius of R_S. According to this hypothesis, the contraction is still going on though the decrease in R_S from year to year would be too small to be noticed.

of energy had yet been suggested.) Simple arithmetic shows that the age of the sun can only be about

$$\frac{4 \times 10^{41}\,\text{J}}{4 \times 10^{26}\,\text{J/sec}} = 10^{15}\,\text{sec} \cong 3 \times 10^{7}\,\text{yr}.$$

Thirty million years, it was argued (correctly) is too short a time for all of the evolutionary changes envisioned by the Darwinians to have taken place. This posed a dilemma for Darwin's followers; until a more potent source of energy could be found, there remained a serious reason to doubt the theory of evolution.

§5.8 CONCLUSION

At the beginning of this chapter, Newton's law of universal gravitation was described as a successful scientific law. Surely this is an understatement. From a fairly modest starting point, by assuming that everyday objects have something in common with the heavenly bodies, Newton developed a single general result—one that is beautiful for its simplicity—which provides a unified description of the motions of apples, rocks, moons, and planets. The law of universal gravitation enables us to calculate the masses of the earth and the sun, it has been used to find previously unknown planets, it describes the motions of satellites and comets as well as the motions of rocks and apples. Indeed, we can affirm with complete confidence that—though other types of forces may often be more important than gravity—there *is* a precisely calculable force of attraction between every pair of objects in the universe. As an outstanding example of a successful theory, the law of universal gravitation established an ideal that scientists—and others—still strive to emulate.

Without its record of successes, this law would not have received the attention it has, but its impact on science and on modern thought in general is due not so much to the particular facts it can explain but to its "style." The law of universal gravitation is above all a *universal* law. It asserts that a force acts not between some objects or most objects but between *every* pair of objects. It is also a very *precise* law; it asserts the existence not just of *a* force but a force of a certain definite size. Its most detailed confirmation comes from studies of the solar system, in which only a very few objects are involved, a small enough number that the motion of each one can be predicted and observed with great precision.

Extrapolation of the successes of the law of universal gravitation in describing the solar system leads to the vision of a "clockwork" universe—a "law and order" universe, in which everything moves according to one master plan, and the motion of every object is completely determined, with no spontaneity and no surprises. Laplace, a French physicist and mathematician whose own work helped to confirm the law of gravitation, put it this way:

> All events, even those which by their insignificance do not seem to follow the great natural laws, are indeed consequences of these laws as surely as is the rising and setting of the sun. We must therefore consider the present state of the universe as the result of its earlier state and as the cause of that which is to come. An intelligence which at a given instant knew all the forces acting in nature and the positions of every object in the universe—if endowed with a brain sufficiently vast to make all the necessary calculations—could describe with a single formula the motions of the largest astronomical bodies and those of the lightest atoms. To such an intelligence, nothing would be uncertain; the future, like the past, would be an open book.

Such a hypothetical being, who could know the past and future with absolute certainty, came to be referred to as "Laplace's demon." Laplace's view of the universe is appealing in some ways, but extremely sterile and unappealing in others.* Of course, no such demon can exist—even in the computer age—but like it or not, the success of the law of universal gravitation at least *suggests* that "in principle" all of nature may operate in such a fashion.

The law of universal gravitation is surely an important part of physics, but even in its influence on physics, it was perhaps *too* successful. Most phenomena with which we must deal are vastly more complicated than the solar system, and we are certain to be disappointed if we insist on evaluating other theories or laws by comparing them with Newton's. When we venture outside physics, the ideal established by the law of universal gravitation is even less appropriate. Biologists are concerned with the behavior of organisms composed of many billions of atoms. Can we hope to understand living beings in terms of an absolutely precise deterministic theory? Maybe so if we had one of Laplace's demons available; since we do not, it is probably futile to make the attempt. Alternative paths to the understanding of living organisms have in fact been more fruitful, approaches in which the complexity of the problem is recognized from the start. Historians and economists, perhaps subtly influenced by Newton's law, have sometimes tried to develop unified deterministic theories of history and economics. If the population of the world were, like that of the solar system, 10 or 20 instead of 4×10^9, they might have been more successful. As it is, economists, for instance, have made more progress by admitting that all people act irrationally and emotionally at times, that people are simply different from one another, and that human actions need to be treated statistically rather than deterministically.

The law of universal gravitation is not the only highly successful scientific generalization. Two of the other great laws of science, the law of conservation of energy (or the "first law of thermodynamics") and the second law of thermodynamics (which explicitly deals with the

*"I see no mention of God in this work," Napoleon is said to have remarked upon looking at Laplace's book. Laplace's reply: "I have no need of that hypothesis."

chaos and disorder found in nature) have been just as successful in their own ways as the law of universal gravitation but are very different in style. The laws of thermodynamics have found direct application in other sciences, and—if it is granted that social scientists ought to seek any guidance at all from science—the laws of thermodynamics would probably serve them better as models than would the law of gravitation. None of this should be taken as an attempt to detract in any way from the power and the beauty of this law. It is undeniably one of the most amazingly successful intellectual achievements in human history.

Finally, a discordant note must be added to this description of the triumphs of the law of gravitation: the law is wrong. However, we shall see that there is no need to discard the theory, or even to modify seriously our description of its successes. We have seen how the solar system provides an ideal testing ground for a law of gravitation, how even the perturbations of the planetary orbits were accounted for when Neptune and Pluto were discovered. There are a few small details, though, that do not fit, the most noticeable being the behavior of the planet Mercury. Mercury travels in an elliptical orbit about the sun, and in successive orbits it does not repeat exactly the same path. This behavior is shown—in highly exaggerated form—in Figure 5.10. We describe this by saying that the ellipse itself rotates, though not very rapidly—only through an angle of about 0.15 degree per century. Most of this rotation can be explained as due to the perturbing effects of the other planets, but when all the calculations are completed, a discrepancy remains. The discrepancy is extremely small, only a rotation through an angle of 0.012 degree per *century*, but it is real. The mere detection of such an effect is a tribute to the precision of astronomical observations. Similar but even smaller discrepancies have been detected for other planets.

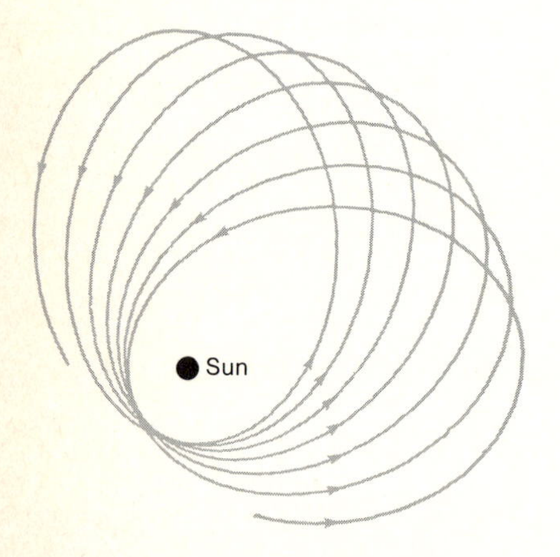

FIGURE 5.10
The orbit of Mercury. The rate of rotation of the ellipse is greatly exaggerated in this figure.

The moral of the Neptune story is obvious: look for another planet and confirm Newton's law once again. This idea was tried, the "planet" was even given a name (Vulcan), but no such planet was found. There seems to be no escape from the conclusion that the law of universal gravitation is not quite right. A theory of gravitation (known as the general theory of relativity), quite different in its basic premises but nearly identical in most of its predictions, was developed by Einstein, a theory that does seem to account for the peculiar behavior of Mercury and for some other effects that are not consistent with Newton's law. At the present time, it is not clear whether Einstein's theory of gravitation or one of several competing theories is the best, but it is clear that Newton's law of universal gravitation cannot be regarded as absolutely and perfectly correct.

It would be a serious mistake, however, to discard Newton's law, to classify it with the multitude of erroneous and obsolete scientific ideas. In a very real sense, Newton's law is *not* wrong. As long as we realize that it does have limitations and that there are some circumstances in which its predictions are in disagreement with observations, we can continue to use it with great confidence for nearly all purposes, and astronauts can continue to stake their lives on its validity. The fact that even after centuries of successes a tiny flaw was found, even in the law of universal gravitation, should serve as a sober reminder of the tentative nature of any scientific law.

FURTHER QUESTIONS

5.21 Estimate the force of attraction between two lead spheres in the Cavendish experiment, if each one has a mass of 0.8 kg and if the separation between their centers is 0.2 m. By what factor does the gravitational force exerted by the earth on each sphere exceed this value?

5.22 Consider two earth satellites, both in circular orbits but orbits of different radii. Which one will have the higher speed? Which one will have the smaller orbital period?

5.23 The smallest imaginable orbit for an earth satellite is one with a radius of about 4000 miles; in such an orbit the satellite would almost graze the surface of the earth. What would be the orbital period of such a satellite? (Ignore all effects such as air resistance.) What would be its speed?

5.24 The acceleration due to gravity has the value 9.8 m/sec^2 on or near the earth's surface but is slightly smaller at the tops of high mountains. At what altitude would it decrease to one-half of this value, to 4.9 m/sec^2?

5.25 The moon takes about a month (actually 27.3 days) to complete one orbit around the earth. Estimate how long it would take if it were half as far away.

5.26 (a) If the earth had a radius of 2000 miles instead of 4000 miles but the same average density, what would be the value of g, the acceleration due to gravity at the earth's surface?

(b) How would this change in the size of the earth affect the orbital period of a satellite such as the one discussed in Question 5.23, one that almost grazes the earth's surface?

5.27 One of Jupiter's moons travels at a speed of 1.73×10^4 m/sec in a circular orbit with a radius of 4.22×10^8 m. From this information, estimate the mass of Jupiter, and compare this with the value in Appendix D.

5.28 The escape velocity (§5.6) is approximately 7 miles/sec. How does this compare with the speed of a jet plane? With the orbital speed of Sputnik I?

5.29 How fast would a rock have to be thrown into the air in order for it to reach a maximum altitude 4000 miles above the earth's surface?

5.30 It is suggested in the text that the laws of thermodynamics might provide a better guide for social scientists than does the law of universal gravitation. It is clear that social scientists need to be aware of the technological impact of science, but should they pay any attention at all to scientific laws in trying to formulate generalizations about their own fields of study?

5.31 (a) Many people have asked themselves what would happen if they were to dig a "hole to China" (a hole straight through the center of the earth) and drop a rock into it. Argue from conservation of energy that the rock would accelerate as it fell toward the center of the earth, attain its maximum kinetic energy at the center, and then slow down as it approached the other end of the hole; if air resistance could be disregarded, the rock would just barely reach the other end, arriving at the level of the ground on the other side of the earth with no kinetic energy left.

(b) If no one were waiting at the other end of the hole to catch the rock, what would happen to it after it had reached the other side of the earth?

(c) How long would it take to get to the other end of the hole after it was dropped? A precise calculation (even for this idealized problem with no air resistance) is beyond the scope of the present discussion, but an estimate can be made in the following way. First show that *if* the law of conservation of energy in the simple form $\frac{1}{2}mv^2 + mgh =$ constant (with $g = 9.8$ m/sec^2) could be used, the speed of the rock when it reached the center of the earth would be about 10^4 m/sec. Its average speed as it fell to the center of the earth would actually be less than this, perhaps half as large, or 5000 m/sec. We would expect the second half of the rock's journey to be just a "mirror image" of the first half; therefore 5000 m/sec would serve as an estimate of its average speed for the whole trip. Using this value, show that the required time would be about 40 minutes.

(d) If air resistance were taken into consideration, how would this affect the behavior of the rock?

6

The First Law of Thermodynamics: The Generalization of the Law of Conservation of Energy

Nancy Richardot

6

The First Law of Thermodynamics: The Generalization of the Law of Conservation of Energy

Now if this motive Energie must be called Heat, let it be so, I contend not. I know not how otherwise to call it.

—Joseph Glanvill, *Scepsis Scientifica,* or *Confest Ignorance, the Way to Science,* 1665.

Heat is work, and work is heat.

—M. Flanders and D. Swann, in "The First and Second Laws of Thermodynamics" (*At the Drop of Another Hat,* Angel Records)

§6.1 HEAT, TEMPERATURE, TEMPERATURE SCALES, AND THERMAL ENERGY

We began the study of energy in Chapter 2 by considering the motions of objects moving under the influence of the earth's gravitational attraction, and we saw that by introducing the concepts of kinetic energy and gravitational potential energy we could formulate a law of conservation of energy applicable to such motions. The generalization of the law of conservation of energy was described in Chapter 4, but the justification for the extension was postponed. It is largely on the basis of careful experiments carried out by James P. Joule in the middle of the 19th century that we are able to include *thermal* energy as one of the many forms of energy and to include a flow of *heat* as one type of energy transfer. We now know that thermal energy can be understood primarily as the manifestation of the rapid random motions of the atoms and molecules of which a substance is composed, that is, that thermal energy is for the most part just disorderly kinetic energy on a microscopic scale. The evidence for this interpretation of thermal energy did not come until after Joule's experiments; Joule was able to show how thermal energy could be included as a form of energy, but the understanding of what thermal energy "really is" came later, and we shall consider this interpretation in §6.5.

The motivation for studying heat and temperature and thermal energy is two-fold. First, the necessary measurements and thinking typify the procedure that should be followed whenever it appears necessary to broaden the concept of energy—to "save" the law of conservation of energy when it appears to have failed. Second, of all the extensions of the energy concept that have been made, it was the incorporation of heat and thermal energy into the picture that was the crucial step, the really significant extension that transformed the law of conservation of energy from a rather specialized result, useful in describing some types of mechanical phenomena, into a universal law of nature. Until the mid-19th century the subjects of heat and mechanics were two distinct branches of science, each with its own experts, its own techniques and units of measurement, its own laws. It was out of the marriage of these two disciplines that the law of conservation of energy emerged as a universally valid law. Before going into the experiments that brought these two areas together, it is necessary to discuss the subject of heat and the measurement of heat and temperature.

Three different temperature scales are used in this book. The first is the familiar Fahrenheit scale, on which the temperature at which water freezes is 32 degrees (32°F) and the temperature at which water boils is 212°F. The Fahrenheit scale is the one to which Americans are accustomed, and it is still used for many engineering purposes as well as in weather reports, but almost everywhere else in the world and for almost all scientific purposes, the Celsius (or centigrade) scale is used instead. On this scale, the temperatures at which water freezes and boils are 0°C and 100°C; a normal room temperature of 70°F is about 21°C.* The most fundamental temperature scale is the *absolute* or *Kelvin* scale. The temperature of zero on the Kelvin scale (0°K, "absolute zero"), is equal to a temperature of −273°C (more precisely, −273.15°C). Except for this important difference, the Kelvin and Celsius scales are the same; any temperature given in °C can be converted into °K simply by adding 273. The normal freezing point of water, then, is 273°K, and the boiling point 373°K. In giving temperatures on the Kelvin scale, the degree symbol (°) is often omitted; thus the temperature of the freezing point of water may be written as 273 K and read as 273 kelvins. We will retain the degree symbol as a reminder that a *temperature* is being referred to. The Kelvin scale is also called the *ideal-gas* temperature scale, and we will describe in §6.5 the properties of gases that are used in establishing this scale; its significance as an "absolute" scale and the significance of a temperature of zero on this scale can only be understood on the basis of the *second* law of thermodynamics. Values of various temperatures are given in Table 6.1. Some of these are familiar ones; others are far beyond the range of ordinary experience. Notice that room temperature is about 294°K; 300°K is an easy round number to remember for the temperature of the environment in which we live.

*For more discussion of conversions of temperatures from one scale to another, see Table A.11.

TABLE 6.1 Values of various temperatures on the Fahrenheit, Celsius, and Kelvin scales.

	°F	°C	°K
Absolute zero	−460	−273	0
Liquid helium	−452	−269	4.2
Liquid hydrogen	−423	−253	20
Liquid nitrogen	−320	−196	77
Liquid oxygen	−297	−183	90
Lowest recorded weather temperature	−125	−87	186
"Dry ice" (solid CO_2)	−110	−79	194
Mercury freezes	−38	−39	234
Water freezes	32	0	273
Room temperature	70	21	294
Body temperature	98.6	37	310
Highest recorded weather temperature	136	58	331
Water boils	212	100	373
Lead melts	621	327	600
Steam temperature in a nuclear power plant	660	350	620
Steam temperature in a fossil-fuel power plant	930	500	770
Uranium fuel rod (interior temperature)	4000	2200	2500
Light bulb filament	4600	2500	2800
Tungsten melts	6170	3410	3683
Sun—surface	10,000	5500	5800
Sun—interior	2.7×10^7	1.5×10^7	1.5×10^7
Deuterium-Tritium fusion (ignition temperature)	7×10^7	4×10^7	4×10^7
Deuterium-Deuterium fusion (ignition temperature)	7×10^8	4×10^8	4×10^8

Temperature and heat are two important concepts; although there are relationships between them, these two concepts are by no means identical. A simple example may make this point clear. Consider a stove, with two identical burners. On one burner place a pan with a small amount of water, which comes to a boil in about two minutes. On the other burner, place a large pan with several gallons of water. It takes a long time to raise the temperature of the larger amount of water. After ten minutes, perhaps, the water in the large pan is still barely warm to the touch. In each case, heat flows from the stove into the water, but more heat flows into the large pan during its ten minutes on the stove than into the small pan during the two minutes it is on the stove. Nevertheless, the temperature of the water in the small pan is higher than the temperature of the water in the large pan. Heat and temperature are related concepts, but the two terms are not synonyms. (Perhaps one reason for the occasional confusion is that the adjective "hot" usually refers to temperature and only indirectly to heat. In this example, the water in the small pan is hotter than the water in the large pan, that is, its *temperature* is higher, even though it has received a smaller amount of *heat* from the stove.)

Whenever two objects at different temperatures are put in contact, there is a tendency for the warmer one to cool off and for the temperature of the cooler one to rise. Eventually the two objects reach an equilibrium temperature, intermediate between the two initial temperatures. We describe this as the result of a flow of heat from the hotter object to the cooler one. A flow of heat is usually described as occurring by conduction, convection, or radiation. Heat flows by conduction through *something*, along a silver spoon placed in a cup of hot coffee, or through the walls of a house or a pane of window glass from the interior of the house to the outside. Convection refers to a flow of heat produced by the actual motion of a liquid or a gas. When a window is fitted with a storm window, heat flows from the warm pane of glass on the inside to the cooler glass on the outside primarily by convection, by means of circulating currents of air, which receive heat from the warm pane, circulate within the air space, and deliver heat to the cooler pane. Heat can also be transferred, even through a vacuum, by radiation. Any object "radiates" to some extent and receives radiation from its surroundings, though the effect is much larger if the temperature is high. If the temperature is high enough, this radiation can be seen as visible light, as is the case for a "red-hot" heating element on an electric stove; at lower temperatures, the same phenomenon of radiation occurs, but the radiation cannot be detected by the eye.*

A more subtle distinction than that between heat and temperature is the one between heat and *thermal energy*. Thermal energy is a form of energy that an object "has"; the higher its temperature, the greater its thermal energy. The term *heat* should be used only to refer to some sort of process—a "flow of heat." Heat is energy in *transit* from one place to another. In §3.4 we made a similar distinction between *work* and *energy* (kinetic energy and gravitational potential energy in that context). Work is a measure of energy transfer: Do work on a baseball and the result is an increase in its energy, a transfer of energy from you to the ball.

Put two objects with different temperatures in contact, put a cold bottle of beer in a pan of warm water, for instance. There is a flow of heat from the water to the beer; the result is an increase in the thermal energy of the beer and a decrease in the thermal energy of the water. This description seems to imply that "heat" *can* be regarded as a synonym for "thermal energy": "Heat flows from the water to the beer;

---

*A flow of heat is just one of the ways of transferring energy; it is something of a semantic problem to decide whether the radiation emitted by a hot object should be called "heat," "light," or "electromagnetic radiation." Whatever its name, the process of energy transfer by radiation is an important one; it is by this process that the earth receives energy from the sun and gives off energy into space, and we shall consider this more carefully in Chapter 11. For now we can simply regard radiation as one of the possible types of heat flows.

the water *loses heat* while the beer *gains heat*." Is there anything wrong with such a statement? If we were to consider only experiments of this sort, such language would be permissible. However—and this is extremely important—thermal energies can be changed even when *no* flow of heat occurs. Rub two sticks together (do *work* on them), and their temperature goes up, their thermal energy increases. Hammer vigorously on a nail, or drill a hole in a block of wood; an increase of thermal energy results even though there is no flow of heat, even though no other object is "losing heat." Thermal energies can be changed by a flow of heat *or* by doing work. Joule's important contribution was to perform careful experiments in which changes in thermal energy were produced by doing work. By measuring the amount of work done and the resulting temperature changes, he showed that a certain amount of work always produced the same effect as a certain amount of heat. In order to describe these experiments properly, we must first consider the simpler experiments in which thermal energy is transferred from one object to another by a flow of heat. Units for measuring heat will be described, and we will return in §6.2 to Joule's experiments, which demonstrated an equivalence between heat and work.

One of the basic units for measuring heat is the *calorie* (cal), originally defined as the amount of heat necessary to raise the temperature of 1 g of water by 1°C. We now know that a flow of heat is one kind of energy transfer, and therefore heat can be measured in joules or kilowatt-hours or any other energy units. (After this had been realized, it was possible to redefine the calorie as equal to a certain number of joules; the new definition was chosen so that it is still true that 1 cal of heat will raise the temperature of 1 g of water 1°C. It was not possible to do this until the subject of heat had been studied and until Joule's crucial experiments had been done.)

QUESTION

6.1 The *BTU* (British Thermal Unit) is defined as the amount of heat necessary to raise the temperature of 1 lb of water by 1°F. Show that 1 BTU $\simeq$ 252 cal.

Consider experiments in which heat is allowed to flow from one object to another, in which we determine the initial and final temperatures and calculate the amount of heat that flows out of the warmer object and into the cooler object. Consider, for instance, two containers, each containing 400 g of water, one of them initially at a temperature of 60°C and the other at 20°C (Figure 6.1). If they are placed in contact, the final common temperature turns out to be 40°C. We can describe this result by saying that 8000 calories of heat have flowed into B. One calorie produces a temperature rise of 1°C if the mass of the water is

1 g. More generally, if the mass is m and the temperature change ΔT, the amount of heat, H, is given by:

$$H \text{ (cal)} = m \text{ (g)} \times \Delta T \text{ (°C)} \quad \text{(for water)}; \qquad (6.1)$$

thus, in the preceding example, $H = 400 \times 20 = 8000$ cal. At the same time, the 400 g of water in A have undergone a temperature decrease of 20°C; 8000 cal of heat have flowed out of A.

Two reservations must be made about this sort of "calorimetry" experiment. The first can cause serious experimental problems but is not a serious matter of principle: care must be taken to make sure that extraneous heat flows to or from the surroundings are small enough that they can be ignored. In addition, the containers themselves get cooler and warmer as do the thermometers used to measure the temperatures. Unless the heat flows to and from the containers, the thermometers, and the surroundings are negligible in comparison to the heat flow from one sample of water to the other, the observed final temperature may be different from that stated. The second reservation is of fundamental importance. If we are to use the observation that the temperature of sample B rose by 20°C to infer that 8000 cal of heat flowed into it, we must be certain that nothing else has been going on which might have increased the temperature of B. It is possible to increase the temperature of a container of water by vigorous stirring, something completely different from a flow of heat. The essential step in the formulation of the first law of thermodynamics, the generalized law of conservation of energy, is precisely the determination of how large a rise in temperature is produced by a certain amount of stirring; for the present, in order to get good "clean" results, we must stipulate that no such extraneous processes are occurring that could produce temperature changes.

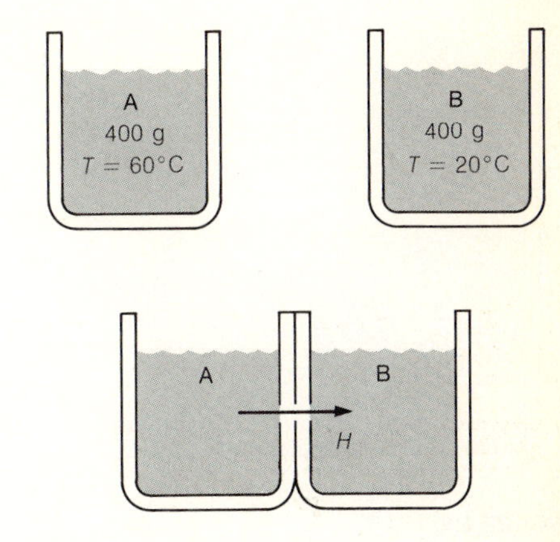

FIGURE 6.1
A flow of heat from A to B results in a common final temperature of 40°C.

What happens in a calorimetry experiment if the two containers have different quantities of water? Consider Figure 6.2. We would certainly expect the final temperature to be closer to 60°C than to 20°C, and in this case it turns out to be about 50°C. Since the temperature of container B increased by 30°C, we infer from Equation 6.1 that the amount of heat flowing into B was $200 \times 30 = 6000$ cal. The temperature of A decreased by 10°C; we infer that the amount of heat flowing out of A was $600 \times 10 = 6000$ cal. Again it appears that the amount of heat leaving A is equal to that flowing into B.

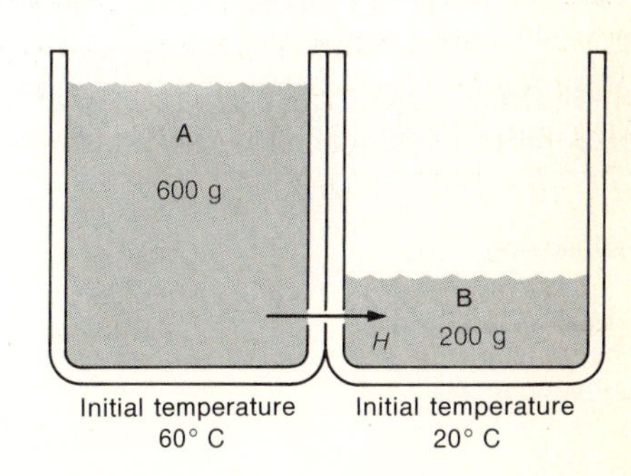

FIGURE 6.2
With different amounts of water in the two containers, the final temperature is closer to 60°C than to 20°C.

One calorie of heat will raise the temperature of 1 g of water 1°C, but for other substances, the amount of heat required to produce this temperature change is a different number of calories. The amount of heat needed to change the temperature of 1 gram of aluminum by 1°C, for instance, is only 0.22 cal. We define the *heat capacity, C,* of any substance as the number of calories required to raise the temperature of 1 g by 1°C. Equation 6.1 can be modified in the following way, to obtain an equation describing all substances:

$$H = C\, m\, \Delta T. \qquad (6.2)$$

With H measured in calories, m in grams and ΔT in °C, the units of heat capacity must be cal/g-°C; for aluminum, $C = 0.22$ cal/g-°C, and for water $C = 1$ cal/g-°C. (Heat capacities can also be given with reference to 1 kg, for instance, rather than 1 g. Thus for aluminum, $C = 220$ cal/kg-°C.) We can determine heat capacities by doing experiments in which heat flows between water and other substances, and by measuring the resulting temperature changes. Once the heat capacity of a material has been measured in one such experiment, other experiments with samples of various sizes and various initial temperatures can be interpreted in a simple way; the amount of heat flowing out of the warmer object is equal to the heat flowing into the cooler one. Consider, for instance, the experiment shown in Figure 6.3. A 500-g piece of aluminum with an initial temperature of 10°C is placed in a jar of water holding 800 g of water at a temperature of 76°C. In this experiment, the common final temperature of the aluminum and the water is about 68°C. The decrease in temperature of the water is 8°C; the amount of heat flowing out of the water is therefore

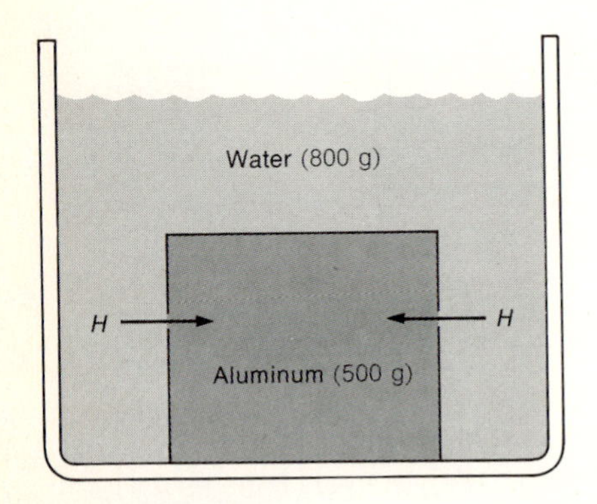

FIGURE 6.3
Transfer of heat from water to a submerged piece of aluminum.

$$H = C\,m\,\Delta T = 1\frac{\text{cal}}{\text{g-°C}} \times 800\text{ g} \times 8\text{°C} = 6400\text{ cal}.$$

The temperature of the aluminum increases by 58°C; the amount of heat flowing into the aluminum is thus

$$H = C\,m\,\Delta T = 0.22\frac{\text{cal}}{\text{g-°C}} \times 500\text{ g} \times 58\text{°C} \simeq 6400\text{ cal},$$

the same amount as that which flowed out of the water.

Heat capacities of a number of materials are given in Table 6.2. Notice that nearly every material has a heat capacity lower than that of water; that is, according to Equation 6.2, a larger amount of heat is required to produce a certain rise in temperature for a given mass of water than for most other substances. A large body of water tends to have a stabilizing effect on the temperature, because a large flow of heat is required to produce a significant change in temperature. This is one reason for the moderate climates usually found in coastal regions.

TABLE 6.2 Heat capacities, C, of various substances near room temperature.

Substance	Heat capacity, C (cal/g-°C)
Water	1.0
Ice	0.5
Aluminum	0.22
Iron	0.11
Copper	0.092
Lead	0.031
Gold	0.031
Brass	0.09
Brick	0.2
Cement	0.16
Rocks	0.2
Dirt (dry)	0.2
Wood	0.3–0.6
Glass	0.1–0.2
Bakelite	0.35
Paper	0.3
Gasoline	0.5
Air	0.17
Hydrogen	2.43
Helium	0.74

It is not always true that a flow of heat results in a change in temperature. Put a bucket of ice with an initial temperature of −10°C out in the room. At first the temperature of the ice increases as heat flows into it, but when the temperature of the ice reaches 0°C, it begins to melt. The temperature of the ice-water mixture remains at 0°C until all the ice has melted, at which point the temperature begins to increase once more. We describe this by saying that melting or freezing involves a "latent heat." A certain number of calories of heat must be added to melt 1 g of ice, and the same number of calories must be removed in order to freeze 1 g of water, in each case with no change in temperature. For the transition from ice to water, this latent heat (the latent heat of melting or the latent heat of "fusion") is 80 cal/g. A similar effect occurs when water boils and turns into water vapor. Add heat to water at a steady rate and its temperature rises until it reaches 100°C; then the temperature remains at 100°C as the water boils away. There is a latent heat of vaporization, which for water is about 540 cal/g.

A mixture of ice and water has a temperature of 0°C, and this temperature will remain fixed even if rather large amounts of heat flow into it or out of it, as long as it is not completely melted or completely frozen. Such a mixture tends to stabilize the temperature at 0°C. Large tubs of water placed in an apple cellar on a cold night are very effective at keeping the temperature from falling below 0°C; the temperature cannot fall below this value until enough heat has been removed from the water so that it is completely frozen.

§6.2 THE NATURE OF HEAT. JOULE'S DETERMINATION OF THE "MECHANICAL EQUIVALENT OF HEAT"

In calorimetry experiments such as those described in §6.1, in which heat flows from one object to another, the amount flowing out of the warmer object is equal to that flowing into the cooler one. In these experiments, heat behaves like a fluid (the "caloric fluid"), whose total amount is unchanged but that can be transferred from one object to another. In the "caloric theory of heat," the flow of heat is considered analogous to the flow of a fluid (Figure 6.4). When the valves are opened to allow flow from one container to another, the fluid levels in the various containers are equalized. The level of the fluid is analogous to the temperature. The wide containers correspond to objects with large masses or large heat capacities, ones for which a large amount of fluid must be added or subtracted to produce a specified change in the level. It is clear that the experiments we have been describing are *consistent* with such a picture. If heat *were* an indestructible fluid that could flow from one object to another, then the experiments we have described would exhibit the regularities which they do in fact exhibit. This is by no means the same thing as proving that heat *is* a fluid, but the idea is nevertheless a useful one; it helps us to see and remember patterns in a large number of different experiments. In such experiments, the term "caloric fluid" could be considered as a picturesque term for "heat," and it would not be necessary to distinguish between the concepts of heat and thermal energy.

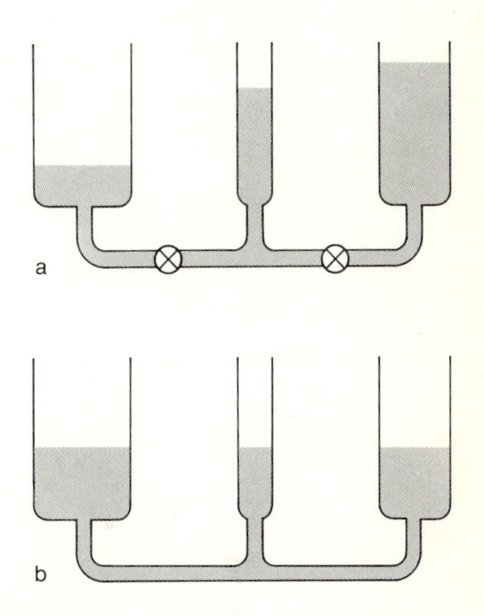

FIGURE 6.4
In the caloric theory, the flow of heat is considered as analogous to the flow of a fluid between various containers, and the fluid level is analogous to the temperature: (a) closed valves isolate the containers from one another and prevent fluid flow; (b) when the valves are opened, the levels in the various containers become equal to one another, just as the flow of heat between objects results in an equalization of their temperatures.

The difficulty with the caloric theory of heat is that, even though it provides a correct summary of a restricted class of experiments, there are many situations in which it is absolutely wrong. We do not need to look very far to find examples of such situations; some were described in §6.1, in which the importance of distinguishing between heat and thermal energy was emphasized. When you rub two sticks together, you are apparently creating new "caloric fluid," and you can continue to do so until you get tired. Count Rumford (an early American scientist whose Tory sympathies caused him to leave the country at the time of the Revolution) pointed out in 1798 that "caloric fluid" could be created in unlimited amounts by doing mechanical work. Observing the way in which cannon were bored by huge drills turned by horses, he called attention to the enormous quantities of heat (caloric fluid?) produced, enough to boil the water in which the cannon were immersed. "Caloric fluid" is not something whose total amount is constant; it can be created at will by doing mechanical work. In some of the simple experiments described in Chapters 2 and 3, "caloric fluid" also seems to be created. As the motion of the swinging pendulum dies down, the pendulum's support and the surrounding air get slightly warmer; in a sticking collision between two air-track gliders, a slight rise in temperature is produced as mechanical energy is lost.

"Caloric fluid" can be created where it did not exist before. It can also be destroyed, and one modern example of this is found in a "steam-electric" power plant fueled by coal or oil or by nuclear fission. In this type of plant (Figure 6.5), water is the "working substance" that circulates constantly through the plant, sometimes as a liquid, in other places in the form of steam. In one part of the plant, heat flows from a nuclear fuel rod, or out of a burner fueled with coal or oil, into the circulating water, turning it into steam. In a different part of the water's journey, it passes through a "heat exchanger," where heat flows from the circulating water into a stream of cooling water from a nearby river or ocean. The important point is that as the water circulates inside the plant, a smaller amount of heat flows out of it (into the river) than flows into it from the coal fire or from the fuel rods. "Caloric fluid has been destroyed!" But something else has been going on: at another point in its passage through the plant, the water does work on the blades of a turbine driving an electrical generator. The eventual effect may be to set a subway train in motion in a distant city (doing work on the subway train,

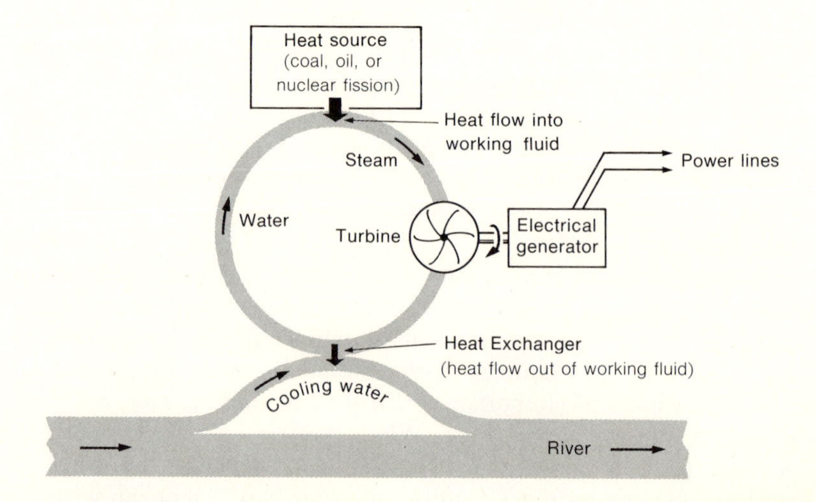

FIGURE 6.5
Schematic diagram of a power plant.

creating kinetic energy), or to turn a motor that raises an elevator in a building (doing work on the elevator, creating gravitational potential energy). "Caloric fluid" has been destroyed, but at the same time mechanical work has been done and kinetic energy or potential energy has been created.

In summary, there is an important class of experiments in which mechanical energy is conserved. In another important class of experiments, the simple calorimetry experiments, heat acts like an indestructible "caloric fluid." If we were never to consider any other phenomena, then the two subjects of heat and mechanics would remain separate from each other, each complete and logically consistent by itself. It is only by putting aside these restrictions and by considering experiments in which work is done—and in which the result is not an increase in kinetic energy or gravitational potential energy but an increase in *temperature*—that we can hope to find a unified theory of heat and mechanics and a universally valid law of conservation of energy.

Performance of mechanical work can produce an increase in temperature, it can have the same effect as a flow of heat. Is there an equivalence between work and heat? Does a certain amount of work (measured in joules) always correspond to the same number of calories, or does it vary from one situation to another? Rarely in the history of science has so much depended on the answers to such apparently innocent and prosaic questions. Though others had realized the importance of these questions and had obtained some preliminary answers, Joule in the 1840s was the first experimenter who had the skill and the persistence to make measurements with enough care and accuracy to produce convincing results. It is highly appropriate that one of the modern units of energy is named for the person who did more than anyone else to provide a firm foundation for the law of conservation of energy.

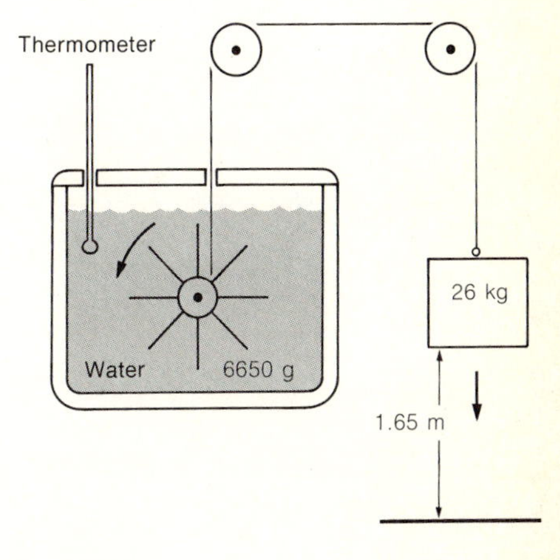

FIGURE 6.6
Joule's paddle-wheel apparatus.

Joule studied a number of situations in which kinetic energy or gravitational potential energy is lost, in which work is done and in which a rise in temperature is produced. One of his most important and most famous experiments was carried out with a paddle-wheel apparatus like that shown in Figure 6.6. When the weight is released at the top, it falls slowly to the ground, losing gravitational potential energy but gaining almost no kinetic energy. The paddle wheel stirs the water as the weight falls; work is done on the water, and the temperature of the water rises. The amount of work done can be calculated from the amount of gravitational potential energy lost; the temperature of the water can be measured, the mass of the water is known, and so we can also calculate the number of calories of heat which would have produced the same rise in temperature. In one of Joule's experiments, a 26-kg weight fell a distance of 1.65 m, stirring 6650 g of water. The experiment is a difficult one because even with this large a weight, only a tiny rise in temperature is produced. Joule found it necessary to put the weight back in its original position and let it fall again, 10 or 20 times altogether, before the temperature of the water was raised by as much as a few tenths of a degree. During all this time (half an hour or more), the water might be losing heat to its surroundings or perhaps gaining heat from the experimenter's warm body.

In spite of all the difficulties, in spite of the necessity of making better thermometers than ever before just to be able to do these experiments, Joule found in his most careful experiments that there was a fixed relationship between the amount of gravitational potential energy lost and the rise in temperature of the water. In the apparatus described above, for example, 20 successive descents of the weight produced a temperature rise of approximately 0.3°C. Since the mass of the water was 6650 g, this is the same temperature rise that would have been produced by an amount of heat equal to

$$C\ m\ \Delta T = 1 \times 6650 \times 0.3 \simeq 2000 \text{ cal.}$$

At the same time, the loss in gravitational potential energy could be calculated as 20 times the amount for one descent of a 26-kg weight through a distance of 1.65 m:

$$20mgh = 20 \times 26 \times 9.8 \times 1.65 \simeq 8400 \text{ J.}$$

(Needless to say, these are not the units used in Joule's own writings.) This is the amount of work done on the water by the blades of the paddle wheel. Doing 8400 J of work has the same effect as does the flow of 2000 cal of heat; *one* calorie is therefore equivalent to about 8400/2000 = 4.2 J. This is what is meant by the "mechanical equivalent of heat," the number of joules equivalent to one calorie.

One experiment is scarcely sufficient, especially since—as skeptics were quick to point out—temperature changes of mere fractions of a degree were produced. Joule's contribution consisted of conducting with great care a wide variety of experiments in which mechanical energy was lost and in which a rise in temperature was produced. The paddle-wheel experiment was repeated with oil and with mercury instead of water; the known values of their heat capacities were used to make new determinations of the mechanical equivalent of heat. He did other experiments, such as ones in which water was heated by forcing it through fine tubes. The results of the various experiments did not always yield precisely the same result, but the most careful experiments all gave a value very close to 4.2 J as the equivalent of 1 cal. Although Joule no doubt hoped that heat and work would prove to be equivalent and that a fixed rate of conversion could be established, it did not "have to turn out" this way. There is nothing in the definitions of heat and work that demands a fixed equivalence between 1 cal of heat and any definite number of joules of work. It might have turned out that in some situations, 1 cal has the same effect as 3 J of work, that in others 1 cal is equivalent to 5 J. It is a wonderfully simple fact about our universe that there *is* a definite factor of equivalence, and on this fact depends the validity of the law of conservation of energy. "Heat is work, and work is heat," goes the song by Flanders and Swann quoted at the beginning of this chapter. This is not literally true; heat is not work, but heat and work are equivalent in that the energy of a system can be increased either by the performance of work or by a flow of heat; these are just two different ways of increasing its energy.

Once the equivalence between heat and work was recognized, on the basis of Joule's experiments, and as the results were confirmed by more and more experiments, it was no longer necessary to have separate units for heat and work. Once we know the conversion factor, heat can be measured in joules, or work can be measured in calories. Calories and joules are both still used, even though one of the two could be dispensed with. But the definition of the calorie now is different from its original definition. Now, by definition,

$$1 \text{ cal} = 4.184 \text{ J}.$$

Heat capacities can be given in joules/g-°C rather than in calories/g-°C. (Joules/kg-°C would be the proper MKS unit.) It is still correct to say that 1 cal of heat raises the temperature of 1 g of water through 1°C, though this statement is no longer the *definition* of the calorie.

QUESTION

6.2 Give the specific heat of water in joules/g-°C. In joules/kg-°C.

For those who want to have an understanding of the logical basis of science and a sense of its historical development, there is an unfortunate aspect to this redefinition of the calorie. Defining the calorie to be a certain number of joules tends to obscure the fact that it is only because of Joule's careful experiments that we have the right to define the calorie in this way; it leads to the possible misconception that the equivalence of heat and work is something tautological, something that holds by definition. To define the inch as precisely 2.54 cm is easy; a great deal of careful work was required to enable us to make the apparently similar statement that 1 cal = 4.184 J. Such a statement contains, by implication, a great deal of information about the world we live in. Though we could dispense with the calorie now, from a historical point of view it seems clear that the existence of a separate unit for heat was unavoidable. The equivalence of heat and work could only be postulated on the basis of a great deal of careful work in two areas of physics which initially seemed to be unrelated: heat and mechanics. It was inevitable that workers in these two areas initially had to develop their own separate systems of measurement.

§6.3 THERMAL ENERGY AND THE GENERALIZED LAW OF CONSERVATION OF ENERGY (THE FIRST LAW OF THERMODYNAMICS). CLOSED AND OPEN SYSTEMS

Consider the water in Joule's paddle-wheel experiment, with a final temperature 0.3°C higher than its initial temperature as a result of the 8400 J of work performed. Shall we say that it has "more heat" than it did in the initial state? This surely makes no sense, for no heat flowed into the water in this experiment; the temperature increase was produced by doing work on the water. Perhaps we could then say instead that the water now "contains more work" than it did initially. But this is not very helpful either, because as far as we can tell from observing only the initial and final states, without knowing how the water got from one state to the other, the change might have resulted from a flow of heat or from the performance of work or from some combination of these effects. If the sample of water is changed from one state to the other, we know that either 8400 J of work were done on it, or that 8400 J of heat flowed into it, or that there was some combination of work and heat such that

$$H + W = 8400 \text{ J},$$

where H is the heat flowing *into* the object of interest and W is the work done *on* the object. (We can allow for the possibility that heat may flow *out* of the object or that the object may do work *on* something else by agreeing to let H or W stand for *negative* numbers where necessary.)

Instead of saying that the water has "more heat" or "more work," we simply say that the water in its final state has a larger value of *thermal energy* (or "internal energy"), a form of energy which may be increased by a flow of heat (as in the calorimetry experiments described in §6.1), or by the performance of work as in Joule's experiments, or both. Thus we can write the conservation of energy equation for these experiments as

$$H + W = \Delta(TE). \tag{6.3}$$

Heat, like work, is a particular type of energy *transfer*. Heat is not itself a form of energy an object can "have"; heat is energy in transit. A flow of heat into an object *produces* a change in the energy of the object, most often a change in the thermal energy of the object. In the same way, work is not energy but a type of energy transfer; in the mechanical context in which the concept of work first appeared, doing work on an object produces a change in its kinetic energy or gravitational potential energy, or in both.

"Heat" is still one of the more confusing terms with which we have to deal. Maximum clarity would be achieved if we were to use the word "heat" only to refer to *flows* of energy (of the type that take place between objects of different temperatures), and if we were always to insist on using the term thermal energy to refer to the kind of energy an object has more of when its temperature is high. Confusion can exist because "heat" is often used as if it were a synonym for thermal energy. We may say that the gravitational potential energy of water at the top of a waterfall is "converted into heat" when it strikes the bottom, when we should say that it is converted into thermal energy. "There is a lot of heat in that tub of water" can be translated as "there is a lot of thermal energy in that tub"; the same statement can be interpreted, using the word heat in its proper meaning, by saying that if the water cools down to room temperature, a great deal of heat will flow out of it in the process.

We have been discussing experiments in which the flow of heat or the performance of work has the effect of producing changes in thermal energy:

$$H + W = \Delta(TE).$$

But there are other forms of energy that may change. Something may acquire kinetic energy by being set into motion as a whole, or it may experience a change in gravitational potential energy by being raised or lowered. The sample of water in the paddle-wheel experiment did not experience any changes in kinetic energy or gravitational potential energy because in that experiment it was prevented from moving as a whole. Other forms of energy may have to be included as well: changes in electrical energy, changes in chemical energy resulting from chemical reactions, and so on. We will not consider the justification for the introduction of each new form of energy as carefully as we did for thermal energy, but the correct procedure is apparent from the method used in introducing thermal energy. In order to conclude, for instance, that the chemical energy content of a gallon of gasoline is 1.2×10^5 BTU, we must burn a known quantity of gasoline in an apparatus designed so that a temperature rise is produced in a surrounding container of water. When this is done, it is found that the rise in temperature is the same as would be produced by a flow of heat, at the rate of 1.2×10^5 BTU per gallon of gasoline burned.

If we allow for the possibility of changes in all forms of energy, Equation 6.3 should be generalized to read

$$\begin{aligned} H + W &= \Delta(KE) + \Delta(GPE) + \Delta(TE) + \Delta(\text{Chemical Energy}) \\ &\qquad + \Delta(\text{Electrical Energy}) + \cdots \qquad (6.4) \\ &= \Delta E, \end{aligned}$$

where E denotes the total energy of the system being considered. Here H represents the total amount of heat flowing into the system, positive if heat flows into the system and negative if heat flows out, and W represents all other types of energy transfer. As pointed out in §4.1.A, this equation must be applied with care. One must be sure that all the forms of energy that may change significantly are taken into account. Any form of energy which does not change can safely be ignored; whatever the chemical energy of the water in Joule's experiment, we did not need to consider it because no chemical reactions occurred in which the chemical energy might have changed. One must have some "system" in mind in applying the energy equation (Equation 6.4), and as discussed in Chapter 4, the same process can legitimately be described in various ways, depending on one's choice of the boundaries of the system. Whatever system may be chosen, this equation says that the overall change in the sum of the energies—kinetic, gravitational potential, thermal, and other—of its parts can be calculated from the total amount of heat and work going into this system from the outside.

One of the most important applications of Equation 6.4 is to a *closed* system, a system chosen so that there is no exchange of heat or work with anything outside the system. In some applications, a single object might be a closed system; more often the systems we consider will include several different parts. In Joule's paddle-wheel experiment, one system which we could study might be just the water. This system is not a closed system; work is done on the water during the experiment. If we choose a larger system to study, one consisting not only of the water, but also of the falling weight, the paddle wheel, the rope, the surrounding air, perhaps even the furniture and walls of the room, then we have, at least approximately, a closed system. For a closed system, Equation 6.4 takes on a particularly simple form:

$$\Delta E = 0, \text{ or } E = \text{ constant.} \tag{6.5}$$

As time goes by, the various forms of energy of all the parts of the system may increase or decrease; Equation 6.5 says that the sum of all these changes is zero. If, for example, the gravitational potential energy of the weight decreases (as it falls), Equation 6.5 says that *something* else must happen. It does not tell us which of the many possible things may happen, but only that, whatever happens, the sum of all the changes of all the other energies will be exactly equal and opposite to the change of gravitational potential energy of the falling weight. Gravitational potential energy may be converted into thermal energy of the water, or into kinetic energy of the falling weight (as it will if we cut the rope instead of letting the weight fall slowly), or into some combination of various types of energies of many parts of the system. It may be converted first into one form of energy, which is then itself converted into some other form. (If we cut the string, the gravitational potential energy of the weight is converted first into kinetic energy of the weight; when the weight hits the floor, this kinetic energy in turn is converted into thermal energy of the weight and the floor.)

The same sort of *result* is obtained when we consider a rather different sort of system, one that is not closed but one for which the net amounts of heat and work flowing into it add up to zero. The earth is very nearly such a system. Think of an imaginary surface drawn around the earth, somewhere above the atmosphere; consider everything inside this surface as the system. The earth does receive a very large amount of heat by radiation from the sun, but at the same time it also loses heat at almost precisely the same rate by radiation in all directions. For the earth as a whole, H is almost precisely zero. We can consider H as being made up of two parts: if H_1 denotes the part of H due to incoming radiation from the sun, and H_2 the outgoing radiation from the earth, then in any one second, $H_1 \simeq 1.7 \times 10^{17}$ J, but $H_2 \simeq -1.7 \times 10^{17}$ J. The negative sign indicates a flow of heat *out* of the system. The total heat flow, $(H_1 + H_2)$, is almost exactly zero. (Although H_1 and H_2 are equal in magnitude, they differ in that H_1 consists mostly of radiation in the form of visible light, whereas H_2 is invisible "infrared" radiation.* Although the important point now is that the numerical values of H_1 and H_2 are almost exactly equal and opposite, the difference in "character" of these two heat flows is absolutely vital to the continued existence of life on earth. We shall return to this point in Chapter 7.)

For the earth as a whole $H + W = 0$, so, just as for a truly closed system, $\Delta E = 0$, the total energy of the earth is constant. The various parts of the earth's total energy may individually change, but always in such a way that the individual changes compensate for one another, such that the sum remains always the same: $E_{\text{Earth}} = \text{constant}$. What a marvelous wealth of complexity lies behind such a simple looking equation, and what a credit it is to the human mind that we have been able to make the measurements and develop the concepts which allow us to write such an equation. Few authors have grasped the implications of these ideas as well as Joule himself; surely no one has a better right than Joule to exult over the significance of these discoveries, as he did in the following words.

> Behold, then, the wonderful arrangements of creation. The earth in its rapid motion round the sun possesses a degree of living force so vast that, if turned into the equivalent of heat, its temperature would be rendered at least 1000 times greater than that of red-hot iron, and the globe on which we tread would in all probability be rendered equal in brightness to the sun itself. And it cannot be doubted that if the course of the earth were changed so that it might fall into the sun, that body, so far from being cooled down by the contact of a comparatively cold body, would actually blaze more brightly than before in consequence of the living force with which the earth struck the sun being converted into its equivalent of heat. Here we see that our existence depends upon the *maintenance* of the living force of the earth.
>
> Descending from the planetary space and firmament to the surface of our earth, we find a vast variety of phenomena connected with the conversion of living force and heat into one another, which speak in language which cannot be misunderstood of the wisdom and beneficence of the Great Architect of nature. The motion of air which we call *wind* arises chiefly from the intense heat of the torrid zone compared with the temperature of the temperate and frigid zones. Here we have an instance of heat being converted into the living force of currents of air. These currents of air, in their progress across the sea, lift up its waves and propel the ships; whilst in passing across the land they shake the trees and disturb every blade of grass. The waves by their violent

*The outgoing radiation, H_2, includes a fraction of the incoming light that is simply reflected back into space.

> motion, the ships by their passage through a resisting medium, and the trees by the rubbing of their branches together and the friction of their leaves against themselves and the air, each and all of them generate heat equivalent to the diminution of the living force of the air which they occasion. The heat thus restored may again contribute to raise fresh currents of air; and thus the phenomena may be repeated in endless succession and variety.
>
> When we consider our own animal frames, "fearfully and wonderfully made," we observe in the motion of our limbs a continual conversion of heat into living force, which may be either converted back again into heat or employed in producing an attraction through space, as when a man ascends a mountain. Indeed the phenomena of nature, whether mechanical, chemical, or vital, consist almost entirely in a continual conversion of attraction through space, living force, and heat into one another. Thus it is that order is maintained in the universe—nothing is deranged, nothing ever lost, but the entire machinery, complicated as it is, works smoothly and harmoniously. And though, as in the awful vision of Ezekiel, "wheel may be in the middle of wheel," and every thing may appear complicated and involved in the apparent confusion and intricacy of an almost endless variety of causes, effects, conversions, and arrangements, yet is the most perfect regularity preserved—the whole being governed by the sovereign will of God.*

Needless to say, such an enthusiastic description of the significance of the law of conservation of energy requires considerable imagination and faith. We cannot logically *deduce* the universal validity of this law, but the experiments that have been described gave Joule the confidence to put forth the law as an inductive generalization. Joule is expressing here his belief that if any phenomena were studied in the same painstaking way in which he carried out his paddle-wheel experiments, the law of conservation of energy would be found to be just as valid in any new situation, no matter how complicated. The "inductive leap" we make in applying the law to all phenomena is a truly bold one; even Newton's guess that there is a gravitational interaction between any two objects in the universe seems timid by comparison.

We shall refer to Equations 6.4 and 6.5 by two terms that are almost synonymous: the law of conservation of energy, and the first law of thermodynamics. Since we now see that thermal energy is just one of many forms of energy, and that a heat flow is just a particular type of energy transfer, we could perhaps do without the term "first law of thermodynamics." The continued use of this name (with the word "thermodynamics" combining references to both heat and mechanics) reflects the historical fact that it was the understanding of the relation-

---

**The Scientific Papers of James Prescott Joule.* (Published by the Physical Society of London. London: Taylor and Francis, 2 vols., 1884–87.) Joule is describing primarily the interconversion of kinetic energy, gravitational potential energy, and thermal energy. Joule refers to "living force" (a more descriptive term than its rather sterile modern counterpart, kinetic energy) and "attraction through space" instead of gravitational potential energy. Joule is using the term "heat" here as a synonym for thermal energy; as pointed out earlier, this is a common practice, but one that is subject to possible misinterpretation.

ship between these two apparently distinct subjects that had to be achieved before one could even contemplate a universally valid law of conservation of energy. The synthesis of these two subjects, heat and mechanics, with the postulation of the general law of conservation of energy, is one of the great intellectual achievements of science. In its impact on various areas of science and technology, and on many other fields as well, this law has been at least as influential as the Newtonian law of gravitation which preceded it. Joule had a key role in the development, but he was by no means alone. The first suggestions that a synthesis of heat and mechanics might be possible came long before Joule's experiments, and the study of the implications of the law of conservation of energy, and of its companion, the second law of thermodynamics, continues to this day.

Especially in the form of Equation 6.5, the law of conservation of energy has occasionally been referred to as the "accountant's delight." Within a closed system, all sorts of changes from one form of energy to another are possible, as long as any one change is canceled out by other changes. The sum of all the energies will always be the same, the "account books" will always balance if all forms of energy are accounted for. We can see a very important limitation on the usefulness of the law of conservation of energy, however. It tells us one very specific thing about any system: the total energy (of a closed system) remains constant. But by itself it tells us almost nothing about the individual changes. It tells us, for instance, that in a process in which only gravitational potential energy and thermal energy change, a decrease in gravitational potential energy of 50 J will be accompanied by an increase in thermal energy of 50 J; this first law of thermodynamics provides no clue about whether such a conversion *will* occur in a particular system, or how rapidly it may occur if it does. Although it is an extremely valuable tool, then, other laws are needed as well. In particular, the *second* law of thermodynamics (Chapter 7) describes the fact that many processes do not occur even though they would be allowed by the first law. The two laws of thermodynamics together limit the possibilities; it is still necessary to consider the details of a particular situation to see what will in fact happen.

It is worth reiterating that in asserting that the law of conservation of energy in its general form (with all forms of energy included) applies to *any* process and *any* system, we are going far beyond anything which we can possibly *deduce* from a few measurements on paddle wheels and tanks of water. This is inductive science at its best. We put forward the law as a generalization to be tested and to be used. We will use it as long as it works. Every new application is one more test. In principle, the law of conservation of energy is still tentative, but it has been used successfully now in countless situations, has allowed us to correlate vast amounts of data and has guided us to the discovery of new effects. More so than for almost any other generalization, we have reason to believe that the law of conservation of energy is "true."

§6.4 LOGICAL STATUS OF THE LAW OF CONSERVATION OF ENERGY

The total energy of a system, the quantity whose numerical value does not change if the system is a closed one, is defined as the sum of the various types of energy:

$$\begin{aligned} E = {} & KE + GPE + \text{Thermal Energy} + \text{Electrical Energy} \\ & + \text{Nuclear Energy} + \text{Chemical Energy} + \cdots \\ & + \text{"possible other forms of energy yet to be discovered."} \end{aligned} \tag{6.6}$$

One may well worry that we have now gone too far, especially in including the final term: "forms of energy yet to be discovered." If we permit this, have we perhaps fallen into the very trap which we were hoping to avoid? Has the law of conservation of energy been reduced to an empty tautology, something that can always be preserved by labeling any discrepancy as a "new form of energy"?

The question of the logical status of the law of conservation of energy is a difficult one. Certainly if *any* discrepancy in *any* experiment were to be routinely explained by inventing a new form of energy, the law would be a mere tautology. This would be no better than the procedure suggested in §2.5 for making a checkbook balance, inventing a category called "unexplained." If it were this easy to "discover" new forms of energy, there could be no conceivable experimental results that could disprove the law of conservation of energy; the law would not be *falsifiable.* A "law" of nature that is not falsifiable is no law at all. In order that a law tell us anything about the world, it must invite repudiation, it must "stick its neck out" by providing explicit predictions about the outcomes of experiments, experiments that could conceivably turn out in some other way. To say that a law is falsifiable is quite different from saying that it *will* be proven false, only that we can describe experiments that *would* prove the law false if they turned out in a certain way. Every successful prediction helps to confirm the law, but every law must in principle be regarded as only tentatively correct. Tomorrow's experiment may disprove it, and one genuine failure is enough to cause its rejection in spite of its past successes, or at least to require that we admit its lack of universal validity. Such was the fate of the law of universal gravitation: again and again it stuck its neck out with predictions of planetary motions that proved to be correct, but the day came when it failed to explain the details of Mercury's orbit. Newton's law of gravitation is still extremely useful, but it is not absolutely and precisely correct.

Suppose that an apparent exception to the law of conservation of energy were reported in a scientific paper, results that could not be explained by any of the presently known forms of energy, and that the author proposed to save the law by inventing a new form of energy to make things come out right. Such a proposal would be received with extreme skepticism by the scientific community. This is not to say that scientists have closed minds and are unreceptive to new ideas, but rather that they are conservative, that they are extremely reluctant to discard or even to modify a highly successful law unless the evidence is absolutely compelling. The experiment would surely be repeated by many others, the procedures used would be subjected to the most searching analysis and criticism. If all else failed, the idea of a new form of energy would then be considered more seriously. How this might be done we cannot say without knowing the particular details of the hypothetical situation. We can, however, describe what was done in one very notable case.

The phenomenon of radioactivity will be discussed in more detail in Chapter 13. For now, it suffices to say that in one type of radioactive decay ("beta decay"), one nucleus changes into another kind of nucleus, emitting an electron in the process. When the energies were measured (in the late 1920s), they did not add up correctly. The new nucleus and the electron simply did not have as much energy between them as they should have had; the "missing energy" was nowhere to be found. The importance of these results was quickly realized. The experiments were repeated, and the discrepancy persisted. Various suggestions were made. One was that the law of conservation of energy had been put to the test and had failed; another, of course, was the presence of a "new form of energy." Those who proposed the existence of a new form of energy made explicit suggestions about the nature of this energy, so that the hypothesis might be tested. It was suggested that a new sort of particle (christened the "neutrino"), previously unknown, was emitted in beta decay, in addition to the electron, and that the neutrino carried away just the right amount of energy to save the law of conservation of energy. The fact that no such neutrinos had been detected was "explained" by supposing that neutrinos have almost no effect on the apparatus used to detect electrons and other known particles. (Perhaps "new form" of energy is not quite the right term, if the missing energy is to be accounted for by the kinetic energy of the neutrinos. However, if neutrinos had no function except that of saving the conservation law, then "neutrino kinetic energy" would really be a new and different kind of energy.)

If this were all there were to the neutrino hypothesis, it would be essentially equivalent to using the word "neutrino" as a synonym for "unexplained"—not a very satisfying procedure. As the theory of beta decay was further developed, the still undetected "neutrino" became increasingly important for explaining a variety of phenomena. Not only was it an invisible carrier of the missing energy, but it had other attributes as well. For example, it was supposed to have *momentum* as well as energy, since momentum conservation also seemed to be violated in beta decay. For twenty years, however, the neutrino remained undetected, because one of its properties is that its interaction with other forms of matter is so weak that the chance of a single neutrino's producing a response from any sort of "particle detector" is extremely small.

According to the theory of the neutrino that had been worked out, neutrinos should interact once in a great while with other kinds of matter, and if the number of neutrinos were large enough, it should be possible to detect a few of them. Eventually, in 1953, in an experiment performed with very sophisticated electronic techniques at a nuclear reactor (which was expected to be a copious source of neutrinos), a few neutrinos were finally detected, exactly as predicted by the theory. Few scientists were surprised, but it is fair to say that everyone was a bit relieved.* The faith in the law of conservation of energy expressed by those who had suggested the existence of the neutrino was vindicated; the law had not only survived but had been successfully used to predict the existence of a previously unknown kind of particle.

We have looked at two examples of how the problem of rescuing the law of conservation of energy has been handled: Joule's investigations of the problem of heat, and the beta-decay problem. It is clear that scientific opinion is not satisfied by simply labeling any discrepancy as a new form of energy; something more is required. By thinking about how the law of conservation of energy might have been falsified in these two cases, we can see how it might fail at some future time. If Joule had found that 1 cal of heat was sometimes equivalent to 3 J of work, sometimes to 4, sometimes to 27, the law of conservation of energy would have been discarded in the 1840s.† If the sole function of the neutrino concept had been to explain the missing energy in beta decay, if the experiments done in 1953 to detect neutrinos had failed, and if no other explanation had been found, the law of conservation of energy would probably be regarded as incorrect, as a law that works most of the time but is not universally valid. (We would not throw it away; it would still be useful, just as Newton's law of gravitation is useful because it is almost precisely correct in almost all circumstances.) A new situation may arise some day in which the law of conservation of energy simply does not work and cannot be saved by the recognition of a new form of energy. The way in which the law survived the beta-decay crisis is just one of the reasons for feeling that it is likely to survive indefinitely.

There is another way of stating the law of conservation of energy, which brings out more clearly the fact that the law *could* be falsified. The statement is that there are no "perpetual motion machines" (perpetual motion machines "of the first kind"**) and that there never

---

*Some scientists were secretly disappointed, for science is especially exhilarating when a solidly established law fails and brand new ideas are needed.

†The equivalence between heat and work is sometimes described as analogous to the equivalence between different kinds of currency. An American company may receive checks in yen and dollars and deposit them all in the same account, just as a physical system receives deposits of both work and heat. Fortunately, the analogy is not a perfect one. If the exchange rate between joules and calories fluctuated from day to day as does the exchange rate between yen and dollars, the law of conservation of energy would have to be rejected or at least modified.

**A perpetual motion machine of the *first* kind would violate the *first* law of thermodynamics; equally interesting and equally impossible devices that would violate the *second* law of thermodynamics are called perpetual motion machines of the *second* kind.

will be. A perpetual motion machine of the first kind would be a device that could deliver energy to the outside (by doing mechanical work, for instance) without itself experiencing *any* internal change, as shown schematically in Figure 6.7. It would be a perpetual source of useful energy because if it can deliver one joule of work without undergoing any internal change, it can also deliver a second, a third, and so on. According to the law of conservation of energy, no such device is possible. Energy can be delivered to the outside only at the expense of the energy of the device itself. *Something* must change inside if energy is extracted, and any actual device must eventually run down unless its energy is replenished. (It is a perpetual motion *machine* which can deliver energy which is forbidden by the first law of thermodynamics; perpetual *motion* is not ruled out. The earth may continue to orbit around the sun with no change in energy. A wheel on a completely frictionless axle could spin forever. There are no frictionless axles, but their existence is not forbidden by the first law of thermodynamics; what the law proclaims to be impossible is the delivery of energy by a spinning wheel without a simultaneous decrease in the energy of the wheel.)

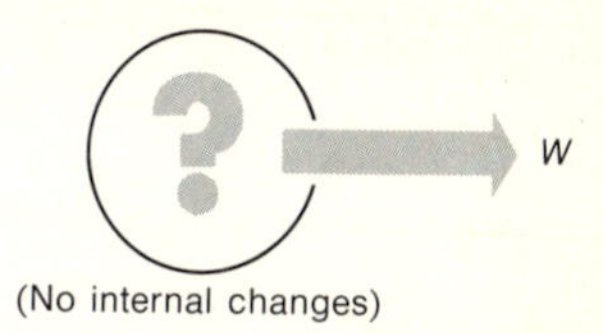

FIGURE 6.7
Schematic diagram of a perpetual motion machine of the first kind.

Countless inventors have tried to build perpetual motion machines, and the first law of thermodynamics (the law of conservation of energy) can be viewed as summarizing their failure and the failure of those who will try in the future. As an example of the thinking that goes into the design of a perpetual motion machine of the first kind, consider the water-driven wheel in Figure 6.8a. What if no waterfall is available? Why not use some of the output of the wheel to operate a pump as shown in Figure 6.8b, to pump the water back to the top? Such a device could run for a while, but even with a perfect pump, all of the output derived from the gravitational potential energy of the water as it descended would be needed to pump it back up again; nothing would be left over for delivery to the outside. With no friction anywhere, the output of the pump would be just sufficient to keep the water circulating. Acceptance of the law of conservation of energy is equivalent to saying that there is no need to examine the insides of the pump, to look at all its gears and pulleys; no matter how ingenious its design, this device will not work. We might be temporarily fooled by an inventor who puts in a battery-powered pump or a mouse-powered pump, but if the machine delivers energy, the batteries will run down or the mice will get hungry, and the machine will stop unless new batteries or new mice are installed.

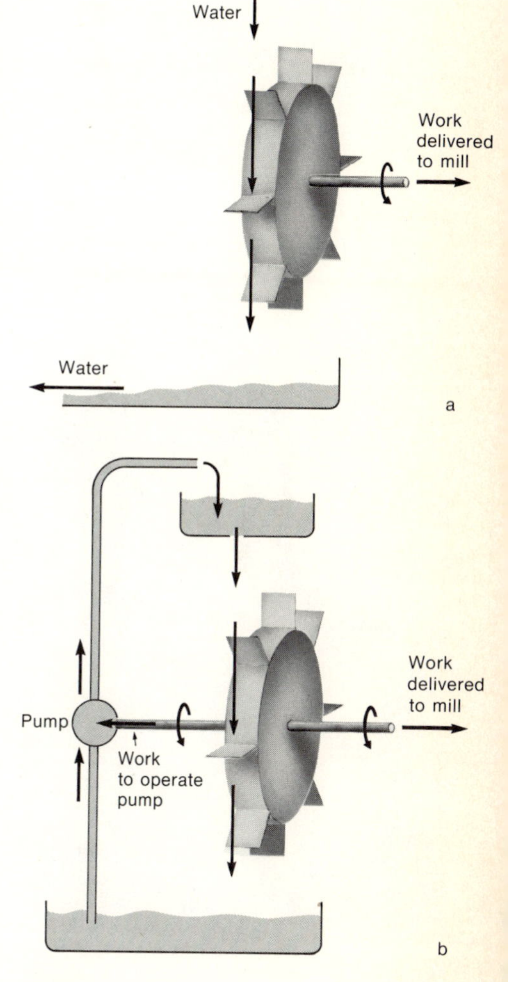

FIGURE 6.8
(a) Conventional water-powered mill wheel. (b) Mill wheel modified for use in desert regions; a perpetual motion machine of the first kind?

Figure 6.9 shows another proposed perpetual motion machine, a wheel with several compartments, each of which contains a heavy ball that is free to roll within the compartment. The idea is that the balls on the left, farther from the axis, turn the wheel in the counterclockwise direction. As the wheel turns, balls passing over the top roll to the outer edge of the wheel, and the rotation supposedly continues; the axle could then be used to raise weights or to drive an electrical generator. If the law of conservation of energy is correct, such a device cannot work. This conclusion is confirmed both by the failure of the wheels of this sort that have actually been built and by a detailed analysis of the motions of the balls.

The law of conservation of energy is put forth as a truly universal law of nature. Its claim is not simply that perpetual motion machines based on the principles of gravity or electricity or magnetism or light will not work, but that it will not be possible to construct perpetual motion machines based on physical phenomena that are not now even imagined, the phenomena represented by "possible other forms of energy yet to be discovered" in Equation 6.6. If a genuine perpetual motion machine of the first kind is ever built, the law of conservation of energy will have been falsified. This might happen, but few scientific beliefs are more widely shared than the conviction that this will never happen.

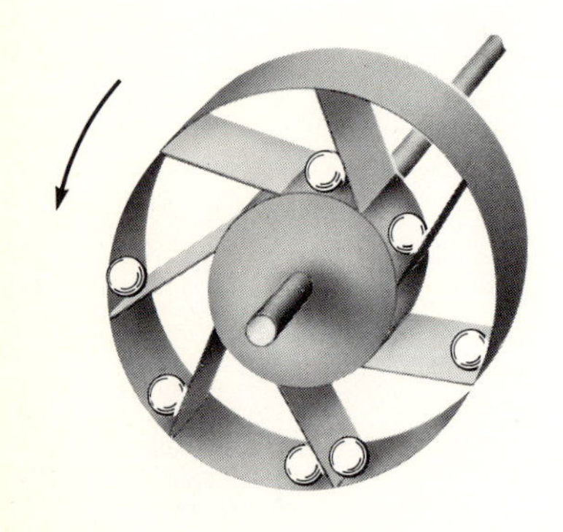

FIGURE 6.9
Another proposed perpetual motion machine.

The law of conservation of energy is *not* tautological; it *is* falsifiable. Experiments had to be carried out before the law could be accepted, and even now we must in principle regard it as only tentatively valid. Curiously enough, Joule himself came very close to saying that it was *a priori* obvious that energy was something whose total amount could not change, that the law of conservation of energy *must* be correct, in the following remarkable statement: "It is manifestly absurd to suppose that the powers with which God has endowed matter can be destroyed any more than that they can be created by man's agency." This is a strange statement coming from a man who spent his life trying to find out *whether or not* the law of conservation of energy was a valid principle. Joule, however, went on to say, "But we are not left with this argument alone, decisive as it must be to every unprejudiced mind. *Experiment* has shown that wherever living force is apparently destroyed or absorbed, heat is produced." It appears that Joule did not really mean it when he said that it would be "manifestly absurd" not to believe in the law of conservation of energy and that in the final analysis, he knew that the decision had to be based on experimental results.

FIGURE 6.10
"The Waterfall", M. C. Escher. A classic example of a perpetual motion machine which looks good on paper but which cannot be built. [Escher Foundation, Haags Gemeentemuseum, The Hague.]

§6.5 WHAT IS "THERMAL ENERGY"? THE BEHAVIOR OF GASES, THE IDEAL-GAS TEMPERATURE SCALE, AND "HEAT AS A MODE OF MOTION"

What *is* thermal energy? From one point of view, it is an abstraction, a number calculated in certain ways, a number that is larger when something is at a high temperature than when it is at a lower temperature. Kinetic energy and gravitational potential energy are abstractions too, ideas that we have invented to help us understand the world. On strictly logical grounds there is no need to go further. Yet somehow one feels a compulsion to ask for further interpretation, not to replace anything that has been said earlier but to supplement it. Long before Joule's experiments, it had been speculated that thermal energy (or "heat") might simply be the manifestation of disorderly motions of atoms and molecules, motions on a microscopic scale, that the atoms and molecules are jiggling back and forth with greater kinetic energies at high temperatures than at low temperatures. Perhaps a "flow of heat" between two bodies is actually a transfer of this random kinetic energy by collisions between their molecules. Let us look more carefully at this idea, "heat as a mode of motion" as it came to be called. Is it possible that thermal energy can be interpreted as the sum of all the microscopic energies of the atoms and molecules? To use the modern term, can we develop a "kinetic theory" to interpret the concept of thermal energy?

If such an interpretation is correct, then a sample of matter (a glass of water or a balloon full of air) is not just an inert lump but, on a microscopic level, a beehive of chaotic activity. Molecules may be rotating, traveling in various directions with various speeds, bouncing off each other, and bouncing off the walls of the container. Such motions must be highly random; at any given moment, if the sample as a whole is not moving, almost exactly as many molecules must be moving in one direction as in the opposite direction. The still smaller particles of which molecules are made may also be in motion within the molecules. *Potential* energies should be included too, for if two molecules attract or repel one another, there will be a potential energy* associated with this interaction. Whatever the sum of all these microscopic kinetic and potential energies may be, perhaps this is what thermal energy "really is."

There are formidable difficulties in trying to verify this idea. (1) These hypothetical energies are those of microscopic objects, molecules, or even parts of molecules. (2) There are vast numbers of objects involved. Even if we can ignore the internal structure of a water molecule, there are about 10^{25} molecules in a glass of water. (3) Everything is changing from moment to moment as the molecules collide with one another and with the walls. (4) We do not know the precise way in which two molecules interact with one another, and therefore it will be difficult, if not impossible, to take account of the potential energies resulting from the interactions between the molecules. The last difficulty is perhaps the most serious, and for this reason we will not attempt to test the kinetic theory idea for all substances. Instead, let us look specifically at gases. Potential energies arise from interactions of molecules. In a gas, the molecules are, on the average, much farther apart from each other than they are in liquids or solids, and we may hope to make some progress by simply ignoring potential energies, and concentrating on

---

*Potential energy is used as a general term to describe any sort of interaction. A rock interacts gravitationally with the earth; there is an associated gravitational potential energy. Molecules do exert gravitational forces on one another, but these forces are extremely weak. The major interaction between molecules is an electrical one; the interaction can be described in terms of a potential energy, which depends in a complicated way on the distance between the molecules. We will not attempt to treat this potential energy quantitatively, but we know that there *is* an interaction and so there is some associated potential energy.

kinetic energies, which we know how to handle. The more "dilute" the gas, the lower its density, the better this approximation should be.

We must realize, though, that in restricting our attention to gases, we are losing a great deal of generality. The first law of thermodynamics applies to *all* substances. Success in constructing a microscopic interpretation of thermal energy for the very special case of dilute gases may suggest, but certainly cannot prove, that this interpretation can be extended to all substances. Even with gases, we have the problem of dealing with enormously large numbers of molecules. It would be a hopeless task to try to analyze the separate motions of all the individual molecules. However, the very fact that the numbers are so large means that we need consider only the *average* kinetic energies of the molecules.

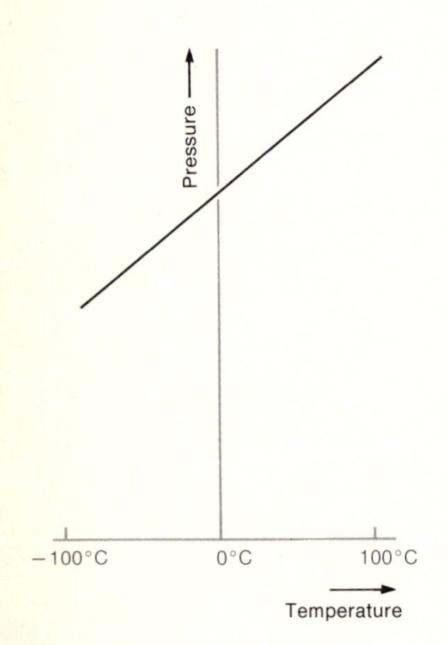

FIGURE 6.11
Relationship between pressure and temperature for a fixed volume of gas.

A gas is in many ways a simpler kind of substance than a liquid or a solid, and gases exhibit striking regularities in their behavior. If one takes a fixed quantity of gas at a fixed temperature and varies the volume, it is found that the pressure* and the volume are inversely proportional to one another. For example, if the volume is doubled, the pressure decreases by a factor of two. Furthermore, if the volume is held fixed and the temperature varied, the relationship between pressure and temperature is a linear one (Figure 6.11). If one carries out the same sort of measurement with a different quantity of gas or with a different *kind* of gas, a very striking regularity emerges, as shown in Figure 6.12. If these lines are extrapolated to lower temperatures, they all cross the horizontal axis at the same point, at a temperature of approximately −273°C. (The more precise value is −273.15°C.)

It is this fact which leads to the definition of the *ideal-gas* temperature scale (or the *absolute* or *Kelvin* scale). If we choose the zero point on this scale (0°K) to be at −273°C, then on this scale the pressure of a fixed quantity of gas kept at a constant volume is not just linearly related to the temperature but is directly proportional to the temperature (Figure 6.13). The pressure itself serves as an indicator of temperature; it is for this reason that this temperature scale is often called the ideal-gas scale. With temperatures measured on this scale (in °K), pressure, p, is directly proportional to temperature, T, if the volume is held constant:

$$p \propto T \text{ (at constant volume).}$$

It was stated above that if the temperature is held constant, pressure and volume, V, are inversely proportional to one another:

$$p \propto \frac{1}{V} \text{ (at constant temperature).}$$

*The pressure is the force per unit area that the gas exerts on the walls of its container, or the equal inward force per unit area required to contain it. In MKS units, pressure is measured in newtons per square meter (N/m^2). Pressure is often measured in other units, for instance in pounds per square inch (abbreviated "psi"), or in atmospheres (atm), with 1 atm equal to the average pressure at sea level exerted by the earth's atmosphere. 1 atm = 1.013×10^5 N/m^2 = 14.7 lb/in.^{2}

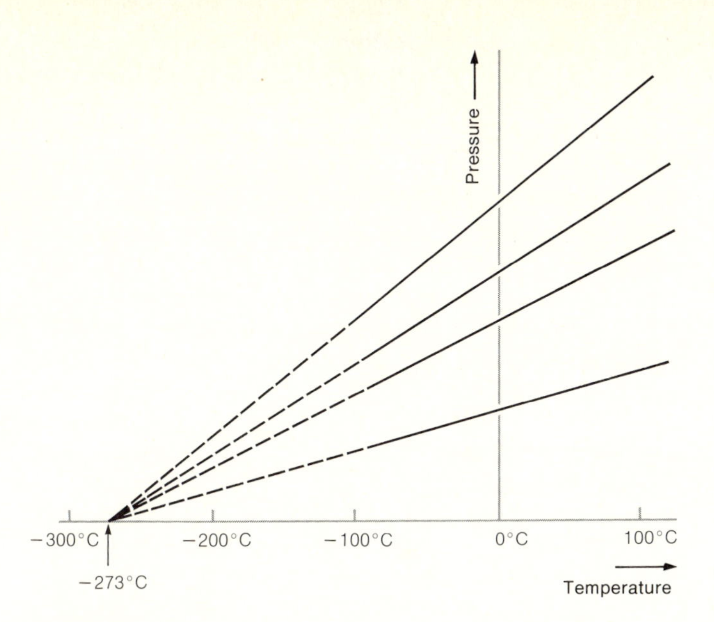

FIGURE 6.12
For various quantities of various gases, the pressure-temperature relationships are similar to those shown in Figure 6.11, and the extrapolated straight lines all cross the axis at the same point, $T = -273°C$.

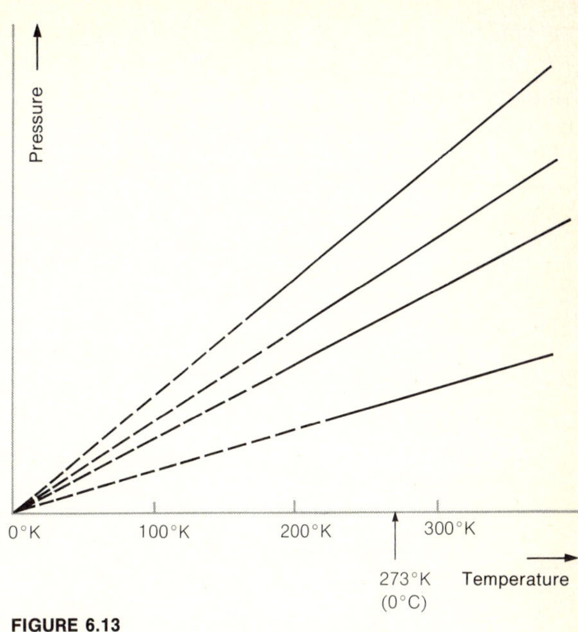

FIGURE 6.13
Data of Figure 6.12 replotted, with 0°K put at a temperature of −273°C.

When neither the volume nor the temperature is held fixed, these proportionalities can be combined into a more general one: $p \propto T/V$. An equivalent statement is that the combination pV/T is a constant:

$$\frac{pV}{T} = \text{constant (for a fixed amount and type of gas).} \qquad (6.7)$$

To a good approximation, all gases have the properties described, but precise measurements show slight deviations. The more dilute the gas, the lower the pressure, the more accurately it obeys these relationships. An *ideal* gas is defined as one having precisely the properties described; in practice one can approach this ideal-gas behavior as closely as desired by using a sufficiently low pressure.

What is the significance of the temperature of "absolute zero," $T = 0°K$? It is tempting to look at Figure 6.13 or Equation 6.7 and to say that absolute zero is the temperature at which the pressure of a fixed volume of gas is reduced to zero. However, as the temperature is made lower and lower, all real gases sooner or later begin to deviate from ideal-gas behavior, and eventually they either liquefy or solidify. In the present context, the significance of absolute zero is simply that it is the temperature at which the extrapolated lines in Figures 6.12 and 6.13 cross the horizontal axis; if temperatures are measured with respect to this point, then as long as we do not approach this temperature too closely, the relationship between pressure, volume and temperature (in degrees

Kelvin) is accurately described by Equation 6.7. The significance of a temperature of 0°K as a true zero emerges from the second law of thermodynamics. We will see in Chapter 7 how this law leads us to regard the temperature scale determined by the behavior of dilute gases as a truly "absolute" scale.

Equation 6.7 describes the relationship between pressure, volume, and temperature for a fixed amount of gas. The value of the "constant" in this equation surely depends on how much gas is present. For a given value of pressure and temperature, twice as much gas will occupy twice as much volume. A simple regularity appears when we compare various quantities of various kinds of gases. All cases can be described by writing Equation 6.7 in the following way:

$$\frac{pV}{T} = nR, \qquad \text{or} \qquad pV = nRT, \tag{6.8}$$

where R is a quantity that has the same value for all gases, and n is the quantity of gas measured in *moles.* Let us digress briefly to explain the term "mole."

The "atomic weight" of an element is the weight (or mass) of one of its atoms relative to that of hydrogen, the lightest atom: the atomic weight of hydrogen is 1, that of helium is 4, that of oxygen is 16, and so on.* A "gram-atomic-weight" of an element is a macroscopic amount of that element, an amount such that its mass in *grams* is equal to the atomic weight: 1 g of hydrogen, 4 g of helium, 16 g of oxygen, etc. It follows from these definitions that no matter what element is considered, a gram-atomic-weight contains the same number of atoms. If we know the mass of any one atom, we can determine the number of atoms in a gram-atomic-weight. The determination of the mass of a single atom is easier said than done, but it is now known that the mass of a hydrogen atom is approximately 1.66×10^{-24} g. The number of hydrogen atoms in 1 g of hydrogen is therefore the reciprocal of this number:

$$\frac{1}{1.66 \times 10^{-24}} = 6.02 \times 10^{23}.$$

One gram-atomic-weight of any element contains this number of atoms, a number denoted by N_A and called Avogadro's number:

$$\text{Avogadro's number} = N_A = 6.02 \times 10^{23}.$$

This is an important number because it allows us to relate a macroscopic description, where quantities are measured in grams or kilograms, to a microscopic description in terms of the number of atoms or molecules.

---

*This statement is not quite exact. For various reasons, the atomic weight of the most common form of carbon is defined to be exactly 12; then the atomic weights of other elements have values that are very close to exact integers. That of hydrogen, for instance, is slightly larger than 1.

QUESTION

6.3 A number such as 6×10^{23} is staggeringly large. Try to get a feeling for its magnitude by comparing it with other large numbers that you may have occasion to deal with. How does N_A compare with the population of the world? With the gross national product of the United States, expressed in dollars?

Many elements in the gaseous state exist not as isolated atoms but as molecules, often diatomic molecules. This is true of hydrogen (H_2), nitrogen (N_2), oxygen (O_2), and others, but not for the "rare gases" (helium, neon, argon, and so on). Their "molecules" are simply single atoms. The "molecular weight" is defined as the mass of a molecule relative to that of a hydrogen *atom*. The molecular weight of hydrogen is 2 and of oxygen 32, whereas for helium the molecular weight is 4, identical to the atomic weight. A "gram-molecular-weight" (abbreviated by the term *mole*) of any substance is a quantity whose mass in *grams* is equal to the molecular weight: 1 mole of hydrogen is 2 g of hydrogen; 1 mole of helium is 4 g of helium; 1 mole of oxygen is 32 g of oxygen. The chemical formula of water is H_2O; 1 mole of water is 18 g of water. Just as 1 gram-atomic-weight of any element contains the same number of atoms (N_A atoms), so 1 mole of any substance contains N_A *molecules* of that substance.

With this definition of the meaning of the term mole, let us return to the general equation (6.8) for describing the behavior of gases,

$$pV = nRT,$$

where n is the quantity of gas measured in moles, and R is a constant that has the same value for all gases, the "gas constant":

$$R = 8.314 \text{ J/mole-°K}.$$

This is the numerical value of R, with p and V expressed in MKS units (N/m^2 and m^3) and with T in °K.

QUESTION

6.4 Check these units. It looks as if the combination (pressure $\times$ volume) has the same units as does energy. Is this right?

Equation 6.8 can be written in terms of the number of molecules. Since n is the number of moles, and since each mole contains N_A molecules, the total number of molecules, N, is $N = nN_A$. Then

$$pV = nRT = \frac{NRT}{N_A}. \tag{6.9}$$

The combination (R/N_A) is a useful one, which has been given a name and a symbol of its own:

$$\frac{R}{N_A} = k \text{ (Boltzmann's constant)} = 1.38 \times 10^{-23} \text{ J/°K}.$$

With this notation, Equation 6.9 can also be written:

$$pV = NkT. \tag{6.10}$$

We have now expressed the ideal-gas equation in various forms. In the last form (Equation 6.10), we have both macroscopic quantities (p, V, and T) and a quantity that refers to the microscopic composition of the gas (N, the number of molecules), together with Boltzmann's constant, k. We should now be in position to construct a microscopic model of a gas. We will see that Boltzmann's constant, k, provides the link between the microscopic and macroscopic properties of gases. Though our model will be a simple one, and though we will have to make a number of simplifying assumptions, this model will be a useful device for helping us to understand the macroscopic behavior of a gas in microscopic terms.

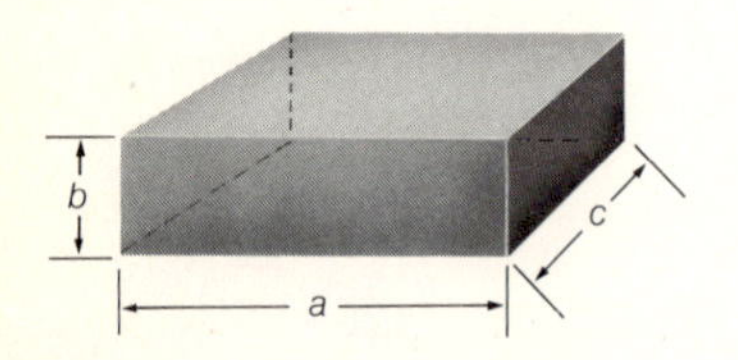

FIGURE 6.14
It is easiest to calculate the pressure exerted by a gas on the walls of its container if we choose a box of this simple shape.

We will suppose that the molecules are in motion, that they have kinetic energies, and we wish to see how, as a result, the gas exerts a pressure on the walls of its container. Is it possible that we can derive anything resembling the ideal-gas equation (Equations 6.9 and 6.10)? Suppose that the gas has a total number of molecules, N, each of which has a mass of m. We want to calculate the pressure that this gas exerts on the walls of its container. Let us assume for simplicity that the container is a box (Figure 6.14) whose edges have lengths a, b, and c. At any given instant, presumably, the molecules are distributed quite uniformly throughout the box and are moving in various directions with various speeds. A particular molecule may move in a straight line for a while, then collide with another molecule, and set out in a new direction with a new speed. Occasionally it will bounce off one of the walls, and as it bounces it will momentarily exert an outward force on the wall. It is in these outward forces exerted by all the molecules that we hope to find an explanation of the pressure exerted by the gas.

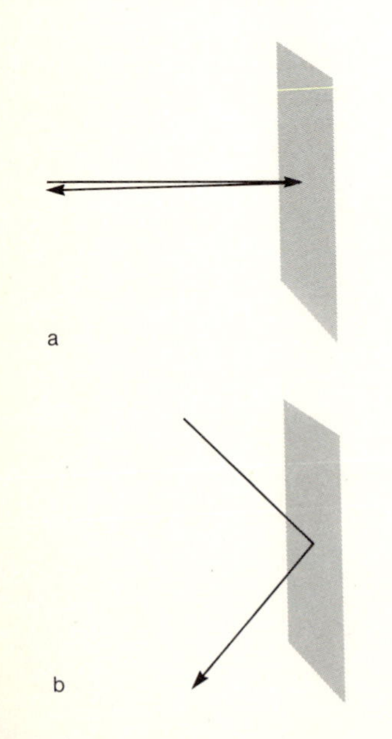

FIGURE 6.15
A head-on collision of a molecule with the wall (a) is easier to analyze than one in which the molecule strikes the wall at an angle (b).

A collision of a molecule with a wall is thus an event that we must consider with care. Such a collision is easier to treat if it is one in which the molecule strikes the wall head-on rather than at an angle (Figure 6.15). In spite of the fact that the molecules move in all directions and therefore may hit the wall at any angle, let us make the obviously incorrect assumption that of the N molecules, $N/3$ of them are traveling in each of the three directions parallel to the edges of the box, and that none of them is traveling in any other direction. This assumption makes every collision a head-on collision and greatly simplifies the calculation, yet one might reasonably suppose that such an oversimplification would at least affect some of the details of our results. One might also guess, though, that the general features of the results would not be profoundly changed. It is actually not a great deal more difficult to make the calculation without this simplifying assumption, and the results turn out to be precisely the same as those we shall derive here.

Since we are interested in the pressure on the walls, and therefore in the collisions of the molecules with the walls, let us completely ignore the collisions of the molecules with one another. Of course the molecules do collide with one another, but a little thought may make it clear that this is probably not a damaging oversimplification, as far as the force exerted on the wall is concerned. The wall does not "care" whether the molecule comes on a straight-line path from the opposite wall, or whether it acquired its velocity in a collision with some other molecule in the vicinity.

In our highly simplified model, each molecule just bounces back and forth between a pair of opposite walls (Figure 6.16), always with the same speed, v, except during the brief moments of collision. Let us focus attention on one such molecule, moving parallel to the edge whose length is a, and consider its interactions with the right-hand wall. As the molecule bounces off the wall, the wall must briefly exert a force to the left on the molecule in order to reverse its momentum; simultaneously the molecule exerts an outward force (to the right) on the wall. The clue needed for calculating this force, at least its average value, is that every time the molecule collides with the wall, its momentum changes by the amount $2mv$.

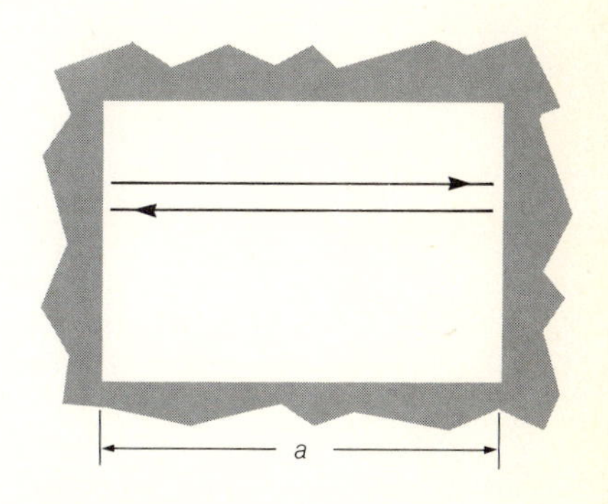

FIGURE 6.16
In our simplified model of a gas, each molecule bounces back and forth between a pair of opposite walls.

QUESTION

6.5 Why is it $2mv$? Where does the factor of 2 come from?

This collision lasts only a very short time, but while it is going on, the wall exerts a leftward force on the molecule, and the molecule exerts an equal and opposite force (to the right) on the wall. After each such collision, the molecule bounces off to the left, but if we wait until the molecule has had time to go over to the left wall, collide there, and come back again, another collision with the right-hand wall takes place. If the time between these collisions of a particular molecule with the right-hand wall is denoted by τ, then

$$\tau = \frac{2a}{v}.$$

The molecule exerts its force on the right-hand wall in a series of brief hammer blows; on the average it delivers a momentum of $2mv$ every τ sec. If there were no other forces acting on the wall, the wall's momentum would change by this amount ($2mv$) every τ sec. This is the same effect as would be produced by a steady force whose size is

$$F = \frac{\text{change in momentum}}{\text{time}} = \frac{2mv}{\tau}.$$

In other words, the average force exerted on the wall by one molecule through its successive collisions is just

$$F = \frac{2mv}{\tau} = \frac{mv^2}{a}. \tag{6.11}$$

Equation 6.11 looks promising. Notice the appearance of mass and velocity in the combination mv^2, just as in the expression for kinetic energy.

So far we have considered just one molecule, but in our model we suppose that a total of $N/3$ molecules are doing the same thing, hitting the right-hand wall, and so the total average outward force on this wall is $N/3$ times as big:

$$F = \frac{Nmv^2}{3a}.$$

The pressure on the wall is just the force per unit area:

$$p = \frac{F}{\text{area}} = \frac{F}{bc} = \frac{Nmv^2}{3abc}.$$

But since (abc) is just the total volume of the box $(V = abc)$, we can rewrite this equation as

$$p = \frac{Nmv^2}{3V}. \tag{6.12}$$

Simply by rewriting Equation 6.12, we can put it in a form that looks quite similar to the ideal-gas equation:

$$pV = \frac{Nmv^2}{3}. \tag{6.13}$$

QUESTION

6.6 We considered one particular side of the box, the right-hand wall; surely our results ought not to depend on which side we chose to consider. Analyze the problem again (this time considering collisions with the top of the box, for example), and again derive Equation 6.13.

Since $\frac{1}{2}mv^2$ is the kinetic energy of a single molecule, it is useful to rewrite Equation 6.13 as

$$pV = \tfrac{2}{3}N\left(\tfrac{1}{2}mv^2\right) = \tfrac{2}{3}N \times (KE \text{ of one molecule}). \tag{6.14}$$

We can take account of the fact that the molecules do not all have the same speed by modifying Equation 6.14 slightly, so that it reads:

$$pV = \tfrac{2}{3}N \times (\text{average } KE \text{ of a molecule}) = \tfrac{2}{3}N \times \overline{\text{KE}}. \tag{6.15}$$

(The bar above a quantity will be used to indicate an average value.)

Since there are N molecules altogether, we can also write

$$pV = \tfrac{2}{3} \times (\text{total } KE \text{ of all the molecules}). \qquad (6.16)$$

Now let us compare these results with the equations that describe the known behavior of ideal gases. In Equations 6.9 and 6.10, the ideal-gas equation was expressed in two different ways: $pV = nRT$, and $pV = NkT$. These equations are closely similar, but not identical, to Equations 6.15 and 6.16, derived from the microscopic kinetic theory. The ideal-gas equations would be identical to the kinetic theory equations if we could make the identification:

$$\text{average } KE \text{ of one molecule} = \tfrac{3}{2}kT, \qquad (6.17)$$

or, in other words,

$$\text{total } KE \text{ of the molecules} = \tfrac{3}{2}NkT = \tfrac{3}{2}nRT. \qquad (6.18)$$

At first this identification is just a reasonable guess as to how temperature and molecular kinetic energy might be related, an idea which we will try, to see what the consequences are. If this interpretation holds up (and it does), then—at least for gases—the macroscopic quantity that we sense as temperature has a microscopic meaning; it is through Boltzmann's constant, k, in Equations 6.17 and 6.18 that the precise connection is established. The molecules of a gas at high temperature have, on the average, more kinetic energy than do molecules of a gas at lower temperature; specifically, for instance, twice as much kinetic energy if the temperature (on the absolute scale) is twice as large.

The thermal energy, TE, of the gas, then, ought to be just the sum of all the molecular kinetic energies, and thus by Equation 6.18,

$$TE = \tfrac{3}{2}NkT = \tfrac{3}{2}nRT. \qquad (6.19)$$

Before we began our investigation of a detailed kinetic theory, we were already hoping to interpret the thermal energy of a gas as the sum of the molecular energies. What we have obtained from the kinetic theory is the result that thermal energy ought to be related to the macroscopic variable, T, in the specific way expressed by Equation 6.19. This is something we can test by experiment, for thermal energies can be measured by determining heat capacities, and we can use Equation 6.19 to predict the numerical values of the heat capacities of gases. The heat capacity of a gas is the amount of heat needed to raise the temperature by 1°K (or 1°C). When a gas is heated, then according to the first law of thermodynamics as applied to this situation,

$$H + W = \Delta(TE).$$

If the gas were free to expand while being heated, then it could do work on its surroundings. To avoid this complication, let us heat the gas while keeping the volume constant, and make sure that no types of energy transfer occur except for the heat flow. Then $W = 0$, so that $H = \Delta(TE)$, and if we use Equation 6.19:

$$H = \tfrac{3}{2}nR\,\Delta T.$$

If $\Delta T = 1°K$ and if we consider one mole of gas ($n = 1$), then $H = \tfrac{3}{2}R$. Thus we can say that if our kinetic theory is correct, the heat capacity* per mole of any ideal gas should be:

$$C = \tfrac{3}{2}R = 12.5 \text{ J/mole-°K}.$$

This is a theoretical prediction that we can compare with experimental data, as shown in Table 6.3. For the monatomic ideal gases, the prediction works beautifully, and provides some numerical evidence that we are on the right track. For the diatomic gases (H_2, N_2, O_2), however, something is clearly wrong. Even for the diatomic gases, a slight elaboration of the kinetic theory leads to an understanding of these data as well. We will not go into the details, but the gist of the idea is as follows. In our model the molecules were treated as if they were simply "point masses." An extended object can have kinetic energy as a result of rotation about its center (like a flywheel), in addition to the ordinary kind of kinetic energy ($\tfrac{1}{2}mv^2$) that results from the translational motion of the object as a whole. To make this distinction, we refer to the usual sort of kinetic energy as "translational" kinetic energy. The contribution of the additional kinetic energy (rotational kinetic energy) might well be important for an extended diatomic molecule such as oxygen, more so than for a single atom. It is only the translational kinetic energy which matters as far as the forces exerted on the walls during collisions are concerned, so we might expect that Equation 6.18 should be modified to read:

$$\text{total } \textit{translational } KE \text{ of the molecules} = \tfrac{3}{2}nRT.$$

TABLE 6.3 Heat capacities (at constant volume) per mole of various gases ($p = 1$ atm; $T = 300°K$).

Gas	Heat Capacity (J/mole-°K)
He	12.5
Ne	12.5
Ar	12.5
Kr	12.5
Xe	12.5
H_2	20.5
N_2	20.8
O_2	21.1

The total molecular kinetic energy, rotational kinetic energy included, ought to be larger than $\tfrac{3}{2}nRT$, and so the heat capacity per mole of a diatomic gas might well be larger than $\tfrac{3}{2}R$, as indeed it is.

There is another and very different test that we can make. We made the initial hypothesis that the molecules are in motion; now we can say how fast they are moving if the theory is right. From Equation 6.17:

$$\overline{\tfrac{1}{2}mv^2} = \tfrac{3}{2}kT, \tag{6.20}$$

so the average speed can be calculated. The speed obtained by simply solving Equation 6.20 for v (as if the bar indicating an average were not there) is called the root-mean-square speed, or "rms" speed, denoted by v_{rms}:

$$v_{rms} = \sqrt{\frac{3kT}{m}}. \tag{6.21}$$

*Strictly speaking this is the "heat capacity at constant volume," usually denoted by C_v. If the gas were allowed to expand and do work on its surroundings, more heat would be needed to raise the temperature by 1°K, so the term "heat capacity" requires qualification.

For example, for helium ($m \simeq 6.64 \times 10^{-27}$ kg) at room temperature ($T \simeq 300°K$),

$$v_{rms} \simeq \sqrt{\frac{3 \times 1.38 \times 10^{-23} \times 300}{6.64 \times 10^{-27}}} \simeq 1370 \text{ m/sec.}$$

QUESTIONS

6.7 In this calculation, the mass of a helium atom was expressed in kilograms; the result came out in meters per second. Check to make sure that the units have been treated correctly.

6.8 Calculate the approximate rms speed of oxygen molecules at room temperature.

6.9 How do these molecular speeds compare with other high speeds such as the speed of an airplane, the speed of sound, the speed of light, or the "escape velocity" (§5.6)?

We can hardly test these predictions for the molecular speeds directly because atoms are too small to watch. (If this were not so, "heat as a mode of motion" would doubtless have been accepted many centuries ago.) By various indirect methods, molecular speeds have been measured, and the predictions of Equation 6.21 have been amply confirmed. It is found, though, that as we would expect, the molecules do not all have the same speed. The speed of a molecule changes rapidly from moment to moment as it collides with other molecules. At any given instant, there is a distribution of speeds, as shown in Figure 6.17, some molecules moving more slowly and others faster, and the rms speed of the whole collection of molecules is correctly predicted by Equation 6.21. This equation is valuable for calculating a typical round number for the speeds of the molecules. In some cases, it is important to remember that the molecules do not all have the same speed. To cite one contemporary example, present plans to generate energy by means of nuclear fusion (Chapter 15) depend on the fact that a few molecules in a hot gas have speeds much greater than the rms speed.

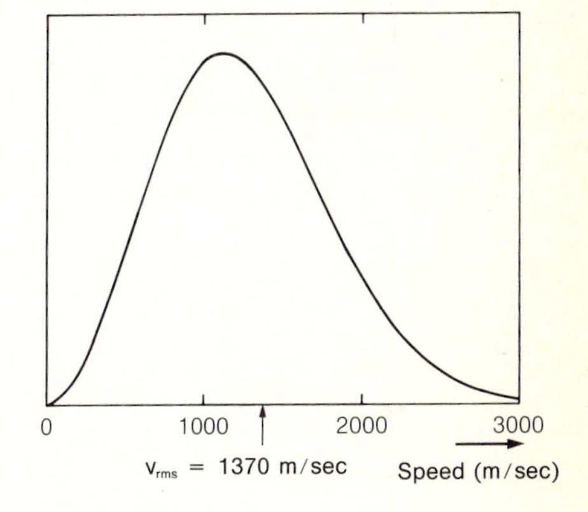

FIGURE 6.17
Distribution of speeds of helium atoms at a temperature of 300°K.

§6.6 SUMMARY

Before the synthesis of heat and mechanics described in this chapter, both subjects had their own conservation laws, each useful but of limited validity. Mechanical energy is a quantity which is often conserved but frequently is not; heat (or "caloric fluid") is something which acts like a conserved quantity in certain experiments but not in all. Both of these conservation principles of limited validity turned out to be special cases of a universally valid law. According to the generalized law of conservation of energy, mechanical energy *is* conserved in those experiments in which other forms of energy (notably thermal energy) do *not* change; the predictions of the caloric theory (heat lost by one object is equal to the heat gained by another) *are* correct in just those experiments in which energy is transferred *only* by heat and in which thermal energies are the only ones which change. These simple processes, the more complex ones investigated by Joule, and indeed all processes, are correctly described by the generalized law of conservation of energy. The first law of thermodynamics, the law of conservation of energy, gives no clues as to what types of energy conversions *do* occur, which kinds are possible and which are impossible. This important matter is dealt with by the second law of thermodynamics, to be discussed in Chapter 7. In addition to seeing how the unification of heat and mechanics led to a general law of conservation of energy, we have also seen in §6.5 that "thermal energy" can be understood as the sum of the energies of the randomly moving atoms and molecules of which matter is composed.

FURTHER QUESTIONS

6.10 If a block of copper at an initial temperature of 100°C and a block of aluminum at 0°C (each of mass 500 g) are brought into contact, what will be the final equilibrium temperature? (Assume that no heat flows to or from the surroundings.) How much heat flows out of the copper and into the aluminum?

6.11 One quart of boiling water ($T = 100°C$) is poured into an aluminum saucepan ($T = 20°C$). Predict the final common temperature of the pan and the water, on the assumption that no heat flows to or from the surroundings. (Make your own estimate of the mass of a saucepan, and state the value you use.)

6.12 A block of ice ($m = 10$ kg) has an initial temperature of $-10°C$. If it is set out in the room ($T = 20°C$), heat flowing into the ice first raises the temperature to 0°C, then causes the ice to melt, and then raises the temperature to 20°C. How much heat is required for each of these three steps? Which of the three amounts of heat is the largest?

6.13 (a) A glass of water (a half pint, about half a pound) has an initial temperature of 20°C. How much ice (at 0°C) should be added so that just as the last ice melts, the temperature of the water reaches 0°C? (Assume that there is no flow of heat from the surroundings.)

(b) About how many ice cubes is this? (The density of ice is approximately the same as that of water.)

6.14 A calorie-conscious weight-watcher might well prefer cold soup to hot soup; equal amounts of hot and cold soup have the same "food energy" (chemical energy), but the hot soup has more thermal energy. Is this a significant effect? Consider a 1-cup serving of tomato soup (about 0.5 lb). If the heat capacity of the soup is approximately equal to that of water, how much greater is the thermal energy of a serving of hot soup ($T \simeq 150°F$) than of cold soup ($T \simeq 50°F$)? How does this difference compare with the food energy of the soup (about 100 kilocalories)?

6.15 (a) If you drink cold water, it will soon warm up to body temperature (98.6°F), and this takes energy. Therefore you ought to be able to keep your weight down even if you eat all the rich food you wish, provided that you also drink lots of cold water. How much ice-cold water should you drink to counteract the effects of eating one chocolate sundae (about 500 kcal)?

(b) An even better plan is to eat ice cubes, to take advantage of the latent heat of melting. How much ice should you eat to offset one sundae?

6.16 The "ton" is an old-fashioned unit used to describe the capacity of an air conditioner. A "1-ton" air conditioner can remove heat at the same rate as would the melting of 1 ton of *ice* per *day*. What is this rate in watts?

6.17 Joule pointed out that the water at the bottom of a waterfall should be hotter than the water at the top. The water's initial gravitational potential energy is first converted to kinetic energy as it falls, but at the bottom it is "lost," while the temperature of the water increases. Is this effect big enough to measure? Estimate what the temperature rise should be at Niagara Falls. (What do you have to know about Niagara Falls in order to make this calculation? Its height? The amount of water going over the falls every second? Are there any special assumptions you have to make in order to make such a calculation? Are your assumptions likely to be correct in practice?)

6.18 In a laboratory experiment, a student is given a long hollow tube, closed at both ends, containing a handful of lead pellets. The tube is held vertically, and then suddenly inverted so that the lead falls the length of the tube. This process is repeated until the student gets tired, the lead is then removed and its temperature measured. Discuss qualitatively the various energy conversion processes involved in this experiment. Estimate how much the temperature of the lead will rise if the tube is 100 cm long and if the tube is inverted 50 times. Could you *melt* lead this way?

6.19 Hammering on a nail produces an increase in its temperature. Estimate the maximum increase in temperature of an iron nail that could be caused by hitting it once.

6.20 Estimate the amount by which the temperature of a 2- or 3-pound chunk of iron might rise, if you hammer on it as hard as you can for about 5 minutes. Does your result seem reasonable? Discuss the simplifying assumptions that must be made in order to answer this question.

6.21 When a lead bullet smashes into a wall, all of its kinetic energy is converted into thermal energy. Might there be enough thermal energy to melt the lead? Estimate the minimum speed that a bullet must have in order that it might be melted in this way. Does the result depend on the mass of the bullet?

6.22 Suppose that you put a pan of water on the stove and that the power goes off before the water boils, when its temperature has only reached 80°C. Perhaps you can still get the water to boil by stirring it vigorously. Try to make a rough estimate of how long you might have to stir in order to bring its temperature up to 100°C (assuming that no heat is lost to the room in the meantime). Do you think this idea will work?

6.23 Speculate on the ways in which our daily lives might be altered if instead of 1 cal = 4.184 J, the mechanical equivalent of heat were:

(a) 1 cal = 4000 J;

(b) 1 cal = 0.004 J.

6.24 It was speculated in the text that Joule hoped that his experiments would reveal a fixed equivalence between amounts of heat and work. Is this a risky attitude for scientists to take toward their work? Would it be better if all scientists were completely open-minded and had no such preconceptions?

6.25 (a) The Rankine (R) temperature scale is an "absolute Fahrenheit" scale, just as the Kelvin scale is an "absolute Celsius" scale; that is, a Rankine degree has the same size as a Fahrenheit degree, but absolute zero is given the value 0°R. What are the temperatures of the freezing and boiling points of water on the Rankine scale?

(b) Table A.11 provides equations that can be used to convert temperatures from one scale to another. Derive similar equations for converting temperatures from °K to °R and from °F to °R. Test your equations by trying them for the temperatures of absolute zero and of the freezing point of water.

6.26 Equation 6.10 can also be written in the form $p = (N/V)kT$. (N/V) is the "particle density," the number of molecules per unit volume. Show that at common temperatures and pressures (say, $T = 300$°K and $p = 1$ atm), the particle density of a gas is approximately 2.5×10^{19} molecules/cm^3.

6.27 A room with a volume of 30 m^3 is completely filled with oxygen, at a pressure of 1 atm and a temperature of 300°K. How many moles of O_2 are there in the room? How many O_2 molecules? How many oxygen atoms? What is the mass of this much oxygen in kilograms? in pounds?

6.28 By what factor does the average kinetic energy of helium atoms increase as the temperature is increased from 300°K to 1200°K? By what factor does the rms speed increase?

6.29 For a gas of helium atoms at 300°K in a cubical box 10 cm on a side, how large is the average force exerted by a *single* atom, as given by Equation 6.11?

6.30 One mole of water has a mass of 18 grams. What is the heat capacity of water in joules/mole-°K? How does this value compare with the heat capacities of gases given in Table 6.3?

6.31 Another unsuccessful explanation of the source of the sun's energy. The possibility that the energy radiated by the sun comes from the gravitational potential energy lost during a gradual decrease in the size of the sun was discussed in §5.7. This idea does not work; too little energy is available from this source. Another possibility is simply that the sun was once hotter than it is now, that it is steadily getting cooler, and that the energy it emits is accounted for by a decrease in its thermal energy. We can test this idea by estimating how much hotter the sun must have been at a time—say, 500 million years ago—when life on earth had already been established.

(a) The sun is primarily composed of hydrogen, and the heat capacity of hydrogen is given in Table 6.2. Using this value and the mass of the sun (2×10^{33} g), show that if the sun is cooling off and losing energy at the rate of 4×10^{26} W, its average temperature is decreasing at the rate of about 2×10^{-8}°C/sec.

(b) Show that, if this decrease has been going on at a steady rate, the sun's average temperature 500 million years ago was about 3×10^{8}°C (or °K) higher than it is now.

Note: The interior temperature of the sun is now "only" about 1.5×10^{7}°K. (The surface temperature is much lower, about 5800°K.) It is difficult to believe that life could have existed if the sun's temperature were 3×10^{8}°K higher than it is now. It is true that heat capacities do vary with temperature, and it is the room temperature value of the heat capacity of hydrogen that has been used in this calculation, but this is not a large enough effect to alter the essential conclusion. We cannot reconcile the proposed explanation of the sun's energy with the geological evidence for the early existence of life.

7
The Second Law of Thermodynamics

T. W. Tenney

7
The Second Law of Thermodynamics

Humpty Dumpty sat on a wall.
Humpty Dumpty had a great fall.
All the king's horses,
And all the king's men,
Couldn't put Humpty together again.

§7.1 ORDER AND DISORDER. THE DIRECTION OF TIME

The total energy of a closed system is conserved. This is the essential content of the first law of thermodynamics, the law of conservation of energy: it is possible to define various forms of energy in such a way that their sum does not change if the system is closed. Alternatively, the first law of thermodynamics asserts that "perpetual motion machines of the first kind" cannot be constructed, that there is no conceivable device from which a net amount of energy can be extracted without a compensating change in the internal state of the device. *Thermal* energy was especially important in the development of the first law of thermodynamics, because it was the recognition of an equivalence between thermal energy and other forms of energy which was the key step in arriving at the law. Within the framework of the first law, though, all forms of energy are equivalent. Conversions of energy of all sorts are allowed by the first law. The first law tells us only that, in a closed system, if any one form of energy increases, there must be a corresponding decrease in some other form of energy. For a system that is not closed, a transfer of energy to or from the system (heat or work) is accompanied by a corresponding change in the total energy of the system.

§7.1 ORDER AND DISORDER. THE DIRECTION OF TIME

The second law of thermodynamics is a rather different sort of law. The second law distinguishes between different forms of energy and establishes very important restrictions on the types of energy conversions that can occur. That is, of all the conceivable energy conversions permitted by the first law, only certain ones actually take place. As an introduction to the second law, consider a number of statements (some quite commonplace), which may not seem, at least at first, to be interrelated.

(1) Almost any movie appears absurd if it is run backwards. Think of something as simple as a movie of a ball striking a pane of glass. As seen in reverse, all the various scraps of glass rise up and join together, and the glass is miraculously repaired as the ball emerges. Think of a diver emerging feet first from a lake, rising smoothly to the diving board, leaving a perfectly calm surface behind. Even a movie of a pendulum is ridiculous if viewed backwards; a pendulum at rest spontaneously begins to move, and swings through a larger and larger angle as time goes on. If we could measure the temperature of the surrounding air, we would observe the air getting slightly cooler as the pendulum's motion increases. There would be no violation of the law of conservation of energy in such a process—only a conversion of thermal energy of the air into kinetic energy and gravitational potential energy of the pendulum—but this process simply does not happen.

(2) It is easy to mix whiskey and water or to dissolve salt in water, but unmixing, turning salt water into fresh water, is much more difficult.

(3) Why is it so easy to break an egg, but so difficult, even for all the king's horses and all the king's men, to put it together again? No sophisticated technology is needed to destroy life, that we have been able to do for millennia. Why have we not been able to improve much on nature's methods for creating life? Is it that there are so many ways of being dead and so many ways in which an egg can be broken, but only a few highly organized ways in which something can be alive and in which an egg can be whole?

(4) Put two objects with different temperatures in contact; heat will flow from the hotter one to the cooler and never the other way around. As expressed by Flanders and Swann in the song quoted at the beginning of Chapter 6: "Heat won't pass from the cooler to the hotter; you can try it if you like, but you'd far better notter." Why not? A flow of heat in either direction would be perfectly consistent with the first law of thermodynamics.

(5) It is quite easy to make sure that a cold bottle of beer and the air in the kitchen reach the same temperature: unplug the refrigerator and wait. The reverse is not true; in order to produce a temperature difference between a bottle of beer and the air in the kitchen, you must plug in the refrigerator, using electrical energy for which you must pay.

(6) If energy is conserved, why is there an energy crisis? There is a lot of thermal energy in the oceans; why not let them cool off a bit and use the energy to run our cars and elevators?

(7) For that matter, since energy is conserved, why don't we simply "recycle" energy the way we recycle steel and aluminum? How can there possibly be an energy crisis?

(8) It is easy to convert mechanical or electrical energy into thermal energy, and to do so with an "efficiency" of 100%; just drop a rock or use electrical energy to run a stove. Conversion in the opposite direction is much more difficult. Steam-electric power plants are designed to convert thermal energy into electrical energy, but they are complicated devices, and they only produce a partial conversion of thermal energy into other forms. Is this simply the result of friction or poor design, or is there some fundamental reason for the fact that their efficiencies are less than 100%?

There are two themes which are common to many of the observations just made. First, time is not symmetrical. There is a profound difference between past and future. A wide variety of very ordinary occurrences would be astonishing if they were seen going backwards. In particular, energy conversions that take place easily and spontaneously in one direction just do not go in the reverse direction. A second theme, closely related to the first, is that of all the forms of energy, thermal energy is different from all the rest. Conversion of other forms of energy into thermal energy goes on all the time; conversion of thermal energy into other forms can be accomplished only with difficulty, and never completely. The first law is of no help in understanding these facts. On the basis of the first law, there is no reason why a pail of water set outdoors on a winter night should not extract thermal energy from its surroundings and get hotter, but this does not happen. As a book slides along a table, its kinetic energy is converted into thermal energy by friction and it slows down. It would be perfectly consistent with the first law if thermal energy were to be converted into kinetic energy, the book increasing in speed and the table cooling off at the same time.

We can begin to understand these facts by remembering that thermal energy is the manifestation of the random disorderly motions of vast numbers of atoms and molecules, that thermal energy is nothing but the sum of these microscopic energies. By contrast, the kinetic energy of a macroscopic object, such as a ball or a car, is the result of the *orderly* motion of all the atoms, a common motion with the same velocity for all, a common motion that is of course superimposed on the random thermal motions of the atoms within the car. Any large collection of things, whether books in a library or atoms in a ball, has a strong tendency to change in the direction of increasing disorder, increasing chaos. There are so many more ways of achieving disorder. In a collection that is in nearly perfect order, almost any random change will increase the amount of disorder. If the books on the shelves of a library are totally mixed up, it is very unlikely that a blindfolded person moving books at random will get them arranged in any sensible orderly way, because there are so few ways for them to be in order and so many ways for them to be disordered.

In qualitative terms, the second law of thermodynamics is simply a recognition of the universal tendency of order to degenerate into disorder. A quantitative measure of disorder has been developed, called *entropy*. A more precise definition of this term is unnecessary for the purpose of this discussion, but in many cases we can readily distinguish one state as being more disorderly than another, and we shall describe the more disorderly one as having the greater entropy. A warm ball at rest is more disorderly on a microscopic scale than a cool ball that is moving. Although the total energies of the two balls (kinetic energy plus thermal energy) may be equal, the atoms in the moving cool ball are all moving in the same direction, whereas there is no pattern to the chaotic internal motions of the atoms in the stationary warm ball. The warm ball has the greater entropy. The second law states that there is a universal tendency for entropy to increase, that in a closed system all the changes, all the energy conversions, are ones that increase the entropy, and that a closed system tends toward a condition of maximum entropy.

This is the most fundamental statement of the second law of thermodynamics: the total entropy of a closed system never decreases. There are other statements, which can be shown to be equivalent to this one, which can be used to give the essential content of the second law. These alternative statements describe the impossibility of converting thermal energy completely into other forms and the fact that heat flows from hot to cold, not the other way around.

Energy is an abstraction, created by human minds to help us perceive regularities in natural phenomena. Entropy is even more of an abstraction, more difficult to deal with. It is not far wrong to think of entropy as a synonym for "degree of disorder," a statement that of course begs the question of how to define order and disorder in each particular case. Entropy is definitely not the same thing as energy; in a closed system, the total energy neither increases nor decreases, but entropy obeys no such law. It can and does change, and its changes are always increases, never decreases. Entropy has often been called "time's arrow," because with its aid we can distinguish past and future. If we are given descriptions of two states of a closed system, we can tell which state preceded the other in time; the state with the smaller entropy, the state with the smaller amount of disorder, is the earlier one. In a closed system, the entropy tends to increase toward a maximum value. Then everything will be at the same temperature, and no further energy conversions will take place. This is the so-called "heat death" of a closed system.

QUESTION

7.1 Think of some other examples of phenomena that would appear very strange if seen in a movie being run backwards. Is it possible to describe some of these strange events as ones in which there is a decrease in disorder (a decrease in entropy)? Can you think of any phenomena that would *not* seem peculiar if observed in reverse?

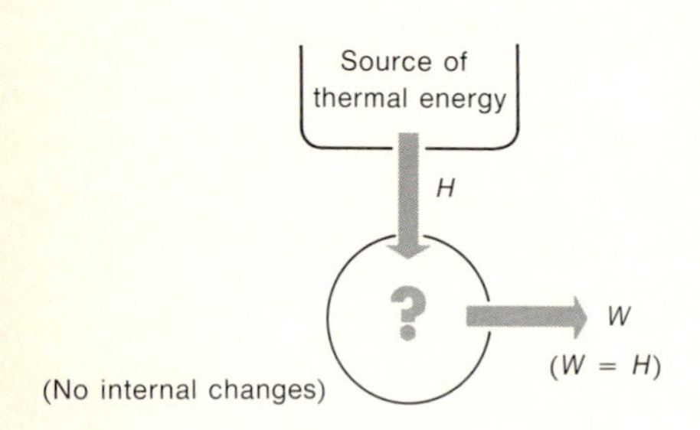

FIGURE 7.1
A perpetual motion machine of the second kind.

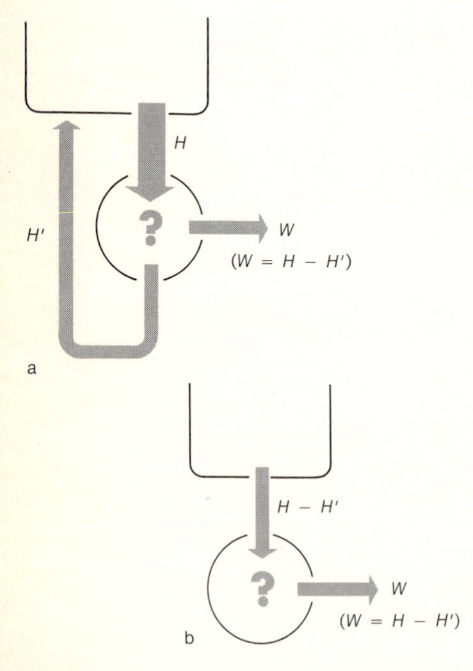

FIGURE 7.2
(a) An impossible heat engine in which the waste heat is put back into the original heat source. (b) Rearrangement of part (a), demonstrating that as far as flows of energy are concerned, it is equivalent to the impossible machine shown in Figure 7.1.

§7.2 HEAT ENGINES. THE THERMODYNAMIC LIMIT ON THE EFFICIENCIES OF HEAT ENGINES

Thermal energy is different from other forms of energy. It is the manifestation of disorderly chaotic atomic and molecular motions, but we do not need even to be aware of the existence of atoms to realize that thermal energy has its own special features. It is easy to convert kinetic energy or gravitational potential energy completely into thermal energy; just drop a rock. The reverse is not possible; there is simply no way of converting thermal energy completely into other forms. Once energy has been converted into thermal energy, it cannot be completely "recycled." The conversion of other forms of energy into thermal energy is in fact a "degradation" of energy.

It would be wonderful if this were not true. Suppose we had a device that *could* convert thermal energy completely into other forms. We could then take advantage of the thermal energy of the oceans in the way shown schematically in Figure 7.1. Heat would flow into the device, the oceans would cool off slightly, the thermal energy entering the device would be converted into some other form and delivered to the outside world as a flow of nonthermal energy, indicated by W. Such a device, which could completely convert thermal energy into nonthermal energy (or "completely convert heat into work"), would be a perpetual motion machine of the *second* kind; it would not be a perpetual motion machine of the *first* kind because as long as $W = H$, the *first* law of thermodynamics would be satisfied.* It would provide a means of decreasing the total amount of disorder, decreasing the total entropy, by converting disorderly thermal energy into more orderly forms. Although many ingenious designs for such machines have been proposed, none has been successful, and with a high degree of confidence, we can assert that none ever will be successful. An alternative way of stating the second law of thermodynamics is simply to say that perpetual motion machines

*There are two ways of indicating the *direction* of an energy transfer. One is to let W and H represent positive quantities for energy transfers into the system and negative quantities for energy flowing out of the system; this procedure has been used in previous chapters in writing the general conservation of energy equation: $W + H = \Delta E$. The second is to let W and H simply represent the *magnitudes* of the energy transfers and to indicate the direction with a diagram; the second method is followed in this chapter.

of the second kind cannot be built. It is impossible to build a device that will have the sole effect of extracting heat from an object and converting it all into "useful work," that is, into nonthermal energy delivered to the outside.

QUESTION

7.2 Estimate the "environmental impact" of such a device. The world uses energy at the rate of about 240 mQ/yr. At what rate would the average temperature of the oceans of the world decrease if energy were extracted at this rate. (Actually, there would probably be only a temporary local cooling, since the energy could be put back into the oceans after we had used it.)

A "heat engine" is a device for extracting thermal energy from something and converting part of this energy into other forms; a steam-electric power plant is an example. According to the second law, no heat engine can have an efficiency of 100%; a fraction of the energy can be converted into other forms, but some must be left over. This cannot be put back into the original source of heat, a possibility suggested in Figure 7.2, for if this were possible, we would have a device just like the impossible one shown in Figure 7.1, one in which a net amount of energy equal to $(H - H')$ is extracted and completely converted into other forms. The energy left over must be put into some other object at a different temperature, as shown in Figure 7.3. In order to convert thermal energy even partially into other forms, *two* "heat reservoirs" at *different* temperatures are required. A single reservoir may contain a great deal of thermal energy (as do the oceans), but it is energy that is unavailable for such purposes as generating electrical energy or doing mechanical work. "Energy," according to the dictionary, is "the capacity for doing work"; at best this is a half-truth, one which is incorrect in this case.

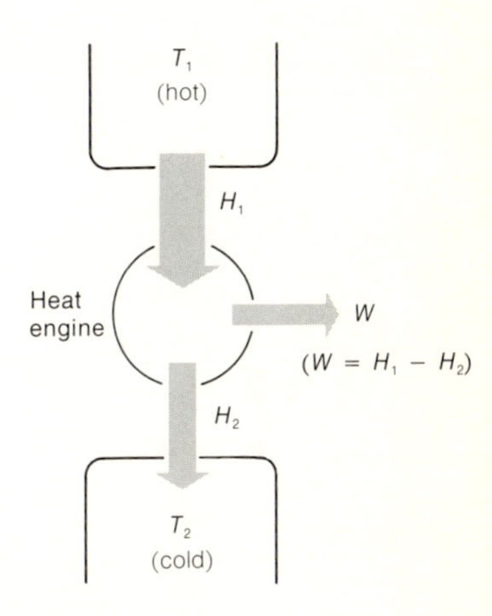

FIGURE 7.3
A possible heat engine (a steam-electric power plant, for example) with two reservoirs at different temperatures.

If we do have two reservoirs available, it is then important to ask to *what extent* it *is* possible to convert heat into useful work, and we shall explore this aspect of the second law of thermodynamics in this section. Even in an ideal heat engine with no friction at all, H_2 (in Figure 7.3) cannot equal zero; that is, W must be less than H_1. H_1 represents a flow of heat out of the hot reservoir. In a steam-electric power plant, this energy is originally derived from the chemical energy of the fuel being burned or (in a nuclear plant) from the nuclear energy of a nuclear "fuel". How much of this can be obtained in the form of useful nonthermal energy? That is, for a specified value of H_1, what is the largest possible value of W? The answer must depend on the temperatures of the two reservoirs. If they are at the same temperature $(T_1 = T_2)$, then the two reservoirs are in effect just one large reservoir; in this limiting case, W must be zero, because if it were not we would have just the perpetual motion machine of Figure 7.1. The greater the difference in temperature, the greater the fraction of H_1 that can be converted into work.

Using the second law of thermodynamics, we can show, by an argument to be outlined below, that the best we can possibly do is given by the following expression:

$$\text{Maximum possible value of } W = H_1\left(1 - \frac{T_2}{T_1}\right) = H_1\frac{(T_1 - T_2)}{T_1}, \quad (7.1)$$

where T_1 and T_2 are the *absolute* temperatures of the two reservoirs. We can express this as an efficiency (denoted by ϵ) by writing the ratio of the work to the heat delivered by the hot reservoir:

$$\epsilon = \frac{W}{H_1};$$

$$\text{Maximum possible value of } \epsilon = 1 - \frac{T_2}{T_1} = \frac{T_1 - T_2}{T_1}. \quad (7.2)$$

Equation 7.2 gives the "thermodynamic limit" on the possible efficiencies of heat engines, or the "Carnot efficiency" of an *ideal* heat engine (after the man whose work played a major role in this development). According to these equations, the efficiency is zero if $T_1 = T_2$, as anticipated. For the best possible efficiency, T_2 (the temperature of the cold reservoir) should be as low as possible, and T_1 should be as high as possible, in order to minimize the ratio T_2/T_1. However, unless T_2 is equal to 0°K (and no such reservoirs are available), the maximum possible efficiency is always less than 1. Of the heat extracted, only a fraction, $\epsilon = 1 - T_2/T_1$, is available for doing work. The temperature of the cold reservoir, T_2, is not likely to be less than the temperature of the environment, 300°K or so (unless we refrigerate the reservoir, a process that itself requires energy). The usefulness of a certain amount of heat H_1, then, depends on the temperature, T_1, of the object from which it comes: the higher the better. Heat from a high-temperature reservoir is "high-grade" energy, a significant fraction of which can be converted into, say, electrical energy. The same amount of energy coming from a low-temperature reservoir, perhaps only slightly above the temperature of the environment, is low-grade energy. A flow of heat is inevitably accompanied by a flow of entropy, but heat from a high-temperature reservoir carries a smaller amount of entropy than does an equal amount of heat from a low-temperature reservoir. This is the fundamental sense in which the terms "high-grade" and "low-grade" have been used. In this sense the performance of work (by mechanical work or by an input of electrical energy, for example) represents transfer of "pure energy," uncontaminated by any accompanying transfer of entropy; this is the fundamental distinction between heat and work, the two general ways in which energy is transferred. Notice that the normal processes of heat flow are always ones that tend to reduce temperature differences and thus to degrade energy, to make energy less available for doing work. (Sometimes low-grade thermal energy is all that we want. A reservoir of water at 150°F would not be very good for generating electrical energy but might be very convenient for keeping a house warm.)

Equations 7.1 and 7.2 are extremely important for understanding the implications of the second law of thermodynamics. The line of reasoning which leads to these equations will not be given in detail; in brief, the argument is as follows. We assume to begin with that two temperature reservoirs are available, with temperatures T_1 and T_2. Since we are interested in the maximum possible efficiency of a heat engine operating between these two reservoirs, we assume that the device itself, whatever pistons, valves, and so on it may contain, has no imperfections, that its parts are completely frictionless, for any friction in the engine would generate additional amounts of thermal energy and cause a corresponding decrease in the useful output. Then it is shown that if there were two such engines with different efficiencies, we could use them together to make a single device that would be a perpetual motion machine of the second kind; according to the second law, this is not possible, and we therefore conclude that all such engines, when operated between the same two temperatures, have the *same* efficiency. This means that it is only necessary to perform the calculation for any one such engine.

The simplest engine to analyze is one that uses an ideal gas as its "working substance," because the behavior of such a gas is well known. The operation of such a heat engine is shown in Figure 7.4. The gas is first brought into thermal contact with the hot reservoir (T_1). As heat (H_1) flows into the gas, the gas expands (Figure 7.4a), doing work on the piston. In a further expansion (Figure 7.4b), the gas does more work on the piston. In the process, the thermal energy of the gas decreases; the amount by which the gas expands can be chosen so that its temperature drops until it is exactly equal to T_2. To complete the cycle, the gas must be returned to its original state. First the gas is compressed (Figure 7.4c); work is done *on* the gas by the piston while heat, H_2, flows from the gas into the cold reservoir. The temperature of the gas during this compression is lower than it was originally; thus the pressure it exerts is smaller, and the work that must be done on it is less than the work the gas did on the piston during the initial expansion. The gas is then decoupled from the cold reservoir, and in a final compression (Figure 7.4d) its temperature is raised to T_1, and the cycle is complete. The work done on the piston during the expansion stages (Figure 7.4a,b) is greater

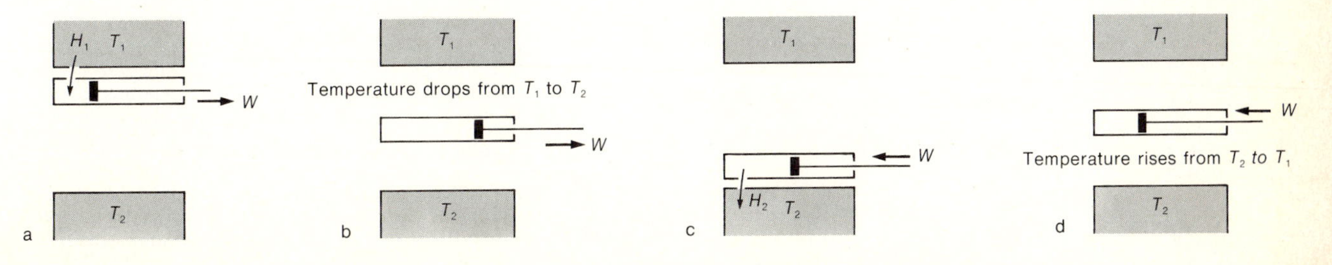

FIGURE 7.4
The sequence of steps in one full cycle of a simple heat engine, which uses an ideal gas as the working substance. A net amount of work is delivered to the outside by means of the piston rod.

than the work the piston must do on the gas to compress it (c,d). Consequently, in the course of a complete cycle, a net amount of work, W, is done *on* the piston, and this energy can be used to raise a weight or drive a motor. The behavior of an ideal gas is simple enough that the value of W can be calculated; Equation 7.1 gives the result of this calculation.*

The efficiency given by Equation 7.2 is the maximum efficiency with which thermal energy from a heat source at a temperature T_1 can possibly be converted into other forms of energy, when a cold reservoir at temperature T_2 is used. Until the middle of the 19th century, when these ideas were developed, it was thought (by Joule, for one) that with sufficient ingenuity and sufficient care in reducing friction, efficiencies close to 100% might be obtained. As a matter of fact, it was the very practical goal of making better heat engines that motivated the analysis described above. Now we know that the second law of thermodynamics establishes definite limits on the efficiencies that can be achieved. Although the Carnot efficiency is one that cannot be exceeded, even with perfect and completely frictionless machinery, it is all too easy to achieve lower efficiencies. A "heat engine" with an efficiency of 0% can be obtained by connecting two reservoirs with a metal bar and allowing heat to flow by conduction; all the heat removed from one reservoir flows into the other and none is converted into work. Any efficiency between 0% and the Carnot efficiency is possible; any departure from ideal conditions in a heat engine will give an efficiency smaller than the Carnot efficiency.

The thermodynamic limit on the efficiencies of heat engines is highly relevant to the operation of a modern power plant. Figure 7.5 shows, in schematic form, the operation of a steam-electric power plant fueled by coal, gas, or oil. The fuel is burned in the combustion chamber, its chemical energy being converted to thermal energy, thermal energy that maintains the temperature of the combustion chamber at T_1. Water is the working substance that circulates through the plant, as liquid in some

---

*It is now possible to see why the ideal-gas temperature scale is also an "absolute" scale. The Carnot efficiency of an ideal heat engine is expressed by a very simple equation if temperatures are given on the ideal-gas scale: $\epsilon = 1 - \frac{T_2}{T_1}$. We could use this equation itself to establish a temperature scale by determining the efficiency of an ideal engine operated between the two reservoirs and defining the temperatures of the two reservoirs so that $\epsilon = 1 - \frac{T_2}{T_1}$. Since the Carnot efficiency does not depend on the properties of the working substance but only on the temperatures of the reservoirs, such a temperature scale can properly be called an "absolute" scale, and it is identical to the ideal-gas scale. The temperature of absolute zero is unobtainable, but its thermodynamic significance can be understood from the same equation. It is the temperature of a hypothetical cold reservoir that would permit us to design a heat engine with an efficiency of 100%. $\left(\text{If } T_2 = 0, \text{ then } \epsilon = 1 - \frac{0}{T_1} = 1.\right)$ From a microscopic point of view, atomic and molecular motions become progressively smaller as the temperature is lowered. At a temperature of absolute zero, these random motions would have their smallest possible values. We now know, however, that according to quantum mechanics, these motions would not completely cease even if this temperature could be reached.

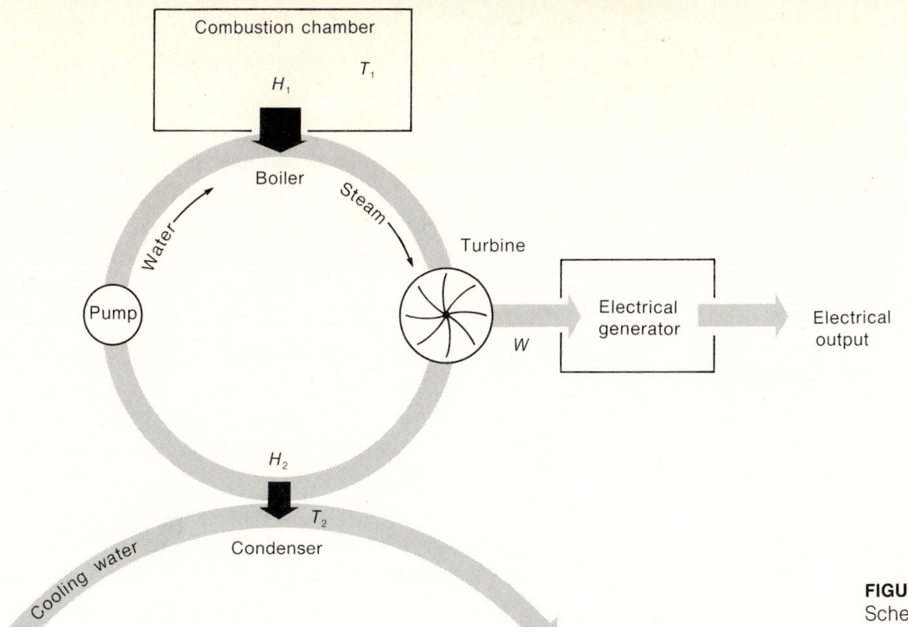

FIGURE 7.5
Schematic diagram of a steam-electric power plant.

parts of the system and as steam in other parts. In the boiler, the water receives heat from the combustion chamber and is converted into high-pressure steam. In the turbine, the steam pushes against the blades of the turbine, doing work on the turbine (Figure 7.6). This work is delivered to an electrical generator, where electrical energy is produced for delivery to the consumers. The water leaves the turbine as very low-pressure steam, at a temperature much lower than when it entered the turbine. In order for the water to complete the cycle, it must be condensed into liquid, and this is done by allowing heat H_2 to flow out into an auxiliary stream of cooling water from a nearby river. This is the "waste heat," the local "thermal pollution," produced by any steam-electric power plant. (The waste heat, H_2, could be delivered to a stream of air, but there must be *some* cooler substance into which this heat can be put.)

For the best possible efficiency, T_2 should be as low as possible and T_1 as high as possible. We have little control over T_2; we have to take whatever water or air temperature is available, and T_2 will probably be in the vicinity of 20° or 25°C (about 300°K). In a modern power plant, T_1 might be as high as 500°C (773°K), and therefore the maximum possible efficiency is

$$\epsilon = 1 - \frac{T_2}{T_1} \simeq 1 - \frac{300}{773} \simeq 0.6,$$

that is, a maximum possible efficiency of about 60%. Thus for every joule of chemical energy released by combustion, only about 0.6 J at the most can be converted into electrical energy. The remaining 0.4 J is not available as electrical energy or any other form of nonthermal energy, and—

FIGURE 7.6
Installation of a steam turbine at the Vermont Yankee nuclear power plant. Several bladed wheels are mounted on the same axle and are spun by a jet of high-pressure steam. [Photograph by Edward Lee, New England Power Service Company.]

almost equally important—this waste heat must be put into the river or into whatever is used for cooling.

Waste heat is an inevitable by-product of the generation of useful forms of energy from thermal energy. The effect of this thermal pollution on the ecology of rivers and oceans is a matter of increasing concern. In some locations, the environmental impact of this waste heat on a river or a shallow bay would be unacceptable, or an adequate supply of cooling water may simply be unavailable. An alternative is to use "cooling towers" (Figure 7.7), where the waste heat can be transferred to the air instead. There are two general types of cooling towers, "dry" towers in which heat is simply transferred from the water to the air, and "wet" towers in which some of the water is evaporated into the air. As water is evaporated, the latent heat of vaporization must be supplied, and the water left behind is cooled. By using cooling towers, we can avoid the direct heating of a river, but these towers nevertheless affect the environment by releasing large quantities of heat into the atmosphere and—if they are wet cooling towers—by releasing large amounts of water vapor.

The maximum possible efficiency of a steam-electric power plant is the Carnot efficiency, determined by the temperature of the combustion chamber, T_1, and of the auxiliary cooling water, T_2. Real power plants have efficiencies significantly lower than the theoretical maximum. Efficiencies of about 40% are typical of modern fossil-fuel power plants, whereas older plants may have efficiencies of 30% or less. (See Figure 7.8.) The nuclear power plants built during the 1960s and early 1970s are also steam-electric plants, which operate in much the same way as fossil-fuel plants; the major difference is that the heat H_1 comes from fuel rods in which a fission chain reaction is occurring rather than from a combustion chamber in which fuel is burned. (The nuclear plant is different also in that it produces no smoke or ashes, but does produce dangerously radioactive materials.) In order to prevent damage to the fuel rods, the temperature T_1 in a nuclear plant has to be kept somewhat

FIGURE 7.7
Two coal-fueled steam-electric power plants—each 750 MW(e)—near Washingtonville, Pennsylvania, each with a 370-ft high cooling tower. [Pennsylvania Power and Light Company.]

lower than in a fossil-fuel plant; largely because of this fact, the efficiencies of nuclear plants are only about 30%, lower than those of modern coal or oil plants. Efficiencies of actual power plants are determined only in part by the fundamental laws of thermodynamics. Other more practical considerations are important as well, but the strategy for improving the efficiency suggested by Equation 7.2 can be effective; other things being equal, the efficiency will be improved either by raising T_1 or by lowering T_2.

The size of an electrical power plant can be described by stating either the electrical power it can deliver or the rate at which heat is delivered from the hot reservoir, a rate equal to that at which chemical energy or nuclear energy is being used up. A power plant that can deliver 300 megawatts (MW) of electrical power is often described as having a capacity of 300 megawatts-electrical [300 MW(e)]. If its efficiency is 30%, it is using thermal energy at the rate of 1000 MW, and the plant may be described as one whose size is 1000 megawatts-thermal [1000 MW(t)].

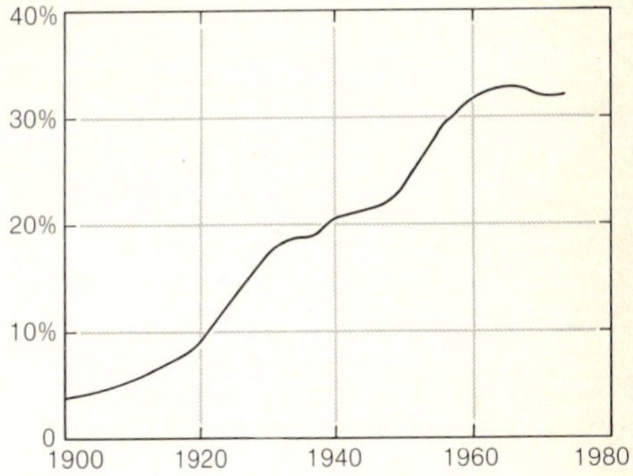

FIGURE 7.8
Average efficiencies of American steam-electric power plants.

§7.3 REFRIGERATORS AND AIR CONDITIONERS

Thermal energy cannot be completely converted into other forms. Closely related to this is the familiar fact that heat normally flows from hot objects to cooler ones and not the other way around. This is another way of stating the second law of thermodynamics. Heat does not by itself flow from a cool body to a hotter body, and moreover there is no way of building a device of any kind whose *sole* effect would be to extract heat from one object and deliver it to a hotter object. A device that could indeed do so (Figure 7.9) would be another variety of a perpetual motion machine of the second kind. As long as the two heat flows were equal, such a device would not violate the first law; the nonexistence of such machines is part of the content of the second law of thermodynamics. If we have two reservoirs at different temperatures, the atoms are to some extent sorted out according to their speeds, fast ones in the hot reservoir and slow ones in the cold reservoir. This state is more ordered than one in which the two reservoirs have the same temperature. A perpetual motion machine of the type shown in Figure 7.9, which could increase the difference in temperature without any other change, would produce a net reduction in the total amount of disorder. If such a perpetual motion machine existed, it would be extremely useful, because it would be a perfect refrigerator or air conditioner. The function of a refrigerator or air conditioner is to extract thermal energy from a cool object (a bottle of beer or the interior of a house), making the cool object still cooler, and to deliver thermal energy to the outside of the refrigerator or to the outside of the house.

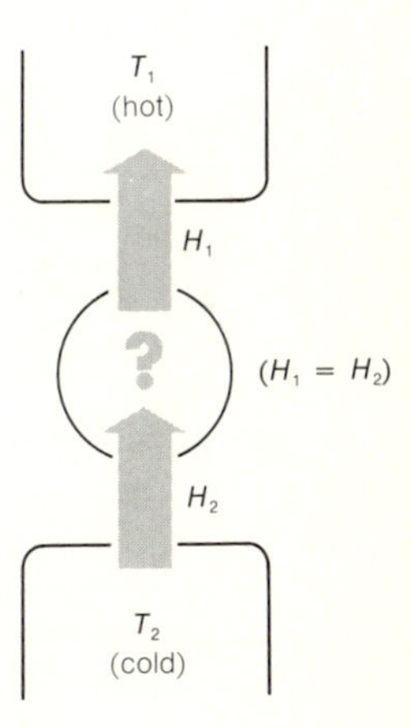

FIGURE 7.9
Another perpetual motion machine of the second kind.

Although machines of the type sketched in Figure 7.9 are only figments of our imagination, refrigerators and air conditioners do exist and are used to transfer heat from cold objects to hotter ones. They function only if some additional energy, usually electrical energy, is supplied. In other words, they work against the natural tendency for heat to flow from hot to cold, but only at the expense of some high-grade energy. A schematic diagram of an actual refrigerator is shown in Figure 7.10. The circles in these diagrams actually represent a complex array of wheels, pistons, and so on. One of the most impressive results of the laws of thermodynamics is that we can draw important conclusions about the operation of refrigerators and air conditioners, as well as about power plants, that are valid no matter how they may be constructed.

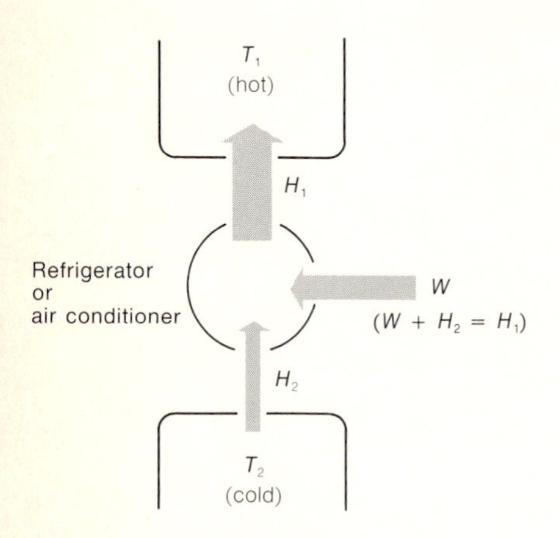

FIGURE 7.10
A real refrigerator or air conditioner.

The second law of thermodynamics enables us to determine the maximum possible amount of work that can be obtained from a given amount of heat. It can also be used to find the *minimum* amount of work required to transfer heat from a cool object to a warmer one. A refrigerator or air conditioner is essentially a heat engine operated backwards, as a comparison of Figures 7.10 and 7.3 will demonstrate. In a refrigerator, an amount of heat equal to H_2 is removed from the cool interior. An input of energy denoted by W (usually in the form of electrical energy) results in a flow of heat H_1 to the outside. By the first law of thermodynamics, H_1 must be equal to the sum of W and H_2:

$$H_1 = W + H_2.$$

The heat H_1 usually flows to the air in the room. (Most refrigerators have coils of tubing on the back that are slightly warm to the touch. This is where H_1 is being delivered to the room.) An air conditioner operates in the same way, except that here the cold reservoir is the interior of the house, and the hot reservoir is the air outdoors. In transferring heat from a cold body to a hotter one, we are operating against the natural tendency towards maximum disorder. As we might anticipate, the greater the existing temperature difference, the more difficult this transfer will be, and the greater the amount of work that will be required. A thermodynamic analysis similar to the one leading to Equation 7.1 shows that the minimum amount of work, W, required to remove an amount of heat, H_2, from the cooler object is given by the following expression:

$$\text{Minimum amount of } W = H_2\left(\frac{T_1 - T_2}{T_2}\right) = H_2\left(\frac{T_1}{T_2} - 1\right). \quad (7.3)$$

As anticipated, the larger the temperature difference, the larger the minimum amount of work required.

In operating a refrigerator, the amount of heat removed from inside, H_2, is what we are most interested in, whereas W is what we have to buy from the electric company. The best refrigerator is one for which a large value of H_2 can be obtained for a small value of W. That is, a refrigerator can be characterized by the ratio H_2/W, a ratio that is similar to an efficiency but is not identical and is called the "coefficient of performance," CP:

$$CP = \text{coefficient of performance} = H_2/W. \qquad (7.4)$$

From this definition and Equation 7.3, it follows that the best possible coefficient of performance is given by

$$\text{Maximum possible } CP = \frac{T_2}{T_1 - T_2}. \qquad (7.5)$$

The smaller the difference in temperature, the larger the possible coefficient of performance. Notice that the expression on the right-hand side of Equation 7.5 can be greater than 1. It is theoretically possible (and indeed possible in practice) to obtain a coefficient of performance larger than 1.

Consider some actual numbers. Suppose the interior of the refrigerator is at a temperature equal to 40°F (T_2 = 40°F = 4°C = 277°K) and the outside at 75°F (T_1 = 75°F = 24°C = 297°K). Then the best possible coefficient of performance is

$$\frac{T_2}{T_1 - T_2} = \frac{277}{297 - 277} = \frac{277}{20} \simeq 14.$$

With such a coefficient of performance, for every one unit of energy fed in from the power line, 14 units of thermal energy would be extracted from the inside of the refrigerator and a total of 15 units of thermal energy delivered to the room. (With 14 units of energy being removed from the inside for every unit of electrical energy used, it is tempting to say that the "efficiency" of this process is 1400%! This is certainly one situation in which the term efficiency would be misleading at the very least, and it is therefore preferable to use the more neutral term "coefficient of performance.")

It is a consequence of the second law of thermodynamics that the efficiency of a heat engine cannot exceed

$$\epsilon = 1 - \frac{T_2}{T_1},$$

and that the coefficient of performance of a refrigerator cannot exceed

$$CP = \frac{T_2}{T_1 - T_2}.$$

Although these theoretical maximum values cannot be achieved in practice, we can approach them fairly closely with the expenditure of sufficient care and money. Since a refrigerator is in a sense a magical device that counteracts the tendency of heat to flow from hot to cold, and since the coefficient of performance of a refrigerator can be greater than 1, it might occur to us that we could evade the consequences of the second law by using a refrigerator and a heat engine together. Could we avoid the production of waste heat in the way shown in Figure 7.11,

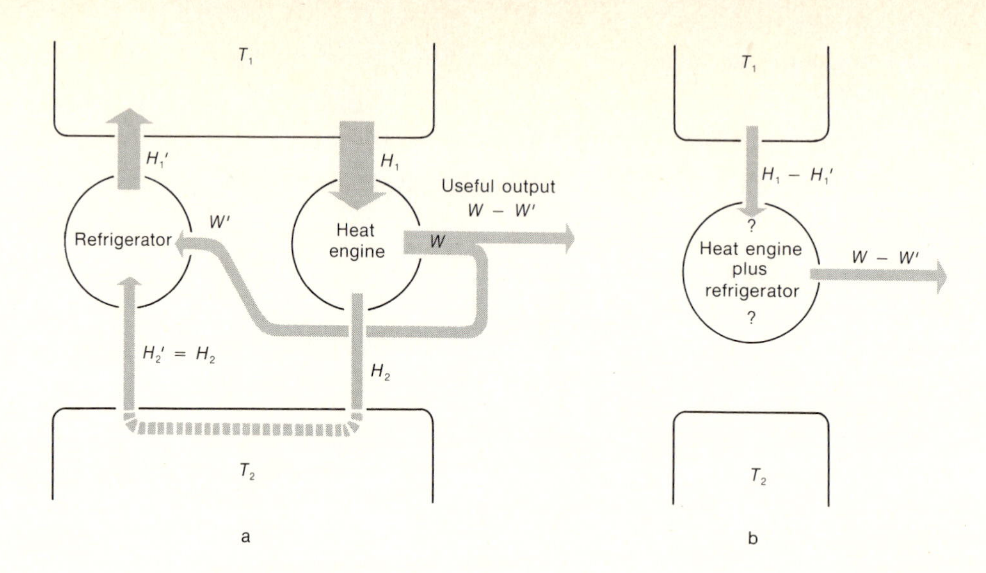

FIGURE 7.11
(a) Another attempt to build a perpetual motion machine of the second kind. *W′* (part of the engine's output) is used to run a refrigerator to remove the engine's waste heat from the cold reservoir. (b) The composite device would be equivalent to a perpetual motion machine just like that shown in Figure 7.1.

by taking a small part of the output of a heat engine to run a refrigerator, which would put the engine's waste heat back in the hot reservoir, thus making a composite device that would be a heat engine with a net useful output but no waste heat? This would be a complicated perpetual motion machine of the second kind, but it would not violate the *first* law of thermodynamics.

To simplify the arithmetic, suppose that $T_1 = 400°\text{K}$ and $T_2 = 300°\text{K}$. Then the maximum possible efficiency of the heat engine is

$$1 - \frac{300}{400} = \frac{1}{4}$$

and the maximum possible coefficient of performance of the refrigerator is

$$\frac{300}{400 - 300} = 3.$$

Now suppose that an amount of heat H_1 equal to 1 joule is delivered to the heat engine. Because the efficiency is no greater than $\frac{1}{4}$, $W \leq \frac{1}{4}$ J, and so H_2 is at least $\frac{3}{4}$ J: $H_2 \geq \frac{3}{4}$ J. The coefficient of performance of the refrigerator is no greater than 3, so the amount of work W' needed to operate the refrigerator, to remove $\frac{3}{4}$ J of thermal energy from the cold reservoir, is at least one-third as large: $W' \geq \frac{1}{4}$ J. Even if both devices are the best possible ones, the *full* output, *W*, of the heat engine is needed to operate the refrigerator; the two devices operate in tandem as shown in Figure 7.12, accomplishing absolutely nothing. This is a complicated device, but it cannot be used to violate the second law of thermodynamics. At any rate, it does leave the two reservoirs unchanged and causes no degradation of energy, as long as both the engine and the refrigerator are ideal ones.

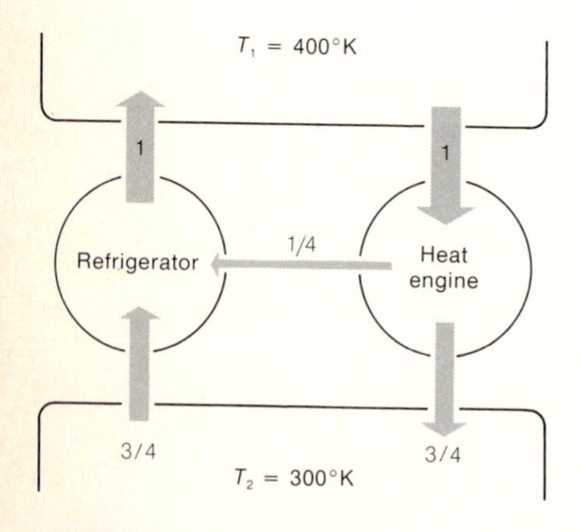

FIGURE 7.12
Even if both the heat engine and the refrigerator are the best possible ones, the method proposed in Figure 7.11 does not succeed; no work is left over for external use.

QUESTION

7.3 Show that the preceding result does not depend on the particular temperatures chosen. That is, consider an ideal heat engine and an ideal refrigerator operating between the same two temperature reservoirs (T_1 and T_2), and show that the energy flows of each device are identical, except that they operate in opposite directions. In other words, show that if the quantity H_1 in Figure 7.10 is equal to H_1 in Figure 7.3, then the values of H_2 and W in these two figures are also equal.

§7.4 A SUMMARY VIEW OF THE SECOND LAW OF THERMODYNAMICS. CLOSED AND OPEN SYSTEMS

The second law of thermodynamics states that in every process, there is a universal tendency toward increasing disorder, increasing entropy, a degradation of energy. In a *closed* system, entropy can at best remain constant; in almost every process, the total entropy of all the things involved increases to some extent, and it never decreases. As time goes on, the entropy tends to increase, and this serves to distinguish the future from the past. This is not to say that the entropy, the disorder, of *part* of the system cannot decrease, but that any local decrease of entropy must be compensated for by at least an equal increase somewhere else. In the long run, every closed system approaches a state of maximum entropy, maximum disorder, in which all other forms of energy have been degraded as far as possible into thermal energy, in which everything is at a single uniform temperature. This is the "heat death" of a closed system.

It follows that, in order for a system to continue to operate—to avoid this heat death—it cannot be a closed system; it must be *open,* it must be able to carry out exchanges of energy with its surroundings. There are two extremely important examples of open systems that continue to function in spite of the second law of thermodynamics, in spite of the universal tendency toward increasing disorder: first, the earth as a whole, and, second, the body of a living organism such as a human being. As described in §6.3, the earth as a whole is *similar* to a closed system in that its total energy is very nearly constant, but it differs in a very important way. Its total energy is constant not because it undergoes *no* energy exchanges with other things but rather because the amounts of energy it receives and emits are in almost perfect balance. The earth receives heat from the sun, heat that is high-grade (low-entropy) heat because of the sun's high temperature. The energy emitted by the earth is heat at a much lower temperature and is of lower quality, carrying more entropy for the same amount of energy. Even though the energy received and the energy emitted are equal in *energy* content, they are not equal in *entropy* content: the entropy associated with the outgoing energy is larger. Thus the earth can remain in a nearly steady condition by exporting more entropy than it receives, the excess amount being

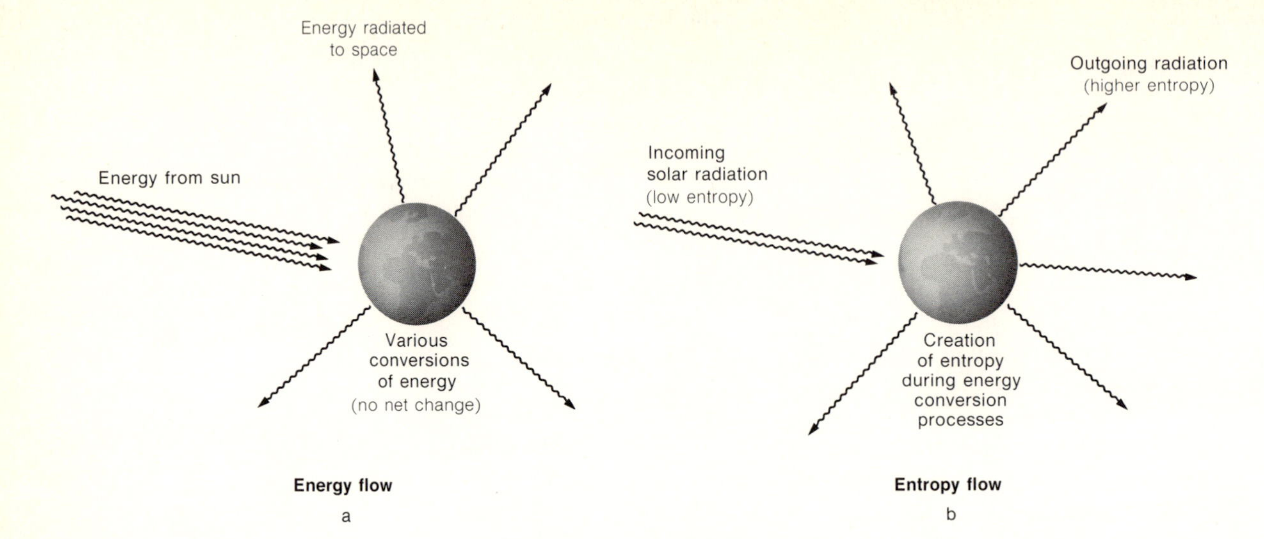

FIGURE 7.13
The earth's nearly unchanging condition is maintained by (a) keeping input and output of energy nearly equal, and (b) by emitting enough *extra entropy* to account for the entropy created on the earth.

just enough to get rid of the entropy produced by all the processes occurring on the earth. The flow of energy and entropy to and from the earth is shown in Figure 7.13. The two laws of thermodynamics were concisely summarized by the 19th-century German physicist Clausius in the following way: "Die Energie der Welt ist constant; die Entropie der Welt strebt einem Maximum zu." (The energy of the world is constant; the entropy of the world *strives toward* a maximum.) This does not mean "the entropy of the world increases to a maximum," but rather that it "strives toward" a maximum; the entropy of the earth does not increase in spite of all the entropy-producing processes which occur.

The preceding thermodynamic view of the earth can be applied to another open system, the body of a living creature. Energy is received in the form of high-grade chemical energy of food and oxygen. Various conversions of energy occur within the body (conversion of one type of chemical energy into another, conversion of chemical energy into thermal energy, and so on), but no creation or destruction of energy takes place. Energy is given out in the form of smaller amounts of chemical energy—in the air that we exhale and in the waste products that we excrete—in the form of mechanical work that we do, and in the form of the heat we give off to our surroundings. Like the energy flow to and from the earth, the total energy received and the total energy given out are in almost perfect balance. Similarly, all the energy conversions that go on in our bodies produce entropy. We get rid of this entropy by emitting low-grade energy in the form of heat, so that the energy emitted, though equal to the amount of energy received, carries off the excess amount of entropy.

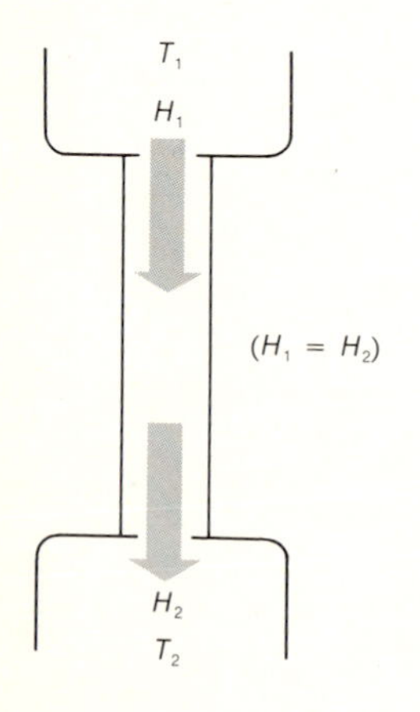

FIGURE 7.14
A metal bar between two temperature reservoirs is an open system similar to the earth.

A much simpler example of such a system is a metal bar connecting two bodies at different temperatures (Figure 7.14). As long as the two bodies are large, the bar remains in nearly a steady-state condition for a long period of time, one end at temperature T_1 and the other at temperature T_2. Since the bar receives just as much energy from the hot object as it delivers to the cold object, its total energy remains constant. The process of conduction of heat through the bar is one that produces entropy. Since the energy the bar receives is high-grade energy (from a hot object) and the energy it loses is of lower grade, and since these two amounts of *energy* are equal, the amount of *entropy* delivered to the cold object is greater than the amount of entropy received from the hot object. Thus the entropy of the bar does not increase in spite of the production of entropy within it.

Processes such as the three described above, in which an open system is maintained in a steady condition by exporting more entropy than it receives, do not constitute in any way violations of the second law of thermodynamics. On the contrary, these processes are controlled by the second law just as by the first. If entropy is created on the earth, in a living body or in a metal bar, but not accumulated there, there must be an increase in entropy somewhere else. In the case of the bar, if the hot object, the cold object and the bar together constitute a closed system, then the entropy of the whole system *does* increase.

What if we consider the earth as just a part of a much larger system, the universe? Is the whole universe a closed system to which the second law of thermodynamics applies? Is the universe therefore approaching a condition of maximum disorder, a heat death, as suggested by the following lines from *The Garden of Proserpine,* by the 19th-century British poet, Swinburne?

We are not sure of sorrow,
  And joy was never sure;
To-day will die to-morrow;
  Time stoops to no man's lure;
And love, grown faint and fretful
With lips but half regretful
Sighs, and with eyes forgetful
  Weeps that no loves endure.

From too much love of living,
  From hope and fear set free,
We thank with brief thanksgiving
  Whatever gods may be
That no life lives for ever;
That dead men rise up never;
That even the weariest river
  Winds somewhere safe to sea.

Then star nor sun shall waken,
  Nor any change of light:
Nor sound of waters shaken,
  Nor any sound or sight:
Nor wintry leaves nor vernal,
Nor days nor things diurnal;
Only the sleep eternal
  In an eternal night.

Although we know that eventually—perhaps 10 or 100 billion years from now—the sun will cool off (after perhaps having first gotten much hotter for a time) and that life on earth will end, it is not certain that we can extrapolate the second law of thermodynamics to the whole universe. The question of the ultimate fate of the universe is an intriguing one, a question to which the answer is not now known.

The second law can be regarded as a formal statement of the well-known fact that heat tends to flow from hot to cold and not the other way around, but the consequences of this obvious fact are profound. The very practical aspects of the second law of thermodynamics, as they affect the "energy problem," are that thermal energy is a special kind of energy and that there are very definite limitations on the types of energy conversions which are possible. All forms of energy are eventually degraded into thermal energy; energy cannot simply be "recycled." An ideal electrical power plant would not itself cause any degradation of energy, but once the electrical energy is converted into thermal energy, either directly as in a toaster or indirectly in a motor (where it may be converted into kinetic energy first and then into thermal energy), it is no longer available for conversion into other forms of energy. The energy has disappeared as *useful* energy, but it has not been destroyed; it is still present as thermal energy, heating up our cities and perhaps affecting the earth's climate. The energy crisis is a result not so much of the amount of energy we use but of the ways in which we use it, of the degradation of energy that inevitably accompanies what we do; indeed the term "entropy crisis" would probably be a more accurate one than "energy crisis."

The thermal energy of a *single* heat reservoir, large though it may be, is unavailable for conversion to other forms of energy. To convert thermal energy to nonthermal forms, we need two reservoirs, and even then we can convert only part of the energy extracted from the hot reservoir. Once chemical energy, for instance, has been converted into thermal energy, the fraction that can be converted into another form (into electrical energy or back into chemical energy, for instance) is, at most, equal to the value given by the Carnot limit. Thus whenever energy is obtained by first converting it into thermal energy, a significant part of it is inevitably wasted, and so this step should be avoided if possible.* An important corollary to the inefficiency resulting from the conversion of chemical energy into electrical energy with thermal energy as an intermediate step is that the thermal energy *not* converted into electrical energy must go *somewhere;* this is the "waste heat," the "thermal pollution" generated by every steam-electric power plant.

*Sometimes it is only the thermal energy we want, as when we burn oil in a furnace and immediately use the thermal energy to heat a house, but even then there is a better way of using the oil than simply burning it in a furnace, as we shall see in the next section.

§7.5 APPLICATIONS OF THERMODYNAMICS TO HOME HEATING

What guidance can thermodynamics provide us for choosing the best way to keep a house warm in the winter? Of course, as pointed out in Chapter 4, the process of keeping a house warm is an inherently wasteful one, since energy must be continuously supplied simply to replace the heat flowing to the outside. Good insulation and storm windows are important, but let us consider some of the other factors. Consider a house that on a winter day needs 10^6 BTU of energy to keep it warm and let us suppose that oil is accepted as the source of the energy. How can we obtain these 10^6 BTU with the least amount of oil? To simplify the numbers, we shall measure energy in millions of BTU's (MBTU's).

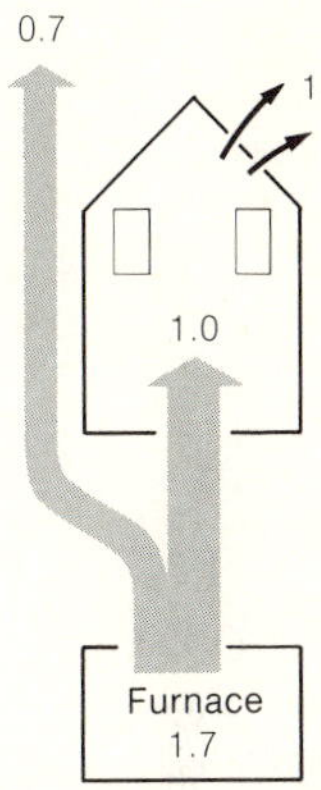

FIGURE 7.15
Home heating with an oil furnace. Energy required: 1.7 MBTU.

THE SIMPLE OIL FURNACE. The typical home furnace has an efficiency of about 60%. Therefore, in order to deliver one unit of energy (1 MBTU) to the house, the chemical energy of the oil used must be larger by the ratio 100/60. That is, we need about 1.7 MBTU, as shown in Figure 7.15. Of the oil's chemical energy, 0.7 MBTU is wasted, either as thermal energy going up the chimney or as the chemical energy of unburned oil.

ELECTRIC HEATING. Electrical energy can be converted into thermal energy at an efficiency of 100%. Therefore, if we purchase 1 MBTU of electrical energy from the power company,* we can do the job, as shown in Figure 7.16. This is appealing at first glance, since an energy input of only 1 MBTU is needed, but the 1 MBTU of electrical energy comes from an oil-fueled power plant with an efficiency of only 33% or so. Thus a more complete view of electric heating is that shown in Figure 7.17. Electric heating is not as desirable after all, since oil with

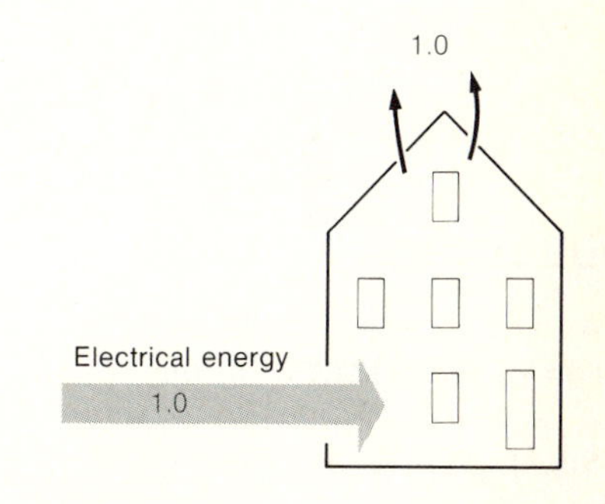

FIGURE 7.16
A superficial diagram of the process of electric heating. Energy required: 1.0 MBTU?

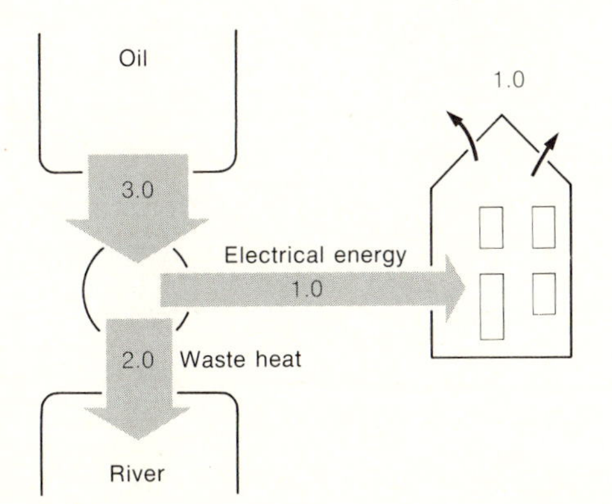

FIGURE 7.17
A more complete diagram of electric heating. Energy required: 3.0 MBTU.

*Power companies usually sell energy by the kWh, not by the BTU. However, because both are energy units, we can express any form of energy in either unit if we wish.

a chemical energy of about 3 MBTU is needed—almost twice the amount required if an oil furnace is used. Furthermore, 2 MBTU of waste heat must be put into the river, affecting the water temperature and the fish. From a thermodynamic point of view, electric heating is a very silly procedure. Since in this case thermal energy is what we want, it is absurd to include the step of converting thermal energy from the oil into electrical energy at an efficiency of 33%, even though the subsequent step can be carried out at an efficiency of 100%. We are using very high-grade energy (electrical energy) when we need only low-grade energy (the thermal energy of warm water or warm air); the "efficiency" may be 100%, but this number is misleading because the conversion from electrical energy to thermal energy is accompanied by a very significant degradation of energy, a large increase in entropy.

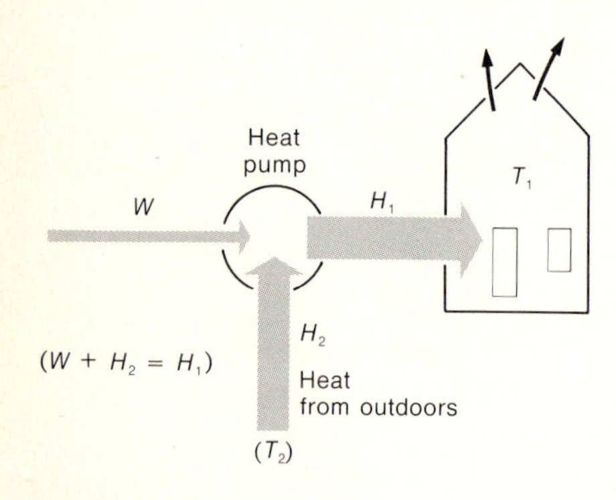

FIGURE 7.18
Home heating with a heat pump.

HEAT PUMPS. A third way of heating a house is suggested by thermodynamics. Why not refrigerate the outdoors—that is, run a refrigerator that removes heat from the cool outdoors and delivers heat to the warmer interior of a building? In this application, such a device is called a heat pump (Figure 7.18), though it is really no different in principle from an air conditioner or a refrigerator. Operation of a refrigerator does require an external source of energy, but as we saw in §7.3, a good deal of heat can sometimes be transferred with a fairly modest expenditure of energy. For an ideal refrigerator (the best allowed by the laws of thermodynamics), it was shown in §7.3 (Equation 7.3) that H_2 and W are related by the equation

$$H_2 = \frac{WT_2}{T_1 - T_2}. \tag{7.6}$$

QUESTIONS

7.4 Our interest here is in delivering a particular value of H_1 (1 MBTU) to the house, which now plays the role of the "hot reservoir." From the first law, we know that $H_1 = H_2 + W$. Using this fact and Equation 7.6, show that therefore

$$H_1 = \frac{WT_1}{T_1 - T_2}. \tag{7.7}$$

7.5 Show that if $H_1 = 1$ MBTU and that if $T_1 \simeq 70°$F and $T_2 \simeq 20°$F, it follows from Equation 7.7 that

$$W = \frac{H_1(T_1 - T_2)}{T_1} \simeq 0.1 \text{ MBTU}.$$

(Observe that the closer together T_1 and T_2 are, the smaller is the amount of externally supplied energy, W, required to deliver the same amount of heat.)

Thus an energy input to the heat pump of only 0.1 MBTU can do the job. A heat pump provides a much better way of taking advantage of the high quality of electrical energy than the wasteful practice of simply converting electrical energy into thermal energy. An energy of 1 MBTU can be delivered to the house with an input of only 0.1 MBTU of electrical energy. This may seem too good to be true, but heat pumps do

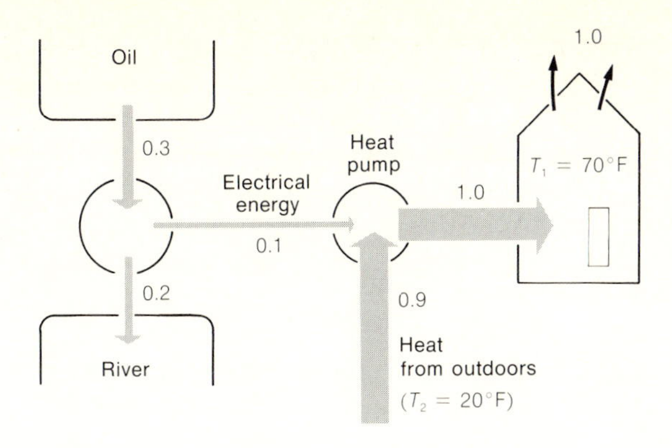

FIGURE 7.19
Home heating with an ideal heat pump. Energy required: 0.3 MBTU.

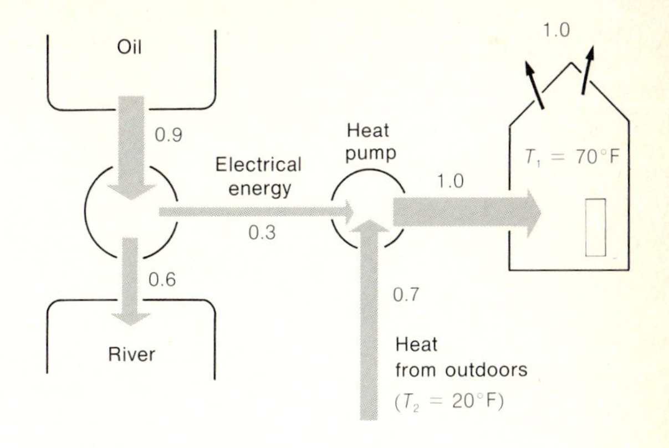

FIGURE 7.20
Home heating with a less than perfect heat pump. Energy required: 0.9 MBTU.

not violate the laws of thermodynamics (neither do refrigerators, and many refrigerators have coefficients of performance that are greater than 1). Most of the heat delivered to the house comes from the outdoors; the electrical energy has the vital function of pumping this energy from a lower to a higher temperature. Even when we take into account the inefficiency at the generating plant (Figure 7.19), the overall picture looks very favorable. However, we have been describing an ideal heat pump, and real heat pumps are not this good. For an actual heat pump, the work W required to deliver 1 MBTU might be about 0.3 MBTU, but even with this figure, the overall energy requirement for a heat pump (Figure 7.20) is smaller than that for the other methods. Even with a less than perfect heat pump, we can deliver 1 MBTU to the house with the expenditure of less than 1 MBTU of chemical energy at the generating plant. (Whether this will be true in any specific case depends on the particular temperatures and on how closely the heat pump resembles an ideal one.)

It appears that the chemical energy of oil required would be about 1.7 MBTU if a furnace were used, 3.0 MBTU with electrical heating, but only 0.9 MBTU with a heat pump. Why, then, are any homes heated by electricity, and why are not all homes heated with heat pumps? Thermodynamics alone is not adequate to answer these questions. Among the many reasons for the ways in which homes are actually heated are that the capital investment required of the home owner is greatest for a heat pump and least for electrical heating, that reliable heat pumps are not readily available, and that electrical heating *is* "cleaner" than an oil furnace. At any rate, electrical heating appears cleaner to the user who is not aware of the smoke and the waste heat produced at the power plant, or the even greater amounts of pollution that may be produced in the future when supplies of clean oil have been depleted. Even if these problems are recognized, it might be advantageous to have the pollution concentrated at the power plant rather than being distributed throughout an already polluted city.

In spite of the complicating social and economic questions, the message of thermodynamics is clear. The energy requirements for heating with heat pumps are less than for heating with conventional furnaces, which in turn are better than electrical heating. Furthermore, a heat pump can be designed so that by opening and closing a few valves, it can be used in the reverse direction during the summer, to pump heat from indoors to outdoors, that is, to air-condition the house. Since a heat pump works best when the indoor and outdoor temperatures are not too far apart (see Question 7.5), heat pumps are most promising in those parts of the country where the winters are fairly mild, just where air conditioning is most desired. The first and second laws of thermodynamics strongly suggest that any energy conservation program should include the development of reliable economical heat pumps, the implementation of measures to discourage electric heating, and of course the installation of better insulation. There is another way of heating houses that requires even less oil, ideally none at all: the use of solar energy, a possibility that will be discussed further in Chapter 16.

§7.6 *WHY* IS THE SECOND LAW CORRECT? THE OPERATION OF CHANCE AT THE MICROSCOPIC LEVEL

All our experience confirms the second law of thermodynamics. Heat does not spontaneously flow from a cool object to a hotter object. Why not? Why is there a universal tendency toward increasing disorder? Consider from a microscopic point of view what happens when heat flows from one object to another. Suppose that a box is divided into two compartments by a sheet of metal (Figure 7.21) and that compartment A is filled with helium gas at a low temperature and compartment B with helium gas at a higher temperature. This is a state with a relatively high degree of order; most of the fast atoms are segregated in B and most of the slow ones in A. When a rapidly moving atom in B strikes the barrier, some of its kinetic energy will be transferred to atoms within the barrier. When a slow atom in A strikes the barrier and interacts with the vibrating atoms of the barrier, the chances are that its kinetic energy will be increased. In this way, atomic kinetic energy can be transferred from B to A through the barrier, increasing the temperature of A and decreasing the temperature of B. Eventually the two temperatures will become equal. Atoms still strike the barrier, energy is still exchanged between the atoms of gas and the atoms in the barrier, but on the average as much energy goes one way as the other. This is a more disordered state; the atoms are no longer sorted out (slow ones in A and fast ones in B), but instead the numbers of slow and fast atoms in each compartment are

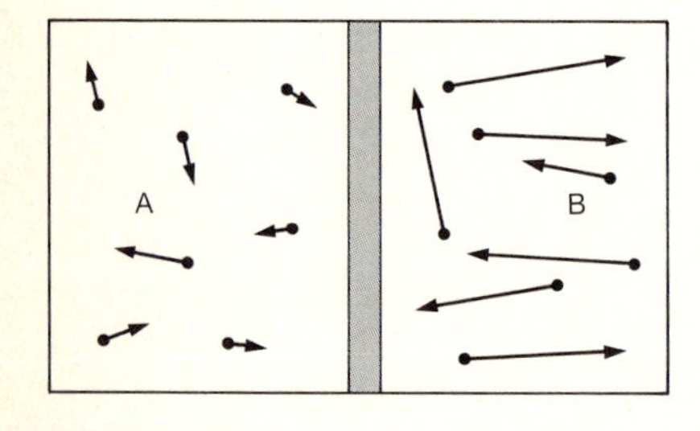

FIGURE 7.21
A box with low-temperature gas on the left side of the barrier, high temperature gas on the right.

approximately equal. From a macroscopic point of view, A and B are in thermal equilibrium, although on a microscopic level this thermal equilibrium is not a static situation but a dynamic fluctuating one in which small amounts of energy are transferred back and forth between A and B.

But if the approach to thermal equilibrium is governed by chance, could it not once in a while go the other way around? Could rapidly moving atoms in B bounce off the barrier, experience head-on collisions with rapidly moving atoms of the barrier, and come back with still greater kinetic energies? Could slow atoms in A hit the barrier and come back with still smaller kinetic energies, thus producing what would be observed as a flow of heat from the cool region, A, to the hotter region, B? Certainly *some* collisions do have this effect, but collisions having the opposite effect are more *likely*. We perhaps should revise the second law of thermodynamics to say that it is more *probable* that heat will flow from hot to cold than vice versa. The circumstance that makes this qualification unnecessary is that we are dealing with such enormous numbers of atoms that probabilities become virtual certainties; the probability that heat will not flow from a cold body to a hotter one is absolutely overwhelming. Perhaps we should qualify all statements based on the second law with terms such as "almost always" or "with a high degree of probability"; perhaps so, but if we insist on such qualifications, then it is hard to think of any statement ever made about any subject that ought not to be similarly qualified.

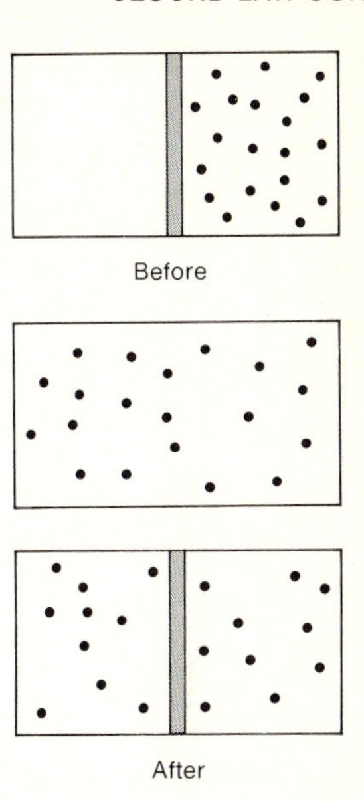

FIGURE 7.22
The natural tendency is for processes to proceed in the direction of increasing disorder.

A related example may indicate how safe we are in treating the predictions of the second law of thermodynamics as definitive rather than just as statements that are true "more often than not." Consider a box with a removable partition (Figure 7.22): one half of the box is initially filled with gas and the other evacuated. This is an ordered state, one of low entropy; all the atoms are on one side of the barrier. Now remove the barrier and the random motions of the atoms quickly produce a state in which the atoms are almost uniformly distributed. This is a state of greater disorder, higher entropy. If we put the partition back in, we find about half the atoms in each compartment. This result is what the second law of thermodynamics predicts: an ordered state has degenerated into a disordered one.

Can it happen the other way around, as sketched in Figure 7.23? In principle, yes, but we can readily calculate the chances of finding the atoms separated in this way. The chance that any one atom is in the right half of the box is $\frac{1}{2}$, the chance that two particular atoms are both in the right half is $\frac{1}{2} \times \frac{1}{2} = \frac{1}{4}$, and so on. If we have N atoms altogether, the probability that all N are in the right half of the box is

$$P = \frac{1}{2} \times \frac{1}{2} \times \frac{1}{2} \times \frac{1}{2} \cdots = \frac{1}{2^N}.$$

If there were only four atoms in the gas, this probability would be $1/2^4 = 1/16$. About one time in every 16, we *would* find all of the atoms in

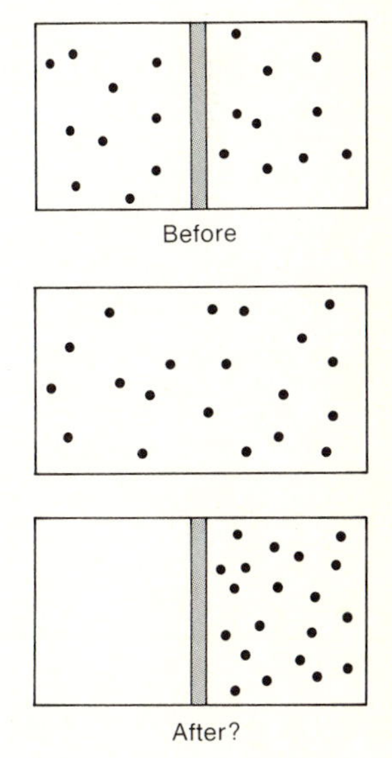

FIGURE 7.23
The gas *might* spontaneously change from a disordered state to a more ordered one, but such a reversal has never been observed.

the right-hand compartment. Any measurable amount of gas, though, consists of many more than four atoms. If $N \simeq 10^{19}$ (approximately the number of atoms in 1 cm^3 of gas under ordinary conditions), the probability of finding all of them on the right is

$$P = \frac{1}{2^{(10^{19})}}, \tag{7.8}$$

and this probability is so small that it is altogether negligible. Because there is only one way in which all the atoms can be in the right-hand compartment and so many ways in which they can be distributed so that about half are in one and half in the other, the chances are overwhelming that we will find the atoms approximately uniformly distributed.*

Ordinary objects, even very small ones, have extremely large numbers of atoms. This is the fact which makes the predictions of the second law of thermodynamics absolute certainties for all practical purposes, not just "practical" in the technological sense but for the pure scientist as well. The probability calculated above of finding all the atoms on one side of the box (Equation 7.8), is truly insignificant. If we tried to write it out in decimal notation, it would look something like this:

$$P = 0.\underbrace{00000000000000000000 \cdots 000000000000000000000000000}_{\text{about } 10^{19} \text{ zeroes}}1.$$

Not just 19 zeroes, 10^{19} zeroes, absolutely impossible to write! If we did this experiment once every second for a billion billion years, we could still be certain that we would not once find all the atoms in the same half of the box.

It is much more probable that a monkey striking typewriter keys at random will correctly type all of Shakespeare's works on the first try without a single error. Let us call P' the probability that this will happen, and calculate its value. Shakespeare's works take up about 1000 pages with about 6000 characters per page, a total of about 6×10^6 characters. Since a typewriter has about 60 keys, the monkey has about 1 chance in 60 of getting each individual character right, so the chance of getting every one of the first 6×10^6 characters right is approximately

$$P' = \frac{1}{60^{(6\times 10^6)}} \tag{7.9}$$

The quantity P' is certainly small but, though it may not be obvious from the way in which the numbers are written, P' is much greater

*We have only calculated the probability of finding *all* the atoms in the same half of the box. It is somewhat more difficult to calculate, for instance, the probability that at least 55% of the atoms will be in one half of the box and no more than 45% in the other. Calculations show that there is a negligibly small chance of observing any significant deviation from uniformity, any deviation large enough to be detected by macroscopic methods such as weighing the two halves.

than *P*. The monkey has to make "only" 6×10^6 lucky guesses in a row; to find all the atoms on one side of the box, every one of 10^{19} "guesses" has to be right.

It is an intriguing fact that the second law of thermodynamics is fundamentally a statistical law. Do not, however, invest in a scheme for evading the second law which relies on the fact that it is a law which is only valid with a high probability and that once in a while disorder will spontaneously change into order; better to put your money into lucky monkeys instead. The predictions of the second law can be relied on with certainty because probabilistic predictions become certainties for all practical purposes when large numbers are involved. If you think you observe heat flowing spontaneously from a cold body to a hotter one, the chances are excellent—to put it mildly—that you are mistaken or dreaming.

§7.7 THREE IMPORTANT NATURAL LAWS: NEWTON'S LAW OF GRAVITATION AND THE TWO LAWS OF THERMODYNAMICS

Of the physical laws which we have discussed, three are particularly noteworthy as powerful and widely applicable generalizations: Newton's law of universal gravitation, the law of conservation of energy (the first law of thermodynamics), and the second law of thermodynamics. It is worth considering the contrasts between these three intellectual achievements. Newton's law of gravitation is the prime example of what physics is popularly supposed to be. It is a precise statement, an assertion that between every pair of objects in the universe there exists a certain calculable force. Under conditions in which no other forces are of importance, Newton's law permits absolutely precise predictions; it leads to the vision of a "clockwork universe" of the type imagined by Laplace (discussed in §5.8). Its most important limitation is that, although put forward as the definitive statement about gravity, it says nothing about any other sorts of interactions. This law also happens to be not quite correct even in the gravitational context, but this is really a minor blemish which detracts little from its importance.

In contrast, the laws of thermodynamics are universal, applying to all phenomena in which any and all interactions may be of importance. The laws of thermodynamics were clearly formulated more than a century ago. Since that time, in spite of the many new things we have learned, these laws have retained their validity; if anything, they are of more importance now than when first proposed. The price which must be paid for this increased generality is that the laws of thermodynamics do not tell the whole story. These laws tell us some very important things about *all* phenomena, but there are no phenomena about which they tell us everything. The first law of thermodynamics says that *if* kinetic energy increases by a certain amount, a corresponding decrease must occur in other forms of energy, but it does not tell us whether this increase in kinetic energy *will* occur or, if it does occur, which of the other forms of energy will experience decreases. The first law of thermodynamics must be used in conjunction with detailed information about the particular phenomena in question. The second law of thermodynamics goes farther in allowing us to say what may happen and what will not happen, but it is still not enough. Heat does not flow from cold to hot; heat *may* flow from hot to cold, but the laws of thermodynamics do not give us any hint as to how rapidly this will happen. The laws of thermodynamics set very important constraints on natural phenomena, constraints that are extremely important to know about, but we must still consider each situation in detail to find out which of the many processes permitted by thermodynamics will actually occur.

The two laws of thermodynamics have been given as statements about the behavior of two physical quantities, energy and entropy, or alternatively as assertions of the impossibility of constructing perpetual motion machines of either the first or second kind. One should not be misled by the alternate version into thinking that the laws simply reflect human weaknesses; even if life had never evolved anywhere in the universe and even if there were no brains to conceive of the ideas of energy and entropy, it would still be true that "energy" would be conserved and that heat would not spontaneously flow from cold to hot.

The second law of thermodynamics has a different sort of "feel" than do all other scientific laws. Its validity depends on the fact that the number of atoms in any macroscopic object is enormous. (In a "universe" of two or three atoms, energy would still be conserved in the interactions of these atoms with one another, but the concepts of temperature, of order and disorder, of thermal energy as opposed to more orderly forms of energy, would be superfluous.) From time to time, the fact that the second law depends on the operation of chance on a microscopic level has caused some confusion, because improbable events do happen. Although this is a matter of great philosophical interest, it provides no grounds for doubting predictions made with the second law. You may sometime "defy the laws of probability" by being dealt a bridge hand with thirteen spades; you might get two or three such hands in a row, but it is safe to assert that you will *never* observe heat flowing from a cold body to a hotter one.

In a sense, the second law is a highly pessimistic generalization about the universe. The gloom that pervades Swinburne's poetry (§7.4) should be contrasted with the exuberance of Pope's lines about Newton quoted in §5.1. Energy is always degraded, order always tends to degenerate into disorder and chaos. These thoughts have occasionally been taken over, without a great deal of care, by persons who write about history and political and social problems, in the idea that it is a "law of nature" that things can only get worse and in the nostalgia expressed for the "good old days." (Before penicillin? Before abolition of the slave trade?) One may easily question the validity of trying to apply the laws of thermodynamics to social and political questions, but it is surely reasonable to ask that the laws at least be used correctly. In a *closed* system, disorder (entropy) always tends to increase; in an open system, the entropy of the system itself need not forever increase, but may well decrease while the total entropy of the larger system of which it is a part inexorably increases.

FURTHER QUESTIONS

7.6 Efficiencies of electrical power plants are often described in a most peculiar way, in terms of the "heat rate." A statement that the heat rate is 10,000 BTU/kWh means that for every kWh of electrical energy generated (W), 10,000 BTU of heat (H_1) must be delivered from the hot reservoir. The higher the heat rate, the lower the efficiency. In fact the heat rate is precisely the *reciprocal* of the efficiency except for the fact that W and H_1 are expressed in different units. What is the efficiency (ϵ) of a power plant whose heat rate is 10,000 BTU/kWh? What is the heat rate of a power plant whose efficiency is 40%?

7.7 Figure 7.8 shows how the average efficiency of American steam-electric power plants has varied since 1900. Using this figure, sketch a graph showing the variation in heat rate for the same period of time.

7.8 Consider two power plants, one with an efficiency of 40% ($\epsilon_A = 0.4$) and the other 30% ($\epsilon_B = 0.3$). The difference between 30% and 40% does not seem very large ("only 10%", or perhaps "only 25%"). Compare these two power plants on the following basis: if each generates the same amount of electrical energy, by what factor will the waste heat from plant B exceed that from plant A?

7.9 By how much does a large steam-electric plant raise the temperature of a river? (If it were 10^{-6}°C, then "thermal pollution" would not be a serious problem.) Consider a 1000 MW(e) plant with an efficiency of 33% [3000 MW(t)] on a river whose average rate of flow you know, for instance the Colorado River discussed in §4.2. (If possible, take an actual power plant in your area and use data for the river on which it is located, estimating the rate of flow if necessary.) What would be the rise in the average temperature of the river? What idealizations and approximations are necessary in order to make this estimate?

7.10 The waste heat produced by steam-electric power plants is an increasingly important environmental problem. If the efficiency is 33%, then for every kilowatt-hour of electrical energy used, 3 kWh of chemical or nuclear energy must be used up and 2 kWh of heat must be put into the environment. Often this results in the warming of a river, producing a large quantity of lukewarm water at a temperature somewhat above that of the environment. Not only may this upset the ecology of the river, but it seems wasteful just to throw this energy away. Can you think of uses to which large quantities of warm water might be put? What kinds of side effects might result from such uses?

7.11 In 1973, the average electrical output of all the steam-electric plants in the United States was about 2×10^{11} W. Suppose that all of this power were generated by one enormous power plant with an efficiency of about 33% and that the Colorado River were used to provide the necessary cooling. How much warmer would the river be after emerging from this plant?

7.12 The efficiency with which thermal energy can be extracted from a hot reservoir and converted into other forms depends on the difference in temperature between the hot reservoir and the cold reservoir into which the waste heat must be put. The maximum possible efficiency can be increased either by using a hotter hot reservoir or a colder cold reservoir. Which change would have the most effect on efficiency, raising the temperature of the hot reservoir by 100°C or lowering the temperature of the cold reservoir by 100°C? What if the choice were between doubling the absolute temperature of the hot reservoir and halving the absolute temperature of the cold reservoir?

7.13 The founders of thermodynamics were motivated not only by scientific curiosity but also by very practical considerations, by the desire to improve the efficiency of steam engines. Joule pointed out that efficiencies of existing steam engines were less than 10% or so and wrote, "at least ten times as much power might be produced as is now obtained by the combustion of coal." Discuss Joule's statement in the light of the second law of thermodynamics.

7.14 The maximum efficiency of a heat engine is given by Equation 7.2; $\epsilon = 1 - \frac{T_2}{T_1}$. Can we improve on this by running two ideal heat engines in series, as shown in the figure (top right), in which the waste heat from the first engine (at a temperature T') provides the high-temperature heat for the second engine? ($T_1 > T' > T_2$). Try this first for a specific case (e.g., $T_1 = 500°\text{K}$, $T' = 400°\text{K}$, $T_2 = 300°\text{K}$), and then show in general that if A and B are both ideal heat engines, the overall efficiency of the combination $\left(\epsilon = \frac{W_A + W_B}{H_1}\right)$ is still given by $\epsilon = 1 - \frac{T_2}{T_1}$.

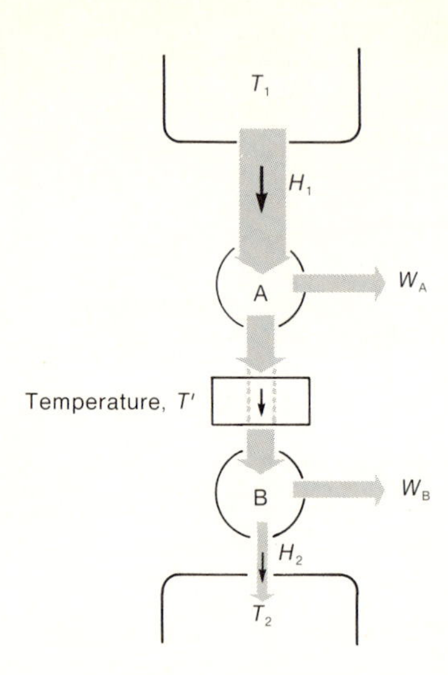

7.15 Could you cool off a room by plugging an air conditioner in and then setting it down in the middle of the floor instead of installing it in a window?

7.16 The coefficient of performance (Equation 7.4) is a convenient quantity for describing how good a refrigerator or air conditioner is at removing heat, H_2, for a given expenditure of electrical energy, W. For some strange reason, manufacturers often provide this information in disguised form, as an "energy efficiency ratio" (E.E.R.), a number that is the ratio of the rate at which heat, H_2, is removed from the room (in BTU/hr) to the electrical power required for operation (in watts), the rate at which work, W, must be done. From this E.E.R., the coefficient of performance can be calculated.

(a) One list of air conditioners shows E.E.R.'s ranging from 5 to 12. Show that an E.E.R. of 12 is equivalent to a coefficient of performance of about 3.5.

(b) The strange combination of units used in the definition of the E.E.R. is similar to that used in the definition of the heat rate of a power plant (Question 7.6). For a power plant, the *lower* the heat rate, the higher the efficiency. Is this also correct for the E.E.R.? (Which is better, an E.E.R. of 5 or 12?)

7.17 Visit a local appliance store, or look at advertisements for air conditioners in a current catalog of a mail-order house, and see whether information about their coefficients of performance is available. If so, is there any correlation between price and coefficient of performance?

7.18 Consider Equation 7.5. Is it possible to choose T_1 and T_2 in such a way that the maximum possible coefficient of performance is equal to 1? less than 1?

7.19 Various methods of home heating can be compared in terms of *cost* as well as in terms of the amounts of energy used. Using cost data from Appendix P (or, preferably, more recent data for your own vicinity), estimate the cost of providing 1.0 MBTU to the house by the following means:

(a) an oil furnace with an efficiency of 60% (Figure 7.15);

(b) a furnace fueled by natural gas, also burned at 60% efficiency;

(c) electrical heating (Figures 7.16 and 7.17);

(d) a heat pump (Figure 7.20).

In making such comparisons, only the cost of the *fuel* or the electrical energy is being considered. The very important differences in capital investment and in maintenance costs are not taken into account, nor are the hidden environmental costs.

7.20 Discuss the following argument that is sometimes presented (though not in such a naked form) to show that electrical home heating uses about the same amount of energy as does heating with an oil furnace. Most electrically heated homes have been built quite recently and therefore usually have better than average insulation. Furthermore, since electrical energy is so expensive, there is an added incentive for builders to spend money on insulation and for occupants to buy storm windows and to keep their thermostats turned down. Therefore on a day when the typical home heated by an oil furnace needs 1 MBTU of heat, the average electrically heated home of the same size in the same town probably needs only about 0.5 MBTU. We can use Figure 7.17 here, if we simply divide all the numbers by two; we see that the total energy required from the oil at the power plant is 1.5 MBTU, about the same as the 1.7 MBTU shown for the conventionally heated home in Figure 7.15.

7.21 Show that P′ (Equation 7.9) is really vastly larger than P (Equation 7.8). Suggestion: Because 64 is an exact power of 2 (2^6), the problem can be simplified by making a slight change and considering a typewriter with 64 keys. Then the denominator of the expression for P' becomes $64^{(6\times10^6)} = (2^6)^{(6\times10^6)}$. Now use the fundamental laws of exponents, and see which is the larger, P or P'.

7.22 If you flip a coin four times and record the result as heads (H) or tails (T), there are 16 possible outcomes, for example:

TTTT
TTTH
TTHT
. . . .
. . . .
. . . .

Of the 16, how many result in all heads and how many in two heads and two tails? How does the likelihood of getting all heads rather than a roughly equal mixture of heads and tails change as you increase the number of tosses in the sequence?

7.23 In §7.6 we considered a gas of 10^{19} atoms. This is approximately the number of atoms in 1 cm^3 of gas under ordinary conditions, but perhaps it was unfair to consider such a large volume of gas. Consider instead a volume of 10^{-6} cm^3; then the number of atoms, N, is only about 10^{13}. In this case, is there a reasonable chance of finding all the atoms on one side of the box?

7.24 It is much more probable that a monkey will type all of Shakespeare's works correctly on the first try than that we will find all 10^{19} atoms in the same half of the box. Perhaps the monkey was given too easy a task. How would the comparison between the two probabilities come out if we asked the monkey to type correctly all the books in a large library?

7.25 Imagine the following demonstration, one often performed by physics professors to dramatize their confidence in the laws of nature. Make a pendulum by attaching a cannonball to the end of a wire attached to the ceiling, stand at one side of the room with the back of your head against the wall, hold the ball against the end of your nose, and release it. Although the return swing of the ball is frightening to watch, no professor has yet been injured in such an experiment. To what scientific principles do they owe their survival? The law of conservation of energy? The second law of thermodynamics? Newton's law of gravitation? All of the above?

8
Electrical Energy and Its Uses

Nancy Richardot

8
Electrical Energy and Its Uses

Welcome to the Joy of Total Electric Living

§8.1 INTRODUCTION

It is hardly necessary to review here the many ways in which electricity provides conveniences and services which would otherwise be difficult to obtain. Those who may not be aware of the blessings of abundant electrical energy are often reminded by the advertisements: "Live Better Electrically," "Use Electric Heat for All It's Worth," etc. Nor is it necessary to be reminded of the problems that have resulted from our increased reliance upon electrical energy. The newspapers all too frequently bring us the news of power shortages and brown-outs, of controversies about the effects on the environment and on the public health of generating electricity from coal or from nuclear energy, and so on. Electricity is more than a convenient source of energy. Virtually all of the technology that, together with the increased use of energy, makes the 20th century so different from preceding centuries depends on electricity for its operation: electronic computers, airport traffic control systems, telephone, radio and television communication systems, "bugging" devices and tape recorders—these are but a few of the features of modern life made possible by electricity. The *amounts* of energy needed for purposes such as these are not very large, but the energy used must be supplied in the form of *electrical* energy.

Electrical phenomena are also of basic scientific importance. Whereas gravitational interactions control the motions of planets and galaxies, and nuclear interactions hold the constituents of atomic nuclei together, electrical interactions are the dominant ones for a vast range of phenomena between these extremes. Electrical interactions are responsible for the structure of atoms, for the interactions between atoms that lead to the formation of molecules, for the whole subject of chemistry, and for the very existence of macroscopic objects such as rocks, trees, houses and people.

The production and use of electrical energy are important aspects of the energy problem. Of all the energy used in this country, 25% is used for generating electricity, and growth in electricity consumption has been especially rapid, with a doubling time of about 10 years. The rapid rate of growth is one of the reasons for the special attention devoted to electrical energy in discussions of the energy problem. Another is the "high visibility" of the problems associated with the generation and transmission of electrical energy. An electrical power plant, no matter what its source of energy, is a large and expensive piece of machinery. A decision to build a new plant is not taken lightly; an investment of several hundred million dollars is required, land must be acquired, public hearings are often held, and innumerable licenses and permits must be obtained. By contrast, when the individual citizen decides to buy a clothes dryer or a dishwasher, no permit is required, no hearings are held, no "environmental impact statement" is filed, no press releases are issued. Perhaps hearings *should* be held, perhaps press releases *should* be issued ("Sierra Club Seeks Injunction to Stop Installation of Clothes Dryer"), for it is the vast number of such individual decisions that produces the need for new power plants.

We shall begin by describing the nature of electrical interactions at the microscopic level. This is not the way in which the development of the subject actually took place. Historically, electrical phenomena were first studied at the macroscopic level, and it was only after a long period of investigation in which many scientists participated that electrical phenomena could be understood from a microscopic point of view.

§8.2 ELECTRICAL PHENOMENA AT THE MICROSCOPIC LEVEL

Ordinary matter is composed of vast numbers of atoms of various types, and atoms are composed of still smaller particles. Virtually all of the mass of an atom is concentrated in a tiny central *nucleus* (whose diameter is about 10^{-15} m), but the nucleus is surrounded by one or more electrons whose orbits extend to distances of 10^{-10} m or so and give the atom its characteristic size, a size that is very small but enormously greater than that of a nucleus. The atom of each chemical element has its own characteristic number of electrons, one for hydrogen, two for helium, three for lithium, and so on. Between the constituents of an atom there exist strong interactions, and it is these interactions that are called "electrical." There is an attractive force between the nucleus and each electron that holds the atom together, and in addition there are repulsive forces between any two electrons. There are also repulsive forces between any two nuclei that come into play when a number of atoms interact with one another.

Long after atoms were first studied, in fact long after nuclei were first studied, it was learned that nuclei are composed of two kinds of particles: *protons*, which are responsible for the electrical properties of nuclei, and *neutrons*, which show no electrical effects. Electrical forces result from the combined interactions of protons with electrons, of protons with other protons, and of electrons with other electrons.* These forces are not gravitational. Presumably there are gravitational forces between electrons and protons, but these gravitational forces are much weaker than the electrical forces, and are therefore not noticeable. When we call these interactions "electrical," we are not explaining their origin; we are merely admitting the existence of these forces. We say that electrons have an electrical "charge" and that protons also have electrical charge, but again, in doing so we are simply giving a name to whatever property it is that electrons and protons have which cause them to attract or repel one another. There are two kinds of charge: that of the protons and that of the electrons. The rule that determines whether the force between two charged objects is attractive or repulsive is simply this: two *like* charges repel one another, whereas two *unlike* charges attract one another.†

We could use any pair of labels to refer to the two varieties, "red" and "green" for instance, but it is convenient to distinguish the two kinds of charge by labeling one *positive* and the other *negative*. According to the universal convention, the charge of the electron is negative and that of the proton positive, though it would make little difference if the opposite convention had been adopted. The basic fact that makes it reasonable to use the algebraic signs (+ and −) as labels for the two kinds of charge is that the electrical properties of an object depend on whether it has equal numbers of protons and electrons or whether it has more of one or the other. A "neutral atom" has as many electrons as it has protons. An atom that has lost one electron has a net positive charge; it is a positively charged ion. An atom with one extra electron, a negatively charged ion, acts electrically just like a single electron.

---

*There also exist other particles that exhibit electrical effects, particles found in cosmic rays, for example, but all the electrical properties of ordinary matter can be attributed to protons and electrons.

†Nuclei contain protons which repel one another electrically and neutrons which are electrically inert; there must therefore be an additional interaction, a nuclear interaction, which serves as the "glue" to hold a nucleus together.

The convention used, in which the electron's charge is *negative*, can be a source of some confusion in describing a *flow* of charge, a current. When a current flows through a wire, it is the negatively charged electrons that move, whereas the positively charged nuclei remain fixed.* It is conventional to describe such a current *as if* it were the result of positive charge flowing in the opposite direction; that is, the "direction of the current" is the direction opposite to that in which the negative electrons are moving. If we were at liberty to revise the terminology of physics, it might be advantageous to avoid this problem by calling the electron's charge positive and that of the proton negative. As it is, we are committed to the opposite convention, and we must be aware of the choice which has been made.

The magnitude of the attractive or repulsive force between two charged objects decreases if the distance between them increases; the electrical force varies with distance in exactly the same way as does the gravitational force, that is, in proportion to the *inverse square* of the distance, as long as the charges are far enough apart to be treated as "point charges." We introduce the symbol q to represent the magnitude of the electrical charge and write the following equation to represent the force between two objects with charges q_1 and q_2, separated by a distance r:

$$F_{\text{elec}} = \frac{k_e q_1 q_2}{r^2}. \tag{8.1}$$

This is Coulomb's law, the basic law governing electrical interactions. Here k_e is a proportionality constant (the "Coulomb's law constant") whose value depends on the units in which force, distance, and charge are measured. The MKS unit of charge is the *coulomb*, and if charge is measured in coulombs (C), force in newtons (N), and distance in meters (m), k_e has the rather awkward value of 9×10^9.

QUESTION

8.1 Show that in order for Equation 8.1 to be dimensionally correct, k_e must have units and can be written as either

$$k_e = 9 \times 10^9 \text{ N-m}^2/\text{C}^2$$

or

$$k_e = 9 \times 10^9 \text{ kg-m}^3/\text{sec}^2\text{-C}^2.$$

---

*A current in a metallic wire is the result of the motions of electrons, but other currents may result from the motions of positive charges, or from motions of both positive and negative charges in opposite directions.

The charge of the electron and the charge of the proton are, as far as we know, precisely equal in size, both having a magnitude of approximately 1.6×10^{-19} C, a value of charge called the electronic charge and given the symbol e:

$$e = 1.6 \times 10^{-19} \text{ C}.$$

A piece of ordinary matter contains a very large number of both protons and electrons, and its net charge is positive, zero, or negative, depending on whether the number of protons is greater than, equal to, or less than the number of electrons. Its net charge must be equal to some integral multiple of the electronic charge (0, -23, or $+128$, for instance).* In many circumstances the fact that electrical charge comes in discrete units is of little importance. The reason is that the electronic charge is so small that a macroscopic object for which the number of protons differs from the number of electrons by less than a million or so appears to be electrically neutral for most purposes. The difference between a net charge of 10^6 electronic charges and $(10^6 + 1)$ is undetectable, so electrical charge can usually be regarded as if it were a continuously variable quantity.

The similarities between the Coulomb law of force (Equation 8.1) and the law of universal gravitation,

$$F_{\text{grav}} = \frac{Gm_1m_2}{r^2}$$

are apparent. Because both types of force vary with distance in precisely the same manner, many of our previous results about gravitational interactions can be readily adapted to the description of electrical interactions. There are some extremely significant differences between the two, however: (1) the constituents of ordinary matter have only one kind of mass, but there are two kinds of charge, positive and negative; (2) although gravitational forces are always attractive, electrical forces can be either attractive or repulsive; (3) at the microscopic level, electrical forces are tremendously stronger than gravitational forces. Even though the gravitational forces between the fundamental constituents of atoms are much weaker than the electrical forces, gravitational forces between large objects can be stronger because of the cancellation of repulsive and attractive electrical forces. If the atoms in a baseball and the atoms in the earth were stripped of their electrons, their protons would repel each other with a fantastically large force. As it is, both the ball and the earth have almost precisely the same number of electrons as protons, the electrical force between the ball and the earth is nearly zero, and the gravitational attraction is dominant.

---

*All of the known "fundamental particles" (not only the electron, proton and neutron, but also those found in cosmic rays, emitted in radioactive decay or created with the aid of high-energy accelerators) have a charge that is either zero or equal to $+e$ or $-e$. Recently it has been speculated that particles such as protons may themselves consist of still smaller units, hypothetical particles called "quarks" with charges equal to $\pm\frac{1}{3}$ or $\pm\frac{2}{3}$ electronic charges.

§8.3 ELECTRICAL PHENOMENA AND ELECTRICAL ENERGY AT THE MACROSCOPIC LEVEL. CONDUCTORS AND INSULATORS, THE CONCEPTS OF ELECTRIC FIELD STRENGTH AND VOLTAGE

Any object of macroscopic dimensions contains an enormous number of atoms. If every atomic nucleus has associated with it just the right number of electrons to make the atom as a whole electrically neutral, then the entire object as well will have a net charge of zero. The electrons contribute a very large negative charge, a charge that is precisely canceled out by the positive charge carried by the nuclei. The electrical effects observed with macroscopic objects arise when this cancellation is not perfect, when an object carries a slight excess of one type of charge or the other. One way of producing electrical effects is to put two objects in intimate contact by rubbing them together. When two objects are pressed together, there is a tendency for electrons to be transferred from one to the other. The result is that one object ends up with a net positive charge and the other a net negative charge. Rub a balloon against the wall and some electrons will be transferred from the balloon to the wallpaper; then the balloon will stick to the wall as a result of the attractive force between the negatively charged wallpaper and the positively charged balloon. Stroke a cat on a dry day, and some of his electrons will be transferred to you; then if you touch your finger to his wet nose, your excess electrons will be attracted to the positively charged cat, and you and he feel a shock as the electrons flow back into the cat. Walk across a carpet and the contact between your shoes and the carpet results in a transfer of electrons from your body to the carpet; you will not be aware of your net charge until you lose it when you grasp a metal doorknob.

One of the most important facts about electrical phenomena is that different materials vary widely in their ability to conduct electrical charge. If, for example, two metallic balls are supported on a plastic stand (Figure 8.1) and given opposite charges, the two balls will retain their net charges for a long period of time, even though the excess electrons on the left-hand ball are attracted by the net positive charge on the right-hand ball. If the two balls are connected by a metallic wire, even a very thin one, then with a speed far too great for measurement by any ordinary methods, the left-hand ball will lose its excess negative charge and the right-hand ball will lose its positive charge (actually by gaining negative electrons), as electrons pour through the wire from left to right. It is possible to group materials into two general categories: (1) electrical conductors, through which charge can flow quite readily, and (2) insulators, which are really extremely poor conductors, and through which charge can flow only with difficulty. Glass, plastics, rubber, paper, rocks, and wood are insulators; all metals are conductors. In §8.5.C we will establish a quantitative measure of a material's ability to conduct electricity. The important point is that there is an enormous difference between good insulators and good conductors. This is the crucial fact

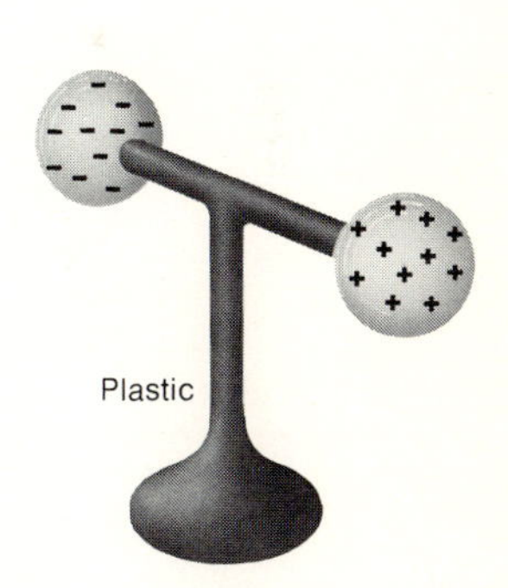

FIGURE 8.1
Two metallic balls with opposite electrical charges, on an insulating stand.

which makes possible the convenient manipulation of electricity for whatever purposes we have in mind. We can construct devices out of appropriate pieces of conducting and insulating materials so that electrical charge will flow when and where we want it to flow. When we plug a toaster into a wall socket, we do so with confidence that charge will flow through the wire in the toaster—not through the air, out across the kitchen table or anywhere else. This is truly an amazing fact, though familiarity makes it seem less astonishing than it really is. It is only because nature has provided us with materials which differ so widely in their ability to conduct electricity that a prosaic device such as a toaster can be made to work and to work reliably.

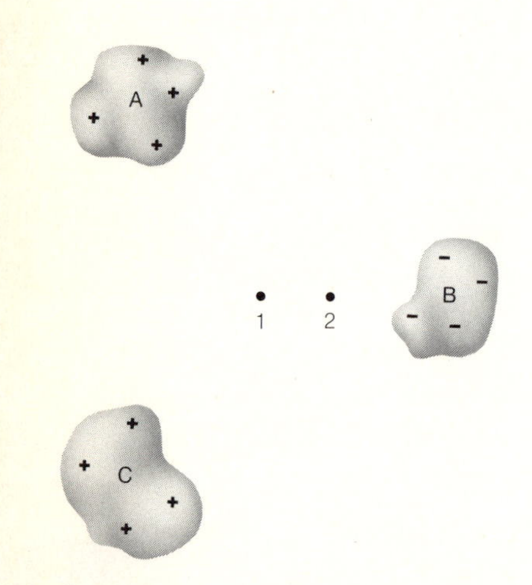

FIGURE 8.2
A positive charge placed at point 1 will experience a force to the right; if released, it will lose electrical potential energy as it moves from point 1 to point 2.

Coulomb's law, the basic law giving the force between two point charges (or between two objects small enough to be treated as point charges) is a very simple one. However, discussion of electrical effects is complicated by the fact that charged objects come in all shapes and sizes. Consider, for example, the arrangement of three objects shown in Figure 8.2. There are forces of attraction or repulsion between these various objects, resulting from their distributed charges. Furthermore, if we explore the region around them by putting another charge, q (a small "test charge"), at various points, this test charge experiences forces from all the charges on the various objects. A positive test charge placed at point 1 will experience an electrical force toward object B, for instance, but even if we know exactly how the charges are distributed, it is difficult to make an exact calculation of this force.

We also want to broaden the energy concept to include electrical energy. A positive test charge released at point 1 in Figure 8.2 will "fall" toward point 2, gaining kinetic energy. Just as a rock falling toward the earth loses potential energy, a moving charge loses *electrical* potential energy. We can immediately see that the discussion of electrical potential energy will be complicated by the fact that there are two kinds of charge; a *negative* charge released at point 1 will "fall" in the opposite direction, and we must be careful about the algebraic signs of electrical potential energy and of changes in electrical potential energy.

It is very useful to consider in detail one particularly simple example, an arrangement of two parallel metal plates as shown in Figure 8.3. If the two plates are separated by a short distance, and if one plate has a positive charge and the other a negative charge of equal magnitude, the electrical effects produced on test charges placed between the plates are exceptionally easy to describe. Such a pair of plates has a "capacity" for storing electrical charge; we refer to the pair of plates as a "parallel-plate capacitor" or, simply a "capacitor."

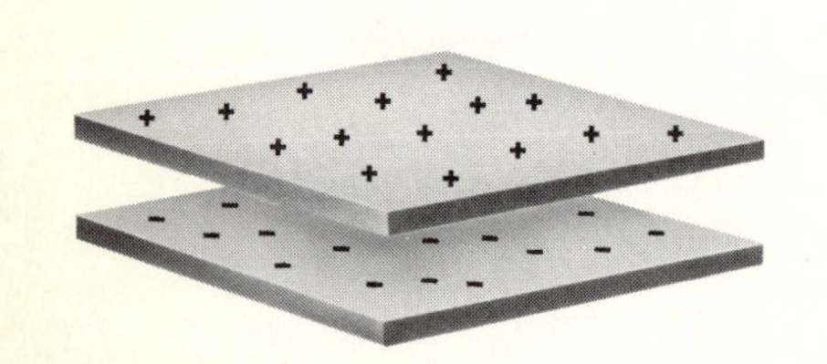

FIGURE 8.3
A parallel-plate capacitor.

If a positive test charge, q, is placed between the plates of this capacitor, it will feel both an attractive force from the lower plate and a repulsive force from the upper plate. Both these forces are in the downward direction and therefore combine to produce a large downward force. As long as the plates are quite close together and as long as the test charge is not out near the edges of the plates, the downward force is almost exactly the same no matter where the test charge is (see Figure 8.4). It is this fact which makes the capacitor a convenient and simple case to study.*

If we put a test charge with a different value of q inside the capacitor, the resulting electrical force increases or decreases in direct proportion to the value of q. This important fact follows from the fundamental equation describing Coulomb's law, Equation 8.1: the electrical force between two charges is directly proportional to the magnitude of each charge. The force on a test charge q is the sum of the forces between q and each bit of charge on the plates; if q is changed by substituting a different test charge, every contribution to this force changes in proportion to q and therefore so does the total force experienced by q.

This observation leads to the introduction of a useful concept. Any test charge placed between the plates of the capacitor experiences an electrical force, and we describe this by saying that the charges on the capacitor plates produce an *electric field* in the space between them. At each point, this field can be described by an electric field vector, $\boldsymbol{\mathcal{E}}$, defined as the force *per unit charge* exerted on a positive test charge placed at that point. In Figure 8.4, the electric field in the space between the plates is in the downward direction; its magnitude depends on the size and spacing of the plates and the amounts of charge on the plates. Once we know the electric field, we can calculate the electrical force that will be exerted on any test charge placed in this field. Since the electric field is the force per unit charge, the electrical force on a test charge q is simply:

$$\boldsymbol{F} = q\boldsymbol{\mathcal{E}}. \tag{8.2}$$

We can even use this equation to calculate the force on a *negative* test charge. If q is negative, then Equation 8.2 tells us to calculate the force by multiplying the vector $\boldsymbol{\mathcal{E}}$ by a negative quantity. We define multiplication of a vector by a negative quantity as an operation that gives a vector of the same *size* as if we had multiplied by a positive quantity but in the opposite direction.

What is the purpose of introducing the concept of electric field? When we describe the charges on the two plates as producing an electric field and then calculate the force on a test charge by using Equation 8.2, what we are really doing is to divide the calculation of the force on the test charge into two parts: first, the calculation of the field, and, second, the calculation of the resulting force. In this example, it is not necessary to use the concept of an electric field, but it does simplify the problem to divide it into two parts in this way. The "field" idea (first electric fields, then magnetic fields, later still other kinds of fields) has turned out to be more than just a computational device, however. In particular, light

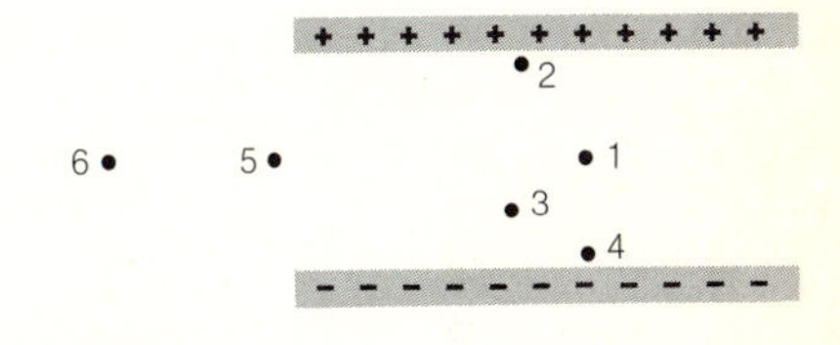

FIGURE 8.4
Side view of a capacitor. The force exerted on a test charge is approximately the same at points 1 through 4; the force will be different, however, on a test charge located at points 5 or 6.

*The total electrical force on a test charge between the capacitor plates results from the combined attractions and repulsions of all the charges on both plates. Vector addition of all these forces is called for. This calculation is omitted here, and we shall merely use the result, that the electrical force on a test charge is almost the same for all points that are not too near the edges of the plates.

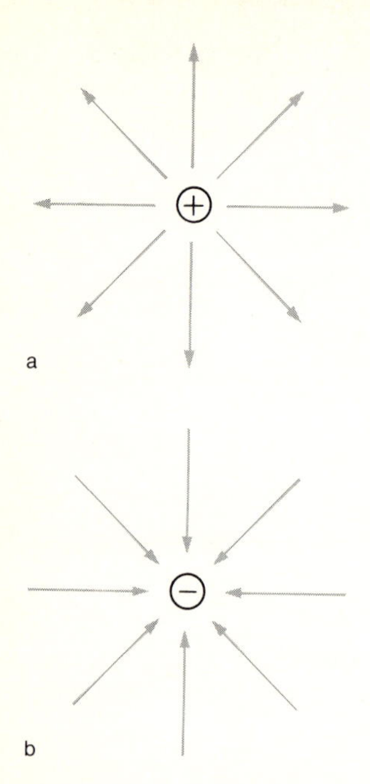

FIGURE 8.5
Electric fields produced by (a) a positive charge, and (b) a negative charge.

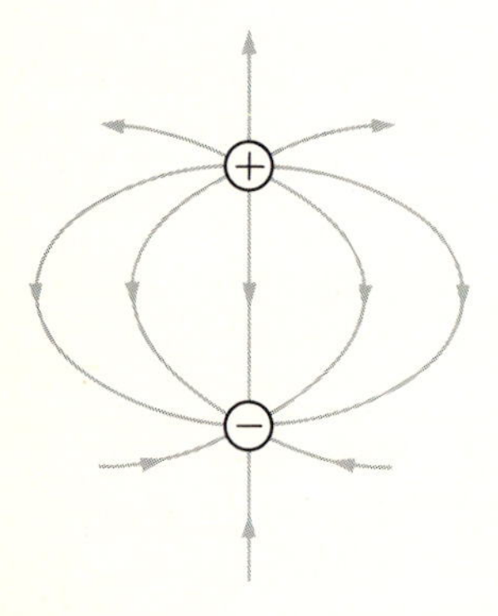

FIGURE 8.6
The electric field produced by a pair of charges of opposite sign.

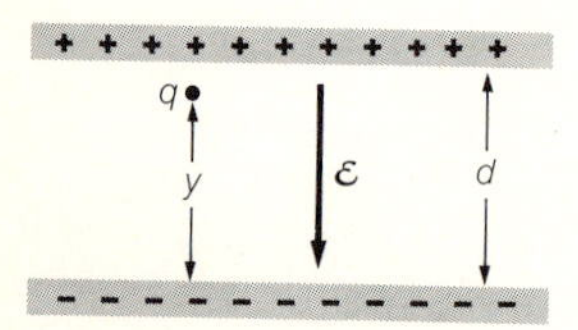

FIGURE 8.7
A charge q inside a capacitor.

waves and radio waves can be described as waves of electric and magnetic fields that propagate through space, and here the field concept is indispensable.

The units of $\boldsymbol{\mathcal{E}}$ are those of force divided by charge, newtons per coulomb in MKS units. In Figure 8.4, $\boldsymbol{\mathcal{E}}$ points downward; if its magnitude is, say, 1000 N/C and if q is $+10^{-6}$ C, then according to Equation 8.2, the force on q is in the downward direction, and its magnitude is

$$10^{-6}\ \text{C} \times 10^{3}\ \text{N/C} = 10^{-3}\ \text{N}.$$

If q is negative, say $q = -10^{-7}$ C, then although the electric field vector is the same (pointing down), the force is in the upward direction and has a magnitude of

$$10^{-7}\ \text{C} \times 10^{3}\ \text{N/C} = 10^{-4}\ \text{N}.$$

The concept of electric field is not restricted to the case of a capacitor. Any charged object or combination of charged objects produces an electric field, a field defined as above by the electrical force per unit positive test charge. A single positive point charge produces an electric field that is at every point directed away from the point charge (Figure 8.5a), and that has a magnitude that decreases as the inverse square of the distance. A negative point charge produces an electric field that is everywhere directed toward the point charge (Figure 8.5b); a positive test charge placed in this field experiences a force in the same direction as $\boldsymbol{\mathcal{E}}$, toward the center, whereas a negative test charge experiences a force in the opposite direction, away from the center. A positive charge and a negative charge together produce the more complicated field shown in Figure 8.6. The field can be calculated at each point by considering a test charge placed at that point and calculating the vector sum of the forces exerted on it.

A test charge between the plates of a capacitor will experience a force that is the same no matter where it is within the capacitor, as long as it is not too near the edges. This force is just like the gravitational force exerted on an object near the surface of the earth—everywhere downward and with a magnitude that does not depend on its position as long as we stay close to the surface. This similarity will be exploited below; the most important difference between the two cases is that, although all rocks experience a downward gravitational force, the direction of the force on a test charge depends on whether it has a positive or a negative charge. What can we say about the electrical potential energy of a charge in the uniform electric field of a capacitor? Suppose that the electric field in Figure 8.7 has a magnitude $\mathcal{E}$. If q is a positive charge, it experiences a downward electrical force whose magnitude is $q\mathcal{E}$. An object of mass m, moving under the influence of gravity near the earth's surface, experiences a downward force of magnitude mg; its gravitational potential energy is equal to mgh, where h is the height above some arbitrarily chosen zero-level. The gravitational potential energy is equal to the force, mg, multiplied by the height, h. In the electrical case, then, we again choose a zero-level, this time for *electrical* potential energy (EPE). If the zero-level is chosen to be *at* the negative plate, then the electrical potential energy at any other point is the magnitude of the force ($q\mathcal{E}$) times the distance from this zero-level. We will denote this distance by y rather than h, because it is not necessarily an actual height. If the capacitor is oriented as in Figure 8.7, this distance *is* measured in the upward direction, but it could just as easily be oriented as in Figures 8.8a or 8.8b; for all these positions, if we choose the zero-level for electrical potential energy to be at the negatively charged plate, the electrical potential energy of the charge, q, at some other point is given by

$$EPE = q\mathcal{E}y.$$

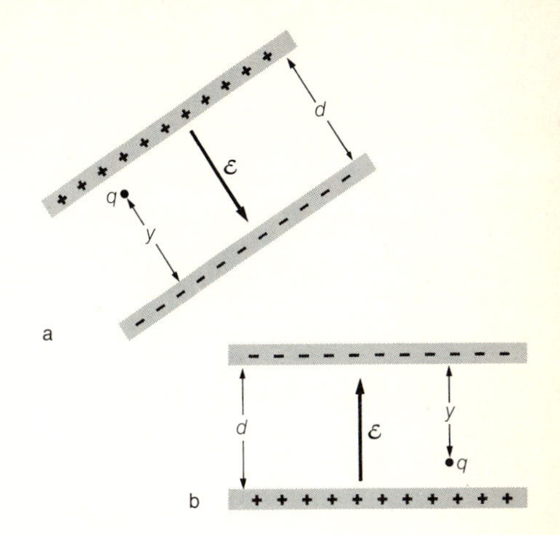

FIGURE 8.8
The capacitor shown in Figure 8.7, with different orientations. The distance y is still measured from the negative plate.

All of our previous results for the motions of rocks and baseballs can be adapted to the electrical case. A ball of mass, m, moves in such a way that the sum of its kinetic energy and its gravitational potential energy, ($\frac{1}{2}mv^2 + mgh$), keeps a constant value. A positive test charge (an object of mass, m, with a charge, q) between the plates of a capacitor moves in such a way that the sum of its kinetic energy and electrical potential energy, ($\frac{1}{2}mv^2 + q\mathcal{E}y$), keeps a constant value. Release such an object at the positively charged plate, for example, where $y = d$ and $EPE = q\mathcal{E}d$, and it will "fall" to the negative plate, losing an amount of electrical potential energy equal to $q\mathcal{E}d$ and therefore gaining an equal amount of kinetic energy.*

If the test charge is negative, then the direction of the electrical force is reversed. A negative charge will "fall" from the negative plate to the positive plate; if we continue to define the electrical potential energy of the charge to be zero when it is at the negatively charged plate, the electrical potential energy of a negative charge must be less (that is, negative) at the other plate. Electrical potential energy diagrams for the two kinds of test charge are shown in Figure 8.9. In each case, the change in electrical potential energy as the test charge moves from one plate to the other has a size equal to $q\mathcal{E}d$, but, for a positively charged object, motion from the negative plate to the positive plate produces an *increase* in electrical potential energy, whereas for a negatively charged object, the same motion results in a *decrease* in electrical potential energy.

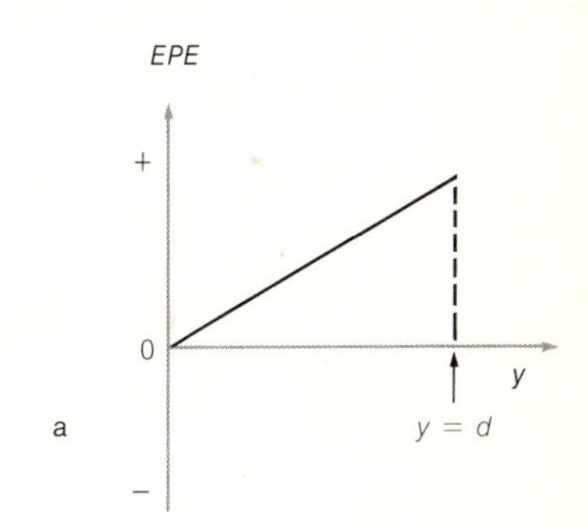

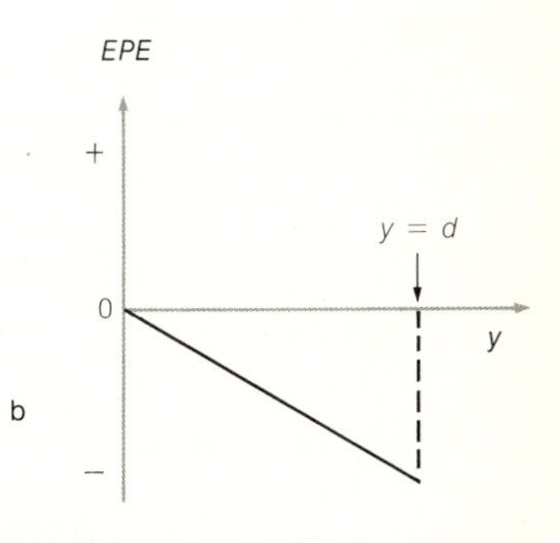

FIGURE 8.9
The electrical potential energy of a test charge in the capacitor of Figure 8.7 as a function of its distance from the negative plate: (a) a positive test charge; (b) a negative test charge.

In order to take a positive test charge, q, from the negative plate to the positive plate, its electrical potential energy must be increased by an amount $q\mathcal{E}d$. Some other form of energy must decrease in order for this to occur. This energy may come from the kinetic energy of the test charge itself if it is projected "up" from the negative plate, or it may come from some other form of energy. Whatever the source of energy, an amount equal to $q\mathcal{E}d$ is required to take it from the negative plate to

*For the sake of simplicity, we have been assuming that the gravitational force itself is much smaller than the electrical one. If it is not, then gravitational effects can be included by describing the motion of the test charge as one in which the sum of three forms of energy, ($KE + EPE + GPE$), keeps a constant value throughout the motion.

the positive one. If the size of q is doubled, twice as much energy is required and twice as large a change in electrical potential energy results, since the electrical potential energy at the positive plate, $q\mathcal{E}d$, is directly proportional to q. The energy required *per unit charge* is simply

$$\frac{q\mathcal{E}d}{q} = \mathcal{E}d,$$

a quantity that depends on the strength of the field and on the separation between the plates but *not* on the size of the test charge itself. This suggests the definition of a new concept, closely related to electrical potential energy, the electrical potential energy *per unit charge*, called the "voltage." Voltage is related to electrical potential energy in the same way that the electric field strength is related to the electrical force; voltage is electrical potential energy per unit charge, and electric field strength is electrical force per unit charge. (The voltage is often called the "electrical potential," but use of this term invites confusion because it sounds very similar to electrical potential *energy*. To avoid some of the confusion between these two closely related ideas, we will use the term voltage and the symbol V.) Just as for electrical potential energy, only *differences* in voltage are significant; to refer to "the" voltage, some zero-level must be chosen. If we choose the zero-level for voltage to be the same as that for electrical potential energy (at the negative plate), then the voltage of the positive plate (the voltage of the positive plate "with respect to" the negative plate) is just equal to $\mathcal{E}d$. Voltage has the units of energy per unit charge, joules per coulomb in MKS units; indeed, it is such an important quantity that this combination is given its own name, the volt (abbreviated as V):

$$1 \text{ J/C} = 1 \text{ V}.$$

QUESTION

8.2 Show that 1 volt can also be considered as 1 N-m/C, and that the units of $\mathcal{E}$ (newtons per coulomb) can also be expressed as volts per meter.

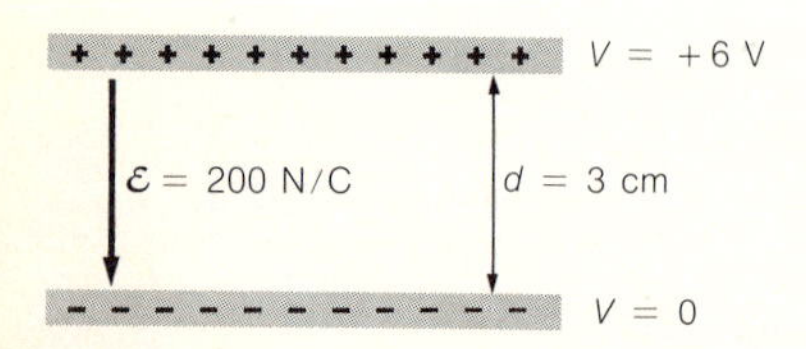

FIGURE 8.10
A capacitor in which the separation between the plates is 3 cm; in this example, the electric field has a magnitude of 200 N/C.

Consider a numerical example. Take a capacitor in which the separation, d, is equal to 3 cm (0.03 m) and the electric field strength, $\mathcal{E}$, has a magnitude of 200 N/C = 200 V/m (Figure 8.10). The voltage difference between the two plates is equal to $V = \mathcal{E}d = 200 \text{ V/m} \times 0.03 \text{ m} = 6 \text{ V}$ (or 6 J/C). If we define the voltage to be zero at the negative plate, the voltage at the positive plate is equal to +6 V. (The voltage *between* the plates is 6 V.) If a positive charge is moved from the negative plate to the positive plate (Figure 8.11a), its electrical potential energy will be increased. If the voltage difference is 6 V, then 6 J of energy must be supplied for every coulomb moved; conversely, if a positive charge moves from the positive plate to the negative plate (Figure 8.11b), 6 J of electrical potential energy are lost (converted into some other form of energy) for every coulomb. But if a *negative* charge is moved from the negative plate to the positive plate (Figure 8.11c), 6 J of electrical potential energy are lost for every coulomb moved; a negative charge can "fall" from the low-voltage plate to the high-voltage plate. In order to take a negative charge from the positive plate to the negative plate (Figure 8.11d), its electrical potential energy must be increased and energy must be supplied.

The concept of voltage can be used to define another unit of energy with an odd name, the *electron-volt* (eV), which is especially useful in describing atomic and nuclear phenomena. Suppose that the voltage difference between two points is 1 V (1 J/C), and consider a particle whose charge is equal to the electronic charge ($e = 1.6 \times 10^{-19}$ C). If it moves from one point to the other, its change in electrical potential energy is, by definition, equal to 1 electron-volt. That is,

$$1 \text{ eV} = 1.6 \times 10^{-19} \text{ C} \times 1 \text{ J/C}$$

$$= 1.6 \times 10^{-19} \text{ J}. \quad (8.3)$$

Suppose that a proton (which has a charge equal to the electronic charge) is released at the upper plate of the capacitor shown in Figure 8.10. As it falls to the lower plate, its electrical potential energy will decrease by 6 eV; we can therefore immediately say that its kinetic energy when it hits the lower plate will be equal to 6 eV. If desired, we can convert this into joules: 6 eV $= 9.6 \times 10^{-19}$ J. The apparatus described is a rudimentary "particle accelerator," a device for increasing the kinetic energy of a particle by a known amount.

To summarize this discussion of voltage, a charged object (or a set of charged objects, such as the two plates of a capacitor) produces an electric field, and the concept of voltage provides a useful way of describing the energy changes of other charged objects moving in this field. If a positive charge moves from a point of low voltage to a point of higher voltage, its electrical potential energy must increase, by an amount equal to its charge times the change in voltage. If it moves in the opposite direction, it loses a corresponding amount of electrical potential energy. Except for a reversal in sign, the same is true for a *negative* charge. For a negative charge, there is a decrease in electrical potential energy as it moves from a point where the voltage is low to a point of higher voltage, and an increase when it moves in the opposite direction.

The concept of voltage is useful primarily because it provides a convenient way of describing the changes of electrical potential energy of charges moving in an electric field. There are two other reasons for the usefulness of the concept. First, in practice we do not have to determine voltage differences by taking a test charge and measuring energy changes: there are instruments (voltmeters), which can simply be connected to the two plates of a capacitor and which give a reading equal to the

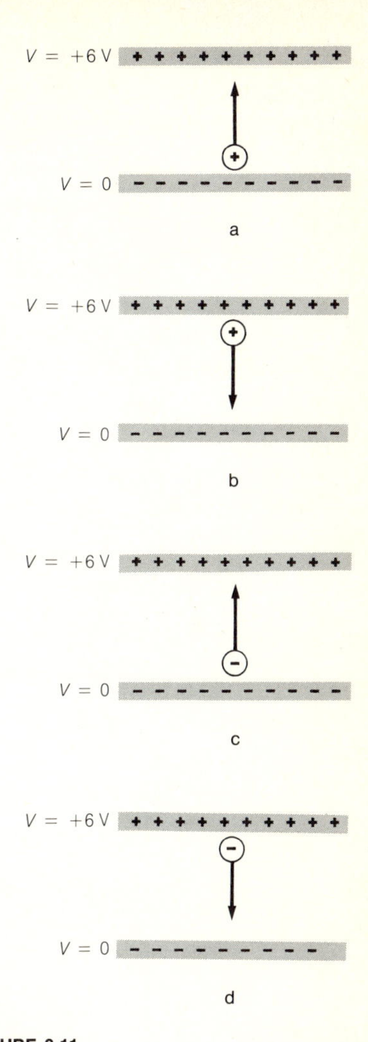

FIGURE 8.11
As a charge moves from one plate to the other, there is a change in electrical potential energy: an increase in (a) and (d), a decrease in (b) and (c).

voltage difference. Second, as we will see in §8.4, a *battery* is a concoction of chemicals with two protruding terminals, that acts in such a way that, under a range of conditions, it maintains a definite voltage difference between the two terminals. Similarly, the common household electric outlet contains the ends of two wires with a voltage difference between them supplied by the power company. In this case, the voltage difference is an *alternating* one, but nevertheless it is the voltage difference between these two wires that we rely upon to run our appliances.

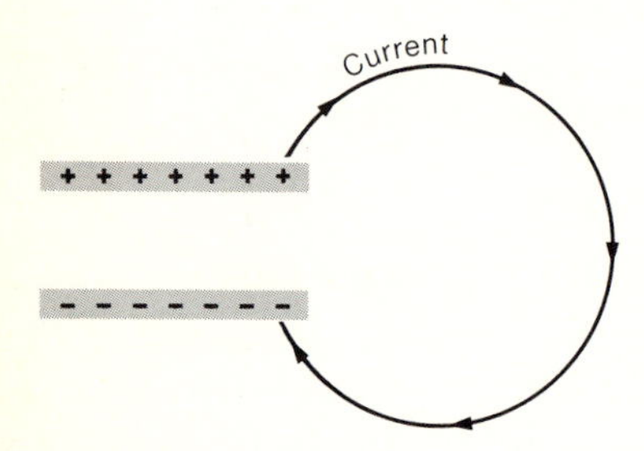

FIGURE 8.12
When the two plates are connected by a wire, there is a momentary flow of positive charge in the clockwise direction.

A charged capacitor contains stored electrical energy. If a positively charged object were removed from the positive plate and released, it would "fall" to the negative plate, losing electrical potential energy and gaining kinetic energy. Is there some other way of converting electrical energy into another form? Suppose that the two plates are connected by a wire (Figure 8.12). The instant that contact is made, the excess positive charge on the upper plate flows through the wire to the lower plate, there is a momentary *current* in the clockwise direction, and—almost instantaneously—the capacitor is discharged, and both plates are then electrically neutral. Where has the electrical potential energy "gone"? Our experience with conservation of energy leads us to expect that the missing energy has not simply vanished but has been converted into some other form. The electrical energy lost has suffered the same fate as the missing energy in many other examples; it has been transformed into thermal energy. The wire gets slightly warmer during the discharge of the capacitor. We have here the elements of an electric hot plate or stove or toaster. It is not a very practical toaster, to be sure, but the essential energy conversion ($EPE \longrightarrow TE$) is the same in a real toaster as in this example of the discharge of a capacitor through a wire. The current is of course only a momentary one in this example, since the capacitor is quickly discharged and the voltage difference between the plates drops to zero. If there were a way of maintaining the voltage difference, of producing new electrical potential energy to replace that which is used up, we could produce a steady current in the wire and continuous heating—a continuous conversion of electrical energy into thermal energy.

There is still one important question about currents that should be raised: is it not true that it is really the negative electrons which flow through the wire? If so, what really happens in the discharge of the capacitor is that the excess electrons on the lower plate flow through the wire; that is, there is a flow of *negative* charge in the *counter*clockwise direction in Figure 8.12. True enough, and the thermal energy is actually generated by collisions between the moving electrons and the atoms of the wire, collisions that cause an increase in the random motions of these atoms. There are two facts, however, that make it legitimate to describe the current as if it were due to the motion of positive charge. First, when negative charge moves from the lower plate to the upper plate, electrical potential energy is lost, just as it is when positive charge moves in the opposite direction (see Figure 8.11). Second, just as it is impossible to tell from macroscopic observations whether a "positively charged" object has an excess of positive charge or a deficiency of negative charge, it is also extremely difficult to distinguish between a flow of positive charge in one direction and a flow of negative charge in the opposite direction. (In some situations, positive charges *do* move, and in some a current is due to the motion of both positive and negative charges in opposite directions.) None of our important results depend on whether a current is considered as a flow of positive charge in one direction or as a flow of negative charge in the opposite direction. Unnecessary confusion would be produced if all the possibilities were discussed in each instance, and so in most cases, we will use the conventional description of a current as a flow of *positive* charge.

§8.4 BATTERIES

A battery, found in any flashlight, portable radio, or automobile, does just what was called for in the previous section. A battery has two terminals between which is maintained, under most circumstances, a definite voltage difference—a voltage called the "electromotive force" of the battery. (This is not a good term because this is a voltage, not a force, but the name has stuck, and we will simply refer to this voltage as the "emf" of the battery.) One can study batteries in two different ways. One approach, a "phenomenological" approach, is to take a battery as it is and find out how it behaves under various circumstances, see what phenomena result when we use it. On the other hand, in order to understand why a battery behaves as it does, it is necesary to examine the chemical reactions going on in its interior, to "cut it open" to see how it works.

From a phenomenological point of view, a battery is simply a device with two metal terminals, between which there is some voltage difference, a voltage difference that is somehow maintained at an approximately constant value even when charge is drained off by connecting a wire from one terminal to the other. *Something* must be going on inside the battery to replenish the charge drained off, to maintain the voltage difference. The two terminals need not be shaped like the two plates of a capacitor. The important thing is that there are two metallic terminals with a known voltage between them. Thus if a known amount of charge moves from one terminal to the other, the change in electrical potential energy can be calculated. We represent a battery by a symbol such as this:

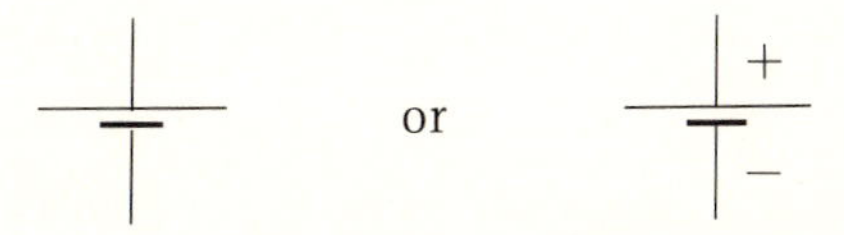

The longer of the two bars is the "positive terminal"; the voltage of this terminal is positive with respect to the other terminal. If a wire is connected between the terminals to provide a conducting path, charge will flow through the wire (losing electrical potential energy and generating thermal energy in the wire), and this tends to reduce the voltage difference between the terminals. If a battery does operate as described, maintaining a fixed* voltage difference between the terminals even as charge is constantly drained off through the wire, then positive charge must be continually transported within the battery from the negative terminal to the positive terminal, with a continuous production of electrical potential energy, and we can thus postulate that a battery cannot continue to supply current indefinitely. A device that could continue to produce electrical potential energy forever, accompanied by no permanent changes within the device itself, would be a perpetual motion machine of the first kind. The production of electrical potential energy must occur at the expense of the chemical energy of the battery, and sooner or later the supply of chemical energy will be depleted; the battery will run down and will no longer be able to maintain a voltage difference.

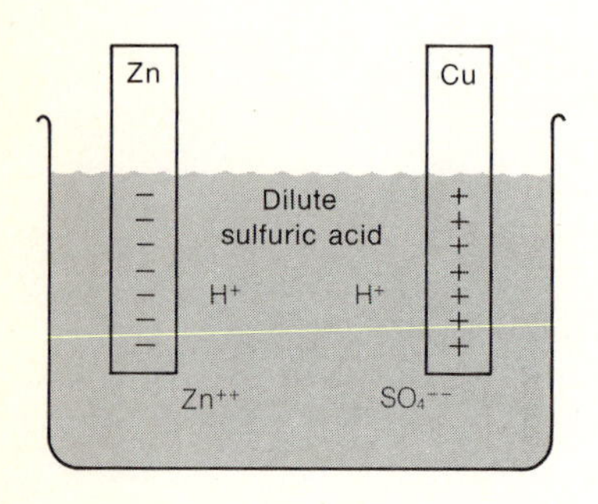

FIGURE 8.13
A copper-zinc battery.

Although a phenomenological approach is very convenient for describing how batteries are used, let us now take a brief look at the inside of a battery. Batteries can be made in many different ways. One simple type of battery (though not a very practical one) consists of two metal "electrodes," one of copper (Cu) and the other of zinc (Zn), dipped into a jar of dilute sulfuric acid (Figure 8.13). The conventional formula for sulfuric acid is H_2SO_4. Actually there are no H_2SO_4 molecules in the water, but instead there are doubly charged negative sulfate ions (SO_4^{--}) and singly charged positive hydrogen ions (H^+). A sulfate ion consists of one sulfur atom and four oxygen atoms plus two extra electrons, so that a sulfate ion has a net charge equal to twice the charge of an electron. Similarly, a hydrogen ion is a hydrogen atom that has lost its electron, in other words simply a proton, with a net positive charge equal in magnitude to the electronic charge.

Zinc tends to dissolve in sulfuric acid, and when a zinc atom enters the solution, it does so, not as a zinc atom but as a doubly charged positive ion (Zn^{++}), leaving two of its electrons behind, giving a negative charge to the left-hand electrode. This is a chemical reaction that can be described as follows:

$$\underset{\text{(from electrode)}}{Zn} \longrightarrow \underset{\text{(left behind in electrode)}}{2e^-} + \underset{\text{(in solution)}}{Zn^{++}}. \tag{8.4}$$

Although copper, like zinc, has some tendency to dissolve in sulfuric

*We will assume that the wire connected between the terminals is a long thin one, which allows only a small current to flow. If the wire is short and thick, a very large current will flow, and no battery can keep the voltage difference truly constant.

acid, it does so to a lesser extent than does zinc. With both a zinc and a copper electrode in the same jar, there is a more important process that takes place at the copper electrode; copper gives electrons to the solution, and these electrons combine with hydrogen ions to form hydrogen gas (H_2), which is released:

$$\underset{\text{(from solution)}}{2H^+} + \underset{\text{(from electrode)}}{2e^-} \longrightarrow \underset{\text{(released as a gas)}}{H_2}. \tag{8.5}$$

This reaction leaves a deficiency of electrons on the copper electrode and gives the copper electrode a net positive charge. After some time these reactions are suppressed. The zinc does not continue to dissolve indefinitely, because as the negative charge on the zinc electrode builds up, it attracts previously dissolved positive zinc ions and inhibits more zinc ions from entering the solution. Similarly, the positively charged copper electrode tends to keep positively charged hydrogen ions away and thus reaction 8.5 is also suppressed. An equilibrium state is reached in which the copper electrode has a voltage of about +1.1 volts with respect to the zinc electrode. This voltage, 1.1 V, is the emf of this particular kind of battery. The two electrodes are the terminals of the battery, the copper being the positive terminal and the zinc the negative terminal.

Now suppose that the circuit is completed by connecting a piece of wire between the two electrodes (Figure 8.14). Excess electrons in the zinc electrode flow through the wire to the copper electrode. This flow reduces the negative charge on the zinc electrode and the positive charge on the copper electrode; the conditions described above for suppression of the chemical reactions are no longer as effective, and so further reactions (8.4 and 8.5) take place. As long as the rate at which electrons flow through the wire is not too large, a dynamic steady state is maintained, in which there is a constant voltage difference of 1.1 V between the two electrodes. Both chemical reactions proceed at a steady rate, zinc steadily enters the solution at the left, positive hydrogen ions move across the jar from left to right, hydrogen gas is evolved at the right-hand electrode, and there is a steady flow of electrons in the wire. Once the process has begun, there is no change in the net amount of charge accumulated anywhere. For every zinc ion that goes into solution (taking two units of positive charge), two electrons leave the zinc electrode and enter the wire; there is no net change in the charge on the zinc electrode. Likewise for every pair of electrons leaving the copper electrode to combine with hydrogen ions (8.5), two electrons enter the copper electrode from the wire, and so the charge on the copper electrode also is unchanged.

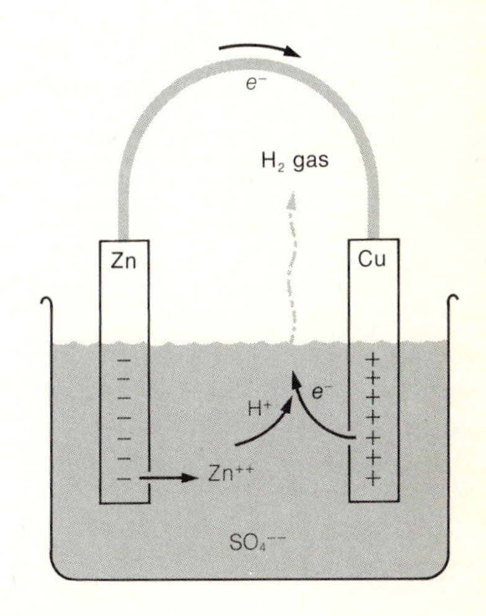

FIGURE 8.14
A battery produces a current in a wire connected between the terminals.

Although there is no continuing buildup of either positive or negative charge anywhere, there is a steady current, a steady flow of charge around the circuit. The current (denoted by the symbol I) is the rate of flow of charge. Its MKS unit is the ampere (A), which is defined as a flow of one coulomb of charge per second:

$$1 \text{ A} = 1 \text{ C/sec.}$$

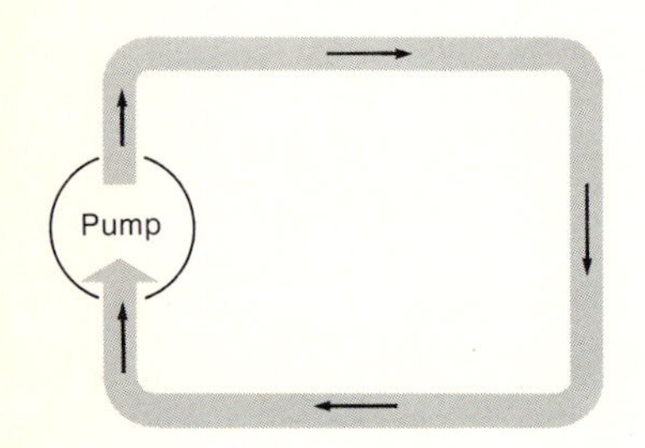

FIGURE 8.15
The flow of water through the cooling system of a car is very similar to the flow of charge in an electrical circuit.

A flow of charge is very much like the flow of water through a closed system of pipes (Figure 8.15), as in the cooling system of a car, for example. There is a complication in the electrical case, because electrical charge is of two types, positive and negative, and some care is necessary in describing the direction of the current. The essence of the actual motion of charge is that negative charge flows from left to right through the wire (Figure 8.14), and positive charge moves from left to right within the battery. However, a flow of one type of charge has the same effect as a flow of the opposite type of charge in the opposite direction, so that the electrons moving from left to right through the wire are equivalent to positive charge flowing in the opposite direction. Thus the overall effect is that of a circulating current (a continuous flow of positive charge) around the circuit in the counterclockwise direction (left to right within the battery, right to left within the wire). Describing this as a continuous flow of positive charge does not provide a complete description of what is occurring at the atomic level, but it is perfectly adequate for describing the processes of energy conversion.

There is no continuing accumulation of charge anywhere in the circuit; charge flows continuously around and around in a closed path. However, important energy conversions are going on: (1) Inside the battery, chemical energy is being converted into electrical potential energy, and (2) electrical energy is being converted into thermal energy in the wire. Consider first the chemical reactions. Each time reaction 8.4 occurs, so does reaction 8.5. The net effect of these two reactions is that a zinc atom goes into solution as a zinc ion, and two hydrogen atoms leave the solution as hydrogen gas:

$$\underset{\text{(from electrode)}}{Zn} + \underset{\text{(from solution)}}{2H^+} \longrightarrow \underset{\text{(in solution)}}{Zn^{++}} + \underset{\text{(released as a gas)}}{H_2}. \qquad (8.6)$$

As this reaction occurs, two units of positive charge move through the battery from left to right. The direction of the electric field, $\boldsymbol{\mathcal{E}}$, within the battery is from right to left; positive charge is being carried *against* the direction of the field, just as if we were physically carrying it from one capacitor plate to another, as in Figure 8.11a. Thus electrical potential energy is being increased, and this increase must occur at the expense of chemical energy; that is, the net reaction (8.6) is one in which chemical energy decreases as a result of the rearrangement of zinc nuclei, hydrogen nuclei, and electrons. (A similar decrease of chemical energy occurs during ordinary processes of combustion, although the specific chemical reactions are different.) It is clear that this cannot continue forever; the zinc electrode will eventually be totally dissolved, and the reactions will stop. That is, the battery will "run down."

At the same time that chemical energy is being converted to electrical potential energy within the battery, electrical potential energy is being converted to thermal energy within the wire, as positive charge flows through the wire from right to left, from the high-voltage terminal to the low-voltage terminal. (Alternatively, what is really happening in the wire is that electrons move from the low-voltage terminal to the high-voltage terminal; whether the current is treated as a flow of positive charge in one direction or as a flow of negative charge in the opposite direction, electrical potential energy is lost.)

Consider the motion of 1 C of positive charge all the way around the closed circuit. The voltage between the two terminals is 1.1 V, and so, as 1 C of charge moves across the battery from left to right, its electrical potential energy increases by 1.1 J. Then as it moves through the wire (from the copper electrode back to the zinc), its electrical potential energy decreases by 1.1 J, and this lost electrical energy appears as thermal energy ("heat") in the wire. We can symbolize the overall energy conversion processes by writing

$$CE \longrightarrow EPE \longrightarrow TE.$$

Chemical energy is being used up within the battery to produce electrical potential energy, which in turn is converted into thermal energy.

There are many other ways of making batteries, all based on the same general principles. Almost any pair of metals can be used as electrodes, and they can be immersed in a wide variety of solutions. It is important that the two metals be different from one another. If both electrodes were made of copper or both of zinc, the battery would be symmetrical and no voltage would be developed. Sometimes the electrodes are immersed in a paste of chemicals instead of a solution, and then we have a "dry cell" rather than a "wet cell." Each set of electrodes and chemicals has its own set of chemical reactions and makes a battery with its own characteristic emf. A number of units are often packaged together to make a single battery two, three, or more times as large. We could construct a 3.3-V battery, for instance, by connecting together in *series* three copper-zinc batteries (Figure 8.16).* The common dry cell, which comes in various sizes (for instance as a flashlight battery), has

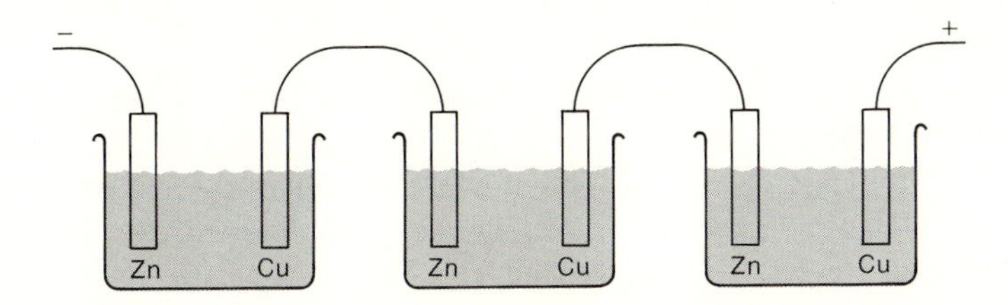

FIGURE 8.16
A combination of three copper-zinc cells to make a 3.3-volt battery.

*Sometimes the term "battery" is reserved for such a composite device, and a single unit is referred to as a "cell." Thus the arrangement shown in Figure 8.16 might be referred to as a 3.3-V battery consisting of three 1.1-V cells.

an emf of 1.5 V. These are often packaged together to form batteries with larger voltages, six small ones, for example, to make a 9-V battery often found in transistor radios. Car batteries are made of cells each with an emf of 2 V; three of these are connected to make a 6-V battery and six of them to make a 12-V battery. We will not try to give a catalog of all the possible types of batteries. Batteries differ from one another not only in the materials from which they are made and the emf they produce, but also in size, in weight, and in the amount of chemical energy they contain and therefore in the amount of electrical energy they can deliver before they run down.

There is another important difference between various kinds of batteries: some batteries can be "recharged" by making current flow through them in the "wrong" direction. The chemical reactions then proceed in the opposite direction, and the chemical composition of the battery can be restored almost to its original state. The chemical energy of the battery is replenished at the expense of electrical energy, which must be supplied from some outside source. The copper-zinc battery is not a good example of a rechargeable battery, because the hydrogen gas produced (reaction 8.5) is given off as a gas and is no longer available. If hydrogen were available, then while recharging such a battery, the overall reaction would go in the opposite direction:

$$\underset{\text{(gas)}}{H_2} + \underset{\text{(in solution)}}{Zn^{++}} \longrightarrow \underset{\text{(in solution)}}{2H^+} + \underset{\text{(building up on the zinc electrode)}}{Zn.}$$

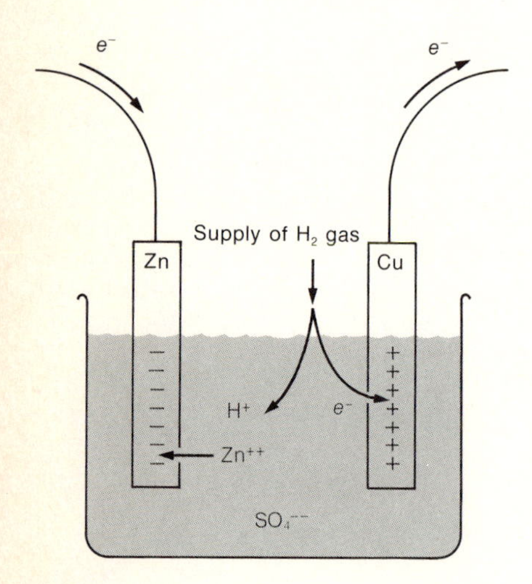

FIGURE 8.17
A battery being recharged.

Inside the battery, positive charges move from the copper electrode to the zinc electrode; the electrical potential energy decreases by 1.1 J for every coulomb of charge that flows through the battery (Figure 8.17). Electrical potential energy is lost, but the chemical energy of the battery is restored. Hydrogen ions go back into solution, and zinc comes out of solution and builds up on the zinc electrode. Ideally this process could be continued until the battery was restored to its original condition. It is not quite correct to speak of this process as "recharging"; whether the battery is being recharged (Figure 8.17) or being used in the normal way (being "discharged") (Figure 8.14), its electrical *charge* is not changing, since the same amount of charge goes into one terminal as comes out of the other.

Some types of batteries, such as dry cells, can be recharged only to a very slight extent; others such as automobile "storage batteries" can be almost completely recharged. When one starts a car, the battery is used to deliver electrical energy at a high rate at the expense of chemical energy; once the engine has started, the battery is slowly recharged by electrical energy produced in a generator or "alternator" driven by the engine.

We have examined here the processes that take place inside a battery. In seeing how batteries are used, we will return to a phenomenological approach and consider a battery as a two-terminal device that "tries" to maintain a definite voltage difference between its two terminals. Even when current is being supplied by the battery, this voltage stays at a nearly constant value (unless too large a current is drawn or if the battery has been run down); electrical potential energy is created as charge is transported within the battery, but at the expense of the chemical energy of the battery.

§8.5 SIMPLE BATTERY-OPERATED CIRCUITS

§8.5.A Energy Relationships

Let us now look more carefully at the simplest and therefore perhaps the most important type of electrical circuit, a battery connected to a wire. (We will assume that the wire is long enough and thin enough so that the current is not so large as to use up the chemical energy of the battery too rapidly.) We denote the wire by the symbol —ww— and draw a diagram of the circuit as shown in Figure 8.18, a diagram that might be used, for instance, as the circuit diagram of a toaster or a hot plate. In order to measure voltages and currents, voltmeters and ammeters* can be added, as shown in Figure 8.19. In this circuit, the reading of the voltmeter is simply equal to the emf of the battery. The two ammeters shown in Figure 8.19 are found to indicate equal currents; the same current flows in all parts of this circuit. (It is really the great difference between conductors and insulators mentioned earlier which is responsible for this fact. It is so much easier for current to flow through the wire rather than through the insulating materials surrounding it that we can be sure that the charge going into the wire at one end all comes out at the other end.)

Imagine that we could watch some definite amount of positive charge, q, as it travels around the circuit, from A through the battery to B, back through the wire to A, and so on (Figure 8.18). What can we say about

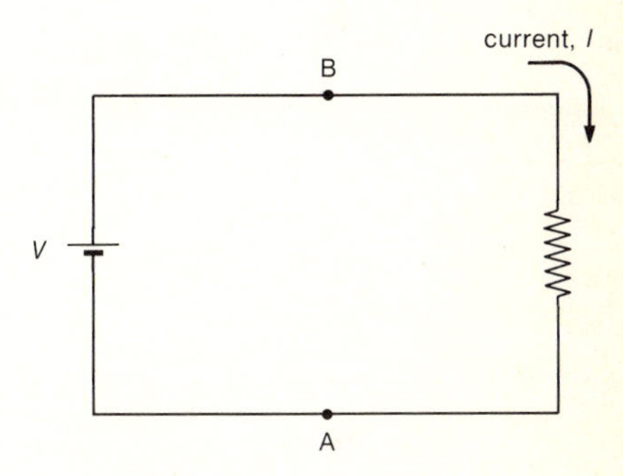

FIGURE 8.18
A simple battery-powered circuit, a battery with an emf equal to V producing a steady current I.

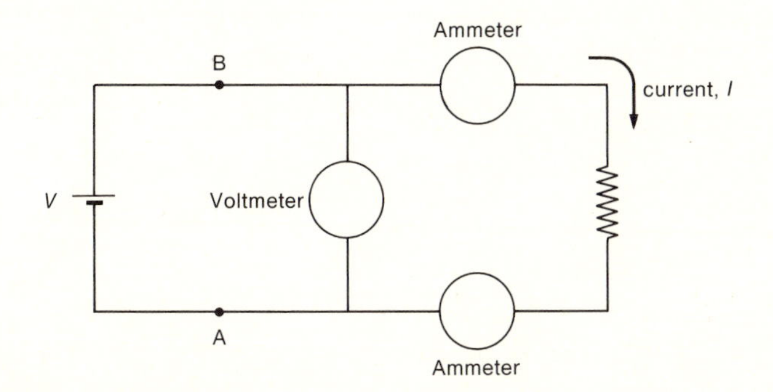

FIGURE 8.19
Meters can be added—voltmeters to measure the voltage, and ammeters to measure the current.

*The principles governing the operation of ammeters and voltmeters will be discussed in Chapter 9.

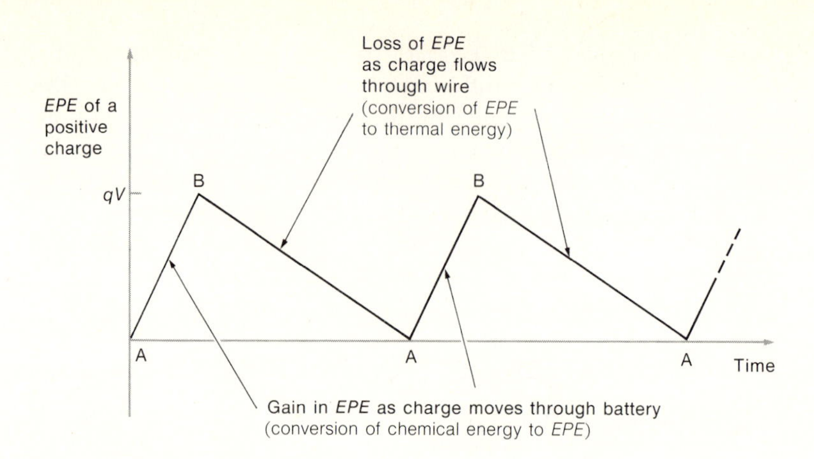

FIGURE 8.20
Electrical-potential-energy history of a positive charge flowing around the circuit shown in Figures 8.18 and 8.19.

its energy? We will define the voltage to be zero at point A, and therefore at A the electrical potential energy of q is zero. As q moves through the battery to B, where the voltage is V, its electrical potential energy increases by qV. (The chemical energy of the battery simultaneously decreases by the same amount.) Then as q moves through the wire, it loses this electrical potential energy. Much like a leaf fluttering down through the air, its kinetic energy does not steadily increase; rather, any kinetic energy it acquires is almost immediately dissipated through collisions with atoms in the wire, producing thermal energy. Figure 8.20 shows the electrical-potential-energy "history" of such a charge as it flows around and around the circuit.

In this circuit, it is not just a single charge that flows, but rather there is a steady flow of charge past every point in the circuit, a continuous conversion of chemical energy to electrical potential energy in the battery, and a corresponding continuous conversion of electrical potential energy to thermal energy in the wire. Because there is a continuous conversion of energy from one form to another, it is appropriate to relate the *rates* of energy conversion to the current, I, and to the voltage of the battery, V. At any point in the circuit, charge is flowing past at the rate of I amperes (I coulombs per second). The total amount of charge, Q, passing any point in the circuit during a time interval Δt is just

$$Q = I\,\Delta t.$$

In effect, an amount of charge equal to Q coulombs has gone completely around the circuit, and therefore the amount of chemical energy converted to electrical potential energy is:

$$QV = I\,\Delta t V.$$

During the same period of time, an equal amount of electrical potential energy is converted into thermal energy. Thus the rate at which chemical energy is being converted into electrical potential energy, and the equal rate at which electrical potential energy is being converted into thermal energy, is

$$\frac{QV}{\Delta t} = \frac{I\,\Delta t V}{\Delta t} = IV.$$

(Notice that electrical potential energy is being continuously produced, but is equally rapidly converted into thermal energy. No net increase or decrease of electrical potential energy takes place while the current flows. Electrical energy is an intermediate in the process of conversion of chemical energy into thermal energy.) The units of IV are those of energy per unit time, that is, of *power*, watts in the MKS system. In this simple circuit, we can write

$$P = IV, \tag{8.7}$$

and refer to P as the rate at which chemical energy is being converted into electrical potential energy, the rate at which electrical potential energy is being converted into thermal energy, or (for the overall process) the rate at which chemical energy is being converted into thermal energy.

QUESTION

8.3 Are the units correct in Equation 8.7? Is it true that (amperes) × (volts) = watts?

Consider a specific numerical example (Figure 8.21). Suppose that we have a 6-V battery and that the current is 2 A (2 C/sec). For every coulomb passing through the battery from the negative terminal to the positive terminal, the increase in electrical potential energy (and the decrease in chemical energy) is 6 J, and so in every second, the amount of chemical energy converted to electrical energy is 2 × 6 = 12 J. At the same time, during each second 2 C of charge flows through the wire, each coulomb losing 6 J of electrical potential energy, and thus 12 J of thermal energy is produced every second. The power being delivered to the wire by the battery is 12 W:

$$P = IV = 2 \times 6 = 12 \text{ W}.$$

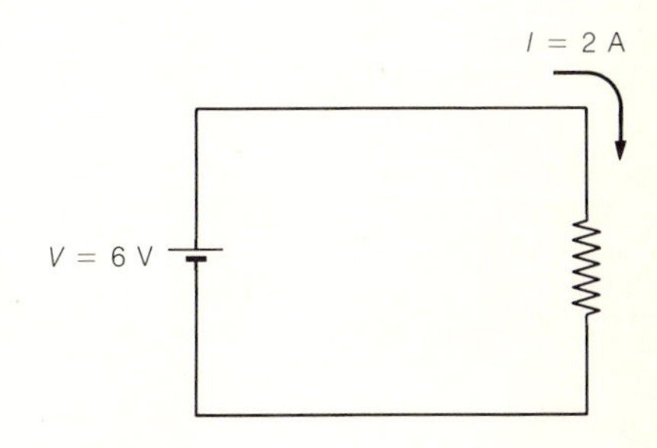

FIGURE 8.21
A particular example of the circuit shown in Figure 8.18.

We have arrived at some extremely important results about the relationship between current, voltage and rates of energy conversion in electrical circuits, results that are applicable to circuits much more complicated than the example that has been discussed thus far. The central result is expressed by Equation 8.7:

$$P = IV \quad \text{(power = current × voltage)}.$$

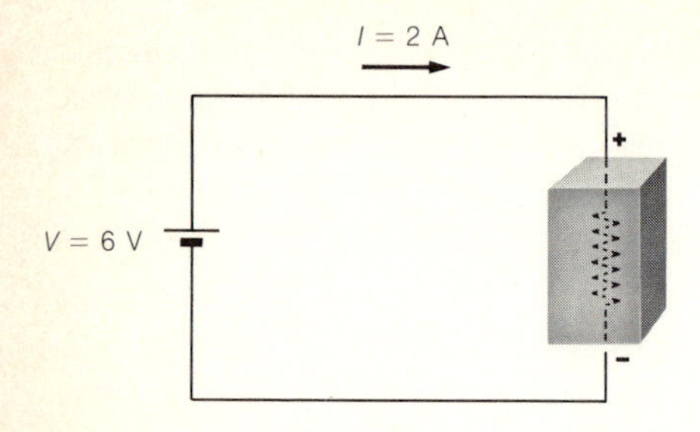

FIGURE 8.22
Even if the wire is hidden inside a box, we can still measure the voltage and current, and thus determine the rate at which electrical energy is going into the box.

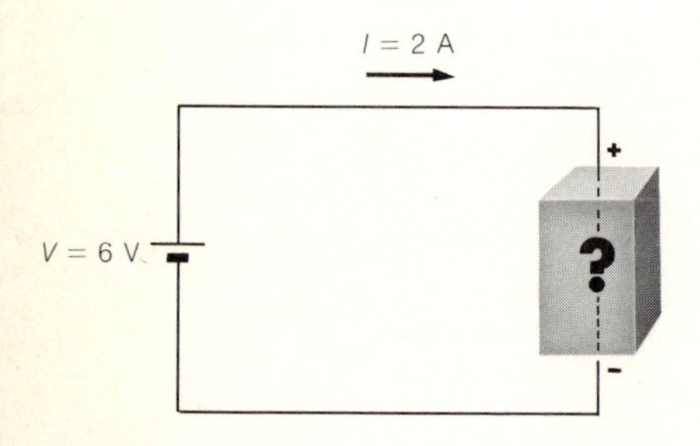

FIGURE 8.23
No matter what may be in the box, we can conclude that electrical energy is being delivered at the rate of 12 watts.

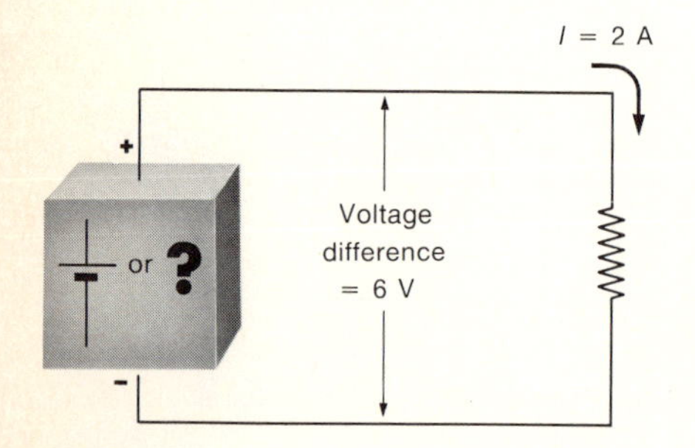

FIGURE 8.24
Whether the left-hand box contains a 6-volt battery or something else, if the current and voltage have the values shown, then electrical energy is being produced within the box at the rate of 12 watts.

Consider once more our example of the battery and the piece of wire, with $V = 6$ V and $I = 2$ A, but imagine that we cannot see the wire, that the wire is enclosed in a box and that we can see only the two protruding terminals (Figure 8.22). We can still *measure* the voltage difference between the terminals, and with an ammeter we can still measure the current flowing through the box, going in at the high-voltage terminal and coming out at the low-voltage terminal. If someone were to get into the box and substitute something for the wire (a different wire, an interconnected set of wires, or perhaps an electric motor), then as long as the voltage and current remained the same, we could not tell from outside that any change had been made. We could still deduce, though, that electrical energy was going into the box at a rate of 12 W, that electrical energy was being used up at this rate. The point is that we know that the current is 2 A and therefore that, in any one second, 2 C of charge flows through the box; because charge flows from the point of higher voltage to a point whose voltage is 6 V lower, each coulomb loses 6 J of electrical potential energy as it goes through the box. Therefore electrical potential energy is going into the box at a rate of $2 \times 6 = 12$ J/sec.

We can generalize this result. For *any* object with two terminals (Figure 8.23), no matter how complicated it may be, if there is a voltage difference, V, between the two terminals and if a current, I, flows in at the high-voltage terminal and out at the low-voltage terminal, then electrical potential energy is being used up at the rate of $P = IV$. If the "object" is simply a piece of wire, all of this electrical potential energy is being converted into thermal energy. If the object includes a motor, then some of this energy is perhaps being converted into kinetic energy of the motor, kinetic energy of the machinery to which the motor is attached, or perhaps to the gravitational potential energy of an elevator that is being raised by the motor. The "object" might even contain another battery that is being recharged, in which case the electrical energy delivered to the box is being partially converted into the chemical energy of this second battery.

We can take a similar point of view toward the other part of the circuit shown in Figure 8.21, the battery. Suppose the battery were enclosed in a box, and imagine that someone were able to substitute something else for the battery without altering the voltage difference or the current. We could still conclude that some form of energy was being converted into electrical energy at the rate of 12 W, because, in any one second, 2 C of charge flows into the low-voltage terminal, and when it comes out at the high-voltage terminal, each coulomb has gained 6 J of electrical potential energy. We can generalize this result as well. Consider any object with two terminals between which there is a voltage difference, V. If a current, I, flows in at the *low*-voltage terminal and out at the high-voltage terminal (Figure 8.24), then electrical energy is being produced and delivered to the outside at the rate of $P = IV$. In our example, the object was a battery, and this energy was produced at the expense of the chemical energy of the battery. It could be that the object contains a tiny person equipped with a collection of balloons and cats, who is frantically rubbing them together and carrying charge from one terminal to the other. Lest this appear frivolous, the point is that from the outside we cannot tell precisely what is going on inside, but we *can* make some definite statements about the rate of production of electrical energy. More realistically, the "object" might be a waterfall driving an electrical generator: then electrical energy would be produced at the expense of the gravitational potential energy which the water loses as it comes down the waterfall. Whatever the object may be, if current flows in at the low-voltage terminal and out at the high-voltage terminal, then electrical energy is being produced and delivered at the rate $P = IV$.

§8.5.B The Electrical Equivalent of Heat

There is one immediate and important application of these results that we can make, an application that serves as a check on the correctness of what has been done in this chapter and as another confirmation of the law of conservation of energy. If electrical energy is being converted into thermal energy in a wire and if the wire is surrounded by a container of water, we can directly measure the amount of thermal energy by observing the temperature rise of the water (Figure 8.25). Since we can also calculate the amount of electrical energy lost, the two figures can be compared to find the number of joules equivalent to one calorie. (The electrical power is $P = IV$; if the experiment is continued for a time Δt, the total amount of electrical energy lost is equal to $IV\,\Delta t$.) Such an experiment is analogous to the experiments of Joule described in Chapter 6, in which gravitational potential energy or other forms of mechanical energy were used up and the resulting rise in temperature of the water observed. Those experiments were described as a determination of the "mechanical equivalent of heat"; the new experiment proposed here is a determination of the "electrical equivalent of heat," and Joule himself performed both kinds of experiments. These experiments do turn out as expected and lead to the same value for the conversion factor between joules and calories: 1 cal = 4.184 J. There is one practical difference between the two sorts of measurements. The mechanical experiments are rather difficult to carry out because the amount of heating produced from reasonable amounts of mechanical energy is quite small, but it is easy to use up enough electrical energy to produce a measurable rise in temperature.

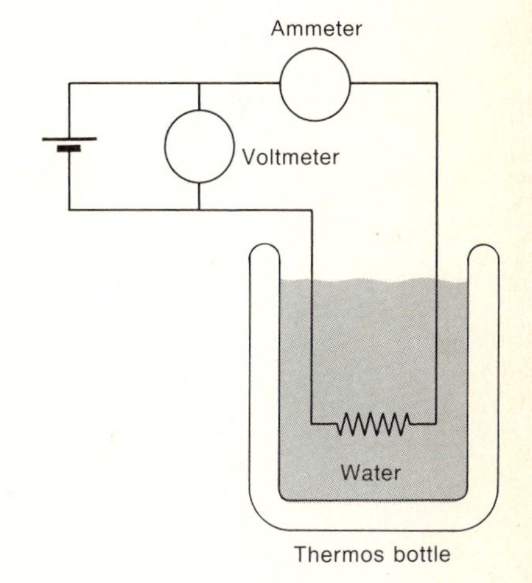

FIGURE 8.25
Apparatus for determining the electrical equivalent of heat.

QUESTION

8.4 In one such experiment, a 6-V battery produced a current of 2 A in a coil of wire in a thermos bottle containing 150 g of water. During a period of 5 minutes, the temperature of the water increased by 5.7°C. On the basis of these data, how many joules are equivalent to one calorie?

---

§8.5.C Ohm's Law and the Concept of Resistance

Our discussion of energy conversion in electrical circuits has, thus far, been completely general. In particular, when a current I flows through something from the high-voltage end to the low-voltage end (the difference in voltage being V), electrical energy is converted to some other form at a rate given by $P = IV$. This is a result based on the fundamental definitions of current and voltage.

Further useful results emerge if we ask what determines the values of current and voltage. For our simplest circuit, a piece of wire connected to a battery, the voltage across the wire is just equal to the emf of the battery. If this voltage is increased by using a battery with a larger emf, the current increases, as we would certainly anticipate. The mathematical relationship between voltage and current turns out to be a simple one: to an extremely high degree of accuracy, the current and the voltage are directly proportional to one another. Actual measurements for two particular wires are shown in Figure 8.26. For each wire, the graph is a straight line through the origin, showing a direct proportionality between current and voltage. This relationship between current and voltage is known as Ohm's law. It is important to realize that the relationship might well have turned out to be of the type shown in Figure 8.27a or 8.27b, and indeed there are some materials which exhibit such "non-Ohmic" behavior. Ohm's law is not a truly fundamental law of nature; it is simply a law that happens to be obeyed almost exactly by a great many materials. When we work, as we almost always do, with materials that follow Ohm's law to within the accuracy of our measurements, this fact makes it possible to express some of our results in a somewhat different form.

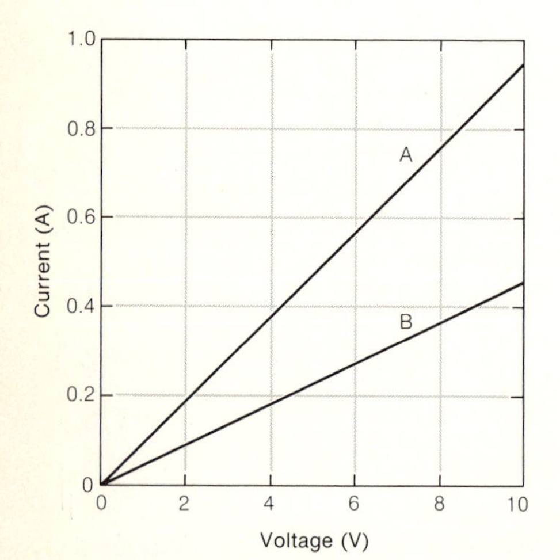

FIGURE 8.26
Relationship between current and voltage for two samples of wire: A—copper, length = 30 m, diameter = 0.25 mm; B—aluminum, length = 25 m, diameter = 0.20 mm.

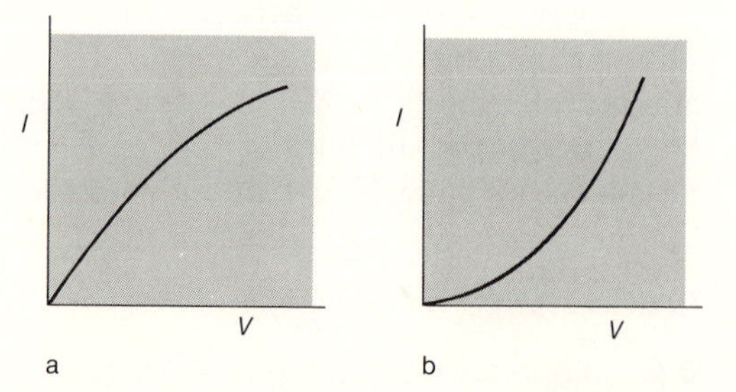

FIGURE 8.27
Relationship between current and voltage for materials that do *not* follow Ohm's law.

For an object that follows Ohm's law, the current and voltage are directly proportional to one another ($I \propto V$ or $V \propto I$). By introducing a constant of proportionality, either of these proportionalities can be written as an equation. It is common practice to use the second one and to call the proportionality constant R, the "resistance" of the wire:

$$V = RI. \tag{8.8}$$

The term "resistance" is appropriate; if R is large, then a large voltage is required to cause some chosen current to flow. From the opposite point of view,

$$I = V/R;$$

for a specified voltage, the current is small if R is large. The units of resistance are those of voltage divided by current, and because the concept of resistance turns out to be very useful, the MKS unit of resistance is given its own name, the ohm:

$$1 \text{ ohm} = 1 \text{ V/A}.$$

With the definition of resistance, $R = V/I$, the general result for the power delivered, $P = IV$, can be written in two useful alternative ways:

$$P = IV = I(IR) = I^2R \tag{8.9}$$

and

$$P = IV = \frac{V}{R}V = \frac{V^2}{R}. \tag{8.10}$$

The resistance of a wire depends on the material of which it is made and also on its dimensions. For any particular material, the resistance is directly proportional to the length, L, and inversely proportional to the cross-sectional area, A:

$$R \propto L/A.$$

We can write this as an equation by introducing a symbol, ρ, for the constant of proportionality:

$$R = \frac{\rho L}{A}. \tag{8.11}$$

The quantity ρ is a property of the material, its "resistivity." For a good conductor, the value of ρ is small; for a poorer conductor, ρ is larger and for a wire of the same dimensions, the resistance will be larger. The resistivity can be determined by making measurements of current and voltage to find the resistance of a wire of known dimensions, and then using Equation 8.11 to calculate ρ.

QUESTION

8.5 Show that the units of ρ are ohm-meters.

TABLE 8.1 Values of resistivity, ρ, for various materials at a temperature of 20°C.

Material	Value, ρ (ohm-meters)
Silver	1.6×10^{-8}
Copper	1.7×10^{-8}
Gold	2.2×10^{-8}
Aluminum	2.7×10^{-8}
Tungsten	5.6×10^{-8}
Brass	7×10^{-8}
Iron	9.8×10^{-8}
Lead	21×10^{-8}
Uranium	26×10^{-8}
Mercury	96×10^{-8}
Nichrome*	100×10^{-8}
Sea water†	0.2
Distilled water	10^{4}
Bakelite	10^{11}
Hard rubber	10^{11}
Nylon	10^{13}
Mica	10^{14}
Glass	10^{14}

*Nichrome is an alloy often used in heating elements of toasters and other appliances.

†The resistivities of sea water, bakelite, rubber, nylon, mica and glass vary considerably from one sample to another; listed values are typical.

Some representative values of resistivity are given in Table 8.1. Notice the enormous differences between the good conductors (the metals) and the insulators, such as bakelite and glass, which are *very* poor conductors. A cautionary note should be added: the resistivity of a material does depend on the temperature. The resistivities of most materials increase with increasing temperature. When a light bulb is first turned on, its resistance is much lower than it is a few tenths of a second later when it has had a chance to get hot. In order even to *measure* the resistance of a wire, a current must be passed through it, producing thermal energy. The resulting rise in temperature may lead to results like those shown in Figure 8.27a, even though the material *is* "Ohmic" and would produce a straight-line graph if the temperature were kept constant. As the temperature decreases, resistivities of most metals decrease, and for many metals a completely new phenomenon appears at temperatures a few degrees above absolute zero, the phenomenon of "superconductivity" in which all traces of resistance abruptly vanish. This happens at a temperature of 1.1°K for aluminum and at 7.2°K for lead. The resistance does not simply become "small"; as far as we know, the resistance of a superconductor is precisely equal to zero. If a current can be started in a closed loop of superconducting wire, it will flow forever! (Admittedly, "forever" is a strong word. In one experiment, a ring of lead wire was kept at a temperature below 7.2°K for a year, and within that time there was no observable decrease in the current; with the same ring at a higher temperature, the current stops within a fraction of a second.) Some alloys have been made that are superconducting at temperatures as high as 23°K. Superconductors may turn out to be of practical importance in the generation and transmission of electrical energy; the higher the temperature at which superconductivity is found, the less the energy needed for refrigeration, and thus the more interesting they will become from a practical point of view.

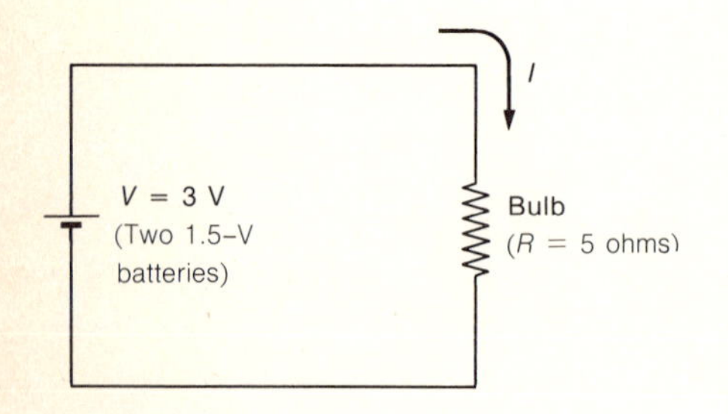

FIGURE 8.28
Circuit diagram of a flashlight.

§8.6 USES OF ELECTRICAL ENERGY: BATTERY-OPERATED DEVICES.

Most of the electrical energy we use is supplied by the power company, but a small amount is obtained from batteries, with the use of circuits like that discussed in detail in §8.5.A. Figure 8.18, for instance, is the circuit diagram of a flashlight; this diagram has been redrawn in Figure 8.28 with numerical values appropriate to an ordinary flashlight.

QUESTION

8.6 Show that the current through the bulb in Figure 8.28 is 0.6 A and that the power delivered to the bulb is 1.8 W. (Check to make sure that Equations 8.7, 8.9, and 8.10 all give the same result for the power.)

We shall consider here a number of practical questions about the use of batteries. How much energy can we get from the battery before it runs down? How much does this energy cost per kilowatt-hour, and how does this price compare with other energy prices? How large and how heavy must batteries be in order to supply a fairly large amount of energy, perhaps enough to drive a car for 100 miles?

Manufacturers often provide information about the quantity of energy obtainable from a battery by specifying a number of "ampere-hours" (abbreviated A h). The common 1.5-V "D-cell" used in many flashlights has a capacity of approximately 3 A h. An "ampere-hour" is not a unit of energy, but this figure can be used to deduce the amount of energy available. To say that a battery has a capacity of 3 A h is to say that it will be discharged in 3 hr if a steady current of 1 A is drawn from it, or in 30 hr if the current is 0.1 A. In an ordinary flashlight, the current is about 0.6 A, and batteries should last about 5 hr.* The total energy supplied in a time Δt is

$$\text{Energy} = P\,\Delta t = VI\,\Delta t.$$

The product $I\Delta t$ is given by the "ampere-hour" figure; if we also know the voltage, the total amount of energy can be calculated. For a D-cell, $V = 1.5$ V, so

$$\text{Energy} = 1.5 \text{ V} \times 3 \text{ A h}.$$

Since the product of volts and amperes is a number of *watts,* this is an energy of 4.5 watt-hours (Wh), or 4.5×10^{-3} kWh. If the price of the battery is 36¢, the cost per kilowatt-hour is

$$\frac{36¢}{4.5 \times 10^{-3}\text{ kWh}} = 8000¢/\text{kWh}$$

$$= \$80/\text{kWh}.$$

Electrical energy purchased from the power company sells at a price more than a thousand times lower; batteries are not cheap sources of energy.

It is also interesting to compare batteries with other sources of energy in terms of energy per unit volume and energy per unit mass. A 1.5-V D-cell has a diameter of 1.3 in. and a length of 2.4 in. (a volume of

*It is an oversimplification to give a single figure for the ampere-hour capacity of a battery. The number of ampere-hours depends somewhat on the size of the current drawn and whether it is used continuously or intermittently. A 3 A h flashlight battery simply cannot supply a current of 300 amperes under any circumstances, though its stated capacity might lead us to expect that it could do so for 0.01 hr. Furthermore, batteries do not *suddenly* run out of energy; instead the emf drops over a period of time. In some applications, a "1.5-V" battery may still be useful even when its emf has dropped to 1.0 V or even less; if so, we can get a greater number of ampere-hours from it than we could if it had to be replaced when its emf had dropped below 1.4 V.

3.2 in.^{3}) and a mass of 0.2 lb. With an available energy of 4.5×10^{-3} kWh, the energy per unit volume is

$$\frac{4.5 \times 10^{-3}}{3.2} = 1.4 \times 10^{-3} \text{ kWh/in.}^3$$

and the energy per unit mass is

$$\frac{4.5 \times 10^{-3}}{0.2} = 22.5 \times 10^{-3} \text{ kWh/lb.}$$

QUESTIONS

8.7 How do these figures compare with the corresponding values for gasoline?

8.8 We saw in Chapter 4 that the propulsive power required at the rear wheels to drive a 2-ton car at a constant speed of 40 miles/hr is approximately 9 kW. Consider the design of an electric car, in which this power would be supplied by flashlight batteries, with enough batteries to drive 100 miles. How many batteries would be needed? How much would they cost? How much mass and volume would be required for the batteries?

Data are given in Table 8.2 for a number of batteries of various types, and figures for the cost per kilowatt-hour can easily be calculated; the CD-21 nickel-cadmium battery, for instance, delivers energy at the astonishing price of $16,700 per kilowatt-hour. For this battery and those listed below it in Table 8.2, it is not really fair to calculate the cost of energy in this way, because these batteries are specifically designed to be rechargeable. We do have to pay for the energy used in recharging, but this energy is usually purchased at a much more modest price, by buying electrical energy from the power company to recharge nickel-cadmium batteries or by burning gasoline in a car to recharge a car's storage battery. The weight and volume data for the automobile storage batteries are extremely relevant in considering the possibilities of electric cars—not the absurd flashlight battery car, but a car with storage batteries which would be used to drive the car for some distance and then recharged.

QUESTION

8.9 Consider any one of the three batteries at the bottom of Table 8.2 and estimate the distance a 2-ton car could be driven at 40 miles/hr on the energy of one such battery. How many batteries would be needed to go 100 miles? How much weight and volume? If the energy for recharging a battery is purchased from the power company at 3¢/kWh, what is the cost of going 100 miles, and how does this cost compare with that of the gasoline for an ordinary automobile?

The results of the preceding question suggest that cars powered by storage batteries are not out of the question, but are still not very practical with conventional storage batteries. A distance of 100 miles is not

TABLE 8.2 Emf, ampere-hour capacity, size, weight, and cost of various batteries.

Battery	Emf (V)	Capacity (A h)	Volume (in.^{3})	Mass (lb)	Cost ($)*
"Pen-light" battery	1.5	0.58	0.47	0.04	0.20
D-Cell (standard flashlight battery)	1.5	3.0	3.2	0.2	0.36
#6 Dry-Cell	1.5	30	33.3	1.9	2.20
#2U6 battery	9.0	0.325	1.15	0.06	1.10
#V-60 battery	90	0.47	17.6	1.27	6.60
#312 Mercury cell	1.4	0.036	0.01	0.0013	0.55
#640 Mercury cell	1.4	0.5	0.13	0.0163	1.40
#1 Mercury cell	1.4	1.0	0.20	0.027	1.50
#12 Mercury cell	1.4	3.6	0.63	0.088	2.70
#42 Mercury cell	1.4	14	3.1	0.365	8.00
CD-21 Nickel-cadmium battery (rechargeable)	6.0	0.15	0.9	0.115	15.00
CD-29 Nickel-cadmium battery (rechargeable)	12.0	0.45	7.9	0.74	43.00
Automobile storage battery A	12.0	40	516	31	21.00
Automobile storage battery B	12.0	96	750	55	45.00
Automobile storage battery C	6.0	84	440	29	30.00

*The costs are based on typical prices for 1974.

very large, and recharging does take time. Research is being done to develop the batteries which are needed to make battery-operated cars more attractive, batteries with higher values of energy per unit mass and per unit volume. As pointed out in Chapter 4, battery-powered cars could be recharged during the night, at a time when power companies have excess generating capacity, so that the use of electric cars would not require the construction of a large number of new power plants. Furthermore, the electrical energy for recharging may come from coal or from nuclear energy, thereby eliminating the need for oil as the particular energy source for automobiles.

In the circuits discussed thus far, one additional factor can be extremely important. In the circuit diagram of the flashlight (Figure 8.28), for instance, we assumed that the wires from the battery to the bulb functioned only as "leads" to conduct current, but these wires do have *some* resistance. A more complete circuit diagram of a flashlight is shown in Figure 8.29, with r_1 and r_2 representing the "lead resistances" and perhaps also the resistance of the switch. The three resistances (r_1, R, and r_2) are connected in *series*. The voltage between the terminals of the battery (its emf) is equal to the sum of the voltages across the three separate resistances, denoted by V_{r_1}, V_R, and V_{r_2}:

$$V = V_{r_1} + V_R + V_{r_2}. \tag{8.12}$$

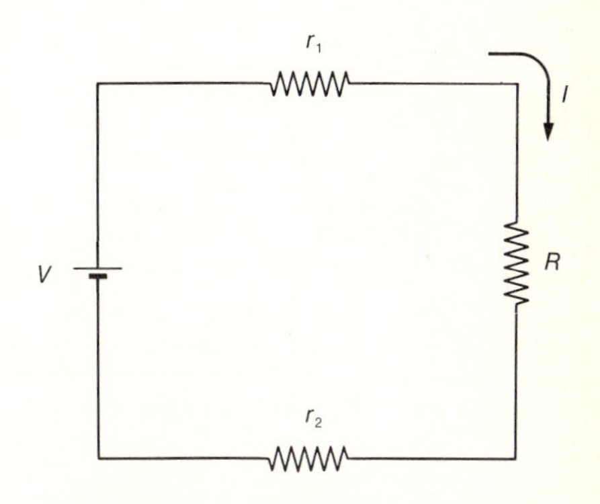

FIGURE 8.29
Circuit diagram of a flashlight, with resistances of the leads (r_1 and r_2) taken into account.

The same current I flows through each of the resistances, and so from Ohm's law (Equation 8.8) we can write:

$$V_{r_1} = Ir_1, \qquad V_R = IR, \qquad \text{and} \qquad V_{r_2} = Ir_2.$$

The voltages across the three resistances are proportional to the respective values of resistance. Substitution of these relationships into Equation 8.12 gives an equation that can be solved for I:

$$V = Ir_1 + IR + Ir_2; \tag{8.13}$$

$$I = \frac{V}{r_1 + R + r_2}. \tag{8.14}$$

The current in a series circuit is the same as if the voltage were applied to a single resistance whose resistance is equal to the sum of the separate resistances. Suppose that $r_1 = r_2 = 0.5$ ohm and $R = 5$ ohms. Then

$$I = \frac{3}{0.5 + 5 + 0.5} = \frac{3}{6} = 0.5 \text{ A},$$

$$V_{r_1} = V_{r_2} = 0.5 \times 0.5 = 0.25 \text{ V},$$

$$V_R = 0.5 \times 5 = 2.5 \text{ V},$$

and the voltages across the various parts of the circuit are as shown in Figure 8.30.

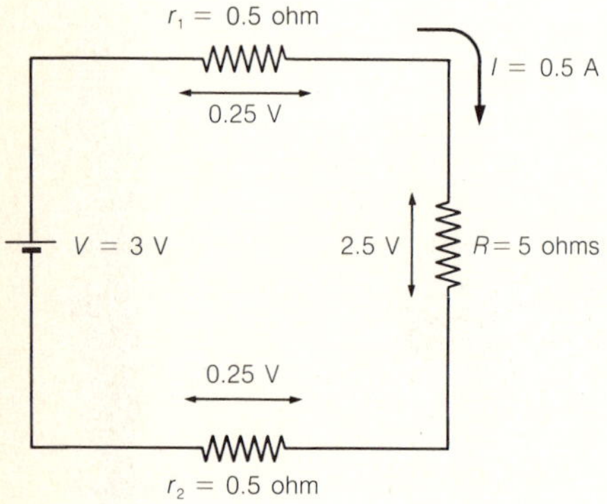

FIGURE 8.30
Circuit diagram of a flashlight, showing voltages across various parts of the circuit.

Another useful way of looking at a series circuit comes from Equation 8.13 and Figure 8.30. The total voltage of the battery, V, is equal to the sum of the "*I-R* drops" across the separate parts of the circuit; in Figure 8.30, 0.5 V of the battery's 3 V is wasted in producing *I-R* drops across the lead resistances and only 2.5 V appears where we want it, across the bulb. In a flashlight, r_1 and r_2 are normally small in comparison with R, but sometimes corroded contacts or a piece of dirt lodged in the switch can in effect make them much larger. Then, according to Equation 8.14, the current will be greatly reduced, and the bulb will not light even though the battery may be perfectly good.

Other series circuits are the same in principle, but changes in the numerical values can make these unwanted *I-R* drops very significant. In starting a car, for instance, a 12-V battery is required to deliver an extremely large current, perhaps 200 A, for a few seconds. If the total resistance of the leads were 1 ohm (as in the flashlight), the current could not possibly be greater than 12 A ($I = V/R = 12$ V/1 ohm $= 12$ A). The lead resistances must therefore be made much smaller than in a flashlight, and we can estimate how small they must be from Figure 8.31. If we require that a voltage of at least 10 V be applied to the starting motor, the sum of the *I-R* drops in the leads cannot exceed 2 V:

$$Ir_1 + Ir_2 \leq 2 \text{ V};$$

$$200(r_1 + r_2) \leq 2 \text{ V};$$

$$(r_1 + r_2) \leq 2/200 = 0.01 \text{ ohm}.$$

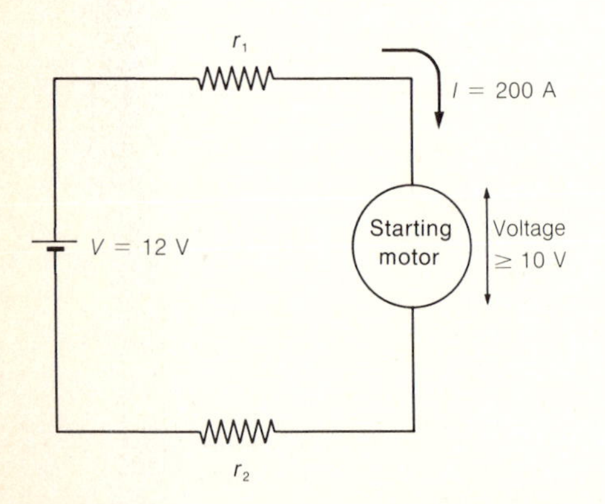

FIGURE 8.31
Circuit diagram of an automobile's battery and starting motor.

QUESTION

8.10 If the two leads together have a length of about 6 ft and are made of copper, estimate the minimum diameter of wire that must be used in order for the total lead resistance to be less than 0.01 ohm.

Whenever large currents must be provided, the effects of the lead resistances may become important, and the ideas just discussed are of significance, both to the person who wants to use an appliance that draws a large current, and to the power company, which must supply a *very* large current from its generating plant to a distant city.

§8.7 RESIDENTIAL USE OF ELECTRICAL ENERGY

Although some of the electrical energy used in the home is derived from the chemical energy of batteries, this is usually only a small fraction of the total. Batteries are expensive, and their use is generally restricted to portable items such as flashlights and transistor radios. Most of our electrical energy is supplied by the power company, and we obtain this energy by plugging appliances into wall sockets, or by screwing bulbs into light fixtures. How the power company generates and delivers this electrical energy will be discussed in the next chapter; here we will be primarily concerned with the user's point of view.

In every wall socket are the ends of two wires with which we make contact when we plug in a floor lamp or a toaster, and leading to every ceiling light fixture are two wires, to the ends of which we make contact when we screw in a light bulb. It is the responsibility of the power company to provide a certain voltage between these two wires. From the user's point of view, it is as if the power company simply had a large battery with its two terminals connected to these two wires. There is one difference between this and the true state of affairs, a difference which is often of little importance but which may sometimes be significant. The power company does not provide a *fixed* voltage difference but rather one that changes rapidly with time. In common household circuits, if the voltage of one of the two wires is defined to be zero, the voltage of the second wire is alternately positive and negative with respect to the first, as shown in Figure 8.32. The power company supplies an "alternating voltage," a voltage that goes through a full cycle of its variations in 1/60 sec. If an appliance is plugged in, the current will not flow steadily in one direction but will instead flow first in one direction and then in the other. This is the origin of the term "AC," alternating current, as opposed to the direct current (DC) provided by a battery. We can describe this as an alternating current with a frequency of 60 cycles per second, or 60 hertz (Hz).*

*The standard power-line frequency in the United States is 60 Hz, though in a few areas direct current is supplied instead of alternating current; a frequency of 50 Hz is standard in Europe.

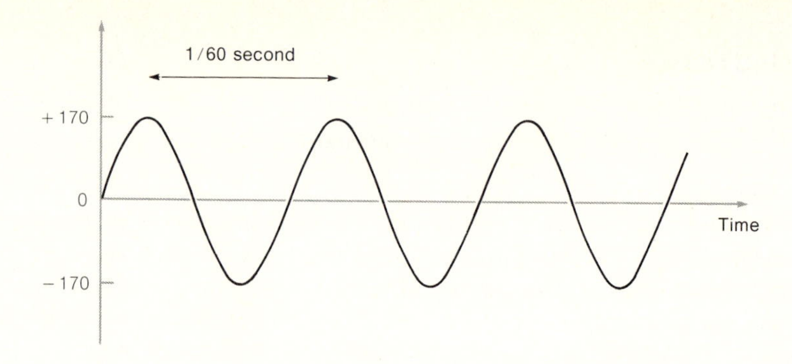

FIGURE 8.32
Variation with time of the voltage of one wire in the power line with respect to the other.

What effect does this complication have for the user? Fortunately, very little. If it is just the heating in which we are interested, there is little difference at all. Thermal energy is produced no matter which way the current flows through a toaster. (More care is needed if the electricity is being used to run a motor, and a motor must be chosen according to whether it is to be run by alternating or direct current.) It is true that the current not only reverses direction every 1/120 sec but also has a larger magnitude at some instants than at others, and thus the rate at which electrical energy is converted into thermal energy also varies. Nevertheless, as long as the variation is quite rapid, whatever it is that is being heated has little chance to cool off during the brief period of time that the voltage is very small, and it never gets as hot as it would if there were a steady voltage equal to the peak value of the applied voltage. (If the voltage varied at the rate of 1 Hz or so instead of 60 Hz, this would not be true; light bulbs would flicker noticeably.) As long as we are concerned with the average amount of energy delivered over periods of time a few sixtieths of a second or longer, then in spite of the fact that we are dealing with AC, we can calculate energies and powers as if a *fixed* voltage difference were maintained between the two terminals. In effect, it is as if there were simply a large battery connected to the two terminals, with an "effective emf" denoted by V_{eff}.

How large is the value of V_{eff}? That is, how large a steady voltage would give the same amount of heating, on the average, as we obtain from the actual varying voltage shown in Figure 8.32? Since the amount of heating is proportional to the square of the voltage (Equation 8.10), we need to find the steady voltage whose square is equal to the *average* value of the *square* of the varying voltage. We will not make the calculation here but will simply give the result. For a varying voltage of the type shown in Figure 8.32, V_{eff} is smaller by a factor of $\sqrt{2}$ than the peak value of the voltage. Thus with a peak voltage of 170 V,

$$V_{\text{eff}} = \frac{170\ \text{V}}{\sqrt{2}} \simeq 120\ \text{V}.$$

The value of V_{eff} varies to some extent from place to place and even from minute to minute; sometimes the standard "line voltage" is referred to as 110 or 115 rather than 120 V. Sometimes in an attempt to save energy, the voltage supplied to the consumers is deliberately reduced by 5 or 10% for a few hours.

Many of the electrical devices used in homes are simply resistances, in which electrical energy is converted directly into thermal energy, and the thermal energy is what we desire; stoves, toasters, electric blankets and electric space heaters are obvious examples. The conversion of electrical energy into thermal energy is a process that is easily carried out at "100% efficiency." (It is all too easy to convert electrical energy, kinetic energy, and other forms into thermal energy; it is the reverse process, the conversion of thermal energy into other forms of energy, which can be done only with difficulty and never completely, as we saw in Chapter 7.) In some household devices, such as electric clocks and mixers, electrical energy is first converted, at least in part, into kinetic energy by means of a motor.

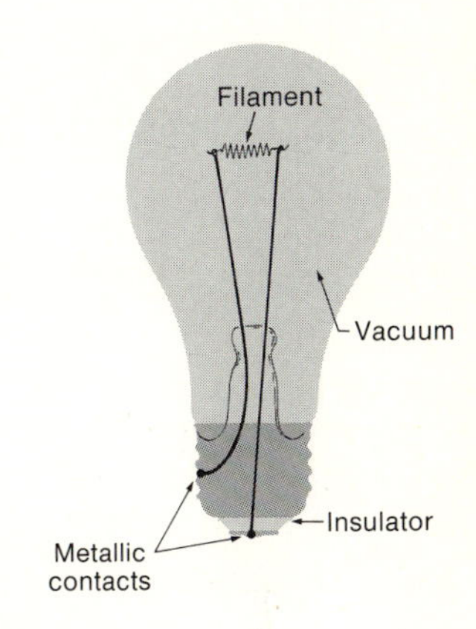

FIGURE 8.33
Common incandescent light bulb.

Perhaps the simplest and surely the most common household "appliance" is the standard incandescent light bulb. This invention has done as much as anything else to change our style of life since its introduction in the late 19th century. The essential part of a light bulb is simply a short piece of very fine wire (a filament) contained in a glass envelope (Figure 8.33). The two ends of the wire are connected so as to make contact with the two wires of the power line when the bulb is screwed into a socket. Thus a circuit diagram of a single lamp in operation is as shown in Figure 8.34, just like the circuit diagram of a flashlight (Figure 8.28). In a practical lamp, there is also a switch so that one can "open" the circuit and stop the current from flowing. Our purpose is to keep the filament hot enough so that it will glow. If the bulb is hot, it will lose thermal energy to its cooler surroundings, and so we must deliver electrical energy to the bulb at a steady rate for continuous conversion into thermal energy. The bulb also transmits energy to its environment in the form of light. This of course is what a light bulb is for. This light energy can be considered as a part of the heat that the bulb transfers to its environment by radiation. The nature of light will be treated in more detail in Chapter 11, but in this discussion we shall focus attention on the rate at which a light bulb or any other device uses electrical energy, whether it converts it all immediately into thermal energy or not. (Not all lights are the same; fluorescent lights give much more useful light for the same amount of electrical energy than do incandescent bulbs.)

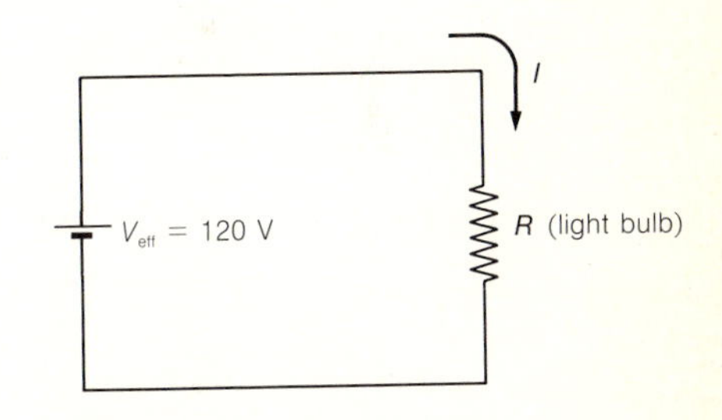

FIGURE 8.34
Circuit diagram of a light bulb in use.

To take a specific example, consider a 100-watt bulb, one which is designed to use electrical energy at the rate of 100 W. A 100-W bulb made to be operated from a 120-V line should have a resistance such that $P = V^2/R = 120^2/R = 100\ W$, and so $R = 120^2/100 = 144$ ohms.

QUESTIONS

8.11 Show that the current drawn by a 100-W lamp is approximately 0.83 A. Thus we can fill in some numbers in the previous circuit diagram, as shown in Figure 8.35.

8.12 Show that a 200-W bulb should have a resistance of about 72 ohms.

8.13 Since $P = I^2R$ (Equation 8.9), shouldn't a 200-W bulb have a *larger* resistance than a 100-W bulb, rather than a smaller one?

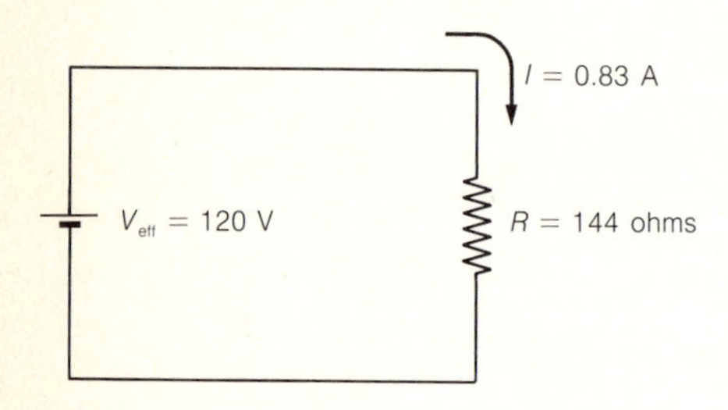

FIGURE 8.35
Circuit diagram of a 100-W light bulb in use.

A number of lights and other appliances can be operated simultaneously from the same electrical power source. Often this is done by means of the familiar double receptacle in the wall, which simply contains two sockets wired in *parallel,* as shown in Figure 8.36a. This circuit diagram can be rearranged as shown in Figure 8.36b; the two lamps are connected in parallel, with the full voltage (120 V) applied to each lamp. Each lamp takes just as much current and uses just as much power as if the other lamp were not in use. The total amount of power being delivered to the two lamps is 300 W. We can check this by looking at this from the power company's point of view. A voltage of 120 V is supplied, and the total amount of current being delivered is $0.83 + 1.67 = 2.5$ A. The power being supplied is thus

$$P = IV = 2.5 \times 120 = 300 \text{ W}.$$

FIGURE 8.36
(a) Diagram showing the operation of a 100-W and a 200-W lamp from a double receptacle. (b) Redrawing of the circuit shown in (a).

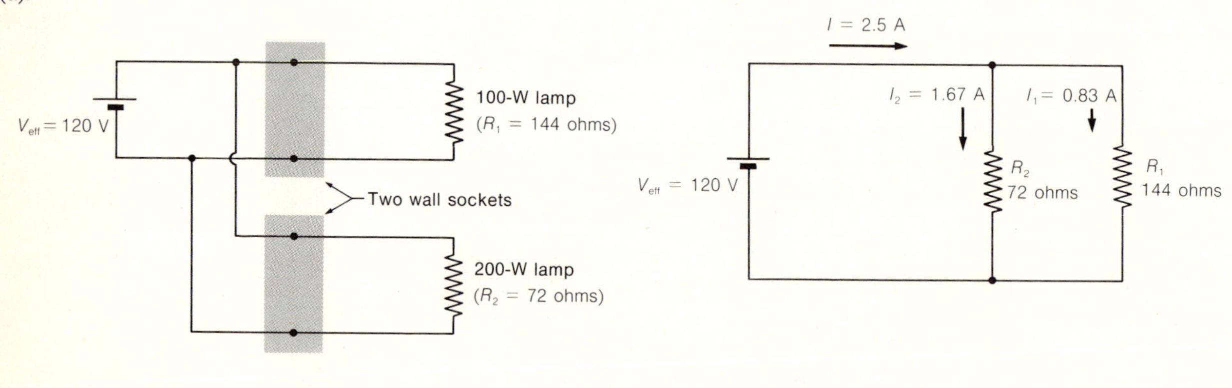

QUESTION

8.14 What would be the resistance of a single lamp which, if connected by itself across the 120-V line, would use a current of 2.5 A and 300 W of power? (Note that this resistance turns out to be *smaller* than the resistance of either the 100-W bulb or the 200-W bulb. Two resistances connected in parallel are equivalent to a single resistance which is smaller than either one.)

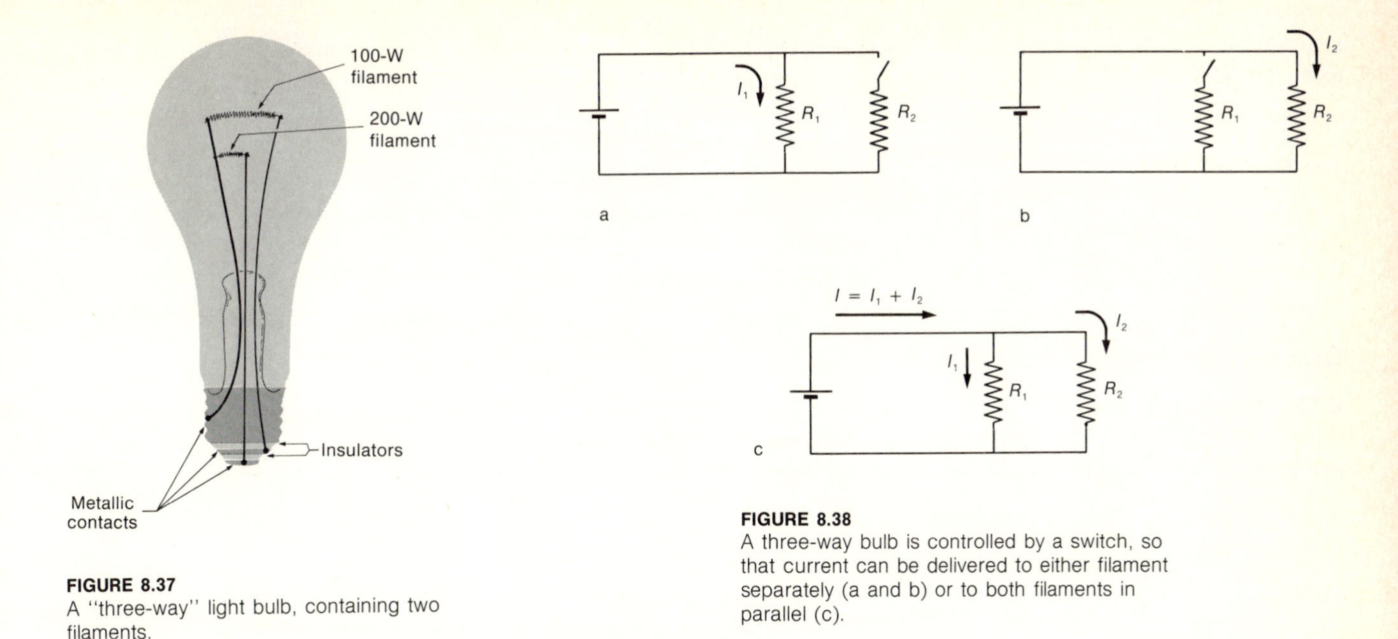

FIGURE 8.37
A "three-way" light bulb, containing two filaments.

FIGURE 8.38
A three-way bulb is controlled by a switch, so that current can be delivered to either filament separately (a and b) or to both filaments in parallel (c).

Occasionally more than one filament is contained in the same bulb. A familiar example is the "three-way" bulb, for example a "100-200-300" bulb, which contains both a 100-W and a 200-W filament (Figure 8.37). Selection of light intensity is made by means of a switch that applies voltage to either filament separately or to both in parallel. Circuit diagrams of the three possibilities are shown in Figure 8.38. We can readily operate two separate appliances from the same 120-V line. But why stop at two? With a "multiple-outlet adapter," we can turn a single electrical outlet into a triple one, and by stacking these adapters we can connect any number of appliances in parallel (Figures 8.39 and 8.40).

It is so easy to operate many appliances simultaneously that we often overlook the consequences of doing so. The energy that is used has to come from *somewhere,* usually from the combustion of a fossil fuel, and the user of these appliances is thus causing the depletion of a natural resource. Since a steam-electric power plant also produces waste heat, use of these appliances also results in thermal pollution at the power plant.

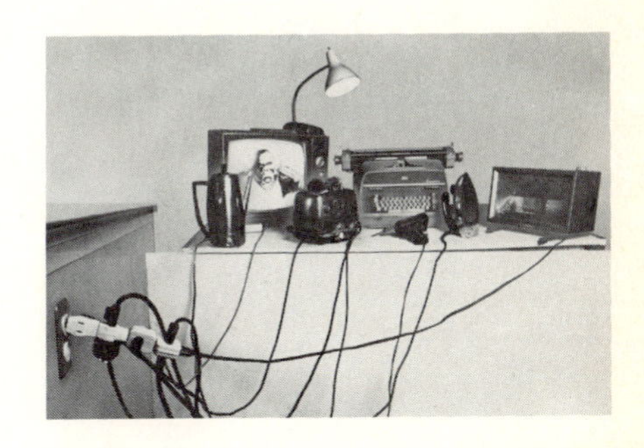

FIGURE 8.39
Many appliances can be plugged into one wall socket. Not recommended!

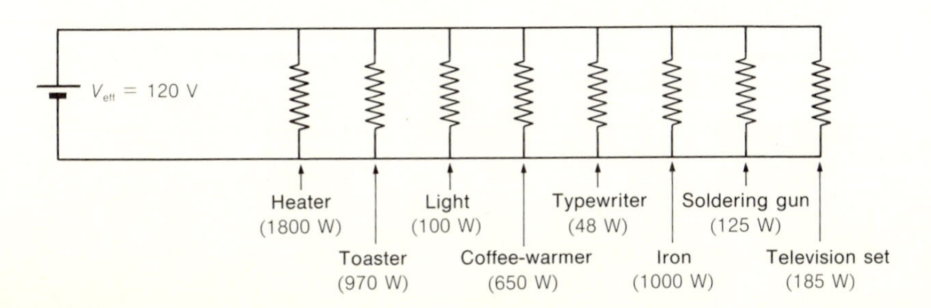

FIGURE 8.40
Circuit diagram of the appliances shown in Figure 8.39.

If a great many customers are using their appliances simultaneously, the power company may be unable to satisfy the demand. It may respond by reducing the voltage supplied or by cutting off the power to some of the area it serves. The customers are usually not directly aware of these problems unless they are subjected to an interruption of service, or the voltage reduction is sufficiently severe so that some of their appliances no longer work properly. They may become aware of their power consumption at the end of the month when they get their electric bills.

QUESTION

8.15 If electrical energy is sold at 3¢/kWh, how much would it cost to leave all the appliances shown in Figure 8.39 on continuously for a whole month? What would be the total current drawn by these appliances?

Even if our hypothetical customers do not care about their electric bills and are not concerned about depletion of the world's resources, there is an immediate practical problem they will have to face. Figure 8.40 is oversimplified because it does not include the resistances of the wires leading from the basement to the wall socket. A more complete diagram, therefore, is the one shown in Figure 8.41. We discussed the same problem in §8.6, using a flashlight as an example, and the same analysis can be used here. The resistances, r, of the wires are quite small, perhaps 0.2 ohm or so, and they play an insignificant role if just a few 100-W light bulbs are in use. But if many appliances are connected in parallel, the total current delivered may be very large. One consequence is that the I-R drops in the wires may be significant, and the voltage applied to each appliance may be reduced. A much more serious consequence is that power will be dissipated in the wires (I^2r), they will get hot, and, if they get hot enough, they can start a fire. Hence every household circuit includes a fuse or a circuit breaker, connected in series with everything else (Figure 8.42) in order to limit the total current to a safe value.

A fuse is simply a very short piece of wire with a fairly low resistance. In normal operation, it has little effect; there is only a very small I-R drop across it, and electrical energy is converted to thermal energy at a small steady rate. The temperature of the fuse rises slightly until the

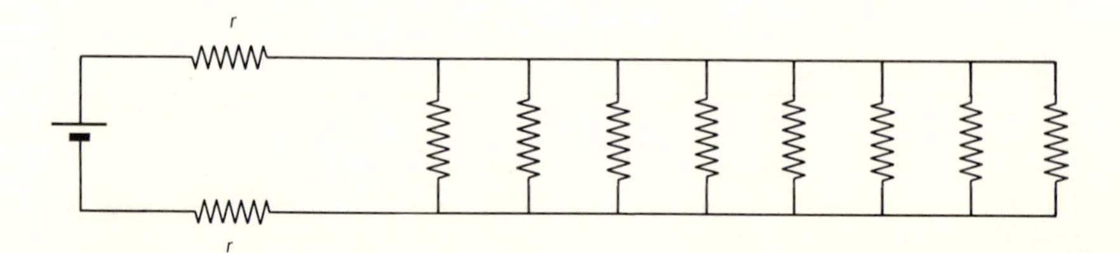

FIGURE 8.41
Circuit of Figure 8.40, with resistances of the leads included.

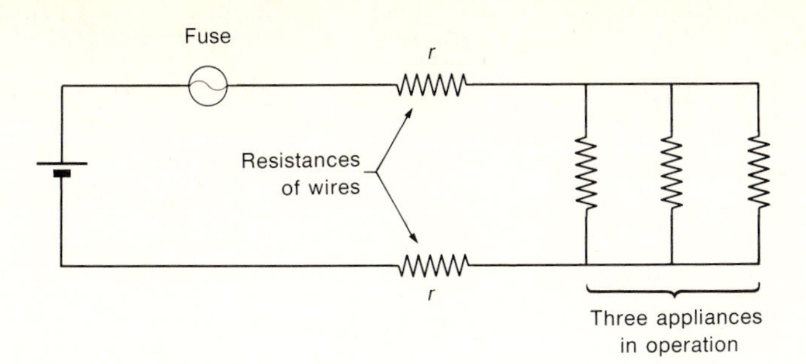

FIGURE 8.42
Use of a fuse to limit current to a safe value.

rate at which heat is produced is equal to the rate at which heat is delivered to the surroundings. The total current used by all the appliances on the circuit goes through the fuse. If this current is increased, the fuse wire gets hotter, and if it gets hot enough, the wire melts (the fuse "blows") and cuts off the current to all the appliances, and a possible fire is averted. The critical current at which the fuse will blow is determined by the design of the fuse. A 15-A fuse is found in many household circuits; a 30-A fuse contains a thicker piece of wire and a larger current can be passed before it melts.

Once the fuse has blown, the only sensible course of action is to turn off or unplug some of the appliances and put in a new fuse.* A little knowledge is a dangerous thing; the person who knows a little physics and understands why a fuse has blown may think of other responses to a blown fuse. One is to put in a larger fuse, for example to substitute a 30-A fuse for a 15-A fuse. This will permit twice as much current to flow and thus increases the possible overheating of the wiring and the possibility of a fire. An extreme response is to "put a penny in the fuse box," that is, to by-pass the protective fuse with a short fat piece of copper. Such a course of action is recommended only to arsonists and pyromaniacs.

QUESTION

8.16 By what factor is the rate of heating of the wiring increased if the current is increased from 15 A to 30 A?

Fuses not only protect us from the consequences of thoughtlessly using all our appliances at the same time, but also provide protection from accidental "short circuits." Suppose that only a single lamp is in

*Circuit breakers are often used instead of fuses. A circuit breaker is a switch designed so that it automatically opens if the current exceeds some particular value. The switch can then simply be closed again ("reset") without replacing anything.

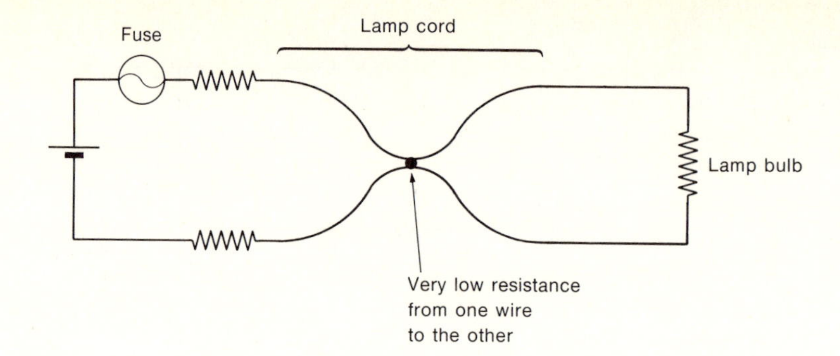

FIGURE 8.43
A fuse provides protection against a short circuit.

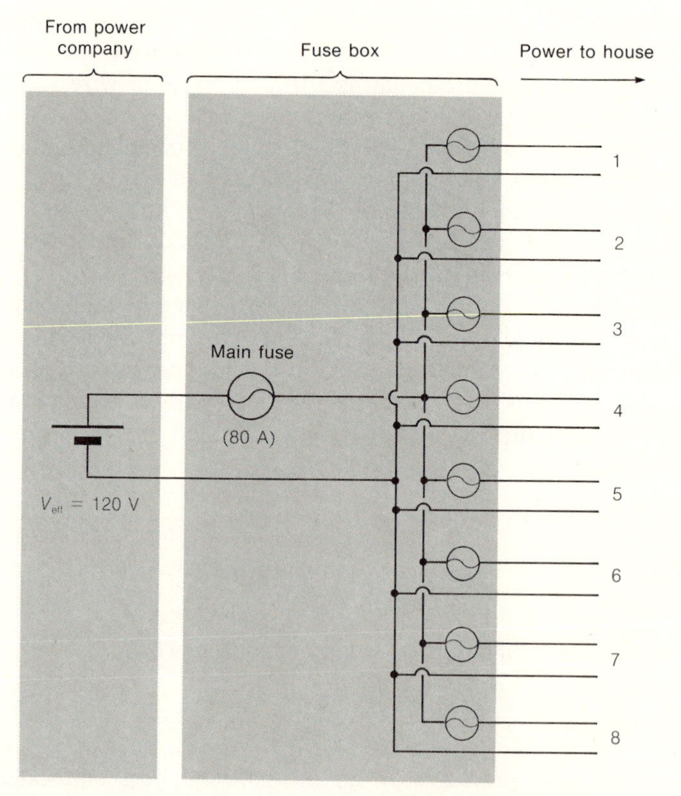

FIGURE 8.44
Power distribution system for a house. Not included in this figure are the kilowatt-hour meter, the main switch, and the 240-V circuits for the stove and clothes dryer.

use but that, because of frayed insulation on the lamp cord, the metallic conductors touch one another. Then there is a very low resistance path in parallel with the lamp (Figure 8.43), a huge current flows, and the fuse will blow.

A single 120-V line with a 15-A fuse can deliver at most $120 \times 15 = 1800$ W of power. Since a single light bulb uses about 100 W, and a single heavy appliance such as a toaster or an electric iron uses about 1000 W, it is clear that one such circuit is not enough for the typical American home. Consequently, heavy-duty lines are normally brought in from the street to a fuse box in the basement; from this point a number of smaller capacity lines lead to various parts of the house. A circuit diagram of one home is shown in Figure 8.44. Power is distributed to various rooms by a total of eight 120-V lines, each protected by a 15-A fuse. Some appliances, such as stoves and clothes dryers, are often designed to be operated with 240 V instead of 120, and special circuits, not shown in Figure 8.44, are included for this purpose. With a variety of lamps, television sets, washing machines, and other appliances connected to the various lines, the whole house is really no different from a giant multiple-outlet adapter; that is, except for the fuses, the circuit diagram of the house is just like that shown in Figure 8.40. Notice the large 80-A fuse that serves as additional protection; the eight lines cannot all be used simultaneously at full capacity; the main fuse will blow if the total current exceeds 80 A.

In order to be able to make intelligent choices about the use of electrical energy, it is important to be aware of approximately how much power each appliance uses. In Appendix M is a list of typical wattages of various appliances, together with estimates of the energy used by each one in the course of a year. This second figure is necessarily based on a guess of the average number of hours per year that an appliance is used. A long list of the electrical devices contained in one particular home is given in Table 8.3 on the next page. It might be noted that this home has no air conditioner, no electric hot-water heater and no permanent electric space heating; any one of these items would add a good many watts to the grand total.

One important general idea should be apparent from these tables. In thinking about ways to cut down on our use of electrical energy, a sense of perspective is important. The heavy-duty appliances such as stoves and clothes dryers are the ones which really eat up the energy, and refrigerators are important too, because they are in use such a large proportion of the time. Although it is certainly true that the energy consumed by light bulbs, toys, small appliances and radios can add up to a significant amount, people who dry their clothes on a clothesline (with solar energy) instead of with an electric dryer are making a much more significant contribution to solving the energy problem than those who devote their efforts to turning off lights and who insist on using mechanical clocks rather than electric ones.

TABLE 8.3 Inventory of electrical devices in one home.

Note: Not included are a number of battery-operated devices. Many of the appliances listed are often operated at less than maximum power; the figure listed is the maximum wattage for each appliance.

Device	Maximum wattage	Device	Maximum wattage
	Appliances		
Stove (two 2400-W heating elements, two 1200-W elements, and two 3800-W ovens)	14800	Hi-fi amplifier (vacuum tubes)	130
Clothes dryer	5200	Soldering gun	125
Portable electric heater	1800	Mixer	120
Dishwasher	1200	Battery charger	120
Waffle iron	1200	Sewing machine	100
Toaster-Oven	1200	Humidifier	85
Deep-fat cooker	1100	Hi-fi tape deck (transistors)	52
Iron	1000	Electric typewriter #1	48
Toaster	970	Electric typewriter #2	48
Coffee-warmer #1	650	Rug sweeper	45
Travel iron	620	Kitchen exhaust fan	40
Vacuum cleaner	600	Toy train set (3 engines)	40
Washing machine	550	Record player (self-contained, vacuum tubes)	40
Electrical starter for charcoal grill	490	AM-FM radio (vacuum tubes)	29
Blender	400	Soldering pencil	29
Vaporizer	400	Hi-fi turntable	21
Coffee-warmer #2	385	Short-wave radio (transistors)	20
Hair dryer	345	FM radio (vacuum tubes)	20
Christmas tree lights (48 bulbs)	340	Intercom (vacuum tubes)	20
Blowers and controls for oil furnace	300	Hi-fi FM tuner (transistors)	16
Electric drill	300	Erector set motor	15
Coffee-cup immersion heater #1	300	Electric razor	14
Coffee-cup immersion heater #2	280	Searchlight	11
Refrigerator	220	Engraving pencil	5
Television (black & white, vacuum tubes)	185	Pocket calculator (with battery recharger)	4
		Doorbell	3
		Electric football game	3
		Clock	2
		Kilowatt-hour meter	2
		Subtotal	36042
	Lights		
1 18-W fluorescent light	18	22 75-W bulbs	1650
1 25-W bulb	25	12 100-W bulbs	1200
1 30-W bulb	30	1 30–70–100 3-way bulb	100
1 40-W fluorescent light	40	1 50–100–150 3-way bulb	150
3 40-W bulbs	120	1 100–200–300 3-way bulb	300
1 50-W bulb	50		
26 60-W bulbs	1560	Subtotal for lights	5243
		Total	41285 W

FURTHER QUESTIONS

8.17 An electron has a mass of 9.11×10^{-31} kg and a charge of -1.6×10^{-19} C. Calculate the ratio of the electrical force of repulsion between two electrons separated by a distance of 10^{-10} m to the attractive gravitational force between them. Does this ratio depend on the distance between them?

8.18 (a) Make an order of magnitude estimate of the force of repulsion between a baseball and the earth that would exist if both the ball and the earth were stripped of their electrons. (From the mass of each object, one can estimate the number of atoms it contains, since every atom has a mass of roughly 10^{-27} kg; an atom that has lost its electrons will have a charge of about 10^{-19} C. Because the electrical force varies as the inverse square of the distance, just like the gravitational force, we can apply the results of §5.2 and calculate the force between the two objects by treating them as two charged objects 4000 miles apart.)

(b) By what factor does this repulsive electrical force exceed the gravitational force of attraction?

8.19 In Chapter 4 the "jelly donut" was defined as an amount of energy equal to 10^6 J. A real jelly donut is about the same size as a flashlight battery. Compare the two objects in terms of their stored chemical energy.

8.20 How long would it take to raise the temperature of one quart of water from room temperature to the boiling point, using a 1200-W heating element, if all the heat were delivered to the water?

8.21 Show that the ohm (1 ohm = 1 V/A) can also be expressed in either of the following two ways:

$$1 \text{ ohm} = 1 \text{ J-sec/C}^2;$$

$$1 \text{ ohm} = 1 \text{ kg-m}^2/\text{C}^2\text{-sec}.$$

8.22 (a) What are the resistances of the two wires described in Figure 8.26?

(b) One of the wires was made of copper and the other of aluminum. From the figure, estimate the resistivities of these two samples, and compare the results with the values in Table 8.1.

8.23 Discuss the following statement. "According to Equation 8.9, power is proportional to resistance, but according to Equation 8.10 it is *inversely* proportional to resistance. At least one of these equations must be wrong."

8.24 The motion of charges around a battery-powered circuit is analogous to the motion of skiers on a mountain equipped with a ski tow. As a skier is lifted up the mountain against the force of gravity, gravitational potential energy is increased at the expense of the source of energy driving the ski tow, just as the electrical potential energy of a charge is increased as it is carried through a battery. The skier then goes down the mountain at a fairly steady speed, losing gravitational potential energy and causing an increase in the thermal energy of the snow and of the skis. A particular ski resort can be described by various quantities such as the height of the mountain, the number of skiers making a complete circuit per unit time, the mass of a skier, and so on. Can you construct a diagram corresponding to Figure 8.20, and can you make the analogy more precise? What quantities in the description of the ski resort correspond to I, to V, to P?

8.25 As far as energy changes are concerned, we may regard a current either as a flow of positive charge or as a flow of negative charge in the opposite direction. Adopting the latter point of view for the moment, consider the electrical-potential-energy history of a negative charge as it moves around the circuit of Figure 8.18 in the counterclockwise direction. Give a verbal description of the energy changes that occur, and construct the appropriate diagram corresponding to Figure 8.20.

8.26 The effective resistance of two resistances in parallel is less than the resistance of either one. Derive a general formula for the effective resistance in terms of the individual resistances, R_1 and R_2.

8.27 (a) How many kilowatt-hours of energy does a 100-W light bulb use in 1 hr?

(b) How much does it cost to run a 100-W bulb for 1 hr?

8.28 Various kinds of 3-way bulbs are on the market: 30-70-100–W bulbs, 100-200-300–W bulbs, and so on. Why are there no 50-75-100–W or 100-150-200–W bulbs?

8.29 Almost all of our electric lights and appliances are designed to be operated in parallel (as in Figure 8.36), with the full 120 V applied to each individual bulb or appliance. An interesting exception is occasionally to be found on a Christmas tree. Part (a) of the figure here shows a circuit diagram of a "series-connected" string of lights. By way of contrast, the more common type, the parallel string, is shown in part (b). Each resistance symbol in these diagrams represents a single bulb, one of ten in the set. A typical Christmas tree bulb is designed to use about 5 W of power. In this respect, the bulbs in the series string and those in the parallel string are similar to each other, but they are otherwise quite different. Show that each bulb in the parallel string should have a resistance of about 3000 ohms, and that each bulb in the series string should have a resistance of about 30 ohms.

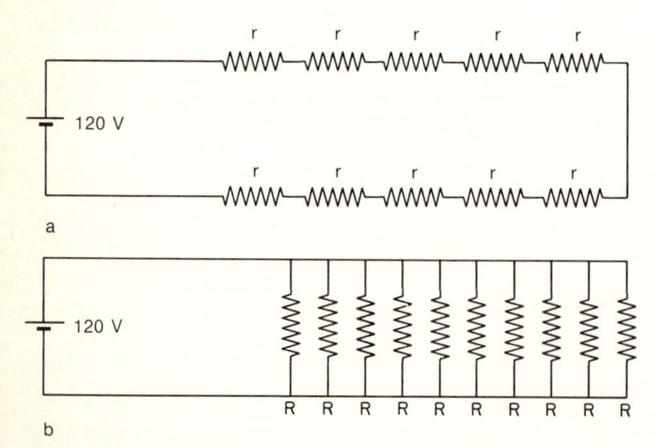

8.30 Suppose that the bulbs discussed in the previous question were accidentally interchanged. How much power would be delivered to each bulb in the series string? to each bulb in the parallel string? What would be the probable results of these mistakes?

8.31 It is often said that, since light bulbs use much more power just after being turned on, one should leave the lights on when leaving the room for less than 5 or 10 minutes. There is some basis for this argument; a 100-W bulb has a resistance of about 144 ohms at its operating temperature but a resistance of only 16 ohms when it is cold, and it therefore does use more power just after being turned on.

(a) How much power does a cold 100-W bulb use when connected to a 120-V line?

(b) A bulb takes about 0.1 sec to reach its operating temperature. Is it a good idea to leave the light on when you plan to be out of the room for 5 minutes?

8.32 When a heavy-duty appliance, such as a heater that requires a current of almost 15 A, is plugged into a circuit with a 15-A fuse, the fuse often blows; this is more likely to happen if several light bulbs connected to the same circuit are also in use. Sometimes one can avoid blowing a fuse by first turning off all the lights, then plugging in the heater and letting it run for a minute or two, and then turning the lights back on. Can you explain this phenomenon?

8.33 Comment on the following advertisement. "No plug! No outlet! No cord! Beat the energy crisis! Save electricity! Our convenient battery-run light fixture. . . ."

8.34 "Blade-shaving uses over 50 times more energy than an electric shaver. Surprising, but factual! Fact: An electric shaver uses under 5¢ worth of energy a year. Fact: It takes over 50 times more energy just to heat the average amount of water a blade shaver uses." Make some rough calculations to test the truth of this advertisement. Be explicit about the assumptions you make. (Remember that if the hot water for the blade shaver is electrically heated, the necessary electrical energy is probably generated at a plant with an efficiency of only 33% or so. However, if the water is heated by gas or oil, the efficiency is probably similar to that of a furnace—approximately 60%.)

8.35 The compilers of the list in Appendix M made some estimates of the number of hours per year that each appliance is used. It should be possible to check their arithmetic by looking at the figures for the electric clock. Are the figures given consistent with the assumption that an electric clock is used continuously?

8.36 In a list of power needs of electrical appliances, an earlier version of the list given in Appendix M, it was stated that a typical electric toothbrush uses 7 W of power and in the course of a year consumes 5 kWh of electrical energy. In the same list, the wattage of an electric razor was given as 14 W and the amount of energy used per year as 18 kWh. If these data were correct, approximately how many hours per day would the toothbrush and the razor be in use? (There is an obvious moral to this question. Don't believe everything that appears in print. Numerical errors are apt to creep into any compilation, no matter how conscientiously prepared. Once a mistake has been made, it may be impossible to correct it, since others may copy the numbers without checking them. This is what has occurred with the two errors mentioned here.)

8.37 In the corrected version of this list (Appendix M), how many hours of use per day are implied for the electric razor and toothbrush? Are these figures reasonable? Test some of the other entries to see whether the numbers are reasonable; are there any obvious mistakes?

8.38 What current is required to operate a 5000-W clothes dryer from a 240-V power line?

8.39 One pound of butter provides an energy of about 3600 kcal. What is the cost per kilowatt-hour of energy obtained from butter? How does this compare with the cost of electrical energy?

8.40 Except for the stove and the clothes dryer, all of the items in Table 8.3 are operated by connecting them to one of the eight 120-V lines shown in Figure 8.44. Is it possible to operate all of them simultaneously?

8.41 How much would it cost per month to operate continuously all the items shown in Table 8.3 (if it were possible to do this without blowing the fuses)?

8.42 If all the items listed in Table 8.3 could be operated continuously, would this provide enough energy to keep that particular house warm? (The house has a thermal conductance of 40,000 BTU/DD; at its location the coldest month—January—has about 1300 degree-days.)

8.43 If you have access to a house, compile a list similar to that shown in Table 8.3. Could the house be kept warm with its inventory of electrical appliances and lights?

8.44 Many towns have a community Christmas tree, a large tree liberally covered with bulbs, which is operated every night for a two- or three-week period in December. Estimate the cost of this annual custom and the amount of electrical energy needed. If this consumption of electrical energy is considered as being shared by all the residents of the town, does this use of power make a significant contribution to an individual's power consumption?

8.45 During a night baseball game in Boston, Fenway Park (with a seating capacity of 33,379) is illuminated by 1120 bulbs, each of which uses 1500 W of power. Make a rough comparison of the energy used in lighting the park with the energy used by the spectators in getting themselves to the game. (This is not an easy calculation. Be explicit about the assumptions you make. How many spectators come by bus, how many by car, from what distance, etc.?)

8.46 Estimate the cost of the electrical energy needed for each of the following activities:

(a) brushing your teeth with an electric toothbrush;

(b) cooking a roast of beef in an electric stove;

(c) leaving five lights on when you leave for a two-week vacation;

(d) drying a load of clothes in an electric clothes dryer;

(e) leaving a 5-W "night light" on continuously, day and night, for a full year;

(f) watching a televised baseball game;

(g) keeping an extra refrigerator operating in the basement during the summer.

8.47 Cars A and B have headlight bulbs designed to use the same amount of power, but in car A the voltage applied to the bulb is 12 V, whereas in car B it is 6 V. For which car should the resistance of the filament be larger and by what factor? Suppose that the filaments of the two bulbs are of the same total length; for which car should the diameter of the filament be larger, and by what factor?

8.48 An electrical heater for melting snow is buried just beneath the surface of a driveway, which is 10 ft wide and 40 ft long.

(a) Estimate the amount of energy needed to melt a 1-ft layer of snow. (The density of snow is about one-tenth the density of water.)

(b) How many watts of power are needed in order to melt this much snow in half an hour?

(c) Compare the cost of the electrical energy for melting this much snow with the cost of hiring someone to do the job with a shovel.

8.49 Owning and operating an electrical appliance involves both a capital investment and an operating cost. Consider two different examples: an electric clock and an electric stove. For each one, estimate the annual cost of the electrical energy used. Estimate the initial cost, and use Appendix Q to convert the capital investment into an equivalent annual amount. Compare the relative importance of initial and operating costs for these two appliances.

8.50 Electric companies do not sell every kilowatt-hour at the same price; large commercial and industrial users usually pay a lower price, and even for residential service, it is common to charge a high price for a certain minimum number of kilowatt-hours each month and to charge for additional energy at a lower rate.

(a) Find out what the residential "rate structure" is in your area.

(b) What is the reason for such a rate structure?

(c) Would it be a good idea to use an "inverted" rate structure? (Perhaps each family might get a reasonable minimum amount of energy for almost nothing, and those wishing to use heavy appliances would have to pay a stiff price.)

(d) Ask representatives of your local electric company to explain the rate structure in use. What is their response to the idea of an inverted rate structure?

9

Electromagnetism: The Generation and Transmission of Electrical Energy

William Tenney

9
Electromagnetism: The Generation and Transmission of Electrical Energy

§9.1 MAGNETISM AND MAGNETIC FIELDS

In the preceding chapter, the fundamental electrical concepts were introduced, and the consumption of electrical energy in the home was discussed in some detail. Little has been said as yet about the large-scale *generation* and *transmission* of electrical energy. Practical methods of generating and transmitting the large amounts of electrical energy on which modern society depends are based on the existence of an intimate relationship between two subjects once thought to be quite distinct from one another: electricity and *magnetism*.

Let us first briefly describe the most important features of this rather complex relationship. First, an electrical current produces a "magnetic field." That is, a current-carrying wire exerts forces on a nearby magnet, forces similar to those which would be produced by a second magnet. Second, a current-carrying wire, which is placed in a magnetic field, experiences a force. This force can cause the wire to move, and this is the principle which makes possible electric motors for converting electrical energy into mechanical energy, and it is the principle on which most ammeters and voltmeters are based. Third, if a wire that is part of a closed circuit is moved through a magnetic field, or if a closed loop of wire is subjected to a *varying* magnetic field, an electrical current is generated ("induced") in the circuit. This fact is fundamental both to the operation of electrical *generators,* which convert mechanical energy to electrical energy, and to the operation of *transformers,* which have a key function in the transmission of electrical energy. The various aspects of the relationship between electricity and magnetism are not independent of one another. The relationship is so close that it is best to treat electricity and magnetism, not as two separate subjects, but as *one* area of science: electromagnetism. We will begin, however, by describing some of the most basic magnetic phenomena.

Long before the relationship between electricity and magnetism was discovered, the behavior of magnets, the interaction of magnets with one another, their ability to attract iron objects and the use of magnets as compasses had been carefully studied. Many of the basic magnetic phenomena are familiar and will be only briefly reviewed. A suitably treated rod of iron constitutes a "bar magnet," one whose ends will attract iron filings or small iron or steel objects. Two similar bar magnets exert forces on one another, attractive or repulsive depending on their relative orientation. If one such magnet is pivoted at its center, it turns so that it is lined up approximately in the north-south direction and thus constitutes a compass. The two ends of a bar magnet are different from one another; if two bar magnets attract one another, simply reversing the direction of either one turns the attractive force into a repulsive one. The distinction between the two ends of a bar magnet is conveniently made by observing its behavior when used as a compass; the end that points toward the north (actually toward the "magnetic pole" in northern Canada rather than to the geographic pole) is called a "north-seeking pole" or, for short, simply a "north pole" and is denoted by the label N. The opposite end is of course referred to as a "south-seeking pole" or "south pole," S.

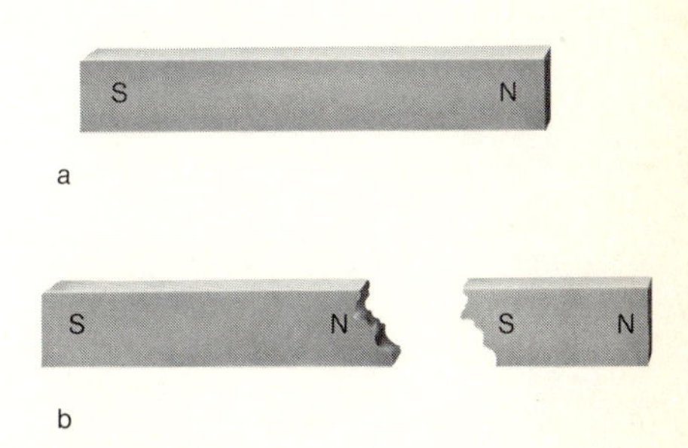

FIGURE 9.1
Breaking a bar magnet in two produces two shorter magnets, each with its own pair of poles.

Once the ends of a number of bar magnets have been identified, the interaction of one magnet with another can be qualitatively described by the statement that two *like* poles, N—N or S—S, repel one another and that two *unlike* poles, N—S, attract one another. This behavior is similar to that of positive and negative charges but with one important difference. North poles or south poles are never found alone but always in pairs, a single magnet always having both a north pole and a south pole. One may attempt to obtain isolated north or south poles by breaking a bar magnet, but the experiment fails; new poles appear on either side of the break, and one obtains simply two shorter bar magnets (Figure 9.1), each with its own pair of poles.

Since any magnetized object has both a north pole and a south pole, the interaction between two such objects is inherently more complicated than the electrical interaction between two charged objects. Whether the net force is one of attraction or repulsion depends on whether the two magnets are oriented so that the attraction between unlike poles is dominant or the repulsion between like poles (Figure 9.2).

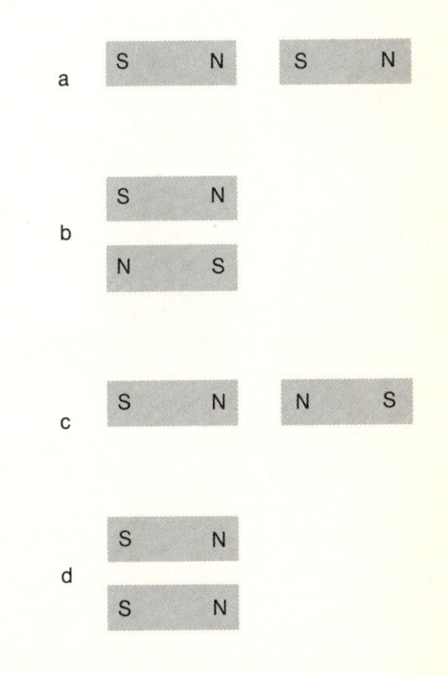

FIGURE 9.2
Two magnets may attract one another (a and b), but if one of the two is reversed (c and d), the net force between the two becomes repulsive.

The behavior of a bar magnet as a compass can be described by considering the earth as a whole to act as a giant magnet, with its "south pole" located at the north magnetic pole in Canada and its "north pole" in Antarctica, near the geographic south pole (Figure 9.3). Observe that, since *unlike* poles attract and since the north pole of a compass needle is by definition the end that points approximately in the northward direction, the supposed magnet within the earth must have its "north" and "south" poles located as shown in Figure 9.3 and not the other way around. At this time, the origin of the earth's magnetism is still not well understood. The earth of course is not simply a large magnet, nor is there a large magnet buried within it, but the effect of the earth on magnetic compasses can be described by saying that the earth acts *as if* it contained a magnet.

In a similar way, we can describe the fact that an unmagnetized object made of iron or steel (such as an iron filing or steel nail) is attracted by a bar magnet and in fact can be picked up by using either the north or south pole of a magnet. Suppose that such an object (a nail, for example) is composed of a large number of microscopic "magnets," which are normally oriented in random directions (Figure 9.4a). If a bar magnet is placed near it, the effect is to magnetize the nail, that is, to remove the randomness and to align the microscopic magnets as indicated in Figure 9.4b. There is a slight attractive force between the bar magnet and each of the microscopic magnets in the nail, and therefore a net attractive force between the magnet and the nail. If the magnet is now removed, the elementary microscopic magnets may return to their random orientations, and the nail loses its magnetism. Sometimes the microscopic magnets within the nail remain partially aligned, in which case the nail itself is permanently magnetized and can be used as a magnet.

It is clear that our description of the behavior of magnets, nails, and so on is only that—a phenomenological description, rather than an explanation. It does not tell us why only a relatively small number of materials (of which iron and alloys containing iron are the most common examples) exhibit strongly magnetic behavior. Indeed, it was not

FIGURE 9.3
The earth acts as if it contained a huge magnet, of which each pole attracts the unlike pole of a compass needle.

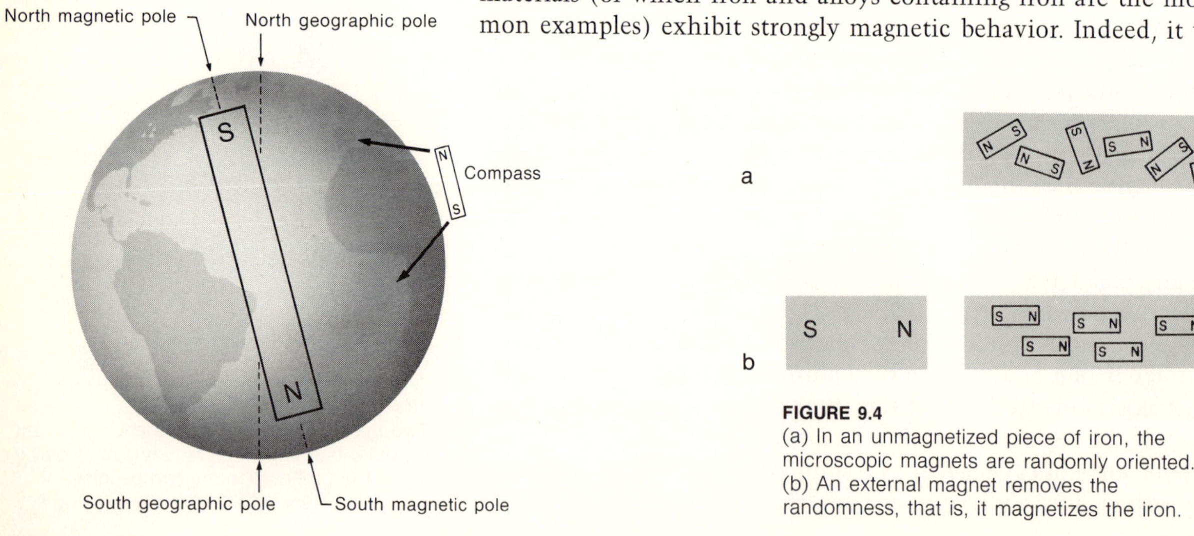

FIGURE 9.4
(a) In an unmagnetized piece of iron, the microscopic magnets are randomly oriented. (b) An external magnet removes the randomness, that is, it magnetizes the iron.

until well into the 20th century, with the development of quantum mechanics, that a fundamental understanding was achieved of the differences between iron, on the one hand, and materials such as copper, silver and paper on the other. Similarly, no explanation was offered in Chapter 8 for the fact that some materials are excellent conductors and others excellent insulators. An understanding of the reasons for these differences is not necessary here, because we can explore the subject of electromagnetism by studying the behavior of magnets and current-carrying wires.

In describing electrical effects, it is useful to describe a charged object or a set of charged objects as producing in the surrounding space an "electrical field," $\boldsymbol{\mathcal{E}}$. Similarly, it is convenient to say that a magnet produces a *magnetic* field, another vector quantity, which is denoted by **B**. The *direction* of the magnetic field at any point in space is defined by the operation of placing a small "test magnet" at that point and noting the direction in which its north pole points. By repeating the operation at a number of points, the direction of the magnetic field can be mapped out, and the result for the field of a bar magnet is shown in Figure 9.5. The nature of the field can also be revealed by sprinkling iron filings around the magnet. Each filing is temporarily magnetized, and then each acts as a tiny test magnet (Figure 9.6).

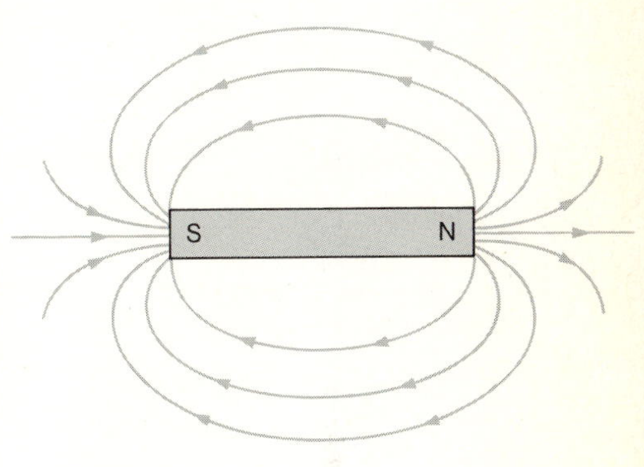

FIGURE 9.5
The magnetic field of a bar magnet.

Although the direction of the magnetic field can readily be determined, defining a unit of magnetic field strength so that numerical values can be assigned to the magnitude of **B** is rather complicated and will not be discussed here. The MKS unit of magnetic field strength is the *tesla*. This is a rather large unit for measuring the magnetic fields which have been discussed in this section. Although magnetic fields of 1 tesla or so are commonly developed in electromagnetic machinery and magnetic fields of 10 to 20 teslas or more can be produced (though with difficulty) in laboratory devices, the field strength in the vicinity of a small bar magnet is typically only about 10^{-2} tesla, and the strength of the earth's magnetic field at most points on the earth's surface is only about 3×10^{-5} tesla. Another unit for magnetic field strength is the *gauss*, much smaller than the tesla, with the conversion factor 1 tesla $= 10^4$ gauss. Thus the strength of the earth's magnetic field is about 0.3 gauss. In any equations containing **B** in this book, it will be assumed that MKS units are used and that **B** is expressed in teslas.

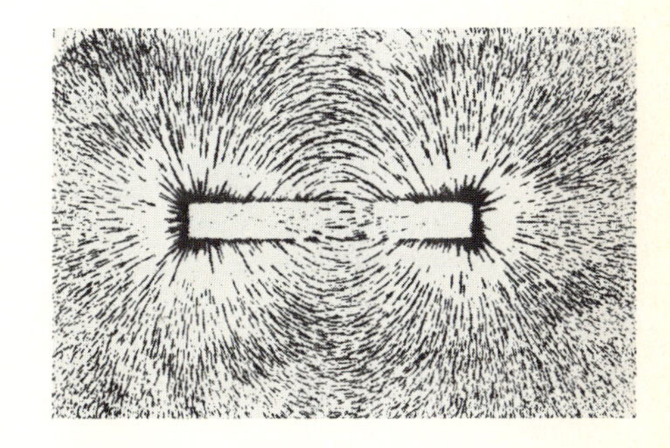

FIGURE 9.6
The magnetic field of a bar magnet, revealed by the alignment of iron filings. [*PSSC Physics*, D. C. Heath, Lexington, Mass., 1965; Education Development Center.]

§9.2 INTERACTIONS OF ELECTRICAL CURRENTS AND MAGNETIC FIELDS. THE OPERATION OF ELECTRICAL METERS AND MOTORS

There are two kinds of magnetic poles just as there are two kinds of electrical charge, and in each case the rules governing attraction and repulsion are that like charges or poles repel one another whereas unlike charges or poles attract one another. Long before any experimental demonstration of a connection between electricity and magnetism, it was speculated that such an interrelationship might exist. It might have turned out that the two types of phenomena are simply similar to one another, just as the fundamental inverse square law of gravitational force is similar to the inverse square law of electrical force. The idea that the relationship between electricity and magnetism might be more than simply one of analogy—that in some way magnetic effects might actually produce electrical effects or vice versa—remained a speculation until 1820 when Oersted carried out a simple experiment that convincingly demonstrated a connection. Oersted discovered that an electrical *current* produces a magnetic field (Figure 9.7). Notice that the direction of the magnetic field produced by a current-carrying wire is not directly toward or away from the wire but is perpendicular to this direction as well as to the direction of the wire itself; the lines representing the field form circular loops around the wire.

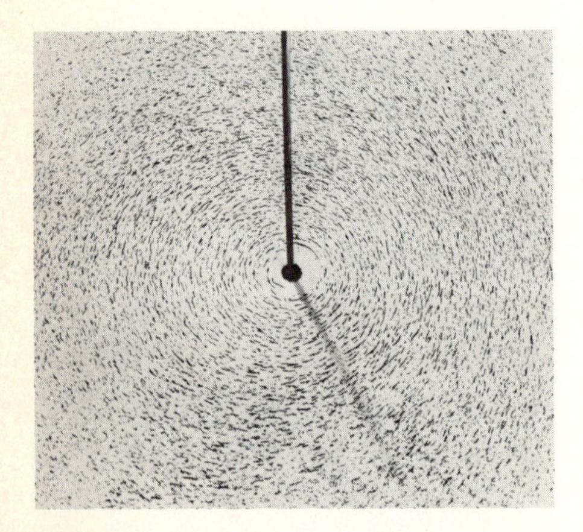

FIGURE 9.7
The magnetic field around a long straight wire, shown by iron filings. [*PSSC Physics*, D. C. Heath, Lexington, Mass., 1965; Education Development Center.]

Diagrams of the magnetic field produced by a current in a wire perpendicular to the page are shown in Figure 9.8. The field shown in part (a) is that produced when the electrical current in the wire is flowing in the direction *out* of the page, toward the viewer. (By the direction of the current, we mean the direction in which *positive* charge would flow to produce these effects; this current usually results from a flow of *electrons* in the *opposite* direction.) If the direction of the current is reversed, the direction of the magnetic field at all points is reversed, as shown in Figure 9.8b. (Many electromagnetic phenomena are difficult to represent on a two-dimensional page. The use of the symbols x and • to show directions of vectors into and out of the page is a useful convention; the dot represents the "point of the arrow," and the cross, the feathers on the tail.) Because the magnetic field produced by a wire is not directed toward or away from the wire but has the more complex structure shown in Figure 9.8, it is useful to have a simple rule for remembering the relationship between the direction of the current and the direction of the magnetic field that it produces. This is provided by the following "Right-Hand Rule" (Right-Hand Rule No. 1): Point the thumb of the right hand in the direction in which the current is flowing; then the fingers of the right hand curl naturally in the direction of the magnetic field (Figure 9.9).

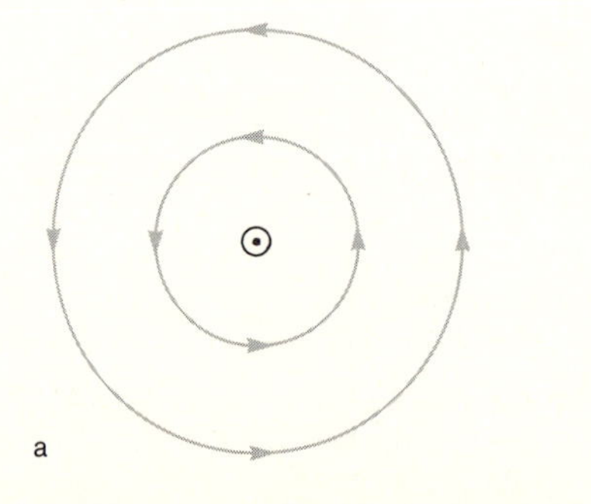

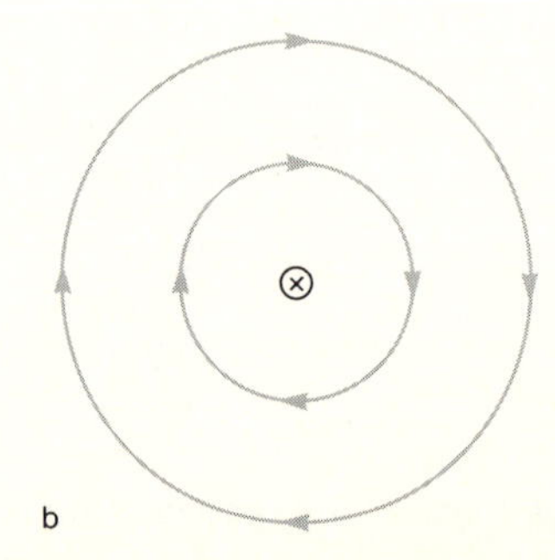

FIGURE 9.8
(a) The magnetic field produced by a current out of the page; the symbol • represents the head of an arrow and shows that the current is directed toward the viewer. (b) The field produced by a current into the page; the x represents the *tail* of an arrow, and shows that the current is directed away from the viewer.

The magnetic field produced by a current in a circular loop of wire is shown in Figure 9.10. The nature of this field can be understood by applying Right-Hand Rule No. 1 to find the field produced by each little piece of wire; the resulting field is the vector sum of the individual contributions. For the loop depicted in Figure 9.11, at the center of the loop each element of the wire makes an upward contribution to the field, leading to a relatively large field. At an exterior point, the magnetic fields contributed by elements on the near and far sides of the loop partially cancel one another; the field is weaker there and in the downward direction, the direction of the field produced by the wire on the nearest part of the loop.

A coil with several turns produces the field shown in Figure 9.12, the superposition of the fields produced by the individual loops. A coil like this is an electromagnet, one whose field can be turned on and off at will by closing and opening a switch in the electrical circuit. The field of a coil can be greatly enhanced by putting a bar of iron inside the coil; the field of the coil magnetizes the iron, and the field produced by the iron adds to that of the coil itself. It should be noted that at points outside the coil, the field produced by a coil of wire (Figure 9.12) is very similar to that of a bar magnet (Figure 9.6). The similarity is not accidental; the modern view is that the magnetic fields of iron bars have their origin in moving charges (currents) or spinning charges.*

Electrical charges that are in motion produce magnetic fields; this is the fundamental fact shown by these various examples. Now if moving charges (currents) produce magnetic fields, that is, if a current-carrying wire exerts a force on a nearby magnet, it is natural to ask whether the reverse is true. Does a magnet exert a force on a current-carrying wire?

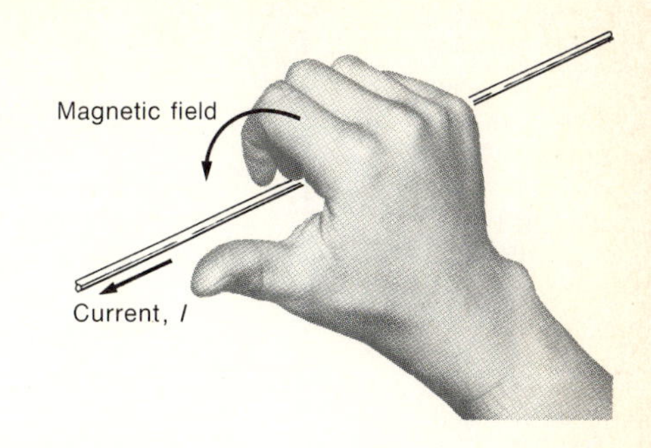

FIGURE 9.9
Right-Hand Rule No. 1, for determining the direction of the magnetic field produced by an electric current.

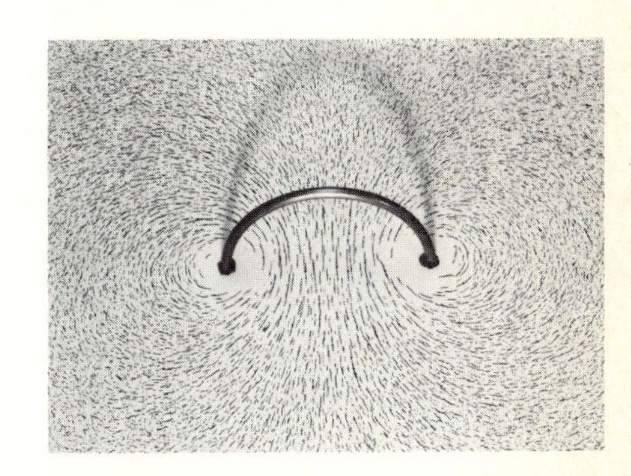

FIGURE 9.10
The magnetic field produced by a current in a circular loop. [*PSSC Physics*, D. C. Heath, Lexington, Mass., 1965; Education Development Center.]

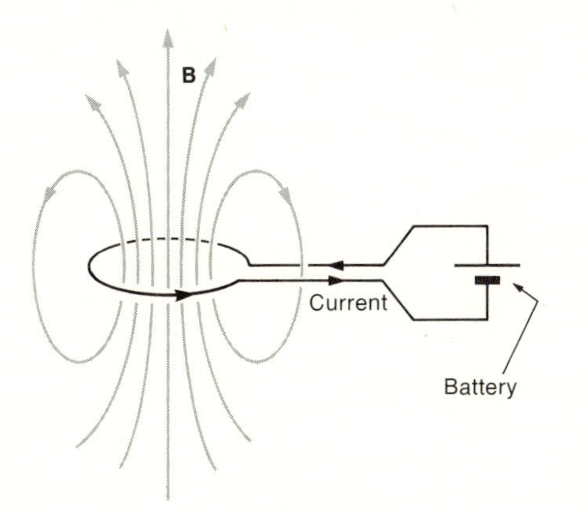

FIGURE 9.11
The nature of the magnetic field produced by a current in a circular loop of wire can be understood by applying Right-Hand Rule No. 1 to each element of wire.

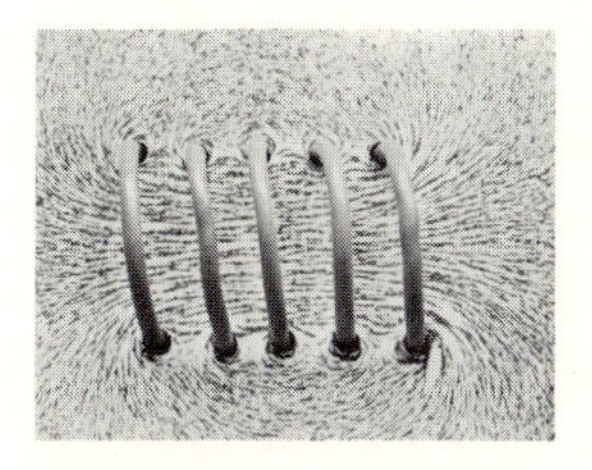

FIGURE 9.12
The magnetic field produced by a coil. [*The Project Physics Course*, Holt, Rinehart and Winston, New York, 1970.]

*As far as is now known, *all* the magnetic fields we normally observe arise from moving or spinning charged particles (electrons or nuclei), not from actual north or south poles. It is conceivable, however, that particles exist that have only an isolated north pole or only a south pole. Most attempts to find these hypothetical particles ("magnetic monopoles") have been unsuccessful, but in 1975 a study of the tracks produced by high-energy particles striking the earth revealed one example of a track which was tentatively identified as that of a magnetic monopole.

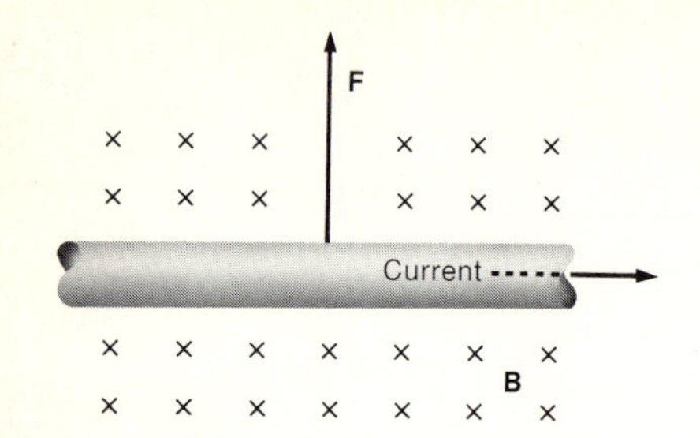

FIGURE 9.13
The electrical current is toward the right, and **B** is into the page; the result is an upward force on the wire.

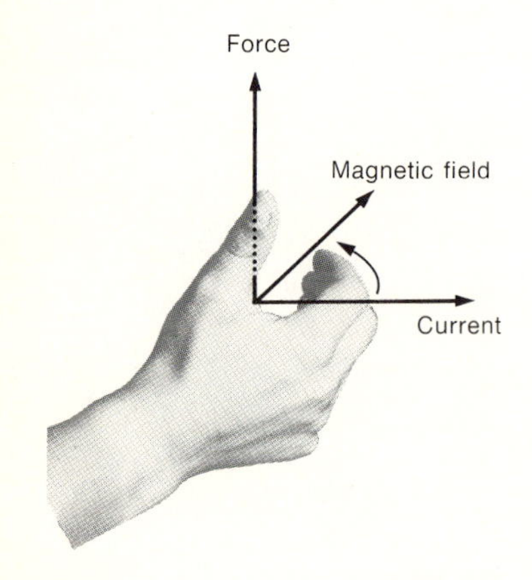

FIGURE 9.14
Right-Hand Rule No. 2, for determining the direction of the force on a wire in a magnetic field.

On the basis of Newton's third law of motion, we might expect an affirmative answer; the gravitational force exerted on a rock by the earth, for example, is accompanied by an equal and opposite force exerted on the earth by the rock. To put the same question in terms of the field concept, we have seen that an electrical current produces a magnetic field; we may then ask whether a current-carrying wire will experience a force when placed in an existing magnetic field. The experiment is not hard to carry out, and it is found that a current-carrying wire placed in a magnetic field does experience a force, though some care is required to describe the magnitude and direction of this force.

An example of such a situation is shown in Figure 9.13. In this case, the direction of the force on the wire is as shown, perpendicular both to the direction of the magnetic field and to the direction in which current is flowing. The direction of this force can be remembered with the aid of Right-Hand Rule No. 2 (Figure 9.14). Put the fingers of the right hand in the direction in which the current is flowing, with the hand oriented so that the fingers curl naturally *from* the direction of the current *toward* the direction of the magnetic field. The thumb then points in the direction of the force on the wire. The force on a wire is a maximum when the current and the field are perpendicular to one another; in this case the force on a segment of wire of length L carrying a current I, is given by:

$$F = BIL.$$

But remember that this equation applies only if the direction of **B** and the direction of current flow are *perpendicular.* If the current and the field are not in perpendicular directions, the force is given by using the *component* of **B** which is perpendicular to the direction of current flow ($B_\perp$):

$$F = B_\perp IL, \qquad (9.1)$$

a force that is zero if the current is either in the same direction as the field or in the opposite direction.

QUESTION

9.1 Apply Right-Hand Rule No. 2 to the situation shown in Figure 9.13. Does this rule allow you to predict an upward force? How is the direction of the force affected if the direction of the current is reversed? If the direction of the field is reversed? If the directions of both the current and the field are reversed?

The facts discussed above are used in the design of electrical meters and electrical motors. The operation of an ammeter is shown in Figure 9.15. When a current is passed through the loop in the direction shown, application of Right-Hand Rule No. 2 shows that the wire on side (b) of the loop experiences a downward force and the wire on side (d) an upward force. The loop will be twisted in the clockwise direction. The loop's motion is restrained by a spring; the greater the current, the greater the force on each side of the loop and the farther the loop is twisted before the spring stops it. A pointer can be attached to indicate the angle through which the loop is twisted, and once the meter has been calibrated, the position of the pointer can be used to measure the current. In practice, the twisting effect is usually enhanced by using a coil of many turns instead of a single loop; by varying the design of the coil and the strength of the magnetic field, ammeters can be constructed to measure currents of various sizes.

The most common *voltmeter* is really an ammeter in disguise, an ammeter containing a large resistance in series with the coil (Figure 9.16). The operation of such a voltmeter is based on Ohm's law. Suppose that the "ammeter" inside the voltmeter gives a full-scale reading for a current of 10^{-3} A and that the value of R is 10^5 ohms. By Ohm's law, if the current is 10^{-3} A, the voltage between the terminals of the voltmeter is $V = IR = 10^{-3} \times 10^5 = 100$ V. This is, then, a voltmeter which gives a full-scale deflection when the voltage between its terminals is 100 V. (The coil itself may have an appreciable resistance; if so, the value of this resistance must be included with that of the series resistance, R.)

An electric motor works in a similar way except that in a motor the coil is allowed to continue turning continuously in the same direction, and more complicated methods of delivering current to the coil are required. Figure 9.17 shows a motor designed for operation with direct current (DC). The two ends of the loop are connected to a split ring (a "commutator"), and the current is delivered through "brushes." At the instant shown, the force on the left-hand side of the loop (d) is upward and that on the right-hand side downward. (Check this by applying Right-Hand Rule No. 2). After the coil turns through 180°, side (b) is on the left. Because of the commutator, the force on the left-hand side is still upward and that on the right side downward, and the loop continues to turn in the clockwise direction.

An AC motor, designed to be supplied with alternating current, is shown in Figure 9.18. Here a pair of "slip rings" is used instead of a commutator. At the instant shown in Figure 9.18a, the force on side (d) is

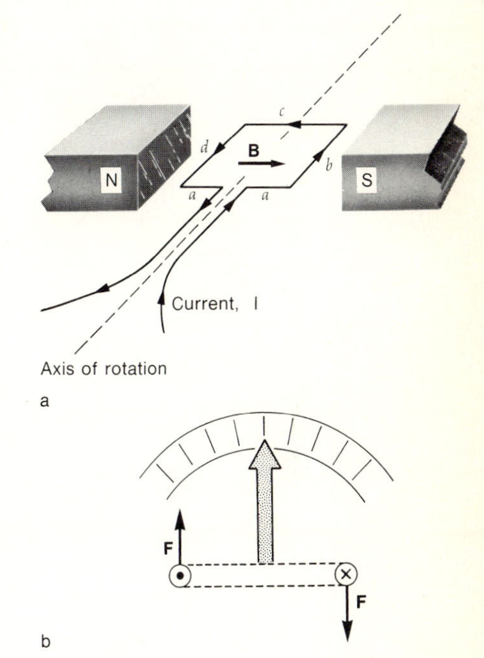

FIGURE 9.15
(a) Principle of operation of an ammeter; (b) front view of the ammeter.

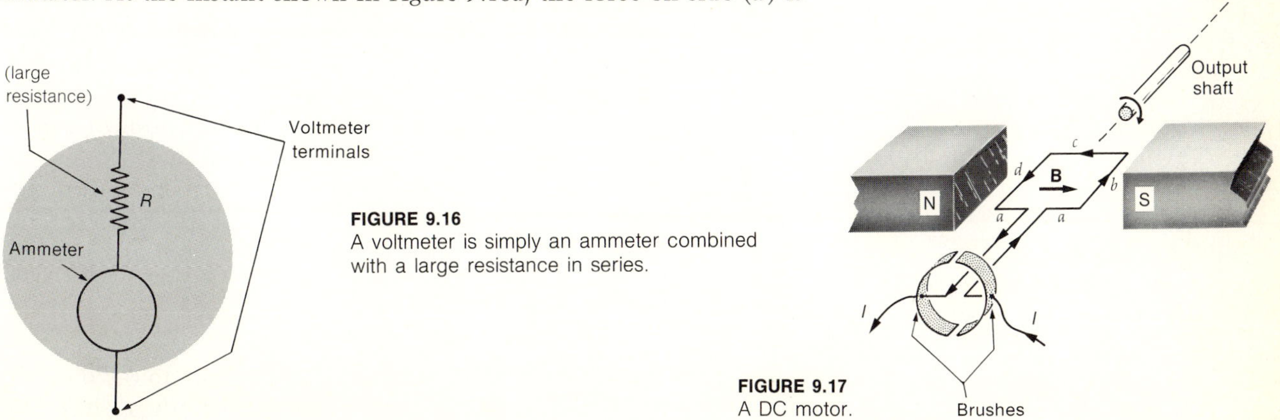

FIGURE 9.16
A voltmeter is simply an ammeter combined with a large resistance in series.

FIGURE 9.17
A DC motor.

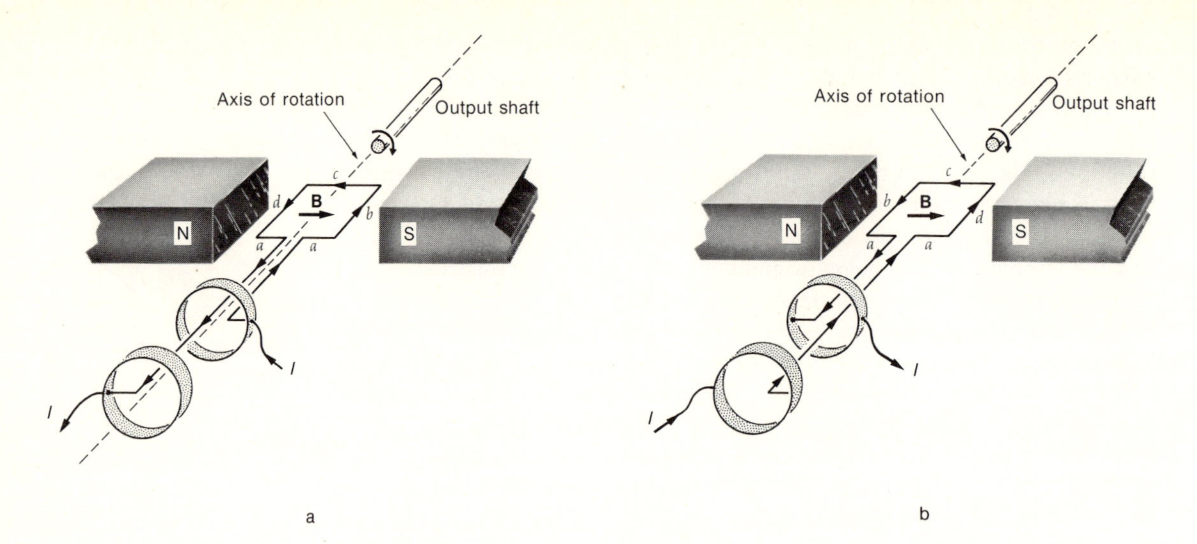

FIGURE 9.18
(a) An AC motor; (b) the same motor after the loop has turned 180°.

upward and that on side (b) is downward, and the motor turns in the clockwise direction. Suppose that the motor is supplied with alternating current with a frequency of 60 Hz, with the current changing from a maximum in one direction to a maximum in the other direction in 1/120 of a second. If the motor turns 180° in this length of time (Figure 9.18b), the forces are still in the direction to continue turning it in the clockwise direction. This is a particular type of AC motor, one designed to turn in precise synchronization with the current that drives it.

The shaft on which the coil of a motor rotates can be used to raise a weight or to turn the wheels of a car—in other words, to deliver mechanical energy. Electric motors come in all shapes and sizes, capable of delivering power over a wide range of values—from the power needed to run a clock to that needed to drive a train. Motors provide the means for using electrical energy for something other than the mere production of thermal energy, by enabling us to convert electrical energy into mechanical energy.

The force experienced by a current-carrying wire in a magnetic field must, from a fundamental point of view, arise from forces on individual moving charged particles. By using Equation 9.1 we can predict the nature of the force on a single charged particle moving through a magnetic field. Consider a section of wire (Figure 9.19) within which there is a current resulting from the motion of positive charges as shown. Let q be the total amount of flowing charge which is in this section of wire at any time. If the charges all move at speed v, then in a time equal to

$$\Delta t = \frac{L}{v}$$

a charge can move from one end of this section of wire to the other. If one were to stand at the right-hand end of this section of wire, during a

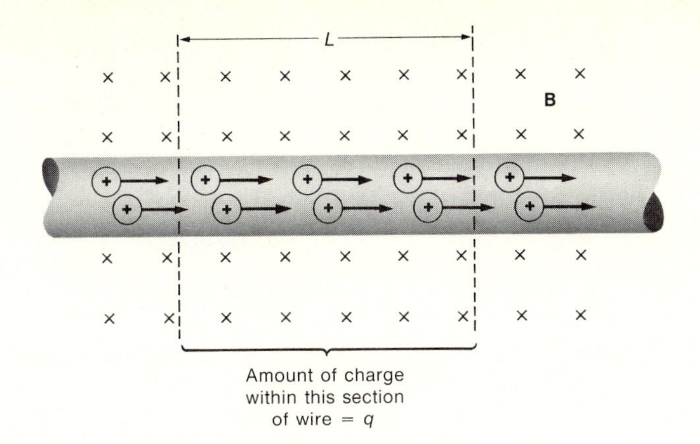

FIGURE 9.19
A current flowing to the right in a magnetic field directed into the page. The resulting upward force on the wire can be interpreted as the sum of the forces acting on individual moving charged particles within the section of wire.

time Δt, *all* the charge initially in this section of wire would pass by; the current, then, is:

$$I = \frac{\text{charge}}{\text{time}} = \frac{q}{\Delta t} = \frac{q}{L/v} = \frac{qv}{L}.$$

From Equation 9.1 the force on this piece of wire is

$$F = B_{\perp}IL = B_{\perp}\frac{qv}{L}L = B_{\perp}qv. \tag{9.2}$$

We can thus predict that, for any charged object in motion through a magnetic field, whether contained within a wire or not, there is a force given by Equation 9.2 with $B_{\perp}$ representing the component of the field perpendicular to the object's velocity. The direction of this force is perpendicular both to **B** and to the velocity and can be found by using Right-Hand Rule No. 2. In applying this rule here, we only need to remember that a moving positive charge constitutes a current in the direction in which the charge is moving. A *negative* charge in motion acts like a "current" in the direction *opposite* to the direction of motion; the force on a moving negative charge is in the direction opposite to that on a positive charge moving in the same direction (Figure 9.20).

a
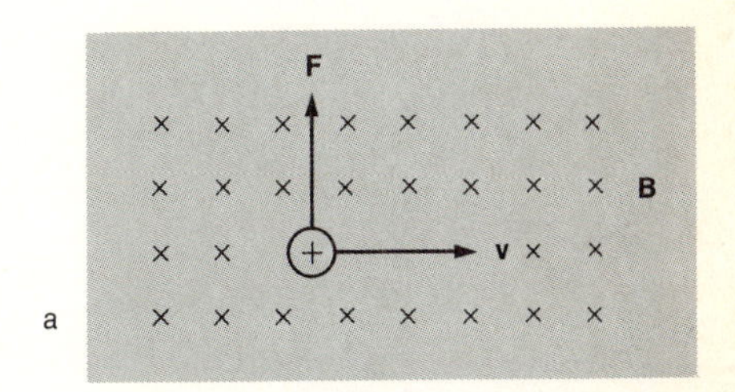

b
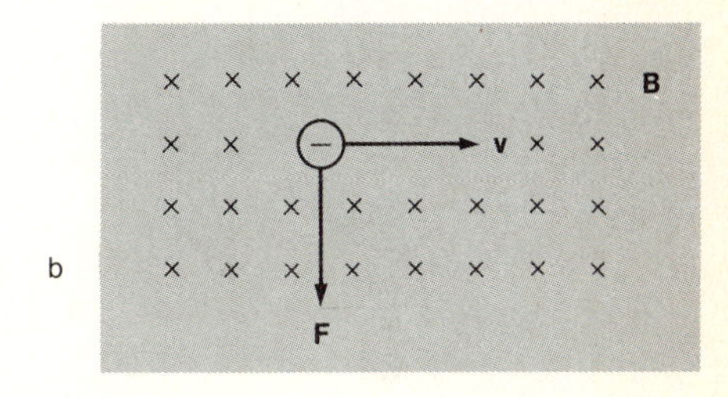

FIGURE 9.20
(a) A positive charge moving to the right acts like a current to the right; Right-Hand Rule No. 2 predicts an upward force when the magnetic field is directed into the page.
(b) A negative charge moving to the right is equivalent to a "current" toward the left; Right-Hand Rule No. 2 predicts a downward force.

§9.3 INDUCED EMF'S, FARADAY'S LAW, AND ELECTRICAL GENERATORS

We have still not come to the most exciting and important aspect of the relationship between electricity and magnetism. If a current can produce a magnetic field, can a magnetic field produce a current? Consider the experimental arrangement shown in Figure 9.21. The circuit on the right contains no battery, and surely no current would flow if the magnet were not there, but does the presence of the magnet produce a current?

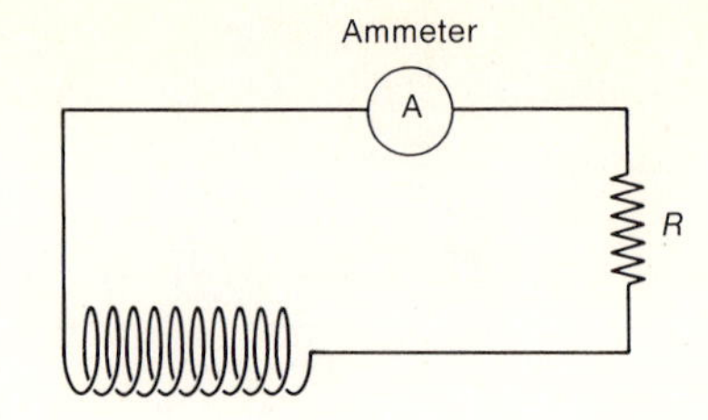

FIGURE 9.21
If the magnet is stationary, no current flows in the circuit.

The answer to this question is definitely negative. No matter how strong a magnet is used, there is no detectable current.

However, if we watch very closely while the magnet is being moved into position, there is a *momentary* current in the circuit while the magnet is in motion. And there is a momentary current in the opposite direction as the magnet is taken away. In fact, almost any motion of the magnet results in a transient current in one direction or the other. A steady magnetic field produces no current, but a *changing* field does result in an "induced" current. This basic fact was discovered by Michael Faraday in 1831 and almost simultaneously by the American physicist Joseph Henry, and it is this fact which is crucial to the generation of electrical energy from mechanical energy. From the experiment described, we can see one way in which we could construct a rudimentary electrical generator. Simply take the apparatus as shown in Figure 9.21 and move the magnet back and forth as indicated in Figure 9.22a. Every time the magnet is in motion to the right, there is a small current in one direction in the circuit and the resistance is heated. As the magnet is moved to the left, a current flows in the opposite direction and again the resistance is heated. The coil and the moving magnet therefore constitute an AC generator, a source of alternating current. By substituting an AC motor (Figure 9.22b), for the resistance, we could drive the motor this way and use it to raise a weight. This simple gen-

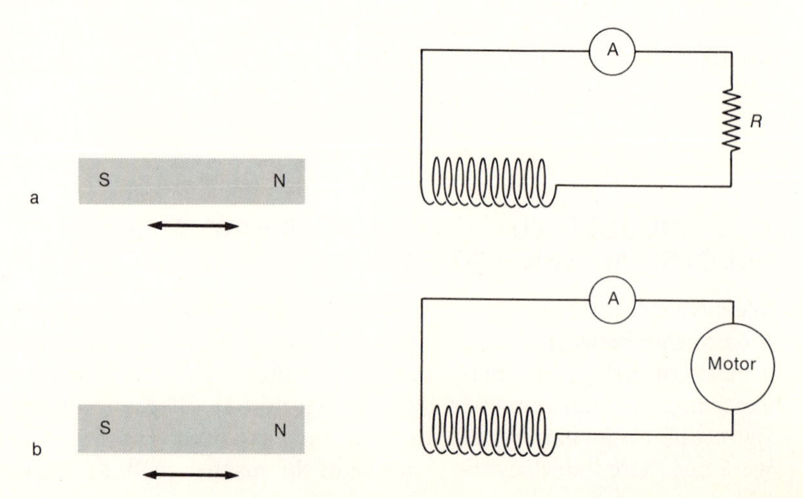

FIGURE 9.22
Oscillatory motion of the magnet induces an alternating current that heats the resistor (a), or operates the motor (b).

erator would not be a very *practical* one, yet the generators found in hydroelectric plants and in nuclear or fossil-fuel steam-electric plants are simply based on elaborations and modifications of this basic plan.

Consider the application of the law of conservation of energy to this situation. Thermal energy is produced in the resistance (Figure 9.22a), and mechanical energy is delivered by the motor (Figure 9.22b). Can we perhaps construct a perpetual motion machine using these ideas, one in which energy is continuously produced without any corresponding decrease in energy elsewhere? In the apparatus shown in Figure 9.22, it was proposed that the magnet be moved back and forth by hand; we can see more clearly the energy problem if we automate this process by making a pendulum out of the magnet, as shown in Figure 9.23. For the sake of this discussion, let us suppose that we have a perfect pendulum, one with frictionless bearings and with no air resistance. The magnet should simply swing back and forth indefinitely, with no decrease in energy, simply a continuous conversion of kinetic energy into gravitational potential energy and back again. This at any rate is what would happen if the magnetic pendulum were left to itself; does the presence of the nearby circuit in any way alter this conclusion?

If the pendulum were to swing with no decrease in amplitude, we would have a perpetual motion machine of the first kind, that is, a violation of the law of conservation of energy. We would have a continuing (though pulsating) production of mechanical energy from the motor with no corresponding decrease in the energy of the pendulum. We must conclude either that the law of conservation of energy does not apply to this situation or (a possibility that is more likely) that some important factor has been overlooked and that the presence of the nearby circuit *does* lead to a decrease in the amplitude with which the magnet swings. There is indeed such a factor. As the magnet moves, its motion results in an induced current in the circuit. But this current itself produces a magnetic field in the surrounding space, and this magnetic field acts on the magnet. Thus the magnetic pendulum is subjected to an extra force because of the presence of the circuit, and this extra force acts to decrease the amplitude of the swinging magnet, to reduce its mechanical

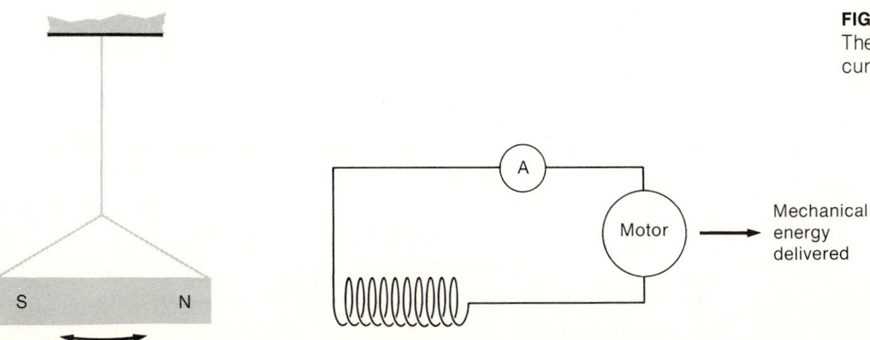

FIGURE 9.23
The magnet swings back and forth, inducing currents which drive the motor.

energy. It is clear that there will be a force; it is not obvious from the development which has been given here that this force acts to *decrease* the energy of the swinging magnet and to do so by the correct amount. If this were not so, the law of conservation of energy would be in severe difficulty. Development of electromagnetic theory in more detail leads to the conclusion that this effect is precisely what is needed to save the law of conservation of energy; the decrease in the energy of the pendulum is of just the right amount to account for the mechanical energy produced by the motor.

We have examined this proposed perpetual motion machine and its failure for two reasons. First, it is often true of proposals for the construction of perpetual motion machines (whether of the first or second kind) that the proposal is correct as far as it goes but that an interesting effect has been overlooked which is of precisely the right size to defeat the scheme. Second, it was pointed out at the beginning of this chapter that the relationship between electricity and magnetism is a complex and many-sided one. We can see from this example how the various aspects of this relationship, together with the law of conservation of energy, form a coherent whole. If changing magnetic fields could produce electrical currents as discussed in this section, and if it were not also true that currents produce magnetic fields, then this proposed perpetual motion machine *would* work and the law of conservation of energy would fail.

If we believe in the law of conservation of energy, we can be sure that no matter how the magnet in the rudimentary AC generator of Figure 9.22 is moved—whether by hand, by attaching it to a pendulum, or by water power, steam or anything else—a continuing input of energy will be necessary to keep it moving, an amount of energy over and above whatever is necessary to overcome friction and air resistance. Work must be done to maintain the oscillatory motion of the magnet, at a rate sufficient to account for the production of mechanical energy by the motor or the production of thermal energy in the resistance. Generators based on the principles discussed here are extremely useful, and provide a convenient way of converting mechanical energy into electrical energy, energy that can then be converted into thermal energy or again into mechanical energy:

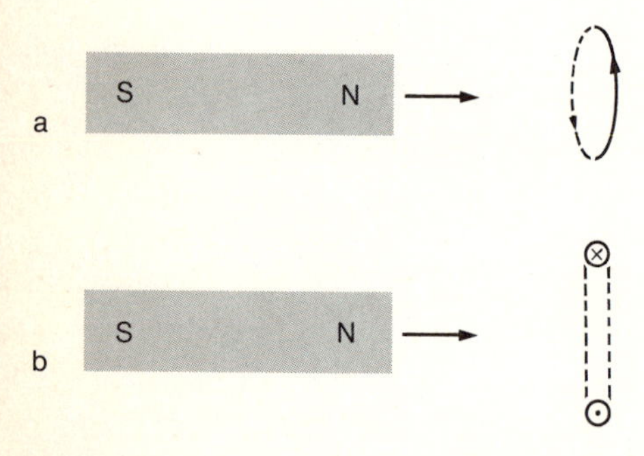

FIGURE 9.24
(a) As the magnet moves toward the loop, an induced current is produced. (b) Side view of the loop: the induced current is into the page at the top of the loop and out of the page at the bottom of the loop.

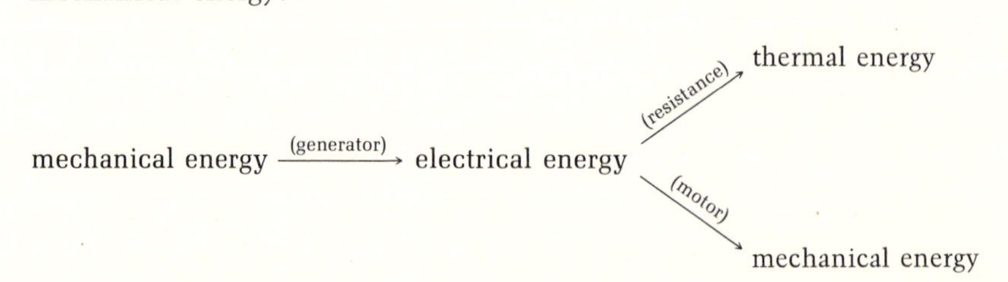

It is necessary to be more precise about the nature of the currents induced by varying magnetic fields. Consider the simple arrangement shown in Figure 9.24. A magnet is being moved toward a circular loop of wire; let us assume that the loop is small enough and far enough away so that the field the magnet produces is nearly uniform over the whole area of the loop. By using magnets of different strengths, or by moving the magnet with varying speeds, we find that the size of the induced current is simply proportional to the rate of change of the magnetic field. By using loops of various sizes, we find that the induced current is also proportional to the area of the loop. We tentatively define a new quantity, the "*flux* of magnetic field through the circuit," as the product of the magnetic field strength and the area of the loop. Then the induced current is proportional to the rate of change of the flux. There is a further complication, however. If the loop is oriented in some other direction (Figure 9.25a), the size of the induced current is smaller, and in the extreme case in which the plane of the loop is parallel to the magnetic field (Figure 9.25b), no current is induced. It is only the change in the *component* of **B** perpendicular to the plane of the loop that is effective in inducing a current. We therefore make a slight modification in the definition of flux:

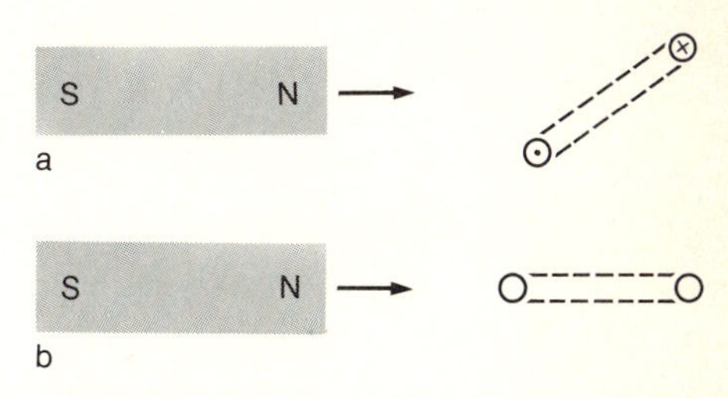

FIGURE 9.25
Two side views of a magnet moving toward the loop: in (a) the orientation of the loop is such that the induced current is smaller than in Figure 9.24; in (b) the loop is parallel to the magnetic field, and no current is induced.

flux = (component of **B** perpendicular to the loop) × (area of the loop).

With this definition of flux, a very simple result emerges from careful studies of the current induced in various situations. The current has the same value as would be produced by a battery with a *voltage* equal to the rate of change of the flux. We say that there is an "induced emf" in the circuit:

$$\text{Induced emf} = \text{rate of change of flux}. \qquad (9.3)$$

QUESTION

9.2 It is not obvious that Equation 9.3 is dimensionally correct, but it is. Remember that an emf is not a force but a voltage and has the units of joules per coulomb, or the dimensions of energy per unit charge. The units of flux are those of magnetic field strength times area (teslas times square meters), and so the units of the right-hand side of Equation 9.3 are those of flux per unit time, that is, tesla-meter^2 per second. From Equation 9.1, we see that 1 tesla can be considered as an abbreviation for 1 N/A-m. From these observations, show that Equation 9.3 is dimensionally correct and that the units are correct, with emf measured in volts, magnetic field strength in teslas, area in square meters, and time in seconds.

Equation 9.3 gives the magnitude of the induced emf and thus—if the properties of the circuit, its resistance for instance, are known—the magnitude of the induced current. The *direction* of the induced current depends on whether the magnetic field is increasing or decreasing and on the direction of the field with respect to the loop. A simple result known as Lenz's law specifies the direction of the induced current: when the flux through a circuit is changing, the induced current is in a direction such that *its* field produces a flux that tends to cancel the change in

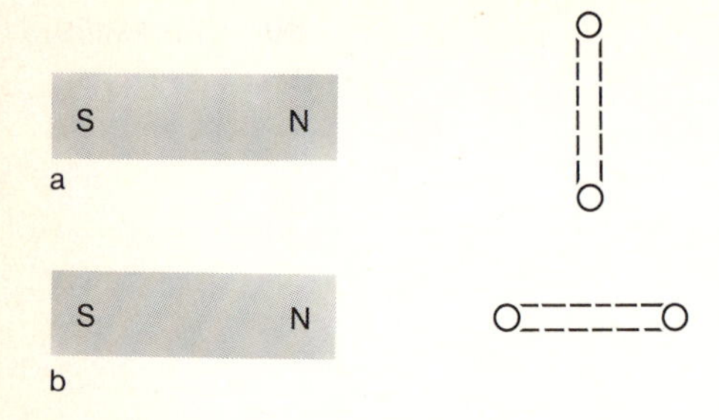

FIGURE 9.26
Changing the orientation of the loop from (a) to (b) results in a change in flux and an induced current.

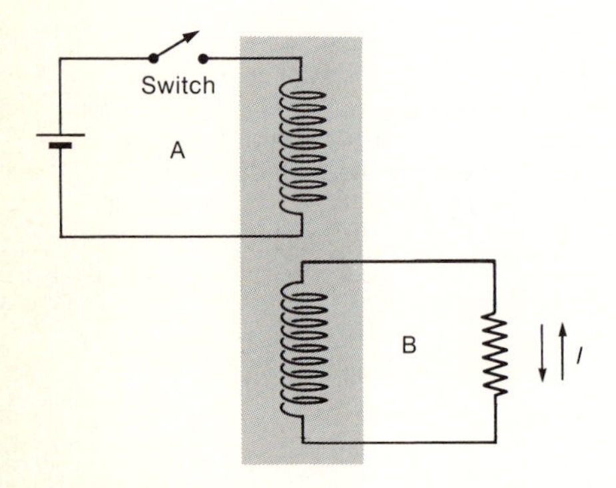

FIGURE 9.27
Opening and closing the switch changes the magnetic flux in circuit B and induces an alternating current.

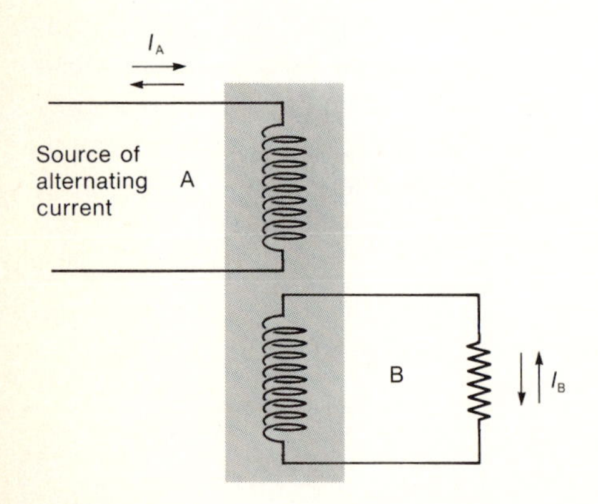

FIGURE 9.28
By connecting coil A to a source of alternating current, an alternating current is induced in circuit B.

flux that was the cause of the current. Let us apply Lenz's law to one example, that of Figure 9.24 in which the magnet is approaching the loop. The magnetic field due to the magnet is to the right and is increasing in size. According to Lenz's law, the induced current in the loop must be in a direction such that it produces a magnetic field through the loop to the *left*. By referring to Figure 9.11 or by using Right-Hand Rule No. 1, we can see that the current induced in the loop must be in the direction shown in Figure 9.24.

Equation 9.3 gives the size of the induced emf, and Lenz's law the direction of the induced current. The two statements together constitute Faraday's law of induced emf's, a law that is more general than one might at first suspect. We have described situations in which the flux through a circuit is changed by moving a magnet toward or away from the circuit. There are other ways in which the flux can be changed, and Faraday's law works equally well in all cases. Instead of moving the magnet, we can equally well move the circuit and keep the magnet fixed; Faraday's law correctly describes the current induced. We can even simply change the orientation of the circuit, as indicated in Figure 9.26. Here too there is a changing flux. The flux is defined in terms of the component of **B** perpendicular to the plane of the circuit, and so the flux is zero in Figure 9.26b whereas it was not zero in Figure 9.26a, and again Faraday's law is applicable. Faraday's law also applies if the circuit is a coil of many turns, provided that the flux is calculated by multiplying the flux through any one loop by the number of loops.

A changing flux can also be produced without any mechanical motion at all. Consider the experiment shown in Figure 9.27. When the switch is closed, the current in coil A increases, and this current produces a magnetic field at the position of coil B. Thus the flux through circuit B changes, and a momentary induced current flows in this circuit. When the switch is opened, the current in coil A stops, the flux through circuit B changes in the opposite direction, and so a momentary current flows in circuit B, one which is in the direction opposite to that produced when the current in A was increasing. Thus by simply closing and opening the switch in circuit A periodically, we can produce successive pulses of current back and forth through circuit B, without any wires between one circuit and the other and without any mechanical motion. The same effect can be produced without the battery and switch if circuit A is connected to a source of alternating current (Figure 9.28). As the current in A changes periodically from a maximum value in one direction to zero, then to a maximum value in the other direction, the corresponding changes in the flux through circuit B induce a similar alternating current in B. There are no wires connecting circuits A and B, yet energy is transferred from one circuit to the other by means of the varying magnetic field. This device is called a *transformer,* and it is essential for the distribution of electrical power, as we shall see in §9.4.

It is surprising that a change in the magnetic flux through a circuit results in an induced current; Faraday himself did not anticipate this result. It is even more surprising that the single generalization called Faraday's law correctly predicts the induced current, no matter how the change in flux is produced, whether by moving a magnet near the circuit, moving the circuit instead, or even with no physical motion at all as in Figures 9.27 and 9.28. With the advantages of hindsight, though, we can see how the induction of a current in a circuit which is in motion through a magnetic field might well have been predicted on the basis of the facts discussed in §9.2, without prior knowledge of Faraday's law. Consider the closed circuit shown in Figure 9.29, a simple rectangular loop of wire that is being moved through a *nonuniform* magnetic field. Suppose that at all points to the right of the dotted line, the magnetic field is directed *into* the paper (perpendicular to the plane of the circuit) and has the magnitude B, and that at all points to the left of this line there is no magnetic field. Let us treat the wire of which the circuit is made as one that contains positive charges that are free to move within the wire, each of magnitude q, and let us consider what happens to one such charge in side (a) of the loop. Such a charge, being inside the wire, moves to the right along with the whole circuit, and since it is a charge in motion through a magnetic field, it experiences a force perpendicular both to the magnetic field and to the direction in which it is moving, as we saw in §9.2. The magnitude of this force is equal to Bqv (Equation 9.2), and from Right-Hand Rule No. 2 the direction of this force is as shown in Figure 9.29. Thus positive charges within side (a) of the loop are pushed toward side (b), they are then free to flow through the rest of the circuit, and thus a circulating current is set up in the counterclockwise direction. (Positive charges in sides (b) and (d) of the loop also experience forces, just as do charges in side (a), but these forces merely push them sideways across the wire and do not tend to produce any current around the loop.) Note that it is important that the magnetic field be nonuniform; if the field were completely uniform, positive charges in both sides (a) and (c) of the loop would be pushed toward side (b), and no circulating current would be produced.

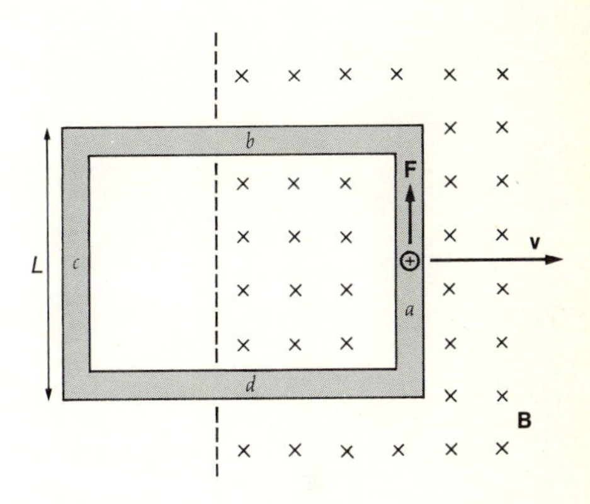

FIGURE 9.29
Motion of a rectangular loop of wire through a nonuniform magnetic field. Positive charges in side (a) experience an upward force; a circulating current in the counterclockwise direction results.

If we accept the law of conservation of energy, we can even predict the size of the induced current, again without prior knowledge of Faraday's law. Denote the magnitude of the induced current, whatever it may be, by I. Now the right-hand section of wire (of length L) carries a current I and is therefore subjected to a force whose magnitude is BIL (Equation 9.1) and whose direction, from Right-Hand Rule No. 2, is to the left. Thus in order to keep the circuit moving at a steady velocity, even in the absence of friction, we must exert a steady force to the right of equal magnitude. We must exert a force on the loop in the direction in which it is moving, that is, we must do work on the circuit, and the *rate* at which we are delivering energy to the circuit is given by:

$$\text{power} = \text{work/time} = \frac{\text{force} \times \text{distance}}{\text{time}} = \text{force} \times \text{velocity}$$

$$= BILv. \tag{9.4}$$

The current in the circuit results in the production of thermal energy; if the total resistance of the circuit is denoted by R, the rate at which thermal energy is produced is given by

$$P = I^2R. \tag{9.5}$$

We are supplying mechanical energy at the rate given by Equation 9.4. Thermal energy is being produced at the rate given by Equation 9.5. These two rates must be equal, and therefore:

$$BILv = I^2R$$

or

$$I = \frac{BLv}{R}. \tag{9.6}$$

Thus without using Faraday's law, we can predict the existence of an induced current in the counterclockwise direction, whose size is given by Equation 9.6. This is the same current that would be produced by a battery with an emf equal to BLv; in other words, the induced emf is equal to BLv.

We can now show that this is precisely what is to be expected from Faraday's law. We must calculate the rate of change of flux through the circuit. Because the field is nonuniform, the flux changes as the circuit moves, as a greater and greater area of the loop is exposed to the field. The rate of change of flux is equal to BLv, as we can see by looking at Figure 9.30. In Figure 9.30a, the flux is equal to BLx_1; during a time Δt the circuit moves a distance equal to $v\Delta t$, and the flux in Figure 9.30b is BLx_2. The change in flux is

$$BLx_2 - BLx_1 = BL(x_2 - x_1) = BLv\Delta t, \tag{9.7}$$

and the rate of change of flux is found by dividing Equation 9.7 by Δt:

$$\text{rate of change of flux} = BLv.$$

Thus the rate of change of flux is equal to the previously calculated value of the induced emf.

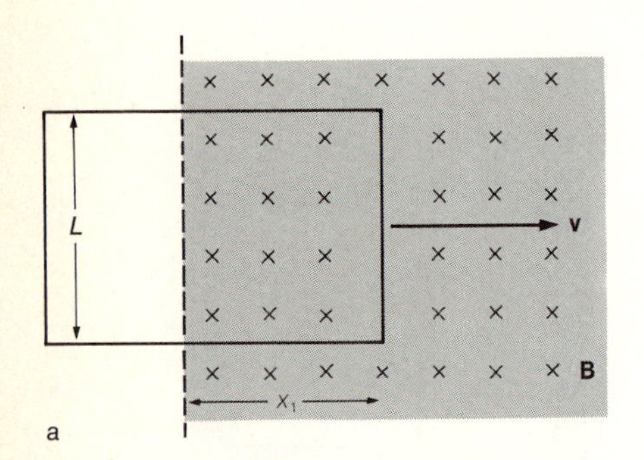

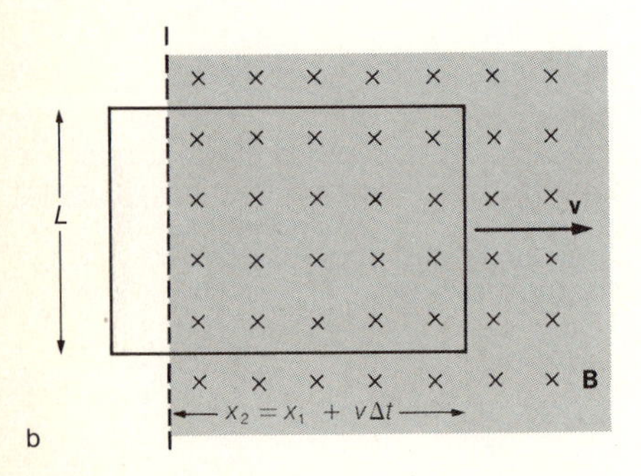

FIGURE 9.30
The circuit of Figure 9.29 is shown here at two different times. As the circuit moves, the flux changes, and thus an emf is induced.

QUESTION

9.3 Show that the induced current is in the direction to be expected from Lenz's law. That is, show that a current in the counterclockwise direction produces a flux through the circuit that tends to cancel the *change* in flux resulting from the motion of the circuit.

At least in this special case, then, the predictions made earlier are in perfect agreement with those made by using Faraday's law. The various relationships between electricity and magnetism are not independent of one another, but instead, together with the law of conservation of energy, they form one coherent and consistent structure.

Using Faraday's law, we can design apparatus to convert mechanical energy into electrical energy, and then with motors or with simple resistances we can convert this electrical energy into mechanical energy or into thermal energy. We do not get something for nothing this way. The law of conservation of energy still applies, and in order to induce electrical currents, energy must be delivered, not only energy to overcome friction but also enough additional energy to account for the electrical energy generated. It is possible to design a generator based on the arrangement shown in Figure 9.22; if some source of energy can be used to move the magnet back and forth continuously (or to move the coil back and forth with the magnet held fixed), a continuing alternating current will be induced in the circuit. It is usually easier in practice to produce a continuous rotational motion than to produce such an oscillatory motion; the fundamental requirement is simply that the motion must result in a change in the magnetic flux through a circuit. One simple way of doing this is shown in Figure 9.31; this device is very similar to modern practical electrical generators. A rectangular loop of wire is fixed to an axle driven by some external source of energy, a stream of water in a hydroelectric plant or a jet of steam in the turbine of a steam-electric plant. As the loop rotates, the perpendicular component of **B** changes from a maximum to zero, and then to a maximum in the opposite direction with respect to the loop; the flux through the loop varies periodically, and a current is induced.

In order that this current can be delivered to an external circuit, one side of the loop is split and wires are brought out to slip rings with which connections are made to an external apparatus (which, in Figure 9.31, consists simply of a resistance and an indicating meter) to form a complete circuit. Notice that Figure 9.31 looks just like an AC motor (Figure 9.18), except that the coil of the generator is driven mechanically and causes a current to flow in a circuit, whereas in a motor a current is supplied from an external source and causes a rotation of the coil, which can be used to deliver mechanical energy. That is, a generator is used to

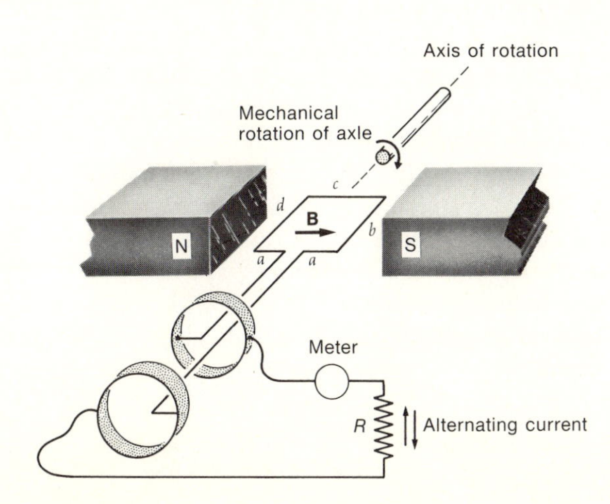

FIGURE 9.31
A simple AC generator.

convert mechanical energy into electrical energy, whereas a motor converts electrical energy into mechanical energy. The same apparatus can be used for either purpose as the occasion demands. As the coil rotates, the magnetic flux through the loop changes periodically. The details will be omitted, but it can easily be shown from Faraday's law that an alternating voltage is induced between the two wires coming from the generator. This device is an AC generator; if the coil is rotated 60 times per second, the voltage between the two wires has the form shown in Figure 8.32 (§8.7), that of the ordinary household voltage.* This is the essence of the modern AC generator, such as that shown in Figure 9.32. Many variations on this basic theme are used. The coil usually has many turns instead of just one, a number of separate coils may be rotated on the same axis with a necessarily more complex arrangement of slip rings, the magnetic field may have a more complicated structure and may be provided by an electromagnet rather than by permanent magnets. Often the circuit in which the emf is induced is held stationary while the source of the magnetic field is rotated. The principles for all such generators are the same. Mechanical rotation results in the production of an alternating voltage between wires leading from the generator, and an alternating current is produced in whatever is connected to these wires. Just as in the case analyzed earlier, mechanical work is required to maintain the rotation; this energy input produces electrical energy, which (in this example) is then converted into thermal energy as the current flows back and forth through the resistance. In a real generator, there is inevitably some friction within the generator and thus some production of thermal energy as well as some generation of thermal energy in the resistance of the coil itself (I^2R). An additional amount of mechanical energy must be fed in to account for these losses. In modern generators, this is not a very large effect; of the mechanical energy delivered to the generator, a small percentage is thus wasted and the remainder is converted into electrical energy for delivery to the external circuit.

FIGURE 9.32
Construction of a large generator. [GEC Turbine Generators Limited, Manchester, England.]

In a hydroelectric generating plant, the mechanical energy needed to drive the generator is derived from the gravitational potential energy of water behind a dam or at the top of a waterfall. The overall efficiency of conversion of gravitational potential energy into electrical energy is quite high, 85% or so. However, in a steam-electric plant, the second law of thermodynamics results in a much more serious inefficiency. Here the mechanical energy to drive the generator is obtained by taking heat from a hot reservoir; the efficiency of this conversion process is limited by the temperature of the hot reservoir and the temperature of the cold reservoir into which the waste heat flows, and the overall efficiency is usually no more than about 33%.

Motors and generators are made in a wide range of sizes, and usually the design is tailored to a particular job. Occasionally it is possible to

*By using a commutator instead of slip rings, one can make a DC generator in which the current in the external circuit is always in the same direction.

take advantage of the fact that the same apparatus can be used either as a motor to convert electrical energy into mechanical energy or as a generator to do the reverse. An important example of this is found in pumped-storage hydroelectric plants (§4.2.B), which employ "reversible pump-turbines." During the night, a motor drives a pump to fill the storage reservoir. During the day—the peak-load period—the pump becomes a turbine and the motor becomes a generator. Water coming down from the reservoir pushes against the turbine blades, causing a rotation of the coil and generating electrical energy, which is delivered to the power lines.

§9.4 THE TRANSMISSION OF ELECTRICAL POWER. THE IMPORTANCE OF THE TRANSFORMER

With generators for the production of electrical energy from mechanical energy, with wires to carry electrical energy from one point to another, and with motors for converting electrical energy into mechanical energy and with simple resistances for converting electrical energy into thermal energy, we have all the elements of a system for the generation, distribution, and consumption of electrical energy as indicated schematically in Figure 9.33. Such a system is all that is needed when the amount of power generated is small and when the place at which the energy is to be used is very close to the point at which it is generated. But when large amounts of energy must be transported over great distances (from Niagara Falls to New York City, perhaps), the simple system shown in Figure 9.33 is inadequate. The problem is a practical one which arises solely from the *quantities* involved. It is caused by the lead resistances of the wires, the same problem discussed in §8.6. The resistances of the wires leading from a flashlight battery to the bulb are small enough that they do not seriously interfere with the operation of a flashlight, but the situation is very different when large amounts of power are involved.

Consider a specific example. Suppose that there is a medium-sized city of 10^4 households served by a power plant 30 miles away. Suppose that at some moment each household is using 1200 W of power. (This is hardly an excessive figure; one toaster or 12 light bulbs will use this much power.) Each household is thus using a current of about 10 A, and thus the current delivered to the whole city is about 10^5 A. From the

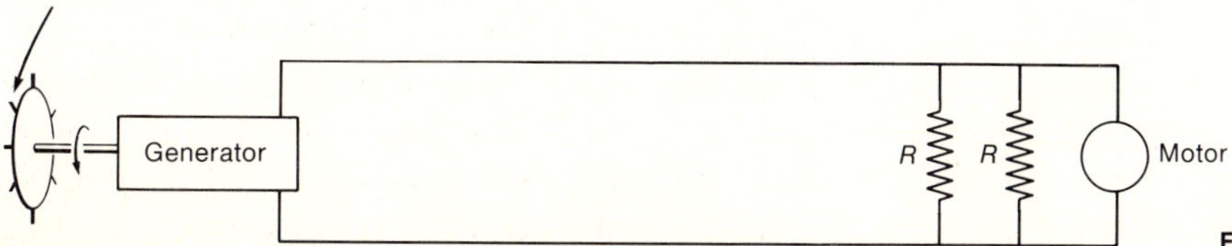

FIGURE 9.33
Diagram of a basic electrical power system.

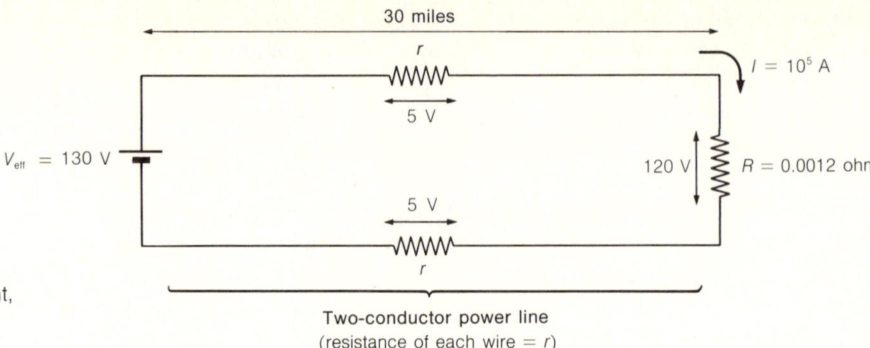

FIGURE 9.34
A simplified circuit diagram of a power plant, transmission line and city (a preliminary version).

power company's point of view, the city is a resistance across which there is a voltage of 120 V and through which a current of 10^5 A is flowing, that is, a resistance equal to $120/10^5 = 0.0012$ ohm. Assume that we will supply power to the city in the same way that power is supplied to the bulb of a flashlight, by running two wires from the source of voltage (the power plant) to the city, and suppose that our criterion is that we waste no more than about 10% of the power generated at the power plant in heating up the wires. If the voltage supplied to the city is 120 V, the voltage at the power plant should be no more than about 10% larger, about 130 V. Then the circuit diagram of the power plant, transmission line, and city will appear as shown in Figure 9.34. Now the question is this: about how thick must the wires in the power cable be?

QUESTIONS

9.4 Show that the value of r, the resistance of each wire, must be 5×10^{-5} ohm.

9.5 Suppose that the wires are made of copper. Show that the diameter of each wire must be about 4.5 m!

Since a wire with a diameter of 4.5 m is obviously out of the question, a serious problem exists. Of course things are not quite as bad if the power plant is 5 miles away instead of 30, and we could run several lines instead of just one, but these modifications do not help very much. We are still faced with the prospect of building fantastically large and expensive power lines. Statewide power transmission networks, even citywide networks, would be out of the question if we were to distribute electrical energy this way.

One possible way of circumventing this problem is to use wires made of superconducting material rather than copper. Then the wires would have zero resistance, and no energy would be wasted in heating the power line. The major difficulties are that superconducting materials are expensive and must be kept at extremely low temperatures; a considerable amount of energy is needed simply for refrigeration. Short experimental superconducting power lines have been built, but further research is necessary before they will be practical and economical for large-scale use.

Another approach is to recognize that since the power is given by the product of current and voltage, we could equally well transmit the same amount of power with a higher voltage and a lower current. Suppose we were to use a voltage 1000 times higher, 120 kV (1.2×10^5 V) instead of 120 V, and a current of only 100 A. This is a possibility worth considering, because for any given wire the rate at which energy is wasted in heating up the wire is proportional to the square of the current (I^2R); with a smaller current, we can use much smaller wires and still not waste an excessive amount of power.

QUESTIONS

9.6 Show that if the city is using a current of 100 A at a voltage of 120 kV, the effective resistance of the whole city is 1200 ohms.

9.7 If we still require that no more than 10% of the power generated at the power plant be wasted in the power lines, the circuit diagram of Figure 9.34 is modified as shown in Figure 9.35.

(a) Show that now the value of r is 50 ohms.

(b) Show that if r is 50 ohms, then the diameter of each wire is about 4.5×10^{-3} m (4.5 mm).

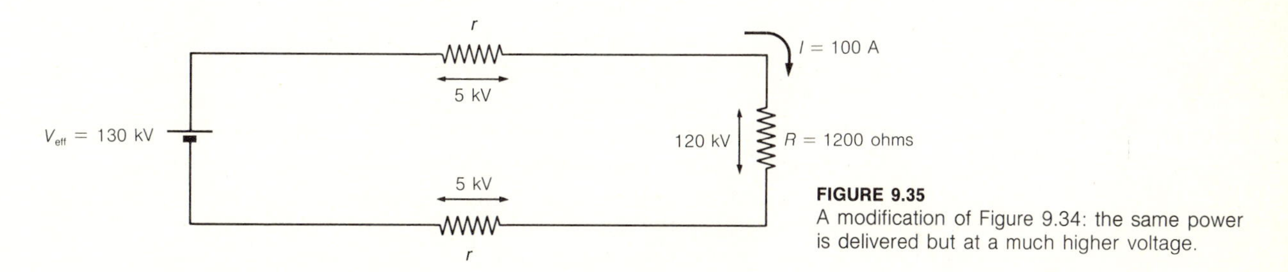

FIGURE 9.35
A modification of Figure 9.34: the same power is delivered but at a much higher voltage.

The results of the preceding question show that the size of the wire needed is much more reasonable if electrical energy can be transmitted at high voltages and low currents, but 120-kV sockets in the home would have to be rejected simply on the grounds of safety. It is bad enough to have an abundance of 120-V outlets around the house; as it is, about 1000 people are accidentally electrocuted every year in the United States. Fortunately, the principles discussed in this chapter provide a means of "transforming" alternating voltages, for "stepping" voltages up or down, so that it is possible to transmit electrical energy at voltages of 100 kV or more and then step the voltages down to 120 V for delivery to the consumer. Consider the arrangement shown in Figure 9.36, the "transformer," which was briefly discussed in §9.3. Suppose that the primary

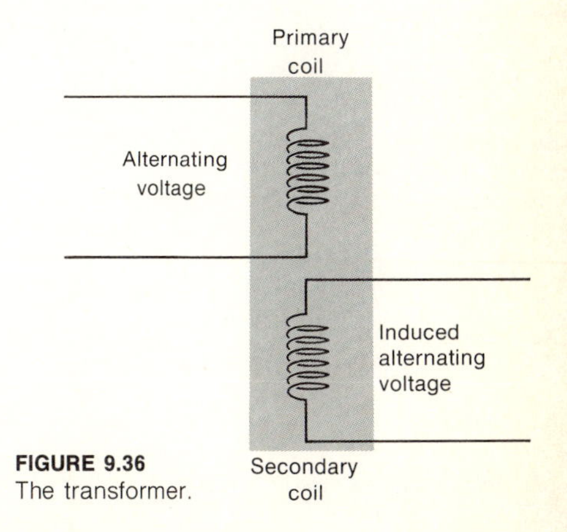

FIGURE 9.36
The transformer.

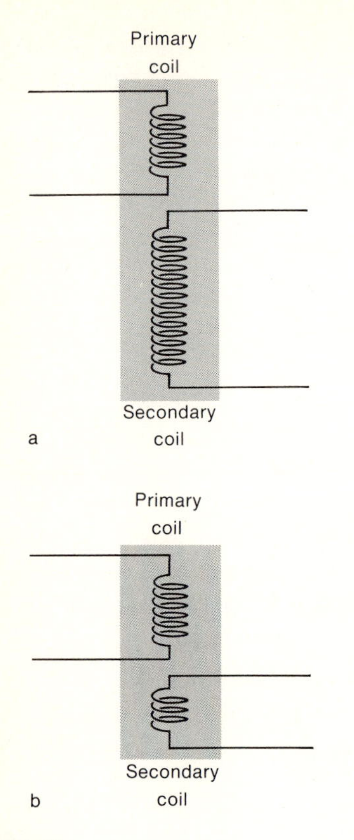

FIGURE 9.37
Transformers for (a) stepping up the voltage in the primary circuit and (b) stepping it down.

coil is fed from a source of alternating current. As current flows back and forth through the primary coil, the magnetic field produced causes a periodic change in the flux through the secondary coil, thus inducing an alternating voltage between the two wires leading from the secondary coil. What happens if we use a secondary coil with more turns of wire, with everything else unchanged (Figure 9.37a)? At every moment the flux through the secondary coil in Figure 9.37a is larger than in Figure 9.36 because there are more turns on the coil, the induced emf is larger, and in this way a relatively low voltage can be stepped up to a higher voltage, or—as shown in Figure 9.37b—stepped down to a lower value.

If the voltage induced in the secondary is greater than that in the primary circuit, the currents must differ in the inverse way. Consider the application of the principles discussed in §8.5.A to the transformer shown in Figure 9.38. Electrical energy is fed into the transformer at a rate given by

$$P_1 = I_1 V_1.$$

Electrical energy is delivered to the secondary circuit at a rate given by

$$P_2 = I_2 V_2.$$

If there is no conversion of electrical energy into other forms within the transformer, these two values of power must be equal:

$$I_1 V_1 = I_2 V_2.$$

Thus if V_2 is 1000 times larger than V_1, I_2 must be 1000 times smaller than I_1. In actual transformers, there are some losses, some conversion of electrical energy into thermal energy within the transformer, and, if so, P_2 may be somewhat smaller than P_1.

The use of transformers provides a tremendous amount of flexibility. Electrical energy can be generated at any voltage that is convenient, stepped up with a transformer to be transmitted at a high voltage and a low current, and then stepped down again for delivery to the consumer at a voltage of 120 V.

Note that the operation of a transformer requires that the primary coil be supplied with *alternating* current. If a steady current were applied to the primary coil, an unchanging flux through the secondary coil would result, and no emf would be induced. This is the major reason for the fact that large-scale systems for generating and distributing electrical energy are almost exclusively AC systems.*

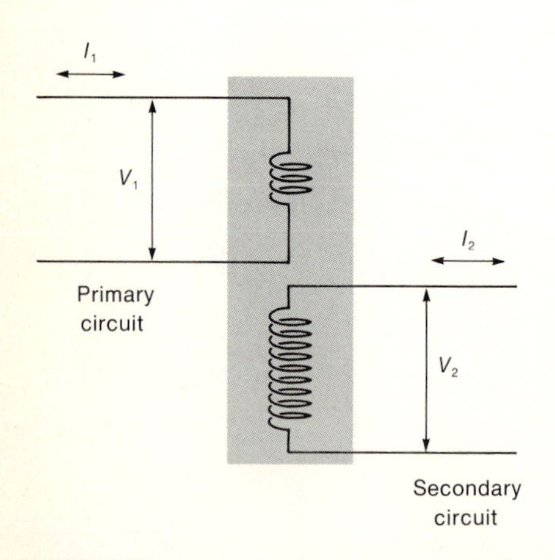

FIGURE 9.38
Power relationships in the transformer.

*It is possible to step a DC voltage up or down, but only by the awkward procedure of using a direct voltage to generate an alternating voltage, using a transformer to change the voltage, and then reconverting to DC. In some special circumstances, there may be factors that make this worthwhile, but at present almost all large systems use AC. Use of direct current would have two advantages. First, the same effective voltage is obtained with a lower peak voltage. To provide an effective alternating voltage of 120 kV, a peak voltage of 170 kV is required; a steady voltage of 120 kV would provide just as much average power. Thus transmission line towers for a DC power line can be somewhat smaller and less expensive. Second, use of DC avoids the necessity of accurate synchronization of alternating voltages generated in various parts of a large power system with many interconnections.

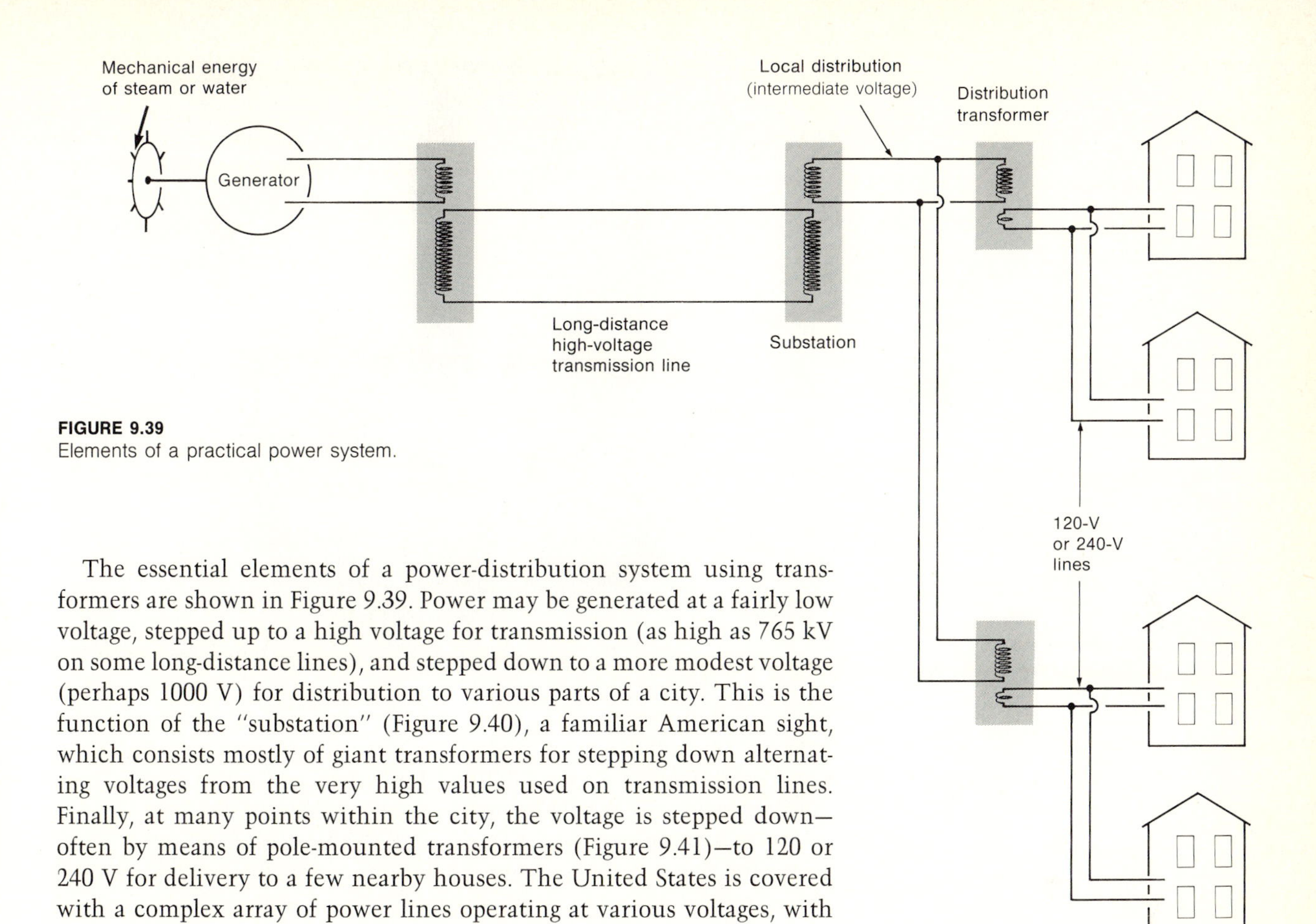

FIGURE 9.39
Elements of a practical power system.

The essential elements of a power-distribution system using transformers are shown in Figure 9.39. Power may be generated at a fairly low voltage, stepped up to a high voltage for transmission (as high as 765 kV on some long-distance lines), and stepped down to a more modest voltage (perhaps 1000 V) for distribution to various parts of a city. This is the function of the "substation" (Figure 9.40), a familiar American sight, which consists mostly of giant transformers for stepping down alternating voltages from the very high values used on transmission lines. Finally, at many points within the city, the voltage is stepped down—often by means of pole-mounted transformers (Figure 9.41)—to 120 or 240 V for delivery to a few nearby houses. The United States is covered with a complex array of power lines operating at various voltages, with

FIGURE 9.40
A large transformer at a substation. [General Electric Company.]

FIGURE 9.41
Pole-mounted distribution transformers.
[Western Massachusetts Electric Company.]

interconnections between power lines that permit power generated in one part of the country to be delivered somewhere else. Sometimes interconnections cause problems or turn local problems into more serious ones (such as the 1965 power blackout in the Northeast), but the system is, for the most part, extremely reliable and convenient; we depend on this electrical generation and distribution system for a great many of our activities.

FURTHER QUESTIONS

9.8 If a charged particle is injected into a uniform magnetic field, **B**, with its velocity perpendicular to the direction of the field, it will experience a force of magnitude qvB in a direction perpendicular to its velocity. As the particle is deflected, the direction of the force changes and always remains perpendicular to the particle's velocity. This is exactly what is required to produce uniform circular motion. Show that the particle will move in a circular path, with a radius given by $r = mv/qB$. (This result has many important applications. One is that it gives us a way of trapping charged particles in a magnetic field, restricting their motion to a limited region. A great deal of the research aimed at obtaining useful energy from nuclear fusion has been directed toward designing "magnetic bottles" based on this idea, to trap charged nuclei so that they have a chance to collide with one another and undergo fusion.)

9.9 Like charges repel one another, whereas unlike charges attract. Do "like" currents (two currents in the same direction) repel or attract? Consider two parallel current-carrying wires. Each wire produces a magnetic field, and the other wire, as a result, experiences a force. Use the right-hand rules to determine whether the force is one of attraction or repulsion. (In practice, the ampere is defined by measurements in this type of experiment; the ampere can be defined as the current in each wire necessary to produce a specified force between the two wires.)

9.10 (a) A positively charged particle is traveling to the right at a speed of 10^6 m/sec through a region in which there is a uniform electric field of magnitude 2×10^5 N/C directed upward. It is possible to superimpose on this electric field a magnetic field, such that the particle simply moves in a straight-line path without deflection. What are the direction and magnitude of the required magnetic field?

(b) Show that in this combination of electric and magnetic fields, charged particles moving at speeds other than 10^6 m/sec would be deflected. (Such a combination of electric and magnetic fields can be used as a "velocity selector," to remove from a beam of particles all but those with a certain speed.)

9.11 Estimate the force on a 1-m length of wire, located in the earth's magnetic field, which is carrying a current of 200 A to the starting motor of a car. Is such a force large enough to be taken into consideration in designing a car?

9.12 If you had an ammeter that would give a full-scale deflection with a current of 10^{-4} A, what series resistance should you use with it to make a voltmeter giving a full-scale deflection for a voltage of 1000 V?

9.13 In analyzing Figure 9.29 and showing how the predictions of Faraday's law might have been anticipated, it was supposed that the wire contained positive charges that were free to move. Repeat the analysis, using the assumption that the wire contains *negative* charges that can flow. Are the conclusions about Faraday's law modified in any way by this change in point of view?

9.14 Suppose that the rotating coil of the generator shown in Figure 9.31 has an area of 2 m^2 and that it is rotated at 60 revolutions/sec in a magnetic field whose magnitude is $B = 0.5$ tesla. Make a rough estimate of the emf that can be induced. (When the coil makes a quarter of a turn, the flux through the coil changes from zero to its maximum possible value.)

9.15 The mechanical output of a motor can be used to drive a generator, and the electrical energy from a generator can be used to run a motor. Discuss the possibility of constructing a perpetual motion machine (as shown in the figure below), in which part of the generator's electrical energy is used to run the motor, with some electrical energy being delivered to the outside.

9.16 Some energy must be supplied to turn an electrical generator. You could supply the necessary energy yourself, by turning a crank. For about how many light bulbs could you provide the energy?

9.17 A farmer has a shed a mile from his house, in which he has electrical equipment requiring a power of about 1 kW. The equipment is designed for a voltage of 120 V but will operate if the voltage is 100 V or more. He plans to run a simple two-wire power line (an "extension cord"), using copper wire with a diameter of 1 mm from his house to the shed. Will this proposal be satisfactory, or will it be necessary instead to install a transformer and transmit the power at a higher voltage?

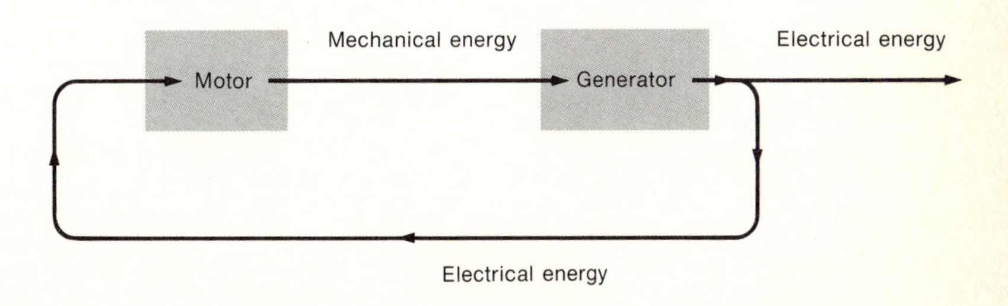

9.18 Find out as much as you can from your power company about where your electricity comes from and how it is transmitted to you. What is the original source of energy: coal, oil, gas, hydropower, nuclear energy, or several of these? At what voltage is it generated? How is the voltage stepped up and down between the place at which it is generated and your 120-V wall socket?

9.19 What plans does your power company have for the future? Are any major new facilities being built or planned? Is there evidence that a real need exists, or is the company just being cautious? What reasons do power company representatives give when asked to explain their reasons for building new plants, the types of plant being built, and their locations?

10

The Energy Problem: Past, Present, and Future Patterns of Energy Supply and Consumption

T. W. Tenney

10

The Energy Problem: Past, Present, and Future Patterns of Energy Supply and Consumption

Man seems to have no function except that of dissipating or degrading energy, . . . of accelerating the operation of the second law of thermodynamics.

—Henry Adams, *A Letter to American Teachers of History,* 1910

§10.1 THE CONSUMPTION OF ENERGY

In this chapter, we will take a closer look at the energy problem. How much energy do we use? Where does it come from? For *what* do we use it? Which of these uses can be singled out as critically important because of their size, and which are insignificant? How much energy will we use in the future, what will we use it for, and where will it come from? How will future patterns of use differ from present ones? For many of these questions, even tentative answers are difficult to find, but it is important to try, if we are to moderate the effects of unchecked exponential growth.

It is difficult to discuss these questions without becoming swamped in a sea of numerical data. Except for a few summary figures, which are given in the text itself, most of the detailed data are contained in the appendixes, to which frequent reference is made in this chapter. The specific numbers will soon be out of date, but the broad patterns, the fractions of the total that are used for one purpose or another, will change less rapidly. We shall concentrate on these patterns and, wherever possible, give figures to show recent annual percentage rates of change.

In discussions of energy supply and consumption on a national or global level, it is convenient to use a very large energy unit, the Q, and its submultiple, the milli-Q (mQ). One Q is by definition equal to 10^{18} BTU, and the milli-Q is 10^{15} BTU. ($1\ Q = 1$ quintillion BTU, and $1\ mQ = 10^{-3}\ Q = 1$ quadrillion BTU.) Unless an explicit statement is made to the contrary, the *year* is used as the unit of time throughout this chapter; thus a statement that we use 13 mQ for space heating means that we use 13 mQ for this purpose per year.

In 1973, the United States used about 73 mQ of energy for all the purposes we recognize as part of the energy problem, virtually everything except for the chemical energy of the food we eat. Worldwide energy consumption (the United States included) amounted to about 240 mQ. It does not require a sophisticated calculation to see that we use far more than our share of the energy; our population is only 5% of the total population of the world, but we use almost one-third of the total amount of energy.

QUESTIONS

10.1 In other words, the per capita rate of energy consumption in the United States is much greater than the worldwide average. Show that on a per capita basis, the United States rate of energy consumption (73 mQ per year) amounts to about 1.15×10^4 J/sec $= 11.5$ kW (roughly 10 kW, an easy figure to remember), and that the worldwide average per capita rate is about 2 kW.

10.2 The amount of *food* energy a person requires is about 2000 kcal per day. Show that this represents an average rate of energy consumption of about 100 W (about 100 times smaller than the rate at which each American uses energy in other forms). A human body is in this sense equivalent to a single light bulb.

The standard of living in the United States is also much higher than the world average, and this is not a coincidence if standard of living is measured in terms of material goods. It does take energy to build and operate automobiles, snowmobiles, air conditioners and printing presses; it also takes energy to provide services, to operate hospitals and universities, for example. Figure K.2 shows per capita gross national product (GNP) and per capita energy consumption in the United States since 1900, and Figure K.3 shows the correlation between per capita rate of energy consumption and per capita GNP for a number of nations. These figures demonstrate only a *correlation* between energy consumption and GNP, not a causal relationship. It might be worthwhile pondering the position of nations such as Sweden, for which the per capita GNP is relatively high in comparison to the amount of energy used.

The validity of using GNP as a measure of standard of living is certainly debatable. GNP measures *something*, something in some way related to prosperity and "quality of life," but whatever the relationship may be, it certainly varies from year to year and from nation to nation. The difficulties of trying to attribute significance to GNP and energy consumption are made obvious by a few examples. Energy consumption includes not only gasoline for pleasure trips but also gasoline for ambulances and electrical energy for hospital emergency rooms. Similarly, the GNP includes the business done by insurance companies and tow trucks as well as by automobile dealers. No one would argue that those who live in the United States have "twice as much happiness" as those who live in Great Britain, but it is probably fair to say that a good life, free of day-to-day fear of disease and hunger, cannot be achieved *without* fairly high levels of GNP and energy consumption, though not necessarily as high as the prevailing levels in the United States. Therefore unless much of the world is to be condemned to perpetual poverty, we must accept the proposition that worldwide energy consumption will and should increase. Whether there is any compelling need for further large increases in the total amount of energy used in the United States is much more questionable.

It is all too obvious that the nature of the energy problem in the United States is very different from that in the developing nations. Still another form of the energy problem is experienced by Japan, a nation almost totally dependent on imported sources of energy. The worldwide energy problem is of the utmost importance and must not be forgotten, but we shall concentrate on the American energy problem, the problem of most immediate concern to us and the one to whose solution individual American citizens have the best chance of contributing.*

The broad patterns of energy supply and consumption that prevailed in the United States during the year 1973 are shown in Figures L.1 and L.2 in the appendix. These two figures together concisely summarize the energy problem except for one crucial factor: the rate of change. It is above all the *growth* in energy consumption that leads us to believe that the present energy problem may develop into a genuine energy crisis during the next few decades. The record of energy consumption in the United States during this century is shown in Figure K.6 in the appendix. Energy consumption has been growing approximately exponentially since World War II. In 1973 we used 73 mQ of energy, and if this pattern of growth continues (as indicated by the extrapolation in Figure K.6), we will use 90 mQ in 1980, 127 mQ in 1990, and 178 mQ in 2000. Where did we get our 73 mQ in 1973, what did we use it for, where can we get 178 mQ in the year 2000? Will we really need 178 mQ per year? What can we do to keep these extrapolations from coming true? A thorough discussion of these important questions would require at least an entire book in itself; this chapter can only serve as an abbreviated guide to some of the important facts and figures and to some of the general conclusions that can be drawn.

For the sake of argument, we will suppose that the projection shown in Figure K.6 is correct. It must be remembered, however, that this is only a projection. There are a number of reasons for believing (or hoping) that the actual rate of growth will be slower. It should be noted, for instance, that the total amount of energy used has been growing more

*The nature of the energy problem is also different for various groups within the United States. Indian reservations in the Southwest, for instance, are being strip-mined to provide coal for generating electricity for Las Vegas and Los Angeles. The Indians bear a share of the environmental burden of the energy problem and receive few of the benefits.

rapidly than has the population. The population of the United States in 1973 was 210 million; extrapolation suggests that it will probably be about 295 million in 2000 (Figure K.1). If, for whatever reason, per capita energy consumption increases less rapidly than it has in the past, the projection shown in Figure K.6 will of course be incorrect.

QUESTION

10.3 If the projected population figure for the year 2000 (295 million) is correct and if per capita energy consumption remains equal to its 1973 value, how will this change affect the projected figure for energy consumption for the year 2000?

---

§10.2 PAST AND PRESENT SOURCES OF ENERGY

Our dependence on various sources of energy since 1850 is shown in Figure K.7. This figure shows how fuel wood,* the dominant source of energy in the middle of the 19th century, has been displaced by coal, which has in turn been displaced by oil and natural gas. Most of our energy now is derived from the chemical energy of the fossil fuels: coal, oil and natural gas.† In 1973, 97% of our energy was derived from these sources, as shown in Figure L.1 and Table L.1. From the late 19th century until after World War II, coal was our dominant source of energy. The amount of energy derived from coal has changed very little since the early years of this century, and since World War II larger amounts of energy have been derived from oil and natural gas. Since the middle 1950s, small amounts of energy have been derived from nuclear energy,

*The total amount of energy derived from wood is still of the same order of magnitude as that supplied by nuclear energy and hydropower, but reliable data on the use of wood are difficult to obtain, and its importance will in all probability become still smaller in future years. Except for this one figure (Figure K.7), which is important in showing the historical record of the past century, energy from fuel wood will not be included in this discussion or in any of the tables and graphs, nor will the small amounts of energy derived from other sources such as the winds.

†In order to simplify the discussion, in most instances only these three categories will be used to describe the fossil fuels. As the term is used here, "oil" includes both crude oil as obtained from oil wells (and the refined products made from it) together with "natural gas liquids" (NGL's), which are fuels such as propane and butane obtained in the processing of natural gas. Some NGL's exist as liquids under ordinary conditions of temperature and pressure; others can easily be liquefied by application of a slight pressure and are conveniently shipped and stored in liquid form. That is, "oil" is used here to refer to the total amount of petroleum *liquids.* (In a few places, separate figures are given for crude oil and for NGL's.) Figures for natural gas refer to "dry" natural gas, the amount remaining after the NGL's have been removed. In some reference works, the NGL's are treated as a separate category; in others they may be included with natural gas. "Coal" includes bituminous coal (the major component), anthracite coal, and lignite. Lignite is a coal with a somewhat smaller fuel value than other coals. Lignite has not been of much importance in the United States in the past, but a significant part of the coal in the western United States is in this form; in coming decades, lignite will probably account for a larger fraction of our coal production.

and the use of nuclear energy has been increasing rapidly. If current projections are borne out, it may be very important by the end of this century.

International trade is an important aspect of the American energy problem. Until about 1950, the United States could supply all of its own energy, but in recent years we have become increasingly dependent on imported oil. Net imports of oil, both crude oil and refined products, constituted 37% of our oil consumption in 1973, as indicated in Table L.1. Figure K.8 shows our increasing dependence on oil imports during recent years; the widening gap between the domestic supply of oil and the total amount of oil we use is a source of much concern. The removal of any one important source of energy can create a short-range "energy crisis" at any time, and as long as we depend on imported energy supplies, we are subject to the actions of other countries. This sort of energy crisis is not the true crisis that may result from the depletion of worldwide fuel supplies, the crisis that we may have to face by the end of this century, but as Americans learned in the fall of 1973 when oil imports from the Middle East were cut off, such a situation can lead to severe disruption at least in the short term. With the exception of oil, foreign energy sources are at present of minor importance and will be omitted from this discussion.

Electrical energy is an extremely important source of energy, particularly from the consumer's point of view, but it is one that must be derived from some other source, as shown in Figure L.1. In 1973, 82% of our electrical energy was generated from fossil fuels, as shown in Table L.2; most of the rest (14%) came from the energy of falling water (hydroelectric power), with about 4% from nuclear energy and about 0.1% (not shown in the tables) from geothermal energy and other sources such as the combustion of waste products.

The fact that electrical energy generation is an important intermediate step in the production and use of energy presents a number of troublesome bookkeeping problems. To the *user,* it is the electrical energy actually delivered that is important, but when we are concerned about the consumption of energy resources, we must consider the losses that occur during generation and transmission. About 10% of the electrical

FIGURE 10.1
A supertanker coming into San Francisco Bay. [Phillips Petroleum Company; American Petroleum Institute Photo Library.]

energy generated is simply wasted in heating up the transmission lines and is never delivered to the customers. The tables in the appendixes refer to the amount of electrical energy generated, but it is worth remembering that the amount of electrical energy delivered is somewhat smaller. A much more important consideration is that most of our electrical energy is generated by thermal processes (in steam-electric power plants), and such processes are highly inefficient. The amount of energy used for generating electricity is about three times as large as the amount of electrical energy generated. For every 1 mQ of electrical energy generated, about 3 mQ of energy is required. The 2 mQ of "waste heat" is a drain on our energy resources, as is the energy generated, and we shall therefore include *all* the energy consumed in generating electricity in considering our "uses of electrical energy." In many of the tables, figures are given to show both the amounts of energy generated and the total amounts of energy consumed in generation. The first row in Table L.2, for instance, shows that the coal used in generation of electricity in 1973 represented an energy of 8.82 mQ; the average efficiency of electrical energy generation in coal-fueled power plants was approximately 33%, and thus the amount of electrical energy actually generated from coal was about 2.95 mQ, and of this energy generated, about 10% was lost in transmission.

Hydroelectric generation of electrical energy can be carried out at an efficiency of about 85%; for purposes of this discussion we will ignore the inefficiency and use the data for the amount of electrical energy actually generated. Thus Table L.2 shows the same value (0.93 mQ) for electrical energy *generated* by hydropower and for energy *used* for this purpose. Notice that the relative importance of hydroelectric power in producing electricity depends on whether we are considering energy used or electrical energy produced. Table L.2 shows that 14% of our electrical energy in 1973 was derived from hydropower, whereas the amount of energy used for hydroelectric generation was only 5% of the total used for generation of electrical energy. In some reference works, this effect is allowed for by treating hydroelectric generation *as if* it were as inefficient as other methods of generating electricity. That is, for bookkeeping purposes, two units of fictitious "waste heat" are invented for every unit of hydroelectric energy generated. This procedure, which will *not* be used here, also yields slightly higher values for the national total "energy consumption."

Tables L.3 and L.4 show how energy consumption is divided between electrical generation and "direct uses" (all uses other than for generating electricity). One reason for the importance of keeping track of the energy used for electrical generation, even though this energy must come from somewhere else, is that energy consumption for this purpose has been growing at a much more rapid rate (6% per year, corresponding to approximately a 10-year doubling time) than has overall energy consumption, which has been increasing at a rate of 3.7% per year. Only one-fourth of our energy in 1973 was used for generating electricity, but this fraction has been increasing. Direct uses of fuels have been increasing too, but more and more new users of energy are inclined to rely on electricity, and, because of the convenience of electrical energy, many conversions take place each year. Some home owners convert their hot-water heaters from oil to electricity, and some old houses are replaced by new ones with electrical heating. Because of the inherent wastefulness of electrical heating discussed in Chapter 7, such conversions produce an increase in the total amount of energy used in the nation.

Another way of looking at the bookkeeping problem is this. Total energy consumption in the United States amounted to 73 mQ in 1973. The amount of electrical energy generated was about 6.6 mQ, and so at first glance it might appear that electrical energy represents "about 10% of the energy problem." However, the amount of energy used for generating electricity was 18.7 mQ, nearly three times as large, and so it would be more accurate to say that electrical energy represents "about 25% of the energy problem."

The information discussed in this section is summarized in Figure L.1, which shows how the energy input is divided between direct uses and use in electrical generating plants, and how the energy used in electrical power plants is divided between the useful electrical output and waste heat. In all of the subsequent discussion of energy uses, we shall include not only the energy that is consumed directly in nonelectrical forms and the electrical energy used, but also the proper share of the energy lost in electrical generation. It would be unfair not to allot this waste heat to the various consumers of electrical energy; if the average efficiency of the nation's power plants is about 33%, then turning on a 100-W light bulb results in an increase of about 300 W in the rate at which energy is used by the power plants.

§10.3 USES OF ENERGY IN THE UNITED STATES

Where does the energy go? Each of us is using energy at the rate of about 10 kW, 10,000 J/sec, and the nation uses about 73 mQ/yr. The second law of thermodynamics tells us where this energy eventually goes; with a very few exceptions, it ends up as "heat" (thermal energy), but our major concern here is how the energy is used between the time it comes from the oil well or coal seam and the time of its eventual degradation.

No one knows in complete detail what all the energy is used for, but information is available that enables us to draw some general conclusions. We will examine the various energy uses with an emphasis on trying to answer two important questions. First, on a national scale, are there specific uses for which large amounts of energy are consumed? If so, these should receive our attention as a nation; these are areas in which legislative action designed to moderate growth or to improve efficiency may have the most impact. Second, of those areas in which the *individual* has some control, which are the most important? Each of us has some influence on the energy problem. We can set the thermostat up or down, we can drive fast or slow or travel by bicycle, we can purchase a new appliance or not purchase it. In which areas can individual choice or individual restraint be most effective, either in alleviating the energy problem or in reducing one's own expenses?

Most of the specific numbers cited refer to the year 1973. In succeeding years, the numbers themselves will be out of date, but the *patterns* of energy production and use change more gradually than do the amounts themselves. For example, in 1973, about 7.7 mQ of energy was used for residential space heating, 10.6% of the national total energy consumption (73 mQ). Whatever the total may be in 1980, about 10% of the energy will probably still be used for residential space heating in spite of national efforts at turning down the thermostat. Patterns of energy consumption do gradually change, though, and some of these trends are shown by the annual rates of change given in Appendix L. Air conditioners, for instance, were rather uncommon in 1950 but now consume a significant amount of energy.

Energy consumption can be divided among four broad "sectors": the residential, commercial, industrial and transportation sectors. It is difficult to define the various sectors in any precise way, and even if this could be done without ambiguity, it would still be somewhat misleading. For example, each of us is responsible for a share of industrial consumption, even though we may be more directly aware of our consumption as members of the residential sector. The role played by *electricity* also complicates the picture. We might choose to regard electrical generating plants as a separate fifth category, but a more accurate view of what the energy is used for is obtained by allotting the energy consumed in the generation of electricity to the sectors in which the electrical energy is eventually consumed. However, because of the special importance of electrical energy, most of the tables show both direct uses of fuels and indirect uses as electrical energy.

The residential sector is fairly easy to define. Energy used in the home is allotted to this sector—largely oil and gas for space heating, cooking, and heating water, and electricity for a wide variety of purposes. Energy consumed in growing, harvesting, or processing food, or energy used in manufacturing appliances, is *not* included, even though these items may end up in the home.

Energy consumption in the transportation sector includes the *fuel* used in transportation and a small quantity of petroleum products used as lubricants (motor oils and greases). Use of oil as a lubricant is described as a "non-energy use," since the oil is not burned. We do not use its chemical energy, but this energy is lost just as surely as if the oil were used as a fuel. The transportation sector does *not* include energy consumed in the manufacture of motor vehicles nor the oil used on roads or the energy used in producing concrete for highways. It does include energy used to transport fuels themselves—to operate oil pipelines, for example, or to move trainloads of coal—in other words the energy used for the transportation *of* energy. This is just one example of the artificiality inherent in any attempt to distribute energy consumption according to use, since the energy used to carry coal to a steel plant might well be assigned to the industrial sector.

The industrial sector is a large one, comprising steel production, petroleum refining, food processing (but not farming), concrete production, and a host of other areas. The manufacture of "petrochemicals" represents an extremely important non-energy use of coal, oil and natural gas. The amount is significant; non-energy uses in all sectors together account for about 6% of the national total. More important than the actual amounts are the many products made from the fossil fuels: various kinds of fertilizers, dyes, medical drugs, synthetic fibers, plastics, paints, pesticides, antifreeze, and so on. The list is almost endless. It would be very difficult to make these items without the fossil fuels as raw materials, as "feedstock." Coal, oil, and gas are not simply "raw energy"; they consist of numerous special molecules on which the petrochemical industry is based. It may turn out that we are being rather foolish to use these precious molecules just to drive our cars and heat our homes.

The commercial sector might be more aptly named "miscellaneous." It includes everything not otherwise accounted for—not only the energy used in heating, lighting, and air-conditioning stores, offices and hotels, but also the energy used in lighting streets and highways, in operating farms, in schools, museums, and hospitals, in all governmental institutions including the armed forces (except for military aviation, which is included in the transportation sector) and so on. It also includes a non-energy use of petroleum for the manufacture of asphalt for highways.

A discussion of the ways in which energy is consumed cannot be used to allocate the "blame" for our energy problems, since each of us is responsible for approximately an equal share of *all* the energy used, whether we use it directly (in the residential and transportation sectors) or indirectly in one of the other sectors. Such an analysis can be helpful, however, for predicting where difficulties are most likely to arise if supplies run short or for discovering where efforts at reducing energy consumption may have the most impact. The overall pattern of energy consumption by sector is shown in Table L.5. It is apparent that the industrial sector is the most important of the four sectors, and it is one in which most individual citizens can take only indirect action to alleviate the energy problem—for example by not making new purchases.

Considerably more information is provided by seeing the purposes for which energy is used within the various sectors, as shown in Tables L.6 through L.9. The information presented in the tables may seem confusing just because so many numbers are given, but some familiarity with these data is necessary if rational plans are to be made for the future. These numbers may provide the essential clues for coping with the energy crisis. For each use shown in the tables, energy consumption is divided into (1) direct uses of fuels and (2) energy consumption in the form of electrical energy; the only area in which this distinction is not made is the transportation sector, in which electrical energy consumption is too small to be significant. Other figures show the percentage contribution of each use to energy consumption within the sector and to the national total. Figures in parentheses show the annual percentage rates of change in recent years.

As an example, consider the first row in Table L.6. Residential space heating accounted for 7.7 mQ of energy in 1973, 55.4% of the energy consumed in this sector, and 10.6% of the national total. Of this energy, 6.4 mQ was obtained directly from the combustion of fuels (oil, natural gas, and a very small amount of coal), and electrical space heating consumed 1.26 mQ. (1.26 mQ represents the total energy used in generating the necessary electrical energy; since the average efficiency of electrical generating plants is about 36%, the electrical energy generated for this purpose was only 36% of this figure, that is, about 0.45 mQ.) Notice that although only about 16% of the energy used for residential space heating is represented by the figure for electrical heating, this type of energy consumption has been increasing rapidly, at a rate of 25% per year, whereas the consumption of energy for direct space heating has been increasing at a rate of less than 1% per year.

A number of important lessons can be learned from these tables. For example, it is clear that space heating is a very important part of the energy problem. Residential and commercial space heating together absorb 18% of the national total, and this figure would probably be increased to about 20% if space heating of factories (included in the industrial sector) were added. Space heating is the largest single item in both the residential and commercial sectors, and the rapid rate of increase in *electrical* space heating is a matter of concern. As we saw in Chapter 7, electrical space heating is a thermodynamically absurd practice. A major national effort to encourage better insulation of homes, to lower thermostats, to discourage electrical heating, and to find alternative ways of heating homes (by solar energy, for instance) could pay rich dividends. Within the residential sector, space heating, water heating, cooking, refrigeration, air conditioning, and lighting together account for 90% of the energy used. Only a very small amount of energy is used for small appliances; electric toothbrushes and carving knives may be silly, but they are not a significant part of the overall problem. The implication of the energy consumption patterns in the transportation sector (Table L.9) is simple and direct. In this large sector, which accounts altogether for 26% of national energy consumption, automobile traffic can be designated as the major problem, with 55% of the energy used in this sector being consumed in this way. Only very small amounts of energy are used to operate buses and trains, even though these modes of transportation are much more efficient ways of moving people than are automobiles.

Energy is used in a great many ways, as these tables show. Solutions to the energy problem would be far easier to find if these tables revealed a small number of unnecessary or frivolous uses of energy that accounted for a major share of the total consumption. If snowmobiles and private planes were responsible for 20% or 30% of the total, we could solve the energy problem, at any rate for a few years, by prohibiting these uses. Unfortunately this is not the case, and the only way to reduce consumption or to moderate the rate of increase is to take a great many steps to conserve energy, each of which may be small but whose total effect may be significant. The concerned citizen who reads these tables and wants to take the most effective possible actions as an individual will notice, for instance, that residential space heating and automobile travel together constitute approximately 25% of the national total. These are areas in which citizens can take action without waiting for their government. A plan of individual action based on these tables would include some or all of the following steps.

(1) In buying or building a house, pay careful attention to the quality of the insulation and avoid electrical heating.

(2) Improve the insulation of a presently occupied house by installing storm windows. Keep the thermostat as low as is reasonably comfortable and do not leave doors and windows open unnecessarily during winter months.

(3) Lower the temperature setting on the hot-water heater, and restrict use of hot water as much as possible.

(4) Travel by foot, bicycle, train or bus if possible. If automobile travel is unavoidable, form car pools.

(5) Keep the old car going as long as possible. (Note the energy consumed in producing steel and other metals, as shown in Table L.8.) In fact, keep all your possessions in working order as long as possible; this is one of the few ways you have of curtailing energy consumption in the industrial sector.

(6) With the lights down low and the thermostat at 62°F, put on a sweater and use your hi-fi to your heart's content. Brush your teeth with an electric toothbrush if you wish; you need not feel guilty about using the small amounts of energy involved.

Another informative way of seeing where the energy goes is to consider what happens to the various fossil fuels. Table L.10 shows how consumption of coal, oil and gas is distributed among direct uses in the four consuming sectors and for generation of electrical energy. (Nuclear energy and hydropower are used almost exclusively for generating electricity and are thus not shown in Table L.10. Table L.5 shows how the total amount of energy used in generating electricity is ultimately allotted to the four sectors.) Two important facts emerge from Table L.10. First, more than half the oil used goes into transportation; as supplies of oil become increasingly tight in coming years, this sector will be affected the most. Second, two-thirds of the coal currently consumed is used for electrical generation. Because coal is the one fossil fuel that is not yet in short supply, this fact alone may stimulate the increased use of electrical energy in this country, in spite of the fact that generation of electrical energy by the combustion of fuels is accompanied by the production of a significant amount of waste heat.

§10.4 FUTURE PATTERNS OF ENERGY SUPPLY

The energy joyride is over.

Is it true that future supplies of energy will be seriously limited? Since most of our energy is now derived from the fossil fuels, the crucial issue is the adequacy of supplies of coal, oil, and natural gas. In Chapter 1, we discussed the serious difficulties involved in making realistic estimates of underground energy resources, but it is better to have the best possible estimates rather than none at all. We saw also that the world appears to be headed for a serious shortage of oil, our most important fuel, within a very few decades. Figures for past consumption, present rates of consumption, and estimates of eventual total production of the various fossil fuels, both for the United States and for the world as a whole, are summarized in Appendix I. A quick survey of the situation is provided in Tables I.1 and I.2, which show the estimated total energy content of the fossil fuels and the number of years these fuels would last at *present* rates of consumption.

The implication of these data is clear. Even if there are no increases in rates of consumption, our two cleanest fuels (oil and natural gas) will become scarce and expensive in the fairly near future. Because consumption of these fuels will undoubtedly continue to increase for some time, the situation is really much worse. We saw in Chapter 1 that world oil resources would last until about the year 2080 at the present rate of consumption, but only until the end of this century if the present exponential growth were to continue without change. A projection of the world history of oil production is shown in Appendix I, Figure I.7; production will probably reach a peak by the year 2000. A similar projection for production of oil in the United States would probably appear quite similar in shape to that shown in Figure I.7 except that the peak has already been passed; the domestic production rate will very probably never again be as high as it was in the early 1970s.

Coal presents quite a different picture. There is enough coal in the United States to provide *all* our energy for 500 years at the 1973 rate (73 mQ/yr) and for 200 years even at the projected rate for the year 2000 (178 mQ/yr). A projected history of world coal production is shown in Figure I.8; a similar graph for the United States would have a similar appearance, the peak appearing at about the same time, roughly in the year 2200. There is a great deal of coal in the ground, but for a number of reasons an energy problem still remains. Coal is a much dirtier fuel than oil or gas, both because of the effect on the environment when it is mined and because of the pollution produced when it is

FIGURE 10.2
A giant shovel for strip-mining coal. [Bucyrus-Erie Company.]

used. Furthermore, coal cannot simply be substituted for oil or gas; one cannot put lumps of coal in the gas tank of an automobile. In some applications—the generation of electrical energy, for instance—coal can be used, and is being used instead of gas or oil. Even here, such a substitution, though technologically straightforward, cannot be done with impunity because of the environmental effects.

Other fossil fuels, not now being used on a large scale, also exist. There are the "tar sands," of which the best known deposit is in western Canada. There is enough energy (about 1.7 Q) in these tars to supply all the world's energy for about a decade (at the 1973 level of energy consumption), *if* it can be economically extracted. Oil shales are another type of fossil fuel, found at many places in the world. It is thought that the energy present in oil shales is about 10^4 Q, a very significant amount since the world now uses energy at the rate of about 0.2 Q/yr. Shale oil is not a true oil but a solid fuel, and the problems in extracting it and refining it are very different from those of processing ordinary oil. Almost all of the oil shales contain so little fuel per ton of rock that they are nearly useless. The best deposits in the United States are thought to contain the equivalent of about 80×10^9 barrels of oil that we can reasonably hope to extract. This represents an energy of about 0.45 Q, and therefore oil shales are obviously worthy of consideration. Development of the necessary technology is underway, but at the present time, it is not yet certain that we can expect a significant contribution from this source, nor is it clear what the environmental effects of using oil shales will be.

Even coal will not last forever. We have as yet used only a small percentage of the world's fossil fuels, but exponential growth is leading us into a period—one of extremely short duration from the perspective of human history—in which the fossil fuels will be rapidly consumed. This is certain to occur if alternative sources of energy are not developed soon. Our descendants may look back on the 20th century and the succeeding few centuries as a brief era in which our heritage of fossil fuels, stored up for hundreds of millions of years, was squandered in a few brief moments of industrial activity as shown in Figure 10.3.

These various numbers suggest a number of trends that are likely to occur in the United States during the remainder of this century. Consumption of oil and natural gas will be lower than would be suggested by simple extrapolation of past trends, because domestic resources are not great, because we will place increasing emphasis on becoming self-sufficient in our energy supply, and because worldwide oil and gas resources will also be getting scarcer. Coal will become increasingly important. "Our civilization is founded upon coal," wrote George Orwell in 1937;* coal may once again become the dominant fuel before the end of this century. It is important to develop practical methods for reclaiming strip-mined land, for making underground coal mining safer,

*George Orwell, *The Road to Wigan Pier* (London: V. Gollancz, Ltd., 1937).

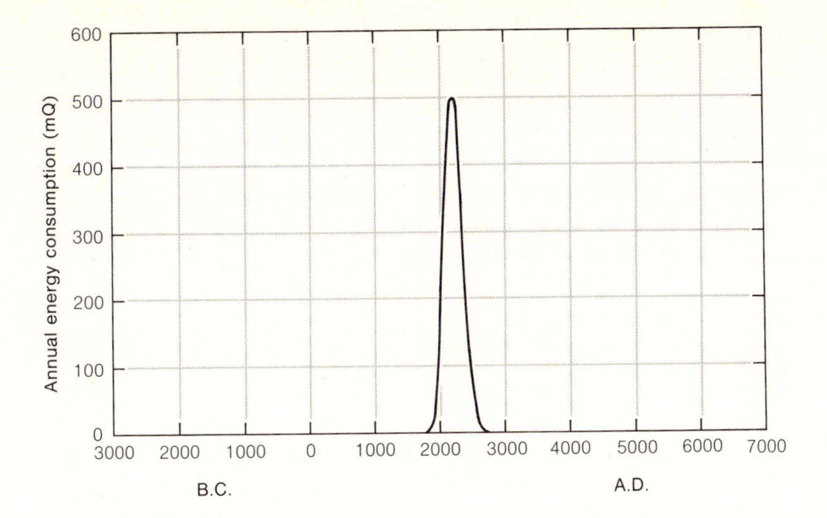

FIGURE 10.3
A long-range look at worldwide fossil-fuel consumption.

for burning coal cleanly, and also for producing clean liquid or gaseous fuels from coal.

Although we could, if necessary, get enough energy from coal alone for quite some time, we will develop alternative sources of energy. Nuclear fission is now of only minor importance, but the use of nuclear energy has been increasing rapidly. In Chapters 13 and 14 we will study the physics of nuclear energy, the adequacy of nuclear energy resources, and some of the environmental problems. A more speculative source of energy is nuclear fusion, to be discussed in Chapter 15. Nuclear fusion may never work, and even if it does, it is not likely to have much of an impact on our energy problem before the end of this century. Geothermal energy is another potential energy resource that is already in use in a few places. A substantial amount of heat is developed in the interior of the earth, largely from the decay of radioactive nuclei. Where the flow of heat to the surface is concentrated in hot springs or geysers, it can be used to generate electrical energy or simply to heat buildings; most of the buildings in Reykjavik, Iceland are heated in this way. At The Geysers in northern California (Figure 10.4), geothermal energy is now being used for electrical generation; in 1973, the installed generating capacity at this plant was about 400 megawatts, 0.1% of the total United States generating capacity. Whether it will be practical to obtain geothermal energy in locations in which the heat is not concentrated, as it is in these sites, remains to be seen. Although geothermal sources may in the future contribute a few percent of the world's energy, it seems unlikely that geothermal energy will ever be as important as coal, oil, or gas.

FIGURE 10.4
The Geysers geothermal generating plant. [Pacific Gas and Electric Company.]

Sunlight is an enormous source of energy, which is clean and—unlike the fossil fuels—renewable. The hydroelectric energy we now use is a form of solar energy, because it is the sun that causes the evaporation of water from the oceans, leading to rainfall over the land. Unfortunately, most of the potential hydroelectric power sites in the United

States have already been developed. Other ways of harnessing solar energy will be discussed in Chapter 16.

Many groups and individuals have been working on the development of small-scale alternative sources of energy, such as windmills to provide the electrical energy for a single house. Those efforts are worthwhile, but for most of us such individual sources of energy are never likely to be very practical. Like it or not, we will continue to have the problem of supplying large amounts of energy to the bulk of the population living in urban and suburban areas. Coal, nuclear fission, nuclear fusion, and solar energy seem to be the only possibilities.

Finally, one very important "alternative source of energy" is energy *conservation*. If the projected energy consumption of 178 mQ in the year 2000 turns out to be false, if the actual value is only 100 mQ, this saving of 78 mQ per year will be just as valuable as the invention of a pollution-free source providing 78 mQ of energy per year.

During the past several decades the generation and use of electrical energy has increased more rapidly than the overall total. For a number of reasons, this trend will probably continue. Electrical energy is convenient to use, and its use produces no pollution at the place it is used (though it may cause a great deal of smoke at the power plant). The pollution produced by the use of coal is most readily handled if coal combustion is concentrated at large power plants, and nuclear energy is used *only* for generating electricity. These facts provide additional incentives for the increased use of electrical energy.

One probable consequence of the increased importance of electricity is that more attention will be paid to developing new methods of storing and transporting energy. Electrical energy is now invariably transported by transmission lines. Underground transmission lines are extremely expensive; conventional lines take up a good deal of space and do not enhance the beauty of the countryside. It may be desirable to generate electrical energy with offshore nuclear power plants or floating platforms carrying giant windmills; if so, transmission of the energy to shore poses new problems. Electrical energy cannot easily be stored. One way of storing electrical energy is to use pumped-storage reservoirs, temporarily converting electrical energy into gravitational potential energy, but suitable sites that are not already in demand for recreational purposes are scarce. Another possibility is to use electrical energy to compress large quantities of air in underground caverns, air that could later be used to drive an electrical generator as the pressure is released. Still another is to convert electrical energy into the kinetic energy of large flywheels, as discussed in Chapter 4. None of these methods of storage is of any help in transporting the energy to another location.

FIGURE 10.5
A liquid hydrogen storage tank. [Chicago Bridge and Iron Company.]

An interesting suggestion for both storage and transportation of energy is that of using electrical energy to produce a fuel such as hydrogen gas. Passage of an electrical current through water breaks water molecules into hydrogen and oxygen by the process of electrolysis. Hydrogen could then be stored and delivered with pipelines like natural gas, or it could be shipped in pressurized containers or as liquid hydrogen if cooled to about 20°K. The energy could then be retrieved by allowing hydrogen to recombine with oxygen as needed. An attractive feature of this idea is that the "ashes" resulting from the combination of hydrogen and oxygen would be simply pure water. Hydrogen could be used as fuel for automobiles and airplanes, or its energy could be reconverted into electrical energy. Of course we could at most get as much energy out of the combination of hydrogen and oxygen to form water as was used to produce the hydrogen in the first place. If we released the energy of hydrogen simply by burning it, thereby producing thermal energy that would then be used to generate electrical energy again, we would get only about one-third of the energy back in useful form. This step should be avoided if at all possible, because it would mean that the overall efficiency of conversion of the original source of energy into electrical energy would be about one-ninth, as indicated in Figure 10.6a.

A possible way around this problem is to use "fuel cells," in which hydrogen and oxygen combine to form water, their chemical energy being directly converted into electrical energy without an intermediate thermal stage. A fuel cell is similar to a battery, in which chemical energy is directly converted into electrical energy, and the efficiency of such an energy conversion is not subject to the limitations of the second law of thermodynamics. No fuel cells are 100% efficient, but if fuel cells with an efficiency of, say, 75% were available, electrical energy could be generated at an overall efficiency of about 25%, as shown in Figure 10.6b.

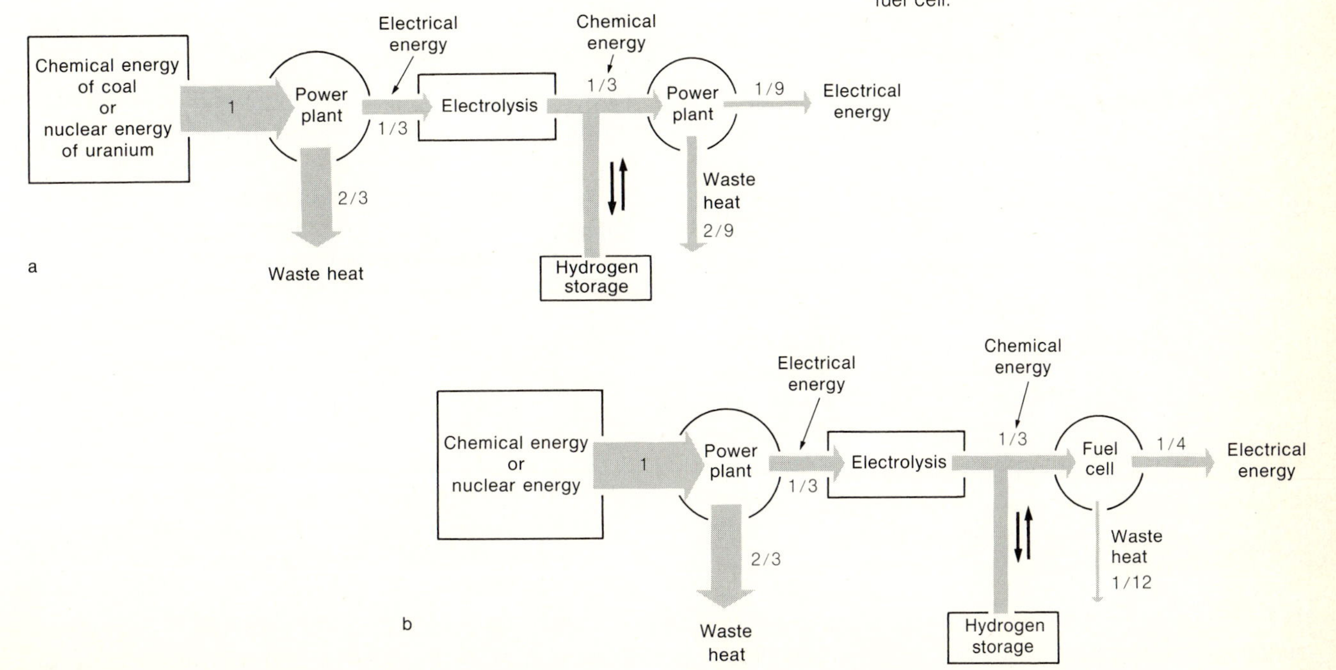

FIGURE 10.6
Transportation and storage of energy in the form of hydrogen gas: (a) at the location where the energy is to be used, electrical energy is generated by burning the hydrogen in a conventional power plant; (b) a higher overall efficiency can be achieved by combining the hydrogen with oxygen in a fuel cell.

Research to develop fuel cells with high efficiencies and large capacities for this purpose is in progress. It appears that if hydrogen is to be an important medium for storing and transporting large amounts of energy, the development of better fuel cells is essential.

Any use of energy has an environmental impact of some kind. If nothing else, virtually all the energy we use is eventually degraded into heat. This heat can alter the ecology of a river or lake; in cities, where energy use is most concentrated, this heat now has a noticeable effect on local climates, and it may produce changes in the global climate if energy consumption continues to grow. Each particular kind of energy use has its own special effect on the environment in addition to the generation of heat, not only on scenery and wildlife but also on the health of the human inhabitants. Some of these effects are well known; there are undoubtedly others of which we are still ignorant. The environmental impact of each type of energy use is not discussed in detail in this book. The reason is simply that a careful discussion of air pollution, for example, would require too much space. It is becoming increasingly obvious, however, that environmental pollution is a matter of serious concern and that all the side effects of our uses of energy must be considered in making decisions.

FURTHER QUESTIONS

10.4 In 1850 the population of the United States was 23 million, and we used, according to Figure K.7, about 2.4 mQ of energy. By what factor has per capita energy consumption increased since 1850?

10.5 About how many cubic miles of coal are used in the United States each year? (The density of coal is about 1.4 times larger than that of water.)

10.6 (a) The planned Alaska pipeline will have a diameter of 48 inches and will carry oil at an average speed of 10 ft/sec; will its completion mean the end of all our energy problems? Estimate the number of barrels per year this pipeline can deliver and the equivalent amount of energy, and compare this quantity with our annual energy consumption.

(b) As of 1974, the known oil resources of Alaska's North Slope amounted to about 10^{10} barrels. How long would this oil last if delivered continuously by the pipeline just described?

10.7 An environmental workbook lists the "environmental costs of producing one megawatt of electricity" in a coal-fueled power plant. Among other items we find "consumption of 683 pounds of coal." What important piece of information is missing? Can you guess what the author probably meant to say?

10.8 Statistical information about energy consumption is sometimes given in informal units such as the equivalent in "tons of coal" or "barrels of oil." By using the energy contents of various fuels given in Appendix H, we can readily convert back and forth between these informal units and "proper" units such as the mQ. (Notice that the energy contents of all the common fuels are of the same order of magnitude when expressed per unit *mass.* Fission and fusion "fuels," however, are quite different; in fission and fusion, it is *nuclear* energy rather than chemical energy that is released.)

(a) In 1973 the United States was said to use energy at the rate of 35×10^6 barrels of oil per day. Show that this is approximately equivalent to 73 mQ/yr.

(b) The United Nations reports energy consumption in terms of equivalent amounts of coal. According to the United Nations, the per capita rate of energy consumption in the United States in 1973 was equivalent to 12,000 kg of coal per person per year. Is this figure consistent with the value given in Question 10.1, an average per capita rate of 11.5 kW?

10.9 According to Table L.6, residential electrical energy consumption was 5.8 mQ in 1973. This figure represents the total amount of energy used for the generation of electricity for the residential sector.

(a) Estimate the total amount of electrical energy actually delivered to residences, taking the average efficiency of generation as about 36%, and convert the result into kilowatt-hours.

(b) Estimate the number of households in the United States and thus the number of kilowatt-hours of electrical energy delivered to the average household per year.

(c) What is the average rate in watts at which electrical energy is delivered to such a household?

(d) Are these reasonable figures for your own household? How many light bulbs, in continuous use, would account for this rate of electrical energy consumption per household?

10.10 As an example in §1.2.B, it was stated that a coal-fueled electrical power plant requires about 3×10^6 tons of coal per year. Is this a reasonable figure for an actual power plant?

10.11 The amount of energy used for generation of electricity has been growing at a more rapid rate than has total energy consumption. This cannot continue indefinitely, but it is interesting to examine recent trends and to "predict" the date at which *all* the energy used in the United States will be devoted to electrical generation.

(a) Plot the data from the two relevant columns of Table K.5 on a single semilogarithmic graph, using the data from 1950 on and leaving room for extrapolation. At what date does it appear that the two lines might intersect?

(b) Such an extrapolation must not be taken as a serious prediction of the future. What are some of the ways in which you can expect the actual course of events to be different?

10.12 Appendix Figure I.1 is a graph of annual worldwide production of coal, and the legend states that the *cumulative* production through 1973 amounted to about 16×10^{10} tons. You can use the graph to make an approximate test of this statement, because the *area* under the plotted curve is a measure of the total produced. Each rectangle on the grid represents a time interval of 20 years and an annual production rate of 500×10^6 tons/yr, and thus an amount of coal equal to $20 \times 500 \times 10^6 = 10^{10}$ tons. Estimate the cumulative production of coal by counting the number of such rectangles underneath the curve.

10.13 (a) The installed electrical generating capacity of the United States is given in Table K.3. If all the generating plants operated continuously at maximum output, how much electrical energy could be generated per year?

(b) According to Table K.4, what percentage of this amount of energy was actually generated during the most recent year listed? (If this number is greater than 100%, at least one of the tables is incorrect; if this number is much less than 100%—less than 5 or 10%, say—we may not be making the best possible use of the generating facilities we have.)

10.14 In the fall of 1973 when oil imports were interrupted, Americans were urged to help by lowering their thermostats from 70°F to 65°F. Let us see whether this might have had a significant impact.

(a) From Figure G.1, estimate the average number of degree-days per year in the United States.

(b) Lowering the thermostat by 5°F would, in effect, decrease the number of degree-days by 5 for each day of the heating season. By about what percentage, then, would the amount of energy needed for space heating be reduced?

(c) By considering the total amount of energy used for space heating, estimate the number of milli-Q of energy per year which could be saved.

(d) How does this amount of energy compare with the energy of the oil we imported during the year 1973?

10.15 It is sometimes said that the fact that energy consumption in the United States is growing more rapidly than the population proves that an increasing population is not the cause of the growth in energy consumption. Although there are almost certainly factors other than population growth, this argument is not itself very convincing. If one person's use of energy produces effects that influence others to use more energy than before (and vice versa), then energy consumption may increase at a more rapid rate than does population even though everyone's life style remains essentially unchanged. An increase in the number of drivers, for instance, leads to more traffic congestion, and everyone ends up using more gasoline to cover the same number of miles as before. Think of some other examples of this type, in which interactions among people may cause energy consumption to grow more rapidly than the population.

10.16 (a) How many large modern power plants would be required to equal the total generating capacity of the United States? (Use the most recent figures in Table K.3 or more recent data if available.)

(b) Does our installed generating capacity appear to be growing exponentially? Estimate the doubling time.

(c) Some of the implications of exponential growth can be seen by asking an absurd question: If this growth continues, by what date will the United States be completely covered with power plants? (To make this calculation, you need a rough estimate of the area required for a single power plant. Think of power plants you have seen, or look, for instance, at Figure 7.7.)

(d) How important is it that the area of a single power plant be accurately known? Recalculate the answer to (c), using a value for the area of a power plant that is half the previously assumed value.

(e) Discuss the results of this calculation. Are there significant conclusions that can be drawn, or is this merely an exercise in arithmetic?

10.17 In late 1973, at the time when many Americans first became aware of the energy problem, the president of the United States was quoted as making the following statement. "There are only seven percent of the people of the world living in the United States, and we use thirty percent of all the energy. That isn't bad; that is good. That means we are the richest, strongest people in the world, and that we have the highest standard of living in the world. That is why we need so much energy, and may it always be that way." Discuss this statement. Are the figures approximately correct? Is this a reasonable attitude? Can these ideas form the basis of a long-range energy policy for the United States?

11

The Nature of Light and Other Types of Radiation, and the Earth's Energy Balance

William Tenney

11
The Nature of Light and Other Types of Radiation, and the Earth's Energy Balance

§11.1 FUNDAMENTAL PROPERTIES OF LIGHT

It is difficult to imagine what life would be like without light or without eyes with which to perceive it. Our knowledge of the world would be rudimentary indeed if all of us were blind, if we were forced to rely on information obtained by touching and listening. Much of what we know about the structure of atoms comes from a study of the light they emit. Light also carries energy. It is the energy of sunlight that keeps the earth warm and that is essential to the growth of plants and hence to all life on earth. It is also solar energy that may provide the best long-range solution to the world's energy problem. For these various reasons, it is important to take up the subject of light. What is it? How fast does it travel? What kind of energy does it carry and how much? What is the difference between one color of light and another?

The speed of light (a speed represented by the symbol c) is approximately 186,000 miles/sec (almost precisely 3×10^8 m/sec or 3×10^{10} cm/sec). Light normally travels through space in a straight line (Figure 11.1); it produces very distinct shadows and does not spread out when passing through a hole in a barrier. Sound and light are quite different in this respect; you cannot see something unless you have an unobstructed "line of sight," but you can hear quite well around corners.

A light beam that strikes a shiny surface is *reflected*, and a light beam is bent, or *refracted*, when it enters or emerges from a piece of glass or a tank of water. The refraction of light was used by Newton in his famous demonstration that what we perceive as white light can be considered as a mixture of lights of various colors. Newton intercepted a narrow beam of sunlight with a prism (a triangular piece of glass) as shown in Figure 11.2. The light was bent as it entered the prism and again as it emerged, but instead of a single white spot on the screen, an elongated *spectrum* was observed with a whole series of colors, varying continuously from red at one end to violet at the other. The interpretation placed on this result was that red light, green light, blue light, and so on are more fundamental than is white light, that white light is really a mixture of lights of various colors, which are refracted by different amounts and can thus be separated in this experiment.

Many of the properties of light are consistent with the view that a beam of light consists of a stream of tiny *particles*, like a stream of bullets from a machine gun, all traveling at the same speed. Such a particle model would explain why light usually travels in straight lines and produces distinct shadows. A stream of particles could bounce off a flat surface, just as light is reflected from a mirror. If light is indeed a stream of particles, the various particles might have different masses and kinetic energies, and this might be the difference between one color and another. Further refinements to the particle model can be made in order to account for phenomena such as refraction, but we shall not pursue this development because there are other phenomena that cannot be reconciled with this particle model (at least not in any reasonably simple way) and that appear to demand that light be interpreted as some sort of *wave*. Before describing these important phenomena, let us discuss briefly the general characteristics of waves.

Shadow

Light beam

a

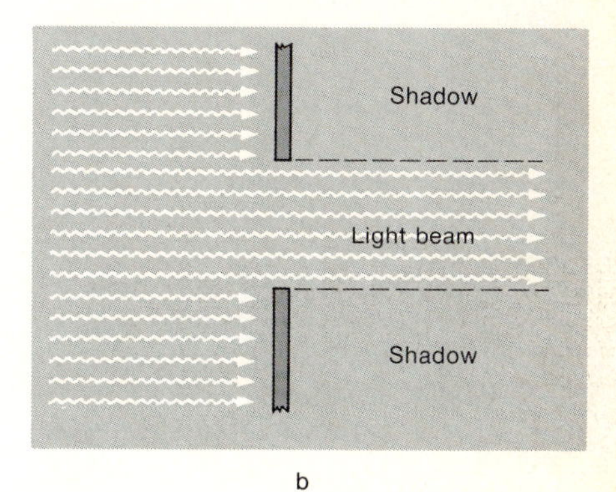

b

FIGURE 11.1
Light normally travels in straight lines and produces distinct shadows.

§11.2 THE CHARACTERISTICS OF WAVES AND THE WAVE NATURE OF LIGHT

A wave can be described as a disturbance that travels from one point to another, carrying energy, even though there is no actual transport of material from one point to the other. Attach one end of a long piece of

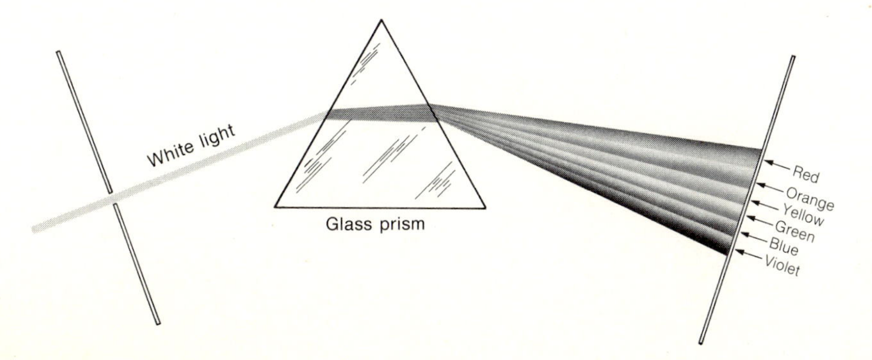

FIGURE 11.2
Newton's demonstration that a beam of white light can be broken up to produce a spectrum. Although the spectrum has been labeled with the names of the various colors, there are no sharp dividing lines, but rather there is a continuous variation from one end of the spectrum to the other.

rope to a post and move the other end rapidly up and down. The neighboring bit of rope responds by also moving up and down with a slight time delay; this in turn sets the next bit of rope into motion, and so on. A wave propagates along the rope toward the post although the motion of all the individual pieces of rope is not toward the post but rather up and down. Water molecules on the surface of an ocean or lake moving up and down produce a similar motion by neighboring molecules; a wave propagates across the surface even though individual molecules remain in very nearly the same place. Sound waves result from oscillatory changes in the density of air at any one point—density changes that cause similar changes in the density at nearby points. Thus a continuous oscillation of density at one point, produced perhaps by the vibration of a loudspeaker, produces subsequent density changes at distant points. By way of contrast, consider the motion of a baseball or a bullet from one point to another; energy is transported by the actual motion of some identifiable object or "particle," by a process quite different from a wave.

The simplest type of wave can be seen by observing a long rope when one end is continuously moved up and down. Figure 11.3 shows a sketch of a section of such a rope, a "snapshot" showing what the rope looks like at some particular instant. It is convenient to characterize a wave by its *wavelength* (denoted by λ), the distance from one crest to the next crest or from one trough to the next trough. A similar snapshot made a very short time later shows that the disturbance as a whole has moved slightly to the right, and Figure 11.4 shows the rope at successive equally spaced instants of time. Each individual crest or trough moves steadily to the right with some definite speed. Figure 11.4g shows the rope at a time when the disturbance has moved exactly *one* wavelength to the right; the rope looks exactly as it did in Figure 11.4a.

The motion of the *wave* is to the right; the motion of any small piece of the rope, however, is up and down. A piece of the rope at A, for example, has a maximum upward deflection in Figure 11.4a. As time goes on, it moves down, reaches its lowest position in Figure 11.4d, and returns to its maximum upward position in Figure 11.4g. This oscillatory time dependence of the disturbance as it is observed at any one point can be described either by giving the *period* of oscillation (the length of time in seconds required for one full cycle of motion), denoted by T, or the *frequency* of oscillation (the number of full cycles executed per second), denoted by ν. Figure 11.4 shows one full cycle of motion for

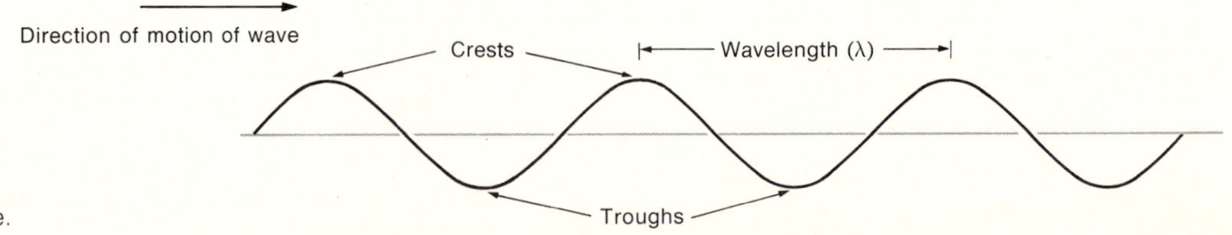

FIGURE 11.3
A wave on a stretched rope.

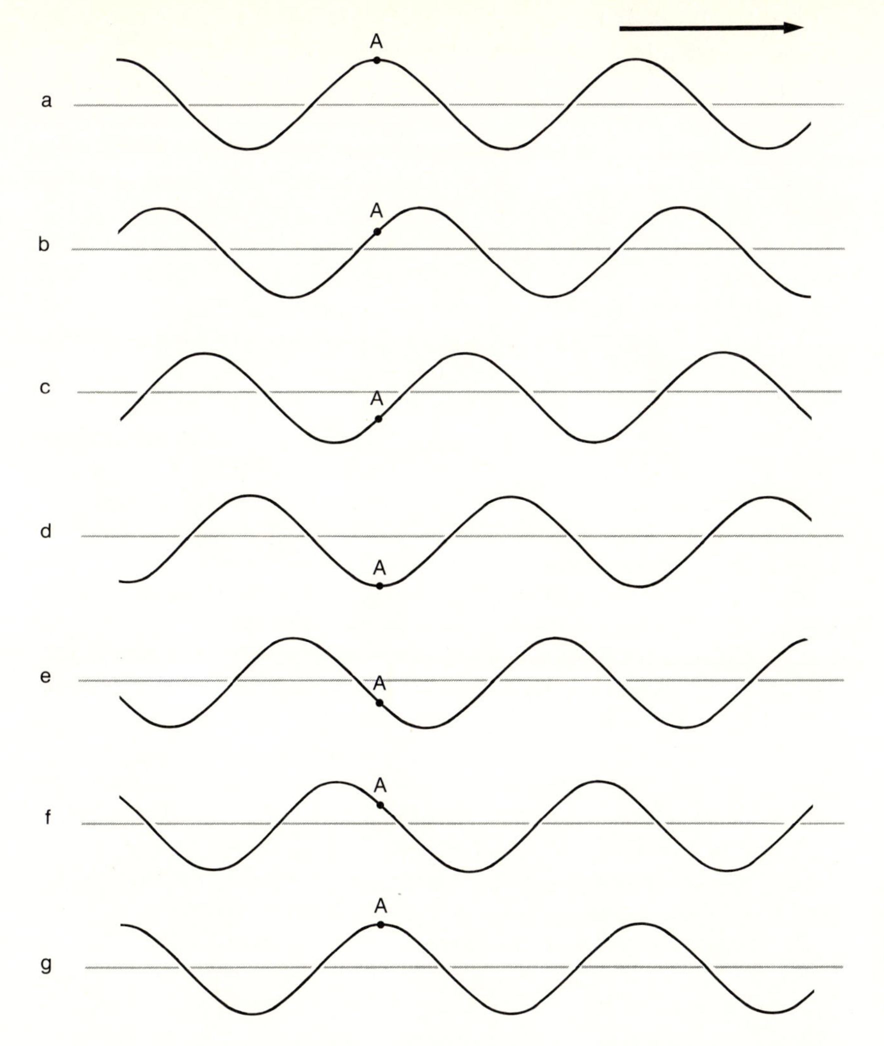

FIGURE 11.4
As the wave moves steadily to the right, every point on the rope moves up and down; in the sequence shown, the wave moves exactly one wavelength to the right.

each piece of the rope. If there are, for instance, three full cycles in one second, the frequency, ν, is 3 cycles/sec, or $\nu = 3$ Hz. (With the second as the unit of time, the units of frequency can be given either as cycles per second or, more concisely, just as seconds^{-1}, since the number of cycles does not depend on the units of measurement.)

QUESTIONS

11.1 The period and the frequency are closely related to one another. Show that $\nu = 1/T$.

11.2 There is an important relationship between the frequency ν (or its reciprocal, T), the wavelength λ, and the velocity, v, with which the wave travels. By considering Figure 11.4, show that these quantities are related in the following way:

$$v = \frac{\lambda}{T}, \qquad \text{or} \qquad v = \lambda\nu.$$

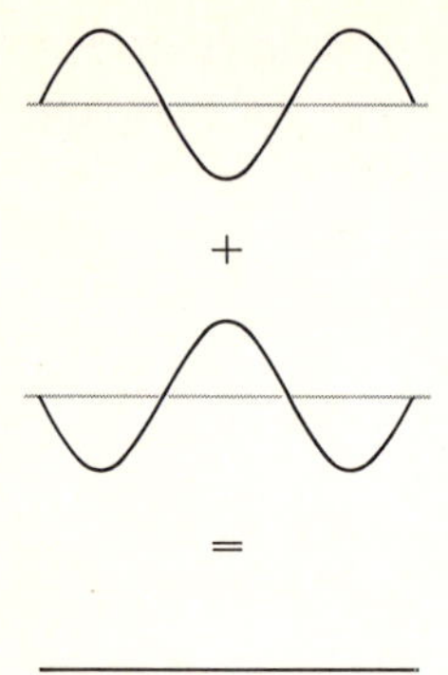

FIGURE 11.5
Destructive interference of two waves that cancel each other.

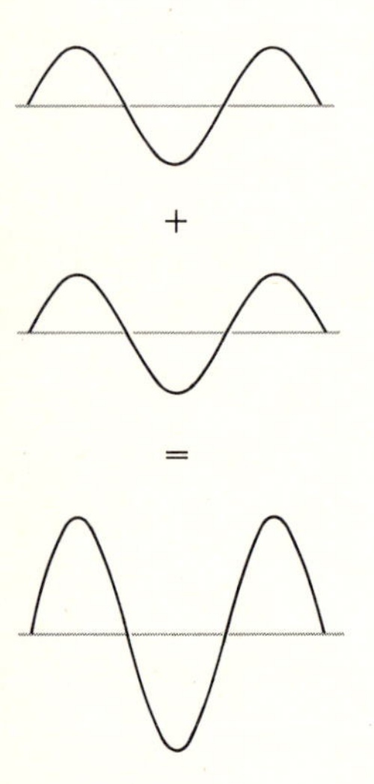

FIGURE 11.6
Constructive interference of two waves to produce a single larger wave.

The most striking feature of waves is the phenomenon of *interference.* Two waves may arrive at some particular point in such a way that crests from one wave coincide with troughs from the other. This is *destructive* interference; the two waves cancel each other out (Figure 11.5). On the other hand, crests from one wave may coincide with crests from the other, and troughs may coincide with troughs; this is *constructive* interference (Figure 11.6), the superposition of two disturbances to produce a wave larger than either one alone. These phenomena can be studied in the laboratory by producing water waves on the surface of a "ripple tank," a miniature ocean provided with a light to project the patterns of crests and troughs on a screen below (Figure 11.7). Repetitive up-and-down motion of a plunger produces crests and troughs that spread out in all directions, as shown in Figure 11.8. Figure 11.9 shows what is observed when *two* sources of waves (S_1 and S_2), which move up and down in synchronism, are used. Regions of predominantly constructive interference are seen, separated by regions of destructive interference. Constructive interference is observed, for instance, at A, B, C, D, and E. Point C is equidistant from the two sources. Two crests emitted simultaneously from S_1 and S_2 require equal times to travel to C, and so they arrive as a single large crest; likewise two troughs emitted simultaneously from S_1 and S_2 are observed at C as a single large trough, and so the up-and-down motion of the surface at C is larger than it would be if either source alone were present. Constructive interference is also seen at other points. Point D, for instance, is one wavelength farther from S_1 than from S_2. Thus a crest from S_2 arrives at D at the same time as does the *previous* crest from S_1, again producing a single large crest. Similarly, point E is *two* wavelengths farther from S_1 than from S_2, and again constructive interference is observed. Between these regions, where there is the maximum amount of constructive interference, are regions of destructive interference. The point midway between C and D, for instance, is one *half* wavelength farther from S_1 than from S_2. Thus at this point, crests from S_1 always coincide with troughs from S_2, and vice

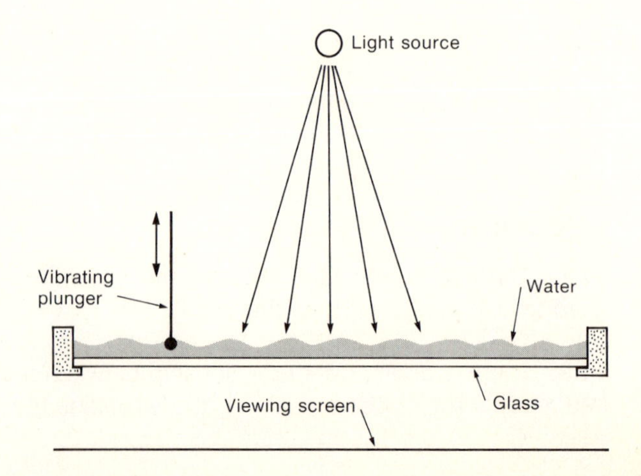

FIGURE 11.7
A ripple tank.

versa, and there is little if any disturbance of the surface of the water.

One very important observation about these phenomena is the following. If we are observing water waves, we can see the water, the "medium" through which the waves travel, and it is apparent how the regions of constructive and destructive interference arise. Suppose, though, that we were to do such an experiment with an invisible liquid ("invisible water") and that we were restricted to drawing inferences from the motions of some indicators, such as a set of corks strung out along a line as in Figure 11.10. What we would observe is that corks at A, B, C, D, and E would move violently up and down whereas corks at intermediate points would scarcely move at all. Having had some experience with the phenomenon of interference as exhibited in ordinary visible water, we would suspect that S_1 and S_2 were producing waves of some kind, and we could deduce their wavelength in the following way. We would interpret the constructive interference seen at D as resulting from the fact that D is one wavelength farther from S_1 than from S_2, and thus we could determine λ by measuring these two distances and subtracting:

$$\lambda = \overline{DS}_1 - \overline{DS}_2.$$

As a check on this result, we would realize that point E must be *two* wavelengths farther from S_1 than from S_2, and thus the wavelength could also be determined by measuring the distances $\overline{ES}_1$ and $\overline{ES}_2$:

$$\lambda = \frac{\overline{ES}_1 - \overline{ES}_2}{2}.$$

FIGURE 11.8
Periodic up-and-down motion of a plunger produces circular crests and troughs that propagate away from the source. [*PSSC Physics*, D. C. Heath, Lexington, Mass., 1965; Education Development Center.]

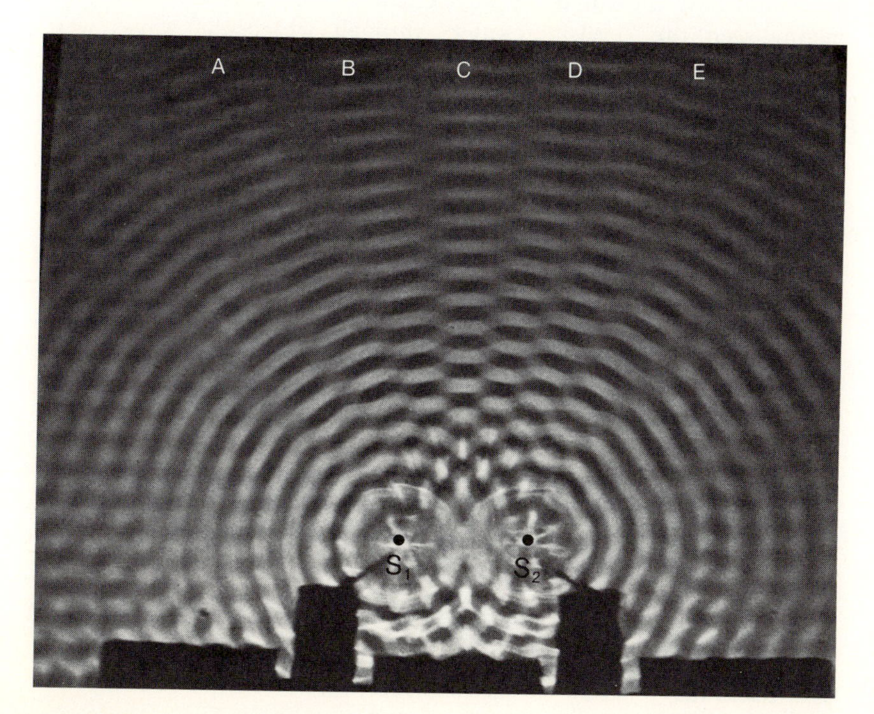

FIGURE 11.9
Regions of constructive and destructive interference produced by two synchronized plungers. [*PSSC Physics*, D. C. Heath, Lexington, Mass., 1965; Education Development Center.]

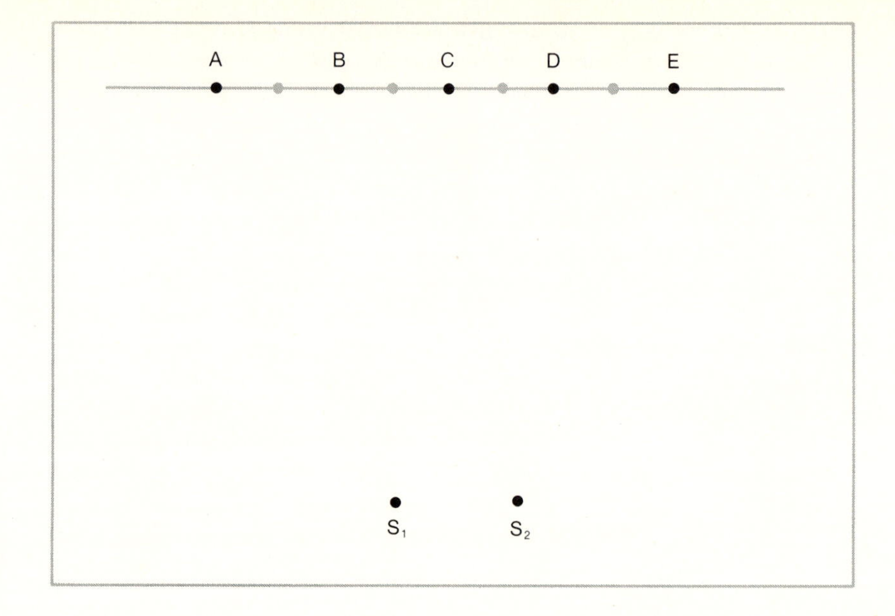

FIGURE 11.10
If the water in Figure 11.9 were invisible, the interference pattern could be revealed by the motions of corks.

If this gave approximately the same value of λ, we would conclude that we were probably correct in interpreting the motions of the corks as the result of two interfering waves. Furthermore (and this is important), we would have measured the wavelength of these waves without ever having seen the "invisible water" through which the waves were traveling.

Our major interest here is in the nature of light, not in that of water waves. The reason for the preceding discussion is that similar interference patterns are observed with light and lead to the inescapable conclusion that light is some sort of wave. The wavelengths of light, however, turn out to be extremely small, and the dimensions of the apparatus needed to observe interference patterns are therefore very different than for water waves. The simplest way of observing interference effects with light is to let a beam of monochromatic light (light of a single pure color) from a laser* strike a plate with two narrow parallel slits (Figure 11.11) and to observe the resulting illumination on a screen. On the screen we see a row of bright lines, parallel to the slits, separated by regions of darkness (Figure 11.12). If the alternating bright and dark regions are understood as the result of constructive and destructive interference, the interpretation of this experiment is straightforward. Light propagates in all directions from each slit, and thus the two slits are sources of waves just like the two plungers in the ripple tank (Figure 11.9). Consider the top view of this experiment

*Lasers are especially convenient sources of light, but interference phenomena were observed 150 years before the invention of the laser in 1960.

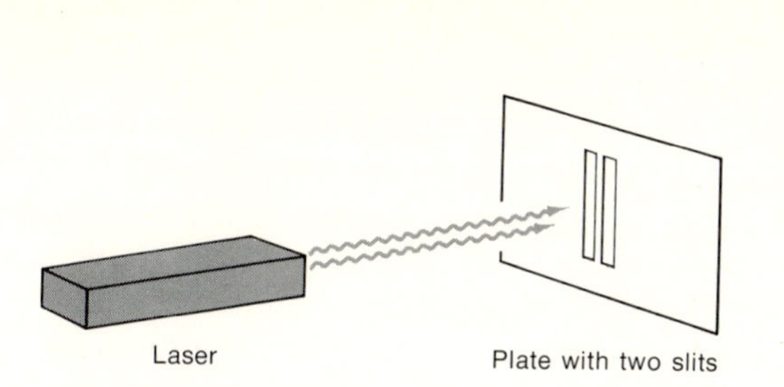

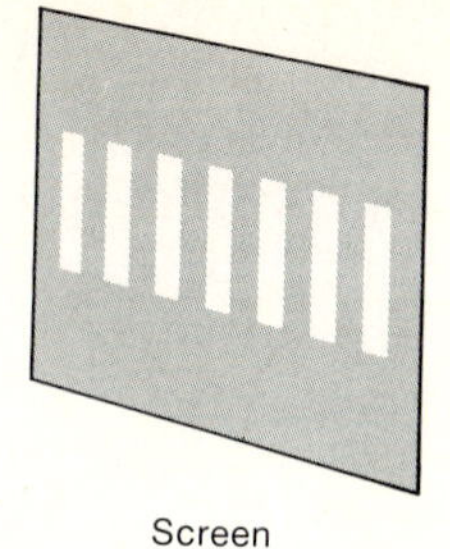

FIGURE 11.11
The two-slit experiment. Light from a laser falls on a plate with two narrow slits, and the pattern on the screen is observed.

shown in Figure 11.13. The bright line at C results from the fact that this point is equidistant from the two slits, the bright line at D from the fact that D is one wavelength farther from one slit than the other, and so on. In this way, the general nature of the observed pattern can be understood. We can go farther and determine the numerical value of λ, as with the "invisible water waves" in Figure 11.10. The wavelengths of light waves are extremely small; for the red light from the most common type of laser, $\lambda = 6.328 \times 10^{-7}$ m. Wavelengths of light are often given in terms of the angstrom unit (Å), where 1 Å $= 10^{-8}$ cm $= 10^{-10}$ m. In angstrom units, this wavelength is 6328 Å. The only difference between this experiment with light and the ripple-tank experiment—and it is an important difference—is that the wavelengths of light are many times smaller than those of water waves.

FIGURE 11.12
Interference pattern observed in a two-slit experiment. [Courtesy Brian Thompson, University of Rochester.]

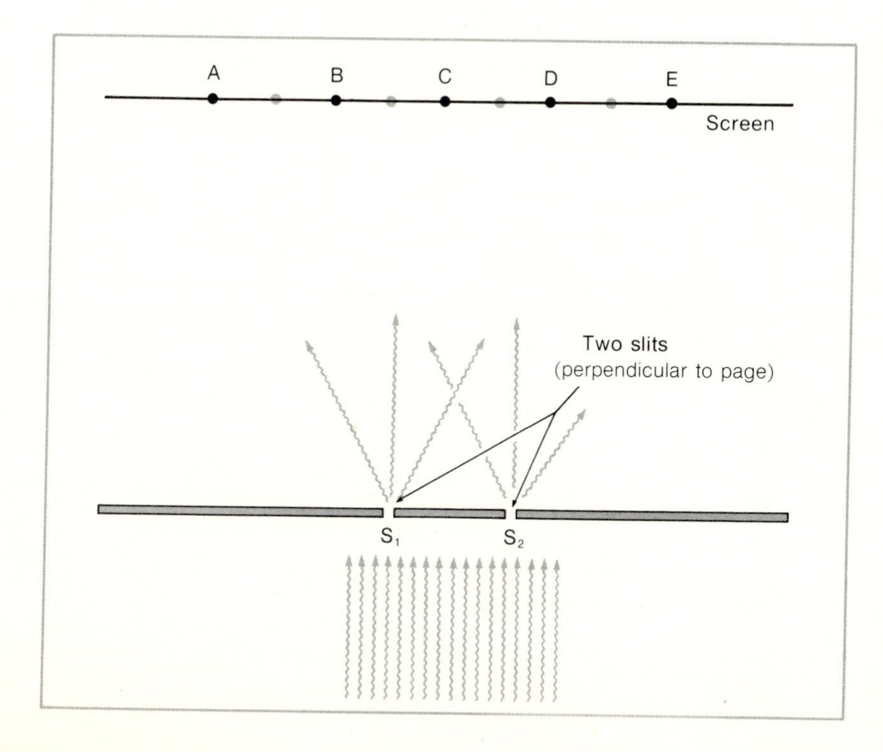

FIGURE 11.13
Top view of the two-slit experiment (not to scale). Brightly illuminated regions—separated by dark regions—are seen at A, B, C, D, E, etc.

If we try to interpret light in terms of a particle model, thinking of a beam of light as being like a stream of bullets, it is impossible to understand the two-slit experiment without making a series of implausible and artificial extra assumptions. On the other hand, the interpretation of this experiment in terms of constructive and destructive interference is so straightforward and direct that it leaves little room for doubt that light is some kind of wave. The wavelength of light from a "red" laser is 6328 Å. When the experiment is repeated, with light of a different color, the spacing between the illuminated regions on the screen changes, and in this way the wavelengths of light throughout the visible spectrum can be determined. These wavelengths range from about 4000 Å at the violet end of the spectrum to about 7000 Å at the red end.

QUESTION

11.3 What is the frequency of a light wave whose wavelength is 6000 Å?

Before wholeheartedly endorsing a wave model of light, we should ask whether some of the other observed facts about light—reflection, refraction, the production of distinct shadows, and other phenomena that are quite consistent with a particle model—can also be explained by a wave model. The details will be omitted, but waves do exhibit reflection when they strike a barrier and also are refracted when they cross a discontinuity; water waves, for instance, are changed in direction in passing from shallow water to deeper water. Waves can often spread out on passing through a hole in a barrier (Figure 11.14a), but this tendency is less pronounced for short wavelengths (Figure 11.14b); if the wavelength is short enough, a rather well-defined beam is observed (Figure 11.14c), with distinct shadow regions to either side. In order to

FIGURE 11.14
When waves pass through an opening, the amount of spreading depends on the width of the opening relative to the wavelength of the waves. In (a) the wavelength, λ, is greater than the width of the opening, in (b) the wavelength and the width of the opening are of about the same size, and in (c) the wavelength is small in comparison to the width of the opening. [Education Development Center.]

a

b

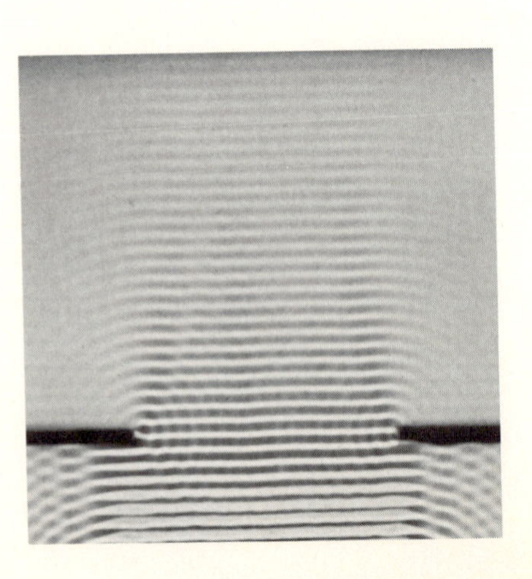

c

observe with light the kind of spreading shown in Figure 11.14a, it is necessary to use an opening whose dimensions are very small, comparable to the wavelength of the waves. Since the wavelengths of light waves are so tiny, it is not surprising that light is normally observed to produce distinct shadows and to travel in well-defined beams when sent through doors and windows.

§11.3 OF *WHAT* IS LIGHT A WAVE? THE MAXWELLIAN SYNTHESIS OF ELECTROMAGNETISM AND OPTICS

Water waves result from the motion of water molecules; waves on a stretched rope arise from motions of the pieces of rope; sound waves represent an oscillatory motion of molecules of the air. Is there a similar medium through which light propagates, analogous to the water, the rope, or the air? What is it that is "waving" up and down or back and forth when a light wave is propagated? Light is different from other sorts of waves in that light can be readily propagated through a perfect vacuum. At one time, the perfectly reasonable (but incorrect) idea was prevalent that there *must* be a medium for light waves, though it was admitted that the medium was a rather strange one if it completely filled the whole of space without having been noticed, without offering any resistance to the motions of planets. The supposed medium for light waves was even given its own name, the "luminiferous ether," a name whose magnificent sound perhaps helped to obscure the fact that no one had any notion of what it was.

One of the great developments in the history of science took place in the mid-19th century, when the British physicist James Clerk Maxwell worked out a unified theory of electricity and magnetism and predicted from his theory that waves of *electric* and *magnetic fields* could be propagated even in empty space and that the speed of these waves should be 3×10^8 m/sec. His electromagnetic theory was essentially an elaboration of the theory described by Coulomb's law, Faraday's law, and so on (Chapters 8 and 9). When Maxwell incorporated these various laws into a single mathematically consistent whole, he had a theory that not only contained Coulomb's law and the other basic facts about electricity and magnetism but also allowed for electromagnetic radiation, electromagnetic waves, waves of oscillating electric and magnetic fields that could transport energy. The speed of the waves could be predicted from his theory using fundamental parameters such as the Coulomb's law constant, and, as stated above, it turned out to be a speed of 3×10^8 m/sec.

Maxwell's theory was intended to be a theory of electromagnetism, not of light, but it could hardly be dismissed as a mere accident that the predicted speed of his electromagnetic waves coincided perfectly with the known speed of light. Light must be an electromagnetic wave, and thus the science of optics, previously a separate field of science, was recognized to be one branch of the science of electromagnetism.* The merging of optics with electromagnetism is a prime example of the sort of scientific synthesis that has often occurred in the past, the type of synthesis that must occur from time to time if science is not to sink under the accumulated weight of apparently unrelated facts and laws. In a similar way, Newton formulated a single theory that described both the motions of planets and the motions of falling rocks and projectiles, and Joule's experiments paved the way for the synthesis of the sciences of mechanics and heat.

According to Maxwell's theory, electromagnetic waves of *any* frequency or wavelength can be propagated through space. Why, then, do the wavelengths of light extend only from about 4000 Å to 7000 Å? The answer is that the limits to what we usually call light are set not by the characteristics of the waves themselves but by the characteristics of the human eye. Other wavelengths may be present, but the eye is insensitive to them. It had been known for a long time that the spectrum of sunlight, for instance, is not bounded by red and violet but also includes radiations of longer wavelength beyond the red end of the spectrum (infrared) and also of shorter wavelength beyond the violet (ultraviolet). This can be demonstrated in Newton's prism experiment (Figure 11.2); detectors sensitive to heat show a response even when placed at parts of the screen beyond the ends of the visible spectrum.

Electromagnetic waves of much longer wavelength and lower frequency are generated by oscillating currents in antennas. Radio waves used in conventional "AM" broadcasting have frequencies between 5×10^5 Hz and 16×10^5 Hz (500 kilohertz to 1600 hilohertz). The numbers on a radio dial indicate the frequency in hundreds of kilohertz: 5, 6, 7 . . . 16. Television is broadcast by waves with frequencies from 54 to 216 megahertz (channels 2 through 13) and UHF television with still higher frequencies. FM broadcasting is done with waves whose frequencies lie between 88 and 108 megahertz, between the television channels 6 and 7. Still higher frequencies, though still far below the frequency of visible light, are used for radar and in microwave ovens. X-rays and gamma rays from nuclear disintegrations are electromagnetic waves with frequencies far greater than those of visible light. The frequencies and wavelengths of electromagnetic waves that have been generated or observed vary over an extremely wide range, and there are *no* definite limits. Figure C-1 gives an overall view of the electromagnetic spectrum. Of all these waves, those we recognize as visible light are only a tiny part, but to us a very important part. All these waves are governed by the same fundamental laws and travel (in empty space)

*By showing that light was an electromagnetic wave and that these waves could travel even through a perfect vacuum, Maxwell really eliminated the need for a special "medium" for light. In retrospect this seems clear, but it was not obvious at the time. Debate about the properties of the "ether" continued, as did unsuccessful attempts to detect the motion of the earth through the ether. The nonexistence of the ether was eventually made a fundamental postulate of Einstein's special theory of relativity.

at precisely the same speed. (In discussing the general properties of electromagnetic waves, the term "light wave" is occasionally used to refer to any such wave, not just to waves in the visible part of the spectrum.) Even though the same basic laws govern all these waves, the enormous differences in frequency and wavelength produce significant differences in behavior. Just to cite one example, a radio wave whose frequency is 10^6 Hz (and whose wavelength is therefore 300 m) can spread around corners far more readily than a light wave whose wavelength is 5×10^{-7} m and which usually casts a very distinct shadow.

§11.4 SOURCES OF ELECTROMAGNETIC RADIATION. BLACK-BODY RADIATION

What are the *sources* of electromagnetic radiation? Aside from radiation such as that produced by currents in a television broadcasting antenna or, toward the other extreme, radiation such as the gamma rays emitted in nuclear events, two quite different types of radiation are commonly encountered. All hot objects (the filaments of incandescent light bulbs or the heating elements of a stove) radiate electromagnetic waves simply by virtue of being hot. The charged particles of which they are composed are engaged in rapid random motions, and it is characteristic of this thermal radiation that it has a *continuous* spectrum, a continuous range of wavelengths, rather than waves of one or more *discrete* wavelengths. On the other hand, discrete spectra are often observed when atoms of a gas are excited by an electrical discharge, in a neon light for example. The discrete spectra produced by electrically excited gases are of fundamental interest in providing clues to the structure of atoms and molecules, and we shall take up this subject in Chapter 12. Here let us examine the continuous spectrum of the thermal radiation emitted by hot bodies. This is the kind of radiation that we receive from the sun and that the earth gives off into space; the temperature of the earth is determined by its "radiation balance." As we saw in Chapter 6, thermal radiation provides a means by which heat can flow from one body to another, even through a vacuum. We now know that "heat radiation" is really a form of electromagnetic energy, which is in transit from one place to another.

It is not necessary that an object be "red-hot" in order to emit thermal electromagnetic radiation, but the amount and nature of the radiated energy do depend on the temperature more than on any other factor. At ordinary temperatures, objects emit energy at a relatively low rate, and the emitted radiation is mostly in the infrared region, with wavelengths longer than those of visible light. As the temperature is increased, the total rate at which energy is radiated increases rapidly, and at the same time the dominant wavelength of the radiated waves decreases. The surface temperature of the sun is approximately 5800°K; at this temperature a large part of the radiated energy is in the visible range.

Temperature is not the only factor that determines the rate at which energy is radiated. Even at the same temperature, two different objects are not necessarily equally good emitters of thermal radiation. There is, in fact, a close connection between the effectiveness of a body as an emitter of radiation and its effectiveness at *absorbing* radiation that strikes it. When radiation strikes an object, some of the incident radiation is reflected, some is absorbed (its energy being converted into thermal energy in the object), and if the object is thin and transparent (a pane of glass, for instance), some may simply be transmitted. Radiation that is transmitted can be considered as not interacting with the object at all; to simplify the discussion we will consider here objects that are thick enough so that all the incident radiation is either reflected or absorbed.*

A *black* surface is one that absorbs most of the light which strikes it, and it is useful to define an ideal "black body" as one that absorbs all of the radiation incident on it, one that is completely "black" for radiation of all wavelengths. As we will see below, at any specified temperature, a black body is the most effective possible emitter of thermal radiation, and it is therefore important to examine the total amount and the spectral distribution of the thermal radiation emitted by such an object. (A cautionary note: Appearances can be deceiving. Something that looks black is a good absorber of light within the visible spectrum, but it may be a poor absorber for radiation of other wavelengths.)

For an ideal black surface, the total rate at which radiant thermal energy of all wavelengths is emitted is proportional to the area of the surface and to the fourth power of the temperature:

$$P_e = \sigma A T^4, \qquad (11.1)$$

where (in MKS units), P_e is the total emitted power in watts, A is the area in m^2, T the absolute temperature in °K, and σ is called the Stefan-Boltzmann constant and has the value

$$\sigma = 5.67 \times 10^{-8}\ W/m^2\text{-}(°K)^4.$$

QUESTIONS

11.4 By considering Equation 11.1, show that the units given to σ are correct.

11.5 The fact that P_e is proportional to a high power of T means that relatively small changes in temperature result in much larger changes in the radiated power. By what factor does P_e increase if T increases by 10%? by 100%?

*The colors seen when objects are illuminated with white light depend on the varying degrees to which different kinds of light are reflected or absorbed. A "black" object absorbs nearly all the light incident upon it. A red object reflects red light but absorbs light of other colors.

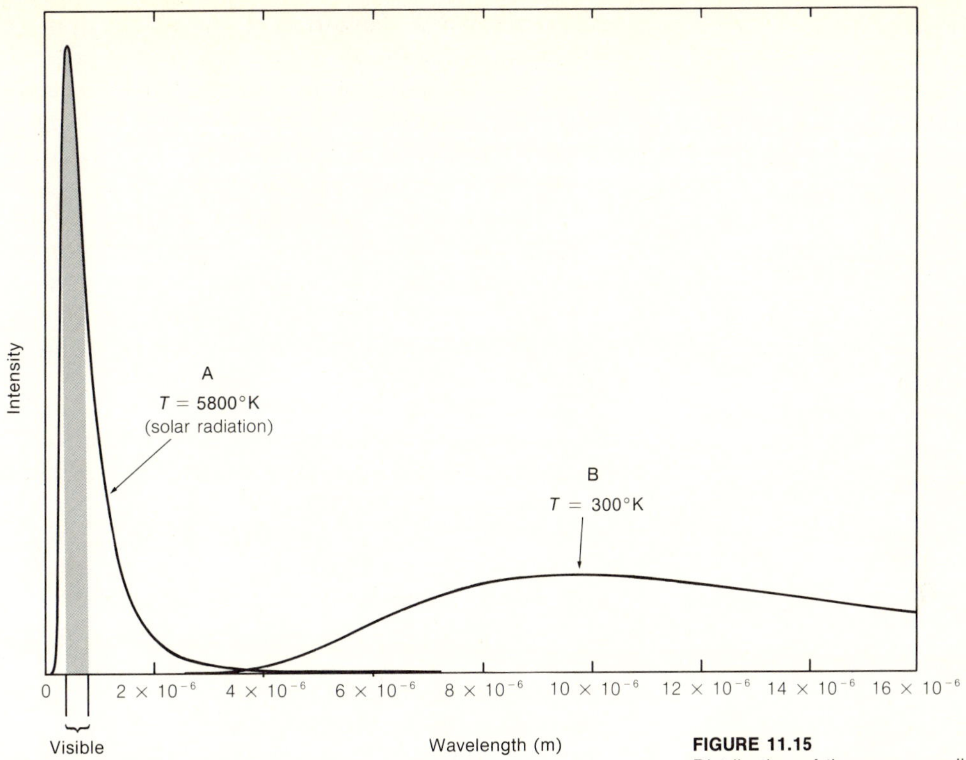

FIGURE 11.15
Distribution of the energy radiated by an object at a temperature of 5800°K (approximately equal to the surface temperature of the sun, curve A) and 300°K (approximately the temperature of the earth, curve B).

This radiated energy is distributed across the spectrum in a way that varies with temperature. The distribution is shown for two different temperatures (approximately the surface temperature of the sun and the temperature of the earth) in Figure 11.15. Notice that a good deal of the radiation from the sun is in the visible part of the spectrum. Objects at ordinary temperatures emit far less energy, and what they do emit is almost all undetectable by the eye; we see them not by their own radiation but by the visible light they reflect. The wavelength at the peak of the spectral distribution (a wavelength denoted by λ_{max}) is inversely proportional to the absolute temperature. If λ_{max} is given in meters and T in °K, this proportionality is expressed by the equation:

$$\lambda_{max} = \frac{2.9 \times 10^{-3}}{T}. \qquad (11.2)$$

QUESTIONS

11.6 What units should be attached to the number 2.9×10^{-3}, the proportionality constant in Equation 11.2?

11.7 What does Equation 11.2 give for λ_{max} for temperatures of 300°K and 5800°K? Compare these results with Figure 11.15.

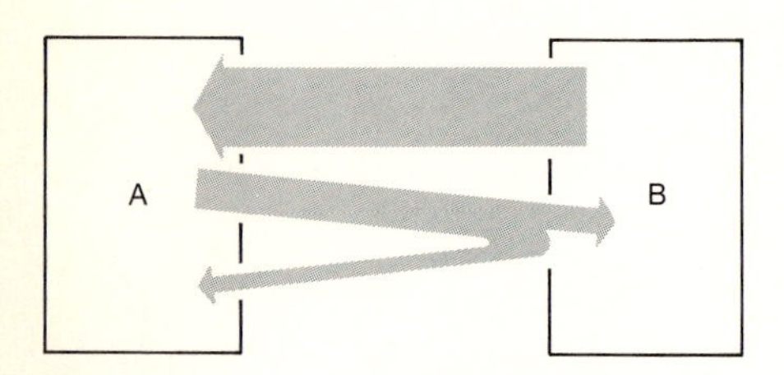

FIGURE 11.16
The second law of thermodynamics would be violated if there were an object (B) which emitted radiation at a greater rate than A (an object with a *black* surface) at the same temperature.

It was mentioned above that at any temperature an ideally black surface (which absorbs all the radiation that strikes it) is the most effective possible emitter of radiation. This fact is a consequence of the second law of thermodynamics, as we can see by imagining what might happen if it were not so. Suppose that two different objects are initially at the same temperature and can exchange energy by means of radiation (Figure 11.16). Assume that object A has a perfectly black surface that absorbs all the radiation that strikes it, and that B (whose surface reflects some of the incident radiation) emits radiation at a greater rate than does A. Since A absorbs all of the radiation from B, A's temperature will rise while that of B will decrease. Such a spontaneous production of a difference in temperature, a continuing flow of heat from a cooler body (B) to a hotter one (A), is contrary to the second law of thermodynamics. We are forced to conclude that object B, which supposedly emits energy at a greater rate than does A, cannot exist, and that therefore a perfectly black object must be a better emitter than any other object at the same temperature. (The qualifying phrase is important; even an ideally black surface emits little radiation if the temperature is low.)

Many objects are good absorbers for radiation of some wavelengths but poor absorbers for radiation of other wavelengths. By extending the argument given above, it can be shown that the properties of absorption and emission go together; a material that is a good absorber for part of the spectrum is also an effective emitter of radiation in the same part of the spectrum. Once again, however, it must be remembered that even a good emitter will give off little radiation unless the temperature is sufficiently high. These ideas have an important application in developing collectors of solar energy (as we shall see in Chapter 16).

§11.5 THE RADIATION BALANCE OF THE EARTH, THE GREENHOUSE EFFECT, AND A GLOBAL VIEW OF ENERGY TRANSFERS

The earth is an extremely interesting object that both absorbs radiation (from the sun) and emits thermal radiation of its own. By considering the earth's "radiation balance," the equilibrium between incoming and outgoing radiation, we can understand some of the factors that determine the average temperature of the earth. The basic idea is that the earth receives radiation from the sun and at the same time emits radiation into space in all directions. If the rate of emission of energy were lower than the rate of absorption, the thermal energy of the earth would increase, its temperature would rise, and the rate at which it emitted radiation would then increase. Equilibrium is reached when, on the average, the temperature of the earth is just high enough so that there is a balance between the rate at which energy is absorbed and the rate at which it is emitted. In other words, if P_a denotes the rate of absorption of energy and P_e the rate of emission, the fundamental equation describing the radiation balance of the earth is

$$P_e = P_a. \tag{11.3}$$

This equation is exactly like the simple equation that describes how a house is kept warm during the winter (Equation 4.8). By separately calculating P_e and P_a and setting them equal to one another, we can predict the average surface temperature of the earth.

First consider P_a, the rate of absorption of energy. The solar radiation striking the earth has the spectral distribution shown by curve A in Figure 11.15. The total intensity varies by a few percent in the course of the year as the distance between the earth and the sun varies, but we will be concerned only with the average value. It is convenient to express the total intensity in terms of the "solar constant," the average power incident per unit area on a surface that is above the earth's atmosphere and oriented so that it is directly facing the sun:

$$\text{Solar constant} = S_o = 1353 \text{ W/m}^2.$$

As seen from the sun, the earth looks like a disk (Figure 11.17) whose area is πR_E^2 (with R_E the radius of the earth), so the total solar power incident upon the earth is given by

$$P_i = S_o \, \pi \, R_E^2. \tag{11.4}$$

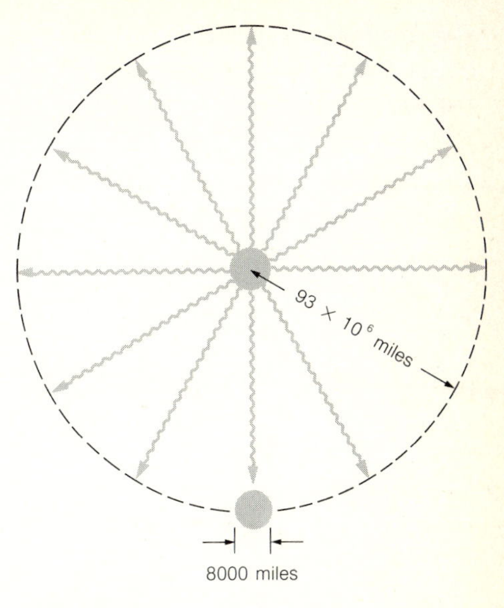

FIGURE 11.17
As seen from the sun (93 × 10^6 miles away) the earth appears as a disk, 8000 miles in diameter (not to scale).

QUESTION

11.8 Show that $P_i \simeq 1.73 \times 10^{17}$ W.

For radiation at the wavelengths that are most important in sunlight, the earth is by no means an ideal black body. Approximately 66% of the incident energy is absorbed in the earth's atmosphere and at the surface, and about 34% is reflected from clouds and from surfaces such as those of snow and water; the average "albedo" of the earth (denoted by α) is approximately 0.34. That is, with an incident power P_i, the reflected power, P_r, is

$$P_r = \alpha P_i = 0.34 \; P_i.$$

The incident power that is not reflected is absorbed:

$$P_a = P_i - P_r = P_i - \alpha P_i = (1 - \alpha) P_i,$$

and so from Equation 11.4,

$$P_a = (1 - \alpha) S_o \, \pi R_E^2. \tag{11.5}$$

Now consider P_e, the rate at which the earth emits energy. The temperature of the earth is in the neighborhood of 300°K, and the energy it emits is therefore concentrated in the infrared (curve B in Figure 11.15).* At these wavelengths, the earth *is* quite a good black body, and we can estimate the rate at which it emits radiation from the black-body formula (Equation 11.1):

$$P_e = \sigma A T^4.$$

This radiation is emitted in all directions, and thus A is the total surface area of the earth:

$$A = 4\pi R_E^2,$$

and so

$$P_e = \sigma 4\pi R_E^2 T^4. \qquad (11.6)$$

On the average, the emitted power and absorbed power are equal (Equation 11.3), and so, using the expressions for P_a and P_e given in Equations 11.5 and 11.6, we obtain the equation

$$\sigma 4\pi R_E^2 T^4 = (1 - \alpha) S_o \pi R_E^2.$$

Solving for T, the average surface temperature of the earth, we obtain:

$$T = \left[\frac{S_o(1 - \alpha)}{4\sigma}\right]^{\frac{1}{4}}. \qquad (11.7)$$

QUESTION

11.9 Show that, with the values given above for S_o and α, the predicted average surface temperature of the earth is approximately 250°K. This value is certainly in the right range but definitely on the low side; the true value is closer to 290°K, well above the freezing point of water. The basic idea behind the preceding calculation is correct; it is largely the atmosphere that makes the temperature higher than it would otherwise be, as will be discussed later in this section.

In our simple model, we have assumed so far that no additional thermal energy is being generated on the earth, that there is a precise balance (Equation 11.3) between incoming and outgoing radiation. This assumption is not quite correct. For one thing, radioactive elements within the earth are a continuing source of thermal energy. In addition, human activities may have some influence. Chemical energy and nuclear energy are converted into thermal energy when fossil fuels are burned and when uranium is used in a nuclear power plant. The earth must get rid of this extra energy, it must radiate at a greater rate, and to do so its temperature must be somewhat higher than previously calculated. Let us see whether this effect can explain the discrepancy between

*At any instant, the incoming radiation is received by just half the earth, whereas the emitted energy goes out in all directions. The earth is somewhat in the position of someone standing in front of a fireplace. The earth's temperature is not uniform, but we will ignore this fact in the following discussion.

the calculated value (250°K) and the observed value (290°K) of the earth's surface temperature. As of 1973, the world was using energy, largely from fossil fuels, at the rate of about 240 mQ/yr. Let us denote this rate at which our society produces thermal energy by $P_{society}$. Equation 11.3 must be modified, because the rate at which the earth emits energy (P_e) must be large enough to equal the sum of the solar radiation absorbed (P_a) and $P_{society}$:

$$P_e = P_a + P_{society}. \tag{11.8}$$

Is the right-hand side of this equation significantly affected by including this new term? That is, how large is $P_{society}$ in comparison to P_a?

QUESTION

11.10 (a) Show that $P_a = (1 - \alpha)P_i \simeq 1.1 \times 10^{17}$ W. (b) Convert the value of $P_{society}$ (240 mQ/yr) into watts, and show that $P_{society}$ is only 0.007% as large as P_a.

Since the value of $P_{society}$ is much less than that of P_a, including this term in Equation 11.8 really has very little effect. The rate at which the earth emits energy, P_e, need only increase by 0.007% in order to get rid of this extra heat, and because the rate of black-body radiation increases as the fourth power of the temperature, a still smaller percentage change in the earth's temperature results from this effect. The calculated value of this temperature increase turns out to be only 0.004°K, less than 0.01°F. We can be thankful that $P_{society}$ is so much smaller than P_a and therefore does not significantly affect the overall radiation balance of the earth, for if this were not so, the temperature of the earth might be significantly altered; the consequences for the earth's climate might be catastrophic.

Although this result suggests that human activities do not now have a large effect on the global climate, this calculation should not be considered as justification for complacency. First, $P_{society}$ is not a constant, but is steadily growing. We should remember that something which is growing exponentially may remain unimportant for a long time and then abruptly become large enough to have an impact. Second, the model we have been using is a very simplified one in which we treated the whole earth as a single object at a uniform temperature. This assumption is surely not correct, and the generation of heat represented by $P_{society}$ is not evenly distributed over the globe; in small areas, such as Manhattan, heat is already generated at a rate greater than the rate at which solar energy is absorbed. Although the global climate has not yet been much affected by our activities, local climates have been affected and would have been altered a great deal more if localized concentrations of heat were not dispersed by atmospheric circulation. We do not have a complete understanding of the climate, and we do not know whether significant climatic changes might be triggered by large local concentrations of heat. Third, human uses of energy have side effects beyond the simple conversion of chemical energy into thermal energy. Combustion of fossil fuels, for instance, produces carbon dioxide (CO_2). Carbon dioxide is a very minor constituent of the atmosphere, but even so it has an important effect on the earth's radiation balance as we will see shortly. Until we understand more fully the effects of CO_2 and other substances we put into the atmosphere, such as particles of soot, we are playing a risky game of tampering with the earth's climate in ways that may turn out to be important.

There is another lesson to be learned from the simple numerical fact that the present rate at which society uses energy is tiny in comparison to the rate at which we receive energy from the sun. If we could learn how to use solar energy for heating buildings and generating electricity and producing fuels to operate vehicles, we could supply our energy needs with only a small fraction of the solar energy that is available to us. This subject is a very important one that will be treated in more detail in Chapter 16.

Our simple model of the earth's radiation balance led to a calculated average surface temperature of about 250°K. This is in the correct range but is definitely on the low side; the actual average surface temperature is more nearly 290°K. Human uses of energy are not nearly large enough to account for this difference; neither is the heat produced by radioactivity within the earth, though this is a larger source of energy than the heat produced by human activities. The simple model is nearly correct, however, and it does contain the essential factors that determine the average temperature of the earth. The temperature calculated with this model would probably be extremely accurate if the earth had no atmosphere, but the presence of the atmosphere modifies the results significantly. The essential fact that has been omitted so far is that the atmosphere can absorb radiation, and it is much more effective at absorbing infrared radiation than it is at absorbing sunlight. The atmosphere is fairly transparent to sunlight; about 50% of the incident sunlight reaches the ground and is absorbed. The earth reradiates infrared radiation, and the crucial fact is that the atmosphere is a very good absorber of infrared. In large part, this atmospheric absorption of infrared radiation is due to relatively minor constituents of the atmosphere: carbon dioxide, water droplets in clouds, and water vapor. The atmosphere is heated by this absorbed radiation, and the atmosphere itself emits infrared radiation, some of it out into space but a great deal of it back to the ground. The atmosphere "traps" a large amount of energy, making the surface temperature of the earth considerably higher than it would otherwise be.

This is called the greenhouse effect because it was once thought that greenhouses were kept warm by a similar mechanism. Suppose that a greenhouse is covered with glass that is completely transparent to visible light but is a perfect absorber of infrared radiation (Figure 11.18). The incident visible light is transmitted through the glass and heats the floor of the greenhouse; the energy radiated from the floor is in the infrared region and is absorbed by the glass. This absorbed energy heats the glass, which then radiates in both directions, toward the floor of the greenhouse as well as toward the outside. The result is that the floor of the greenhouse receives not only the incident sunlight but also half of the infrared radiation emitted by the glass, and so the interior temperature of the greenhouse is higher than if the glass were not there. For the earth, the atmosphere plays the role of the glass, and this is the mechanism that makes the average surface temperature of the earth 290°K instead of 250°K. It so happens that this effect is not of much importance in actual greenhouses; although glass *is* an effective absorber of infrared radiation, its most important role is to reduce the amount of heat lost by conduction and convection. The effect *is* an important one for the atmosphere, though. Even though real greenhouses do not work this way, the term *greenhouse effect* is still used to describe the way in which the atmosphere traps radiant energy, and it is primarily the greenhouse effect that is the cause of the earth's higher surface temperature.

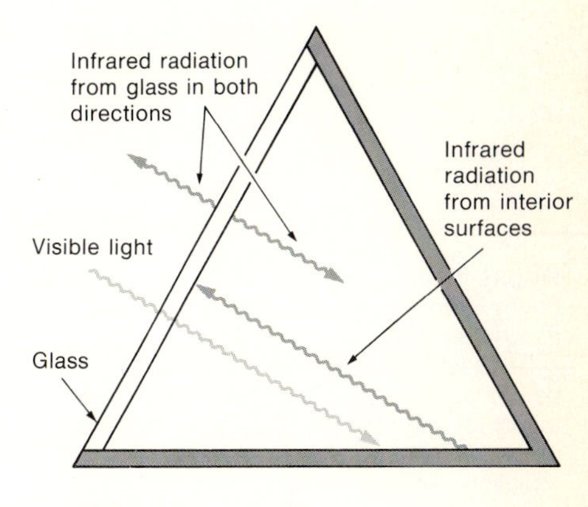

FIGURE 11.18
The greenhouse effect.

When examined more closely, the earth and its atmosphere appear a good deal more complicated than the simple "greenhouse" shown in Figure 11.18. An overall view of the process of energy transfer between the sun, the atmosphere, the earth, and outer space is shown in Figure 11.19. If the solar power incident upon the earth and its atmosphere is taken as 100 units, then of this power, 31 units are reflected from clouds and other components of the atmosphere and 19 are absorbed in the atmosphere. Of the remaining 50 units of solar radiation that reach the ground, most is absorbed and a small amount (3 units) reflected. (Some of this reflected light does not escape but is reflected from the clouds or absorbed in the atmosphere, an effect not shown in Figure 11.19.) There are three important ways in which energy is transferred from the earth to the atmosphere. Air in contact with the earth's surface is heated by conduction at a rate equal to 10% of the incident solar power. Energy is transferred at a larger rate (23 units) through the process of evaporation. Water is evaporated at the surface, and for this to occur the latent heat of vaporization must be supplied, and the water left behind is cooled. The water vapor subsequently condenses in clouds, giving this latent heat to the atmosphere before returning to the earth as rain. The result is a continuing transfer of thermal energy from the earth to the atmosphere. The most important way in which the surface loses energy is by black-body radiation, energy in the form of infrared radiation. Most of the radiation emitted from the surface is absorbed by the atmosphere. The atmosphere then emits infrared radiation, some out into space but a large amount back toward the surface. It is this stream of infrared radiation, delivering energy from the atmosphere to the surface of the earth at a rate equal to 105 units (even larger than the rate at which solar energy is incident at the top of the atmosphere), which is the central feature of the greenhouse effect.

QUESTION

11.11 Check the numbers in Figure 11.19. Observe that the total rate at which the surface of the earth receives energy is equal to the rate at which it loses energy, and that the same is true for the atmosphere.

If we consider the earth and its atmosphere *together*, the complicated picture shown in Figure 11.19 takes on the simpler form shown in Figure 11.20. Of 100 units of solar energy incident upon the earth and the atmosphere, 34 units are reflected into space and 66 units are absorbed. The net amount of infrared radiation given off is also equal to 66 units, and the radiation balance is maintained.

The major energy flows are included in Figures 11.19 and 11.20, but it is of interest to include other flows of energy that are important though of smaller magnitude and also to consider what happens to the energy that is absorbed, between the time it is absorbed and its subsequent emission as infrared radiation. The most important features are shown in Figure E.3. Of the 66 units of solar energy absorbed, 23 units cause evaporation of water, as shown in Figure 11.19, and most of this energy

FIGURE 11.19
Energy transfer between the sun, the atmosphere, the earth's surface and the surrounding space. Numerical values are percentages of the solar power incident at the top of the atmosphere (1.73×10^{17} W). Light gray indicates short-wavelength solar radiation, medium gray shows long-wavelength infrared radiation, and dark gray shows nonradiative energy transfer.

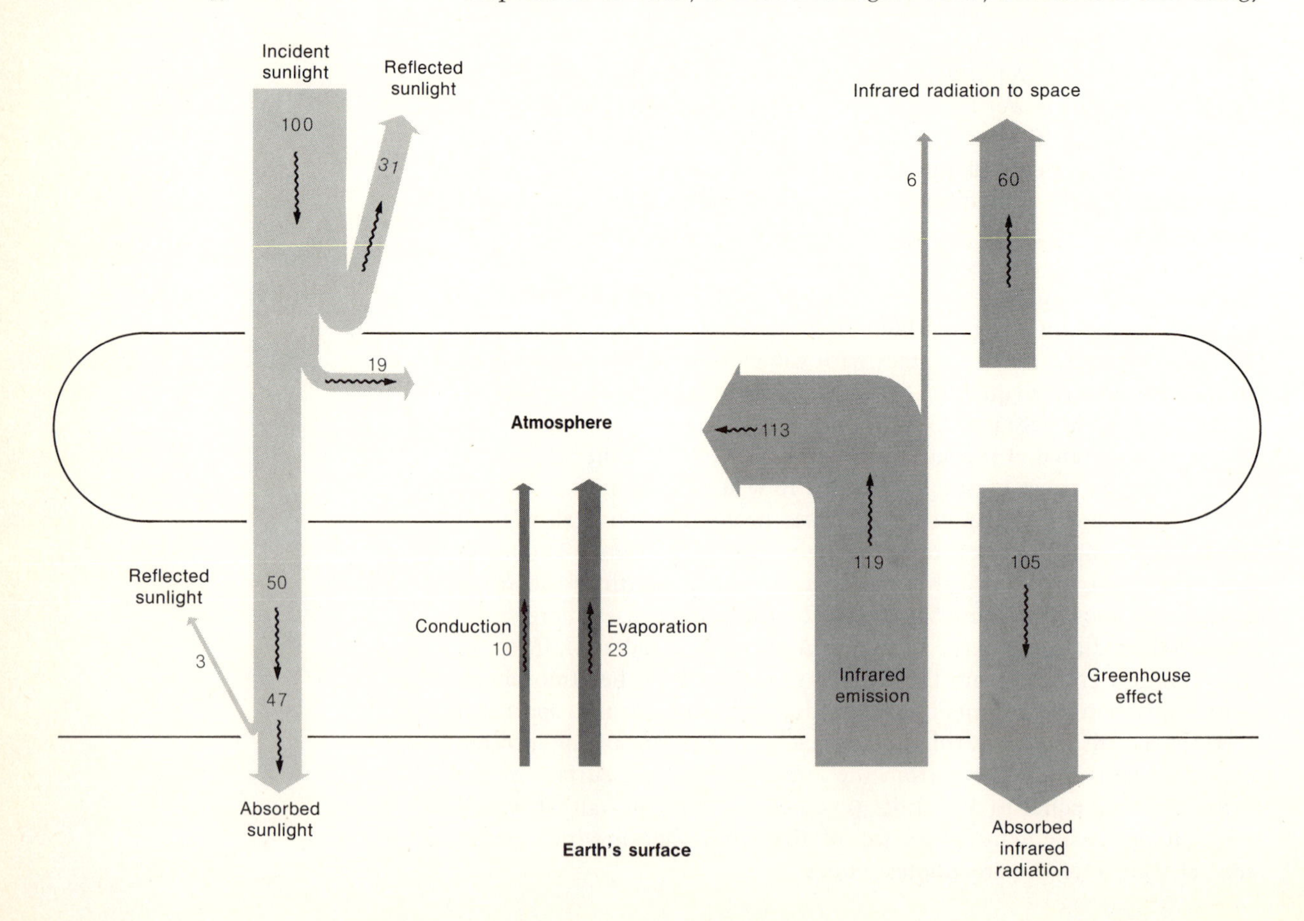

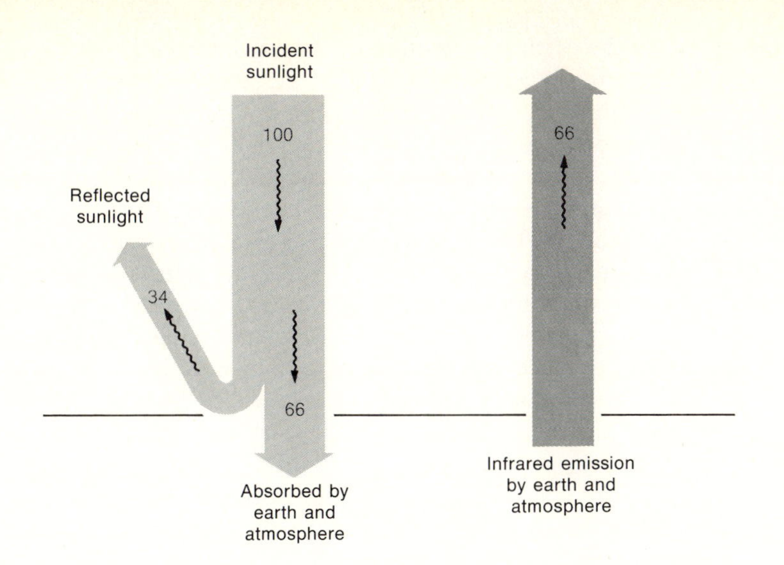

FIGURE 11.20
Energy flows to and from the system composed of the earth together with its atmosphere.

is delivered to the atmosphere as water vapor condenses into droplets, giving off the latent heat of vaporization. (A small fraction of the absorbed energy provides the increase in gravitational potential energy as water is raised from the oceans into the atmosphere; this energy too is converted into thermal energy when falling rain hits the ground and eventually works its way back to the sea.) Most of the rest of the absorbed energy, 42 units, is directly converted into thermal energy at the surface. About one unit is converted into the kinetic energy of winds, waves, and ocean currents; this kinetic energy is converted into thermal energy by frictional processes at the same rate as it is produced. A very small amount, about 0.023 units, is the source of energy for photosynthesis in plants, producing chemical energy, and this is the energy flow on which life depends. Most of this energy, too, is eventually turned into thermal energy as dead plants decay or as animals which have eaten the plants die and decompose. A small fraction of this plant and animal matter does not completely decompose right away; this chemical energy is stored—in peat bogs, for instance—in the form of fossil fuels (or future fossil fuels). Largely as a result of human activities, the chemical energy of the fossil fuels is being used up and converted into thermal energy (some of it being first converted into kinetic energy of automobiles or into electrical energy) at a much faster rate than new fossil fuels are being produced; this is the energy flow that was denoted by P_{society}, and it represents a utilization of solar energy that arrived hundreds of millions of years ago and has been stored as chemical energy of incompletely decayed plants and animals ever since. In addition to these energy flows whose original source is solar energy, we also have a contribution from geothermal energy, about 0.02 units, a flow of heat from the interior of the earth largely due to radioactive materials; this represents a decrease in the amount of nuclear energy stored in the earth, as does human utilization of nuclear energy in nuclear power plants, still an extremely minor contribution. In addition, there is an input of energy from the kinetic energies of the moon and the earth and from their gravitational potential energy into the tides, energy that is also dissipated by friction into thermal energy. Even though a great many different types of energy flows are represented in these figures, the picture presented is still a highly oversimplified one in which the earth is treated as if it were a single homogeneous object and in which all the complexities of variation from place to place and from hour to hour have been ignored.

One final important point (already mentioned in Chapter 7) should be added about the thermodynamics of the earth as a whole. The earth is almost exactly in a steady state, giving off as much energy as it receives,* but *entropy* is continuously being created by all the processes that go on in the atmosphere and on the surface of the earth. The only reason that the earth can be maintained in a steady state is that we export more entropy than we receive. The earth is similar to a closed system in that the amounts of energy received and emitted are very nearly equal, but it is very important to us that the earth is in fact *not* a closed system. The energy we receive from the sun is high-grade, low-entropy energy from a very hot object; the energy we give off is lower-grade energy, with more entropy for the same amount of energy, enough extra entropy to get rid of the entropy generated on the earth. If the earth were a truly closed system (as is perhaps implied by the catchy term "Spaceship Earth"), it would inevitably be approaching a "heat death," a state of thermal equilibrium. As it is, as long as we continue to receive an input of low-entropy energy from the sun, life and all the other interesting processes that go on can be maintained in spite of the second law of thermodynamics.

§11.6 THE PHOTOELECTRIC EFFECT AND THE QUANTUM NATURE OF LIGHT

The evidence in support of the wave model of light is overwhelming. We even know the *wavelengths* and *frequencies* of light and other electromagnetic radiation. But other experiments seem completely inconsistent with this point of view. The clearest evidence against the wave interpretation comes from studies of the "photoelectric effect." Under certain

*The total energy of the earth is not quite constant. The combustion of fossil fuels, for instance, represents a steady depletion of the earth's store of chemical energy. However, as seen in Question 11.10, the rate at which this depletion of chemical energy takes place is extremely small in comparison to the rate at which energy is received from the sun; to a very good approximation, the total energy of the earth is constant, and there is a balance between incoming and outgoing radiation.

circumstances, a light beam hitting a metal plate can knock electrons out of the metal, electrons that can be collected and detected by the current they produce in an electrical circuit (Figure 11.21). This in itself is not surprising, no more surprising than the fact that ocean waves can dislodge grains of sand from a beach. The ability of light to produce electrical currents is already being used on a small scale to extract useful energy from light, especially as a source of energy in earth satellites, and it is one of the ways in which solar energy may be used in the future to provide us with much larger amounts of energy. The surprising features of the photoelectric effect emerge when the effect is studied in more detail, when one investigates under what circumstances light can eject electrons and under what circumstances it cannot. For any particular metal, there is a critical wavelength such that light of *shorter* wavelength (farther toward the violet end of the spectrum) can knock electrons out of the metal but light of longer wavelength, no matter how intense, cannot. The value of this critical wavelength varies from one metal to another, as shown in Table 11.1.

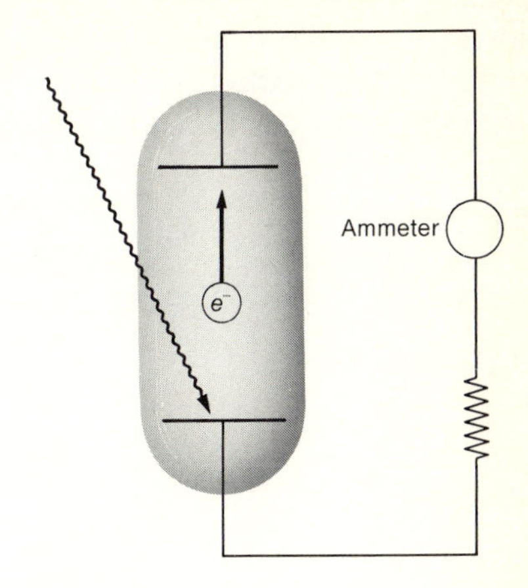

FIGURE 11.21
Apparatus for observing the photoelectric effect.

It is not surprising that the ability of light to eject electrons varies with wavelength, but such a sharp all-or-nothing variation is quite astonishing. Consider lithium, for instance: blue-green light ($\lambda \simeq 5000$ Å), even weak blue-green light, ejects electrons and produces a photoelectric current (the size of which increases as the intensity of the light is increased), but green-yellow light ($\lambda \simeq 5500$ Å), even of great intensity, has no such effect. Attempts to understand these facts on the basis of a simple wave model of light have either been total failures or (for those which seemed to show some promise) required so many *ad hoc* assumptions that they were totally unconvincing. There is an extremely simple explanation of the photoelectric effect, an explanation whose major drawback is that we must forget for the moment our knowledge that light "is" a wave and suppose instead that light consists of a stream of energetic particles called *photons*. No matter how hard we may try, it is impossible to obliterate completely the knowledge that many aspects of light indicate its wave nature, and in the following discussion we will continue to refer to red light, for instance, as having a greater wavelength and a smaller frequency than blue light. Light *does* exhibit interference, and we cannot invoke a simple stream-of-machine-gun-bullets model, not if we want a single explanation for all of the features of light.

TABLE 11.1 Critical wavelengths for various metals.

NOTE: Light of wavelength shorter than the critical wavelength produces a photoelectric effect; light of longer wavelength does not.

Metal	Wavelength (Å)
Potassium	5510
Lithium	5260
Silver	2820
Copper	2580
Gold	2360

It was Einstein who first suggested a particle model as an explanation of the photoelectric effect, a suggestion that can be elaborated in the following way. Suppose that a beam of light consists of a stream of energetic particles (photons) and suppose further that the difference between the various colors lies in the amount of energy *per particle*. Thus each photon in a beam of red light would have the same amount of energy; an intense red light would deliver *more photons* per second, but each photon would have the same amount of energy as one in a weaker red light, and a blue light would contain photons whose individual energies were different from those in a beam of red light. If we assume that the energy per photon is smaller for light toward the red and infrared end of the spectrum than for blue and ultraviolet light, we can immediately explain the existence of a critical wavelength in the photoelectric effect. Presumably, for a particular kind of metal, a certain minimum energy is required to eject one electron. If a photon has this much energy or more (that is, if it is sufficiently far toward the blue end of the spectrum, of sufficiently short wavelength), one photon can eject an electron. If the light is "too red," no individual photon can eject an electron, and so no photoelectric current will be observed (unless two photons happen to deliver their energy to one electron simultaneously, a most unlikely event). It is rather like throwing rocks at a row of bottles on a fence. If each individual rock has too small an energy, bottles may wobble occasionally, but no matter how many rocks one throws, all the bottles will remain on the fence. However, if each individual rock has enough energy, then even with a very small number of rocks per second, once in a while a bottle will be hit and knocked off the fence.

The simple assumption that the energy of a light beam is carried in discrete "lumps" by individual photons accounts immediately for the fundamental observations in a very simple and straightforward way. If we knew nothing else about light beyond the observations of the photoelectric effect, the conclusion would be inescapable that light is a stream of particles. Further indications of the correctness of the particle interpretation and a clue about the relationship between the wave-like and particle-like aspects of light come from an examination of the kinetic energies of the ejected electrons. Suppose that the minimum energy needed to remove an electron from a particular metal is denoted by E_{min}. Let the energy of an individual photon be denoted by E_{photon}. If $E_{\text{photon}} \geq E_{\text{min}}$, then such photons can produce a photoelectric effect, but of the photon's energy, at least an amount equal to E_{min} is required to remove an electron from the metal. Thus the maximum amount of energy left in the form of the electron's kinetic energy is the difference:

$$\text{Maximum } KE \text{ of ejected electrons} = E_{\text{photon}} - E_{\text{min}}. \qquad (11.9)$$

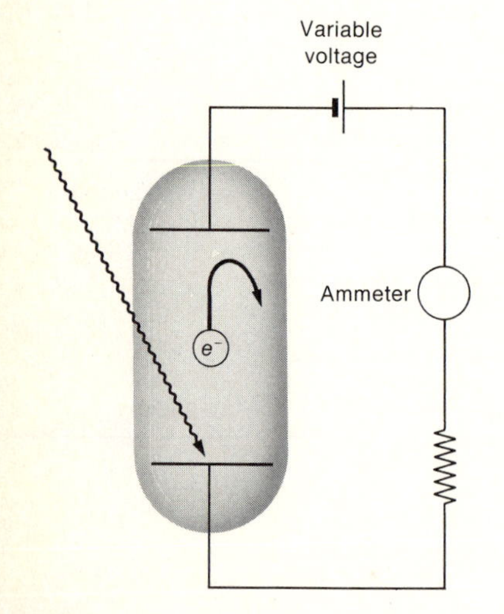

FIGURE 11.22
By adding a battery to the circuit shown in Figure 11.21, the maximum kinetic energy of the ejected electrons can be measured.

The kinetic energies of the ejected electrons can be measured by applying a voltage difference between the two plates. If a battery is connected in the circuit as shown in Figure 11.22, emitted electrons will be retarded; if an electron's kinetic energy is not large enough, it will not be able to reach the collecting plate (even if aimed in precisely the right direction); instead it will be turned around and will not contribute to the current observed on the meter. As the size of the retarding voltage is increased, more and more electrons will be turned around and the observed current will decrease. By adjusting the retarding voltage to the point at which the current is completely stopped, we can determine the *maximum* kinetic energy that any electrons had at the moment they were ejected.

QUESTION

11.12 Consider a simple analogy. Suppose someone at ground level is throwing invisible rocks into the air, all of which have the same mass (0.4 kg) but are thrown in various directions with various speeds. Your job is to study the kinetic energies of these rocks, in particular to determine the energy of the fastest rocks as they leave the ground. Although you cannot see the rocks, you can hear them if they hit the ceiling, and, fortunately, the ceiling is adjustable. As you raise the ceiling, fewer and fewer impacts are heard. The critical height is found to be 6 m; below that height, some rocks are heard but above this height there are none. What was the kinetic energy of the most energetic rocks as they left the ground?

In this way one can measure the maximum kinetic energy of electrons ejected from a particular metal by light of a particular color, and the measurement can then be repeated for lights of various colors. The results of such a series of measurements of the energies of electrons ejected from lithium are plotted in Figure 11.23. It is clear that the energy of a photon is greater, the shorter the wavelength or the higher the frequency; Figure 11.23b in particular, because it appears to be a straight line, suggests a simple interpretation. According to Equation 11.9, the maximum kinetic energy of the ejected electrons is a linear function of the energy of each individual photon. According to the experimental data shown in Figure 11.23b, the maximum kinetic energy of the ejected electrons is a linear function of the *frequency* of the light used. It is hard to avoid the conclusion that the energy of a photon is a linear function of the frequency, and indeed the simplest sort of linear relationship, a direct proportionality, explains all the results. We write this relationship in the form

$$E_{\text{photon}} = h\nu \tag{11.10}$$

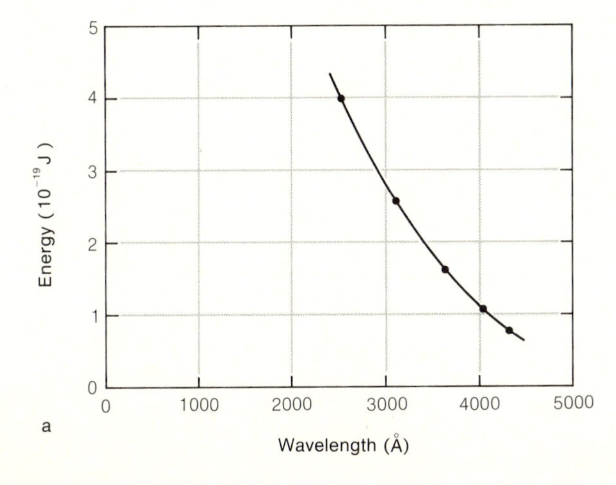

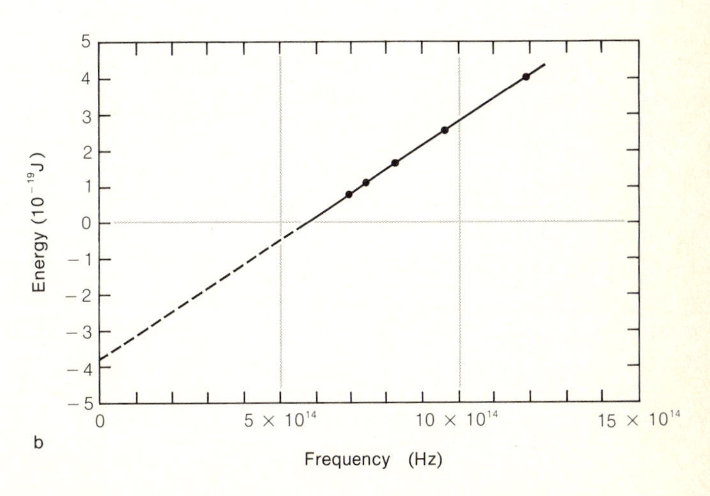

FIGURE 11.23
Maximum kinetic energy of electrons ejected from a lithium surface, plotted versus (a) wavelength and (b) frequency of the incident light.

with h a constant of proportionality. Using this equation, we can rewrite Equation 11.9 as

$$\text{Maximum } KE \text{ of ejected electrons} = h\nu - E_{min}. \quad (11.11)$$

The proportionality constant h is thus the *slope* of the line drawn in Figure 11.23b, and E_{min} must be approximately 3.7×10^{-19} J. That Equation 11.10 really represents a valid conclusion about the energies of photons is confirmed by the fact that similar measurements using other metals yield similar results and give the same value of h but different values of E_{min}. It is expected that E_{min} should vary from one metal to another, since it is the minimum amount of energy needed to eject an electron from the metal. The constant h, relating the energy of a photon to the frequency of the light, is one of the most fundamental of all physical constants and is called Planck's constant. Its units are those of energy divided by frequency:

$$\frac{\text{J}}{\text{sec}^{-1}} \quad \text{or} \quad \text{J-sec}$$

and its numerical value, according to the most recent measurements, is:

$$h = 6.626 \times 10^{-34} \text{ J-sec.}$$

QUESTION

11.13 Estimate the value of h by determining the slope of the straight line shown in Figure 11.23b.

What *is* light? A wave? A stream of particles? Both? Neither? Suppose that our only information about light came from experiments such as the two-slit experiment. The wave interpretation of this experiment is so simple and convincing, and attempts to explain the results in terms of particles so difficult, that there is only one possible conclusion: light is a wave. If, however, our only information about the nature of light came from the photoelectric effect, we would be forced to draw the opposite conclusion. If our knowledge were limited in either way, life would be greatly simplified, but the difficulty is that we have *both* kinds of information. Perhaps the best answer to the question "What is light?" is that it is neither wave nor particle. It is something like a wave and something like a stream of particles, but, in the last analysis, light is light; it behaves in its own strange and complicated fashion, no matter how we may try to make its behavior fit one or another of our models.

During the early part of the 20th century, attempts to understand both the nature of light and the facts of atomic structure led to the development of quantum mechanics, probably the most significant conceptual revolution in science that has ever occurred. Unfortunately, space does not permit an exploration of this fascinating subject. Nevertheless, the experimental facts about the behavior of light are clear. In traveling through space, in going through a pair of slits, light acts like a wave and exhibits interference phenomena, but its energy is delivered in discrete amounts, in the form of photons, and it is Equation 11.10 that provides the essential link between the wave-like and particle-like properties of light.

FURTHER QUESTIONS

11.14 By approximately what factor is the speed of sound less than the speed of light?

11.15 When you listen to a live broadcast of a concert, the music reaches your ears by means of sound waves traveling to a microphone on stage, thence by radio waves traveling at the speed of light, and finally by means of sound waves crossing your living room. Who hears the music first, someone in the back row of the concert hall or you in your living room, 500 miles away?

11.16 Estimate the wavelength of a 400-Hz sound wave. What would be the wavelength of an electromagnetic wave of the same frequency?

11.17 How long would it take light to go 10 ft? Once around the earth? From the moon to the earth? From the sun to the earth?

11.18 Calculate the wavelength of a 100-megahertz FM radio wave.

11.19 What are the possibilities for observing interference with sound waves? If two loudspeakers 10 ft apart are both emitting a 400-Hz tone, can you find a location at which the two waves would nearly cancel one another out?

11.20 Figure 11.9 shows the interference pattern produced in a ripple tank for one particular set of conditions. Make a sketch showing how the pattern would change if the two plungers were brought closer together. What if the two plungers were left at their original position but moved up and down more rapidly? (This would increase the frequency of the waves; the speed of the waves would be nearly the same, and therefore the wavelength would be smaller.)

11.21 Consider a black surface of area 1 m^2 and a temperature of 100°C. At what rate, in watts, does it emit radiation?

11.22 The filament of a light bulb operates at a temperature of about 2800°K. What is the value of λ_{max} for black-body radiation at this temperature? In what part of the spectrum is this wavelength?

11.23 Using the value of the solar constant given in §11.5 (S_o = 1353 W/m^2), estimate the total rate at which the sun emits energy in all directions. (Think of a spherical surface with the sun at the center and a radius equal to the earth-sun distance; each square meter of the surface would receive a power of 1353 W.) Is this result in agreement with the figure given in §5.7 for the total rate at which the sun emits energy, 4×10^{26} W?

11.24 (a) What is the surface area of the sun?

(b) Assuming that the sun radiates like a black body with a surface temperature of 5800°K, calculate the total rate at which the sun emits energy, and compare this with the stated figure of 4×10^{26} W.

11.25 A 100-W bulb uses an electrical power of 100 J/sec and gives off about 3% of this energy in the form of visible light. About how many photons of visible light per second does it emit?

11.26 The human eye is amazingly sensitive; under optimum conditions, a person can see a weak flash of light if about 5 photons of green light act on the retina of the eye. On this basis it has been said that if a penny were dropped from a height of 1 cm and if all its kinetic energy could be converted into photons of green light, there would be enough photons to be seen by every person now living in the world. Do a rough calculation to test this statement. Is it right? Is a height of 1 cm just barely enough, or would a still smaller value suffice?

12
Atoms: Their Structure and Energy

T. W. Tenney

12 Atoms: Their Structure and Energy

§12.1 INTRODUCTION

All familiar materials—water, air, food, our own bodies—are composed of atoms: single atoms, atoms combined in small groups to form molecules, or large collections of interacting atoms that form solids or liquids. A number of atoms or molecules may combine to form a new molecule, or large molecules may be broken apart. These are the chemical reactions that are so important to us—reactions that go on inside our own bodies and in the combustion of a lump of coal, and in which important energy transformations occur. Although the description of atomic structure to be given here is too brief to provide a detailed understanding of the structure of atoms and exact calculations of their energies, we will be able to draw some general conclusions. The first is that each kind of atom or molecule has a particular set of "energy levels"; the internal energy of an atom cannot vary continuously but must have one of a number of "allowed" values. The other important conclusion concerns the numerical values of these energies; when the energy of an atom or molecule changes from one of its allowed values to another, in most cases the change in energy is in the neighborhood of 1 electron-volt (1 eV = 1.6×10^{-19} J). Similarly, in most chemical reactions, the order of magnitude of the change in chemical energy is about 1 eV for each individual occurrence of the reaction.

Atoms were once considered indestructible and indivisible; the word *atom* is derived from a Greek word meaning indivisible. In discussing kinetic theory in Chapter 6, we assumed that the internal energy of an atom had a fixed value, we treated atoms as if they were tiny balls, which could move as a whole and have kinetic energies, but whose "insides" were of no concern. Now let us examine the internal structure of atoms. They are composed of electrons and nuclei, and the electrons may be arranged in many different possible ways, with differing amounts of energy. Changes in atomic structure may lead to a decrease in energy, making energy available in some other form. The reverse process may also occur; addition of energy from outside may cause a rearrangement of the parts of the atom, with an increase in the internal energy of the atom. We will not attempt to present all of the evidence that has led to our present understanding of atomic structure. For the most part, we will merely describe the present view and give some of the most important facts which support this description.

The mass of an individual atom is of course very small. The lightest atom of all (the common hydrogen atom) has a mass of about 1.7×10^{-27} kg, and even uranium atoms, the heaviest atoms found in nature, are only about 240 times heavier. In discussing atomic structure, it is very convenient to measure masses in *atomic mass units* (amu):

$$1 \text{ amu} = 1.661 \times 10^{-27} \text{ kg}.$$

The atomic mass unit is chosen so that the mass of a hydrogen atom is approximately 1 amu; all atomic masses, then, lie approximately in the range between 1 amu and 240 amu.

It is the electrical interaction between the constituents of an atom that determines its structure. At the center of each atom lies a tiny *nucleus*, which has most of the mass of the atom and a *positive* electrical charge. This positive charge is due to *protons* within the nucleus, particles that have a mass of approximately 1 amu and a positive electrical charge whose magnitude is precisely equal to that of the negative charge carried by electrons:

$$e = 1.6 \times 10^{-19} \text{ coulombs (C)}.$$

Also within the nucleus, contributing to its mass, are a number of *neutrons,** particles that also have a mass of about 1 amu but no electrical charge. Around the nucleus are a number of "orbital" electrons, normally at a distance of 1 Å or so (1 Å $= 10^{-10}$ m); it is the electrons that give an atom its characteristic size.

QUESTION

12.1 The sizes of atoms can be estimated from the known densities of solid materials, if we know the masses of individual atoms and assume that the atoms in a solid are in contact with one another. The mass of a copper atom, for example, is about 64 amu, and the density of metallic copper is 8.9 g/cm³. On the basis of this information, how much volume is occupied by a single copper atom? If this were a spherical volume, what would be the radius of the sphere in angstroms?

---

*The only exception is the common form of the hydrogen atom, whose nucleus is simply a proton.

The various elements can be arranged according to the number of elementary units of positive charge on the nucleus, a number given the symbol Z, the "atomic number." For hydrogen $Z = 1$, for helium $Z = 2$, and so on up to uranium ($Z = 92$); the artificial production of elements with still higher values of Z will be described in Chapter 14. The electrons have very little mass compared to the nuclei, the mass of an electron being only 0.0005486 amu, nearly 2000 times smaller than that of a proton, but electrons are of the utmost importance in determining chemical properties. If an atom is electrically neutral, its nucleus is surrounded by a number of electrons equal to the number of protons in the nucleus. Thus the integer Z can be interpreted in various ways: as the value of the positive charge on the nucleus measured in units of the electronic charge, as the number of protons in the nucleus (the "proton number" of the nucleus), or as the number of orbital electrons for an electrically neutral atom. When the elements are arranged according to their values of Z, studies of their chemical properties reveal striking periodicities. Helium, neon, argon, krypton, xenon and radon (the "rare gases") have respectively 2, 10, 18, 36, 54, and 86 orbital electrons. These atoms have their electrons arranged in especially compact and stable ways. Rare-gas atoms combine neither with each other nor—except rarely—with other kinds of atoms. The atoms of the elements lithium, sodium, potassium, rubidium, cesium, and francium have 3, 11, 19, 37, 55, and 87 electrons, in each case one more than for a rare-gas element. These elements exhibit chemical behavior very similar to one another; they each have a strong tendency to lose the single extra electron, forming singly charged positive ions (Li^+, Na^+, and so on). The arrangement of the elements in the periodic table (Table C.2) is based on such chemical properties as these. (Not all the elements were known at the time that the regularities in chemical behavior were first noticed. There were gaps in the early versions of the periodic table, suggesting the existence of elements that had not yet been discovered, but whose chemical properties could be predicted from their location in the table. In some cases, these predictions led to the discovery of the missing elements.)

Curiously, even though uranium atoms have 92 orbital electrons and hydrogen atoms only one, their sizes do not differ widely. All neutral atoms have radii of about 1 or 2 Å. This fact can be qualitatively understood as a result of the larger electrical charge on the nuclei of heavier atoms, a charge that is more effective at pulling in the surrounding electrons.

§12.2 ATOMIC SPECTRA AND ATOMIC ENERGY LEVELS

Our primary concern is with the *energy* involved in atomic structure, and with the *changes* in energy that occur when changes in the arrangements of the electrons take place. What general statements can we make about these energy changes? The most important clue comes from the study of atomic spectra, of the light emitted when a gas is "excited" by an electrical discharge. The striking feature of such a spectrum is that the spectrum is *discrete* rather than *continuous* (§11.4). The atoms of a given element in gaseous form emit light of only certain definite wavelengths or frequencies, as opposed to the continuous spectrum emitted by a hot solid object.

There is a very practical application of this fact. In an ordinary incandescent light bulb, a filament is heated to a temperature of about 2800°K and gives off energy by black-body radiation; even at this temperature, most of the energy is in the infrared region and is useless for illumination. Most of the 100 watts of power required to operate a 100-W bulb is simply wasted in heating up the room. But in a fluorescent light or in a neon light, an electrical discharge results in the emission of light of various discrete wavelengths. Some of these wavelengths are in the visible part of the spectrum; this is the origin of the red color of a neon light. Fluorescent lights contain mercury atoms, which emit radiation largely in the ultraviolet region; these lights are coated with "phosphors," which, when struck by high-energy ultraviolet photons, re-emit a significant fraction of the energy in the form of visible light. As a result, we get more useful light for the same amount of energy delivered to the lamp.

Each kind of atom or molecule has its own characteristic emission spectrum; some examples are shown in Figure 12.1. One way of determining the wavelengths present in a beam of light depends on the fact that lights of various wavelengths are bent through different angles when sent through a prism. If a beam of monochromatic light (with only a single wavelength) is first sent through a narrow slit (as in Figure 11.2) and then bent by a prism, a single narrow line is seen on the screen, parallel to the slit through which the light was sent. When Newton did such an experiment with white light, an illuminated region of oblong shape was observed, with a continuous range of colors from red to violet.

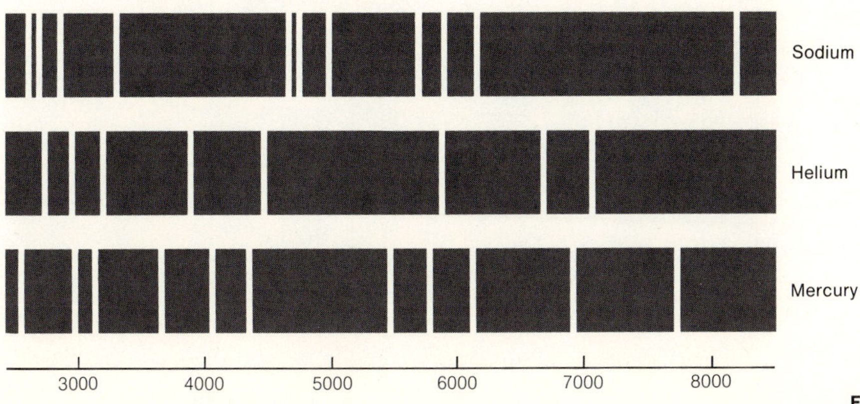

FIGURE 12.1
Portions of some typical emission spectra.

When light from an electrically excited gas is used, most of the colors are missing; a number of lines are observed, each one resulting from light of one particular wavelength. By using photographic films that are sensitive to ultraviolet and infrared radiation as well as to visible light, other parts of the spectrum can be mapped out.

We saw in Chapter 11 that light of a particular wavelength or frequency carries energy in the form of photons, and that the energy of a photon is related to the frequency of the light wave in a very simple way:

$$E = h\nu,$$

where h, Planck's constant, has the value 6.626×10^{-34} joule-seconds (J-sec). An atom excited by an electrical discharge gives off light of certain discrete wavelengths or frequencies; thus the emitted photons have only certain discrete values of energy. The emission of a photon with a certain energy must be accompanied by a decrease of the energy of the atom by the same amount. If the energies of the emitted photons are restricted to certain discrete values, we are almost forced to conclude that the energy of the atom itself must be quantized. The electrons can be arranged only in one of a number of discrete allowed patterns, each with its own corresponding value of energy.

On this basis, we suppose that each atom has its own set of allowed "energy levels," one energy for each possible arrangement of its electrons. We can represent this schematically by an "energy level diagram,"

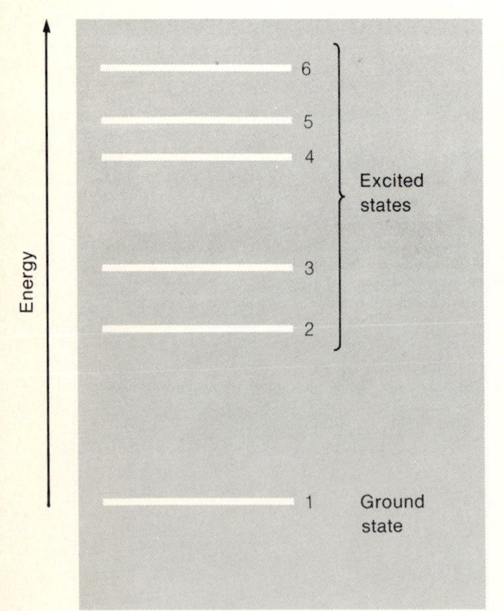

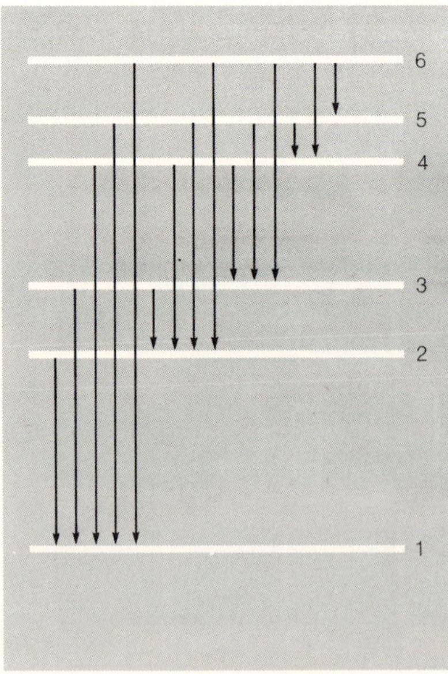

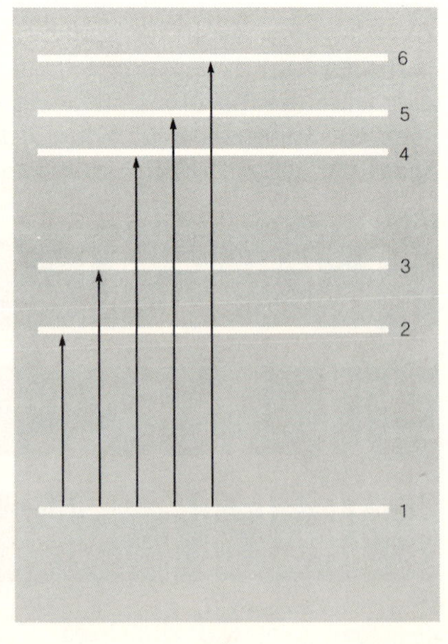

FIGURE 12.2
(a) An atomic energy level diagram;
(b) possible transitions between energy levels;
(c) transitions that may be produced when light is absorbed.

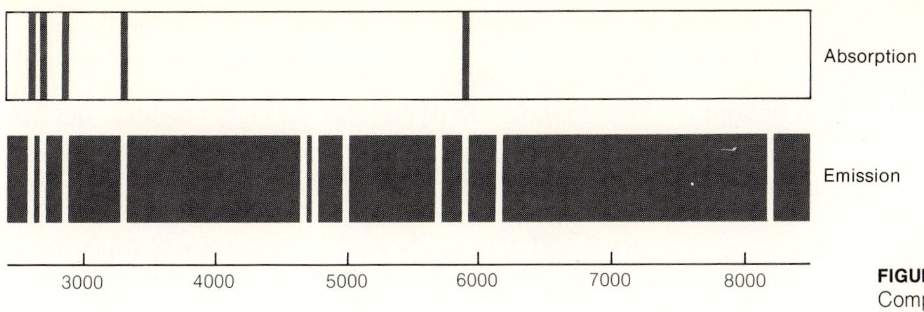

FIGURE 12.3
Comparison of the emission and absorption spectra of sodium.

as shown in Figure 12.2a. An atom left to itself will give off energy and reach its lowest possible energy level, its "ground state" (1). An electrical discharge can provide the energy to raise atoms into one or another of the "excited states" (2, 3, 4 . . .); an excited atom will then fall back to the ground state by one or more jumps, in each jump losing energy and giving off this energy by emitting a photon. Notice the great number of transitions that are possible, as indicated in Figure 12.2b. An atom in state 4, for instance, may get to the ground state either directly, or by way of one or more of the intermediate energy levels.

This description explains another of the observed facts about the interaction of atoms and light. A container of gas, not excited by an electrical discharge, can *absorb* light, but only light of certain discrete wavelengths. If light with a continuous spectrum is passed through a gas, the light emerging on the other side has a nearly continuous spectrum, but certain wavelengths are missing. All of these absorbed wavelengths correspond to wavelengths observed in the emission spectrum of the same gas, but many other wavelengths in the emission spectrum are not observed in the absorption spectrum (Figure 12.3). The explanation is as follows. Suppose that the atom has the energy levels shown in Figure 12.2a. In observing an absorption spectrum, we shine light containing a mixture of all wavelengths (a beam of photons of all possible energies) on the atoms and see which wavelengths are removed. A photon can only be absorbed if its energy has just the right value to cause an atomic transition between one allowed level and another. If all the atoms are initially in the ground state, then the only possible transitions are those indicated in Figure 12.2c; many of the transitions that occur when the emission spectrum is being observed (between levels 2 and 3, 3 and 4, and so on) simply do not take place.

An important *quantitative* conclusion can be drawn from observations of atomic spectra. The wavelengths are typically in or near the visible part of the spectrum; we can take a wavelength of about 6000 Å as characteristic of the wavelengths found in atomic spectra. Thus the energy of a photon of 6000-Å light is typical of the difference in energy between one atomic energy level and another.

QUESTION

12.2 What is the energy of a single photon of light whose wavelength is 6000 Å?

(a) Show that the frequency of this light is $\nu = 5 \times 10^{14}\ \text{sec}^{-1}$.

(b) Show that a single photon of this light has an energy of about 3.3×10^{-19} J.

(c) In discussion of atomic and nuclear energies, it is often convenient to measure energies in *electron-volts* (eV). An electron-volt, as defined in Chapter 8, is equal to 1.6×10^{-19} J. Show that the energy of a photon of 6000-Å light is approximately 2 eV.

Thus the fact that electrically excited atoms emit light of certain discrete wavelengths and the fact that these wavelengths are usually in or near the visible part of the spectrum suggest that atoms have only certain allowed energy levels and that the difference in energy between one level and another is in the neighborhood of one or a few electron-volts. No single number can tell the whole story, of course, but it is important to know approximately the magnitude of the energy changes involved.

There is another and quite different type of experiment which reveals the existence of discrete allowed atomic energy levels. Suppose that an atom in its ground state is struck by an energetic particle, for example an electron that has been given an appreciable kinetic energy. In the collision, the atom may be raised from its ground state to its first excited state, but only if the electron has enough kinetic energy. If the electron's kinetic energy is smaller than the energy difference between the atom's ground state and the first excited state, it cannot cause a transition. Then there will simply be an elastic collision between the electron and the atom, and the atom as a whole will acquire a small amount of kinetic energy. The internal state of the atom will be unchanged, however, and the electron's kinetic energy will be only very slightly reduced.

QUESTION

12.3 Consider a head-on elastic collision between an electron and a hydrogen atom. What fraction of the electron's initial kinetic energy will be transferred to the atom?

The plan of the experiment, then, is to bombard atoms with electrons of known kinetic energy and see whether or not the electrons lose an appreciable amount of kinetic energy as they collide with the atoms. When this experiment is done with mercury vapor (a gas of mercury atoms), it is observed that as long as the electrons have kinetic energies less than about 5 eV, they lose very little energy in colliding with mercury atoms. Electrons with kinetic energies greater than 5 eV, on the other hand, often lose most of their energy in collisions. The interpretation of this result is that for the mercury atom, the separation in energy between the ground state and the first excited state is about 5 eV. This interpretation is confirmed by the observation of the light emitted by the mercury atoms that have been excited. As excited atoms return to the ground state, they emit photons. The energies of the photons can be determined by measuring the wavelength of the emitted light, and the photon energy is found, as expected, to be about 5 eV.

QUESTION

12.4 The "light" emitted by excited mercury atoms can only be seen with the eye if the wavelength is in the visible region of the spectrum. What is the wavelength of a 5-eV photon? Is it in the visible, infrared, or ultraviolet part of the spectrum?

§12.3 THE ROLE OF ELECTRICAL POTENTIAL ENERGY IN ATOMIC STRUCTURE

From a fundamental point of view, there are two forms of energy which contribute to the internal energy of an atom: first, the kinetic energies of the electrons and, second, the electrical potential energy associated with the interaction of the electrons and the nucleus, the interaction that holds the atom together. The discussion of electrical potential energy in Chapter 8 was restricted to the example of a charge in the uniform electric field between the plates of a capacitor; now, however, we must consider an electron subject to the force of the nucleus, a force that varies rapidly with the distance between them, and we need an expression for the electrical potential energy associated with this force. Fortunately, we can use the similarity between the law of gravitational force and the fundamental law of electrical force to derive an expression for electrical potential energy in this situation.

The gravitational force between two objects is given by

$$F_{\mathrm{grav}} = \frac{Gm_1m_2}{r^2}$$

and associated with this, as we saw in Chapter 5, is a gravitational potential energy given by

$$GPE = -\frac{Gm_1m_2}{r}. \tag{12.1}$$

It is understood that in this expression, the zero-level is at "infinity," that is, for an extremely large distance between the two masses. The gravitational force is always attractive; a rock released far above the earth's surface falls toward the earth, its gravitational potential energy becoming more and more negative as it falls.

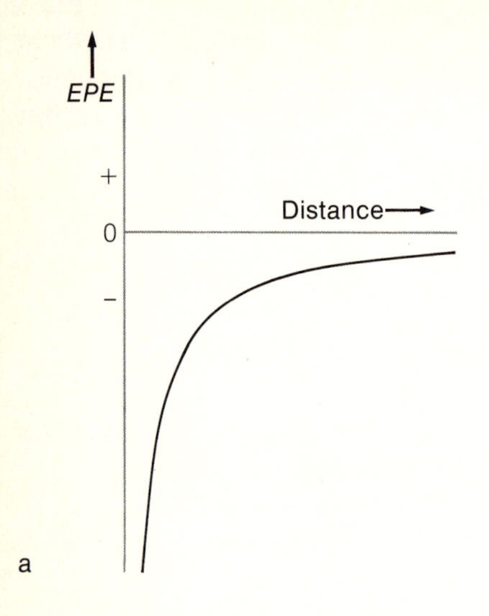

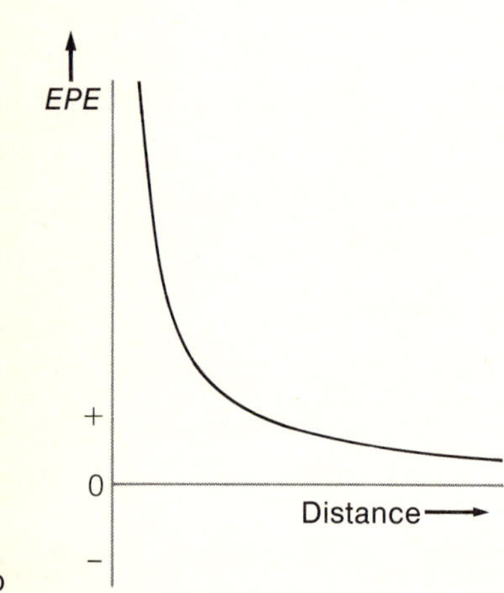

FIGURE 12.4
Electrical potential energy of two charged particles as a function of their separation: (a) for two charges of opposite sign (an attractive force); (b) for two charges of the same sign (a repulsive force).

The Coulomb law of electrical force is

$$F_{\text{elec}} = \frac{k_e q_1 q_2}{r^2},$$

exactly like the law of gravitational force except for the replacement of (Gm_1m_2) by $(k_e q_1 q_2)$, where $k_e = 9 \times 10^9$ N-m^2/C^2, the Coulomb's law constant. If in this case we agree to let the electrical potential energy equal zero when the two interacting charges are very far apart, then we can argue from Equation 12.1 that when they are a distance r apart, the electrical potential energy is $(k_e q_1 q_2/r)$, but we must be careful about the *sign* of this potential energy.

The electrical potential energy is given by the expression:

$$EPE = \frac{k_e q_1 q_2}{r}. \tag{12.2}$$

This equation is correct, but by comparison with Equation 12.1, it seems to have the wrong sign. The point is that if the electrical force is attractive, q_1 and q_2 must themselves be of opposite sign, and so the product q_1q_2 will be a *negative* number. Thus if q_1 represents the positive charge on the nucleus and q_2 the negative charge on the electron, the electrical potential energy calculated from Equation 12.2 is negative and varies with position in the way shown in Figure 12.4a. A negative electron released near a positive nucleus will fall towards the nucleus, losing electrical potential energy, just as a rock loses gravitational potential energy as it falls toward the earth. For future reference, we notice that Equation 12.2 correctly describes the electrical potential energy associated with the repulsive force that acts between two charges of the same sign. In such a case, q_1q_2 is a positive quantity, and Equation 12.2 describes an electrical potential energy that varies with distance in the way shown in Figure 12.4b; a positive charge released near another positive charge will "fall" away, losing electrical potential energy and gaining kinetic energy as the distance increases.

Equation 12.2 gives the electrical potential energy of any two charges at any distance of separation. We can use this equation to estimate the change in electrical potential energy that occurs when an atom makes a transition from one energy level to another. Consider the simplest atom, a hydrogen atom. In this case, we have an electron ($q = -e$) near a hydrogen nucleus (a proton, $q = +e$), and so

$$EPE = \frac{-k_e e^2}{r},$$

where $e = 1.6 \times 10^{-19}$ C.

QUESTION

12.5 Show that if $r = 1$ Å $= 10^{-10}$ m, then $k_e e^2/r \simeq 2.3 \times 10^{-18}$ J, or about 14.4 eV.

That is, the electrical potential energy of an electron at a distance of 1 Å is -14.4 eV; for $r = 2$ Å, this potential energy is -7.2 eV, and so on. Thus we can make a quantitative graph (Figure 12.5) corresponding to the general case shown in Figure 12.4a. The important point is that since atoms have dimensions of a few angstrom units, we expect corresponding changes in the distances between the electrons and the nucleus—from $r = 1$ Å to $r = 2$ Å, for instance—during the rearrangements of electrons which occur as atoms change from one energy level to another. The resulting changes in electrical potential energy amount to a few electron-volts, and since it is known that the difference in energy between atomic energy levels is about this size, this result confirms the idea that electrical potential energy is one of the important contributors to the internal energy of an atom.

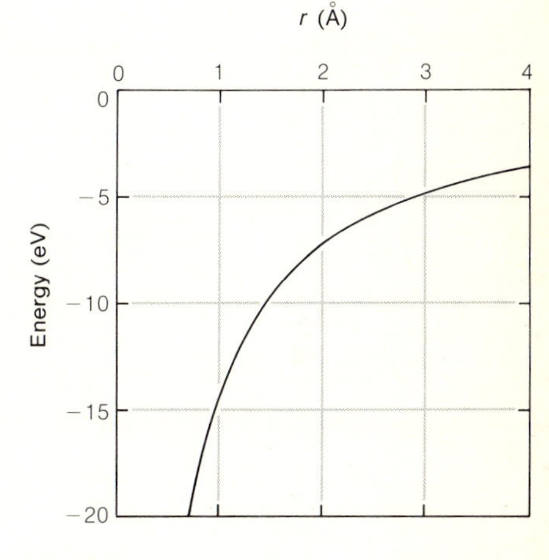

FIGURE 12.5
The electrical potential energy of an electron as a function of its distance from a proton.

§12.4 CHEMICAL ENERGY

In processes such as combustion, chemical reactions occur that lead to the release of "chemical energy." The burning of carbon, for instance, can be described by the reaction

$$C + O_2 \longrightarrow CO_2.$$

An important clue for understanding the nature of chemical energy can be obtained by examining the energy content of various fuels given in Appendix H. These data are often given in practical units appropriate to the particular fuel: energy per ton of coal, per barrel of oil, per cubic foot of natural gas, and so on. A striking regularity is found, however, when various fuels are compared in terms of energy per unit *mass*; the chemical energy per unit mass is about the same for all common fuels, in the vicinity of 10^7 J/kg. We can understand this on the basis of the earlier discussion in this chapter. In the chemical reactions that occur in combustion, rearrangements of the electrons take place. The electrical potential energies and kinetic energies of the electrons change, and we would expect these changes to be of approximately the same size as those which occur when an excited atom emits a photon. Thus, for example, we would anticipate that every time a carbon atom is burned to form CO_2, the amount of energy released would be about 1 eV. A carbon atom has a mass of 12 amu. Within the spirit of this approximation, we expect that energy released in combustion (per unit mass) is about 0.1 eV/amu.

QUESTION

12.6 Show that 0.1 eV/amu $\simeq 10^7$ J/kg, in excellent agreement with the energy content of common fuels.

Chemical reactions also occur in batteries: for instance, the reaction

$$Zn + 2H^+ \longrightarrow Zn^{++} + H_2 \qquad (12.3)$$

takes place in the copper-zinc battery described in Chapter 8. Such a reaction causes the conversion of chemical energy into electrical energy, and another piece of evidence about the nature of chemical energy is provided by the fact that most batteries have emf's of approximately 1 volt: 1.1 V for a copper-zinc battery, for instance, and 1.5 V for a common dry cell. In a copper-zinc battery, charge is transported through the battery, producing electrical potential energy at the expense of chemical energy. Since the emf is 1.1 V, the movement of one unit of charge (one electronic charge) from one terminal to the other results in an increase in electrical potential energy of 1.1 eV. Every time reaction 12.3 occurs, two units of charge are moved from one terminal to the other. That is, for every occurrence of this reaction, there is an increase in electrical potential energy of 2.2 eV and a decrease in chemical energy of the same amount. Again, the evidence is consistent with the view that individual atomic and molecular events involve energy changes of roughly an electron-volt or so.

Even though we have not studied these chemical reactions in detail, we can see that releases of chemical energy are accompanied by changes in the arrangements of electrons that are similar to those occurring when excited atoms emit photons. We saw in §12.3 that the energies involved in atomic structure are, from a fundamental point of view, simply kinetic and electrical potential energies. Therefore chemical energy is not something new and different, but even though this form of energy can be understood in more basic terms, it is still useful to retain the term "chemical energy" in discussing the energy released in chemical reactions.

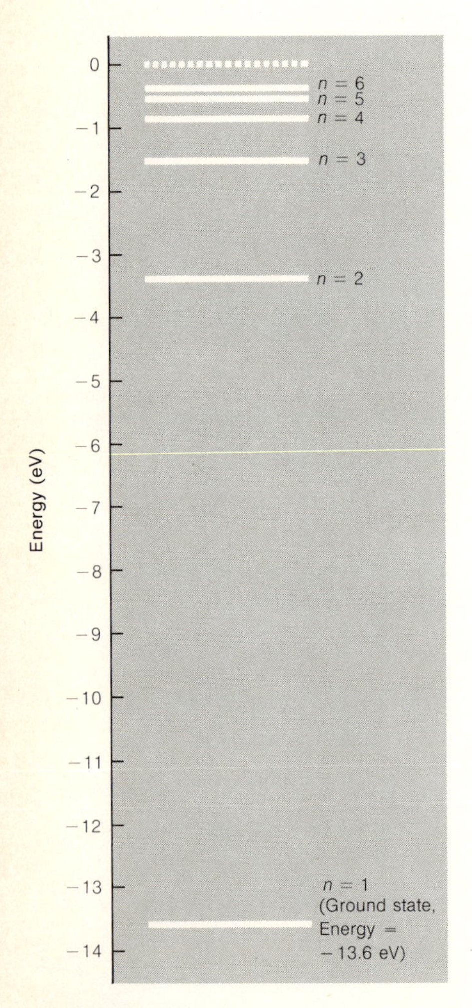

FIGURE 12.6
Energy level diagram of a hydrogen atom. (Numerous levels that lie between the $n = 6$ level and the energy $E = 0$ are not shown.)

§12.5 ATOMIC STRUCTURE AND THE PERIODIC TABLE

Largely from studies of atomic spectra, it has been possible to determine the energy levels of a wide variety of atoms. For the hydrogen atom, the simplest of all, the energy level diagram is shown in Figure 12.6. The ground state has been labeled with the *quantum number* $n = 1$; those labeled $n = 2, 3, 4 \ldots$ are the various excited states. With increasing energy, the spacing between the various excited states rapidly decreases, and the states for which n is greater than 6 are not shown individually. The position of the origin on the energy axis ($E = 0$) has been chosen to correspond to the energy of a hydrogen *ion*, a hydrogen atom whose electron is completely removed. When an electron and a proton come together to form a hydrogen atom, the energy decreases as the electron falls toward the proton, the total decrease in energy being 13.6 eV by the time the ground state is reached. As this happens, an energy of 13.6 eV is released, usually by a whole series of photons whose energies add up to 13.6 eV, as the electron makes a series of jumps from one level to another, losing energy at each jump. Correspondingly, in order to ionize a hydrogen atom, an energy of at least 13.6 eV must be supplied. This amount of energy, 13.6 eV, is the "ionization energy" of a hydrogen atom, or its "binding energy," the energy that must be supplied to break it up into a separated proton and electron.

The various energy levels correspond to various allowed "orbits"* for an electron around a proton, as indicated in Figure 12.7. Each possible orbit has its own particular energy, an energy which is made up of two parts: the kinetic energy of the electron and the electrical potential energy arising from the electrical interaction of the proton and the electron. If an electron jumps from an orbit of large radius to a smaller one, the kinetic energy of the electron actually increases, but this increase is more than outweighed by the decrease in electrical potential energy; the result is that the orbits closest to the nucleus have the lowest total energy.

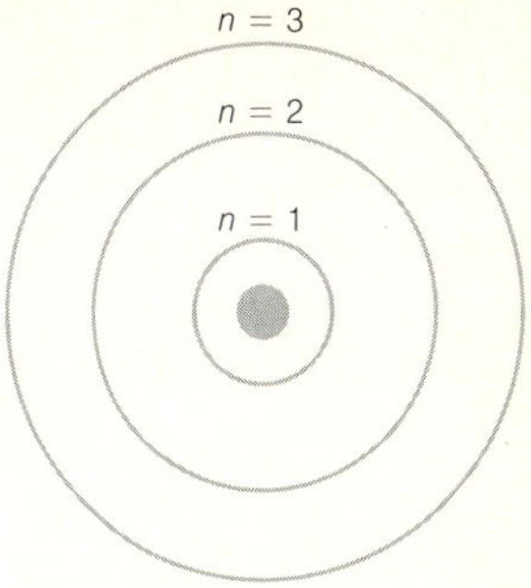

$n = 3$
$n = 2$
$n = 1$

FIGURE 12.7
Some of the possible orbits of the electron in a hydrogen atom (not to scale).

For atoms with more than one electron, similar energy levels are found. Because of the interactions between electrons, the details are more complicated, but to a first approximation the electrons can be considered to be in orbits similar to those of the single electron of the hydrogen atom. A helium atom in its ground state has two electrons, both with $n = 1$. It turns out that no more than two such electrons can be accommodated in a single atom; in the ground state of the helium atom, the $n = 1$ "shell" is full, and this is a particularly stable arrangement.† For lithium, with a total of three electrons, the third electron must go in the $n = 2$ shell (Figure 12.8a). This shell can have as many as eight electrons; by the time we get to neon, with a total of 10 electrons, the $n = 2$ shell is filled, and neon is one of the "rare gases," with a very compact and stable structure. For sodium, with 11 electrons, the last electron must go in an outer orbit (Figure 12.8b). Sodium and lithium both have a single electron outside a filled shell. The chemical behavior and the character of the atomic spectrum of an element are determined primarily by the outermost electrons, and sodium and lithium are very similar in these respects. As we proceed through the list of elements, the behavior gets more complex, various shells can contain various numbers of electrons, but the general pattern continues. A similar diagram for the ground state of the copper atom is shown in Figure 12.9.

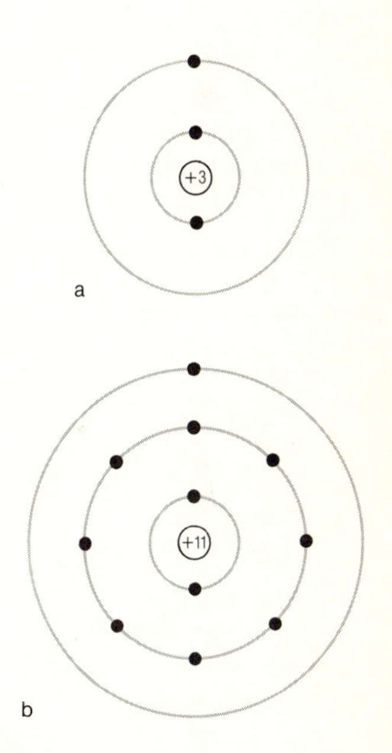

FIGURE 12.8
Ground state of (a) a lithium atom ($Z = 3$), and (b) a sodium atom ($Z = 11$).

As in any atom, the normally observed spectra and the chemical behavior of copper are determined by the outermost electrons, the full

*It is somewhat misleading to refer to electrons as being in orbits, as if they were planets in a miniature solar system, obeying the laws of ordinary *Newtonian* mechanics. Such a description is not altogether wrong, but according to the quantum-mechanical view, which will be briefly described in the next section, electrons do not have well-defined orbits like those of the planets around the sun.

†Why two? Why not just one, or why not five? The limitation on the number of electrons which can occupy each shell is something that can be explained only by quantum mechanics.

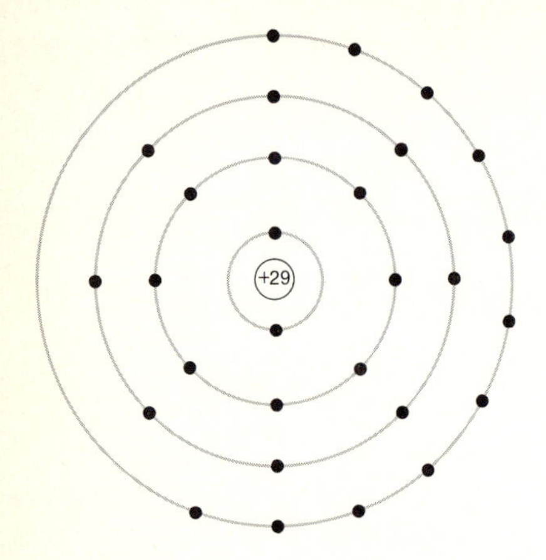

FIGURE 12.9
Ground state of a copper atom ($Z = 29$).

inner shells normally remaining unaffected. It is possible, however, to remove an electron from an inner shell of a large atom, and this is one way in which X-rays are produced. When a high-speed beam of electrons hits a copper target, for instance, it sometimes happens that one of the innermost electrons ($n = 1$) is knocked loose from the atom, leaving the copper atom as in Figure 12.10a. This hole in the $n = 1$ shell will quickly be filled, perhaps as one of the electrons in the $n = 2$ shell jumps into the $n = 1$ shell (Figure 12.10b), emitting a photon in the process, an event that might be followed by another electron jumping from the $n = 3$ shell into the $n = 2$ shell, and so on. Such photons have high energies, for two reasons. First, an electron in the $n = 1$ shell is exposed to the full electrical charge of the nucleus; the nucleus is not shielded by intervening shells of negatively charged electrons. Second, because of the large charge on the copper nucleus, the inner shells are much closer to the nucleus than in the hydrogen atom, and again the effect is to increase the change in energy as the electron jumps from one shell to another. For the transition from the $n = 2$ shell to the $n = 1$ shell of copper, the actual energy change is approximately 8000 eV, far greater than the energy changes of a few electron-volts normally encountered in transitions of the outermost electrons. The emitted photon therefore has a much higher frequency and a shorter wavelength than a photon of visible light. These high-energy photons arising from rearrangements of electrons in the innermost shells are X-rays, photons that because of their high energies can often penetrate materials that are completely opaque to visible light.

FIGURE 12.10
(a) When one of the inner electrons is knocked loose from a copper atom, a hole is left behind. (b) When this hole is filled, a high-energy photon is emitted.

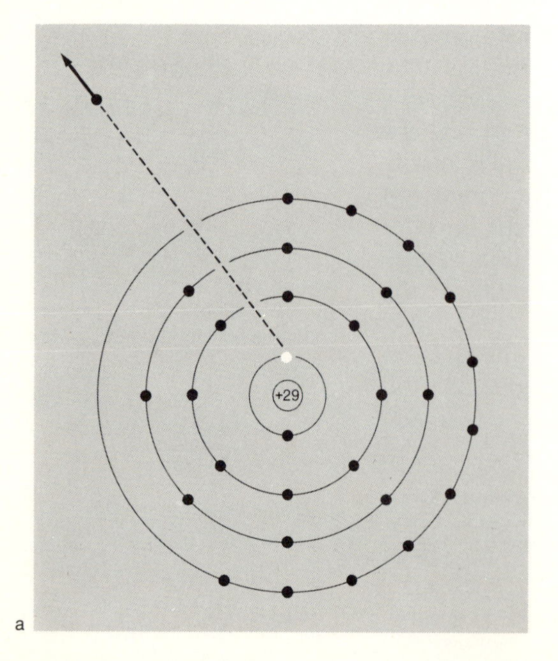

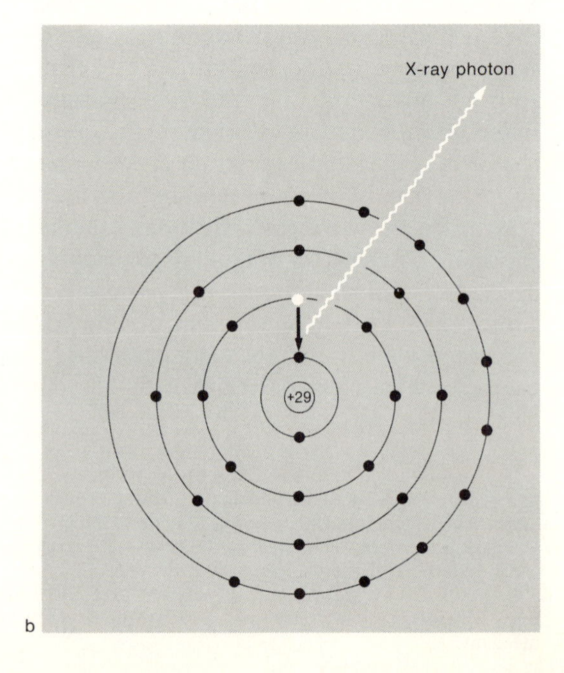

§12.6 QUANTUM MECHANICS, THE EXPLANATION OF DISCRETE ENERGY LEVELS

There is a great deal of experimental evidence demonstrating that atomic energy levels are discrete, and that the energy levels of particular kinds of atoms can be experimentally determined. But the question of "why" remains. Why is there a discrete set of energy levels? Can we deduce from some fundamental theory what the allowed levels of various atoms and molecules should be? It is not possible to understand the facts of atomic structure and atomic spectra on the basis of ordinary Newtonian mechanics. Quantum mechanics is the theory that was developed for this purpose, and also for describing the puzzling wave-particle properties of light. One of the characteristic features of quantum mechanics is that the energy of a microscopic system can have only one of a discrete set of values, though it can change from one to another.

According to quantum mechanics, electrons and other "particles" have some "wave-like" features, just as light waves exhibit "particle-like" features. It is necessary to give up the idea that electrons travel in well-defined orbits around the nucleus, and we can never know precisely both the position and velocity of an electron. (Nor can we know these quantities with perfect precision for *any* object, not even—in principle—for macroscopic objects, though the larger the object the less significant is this limitation.) This idea finds a precise expression as the principle of indeterminacy or the Heisenberg uncertainty principle, a cornerstone of quantum mechanics. Laplace's vision of an all-powerful intelligence who could at some instant know the positions and velocities of all particles and hence calculate the past and future by application of natural laws can no longer be retained, not even as a theoretically conceivable idea. Another consequence of quantum mechanics is that it is not possible, even in principle, to predict *when* an atom in an excited state will emit a photon and change to another state; one can only calculate *probabilities*. At a very basic level, the world is probabilistic and not deterministic, at least not in the naive way in which Laplace thought of it. The question of the interpretation of quantum mechanics and its implications for the ideas of determinism and causality is one that still causes controversy among physicists and philosophers. Einstein himself, whose interpretation of the photoelectric effect helped to introduce quantum mechanics, was reluctant to accept a probabilistic interpretation. Einstein is said to have summarized his objections with the remark: "I refuse to believe that God plays dice with the universe."

The energies of atoms and molecules are made up of the kinetic energies and electrical potential energies of their electrons and nuclei. From the basic laws of quantum mechanics, the allowed energy levels can be calculated for the simplest atoms and molecules. The calculated energy levels of the hydrogen atom, the simplest atom of all, are in excellent agreement with the levels deduced from observation of spectra. For all but the simplest atoms and molecules, though, the theoretical calculation of the allowed energy levels is extremely difficult. Nevertheless, energy levels can be experimentally determined from studies of spectra and from other experiments. Quantum mechanics is also applicable to the interactions of neutrons and protons, and nuclei, too, have discrete energy levels, as we will see in the next chapter. At the atomic and nuclear level, the discreteness of allowed energies is a fact of life, and Newtonian mechanics must be replaced by quantum mechanics. Quantum mechanics is applicable to all phenomena, but for large objects and distances significantly greater than atomic dimensions, the predictions of quantum mechanics are imperceptibly different from those of Newtonian mechanics. One very important point must be added. Although Newtonian mechanics does not provide a correct description of atomic and nuclear structure, the law of conservation of energy is still valid. With the advent of quantum mechanics, the concept of energy in fact became more important than ever before.

FURTHER QUESTIONS

12.7 About how many atoms thick is a page of this book?

12.8 What is the greatest positive charge that a helium ion can possibly have?

12.9 The average atom in the body has a mass of about 10 amu and a diameter of about 1 Å. If all the atoms in your body were placed in a row, how long a line would they form?

12.10 If all the molecules in 1 cm^3 of air were evenly spread out over the United States, about how many molecules per square centimeter would there be?

12.11 The atoms of a solid are closely packed together, but this is not the case for a gas. As shown in Question 6.26, a gas under ordinary conditions contains about 2.5×10^{19} molecules/cm^3. If each molecule has a diameter of about 1 Å, what fraction of the volume occupied by the gas is taken up by the molecules themselves?

12.12 (a) The mass of a water molecule (H_2O) is 18 amu. Approximately how many molecules are there in a glass of water?

(b) Approximately how many molecules of water are there in the world's oceans?

(c) A statement often made in order to dramatize the enormous numbers of atoms contained in macroscopic objects goes something like this: "Every time you drink a glass of water, you drink about __________ water molecules from a cupful of water once drunk by Julius Caesar." Make an estimate of the appropriate number. Assume that the water he drank was eventually returned to the oceans and that the oceans have been thoroughly mixed, so that every glass of water contains a random sample of water molecules.

12.13 As an electron moves toward a proton, starting a great distance away and ending up

1 Å (10^{-10} m) from the proton, does the electron gain or lose electrical potential energy? How much?

12.14 The energy content of gasoline is given in Appendix H as 5.2×10^6 BTU/barrel and also as 38×10^6 BTU/ton. Use these two figures to determine the density of gasoline, in tons per barrel and in grams per cubic centimeter. We know that gasoline is slightly less dense than water, and therefore if the two given numbers are consistent with one another, the result should be slightly less than 1 g/cm^3.

12.15 A "kiloton" is a unit of energy, defined as equal to 10^{12} cal. This is approximately the chemical energy released in the explosion of 1000 tons of TNT, and it is often used in describing the amounts of energy released in the explosions of nuclear weapons. Compare bread and TNT in terms of the chemical energy available per unit mass.

12.16 In §8.6 batteries were described in terms of their "ampere-hour" capacity. We can now estimate the minimum amount of material needed to make a copper-zinc battery with a capacity of 100 A h, a capacity similar to those of the storage batteries listed in Table 8.2.

(a) Show that the ampere-hour is a unit of *charge*, that 1 A h = 3600 coulombs (C), and that 100 A h = 3.6×10^5 C. This is the total amount of charge that flows through a 100 A h battery as it is being completely discharged.

(b) Observe that in a copper-zinc battery, for every zinc atom which goes into solution, a charge equal to twice the electronic charge is carried from one terminal to the other. Show that in the discharge of a 100 A h copper-zinc battery, approximately 10^{24} zinc atoms are dissolved.

(c) Show that 10^{24} zinc atoms have a mass of about 100 g. (Therefore unless the initial mass of the zinc electrode is at least this large, the electrode would be completely dissolved before the battery could deliver 100 A h.)

12.17 There are two potentially disastrous environmental effects associated with the burning of fossil fuels: consumption of oxygen and production of carbon dioxide. Consumption of a significant fraction of the oxygen in the atmosphere would obviously have an effect on life on earth, and a change in the amount of CO_2 in the atmosphere would alter the earth's radiation balance and hence its average temperature. We can make some rough estimates to see whether these two problems might be important.

(a) For purposes of this calculation, assume that coal is pure carbon. Show that the number of carbon atoms in the world's coal resources (Appendix Table I-1) is about 4×10^{41}.

(b) Using Table E.3, show that the number of O_2 molecules now in the atmosphere is about 2×10^{43} and the number of CO_2 molecules about 3×10^{40}.

(c) The basic chemical reaction occurring in the combustion of carbon is: $C + O_2 \longrightarrow CO_2$. For every atom of carbon burned, one O_2 molecule is used up and one CO_2 molecule is produced. What fraction of our oxygen would be consumed in the combustion of all the world's coal, and by what factor would the amount of CO_2 in the atmosphere be increased? What tentative conclusions can you draw from these results? (Calculations such as those suggested here may give hints as to whether these problems may be serious, but the situation is actually more complicated than it seems at first sight. The mechanisms by which oxygen and CO_2 are exchanged between the atmosphere, the oceans and living organisms are not completely understood. CO_2 molecules produced during combustion do not all remain in the atmosphere; some are dissolved in the oceans and others are used by growing plants, whose growth puts oxygen back into the atmosphere.)

12.18 Several possible explanations of the origin of the sun's energy have been discussed earlier. All those discussed so far failed on simple quantitative grounds: not enough available energy to account for the 10^{44} J of energy the sun has given off. Another possible explanation is that the sun is using up chemical energy, that—as seems obvious if we look at it—fuel of some sort is being burned. The sun is largely composed of hydrogen, and the energy released by burning hydrogen is about 10^8 J/kg. Using the known mass of the sun, show that even if the sun were purely composed of hydrogen, the amount of chemical energy released by burning this much hydrogen would be only about 2×10^{38} J. The suggestion that the sun is simply burning hydrogen is therefore completely inadequate as an explanation of the sun's energy. Even if the numerical calculation had produced a different result, this suggestion would still be subject to the serious objection that there is very little oxygen in the sun, and oxygen is required for the burning of hydrogen.

---

13 Radioactivity and Nuclear Physics

13 Radioactivity and Nuclear Physics

He wanted to ask Mme. Curie to invent a motor attachable to her salt of radium, and pump its forces through it, as Faraday did with a magnet.

—Henry Adams, *The Education of Henry Adams,* 1905

§13.1 THE DISCOVERY OF RADIOACTIVITY

The study of nuclear physics began at the close of the 19th century, and for 50 years appeared to be an important area of science but one with few technological applications. This situation changed suddenly in 1945, with the explosion of the first nuclear bombs, and since the end of World War II vast amounts of money and effort have been devoted to the development of peaceful as well as military applications. In surveying the field of nuclear physics, we will concentrate on those features of most significance from a fundamental point of view and those which are important in applying nuclear physics to provide useful forms of energy. Nuclear physics began with the discovery of radioactivity, and this discovery quickly led to an "energy crisis," in which it seemed for a time that the law of conservation of energy would have to be abandoned. A new form of energy, "nuclear energy,"* must be introduced to describe the ways in which nuclei and the constituents of nuclei interact with one another.

In this chapter we will also examine the meaning and significance of what is surely the most famous equation of all time, $E = mc^2$, an equation with important applications in nuclear physics. Then, in the final

*Nuclear energy is often incorrectly referred to as "atomic energy." It is *nuclear* energy that is released in the operation of so-called "atomic bombs" and "atomic power plants"; they should be called nuclear bombs and nuclear power plants.

two sections, we will examine two important special topics in nuclear physics. The first is "beta decay" (already mentioned in Chapter 6), a subject that is important in the story of energy because it led to the postulated existence of a new kind of particle, the neutrino, whose sole function to begin with was to "save" the law of conservation of energy when it seemed that energy was disappearing. The second topic is the discovery of the neutron. The neutron is a key experimental tool of the nuclear physicist and is essential for nuclear bombs and nuclear power plants. A study of the discovery of the neutron is appropriate because the central ideas are simply those involved in the study of colliding air-track gliders in Chapter 3.

Radioactivity was discovered by accident, early in 1896, by the French physicist Becquerel. Becquerel was beginning a study of a "fluorescence" associated with X-rays, then recently discovered, a phenomenon that has no direct relationship to nuclear physics. The accident was that two items he was using in his studies, some photographic film and some crystals of a particular chemical compound (potassium-uranium-sulfate), were left together in a closed drawer over a weekend. When Becquerel developed the film, he found to his surprise that it was not completely blank, as it should have been. Although the film had been in a dark drawer, and no X-rays had been near it, there were dark patches on the film, as if it had been partially exposed to light. Something, apparently, was coming out of the potassium-uranium-sulfate. What happened next was not accidental at all. A lesser scientist might have thrown out that batch of film and gone back to studying fluorescence and X-rays. Becquerel, though, pursued his chance discovery and initiated the study of radioactivity. It was quickly discovered that the element uranium was the cause of the exposed film. Changing the chemical environment of the uranium atoms by dissolving the crystals in water, by using other compounds containing uranium, or even by using pure uranium metal did not alter the effect of uranium on photographic film. These were strong clues that radioactivity did not arise from the outer orbital electrons of the uranium atom, the electrons responsible for its chemistry. Eventually it became clear that the *nucleus* of the atom was the source of the radioactivity, but in Becquerel's time, it was not yet known that there *was* such a thing as a nucleus.

Within a few years, Pierre and Marie Curie took up the subject of radioactivity and showed that this phenomenon was not unique to uranium, discovering two new chemical elements in the process. First, the Curies found that pitchblende, a uranium ore, was more intensely radioactive than pure uranium itself. Chemical separations were carried out, and when the uranium had been removed, an intense radioactivity was still present. Further chemical separations revealed two distinct radioactive materials, whose chemical behavior was different from uranium and different from that of all previously known elements. Two *new elements* had been discovered, which were christened polonium and radium, and which fitted neatly into existing gaps in the periodic table. These were exciting and completely unanticipated discoveries. Perhaps the single most important event in the early history of nuclear physics was the decision of Ernest Rutherford to enter the new field. Significant contributions were made by many scientists, but it was Rutherford who, by means of his own experiments and ideas and those of his colleagues and students, dominated the field for more than thirty years.

Largely through the work of Rutherford and his collaborators, the nature of the emissions from radioactive substances was determined. It was found that the "rays" from such a substance were usually of three different types, which were given the labels α, β and γ (alpha, beta, and gamma), names that have been retained ever since. The three types of radiation differ greatly in their ability to penetrate other materials. α-rays can be stopped by a single sheet of paper or a few centimeters of air. The penetrating power of β-rays is much greater; typical β-rays can pass through a 1-mm thick piece of aluminum or about 10 ft of air. γ-rays are still more penetrating; to reduce the intensity of a beam of γ-rays by a factor of 10 requires several centimeters of lead or several hundred feet of air. By applying electric or magnetic fields to the rays, it is possible to determine their charge, mass and velocity. Alpha-rays (or α-particles, as they are now called) carry a positive electrical charge, and they behave in the same way as a helium atom that has lost its two electrons, that is, a bare helium nucleus. This identification was confirmed in a beautiful experiment performed with the apparatus shown in Figure 13.1. A source of α-particles was contained in a thin-walled glass tube, surrounded by a larger evacuated tube. The nature of any gas in this surrounding tube could be examined by producing an electrical discharge and studying the spectrum of the emitted light. In the course of several days, the wavelengths characteristic of helium gradually appeared in the spectrum. There could be no doubt that the α-particles were penetrating into the evacuated region, where they quickly picked up two electrons to become atoms of helium. Beta-rays (or β-particles) are deflected in the opposite way by a magnetic field and are much less massive than α-particles; they are, in fact, simply electrons with very large kinetic energies.* Gamma-rays are not deflected at all by electric or magnetic fields; they were eventually identified as photons of electromagnetic radiation, just like photons of visible light but with much greater energies, energies comparable to those of X-rays, and even larger. The energies of α-particles, β-particles, and γ-rays are amazingly large, on the order of a million electron-volts (1 MeV) for each one. Such an energy is about a million times larger than the energies of the electrons

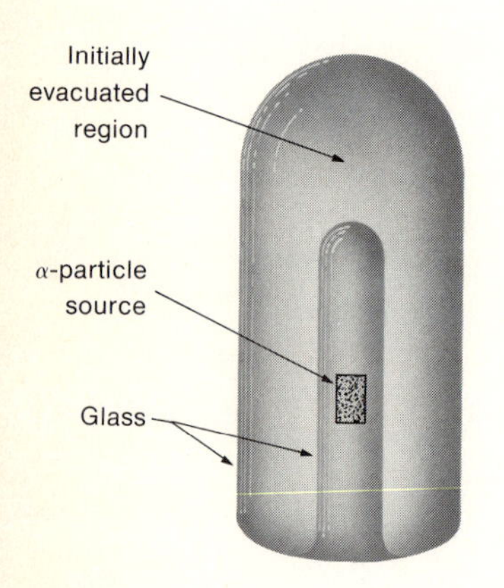

FIGURE 13.1
Apparatus used to show that α-particles are helium nuclei.

*We will denote these electrons by the symbol β^-, the sign indicating their negative charge. The electron has a positive counterpart, the positron (β^+), the electron's "antiparticle," identical to the electron except for the sign of its charge. Positrons are not found in ordinary matter but are produced in cosmic rays and are emitted by some artificially produced radioactive materials. Since β-particles are simply fast electrons, they are sometimes represented simply by the symbol e^-, and positrons by e^+.

in their orbits around an atom, or the energy changes that occur in the chemical reactions of atoms or molecules. This was a clear indication that radioactivity was something new and different.

Needless to say, measurement of the velocities and energies of α-particles, β-particles, and γ-rays—even the simple detection of their presence—is not a simple matter. A great many methods have been developed, and a few of these will be briefly described. The oldest is that used by Becquerel in his initial discovery of radioactivity, a method that depends on the fact that radiation affects photographic films, producing a latent image like that caused by exposure of the film to light. Photographic films give a general indication of the intensity of the radiations and are still used in the "film badges" worn by workers in nuclear installations to monitor potentially dangerous exposures.

Shortly after Becquerel's discovery, it was found that the energetic particles emitted by radioactive substances rip electrons off the atoms of anything through which they pass, leaving in their wake a trail of positively charged ions and detached electrons. This phenomenon makes possible a large number of methods for detection and study of radioactivity, because the ions and electrons carry electric charges and can therefore be readily manipulated by electric fields and collected to produce indications on ammeters and voltmeters. Any one energetic particle can ionize a large number of atoms; since we know about how much energy is needed to ionize a single atom (a few electron-volts or so), the number of ions produced can be used to estimate the number and the energy of the particles causing the ionization. The simplest instrument of this type is the ionization chamber (Figure 13.2). Under normal circumstances, the gas in the chamber is an excellent insulator; no current flows in the circuit. If ionization takes place in the gas, the gas becomes weakly conducting, the circuit is completed by positive ions moving toward the outer wall and electrons moving toward the inner electrode, and the current in the circuit can be measured on the ammeter. The familiar Geiger counter is based on an elaboration of this idea.

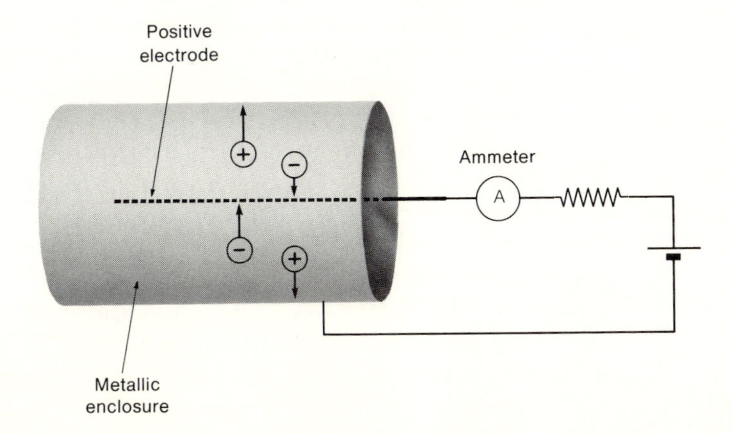

FIGURE 13.2
Schematic diagram of an ionization chamber. Ions and electrons move in opposite directions within the chamber, producing a current in the external circuit.

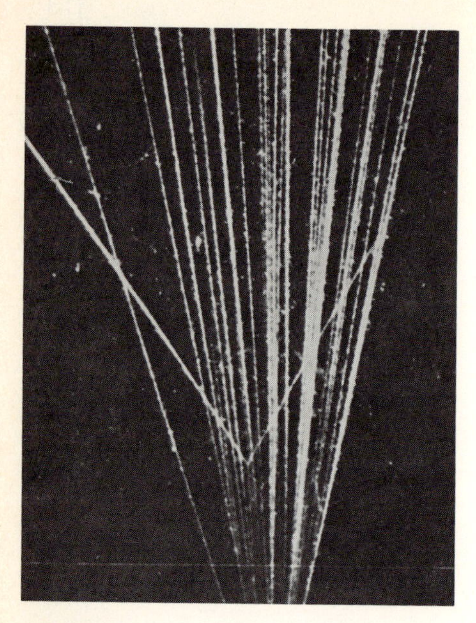

FIGURE 13.3
Cloud-chamber photograph of α-particles; in this photograph one α-particle has collided with a nucleus of one of the atoms of the gas in the cloud chamber. [Courtesy of P. M. S. Blackett.]

Many other types of instruments have been used. A particularly interesting one is the cloud chamber. Meteorologists are familiar with the fact that air may contain so much water vapor that the water is ready to condense into droplets if only something is present to initiate the process. In a cloud chamber the ions left in the path of an energetic particle serve this purpose, enabling us literally to see the tracks of the ionizing particles (Figure 13.3) even though we cannot see the particles themselves.

§13.2 THE "ENERGY CRISIS" OF RADIOACTIVITY AND ITS RESOLUTION. RADIOACTIVE DECAY

Does the discovery of radioactivity require the rejection of the law of conservation of energy? Consider a piece of radium sitting on the laboratory table. There it sits, sending out energetic α-, β-, or γ-rays by the millions, hour after hour, month after month, apparently with no decrease in intensity and no change in the radium itself. Is this a perpetual motion machine of the first kind, the source of an indefinite amount of kinetic energy accompanied by no decrease in any other form of energy? Whether it is a *useful* perpetual motion machine is not the issue; anything that could produce energy forever would present a contradiction to the law of conservation of energy. If a piece of radium is *not* a perpetual motion machine, then where does its energy come from? What is the other form of energy that is diminishing that can account for the kinetic energies of the emitted particles? The discovery of radioactivity posed a strong challenge to the validity of the law of conservation of energy. The claim of the law of conservation of energy is that the total amount of energy is strictly constant; either a loss or a creation of energy would require rejection of the law.

Here was a true *energy crisis*, although it would be another 70 years before this term (with quite a different meaning) would begin to make headlines. If there was in 1900 an energy crisis in the modern sense, it was reflected in a concern about future supplies of wood or coal rather than oil; urban pollution problems were measured in tons of horse dung on city streets rather than in tons of carbon monoxide in the air. If there were any thoughts in 1900 that the energy crisis of radioactivity might have implications for society at large, such thoughts were voiced only rarely and very speculatively. This was an intellectual energy crisis, one to be resolved in the laboratory and in the pages of scientific periodicals. There were a number of possible responses to the challenge which radioactivity posed to the law of conservation of energy. One possibility, which few if any rushed to embrace, was to admit that the law of conservation of energy must indeed be rejected, that radium *was* a perpetual motion machine. Another response was the ingenious suggestion that radium and similar materials somehow had the ability to suck in thermal energy from their surroundings and convert it into the kinetic energy of emitted particles. Such a resolution would have preserved the law of conservation of energy but would have provided damaging evidence against the validity of the second law of thermodynamics; such a conversion of the disorderly thermal energy of the environment into orderly kinetic energy of particles is precisely one of the possibilities denied by the second law. This idea would have been exciting if true, but it was quickly disproved by a new experiment. If the suggestion were correct, then a sample of radium would cool itself and its immediate surroundings by extracting thermal energy; if the radium were surrounded by a container thick enough to absorb all the emitted particles and to convert their kinetic energy back into thermal energy, then no net change in temperature would occur. When the experiment was done, however, it was found that the radioactivity produced a definite *increase* in the temperature of the radium and its surroundings.

Another quite different possibility is worth describing because of the insight it gives into the logical status of the law of conservation of energy. This "resolution" of the energy problem was well described by the French mathematician and philosopher of science, Henri Poincaré:

> It has been supposed that radium is nothing but an intermediary, that it only stores radiations of an unknown nature which fly through space in all directions, passing through all objects—except radium—without any effect either on the objects or on the radiations themselves. Only radium can take a little of this energy and give it to us in various forms. What an advantageous explanation and how convenient! First, it is unprovable and therefore irrefutable. Second, it can account for any exception to [the law of conservation of energy]; it accounts not only for the case of radioactivity but for all the exceptions which future experimenters might accumulate. This new and unknown energy could be used for anything. And now, what have we gained by this stroke? The principle is intact, but what can it be used for? Formerly it allowed us to predict that in such-and-such a situation we could count on a certain amount of energy; it limited us; but now that we have at our disposition this indefinite provision of new energy, we are limited by nothing.*

As Poincaré says, it would be disastrous for the law of conservation of energy if it were necessary to accept such an "explanation." In the act of saving the law, we would have destroyed it by turning it into an empty tautology.

Happily, there is another resolution. It was eventually realized that radium is not a *perpetual* source of energy, that changes do occur in the radium and that the rate at which a piece of radium emits energy slowly but surely decreases. This realization first came from the discovery that at least some forms of radioactivity are only temporary. In particular, the intensity of the radioactivity of a freshly isolated sample of polonium

---

*Henri Poincaré, *La Valeur de la Science* (Paris: Flammarion, 1905). Above translation by Evan Romer.

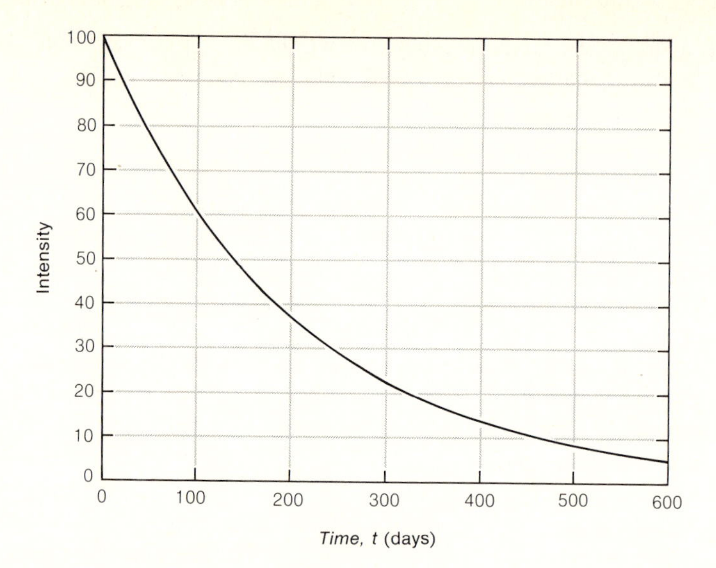

FIGURE 13.4
Decrease in intensity of radioactivity of polonium. (The numbers on the vertical axis give the intensity as a percentage of the intensity at $t = 0$.)

steadily decreases over a period of months (Figure 13.4). Now if one type of radioactivity is temporary, we might suspect that all radioactivities are temporary. Perhaps every radioactive material experiences a change in the process of emitting energetic particles, but the *time scale* may vary from one substance to another. In order to explain the apparently perpetual radioactivity of radium, it was necessary only to hypothesize that the radioactivity of radium decreases extremely slowly. Eventually it was learned that the radioactivity of radium persists for several thousand years; no wonder that any change occurring in the course of a year or two was not noticed. It is still necessary to ask where the energy comes from, what *other* form of energy decreases when a nucleus emits an energetic particle. It is with these subjects that this chapter is largely concerned, but the real "energy crisis" of radioactivity was resolved by the realization that radioactivity, mysterious though it may be, is not perpetual.

Before considering these questions, let us examine more carefully the process of radioactive decay. The experimental facts can be quickly summarized. The rate at which a material emits radiation decreases with time, rapidly at first, then more and more slowly, as shown in Figure 13.4. The radioactive nuclei (polonium nuclei in this example) are apparently changing in some way; a nucleus disintegrates, emitting an α-particle, leaving behind a nucleus which is not radioactive. If we assume that the rate at which a sample emits α-particles is proportional to the amount of material that is still radioactive, this means that the amount of radioactive material, the number of nuclei that have not yet changed, is decreasing in similar fashion. All radioactive materials behave in the same general way, but the time scale varies from one case to another.

There is a curious feature of the experimental results shown in Figure 13.4. After approximately 140 days, half of the original material is still radioactive; after another 140 days, half of that amount remains, that is, one-fourth of the original radioactive material is left after a total of 280 days has elapsed. And so it continues: one-eighth of the original amount remains after 420 days, one-sixteenth after 560 days, and so on. This situation is almost exactly like the exponential growth discussed in Chapter 1, but here it is exponential *decay*. In any given period of time, the same *percentage* decrease takes place. Just as it is useful to describe the growth of an exponentially increasing quantity in terms of its doubling time, it is convenient to refer to the "halving time" of radioactive decay, or, as it is more commonly called, the *half-life*. The half-life of polonium is 140 days. Every pure radioactive material has its own half-life. Half-lives that have been measured (some of them indirectly) range from small fractions of a second to more than 10^{15} years, far greater than the age of the earth. The half-life of the Curies' radium is approximately 1600 years, that of Becquerel's uranium 4.5×10^9 years.

The fact that the number of surviving nuclei (the number which have not yet "died") decreases exponentially with time, that the decay can be described by a half-life, is really extremely strange. Most things that die (literally or figuratively) do not behave this way at all. Automobiles do not behave this way; neither do people. There is no such thing as a "half-life" for a group of automobiles or people.

QUESTIONS

13.1 Make rough sketches with an approximate indication of the time scale showing the fraction of survivors as a function of time for:

(a) a collection of automobiles that came off the assembly line at $t = 0$;

(b) a collection of people, all born at $t = 0$.

13.2 Contrast the shapes of these curves with Figure 13.4. For a group of people, for instance, why cannot one define a "half-life"? (At what time will half of the original number be left? How many will be left after another equal interval of time?)

13.3 Although a person has no half-life, a newborn baby does have a "life expectancy," about 70 years for a 20th-century American baby. What is the difference between a half-life and a life expectancy? Does a polonium nucleus have a life expectancy?

The only successful explanation of the fact that radioactive decay is exponential, that is, that it can be described by a half-life, is the following. Radioactive disintegration is an all-or-nothing event, and the time at which it occurs for any individual nucleus is governed by *chance*, independently of whether or not its neighbors have disintegrated. For each nucleus having the potential for disintegration, there is a certain probability that it will do so in the next second or minute or year. How large this probability is depends on what the radioactive material is, but there is no way to tell whether any one nucleus is on the verge of disintegration. A nucleus has no "memory"; it does not know how long it has existed in its unstable state. There is no such thing as a "middle-aged nucleus"; a brand new nucleus is just as likely to disintegrate as an old one.

This description enables us to make various predictions about radioactive decay. Each nucleus has a certain probability of disintegrating in a particular interval of time; for a polonium nucleus, the chance is 50% that this will occur during the next 140 days. If we have a large collection of such nuclei, we can predict that almost exactly half of them will disintegrate in the next 140 days. Those which are left after the first 140 days are still as good as new; approximately half of them will disintegrate during the subsequent 140 days, and this is the observed behavior. However, if the disintegration of each individual nucleus is strictly a chance phenomenon, there is a possibility that something odd will happen. A collection of 10^{10} polonium nuclei, for example, might *all* disintegrate during the first day. This phenomenon has never been observed, but our failure to observe it is not surprising; such bizarre behavior is, in principle, not impossible, but it is considerably less probable than is a sequence of one million consecutive heads in the tossing of a coin.* When applied to *large* numbers of things, predictions based on probability can be made with an extremely high degree of confidence as we saw in Chapter 7. This is not true when the number of objects is small; no one is especially astonished to flip a coin three times and get three heads. Suppose we have a collection of just 10 polonium nuclei. How many will be left after 140 days? Five? Perhaps, but other results would not be surprising. How many will be left after another 140 days? Two and a half? Definitely not; two perhaps, or three, but not two and a half. In fact it would not be especially surprising to discover that all ten nuclei lasted longer than 280 days or that all ten disintegrated during the first 140 days. A decay curve such as that shown in Figure 13.4 must describe the average behavior, that expected when the number of nuclei is large, but when we have only a small number, we must expect some "statistical fluctuations." It would take us too far afield to consider the details; suffice it to say that calculated statistical fluctuations, based on the probabilistic description given here, are in excellent agreement with observations, thus lending additional support to the description.

A piece of radioactive material is truly an amazing thing. Take a sample of uranium, for example. At some time in the distant past, those uranium nuclei were somehow formed, and almost immediately a few

*The use of the concept of probability in describing the tosses of a coin is necessitated primarily by our ignorance of factors such as the exact value of the speed imparted to the coin as we toss it. By contrast, we have every reason to believe that the disintegration of a nucleus is a quantum-mechanical event and is governed only by chance, that there is no conceivable knowledge we could have that would permit us to make more than a probabilistic prediction.

of them disintegrated. Others survived until the time the pyramids were built, others survived until 1896 and conveniently disintegrated in Becquerel's laboratory drawer. Others will be as good as new hundreds of millions of years from now. No one, not even Laplace's "demon," could have predicted which nuclei would survive until 1896, only that approximately a certain fraction would probably survive that long and that approximately a certain fraction of those would disintegrate during that particular weekend.

§13.3 THE NUCLEUS AND ITS STRUCTURE. TRANSMUTATION OF THE ELEMENTS IN RADIOACTIVE DECAY

Reference has been made on several occasions to the modern view of the atom as consisting of a tiny positively charged central nucleus, containing most of the mass of the atom, surrounded by one or more orbital electrons, whose orbits extend to distances of 1 Å or so. The electrons determine the size of the atom and its chemical properties. The fundamental experiment supporting this view was carried out by Rutherford and his collaborators, who directed a beam of 9-MeV α-particles from a radioactive source at a thin foil of gold and observed the way in which the α-particles were scattered (Figure 13.5). Most of the α-particles go right through the foil with little or no deflection, as if it were transparent. Some are deflected, a few are deflected through quite large angles, and some come almost straight back toward the source (Figure 13.6).

With information about the numbers of α-particles that are scattered through various angles in going through a gold foil, one can try out various models of the structure of the atom and see which is most successful at explaining the experimental results. The model that emerged from these considerations is the following. In the center of a gold atom

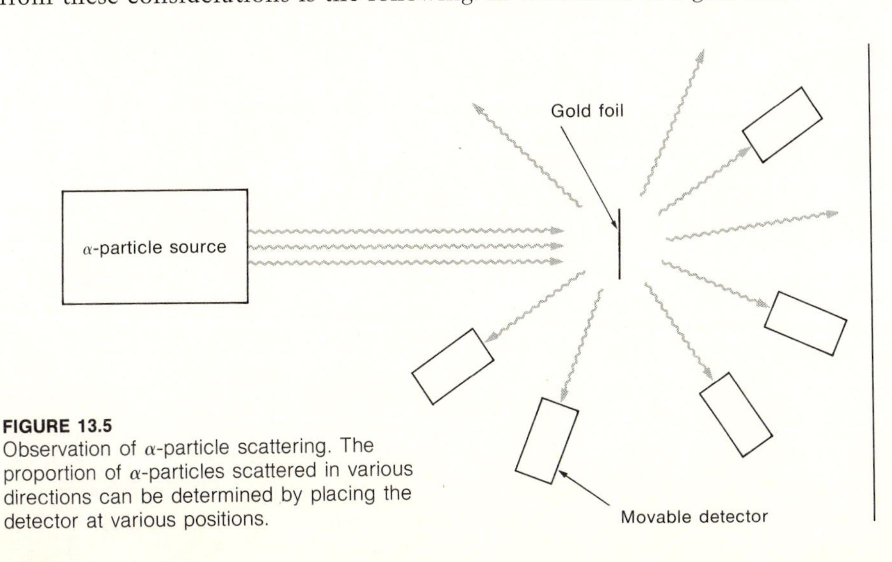

FIGURE 13.5
Observation of α-particle scattering. The proportion of α-particles scattered in various directions can be determined by placing the detector at various positions.

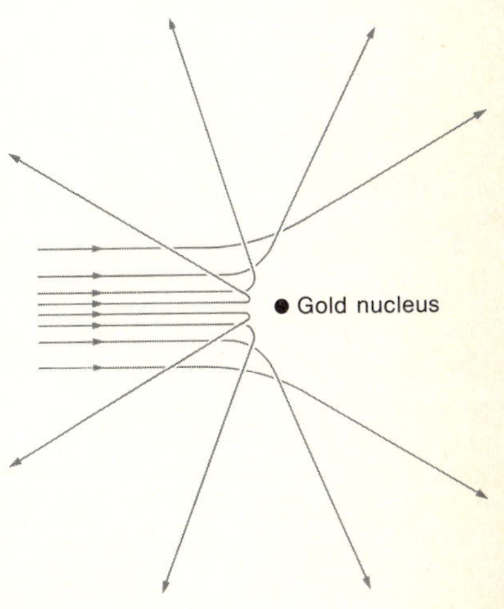

FIGURE 13.6
Trajectories of α-particles deflected by a gold nucleus.

sits a massive nucleus carrying a positive charge, a nucleus so small that it acts like a *point* charge in these experiments. The positive charge on the gold nucleus is 79 times that of the electronic charge; if the gold atom is an electrically neutral one, the nucleus is surrounded by a cloud of exactly 79 electrons. An α-particle entering the cloud of electrons is so much more massive than an electron that it is scarcely deflected at all. It plows through the electrons like a rhinoceros charging through a cloud of flies. As the α-particle nears the center of the atom, there sits the nucleus, exerting a repulsive electrical force on the α-particle. The nucleus is not so easily brushed aside; in fact it is 50 times heavier than the α-particle and scarcely moves at all, but the α-particle *will* be deflected. The angle through which it will be deflected by the nucleus is determined by how close to the nucleus it comes (Figure 13.6). If its trajectory is one that would cause it to miss the nucleus by a wide margin, the force on it will never be large and it will be only slightly deflected. If it happens to be heading almost directly at the nucleus, it will be deflected through a larger angle. If it is headed straight at the nucleus, it will slow down (losing kinetic energy as its electrical potential energy increases) and then stop, turn around, and gain speed again. The detailed analysis is complicated, but the treatment of the nucleus as a point charge exerting an inverse square repulsive force on the α-particle gives predictions precisely in accord with the experimental results.

We cannot conclude that the nucleus is a "mathematical point." All we can conclude from these experiments is that since the nucleus appears to behave like a point charge in these experiments, then whatever its radius may be, none of the α-particles got closer than this to the center of the nucleus. Similarly, orbits of earth satellites are consistent with the hypothesis that the earth is a "point mass." If the orbit of a particular satellite is such that it is always at least 5000 miles from the center of the earth, we could conclude from observing it that the radius of the earth is less than or equal to 5000 miles.

QUESTION

13.4 The α-particles that approach most closely are those that happen to be heading directly toward a nucleus. For these α-particles, simple energy conservation arguments allow us to estimate the distance at which they will stop and turn around. Show that for 9-MeV α-particles projected at gold nuclei, this distance is approximately 2.5×10^{-14} m and therefore that the radius of a gold nucleus is less than or equal to this value.

Most of the mass of an atom is concentrated in the nucleus, which is composed of protons and neutrons. The mass of both the proton and the neutron is about one atomic mass unit ($1 \text{ amu} = 1.66 \times 10^{-27}$ kg). Except for the very important difference that protons have electrical charge and neutrons do not, protons and neutrons are quite similar to one another; the term *nucleon* refers indiscriminately to protons and neutrons. Any particular kind of nucleus can be described by giving the number of neutrons and the number of protons it contains. The proton number, also called the atomic number, is denoted by the symbol Z, and the neutron number by N. Their sum is called the nucleon number and denoted by the symbol A ($A = Z + N$); the nucleon number is also called the *mass* number, since the mass of a nucleus measured in amu is approximately equal to the total number of nucleons in the nucleus.

The electrical charge on the nucleus is contributed by the protons, each carrying one elementary unit of charge (one electronic charge = $e = 1.6 \times 10^{-19}$ coulombs). An electrically neutral atom has as many electrons around the nucleus as it has protons within the nucleus; the proton number, Z, is therefore equal to the number of orbital electrons for a neutral atom. Since the chemistry of an atom is determined by the number of electrons, all nuclei with a particular value of Z are nuclei of the same chemical element. An oxygen nucleus, for instance, has 8 protons ($Z = 8$). It is the eighth element in the periodic table, and a neutral oxygen atom has 8 electrons around its nucleus. There are several kinds of oxygen nuclei, however, with differing numbers of neutrons, several different oxygen *isotopes*. The most common oxygen isotope has 8 neutrons ($N = 8$, $A = 16$); in an ordinary sample of oxygen, 99.76% of the oxygen nuclei are of this type. (The "natural abundance" of this isotope is 99.76%.) There are also two heavier oxygen isotopes, one with $N = 9$ and $A = 17$ (natural abundance 0.04%) and the other with $N = 10$ and $A = 18$ (natural abundance 0.2%). Only these three oxygen isotopes occur naturally, but there are others that have been artificially produced, isotopes that are radioactive and have quite short half-lives.

A convenient shorthand notation for describing any species of nucleus can be illustrated with the example of the heaviest of the three naturally occurring oxygen isotopes:

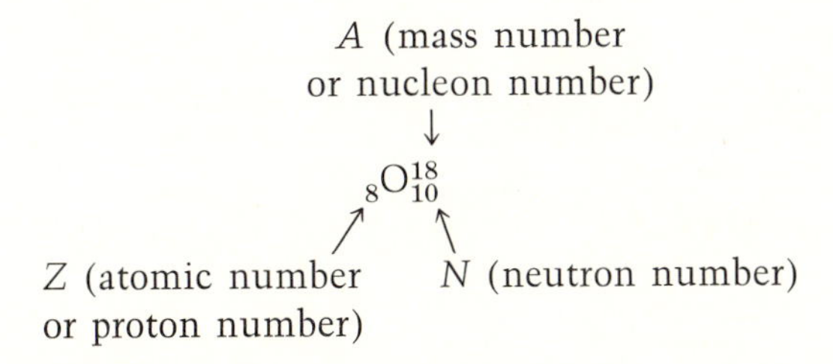

Often this is abbreviated by writing simply $_8O^{18}$ or just O^{18}. The notation O^{18} really tells us all we need to know. If it is oxygen, then Z must be 8, and if $A = 18$ and $Z = 8$, then N must be 10.

Consider a few more examples of this sort of notation. The proton itself is the nucleus of the most common type of hydrogen atom ($_1H^1_0$). There are also two heavier hydrogen isotopes, which are frequently referred to by their own special names (deuterium, D, and tritium, T), even though they are isotopes of hydrogen. Deuterium has one neutron in its nucleus in addition to the proton ($_1H_1^2$ or sometimes $_1D_1^2$). This isotope of hydrogen occurs naturally; of the hydrogen atoms found in nature, one out of every 6500 is of this type. The nucleus of tritium contains two neutrons ($_1H_2^3$ or $_1T_2^3$); tritium is radioactive with a half-life of 12 years. The nucleus of the common isotope of helium is the α-particle, consisting of two neutrons and two protons ($_2He_2^4$). Near the other end of the periodic table we find uranium ($Z = 92$). Most uranium nuclei have a nucleon number of 238 (U^{238} or $_{92}U_{146}^{238}$); there is also a light isotope whose nucleon number is 235 (U^{235} or $_{92}U_{143}^{235}$). It is relatively rare (natural abundance 0.7%), but it is a nuclear material that provides the energy for some nuclear weapons and power plants.

A word of caution about notation is necessary. Does a symbol such as O^{18} stand for a bare nucleus or for a whole atom? It will be used with either meaning as the occasion requires. Sometimes it is not important to be more specific, but frequently it does matter, and then care must be taken to determine which is meant. An O^{18} nucleus and an O^{18} atom differ dramatically in their net electrical charges; their *masses* are also different by an amount equal to eight electron-masses, and though this is not a large percentage difference, even such a small difference can be extremely significant as we will see later. A few particular nuclei are given their own special names and symbols. Thus p denotes a proton, the nuclei of deuterium and tritium are called the deuteron (d) and the triton (t), and the α-particle can be simply represented by the symbol α. Likewise the neutron will be denoted by the symbol n.

How large is the nucleus? From an analysis of Rutherford's scattering experiments (Question 13.4), we concluded that the radius of a gold nucleus is less than 2.5×10^{-14} m, but no more definite conclusions could be drawn. In similar experiments with α-particles of higher energy, some of the α-particles approach the nucleus more closely, and in these experiments the nucleus does *not* act quite like a point charge. From these and related experiments it is possible to estimate the sizes of nuclei. It appears that the volumes of nuclei are approximately proportional to the total number of nucleons (A), as one might have anticipated.* The radii of various nuclei are thus proportional to the cube root of A, $A^{1/3}$, and the numerical values can be expressed by the approximate formula:

$$r \simeq r_0 A^{1/3},$$

where $r_0 = 1.4 \times 10^{-15}$ m. Even for the heaviest nuclei ($A \simeq 240$), the radius is only about 10^{-14} m. Since the electron orbits extend to distances of about 1 Å (10^{-10} m), the atom is a very strange object, almost completely empty except for a few lonely electrons circling around a tiny massive nucleus.

---

*However, as we saw in Chapter 12, the volumes of *atoms* are not proportional to the number of electrons.

QUESTIONS

13.5 Imagine a scale model of a hydrogen atom, with its nucleus (the proton) represented by a sphere whose size and mass is that of a baseball. How massive an object should be used to represent the electron, and how far away should it be placed?

13.6 Estimate the density of a nucleus. How do nuclear densities compare with those of ordinary substances such as water?

What happens to a nucleus when it disintegrates? There is an "obvious" answer to this question, an answer that turns out to be the correct one, though as with many things which seem obvious in retrospect, it was not at all clear during the first hectic years of the investigation of radioactivity. Consider first the process of α-decay. An α-particle has four nucleons and carries two units of positive electrical charge. If an α-particle comes out of a nucleus, the nucleus left behind has four fewer nucleons and two units less of electrical charge. That is, A (the nucleon number) decreases by four and Z decreases by two. But if Z has changed, the new nucleus is a nucleus of a *different chemical element!* The ancient dream of the alchemists, the transmutation of one element into another, is going on under our very noses in any radioactive material.* Take a specific example, ${}_{84}Po^{210}$, which emits an α-particle when it decays:

$${}_{84}Po^{210} \longrightarrow X + \alpha$$

or

$${}_{84}Po^{210} \longrightarrow X + {}_{2}He^{4}. \tag{13.1}$$

What is X, the new nucleus? It must have a nucleon number, A, of 206 and a proton number, Z, of 82. That is, it is two places to the left of polonium in the periodic table; it must be a nucleus of the element lead (Pb), and it is the isotope whose nucleon number is 206. Thus Equation 13.1 can be written:

$${}_{84}Po^{210} \longrightarrow {}_{82}Pb^{206} + {}_{2}He^{4}.$$

The Po^{210} nucleus in this example is referred to as the "parent" nucleus and Pb^{206} as the "daughter." Notice that the total number of nucleons as well as the total electrical charge is the same before and after the disintegration ($206 + 4 = 210$ and $82 + 2 = 84$).

As the nucleus disintegrates, what happens to its orbital electrons? If the original ${}_{84}Po^{210}$ nucleus was the nucleus of an electrically neutral

*Admittedly we have no control over the actual process of radioactive decay. It is now possible to produce artificial transmutations on a limited scale. We will see in Chapter 14 how we can turn gold into lead, for example, a process that was not quite the one the alchemists had in mind.

atom, it had 84 orbital electrons. As it disintegrates, the α-particle comes out with high energy, carrying no electrons with it. The 84 electrons around the nucleus are two more than a lead nucleus ought to have; it is left as a doubly charged negative ion, but the two extra electrons will probably be quickly lost, leaving a neutral lead atom. The changes in energy that result from the rearrangements of electron orbits around the nucleus or from the loss of the two extra electrons are typically an electron-volt or so in magnitude, insignificant in comparison to the kinetic energy released in a nuclear disintegration, which is generally about a million times larger (5.4 MeV in this case).

What happens during β-decay? Again, the straightforward answer is the correct one, though the details are more complicated. A β-particle (an electron, β^-) has a mass of only about $\frac{1}{1822}$ of an amu; its emission will not alter the nucleon number, A. It does, though, carry one unit of negative charge, and therefore the nucleus left behind must have one unit more of *positive* charge. It must be a nucleus of a different chemical element, but this time an element that is one place *higher up* in the periodic table. For example, the β-decay of strontium-90 (${}_{38}Sr^{90}$) leads to an isotope of the chemical element whose proton number is 39 (yttrium) and whose nucleon number is 90, the same as for the original nucleus:

$${}_{38}Sr^{90} \longrightarrow {}_{39}Y^{90} + \beta^-.$$

The total number of nucleons in the daughter nucleus is 90, just as it is for the parent nucleus, but whereas ${}_{38}Sr^{90}$ contains 38 protons and 52 neutrons, ${}_{39}Y^{90}$ has 39 protons and 51 neutrons. In effect, one neutron has turned into a proton, with the emission of an electron. That is, the fundamental process of β-decay is just:

$$n \longrightarrow p + \beta^-.$$

It was not until 1932 that neutrons by themselves—outside the nucleus—were first identified. Before that time it had been speculated that nuclei might be composed of protons and electrons, that an O^{18} nucleus, for instance, might contain 18 protons (to provide the mass) together with 10 electrons within the nucleus (to make the net nuclear charge equal to +8 elementary units). This hypothesis is supported by the fact that electrons are emitted from the nucleus in β-decay, but there are other facts that are much more readily understood if the nucleus is composed of protons and neutrons. We now regard the electrons emitted in β-decay as being created at the instant of disintegration. One might observe that *photons* do not exist inside atoms, yet photons come out of atoms when their electrons are rearranged, and that *sounds* come from a person's mouth that did not previously exist. Photons and sounds, like the electrons emitted in β-decay, are created at the moment of emission.

Most γ-ray emissions follow immediately after emission of an α-particle or a β-particle. Gamma-ray emission changes neither A nor Z; it results from internal rearrangements of the nucleons in the daughter nucleus, changes in their energy levels, in much the same way that emission of a photon by an atom results from rearrangements of the orbital electrons.

It is easy now to say that these processes of transmutation must be going on when nuclei emit α-particles or β-particles. When this idea was first proposed, it was so revolutionary that considerable courage was required to put it forward. If this interpretation is accepted, we can trace what is going on in a whole series of radioactive decays. Some nuclear disintegrations produce daughter nuclei which are stable. In other cases, the daughter nucleus itself is radioactive; if so, there will be a subsequent disintegration, and perhaps a whole chain of disintegrations, one after another. One such chain, which includes the isotope of radium discovered by the Curies, begins with U^{238}, which decays by α-emission with a half-life of 4.5×10^9 years. This leads to an isotope of the element thorium:

$$_{92}U^{238} \xrightarrow{4.5 \times 10^9 \text{ years}} {}_{90}Th^{234} + \alpha.$$

This isotope in turn decays by β-decay with a 24-day half-life:

$$_{90}Th^{234} \xrightarrow{24 \text{ days}} {}_{91}Pa^{234} + \beta^-.$$

A long series of transformations follows, ending with a stable isotope of lead (Pb^{206}).

A convenient way of representing nuclear decays is shown in Figure 13.7. Any nucleus can be located on such a diagram by two coordinates, its neutron number, N, and its proton number, Z. In α-decay, N and Z both decrease by two, so an α-decay is represented by a displacement two units to the left and two units down. In β-decay, Z increases by one unit while N decreases by one, so β-decay is represented by a shorter displacement, this time downward and to the right. The whole series of decays, which begins with U^{238}, is shown in Figure 13.8. At several points, *two* possible paths are indicated. $_{84}Po^{218}$, for example, sometimes emits an α-particle, turning into $_{82}Pb^{214}$, a decay that is followed by β-emission leading to $_{83}Bi^{214}$. Sometimes the β-emission occurs first and then the α-emission; again the result is $_{83}Bi^{214}$, but the intermediate nucleus is $_{85}At^{218}$. Notice that all points on the same vertical line have the same value of Z and are thus isotopes of the same chemical element. Likewise, all points that lie on a diagonal line sloping downward to the right have the same nucleon number, A.

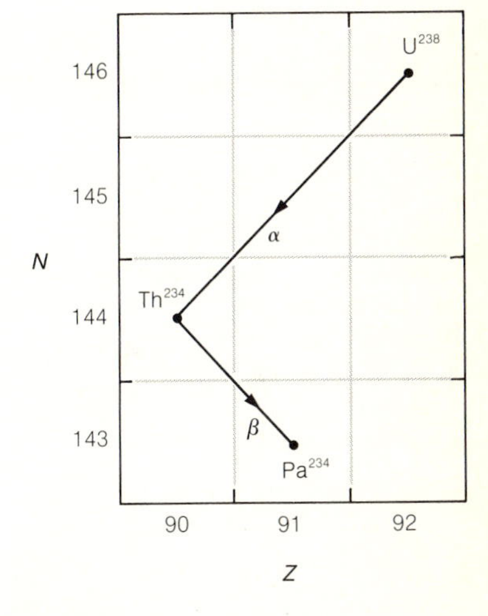

FIGURE 13.7
A graphical method of showing nuclear transformations.

The isotope U^{238} has a half-life of 4.5×10^9 years, comparable to the age of the earth. None of the other half-lives in this chain of decays is anywhere near this long. Nearly all the original representatives of other species of nuclei in this chain disintegrated long ago; these nuclei are found in nature only because new ones are constantly created by the decays of preceding members of the sequence. All of the decays in the sequence discussed are either α-decays (with a change of four units in

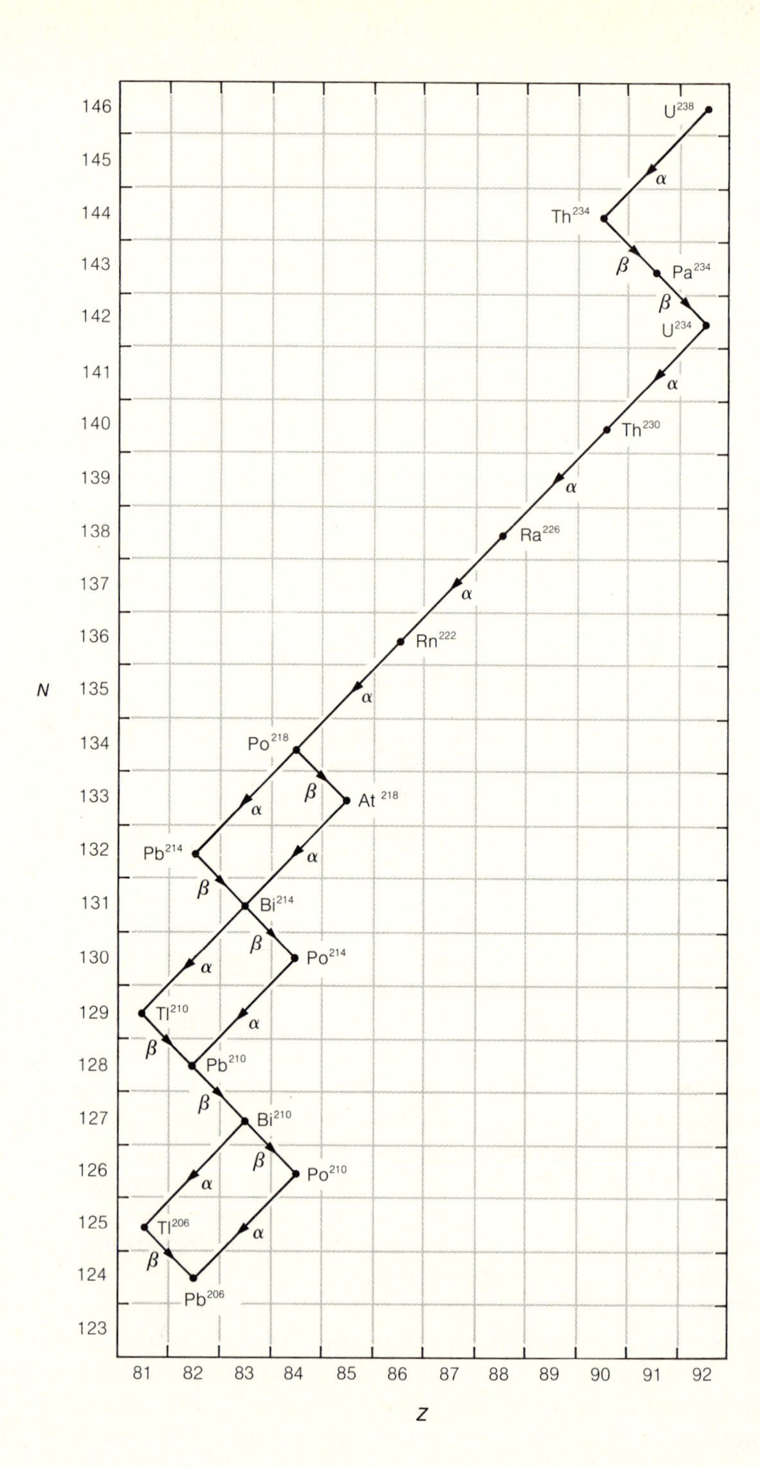

FIGURE 13.8
The chain of decays from U^{238} to Pb^{206}.

A) or β-decays (with no change in A). Thus, beginning with $A = 238$, the only nucleon numbers appearing are 238, 234, 230 . . . 206. There are three similar decay chains terminating in three other stable isotopes: Pb^{207}, Pb^{208} and Bi^{209}. Of these, the one that ends with Bi^{209} contains no nuclei with half-lives longer than 2×10^6 years; this is much less than the age of the earth, and the members of this chain are not found in nature but have been artifically produced. Of the other two, one begins with U^{235}, with a half-life of 7×10^8 years, and the other with Th^{232} whose half-life is 1.4×10^{10} years. (The half-life of U^{235} is long but significantly less than the age of the earth; a billion years ago, U^{235} was much more plentiful than it is now.) A uranium ore deposit consists of a complex mixture of U^{238} and its daughters, granddaughters, and so on, together with U^{235} and all of its descendants. It is only because of the accidental fact that the half-lives of U^{238}, U^{235} and Th^{232} are extremely long that any significant number of radioactive nuclei still exist on earth. If these half-lives had happened to be 1000 times shorter, Becquerel would not have discovered radioactivity, the history of 20th-century science would have been very different, and nuclear fission weapons and nuclear fission power plants would have been produced much later in our history if at all.

QUESTION

13.7 In the decay chain that begins with $_{92}U^{235}$ and ends at $_{82}Pb^{207}$, the types of decay in the order in which they occur are:

$$U^{235} \xrightarrow{\alpha} X \xrightarrow{\beta} X \xrightarrow{\alpha} X \begin{array}{c} \nearrow^{\beta}\ X\ \searrow^{\alpha} \\ \searrow_{\alpha}\ X\ \nearrow_{\beta} \end{array} X \xrightarrow{\alpha} X \xrightarrow{\alpha} X \begin{array}{c} \nearrow^{\beta}\ X\ \searrow^{\alpha} \\ \searrow_{\alpha}\ X\ \nearrow_{\beta} \end{array} X \begin{array}{c} \nearrow^{\beta}\ X\ \searrow^{\alpha} \\ \searrow_{\alpha}\ X\ \nearrow_{\beta} \end{array} Pb^{207}$$

where the X symbols represent the intermediate nuclei. At several points, two possible paths exist as in the chain discussed earlier. Plot this decay chain on a chart like that shown in Figure 13.8, and identify the various species (chemical element and nucleon number) found in this chain.

---

§13.4 NUCLEAR ENERGY. STABLE AND UNSTABLE NUCLEI

What is the source of energy in radioactive decay? A radioactive substance is not a perpetual motion machine. It does run down, though perhaps not for millions of years, but it is still important to ask, about an α-decay for instance: "Where does the kinetic energy of the α-particle come from?" What other form of energy decreases which can account for the kinetic energy of the α-particle? Sometimes it is helpful to attack a problem from the opposite direction. Consider the following simple fact. *Not all* nuclei are radioactive; there are many *stable* nuclei that do not undergo any sort of spontaneous radioactive decay. This may seem like a very obvious remark, until we realize that on the basis of the interactions we have studied in previous chapters, it ought to be false. Consider, for instance, the stable gold nucleus, $_{79}Au^{197}$, with 79 protons and 118 neutrons. The protons all carry positive charge, and thus there is a strong electrical repulsion between them, an extremely strong repulsion at these short distances. The gold nucleus ought to blow itself apart in an instant. Since it does not, we conclude that there is some kind of attractive interaction between nucleons, a *nuclear* interaction, which is important at very short distances and which must be strong enough to overcome the electrical repulsion and to hold the nucleus together.

We gain valuable insight into the strength of this nuclear interaction by asking what the energy of a nucleus would be if there were no nuclear interaction, if the electrical interaction were the only interaction present, and if we could somehow nevertheless construct such a nucleus. It takes energy to push protons together because of their electrical repulsion. If we define the electrical potential energy of a collection of protons to be zero when they are far apart, then the protons in a $_{79}Au^{197}$ nucleus contain a large positive amount of electrical potential energy, ready to be released if they are allowed to fly apart. How *much* electrical potential energy? We have already done an analogous calculation, the results of which we can use to answer this question. Gravitational forces follow an inverse square law just as do electrical forces. In §5.7, in considering whether the energy emitted by the sun might be accounted for by the gravitational potential energy lost as the bits and pieces of the sun came together, we concluded that the gravitational potential energy of the sun itself was approximately:

$$GPE \text{ of sun} \simeq -\frac{GM_{\text{sun}}^2}{R_{\text{sun}}}. \tag{13.2}$$

With a few changes, this formula can be adapted to our present need. The negative sign in Equation 13.2 arose from the fact that the gravitational interaction is always attractive, but the electrical force between positive charges is repulsive, and the sign must therefore be changed. Instead of the combination Gm_1m_2 appearing in the gravitational force law, in the electrical force law we have $k_eq_1q_2$, with k_e the Coulomb's law constant, 9×10^9 in MKS units. Thus we can modify Equation 13.2 to estimate the electrical potential energy of the protons in a nucleus as

$$EPE \simeq \frac{k_eq_{\text{nucleus}}^2}{R_{\text{nucleus}}}. \tag{13.3}$$

QUESTION

13.8 Show that, according to this estimate, the electrical potential energy of the protons in a $_{79}Au^{197}$ nucleus is approximately 1100 MeV. (Be careful with the units. With q in coulombs and R in meters, Equation 13.3 will give a correct result in joules, which can then be converted into MeV.)

Since stable nuclei do not blow themselves apart, there must be a strong attractive nuclear interaction between nucleons. There is a nuclear potential energy associated with this interaction; just as the gravitational potential energy of two attracting objects is negative ($-Gm_1m_2/r$) if the zero-level is chosen to be at infinity, the nuclear potential energy must also be negative for two nucleons that are close together. This attractive nuclear interaction must make a negative contribution to the energy of a $_{79}Au^{197}$ nucleus of at least 1100 MeV.

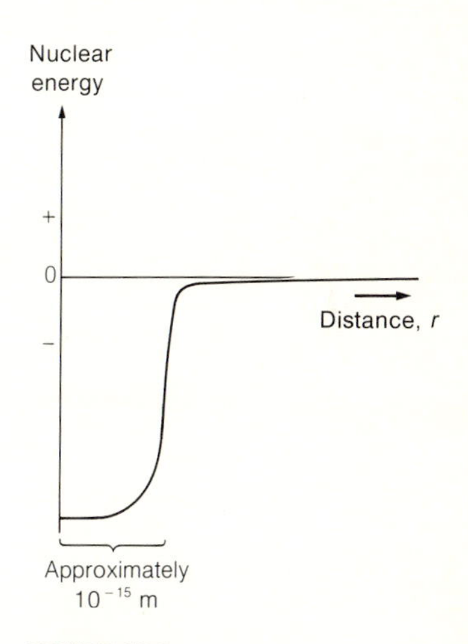

FIGURE 13.9
Variation of nuclear energy as a function of distance, r, between two nucleons.

This energy, 1100 MeV, is fantastically large. This is an energy associated with a single nucleus. It is 10^9 times larger than the energy changes in typical chemical reactions. We know that at distances greater than 10^{-14} m or so, protons do repel one another in accord with Coulomb's law, and so the nuclear interaction must be one that is important only at short distances but is very strong when it does come into play. Thus the nuclear interaction, one that presumably acts between neutrons and neutrons, and between neutrons and protons, as well as between protons and protons, must give rise to a nuclear potential energy that varies with the separation between two nucleons somewhat as shown in Figure 13.9. This is the only kind of interaction between two neutrons or between a proton and a neutron. For two protons, the attractive nuclear interaction together with the repulsive electrical interaction gives rise to an overall energy of interaction between two protons that varies with their separation as indicated in Figure 13.10.

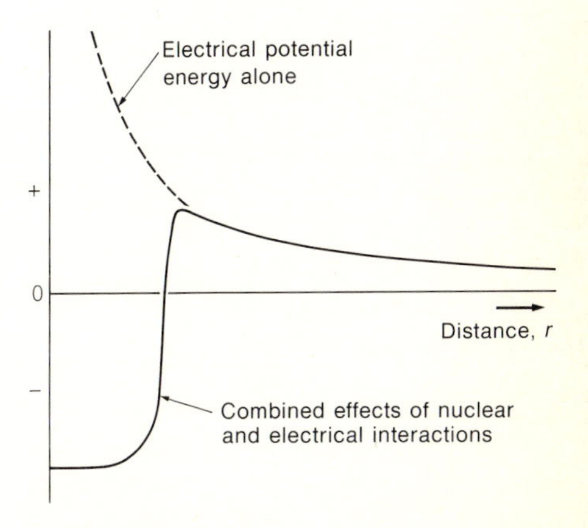

FIGURE 13.10
Combined effects of the electrical repulsion and the nuclear attraction of two protons.

Although many stable nuclei exist, many more nuclei that are unstable are known, not only the naturally occurring radioactive nuclei but many others that have been artifically produced. We can get additional clues about the nature of the nuclear interaction just by looking at the relative numbers of neutrons and protons in stable nuclei and comparing these numbers with similar data about unstable nuclei. Examination of a list of nuclei reveals that stable nuclei tend to have roughly equal numbers of neutrons and protons, that is, equal values of N and Z. This rule of thumb is followed particularly closely among the lightest nuclei. Lithium ($Z = 3$) has two stable isotopes with $N = 3$ and 4. Argon ($Z = 18$) has three stable isotopes with $N = 18$, 20, and 22. As we examine heavier nuclei, we find deviations from this rule. Stable isotopes of heavier elements tend to have a somewhat larger number of neutrons than protons. Mercury ($Z = 80$) has seven stable isotopes with $N = 116$, 118, 119, 120, 121, 122, and 124. We can see this trend

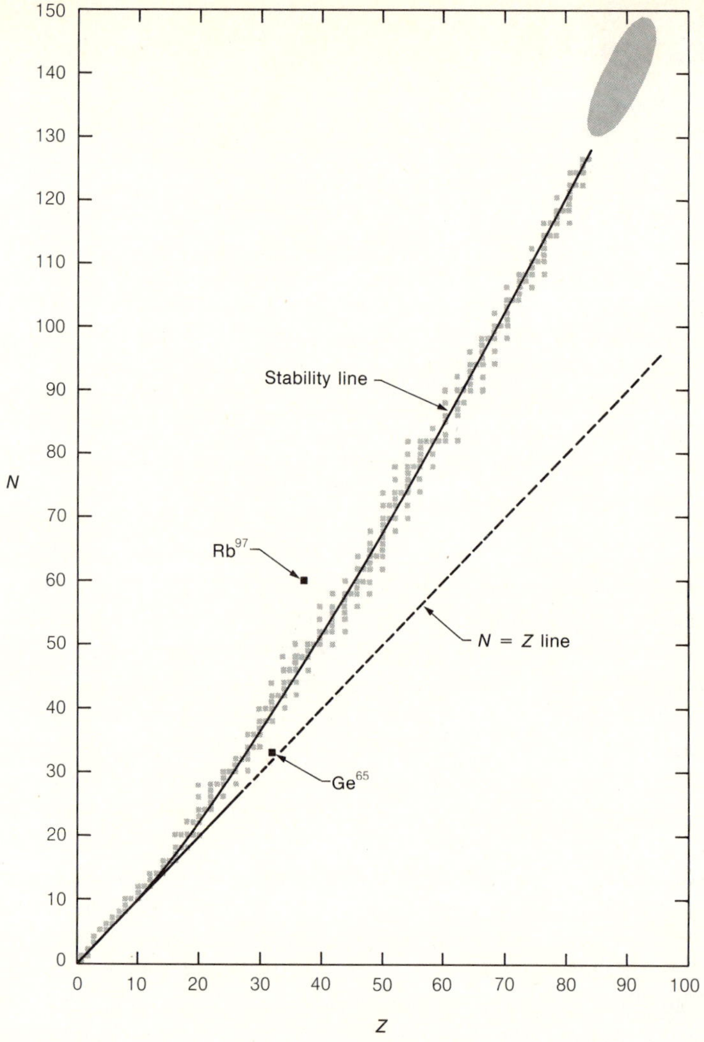

FIGURE 13.11
The stable nuclei. Two unstable nuclei, Rb^{97} and Ge^{65}, are also shown. Heavy radioactive nuclei found in nature such as U^{238} lie in the shaded region.

by locating all the stable nuclei on a graph (Figure 13.11) similar to Figure 13.8. The stable nuclei are clustered along a fairly well-defined line, the "stability line." For the lighter stable nuclei $N \simeq Z$ but for the heavier nuclei N becomes steadily greater than Z.

For the stable nuclei the attractive nuclear interaction is somewhat more effective than it is for the unstable nuclei; that is why they are stable. Thus we may infer, from the approximate equality of N and Z in stable nuclei, that the nuclear interaction somehow produces a lower value for the nuclear energy if neutrons and protons are paired. A good example is provided by the α-particle ($Z = 2$, $N = 2$), which is a stable nucleus in its own right and is also emitted as a unit in many disintegrations of unstable nuclei. If we make the assumption that nuclear energy is lowered if neutrons and protons are paired, we can even understand qualitatively why N tends to be somewhat larger than Z for the heavier stable nuclei. This is due to the competing effects of the attractive nuclear interactions operating between all the nucleons and the repulsive electrical interactions between protons. A gold nucleus has 79 protons. It would have the lowest possible value of nuclear energy if it also had exactly 79 neutrons. However, the repulsive electrical forces between the protons make a positive contribution to the total energy. Adding more neutrons makes the nucleus slightly larger and keeps the protons farther apart from one another. The result is that the total energy (nuclear plus electrical) is lower, and thus nuclei with more neutrons than protons tend to be more stable.

Further support for the preceding ideas is obtained by locating the *unstable* nuclei on Figure 13.11. The radioactive nuclei found in nature belong in the upper right-hand corner of this diagram,* but there are many artificially produced unstable nuclei that lie to one side or the other of the stability line. Two such nuclei are shown in Figure 13.11. The isotope ${}_{37}Rb^{97}$ is radioactive with a short half-life. It decays by β-emission into ${}_{38}Sr^{97}$, which in turn decays into ${}_{39}Y^{97}$, and so on:

$$
\begin{aligned}
{}_{37}Rb^{97} &\longrightarrow {}_{38}Sr^{97} + \beta^- \\
{}_{38}Sr^{97} &\longrightarrow {}_{39}Y^{97} + \beta^- \\
{}_{39}Y^{97} &\longrightarrow {}_{40}Zr^{97} + \beta^- \\
{}_{40}Zr^{97} &\longrightarrow {}_{41}Nb^{97} + \beta^- \\
{}_{41}Nb^{97} &\longrightarrow \underset{\text{(stable)}}{{}_{42}Mo^{97}} + \beta^-.
\end{aligned}
$$

Each step is a displacement downward and to the right on Figure 13.11, toward the stability line, until ${}_{42}Mo^{97}$, a stable nucleus, is reached. The initial nucleus in this sequence, ${}_{37}Rb^{97}$, has too many neutrons; in effect, in each β-decay a neutron is transformed into a proton, making the nucleus more stable. Some artificially produced nuclei have too many protons. They lie to the right of the stability line and decay by emitting positrons (β^+). In positron emission, A does not change, but Z decreases by one while N increases by one, exactly the opposite of ordinary β-decay. The positron emitter ${}_{32}Ge^{65}$ is also shown in Figure 13.11. It decays by a series of positron emissions until it too reaches the stability line:†

$$
\begin{aligned}
{}_{32}Ge^{65} &\longrightarrow {}_{31}Ga^{65} + \beta^+ \\
{}_{31}Ga^{65} &\longrightarrow {}_{30}Zn^{65} + \beta^+ \\
{}_{30}Zn^{65} &\longrightarrow \underset{\text{(stable)}}{{}_{29}Cu^{65}} + \beta^+.
\end{aligned}
$$

*The heaviest stable nucleus has 209 nucleons (${}_{83}Bi^{209}$). Beyond this point, there is no way of adding either neutrons or protons to make a stable nucleus, although some of the very heavy nuclei, such as ${}_{92}U^{238}$, have extremely long half-lives.

†Any positron emitter may sometimes undergo the same transformation by a process called "K-capture," in which instead of emitting a positron, the nucleus captures one of the atom's orbital electrons.

On the basis of the evidence examined, we cannot conclude very much more about the detailed nature of the nuclear interaction. We have, however, reached the important conclusion that in addition to the large repulsive electrical energy, there must be a negative nuclear potential energy, arising from a strong attractive nuclear interaction, whose magnitude must be at least several hundreds of MeV for the larger nuclei. If so, then relatively minor rearrangements of protons and neutrons may well result in energy changes of several MeV and thus may easily account for the energies released in radioactive decay.

Just as the electrons around an atom can exist in various excited states as well as in the ground state, the nucleons in a nucleus have excited energy states as well as a ground state. In addition to the violent rearrangements of nucleons which occur in α or β-emission, transitions between energy states can occur without the emission of a particle; this is what gives rise to γ-rays. ${}_{83}Bi^{211}$, for instance, is an α-emitter, and it is observed that some of the α-particles from a Bi^{211} sample have an energy of 6.28 MeV, whereas others have an energy of 6.63 MeV. A Bi^{211} sample is also a source of γ-rays, whose energy is 0.35 MeV. What must be happening is that the decay of Bi^{211} sometimes leads directly to the ground state of the daughter nucleus (Tl^{207}) but sometimes to an excited state, 0.35 MeV higher in energy, from which there is a subsequent decay to the ground state with emission of a γ-ray. We can represent these two possibilities on an energy level diagram as shown in Figure 13.12. There are many more complex cases in which measurements of the various energies provide similar information about nuclear energy levels.

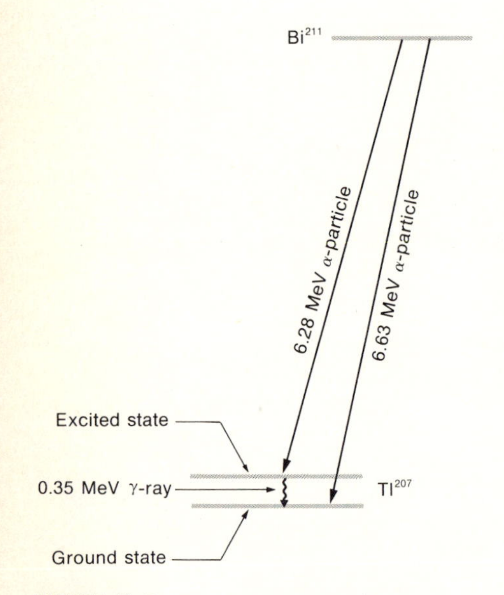

FIGURE 13.12
The α-decay of Bi^{211} leads to the ground state of Tl^{207}, either directly or by way of an excited state of Tl^{207} with emission of a γ-ray.

The fact that of all conceivable nuclei, only a relatively small number are stable—a fact made evident by the decays other nuclei undergo in order to satisfy their "desire" to get to the stability line—has important social implications. In the process of nuclear fission (the splitting of heavy nuclei that occurs in nuclear power plants and nuclear weapons), great quantities of nuclei are produced that (like Rb^{97}) have too many neutrons. These nuclei are radioactive, some with short half-lives and some with very long ones. After an emergency shutdown of a nuclear power plant, nuclei of this type that have already been created continue to release energy as they decay. There is no way to turn them off. All we can do is wait, and, while we are waiting, these nuclei continue to deliver energy to their surroundings. Failure to provide adequate cooling may lead to a disastrous accident. Furthermore, such nuclei, especially those with fairly long half-lives, are removed along with the used nuclear fuel during routine refueling operations. These radioactive nuclei constitute the major part of the radioactive wastes produced by a nuclear power plant, wastes that are extremely dangerous and that must be stored somewhere until their nuclei reach the stability line. The operation of nuclear power plants will be discussed in more detail in the next chapter, but we can already see that some of the hazards posed by the use of nuclear power arise from the fundamental questions of nuclear stability discussed above.

§13.5 NUCLEAR MASS DEFECTS, CHANGES IN MASS, AND THE RELATIONSHIP BETWEEN MASS AND ENERGY

There is something strange about the behavior of *mass* in nuclear physics. First, nuclei have *mass defects*: the mass of a nucleus is smaller than the sum of the masses of its constituent nucleons. Second, there is a decrease in mass in any nuclear disintegration. These peculiarities are of great intrinsic interest. Furthermore, they provide an extremely valuable tool for predicting which nuclei will be stable and which radioactive and for studying the possibility of large-scale conversion of nuclear energy into other forms.*

These effects can be discerned only with extremely precise mass data. Atomic masses can be measured to very high accuracy with a mass spectrometer (Figure 13.13). In this device, a beam of positive ions (atoms that have lost one electron, have a net positive charge, and can therefore be deflected by electric or magnetic fields) is directed into a magnetic field that causes them to travel in circular orbits. From a knowledge of the speed of the ions, their charge, the strength of the magnetic field, and the radius of their circular trajectory, one can deduce their masses. (Thus what is actually measured is the mass of an *ion*; the mass of the neutral atom is calculated by adding a small correction for the mass of the missing electron.) Oxygen, for instance, has three naturally occurring isotopes (O^{16}, O^{17} and O^{18}), and when oxygen ions are studied with this apparatus, the beam of ions entering the magnetic field is divided into three separate beams with different masses. The masses of the three kinds of oxygen atoms are:

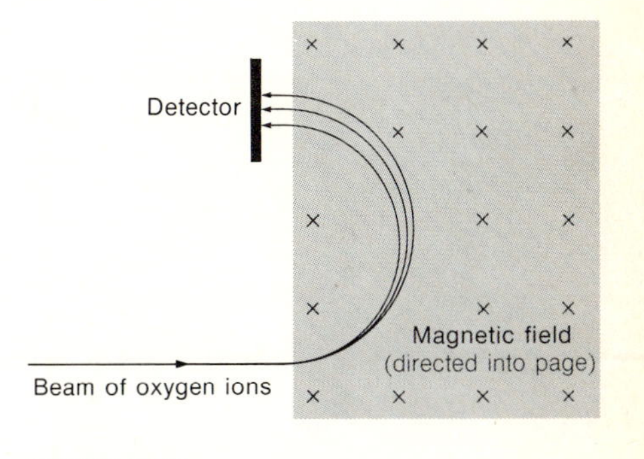

FIGURE 13.13
Schematic diagram of a mass spectrometer. The incoming beam of oxygen ions is divided into three separate beams.

O^{16}: 15.99492 amu
O^{17}: 16.99913 amu
O^{18}: 17.99916 amu.

These results are based on the precise definition of the amu, according to which the mass of the most common form of carbon atom (C^{12}) is defined to be precisely 12 amu; with this definition, the mass of any atom, measured in atomic mass units, turns out to be nearly equal to its nucleon number, A.

As an example of nuclear mass defects, consider the O^{18} nucleus. The mass of an O^{18} nucleus is approximately 18 amu, and the masses of the proton and the neutron are each approximately 1 amu. This makes it reasonable to think of an O^{18} nucleus as a combination of 8 protons and 10 neutrons, but when we examine the situation more carefully, it does not seem to work out quite right. The fact that the mass of an O^{18} atom is not precisely equal to 18 amu is of no importance, because the atomic mass unit is an arbitrarily defined unit. What is extremely significant is that the mass of an O^{18} nucleus is slightly less than the sum of the masses

*It should be noted that there are various ways of treating the subjects discussed in this section. We use the unqualified term *mass* to refer to what in some treatments is called the "rest mass," as opposed to the "relativistic mass."

of the nucleons of which it is composed. The mass of an O^{18} atom is 17.99916 amu; to find the mass of an O^{18} nucleus, we must subtract the masses of 8 electrons:

$$\text{Mass of an } O^{18} \text{ nucleus} = 17.99916 - 8 \times 0.000549$$
$$\simeq 17.995 \text{ amu.}$$

What "should" it be? The mass of a proton is 1.00728 amu and that of a neutron is 1.00867 amu. Thus the mass of 8 protons plus 10 neutrons is equal to:

$$8 \times 1.00728 + 10 \times 1.00867 \simeq 18.145 \text{ amu.}$$

The actual mass of the O^{18} nucleus is *less* than this by 0.15 amu:

$$18.145 \text{ amu} - 17.995 \text{ amu} = 0.15 \text{ amu.}$$

This is not a very large discrepancy, but it is a real one. The mass data are much too accurate for us to write this off as experimental error. The O^{18} nucleus has a "mass defect" of 0.15 amu.

QUESTION

13.9 O^{18} is not an isolated example. Other nuclei, too, have mass defects. Calculate the mass defects of: (a) the α-particle; (b) the U^{235} nucleus.

Now consider the decreases in mass that occur in nuclear disintegrations. We will look at two specific examples of α-decay, that of Po^{210}, one of the substances discovered by the Curies, and Sm^{146}, an artificially produced nucleus:

$$_{84}Po^{210} \longrightarrow {}_{82}Pb^{206} + \alpha + 5.4 \text{ MeV;} \tag{13.4}$$
$$_{62}Sm^{146} \longrightarrow {}_{60}Nd^{142} + \alpha + 2.53 \text{ MeV.} \tag{13.5}$$

The meaning of Equation 13.4, for instance, is that a Po^{210} nucleus is transformed into a Pb^{206} nucleus with the emission of an α-particle, delivering an energy of 5.4 MeV, which appears as kinetic energy shared between the α-particle and the Pb^{206} nucleus.*

In each of these reactions, there is a decrease in total mass. Consider Equation 13.4, for instance. The initial mass is that of a $_{84}Po^{210}$ nucleus; the final mass is the sum of the masses of a $_{82}Pb^{206}$ nucleus and an α-particle (a $_2He^4$ nucleus). We could easily calculate the nuclear masses, for instance by looking up the atomic mass of $_{84}Po^{210}$ and subtracting 84 electron-masses, but the fact that tabulated values of mass are

*In these two examples, all the energy produced appears as kinetic energy. In many other disintegrations, one or more γ-rays are emitted that carry off some of the energy. An example of this type was discussed in §13.4 (Figure 13.12).

usually those of atoms, not of bare nuclei, is a nuisance, and the following labor-saving trick is helpful. No error will be made in calculating the *change* in mass in Equation 13.4 if we use *atomic* masses instead of nuclear masses. This corresponds to adding 84 electron-masses to both sides of Equation 13.4, 84 on the left to make the mass that of a polonium *atom* and 84 on the right (82 for lead and 2 for the α-particle). The change in mass can then be calculated by simply using the various atomic masses. For this reaction, the data are as follows:

Pb^{206}	205.97447 amu
He^{4}	4.00260 amu
Final mass:	209.97707 amu
Initial mass (Po^{210}):	209.98288 amu
Change in mass:	−0.00581 amu

The change in mass is rather small. In most of the calculations in this book, it is not necessary to use so many figures, but in this one, the decrease in mass would not be noticed if we were to round off the original data.

The creation of kinetic energy in this event is no cause for alarm. From the fact the kinetic energy is produced, we may infer that decreases have occurred in one or more other forms of energy. The nucleons in the original Po^{210} nucleus have been rearranged, and changes in the distance of separation of the nucleons result in changes in their nuclear energy and electrical potential energy. The sum of these two forms of energy must have decreased by 5.4 MeV to account for the kinetic energy that was produced. Think of the disintegration of a Po^{210} nucleus from the point of view of the α-particle. Before the event, the α-particle is part of the nucleus. The energy resulting from the specifically nuclear interaction between nucleons is low because the nucleons are fairly close together; the electrical potential energy resulting from the repulsions between protons, though, is very large. However, once the α-particle has been emitted (leaving a Pb^{206} nucleus behind), its electrical potential energy decreases as the distance increases. The overall potential energy (nuclear plus electrical) varies with distance as sketched in Figure 13.14. There is an increase in nuclear energy as the α-particle is emitted, but once it is outside, the electrical repulsion between the daughter nucleus and the α-particle blows them apart; the α-particle rolls "downhill," losing electrical potential energy but gaining kinetic energy.

Without more detailed information about the nuclear interaction, we cannot calculate the total amount of kinetic energy released, but we can make a rough calculation of the change in *electrical* energy. The electrical potential energy of the α-particle just as it is emitted is given by $k_e q_1 q_2 / r$, where q_1 and q_2 are the charges of the α-particle and the Pb^{206} nucleus, and r is approximately the radius of the Pb^{206} nucleus, about 10^{-14} m.

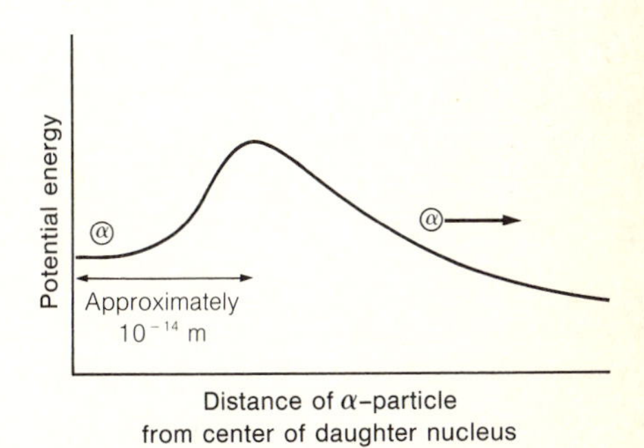

FIGURE 13.14
Once the α-particle has escaped, it is electrically repelled and gains kinetic energy.

QUESTION

13.10 With these assumptions, show that $k_e q_1 q_2 / r$ is approximately 24 MeV.

This is a very rough calculation, but the result is of the same order of magnitude as the kinetic energy produced in this event. We would expect that the amount of kinetic energy produced would be smaller than this value, because the energy associated with the nuclear attraction increases at the same time that the electrical potential energy decreases. Thus it is the *sum* of the electrical potential energy (arising from the repulsions between protons) and the nuclear energy (arising from the attractions of all the nucleons) that decreases as the α-particle is emitted and that accounts for the kinetic energy produced. Since the electrical potential energy arises from the electrical interaction of nucleons (protons), we will refer to the sum of this electrical energy and the energy arising from the attractive nuclear interaction as simply "nuclear energy." It is this nuclear energy that decreases as the α-particle is emitted. In a similar way, when a hydrogen *atom* in an excited state emits a photon, the creation of energy in the form of a photon is evidence for the fact that the internal energy of the atom decreases as the electron jumps into a state closer to its nucleus. What is interesting and apparently different about radioactive decay is that the creation of kinetic energy is accompanied by a loss of mass. A nuclear disintegration is an example of what is often (rather loosely) described as a "conversion of mass into energy." Although it is really a conversion of nuclear energy into kinetic energy, as we examine this process more carefully we shall see why it is not altogether wrong to describe it as a conversion of mass into energy.

QUESTIONS

13.11 Show that the loss of mass in the reaction described by Equation 13.5 is 0.0027 amu.

13.12 By comparing these two reactions, show that the loss of mass is proportional to the amount of kinetic energy created.

There seems to be a connection between loss of mass and creation of kinetic energy. This is reminiscent of Joule's experiments, in which mechanical energy was lost and "heat" created. What is the relationship between the loss of mass in Equation 13.4 and the amount of kinetic energy created? Consider this ratio:

$$\frac{KE \text{ created}}{\text{loss of mass}} = \frac{5.4 \text{ MeV}}{0.00581 \text{ amu}}. \qquad (13.6)$$

QUESTION

13.13 Since energy and mass do not have the same dimensions, this is not a true ratio (such as 2 or π) but a number with *units* and *dimensions*. Show that this ratio must have the same dimensions as the square of a speed, and thus that its MKS units are m^2/sec^2.

We can convert the ratio 13.6 into MKS units, take the square root, and find the numerical value of this speed in meters per second. The result is 8.96×10^{16} m^2/sec^2, and the square root of this number is a speed: 2.99×10^8 m/sec.

QUESTION

13.14 Check this numerical calculation, and make the same calculation for reaction 13.5.

The speed calculated in this way is too close to the speed of light for this result to be regarded as accidental. It appears that we can write:

$$\frac{KE \text{ created}}{\text{loss of mass}} = c^2,$$

where c is the speed of light. This conclusion is borne out by examination of other such reactions. The loss of mass and the amount of kinetic energy produced are always related in this way. Thus we have a very useful result. The loss of mass serves as an indicator of the amount of kinetic energy produced. If we have accurate mass data, we can predict how much kinetic energy will be produced in other reactions:

$$KE \text{ created} = c^2 \times \text{ decrease in mass} \qquad (13.7)$$

or, for short:

$$E = mc^2. \qquad (13.8)$$

Why is there a decrease in mass when kinetic energy is produced? Why should the speed of *light* appear in these equations? The special theory of relativity enables us to understand these facts. Einstein actually predicted that the creation of kinetic energy should be accompanied by a decrease in mass and that the relationship should be that described here, predictions that were later confirmed by measurements. Space does not permit the study of Einstein's theory, but we can nevertheless accept the experimental facts and use the observed changes in mass to predict the amounts of kinetic energy that will be produced in various processes.

Is a process such as α-emission correctly described by the term "conversion of mass into energy"? This is a catchy phrase, but it should be used with great caution, if at all. It is true that a decrease in mass and a creation of kinetic energy take place, but (as stated earlier) there is no change in *total* energy, only a conversion of *nuclear* energy into *kinetic* energy. The change in mass merely indicates the amount of kinetic energy produced as other forms of energy decrease.

It is convenient to measure energies in MeV and masses in atomic mass units. Therefore it is useful to know the amount of kinetic energy in MeV released when there is a decrease in mass of 1 amu. This amount of energy is approximately 931 MeV:

$$1 \text{ amu} \longleftrightarrow 931 \text{ MeV}. \tag{13.9}$$

QUESTION

13.15 Check this result by using Equation 13.8.

This is, in a sense, the "conversion factor" between mass and energy, but it is quite different from an ordinary conversion factor such as 1 inch = 2.54 cm. Mass and energy are not the same. A decrease in mass of 1 amu is accompanied by the production of 931 MeV of kinetic energy and the simultaneous decrease in some other form of energy by the same amount.

Now consider a few applications of these results. How much kinetic energy will be produced in the following β-decay:

$${}_{38}\text{Sr}^{90} \longrightarrow {}_{39}\text{Y}^{90} + \beta^{-}?$$

We need to know the amount by which the mass of a Sr^{90} nucleus exceeds the sum of the masses of a Y^{90} nucleus plus an electron. The same trick used earlier can be employed here to avoid the trouble of calculating nuclear masses. If we add 38 electrons to each side of this reaction, we have 38 electrons on the left, enough for a neutral Sr^{90} atom, and a total of 39 on the right, enough for a neutral Y^{90} atom. Therefore we need only calculate the difference between the *atomic* masses of Sr^{90} and Y^{90}:

Sr^{90}	89.90775 amu
Y^{90}	89.90716 amu
Decrease in mass:	0.00059 amu.

Thus the energy released is $0.00059 \times 931 = 0.55$ MeV.

QUESTION

13.16 ${}_{29}\text{Cu}^{58}$ is a nucleus that decays by *positron* emission:

$${}_{29}\text{Cu}^{58} \longrightarrow {}_{28}\text{Ni}^{58} + \beta^{+}.$$

Show that the amount of kinetic energy produced is approximately 7.5 MeV. (Be careful! In positron emission, it is not sufficient simply to calculate the difference in mass between a Cu^{58} atom and a Ni^{58} atom. The masses of the electrons must be treated more carefully.)

In some nuclear processes, some of the energy is released in the form of γ-rays rather than as the kinetic energy of particles. These processes, too, fit the description of Equation 13.7 if we simply count the energy of a γ-ray as a form of kinetic energy, the "kinetic energy of a photon."

Consider a somewhat different question. Is it possible for a deuteron to disintegrate spontaneously into a proton and a neutron: $d \longrightarrow p + n$? When we calculate the change in mass that would occur in this hypothetical reaction, we find that instead of a decrease in mass, there is an *increase* in mass of 0.0024 amu. This reaction would not produce kinetic energy; on the contrary, it is one in which the total nuclear energy must be increased, and it cannot occur unless some additional energy is supplied. The converse of Equation 13.7 is that in any hypothetical reaction in which there is an increase in mass, some energy must be supplied, and the minimum amount of energy required can be calculated from the change in mass:

$$\text{minimum amount of energy required} = c^2 \times \text{increase in mass.} \quad (13.10)$$

In order to break up a deuteron, there must be an increase in mass of 0.0024 amu, and from Equation 13.10, the minimum amount of energy required is $0.0024 \times 931 = 2.2$ MeV.

Any proposed reaction can be examined in this way. If the total mass decreases, then the reaction may take place spontaneously; if the mass increases, the reaction cannot occur unless energy is supplied. Even if the calculation shows that the reaction may occur, we cannot tell what the probability is that it *will* occur. We can use this method, for instance, to see whether it is possible for a particular nucleus to undergo β-decay, but we get no information indicating whether its half-life will be long or short.

QUESTIONS

13.17 Is it possible for $_8O^{16}$ to decay by α-emission?

13.18 Is it possible for a neutron to undergo β-decay? If so, the reaction would be:

$$n \longrightarrow p + \beta^-.$$

13.19 Show that $_{29}Cu^{64}$ can decay either by ordinary β-decay or by positron emission. What is the daughter nucleus for each type of disintegration? How much energy is released in each process?

It is now easy to understand the fact that nuclei exhibit mass defects. The overall energy of a collection of nucleons is lower when they are together in a nucleus than when they are separated. To break up a nucleus, energy must be supplied. The necessary amount of energy is called the *binding energy* of the nucleus. From Equation 13.10, there would be an increase in mass if a nucleus were to be completely disintegrated into its constituent nucleons; that is, the mass of a nucleus is less than the sum of the masses of its nucleons. The binding energy and the mass defect are related by Equation 13.10:

$$\text{binding energy} = c^2 \times \text{mass defect}. \qquad (13.11)$$

The mass defect of the deuteron is 0.0024 amu, and its binding energy is 2.2 MeV. We saw earlier that the O^{18} nucleus has a mass defect of 0.15 amu. Its binding energy is therefore $931 \times 0.15 = 140$ MeV. This is the amount of energy that would have to be supplied to break up an O^{18} nucleus, to separate its 18 nucleons from one another.

If we know the mass defect of any nucleus, we can calculate the binding energy. The larger the binding energy, the lower the nuclear energy, and the more stable the nucleus is against disintegration. The variation in binding energy from one nucleus to another is of fundamental importance in considering nuclear fission and fusion, and we shall examine nuclear binding energies in more detail in the following chapter.

Any decrease in mass is accompanied by the production of kinetic energy. How far can we push this result? Suppose that a particle such as an electron were to *vanish*. It is easy to calculate the amount of kinetic energy that would be produced as a result. The mass of an electron is 0.0005486 amu. A decrease in mass of this amount would be accompanied by the production of a kinetic energy of $0.0005486 \times 931 = 0.511$ MeV. Since electrons left to themselves do not vanish, this would be idle speculation were it not for the existence of positrons, the antiparticles of electrons. When an electron meets a positron, they annihilate one another. The mass of a positron is equal to that of an electron, so there is a decrease in mass of two electron-masses, and γ-rays are given off whose energies add up to $2 \times 0.511 = 1.022$ MeV.

Where does this energy come from? It cannot be accounted for by a decrease in any of the familiar forms of energy. Instead it is necessary to attribute to an electron (or to any particle, for that matter) an intrinsic energy that it has simply by virtue of its existence, a "rest energy" ($E = mc^2$), which it has even when not in motion. The rest energy of an electron or a positron is just 0.511 MeV. Other fundamental particles also have antiparticles, and similar annihilation events result in the conversion of rest energy into the energies of γ-rays. The annihilation of a particle and its antiparticle is often described as the "complete conversion of mass into energy." This is not a bad description, but what is really happening is the complete conversion of *rest energy* into other forms of energy. With this terminology, other processes such as the emission of an α-particle by Po^{210} (Equation 13.4) can be described as the partial conversion of mass into energy. In that example, the ratio of the decrease in mass to the initial mass is

$$\frac{0.00581 \text{ amu}}{209.98 \text{ amu}} = 0.000028,$$

and so we can describe this as a 0.0028% conversion of mass into energy.

Are changes in mass peculiar to nuclear events, or are these results applicable to other phenomena as well? Consider what happens when an electron encounters a proton and is caught, forming a hydrogen atom in its ground state. In the process, the energy of the electron decreases by 13.6 eV, as we saw in Chapter 12, and some photons whose energies add up to 13.6 eV are emitted. The binding energy of a hydrogen atom is 13.6 eV. According to Equation 13.11, the mass of a hydrogen atom should therefore be less than the sum of the masses of a proton and an electron—not very much less, though, because the binding energy is only a few electron-volts rather than millions of electron-volts.

QUESTION

13.20 Show that the mass corresponding to an energy of 13.6 eV is approximately 1.5×10^{-8} amu.

The calculated "mass defect" of the hydrogen atom is 1.5×10^{-8} amu, too small to be detected even with the most sophisticated apparatus. In earlier calculations, we assumed without question that the mass of a hydrogen atom, for instance, was precisely equal to that of a proton plus an electron. This is not quite correct, but the difference is undetectably small. Similarly, in an ordinary chemical reaction such as the combustion of carbon,

$$C + O_2 \longrightarrow CO_2,$$

energy is released, and so the mass of a CO_2 molecule must be less than the sum of the masses of a carbon atom and an oxygen molecule. Again, the numbers are such that the change in mass is far too small to be measured, but it is just as correct (or just as incorrect, depending on one's taste) to describe a chemical reaction such as this as a conversion of mass into energy as it is for a nuclear reaction; it is only a matter of degree.

The question of degree is very important. We can see why it is that the changes in mass we believe occur in chemical reactions have never been noticed, and we can also see an important practical limitation on the usefulness of Equation 13.7. This result can be successfully used to calculate energy releases only if the available mass data are accurate enough that the decrease in mass (which must be calculated by subtracting one value of mass from another) is accurately known. Only for nuclear interactions or the even more dramatic annihilation events is the percentage change in mass large enough to permit this, and even for nuclear reactions such as α-decay, quite accurate data are required since the changes in mass amount to only a fraction of an atomic mass unit.

§13.6 THE PECULIARITIES OF β-DECAY AND THE DISCOVERY OF THE NEUTRINO

The process of β-decay shows some anomalous features which have not been mentioned thus far. At only one time during the period of more than a century since the law of conservation of energy was clearly formulated have scientists in any significant numbers seriously considered abandoning it; this crisis arose as a result of the anomalies associated with β-decay, and heroic measures were required to save the law. We have touched briefly on this problem and how it led to the postulated existence of a new particle (the neutrino) in Chapter 6; now let us examine the evidence in more detail.

The main problem is that, for a collection of identical nuclei that decay by β-emission, the emitted β-particles do not all have the same energy; instead there is a continuous distribution of energies from extremely low values to some maximum value, as shown in Figure 13.15 for ${}_{83}Bi^{210}$, which decays to ${}_{84}Po^{210}$:

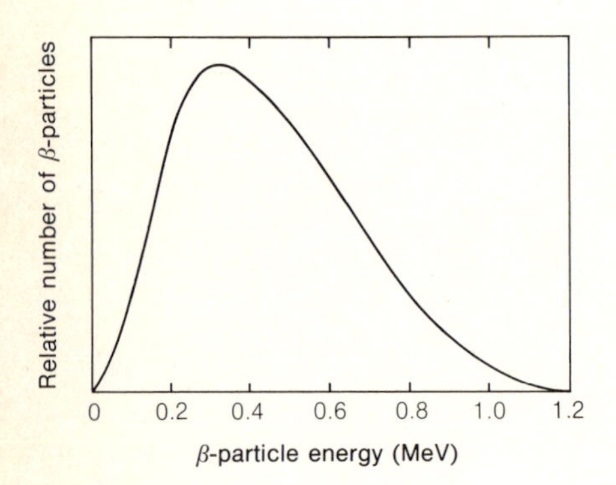

FIGURE 13.15
Distribution of energies of the β-particles emitted by Bi^{210}.

$$ {}_{83}Bi^{210} \longrightarrow {}_{84}Po^{210} + \beta^-. \tag{13.12}$$

The same amount of energy should be released in each decay. However, from Figure 13.15 it appears that sometimes when this decay occurs only a tiny amount of energy is released as kinetic energy, sometimes as much as 1.16 MeV, and in fact the measured kinetic energy of the β-particle can have any value between 0 and 1.16 MeV. We will consider various possible explanations of these results.

First, we have seen other examples of nuclear disintegrations in which the energy of the emitted particle is not always the same. In most α-decays, the emitted α-particle may have one of two or more possible values of energy; the explanation is that the decay can lead either to the ground state of the daughter nucleus or to one of the excited states (Figure 13.12). Beta decay is different in that there is a continuous (rather than discrete) spectrum of energies of the emitted particles. We might suppose that the daughter nucleus has a large number of very closely spaced excited energy states, any one of which might result from β-decay. Thus we would construct a diagram for β-decay as shown in Figure 13.16; there would be a large number of possible β-particle energies, and if they were sufficiently close together, the distribution of energies would appear to be continuous. For an α-decay with two or more possible α-particle energies, decay that leads to an excited state is followed by emission of one or more γ-rays as the daughter nucleus goes to its ground state, as shown in Figure 13.12. Therefore, if the present suggestion is correct, every β-decay that leads to an excited state should be followed by the emission of one or more γ-rays such that the total energy released in the form of kinetic energy of the β-particle plus the subsequent γ-rays would be the same for each event. Thus Figure 13.16 should be completed as shown in Figure 13.17. A nearly continuous spectrum of γ-rays is predicted by this hypothesis.

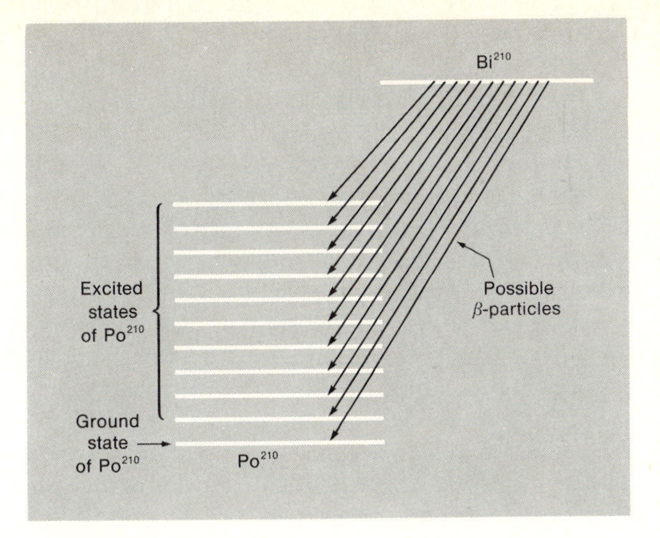

FIGURE 13.16
The seemingly continuous distribution of β-particle energies from Bi^{210} could be explained if the daughter nucleus had a large number of closely spaced excited states.

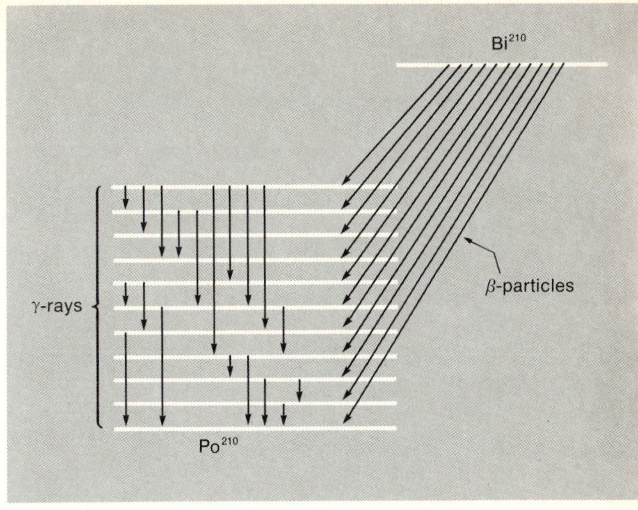

FIGURE 13.17
If Figure 13.16 were correct, γ-rays of various energies should be observed. (Only a few of the supposed γ-ray transitions are shown.) No such γ-rays are observed.

This idea fails because the γ-rays are not observed. Some β-decay processes are accompanied by γ-rays, but these γ-rays have only certain discrete energies instead of a continuous spectrum, and in some β-decays, of which that of Bi^{210} is an example, *no* γ-rays are observed.

A second possibility is that, although energy may not be conserved in each individual decay, it is conserved on the *average.* It would be rather unsatisfying if we had to accept this idea, since a law of conservation of energy that is correct only on the average is not as powerful and general as one that is correct on each and every occasion. In any event, this explanation is incorrect, as shown by a calculation of the total amount of energy released in the β-disintegration.

QUESTION

13.21 From the listed values of the masses of Bi^{210} and Po^{210}, show that the amount of energy released in the β-decay of Bi^{210} is approximately 1.16 MeV.

If energy were conserved on the average, then the average value of the observed β-particle energies should be equal to 1.16 MeV. The average value, however, is about 0.4 MeV (Figure 13.15). The total amount of energy released, 1.16 MeV, corresponds instead to the maximum value of the observed β-particle energies. That is, the few β-particles coming off with energies of 1.16 MeV have the correct amount of energy; for all the rest, some energy is missing.

These results suggest a third explanation. Perhaps Figure 13.15 is not a correct representation of the energies with which the β-particles are emitted. Perhaps all the β-particles are emitted with the correct amount of energy (1.16 MeV), but some of them lose energy by colliding with other atoms before getting to the measuring apparatus. This idea can be tested, since this energy should show up in the form of thermal energy; we can simply enclose the radioactive sample in a container thick enough to stop all the β-particles and measure the rate at which it heats up. Knowing the heat capacity of the container and its contents, we can find the total rate at which energy is being released. If each β-particle carries off an energy of 1.16 MeV, all this energy will be converted into thermal energy, and thus the thermal energy produced should amount to 1.16 MeV for every nuclear disintegration. Such measurements have been made, but the measured rate of heating corresponds to an average β-particle energy of only 0.4 MeV. This value agrees with that derived from Figure 13.15, and so the suggestion that all the β-particles are emitted with an energy of 1.16 MeV is not a valid one.

Perhaps the law of conservation of energy *does* fail in β-decay. Even if it did, we would not have to abandon it completely. We could still say that energy never increases, that in almost all circumstances it neither increases nor decreases, but that in nuclear β-decay a small and variable amount of energy simply disappears. This suggestion conforms to the facts, and for a time it seemed to be the inevitable conclusion.

However, there is another explanation, which permits us to retain the law of conservation of energy without any such qualifications. Let us suppose that another particle, in addition to a β-particle, is emitted in the process of β-decay. We give it a name, the "neutrino," and include it in all our descriptions of β-decay processes. For example, instead of Equation 13.12, we have

$$Bi^{210} \longrightarrow Po^{210} + \beta^- + \text{neutrino}.$$

We will of course say that the energy carried off by the neutrino is just enough to make the books balance. The total amount of energy available, 1.16 MeV, is shared between the β-particle and the neutrino. For those events in which the β-particle comes off with almost no kinetic energy, the neutrino carries off approximately 1.16 MeV of energy; in those events in which the β-particle is emitted with a kinetic energy of 1.16 MeV, the neutrino must be emitted with very little energy. Since electrical charge is already accounted for by the previously known particles, the neutrino must be electrically neutral. (Thus the name: "little neutral one.") We must also assume that neutrinos can pass through matter with very little chance of being absorbed or slowed down; otherwise some of their energy would have been observed as thermal energy in the experiments described above. A neutrino with these properties would explain all the experimental facts.

But are we not deluding ourselves? What have we gained by concocting in our imaginations this invisible undetectable particle with an exotic name? Have we not simply substituted the name "neutrino" for what should more honestly be called "decrease in energy"? The passage by Poincaré quoted in §13.2 can be profitably applied to the neutrino hypothesis as well. Is the neutrino one of those superficially attractive notions that is really *too* successful, an idea that cannot conceivably be disproved and is therefore essentially sterile? It is important to consider further the properties of the supposed neutrinos, to see whether we can imagine *any* experimental results that would lead us to reject the neutrino hypothesis. For one thing, the mass of a neutrino must be zero or nearly so. Why is this? Consider the β-decay of Bi^{210}. The most energetic β-particles have kinetic energies of 1.16 MeV; this fits perfectly with the prediction of the amount of energy released in this decay, a prediction made from the decrease in mass that occurs in the reaction. That calculation was based on the masses of the Po^{210} nucleus and the β-particle. Therefore in order not to create any new problems, we must suppose that neutrinos have no mass. (Neutrinos are similar to photons in this respect, and both photons and neutrinos always travel at the speed of light. "Neutrino" is not just another word for "photon," though; photons with energies of 1 MeV or so are energetic γ-rays and easily detectable.)

The law of conservation of energy is not the only conservation law that appears to be violated in β-decay. Some momentum appears to be missing also, though the evidence has not been presented here. "Angular momentum," a measure of the rate at which something is spinning, is still another quantity that should be conserved, but in β-decay apparently it is not. What could be more natural than to suppose that the neutrino carries off not only just the right amount of energy but also the correct amounts of momentum and angular momentum so that all three conservation laws are preserved? Does the attribution of momentum and angular momentum to the neutrino make its existence any more believable? Or should this, on the contrary, be regarded as a series of increasingly desperate attempts to salvage conservation laws when any rational person would feel obliged to acknowledge the fact that all these laws are violated in β-decay?

The existence of the neutrino was first suggested in 1931. In the following years, a theory of β-decay was developed, a theory based on the assumption that neutrinos do exist. With this theory it was possible to correlate many of the observed facts about β-decay, such as the half-lives of β-emitting nuclei and the shapes of curves such as Figure 13.15. Belief in the reality of the neutrino increased—because of the way in which it fit into this successful theory as well as because of the deep-seated desire to keep the law of conservation of energy—even though no one could exhibit anything like a cloud-chamber photograph of a neutrino's track.

Neutrinos interact only weakly with ordinary matter. It was necessary to make this supposition from the start, for if neutrinos could not pass readily through matter, they would have been detected, and there would have been no problem of "missing energy" to begin with. To say that a particle interacts weakly with matter is not the same as saying that it does not interact at all. With the theory of β-decay referred to above, it is possible to predict how strong this interaction should be. The prediction is that the interaction is so weak that on the average a neutrino can pass through a thickness of about 50 *light-years* of lead. To a neutrino, the earth would be virtually transparent; the chance that a neutrino will be absorbed as it travels through the earth is about one in 10^{11}. As weak as this interaction may be, if the number of neutrinos is large enough, some of them should interact with the matter through which they pass and make their presence known. During and after World War II, a number of nuclear reactors were built, and—if neutrinos do indeed exist—enormous numbers of them are produced in such reactors as an unwanted and unnoticed by-product.

In the early 1950s, an experiment to search for neutrinos was begun. The idea was to let neutrinos from a reactor pass through a large quantity of material containing hydrogen atoms (about 10 tons!) and to look for the predicted reaction:

$$\text{neutrino} + p \longrightarrow \beta^+ + n.$$

That this reaction should occur—not often, but for one neutrino out of a great many—could be confidently predicted on the basis of existing theory. One could even predict about how often such a reaction should occur. It was necessary to use sophisticated apparatus that would respond only to reactions in which both a positron and a neutron were produced, so as to discriminate against all sorts of extraneous events. It was not an easy experiment, but by 1953 the apparatus was working and the experiment was successful. The neutrino-produced reactions did occur at the predicted rate. It had taken 20 years, but the lingering cloud of suspicion attached to the neutrino was finally removed, and this particle, its existence initially proposed to save the law of conservation of energy, became as real as the electron, the proton, and the neutron.

We said earlier that it was important to ask whether the neutrino hypothesis is one that could conceivably be refuted by experiments, whether it is falsifiable. If the experiments just described had turned out differently, the neutrino hypothesis would have been rejected, and we would still be wondering about the validity of the law of conservation of energy. Instead, the law of conservation of energy did work after all. Those who had faith in it had used it to predict the existence of a new particle and they were right. Just as the discovery of Neptune served to dispel lingering doubts about Newton's law of gravitation, so the story of the neutrino makes us more willing than ever to believe in the law of conservation of energy. Perhaps, though, if we remember the eventual failure of Newton's law in describing the motion of Mercury, we ought to be prepared for eventually discovering a true failure of the law of conservation of energy.

§13.7 THE DISCOVERY OF THE NEUTRON

The neutron has been of central importance in the discussion of nuclear structure in this chapter, but it was not until 1932 that we could be certain of the existence of neutrons. Although many people realized that the description of nuclear phenomena would be greatly simplified if such a particle—an electrically neutral particle with a mass approximately equal to that of a proton—did indeed exist, neutrons had never been observed. The discovery of the neutron in 1932 resulted from a neat piece of scientific detective work based on principles discussed in Chapter 3. The first important discovery was that when a sheet of beryllium was bombarded with α-particles, radiation of some kind was produced (Figure 13.18). It eventually turned out that this radiation consisted of neutrons, but additional experiments and some clever thinking were needed before this conclusion could be reached. The radiation was an extremely penetrating one; putting several centimeters of lead between the beryllium and the ionization chamber scarcely reduced the observed intensity at all. The first suggestion was that it was γ-rays that were coming from the beryllium; this was a perfectly reasonable suggestion, although the ease with which they could pass through so much lead made it necessary to suppose that these γ-rays were very energetic, more energetic than any then known.

The next important discovery was that insertion of a sheet of paraffin (which contains numerous protons, that is, hydrogen nuclei) between the beryllium and the ionization chamber (Figure 13.19) caused a considerable increase in the amount of ionization produced. Apparently the unknown radiation was colliding with protons and ejecting them from the paraffin, and it was these protons which caused the increased ionization in the detector. This idea was confirmed by using a cloud chamber instead of an ionization chamber; the tracks had all the characteristics of proton tracks, and it was possible even to estimate the speeds of these

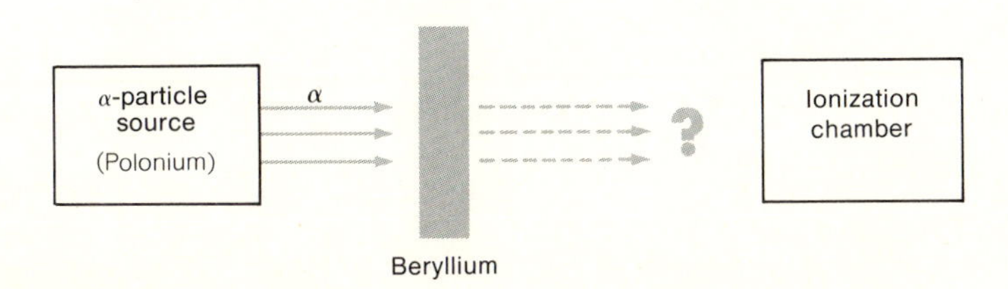

FIGURE 13.18
Alpha-particles striking a beryllium sheet produce a penetrating radiation ("?"), which was eventually identified as neutrons.

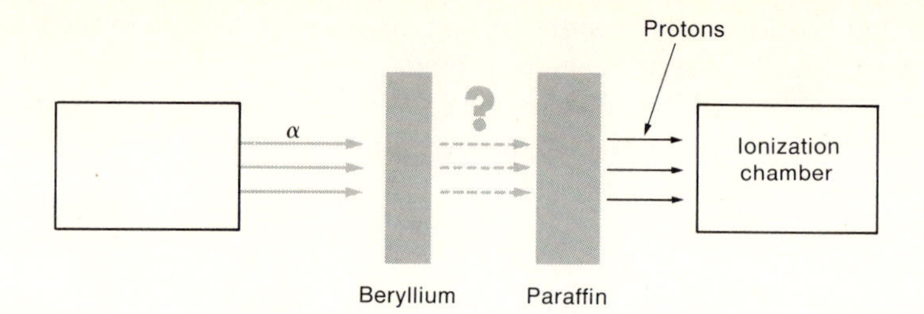

FIGURE 13.19
The unidentified radiation knocks protons out of a paraffin sheet, causing an increase in the observed ionization.

protons. Protons of various speeds were observed, the maximum speed being about 3.3×10^7 m/sec, corresponding to a kinetic energy of approximately 5.7 MeV.

With this information, it is possible to estimate the energies of the supposed γ-rays that knocked the protons out of the paraffin. The hypothesis is that a γ-ray photon has an elastic collision with a proton, transferring some of its energy to the proton. What is the original energy of this γ-ray? We can apply the simple laws of conservation of momentum and energy that apply to collisions of gliders on an air track, with one additional piece of information, the relationship between the energy and momentum of a γ-ray. It can be shown from Maxwell's theory that electromagnetic waves carry momentum as well as energy; a γ-ray or any other photon therefore has a momentum, p, and a photon's momentum is related to its energy by:

$$p = \frac{E}{c}.$$

Now apply the laws of conservation of momentum and energy to the head-on elastic collision of a γ-ray and a stationary proton shown in Figure 13.20. By conservation of energy,

$$E = E' + \tfrac{1}{2}mv^2, \tag{13.13}$$

and by conservation of momentum (remember that momentum is a vector quantity):

$$\frac{E}{c} = mv - \frac{E'}{c}. \tag{13.14}$$

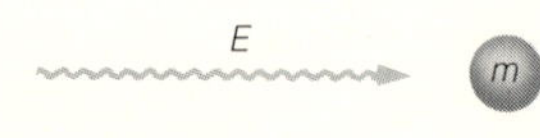

a Before

b After

FIGURE 13.20
Head-on elastic collision of a γ-ray (initial energy, E) with a stationary nucleus.

QUESTIONS

13.22 By eliminating E' from Equations 13.13 and 13.14, show that E, the original energy of the γ-ray, is:

$$E = \frac{mv}{2}\left(c + \frac{v}{2}\right). \qquad (13.15)$$

13.23 In this experiment, m is the mass of a proton ($m \simeq 1$ amu), and v is observed to be 3.3×10^7 m/sec. Show that $E \simeq 54$ MeV. (Be careful with the units. It is probably easiest to do the calculation in MKS units, find E in joules, and then convert to MeV.)

An energy of 54 MeV is very large for a γ-ray, but this is not the real problem with the hypothesis that the unknown radiation consists of γ-rays. The difficulty is that the same "γ-rays" can also be observed in similar collisions with other sorts of nuclei, and they then act as if their energy were quite different. For example if, instead of colliding with the protons in a piece of paraffin, they collide with nitrogen nuclei in the air inside the cloud chamber, nitrogen nuclei are observed coming off with speeds of 4.7×10^6 m/sec. This gives a second piece of data with which we can calculate the energy of the supposed γ-rays.

QUESTION

13.24 By applying Equation 13.15 to this situation, where m = the mass of a nitrogen nucleus ($m \simeq 14$ amu) and $v = 4.7 \times 10^6$ m/sec, show that $E \simeq$ 104 MeV.

We cannot simultaneously attribute two different energies to the same γ-rays; the energy cannot be both 54 MeV and 104 MeV. The γ-ray interpretation was an initially reasonable hypothesis, but it does not survive this numerical test. What if this radiation consists of particles of some kind? If we make this hypothesis (and we may as well call the particles neutrons because this hypothesis turns out to work), we can use the data given above to find the mass of the neutron by adapting some results obtained in §3.2.C. Consider Figure 3.11. In the collision of a neutron with a proton, the neutron plays the role of object A, and the proton that of object B. Using the subscripts n and p to denote the neutron and proton, we have from Equation 3.5:

$$v_f(\text{proton}) = \frac{2m_n}{m_n + m_p}\, v_i(\text{neutron}). \qquad (13.16)$$

Similarly in the collision of a neutron with a nitrogen nucleus, using the subscript "nit" to indicate the nitrogen nucleus:

$$v_f(\text{nitrogen}) = \frac{2m_n}{m_n + m_{\text{nit}}}\, v_i(\text{neutron}). \qquad (13.17)$$

QUESTION

13.25 By eliminating v_i (neutron) from Equations 13.16 and 13.17, show that

$$\frac{v_f(\text{proton})}{v_f(\text{nitrogen})} = \frac{m_n + m_{\text{nit}}}{m_n + m_p}. \qquad (13.18)$$

The values of v_f(proton) and v_f(nitrogen) were measured in the experiments described, and the left-hand side of Equation 13.18 is

$$\frac{v_f(\text{proton})}{v_f(\text{nitrogen})} = \frac{3.3 \times 10^7 \text{ m/sec}}{4.7 \times 10^6 \text{ m/sec}} \simeq 7.$$

Thus Equation 13.18 becomes

$$\frac{m_n + m_{\text{nit}}}{m_n + m_p} = 7. \qquad (13.19)$$

QUESTION

13.26 Solve Equation 13.19 for m_n and show that

$$m_n = \frac{m_{\text{nit}} - 7m_p}{6}.$$

Since the mass of the nitrogen nucleus is approximately 14 amu and that of the proton 1 amu,

$$m_n \simeq \frac{14 \text{ amu} - 7 \text{ amu}}{6} = \frac{7 \text{ amu}}{6} = 1.17 \text{ amu}.$$

The hypothesis that the unknown radiation consists of particles whose mass is approximately equal to that of the proton seems to work. When the beam interacts with nuclei other than protons and nitrogen nuclei, the velocities with which these nuclei are ejected can also be measured, and all of the observations are in accord with this hypothesis. Determination of the mass of the neutron in this way is not very accurate, and more refined methods give a slightly lower value, but the experiments described here gave the first convincing proof of the neutron's existence. (Incidentally, although neutrons bound within nuclei seem to be quite stable, free neutrons are not; they undergo β-decay into protons with a half-life of about 11 minutes.)

Discovery of the neutron had two very important consequences. One was that it was just what was needed to tie together a great many of the known facts about nuclei, and it made possible the description of nuclear structure given in this chapter. Second, in experimental work, neutrons are very convenient projectiles for studying nuclear interactions because of their lack of electrical charge; it is their electrical neutrality which gives them their great penetrating power. Whereas protons and α-particles must be given very high energies if they are to approach other nuclei, even a slow neutron can easily penetrate to the center of an atom, and so neutrons provide an ideal tool for studying nuclear physics.

FURTHER QUESTIONS

13.27 A sample of 10^{20} nuclei with a half-life of 1 hour is prepared on Friday afternoon. How many of these nuclei will probably be left on Monday morning?

13.28 A semilog graph of a quantity that is decreasing exponentially should be a straight line, but with a negative slope. Read numerical values from Figure 13.4, and make a semilog graph of these data.

13.29 One method of dating archeological specimens is based on the properties of the radioactive nucleus C^{14}. This nucleus has a half-life of about 5700 years. Although atmospheric carbon nuclei are mostly the stable form (C^{12}), the supply of C^{14} nuclei in the atmosphere is maintained at a low but steady concentration by cosmic rays. The carbon in a living plant consists of a mixture of C^{12} and C^{14} nuclei, but when the plant dies, the C^{14} nuclei begin to disintegrate and are not replaced. If a sample of charcoal contains only one-eighth as much C^{14} as is found in living trees, approximately how old is the charcoal?

13.30 In the scattering experiment described in §13.3, those α-particles headed directly at gold nuclei approached to within a distance of 2.5×10^{-14} m before being turned around. At this distance, what is the repulsive force (in newtons and in pounds) between an α-particle and a gold nucleus?

13.31 The gravitational effects of ordinary objects are extremely small and difficult to measure. If this book had its present size but a density equal to that of a nucleus, how large a gravitational force would it exert on you? Would this force be large enough to be detectable?

13.32 Tritium (H^3) decays by β-emission. What is the resulting daughter nucleus?

13.33 In the α-decays discussed in §13.5 (Equations 13.4 and 13.5), most of the kinetic energy is delivered to the α-particle, and only a small amount to the daughter nucleus. This fact follows from the principles discussed in Chapter 3. Approximately what fraction of the kinetic energy does the α-particle receive in each of these two reactions?

13.34 Why is it that any nucleus that can decay by positron emission can also decay by capturing one of its own orbital electrons (K-capture), but that the converse is not correct? (Ca^{41}, for instance, can decay by K-capture but never by positron emission.)

13.35 The nuclear species K^{40} is of particular interest to humans; as a constituent of our bodies, it is responsible for much of our exposure to radioactivity. It can decay in three different ways: by β^- emission, by β^+ emission and by K-capture. By considering the relevant atomic masses, explain why all three modes of decay are possible.

13.36 Suppose that a proton and an antiproton were to meet and annihilate each other, producing two identical γ-rays. Estimate the frequency and wavelength of these γ-rays.

13.37 If it were possible to achieve complete "conversion of mass into energy," how much mass would be needed to supply all the energy for the United States for one year?

13.38 Between one meal and the next, the chemical reactions that go on in your body release an energy of about 1000 kilocalories. By how much should your mass decrease as a result of these reactions?

13.39 All photons carry momentum, and therefore a flashlight that emits light ought to recoil in the opposite direction, just as a gun recoils when it is fired. Suppose that a flashlight is turned on for one second and that all the electrical energy used is converted into photons. Make a rough estimate of the speed with which the flashlight would recoil if it were on a completely frictionless surface.

14

Nuclear Fission and Nuclear Fission Power Plants

14
Nuclear Fission and Nuclear Fission Power Plants

We nuclear people have made a Faustian bargain with society. On the one hand, we offer an inexhaustible source of energy. But the price that we demand of society for this magical energy source is both a vigilance and a longevity of our social institutions that we are quite unaccustomed to.

—A. Weinberg, former director of Oak Ridge National Laboratory, 1972.

§14.1 ENERGY AVAILABLE FROM NUCLEAR FISSION AND FUSION

Nuclear energies are very large, at least when compared with other forms of energy on an atom for atom basis. A typical nuclear disintegration releases several MeV in a single event; by contrast the energy changes in the chemical reaction of two atoms or the emission of light by an atom amount only to an electron-volt or so, about one million times smaller. Ever since the turn of the century, when Henry Adams dreamed of "attaching a motor to radium," people have speculated about the practical possibilities of tapping this source of energy. Henry Adams' vision has in fact been realized by using the thermal energy produced by radioactive decay as an auxiliary energy source in artificial satellites, but this is far too expensive for most purposes; something other than simple radioactive decay is needed.

A general view of the possibilities is obtained by considering the *binding energies* of various nuclei. Recall from Chapter 13 that the nucleons that form a nucleus attract one another, having a lower nuclear energy when close together than when separated. The amount of energy that would have to be supplied to break up a nucleus into its constituent nucleons is the binding energy of the nucleus; the larger the binding energy, the more stable is the nucleus. The numerical value of the binding energy can be determined from the mass defect of the nucleus, the

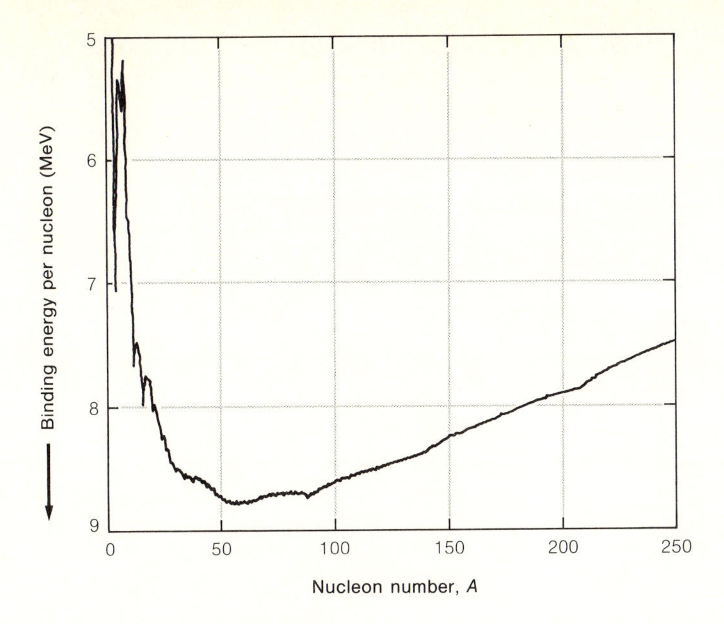

FIGURE 14.1
Binding energy per nucleon, as a function of nucleon number, A. (For many values of A, there are several nuclei; the one with the largest binding energy has been plotted.)

amount by which the mass of a nucleus is less than the sum of the masses of the individual nucleons.

More informative than the binding energy itself is the binding energy *per nucleon*, as we will see below; a graph of this quantity is shown in Figure 14.1. The binding energy is really a measure of *negative* contributions to the energy; the larger the binding energy, the lower the nuclear energy of the nucleons in the nucleus. For this reason, Figure 14.1 is drawn "upside down," so that the most stable nuclei lie lowest on this curve. Although some nuclei (He^4, the α-particle, for example) have exceptionally large binding energies in comparison with their neighbors, the majority of nuclei have a binding energy of about 8 MeV per nucleon. The most significant feature of Figure 14.1, however, is that there is a broad minimum (a *maximum* binding energy per nucleon) in the vicinity of $A = 60$, for elements near the position of iron in the periodic table. Nuclei near this minimum have the lowest possible value of nuclear energy. Somehow in these nuclei, the attractive interactions between nucleons are most effective in binding the nucleons to one another, in achieving a low value of nuclear energy.

If a method could be found either to join two light nuclei to make a heavier one or to split a very heavy nucleus into two approximately equal halves, an overall lowering of nuclear energy would result, and a corresponding amount of energy would become available in some other form. These two processes are called nuclear fusion and nuclear fission respectively, and we can use Figure 14.1 to see about how much energy would be released. Suppose, for instance, that we could join together two light nuclei with $A = 30$ to make a single nucleus with $A = 60$. For $A = 30$, the binding energy per nucleon is about 8.5 MeV; the two nuclei together have 60 nucleons and therefore a total binding energy of 60×8.5 MeV. For $A = 60$, the binding energy per nucleon is about 8.8 MeV. Thus the total binding energy increases by $60 \times (8.8 - 8.5) = 18$ MeV. In other words, the overall nuclear energy of the 60 nucleons decreases by 18 MeV, and an energy of 18 MeV will be released.

QUESTION

14.1 By the same line of reasoning, show that the *fission* of a nucleus with $A = 120$ into two equal halves would release an energy of about 36 MeV.

These two hypothetical examples were used only as illustrations. Almost any fusion reaction between two light nuclei would release energy, as would the fission of any heavy nucleus. It is the fusion of very light nuclei and the fission of the very heaviest nuclei which turn out to provide the most feasible way of obtaining useful energy. Either fission or fusion would lead to a release of energy, but what are the circumstances under which these reactions will actually occur? Why haven't all the light nuclei fused together already, and why haven't all the heavy nuclei undergone fission? In short, why do we not live in a world of nothing but "lukewarm iron"? As with many questions about energy conversions, just because something *can* happen does not mean that it *will* happen. The law of conservation of energy tells us that if a heavy nucleus splits in two, there will be a decrease in nuclear energy and a release of energy in some other form. It does not tell us whether this *will* happen, or if it does happen, how soon. A wood fire burns beautifully when you once get it started; a mixture of hydrogen and oxygen gas can combine explosively, but a spark is needed to get it started; a stalled car can coast downhill, releasing its gravitational potential energy, but an initial push is needed if there is a bump in the road (Figure 14.2). In all these situations, some kind of "energy barrier" (literally a barrier in the last example) keeps the process from starting; some initial push or some sort of "match" is required.

FIGURE 14.2
The energy barrier for a stalled car.

In nuclear fusion, the barrier is an electrical one. Consider two light nuclei. If they are more than about 10^{-15} m apart, they repel one another. If they are to get closer together, their electrical potential energy ($k_e q_1 q_2 / r$) must be increased. Only if they succeed in getting very close together will the attractive nuclear force take over and fuse them, lowering their nuclear energy. Figure 14.3 shows how life looks to one of these two nuclei; the solid line shows the energy resulting from the combination of the electrical repulsion and the attractive nuclear force. If the barrier can be surmounted, the nuclear energy will decrease as the

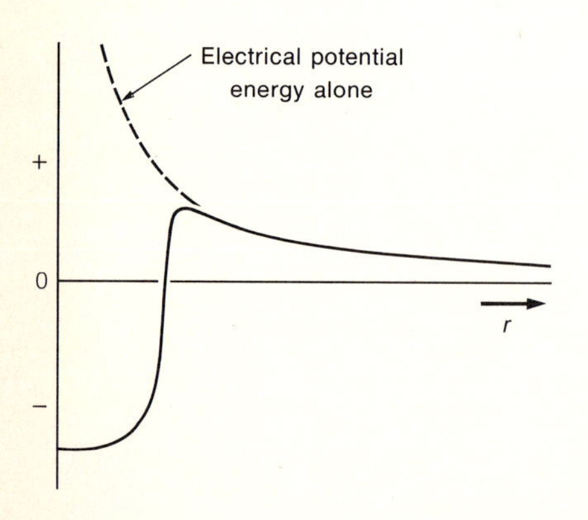

FIGURE 14.3
The energy barrier for fusion.

nuclei come together, but an initial supply of energy, a push of some sort, is required. The problem is similar in fission. Once a nucleus has managed to split into two halves, the electrical repulsion will blow the two segments apart. But just at the instant of fission, there is still an attractive nuclear force acting that tends to pull the two halves back together. Life as seen by such a hypothetical half of a heavy nucleus is shown in Figure 14.4. Its normal abode is to the left of the barrier. If enough energy can be added to get it to the top of the hill, then it can coast the rest of the way. A cow would prefer to be on the other side of the fence where the grass is greener. She needs to be a strong jumper to give herself enough kinetic energy so that she can gain the necessary amount of gravitational potential energy to get over the fence; otherwise she remains where she is. Thus energy barriers exist that inhibit fusion and fission. It turned out that absorption of a neutron makes some heavy nuclei unstable and can lead to fission; we will return to fusion in the following chapter.

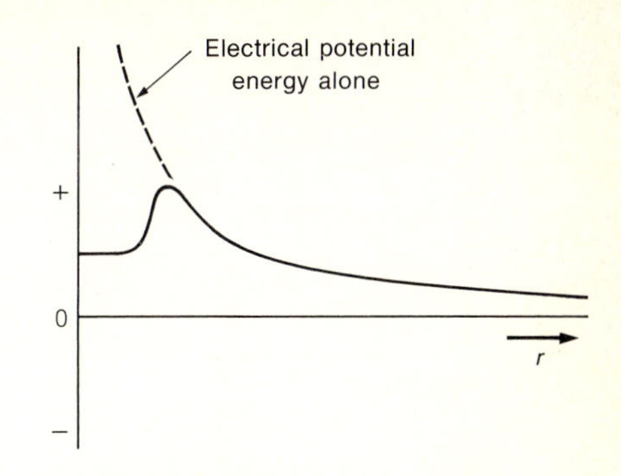

FIGURE 14.4
The energy barrier for fission.

§14.2 NEUTRON ACTIVATION, THE SEARCH FOR TRANSURANIUM ELEMENTS, AND THE DISCOVERY OF NUCLEAR FISSION

Not long after its discovery in 1932, the neutron was being used to study nuclear physics, and in the course of this work, it was discovered that a slow neutron can induce fission of a U^{235} nucleus. This was an accidental discovery, not one made in the course of a "crash program for the harnessing of nuclear energy." The property that makes neutrons useful to the experimental nuclear physicist is that they have no electrical charge. The slowest neutron may wander up to a nucleus and "fall in." When a neutron does fall into another nucleus (or is "absorbed"), a new nucleus is formed, as in the following reaction, for example:

$$_{47}\text{Ag}^{107} + n \longrightarrow {}_{47}\text{Ag}^{108}.$$

Like the example of Rb^{97} discussed in §13.4, Ag^{108} has too many neutrons for the number of protons and decays by β-emission:

$$_{47}\text{Ag}^{108} \xrightarrow{\text{2.4 minutes}} {}_{48}\text{Cd}^{108} + \beta^- + \text{neutrino}.$$

This creation of a new radioactive substance (Ag^{108}) is described as the "neutron activation" of silver.* Observe what has happened. By bombarding silver with neutrons and waiting for the subsequent β-emission, we have changed a few silver nuclei into cadmium nuclei ($Z = 48$), the element one place higher up in the periodic table. By repeating this

*Neutron activation is a very useful method for detecting the presence of small amounts of various elements. Does an old coin contain silver? We can test it by exposing it to neutrons; the subsequent detection of β-emission with a half-life of 2.4 minutes would provide a strong clue.

process, we may be able to move steadily upward through the periodic table.*

These results led to an interesting idea. What if we bombard uranium, then the element at the very end of the known periodic table ($Z = 92$), with neutrons? Can we make nuclei of a brand new element ($Z = 93$) and perhaps go even further, bombarding this element to make nuclei with $Z = 94$, and so on? The idea of making "transuranium" elements in this way is a sound one. In retrospect, it is certain that the physicists of the 1930s did produce a few transuranium nuclei, and measurable amounts of elements as high as $Z = 106$ have since been produced. There was a quite different and unexpected process going on at the same time, though, and to a much greater extent, and that was fission. The realization of this fact was obscured by the experimenters' firm expectation—a reasonable and indeed a correct one as far as it went—that they were producing transuranium elements, and by their understandable failure to anticipate the possibility that fission might be induced by neutrons. When the confusion and controversy had been resolved, it was apparent that when a nucleus of U^{235}, the less common isotope of uranium, absorbs a neutron to form U^{236}, the resulting nucleus is unstable and has enough energy so that it can get over the fission barrier and split into two approximately equal halves. (It is actually the U^{236} nucleus that undergoes fission; we commonly refer to this process, however, as the "neutron-induced fission of U^{235}.") The two parts into which the uranium nucleus splits (the fission fragments) are nuclei of elements near the middle of the periodic table. (Careful attention must be paid to these fission fragments; most of them are radioactive and are the source of most of the health and safety problems arising from the use of nuclear fission power.)

Moreover, in the fission of U^{236} this nucleus does not simply break in two; some *neutrons* are also emitted. This is the key fact that makes a chain reaction possible; the new neutrons can themselves cause the fission of other uranium nuclei, which in turn cause the fission of still more, and so on. The number of neutrons emitted in the fission process is variable, an average of about 2.5 neutrons in each event. The important point is that the average number of emitted neutrons is greater than one, so that a growing chain reaction is possible. The fission of a single uranium nucleus is shown schematically in Figure 14.5, and Figure 14.6 shows how the neutrons produced in one fission event can lead to a growing chain reaction.

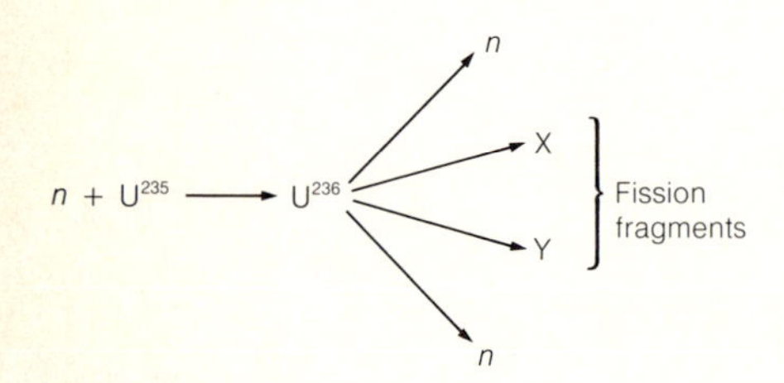

FIGURE 14.5
A fundamental fission event.

The three most important facts about fission are these. (1) If fission occurs, a large amount of energy is released. (2) A neutron can make a nucleus unstable and cause fission. (3) On the average, more than one neutron is emitted in each fission event, and thus a chain reaction is

*It was observed in Chapter 13 that we can turn gold ($Z = 79$) into lead ($Z = 82$). This could be accomplished by a series of neutron activations. However, the small amounts of lead produced would be extremely expensive, and even if the relative prices of gold and lead were reversed, this proposal would not represent a financially attractive investment.

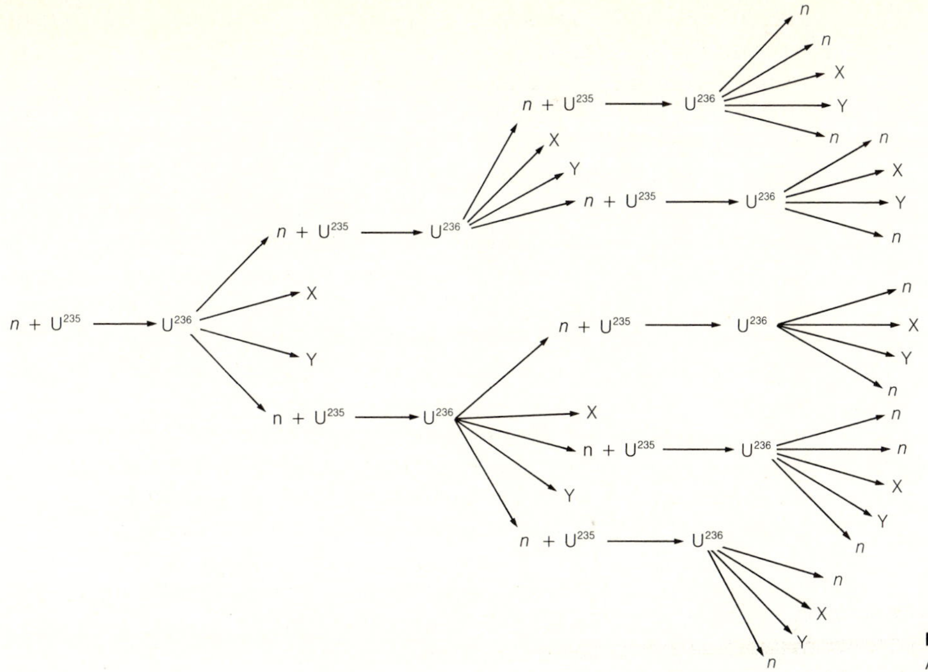

FIGURE 14.6
A growing fission chain reaction.

possible, either an uncontrolled explosive reaction or one that delivers useful energy at a steady rate. One consequence of these facts was the dropping of two nuclear bombs on Japan in 1945; another was the successful operation of the first American commercial nuclear power plant for generating electrical energy at Shippingport, Pennsylvania in 1957.

§14.3 FURTHER DETAILS ABOUT NUCLEAR FISSION

In a sample of uranium as found in nature, only 0.7% of the nuclei are of the species U^{235}. A U^{235} nucleus has the special property that instability leading to fission results from absorption of a *slow* neutron; in fact, a slow neutron is more likely to cause fission than a fast one, simply because a slowly moving neutron has more time to interact with the nucleus. A nucleus whose fission can be induced by a slow neutron is called a *fissile* nucleus, and U^{235} is the only fissile nucleus found in nature. Most uranium nuclei are U^{238} nuclei; fission of U^{238} requires additional energy and occurs only with fast neutrons. Slow neutrons are simply absorbed by U^{238} nuclei, and the chance that even a fast neutron will cause fission is quite low. A fission chain reaction with U^{238} nuclei is impossible, and, in an ordinary sample of uranium, the neutrons absorbed by the U^{238} nuclei make it more difficult to achieve a U^{235} chain reaction. For this reason, a process of "isotopic enrichment" is usually required, to manufacture uranium fuel with an adequate percentage of U^{235} nuclei. The absorption of neutrons by U^{238} nuclei is itself an important process, however—a process used to produce fissile Pu^{239} nuclei, as will be discussed in §14.7.

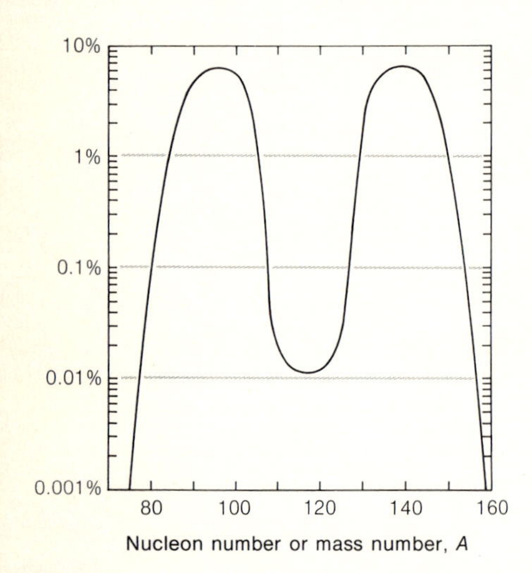

FIGURE 14.7
Distribution of masses of fission products resulting from U^{235} fission. Note the logarithmic scale on the vertical axis.

The fission of the U^{236} nucleus, formed when a U^{235} nucleus absorbs a neutron, results in the emission of several energetic neutrons and two fission fragments. We might have expected that the two fission fragments (X and Y in Figure 14.5) would each have a nucleon number of 117 or so. However, fission is almost always asymmetric, one of the two fragments being significantly heavier than the other. Typically, one has a nucleon number in the vicinity of $A = 95$, and the other about $A = 140$, but the fragments produced are not the same on each occasion. There is a distribution of the masses of the fission fragments (or "fission products") as shown in Figure 14.7. In about 6% of the total number of fissions, for instance, one of the two fragments has a nucleon number of $A = 95$; in about 1% of the events, a fragment is produced with $A = 84$, and so on. Consider just one of these possible fission events in detail. Suppose that the U^{236} nucleus emits three neutrons and breaks up into two fission fragments, one of which has a nucleon number of $A = 90$. The total number of nucleons in the U^{236} nucleus is 236, 93 of which are accounted for by the neutrons and the first fragment, and so the nucleon number of the second fragment must be $236 - 93 = 143$. Thus we can describe the event by writing:

$$_{92}U^{235} + n \longrightarrow {}_{92}U^{236}$$

$$_{92}U^{236} \longrightarrow X^{90} + Y^{143} + 3n.$$

The sum of the proton numbers (Z) of the two fragments must be equal to 92. If they are, for example, 37 and 55, then X and Y are isotopes of the elements rubidium and cesium:

$$_{92}U^{236} \longrightarrow {}_{37}Rb^{90} + {}_{55}Cs^{143} + 3n. \quad (14.1)$$

These two fission products are highly unstable; they have too many neutrons in comparison with the number of protons.

QUESTION

14.2 Where would these two nuclei be located on Figure 13.11? To the left or right of the stability line?

As with other unstable nuclei discussed in Chapter 13, these fission products move toward the stability line by β-emission, usually by a *series* of β-emissions. The sequence of steps for Rb^{90} is shown below, together with the various half-lives:

$$_{37}Rb^{90} \xrightarrow{2.6 \text{ minutes}} {}_{38}Sr^{90} + \beta^- + \text{neutrino}$$

$$_{38}Sr^{90} \xrightarrow{28 \text{ years}} {}_{39}Y^{90} + \beta^- + \text{neutrino}$$

$$_{39}Y^{90} \xrightarrow{64 \text{ hours}} \underset{\text{(stable)}}{{}_{40}Zr^{90}} + \beta^- + \text{neutrino.}$$

The radioactivity of the fission products has two important practical consequences. First, there is the "decay-heat" problem. It is impossible to turn off a nuclear power plant completely. One may stop the fission process itself, but the fission products already produced will continue to disintegrate, liberating energy in amounts sufficient to melt the fuel and perhaps destroy the plant unless steps are taken to keep it cool. Second, some of the resulting nuclei, such as Sr^{90} in the example discussed, have long enough half-lives so that if they are ever released into the environment, they may be incorporated into human food and thus cause biological damage. (Strontium is chemically very similar to calcium, and significant amounts of Sr^{90} produced in bomb tests during the 1950s found their way into milk and hence into children's bones and teeth.) The production of radioactive fission products is an inevitable consequence of the use of nuclear fission. If we insist on using nuclear fission as a source of energy, we must learn to deal with these fission products. The energy problem would be a much easier one to deal with if only it were true that all fission products were stable or had half-lives of a few minutes or less.

QUESTION

14.3 Consider the amount of energy released in the fission of U^{235} and the accompanying decrease in mass.

(a) Consider the fission event described by Equation 14.1. By referring to the binding energy per nucleon curve (Figure 14.1), show that an energy of approximately 200 MeV is released in such an event.

(b) Show that if an energy of 200 MeV is released, the decrease in mass is about 0.21 amu.

A figure of 200 MeV/fission is a convenient round number to use. Most of the energy released is converted into the kinetic energies of the neutrons and of the fission fragments.* This kinetic energy is transferred to surrounding atoms by collisions, thus leading to an increase in the thermal energy of this material, thermal energy that can be extracted as

*Some energy is given to the neutrinos emitted in the β-decay of the fission products, and nearly all of this energy is therefore carried off into outer space. These are the neutrinos detected in 1953 that finally demonstrated the reality of neutrinos (§13.6).

heat and used to generate electrical energy. The decrease in mass in one fission event is less than 1 amu. Most of the mass remains, largely in the form of fission products, and we can therefore say quite accurately that using up 1 ton, say, of uranium nuclei results in the production of nearly 1 ton of fission products.

QUESTION

14.4 It is easy to convert the basic result, 200 MeV per atom used, into more practical terms. If 200 MeV is available from each atom, show that:

(a) The energy available per *ton* of U^{235} is about 7×10^{13} BTU.

(b) An energy of 1 Q (10^{18} BTU) could be obtained from 1.3×10^7 kg of U^{235}.

(c) The total amount of energy used by the average American in one year could be supplied by about 4.5 g of U^{235}.

(d) The total amount of energy used in the whole world in one year could be supplied by about 3400 tons of U^{235}.

(e) An energy of 1000 megawatt-years (MW-yr) could be obtained from 800 lb of U^{235}.

Two important warnings must be appended to these results. First, if the energy released is first converted into thermal energy (and this seems to be the only practical procedure), then the second law of thermodynamics prevents us from converting all of this energy into more useful forms; electrical energy can be generated only at an efficiency of 30% or so. A 1000-MW(e) power plant (1000 megawatts-electrical), a plant with an electrical output of 1000 MW, produces 1000 MW-yr of electrical energy during one year but requires an energy input about three times as large and thus will use up about 2400 lb of U^{235} during a year's operation. (For such a typical large power plant, 1 ton of U^{235} per year is a useful number to remember. Such a plant also inevitably produces about 1 ton of radioactive fission products per year.) A second warning is that of all uranium nuclei, only 0.7% (one in 140) are U^{235} nuclei; if we were to forget this fact, we would be seriously overestimating the value of any given quantity of raw uranium.

§14.4 NUCLEAR FISSION POWER PLANTS

Now that we have discussed the essential physics of a nuclear power plant, let us consider some criteria that must be met in order to put these ideas into practice. We need enough U^{235} (or another fissile material, if available) to maintain a chain reaction. We must be sure that, on the average, the number of neutrons emitted in each fission event that subsequently cause another fission is precisely equal to one. If this number were less than one, the reaction would fizzle out and stop. If this number were greater than one, we would have uncontrolled exponential growth, a runaway chain reaction, releasing a great deal of energy in a short period of time. This is what is desired if we are designing a bomb, but it must be avoided at all cost in a nuclear power plant.

If the amount of uranium is too small, neither a power plant nor a bomb can possibly work; too many neutrons escape and no chain reaction takes place. If the uranium were in the most compact possible shape, that of a sphere, increasing the size would increase the surface area through which neutrons could escape in proportion to the square of the radius, but the amount of uranium would increase in proportion to the *cube* of the radius. There is a critical size, or a "critical mass" of uranium, below which no chain reaction occurs even under the most favorable conditions. The critical mass for pure U^{235} is about 5 kg. All nuclear power plants contain far more than a critical mass of uranium, but even with enough uranium, great care is needed to make a successful bomb, in which a great deal of energy is released in a small fraction of a second. Many hazards are associated with the use of nuclear power, but there is virtually no chance of a genuine nuclear explosion.

We must have the option of stopping the chain reaction in case something goes wrong or simply for maintenance or refueling. (Nuclear power plants are not perpetual motion machines; new supplies of fissile material must be added from time to time.) Because slow neutrons are more effective than fast ones in causing fission, some sort of "moderator" is usually needed to slow down the energetic neutrons released by fission. The nuclear energy liberated by fission is converted into thermal energy. This thermal energy must be removed, not only to keep the reactor from melting but also because this is the energy we want, and provisions must be made for using it to generate electrical energy. Then, since we can convert thermal energy into electrical energy only at an efficiency of about 30%, we must have some place to put the waste heat. Finally, the plant must operate safely and reliably and, if possible, produce electrical energy at an "acceptable" price.

There are many different kinds of nuclear power plants, many different ways of trying to meet these requirements. Whether enough attention has been paid to safety is a matter of great controversy. We will describe the basic principles of one common type of plant, the boiling water reactor (BWR) and some of the details of a particular example. All of the specific details in the following description refer to the Vermont Yankee plant at Vernon, Vermont, a BWR with a rated output of 540 MW [540 MW(e)], which went into operation in 1972. The efficiency of this plant is about 30%, so that the total rate at which nuclear energy must be converted into thermal energy is $540/0.30 = 1800$ MW. It can thus be described as an 1800-MW(t) power plant (1800 megawatts-thermal). A schematic diagram of a BWR is shown in Figure 14.8. The fuel is packed into long thin "fuel rods," 18,000 of them, each 12 ft long and 1/2 in. in diameter. Such a shape is necessary so that the heat generated can be readily removed, keeping the temperature within the fuel rods low enough that they do not melt. Altogether, the reactor core contains 68 tons of uranium, enriched so that 2.7% of the uranium nuclei are

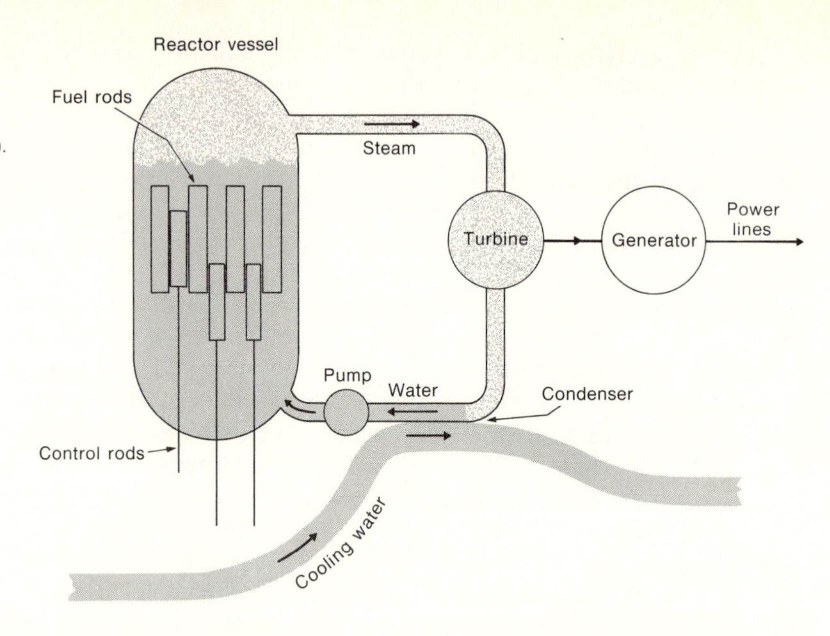

FIGURE 14.8
Schematic diagram of a boiling water reactor (BWR).

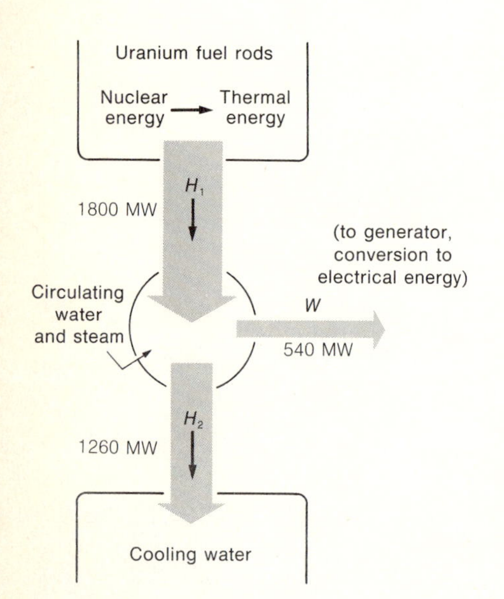

FIGURE 14.9
The essential flows of energy in a nuclear power plant.

U^{235} nuclei. The fission chain reaction produces thermal energy which heats the circulating water to produce steam. Once the steam is produced, a nuclear power plant is little different from any other steam-electric plant. The steam drives turbines to generate electrical energy and is then returned to the reactor vessel, along the way passing through a condenser where the waste heat is extracted before the water, as a liquid, returns to the reactor vessel. An even more highly schematic diagram, showing the essential flows of energy, is given in Figure 14.9.

QUESTION

14.5 How much energy could be obtained from the fission of all the U^{235} in the fuel? About how long could the plant operate at its rated level before refueling? (In practice, fuel rods must be replaced before all the U^{235} is used up, because the accumulated fission products hinder the operation of the chain reaction. When the fuel rods are removed, the remaining uranium is carefully extracted for further use, and the dangerous fission products must somehow be disposed of. The fuel rods are usually not replaced all at one time; at Vernon, the plant is shut down for partial refueling about once a year.)

The water that circulates through the plant in a closed cycle (in the form of both liquid water and steam) is both the moderator that slows down the neutrons and the "primary coolant" that extracts the energy we want and delivers it to the turbines for generating electrical energy. The neutrons produced by fission are slowed down by collisions with protons in the water molecules; this makes the neutrons more effective at causing subsequent fissions. (As we saw in Chapter 3, an ideal way of slowing something down is to let it undergo an elastic collision with an object of equal mass. Since the masses of protons and neutrons are nearly identical, protons are ideal for this purpose.) The same collisions of course heat up the water; the water boils within the reactor core, and the steam thus produced drives the turbines. An unwanted effect is produced by the water: that is, instead of an elastic collision, a sticking collision sometimes occurs between a neutron and a proton. The neutron is absorbed, a new heavier nucleus is formed, and such a neutron is lost as far as the fission process is concerned. After passing through the turbines, the water is in the form of very low pressure steam. In order to get it back to the reactor vessel, it must be condensed into liquid, and in order to do this, heat must be extracted. In the condenser, heat flows from the water within the reactor into a stream of cooling water, at a rate of 1260 MW.

QUESTION

14.6 At the Vernon plant, cooling water flows through the condenser at a rate of 800 ft^3/sec. By about how much is the temperature of the cooling water raised as it passes through the condenser?

The Vernon plant is designed so that the waste heat can be disposed of in two different ways. At some times, a fraction of the normal flow of the Connecticut River passes through the condenser, but during some parts of the year, the river's rate of flow is so low that the resulting rise in temperature of the river would exceed the standards set by the state of Vermont. On these occasions, auxiliary cooling water is used that circulates in a closed loop, receiving heat in the condenser and then delivering its energy to the atmosphere in cooling towers.

FIGURE 14.10
Construction of the containment vessel of the Vermont Yankee nuclear power plant. [Photograph by Edward Lee, New England Power Service Company.]

There are two principal methods of controlling the reaction. The first is an inherent part of the design. If the rate of the reaction should increase, the water within the core would boil more rapidly and contain more bubbles. Thus the amount of water in the core would be smaller, the number of moderating collisions between neutrons and protons would decrease, and, since fast neutrons have a smaller chance of inducing fission, the rate of the reaction would tend to drop to its former level. In addition, control rods containing boron, a good absorber of neutrons, can be inserted into the reactor vessel. The same control rods, when fully inserted, absorb so many neutrons that the chain reaction is stopped. This can be done deliberately, to shut down the reactor for refueling; it will also be done automatically in case the rate of the reaction reaches a dangerously high level.

About half of the plants in operation or being planned as of 1975 were BWR's, but many variations are possible. Many plants are pressurized water reactors (PWR's), in which the water in the core is heated at such a high pressure that it does not boil; an auxiliary water circuit is needed

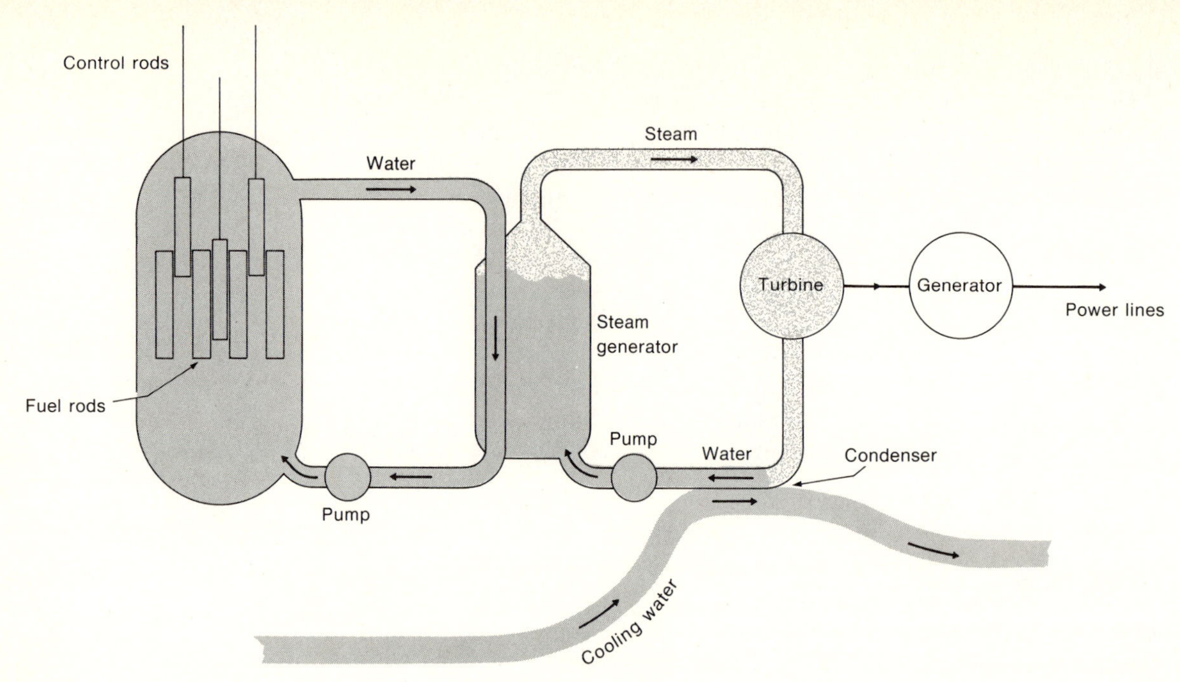

FIGURE 14.11
Schematic diagram of a pressurized water reactor (PWR).

to make steam to drive the turbines (Figure 14.11). Another variation is the high-temperature gas-cooled reactor (HTGR), in which helium gas is used instead of water to extract heat from the core. Nuclear power plants also differ widely in their size. The first commercial plant was a 90 MW(e) plant, which was put into operation at Shippingport, Pennsylvania in 1957; many of the plants scheduled to begin operation in the late 1970s will have electrical outputs of more than 1 GW (1000 MW).

The fundamental physics of a nuclear plant, then, is quite simple and well understood, but it should be noted that the engineering necessary to put these concepts into practice is not easy. A 1000-MW(e) power plant is an impressively large, complicated and potentially dangerous piece of machinery. A capital investment of several hundred million dollars is required, and it takes a minimum of five to seven years from the time the decision is made to build a plant until it goes into operation. It is inevitable that we will have to make long-term plans with less than perfect information, making the best possible estimates of future power needs and trying to make the best possible choices on number, type, and location of plants to be built.

§14.5 SAFETY AND ENVIRONMENTAL PROBLEMS

Even persons who remain blissfully unaware of many environmental problems know that there are risks associated with the use of nuclear power. The most important problems, some of which have been mentioned earlier, will be briefly discussed in this section. While a reactor is operating, large numbers of neutrons are produced, and some sort of "biological shield" must be built to protect the operating personnel against stray neutrons as well as against other types of radiation that are less penetrating. (The neutrons *within* the core are sufficient to deliver a lethal dose in about 10^{-3} sec.) Some neutrons may be absorbed by O^{16} nuclei in the water or by a variety of nuclei in structural materials or in impurities in the water. This absorption of neutrons often results in radioactive products, some with very short half-lives but also some with longer half-lives. This means that the water passing through the core must be kept extremely pure, that any water that leaks out may contaminate the environment, and also that any equipment that has been exposed to neutrons may still be radioactive after it has been removed.

The waste heat that is inevitably produced may result in a serious local thermal pollution problem, as mentioned before. Any power plant in which electrical energy is generated from thermal energy produces thermal pollution, but the operating temperatures in fission reactors are somewhat lower than those in fossil-fuel plants, and nuclear power plants therefore tend to have lower efficiencies. The efficiency of a 1970-model nuclear power plant is about 30%, comparable to that of many existing fossil-fuel power plants but lower than the efficiency of about 40% that can be achieved in modern fossil-fuel plants.

The problems of most concern are those which arise from the fission products. Almost all fission products are radioactive, and their production—in dangerously large amounts—is an inevitable consequence of nuclear fission. Most of the public's experience so far with fission products results from fallout from bomb tests, either fission bombs of the World War II variety or hydrogen bombs (which use a fission trigger to set off a fusion reaction). There are four major ways in which dangerous quantities of fission products *could* be released to the environment.

First, even during routine operation, some of the fission products within the fuel rods may escape into the cooling water. Some of these products inevitably get out into the environment by one means or another. Although this problem is a cause for some concern, it is considerably less serious than others.

Second, reactor accidents are possible. To take the extreme case, any accident that resulted in the demolition of a reactor would be very serious. Accidents might result from external causes such as earthquakes, and therefore reactors are designed to provide safe containment even in the event of severe earthquakes. A direct hit by a large airplane could also do serious damage. Plane crashes are not frequent, and the probability that a reactor will be hit, let alone hit in such a way as to release a large amount of radioactive material, is extremely small. Nevertheless, the possibility cannot be ruled out that a terrorist group might carry out, or threaten to carry out, a suicidal mission directed at a nuclear power plant.

More worrisome are the accidents which might follow from malfunctions in the power plant itself, such as the accidental rupture of faulty pipes. A major cause of concern is that, even though the fission chain reaction itself would almost certainly stop in the event of an accident, the accumulated fission products would make the accident more serious. The point is that fission products are radioactive and give off heat as they decay. Even after a reactor has been completely shut down, the accumulated fission products continue to deliver energy (the decay-heat); immediately after a shutdown, energy is delivered at a rate equal to about 5% of the rate at which energy is generated when the plant is operating. This is a significant amount of power, enough to melt the reactor core in a few minutes unless provision is taken to continue cooling. After a shutdown, the decay-heat steadily decreases, but cooling must be continued for a considerable period of time. It is conceivable that rupture of the pipes carrying the cooling water could lead to a sudden draining of all the water from the reactor core. With the loss of the moderator (the water), the fission chain reaction would immediately stop, but emergency cooling must be provided to keep the core from melting. All contemporary reactors are equipped with an emergency core-cooling system. Unfortunately it is by no means certain that these systems will work. They have never been subjected to full-scale tests, theoretical calculations are extremely difficult, and (fortunately) no accidents have occurred that required their use. Although a serious loss-of-coolant accident is thought to be an extremely unlikely event, the consequences could be so serious that dependable emergency cooling systems are essential. A complete lack of cooling could result in a complete melt-down in which a sizable fraction of the fission products within the plant would be released to the environment while the core of the reactor, much of it in molten form, would slowly sink into the ground. This sort of possibility is sometimes described as the "China syndrome," but the humorous name does not make the problem any less serious. Whether or not emergency cooling systems will work if needed is one of the major unresolved problems about nuclear power plants.

A third way in which fission products could reach the environment arises during reprocessing of the nuclear fuel. A piece of pure uranium, or a uranium fuel rod that has never been in a reactor, is not especially dangerous, but a used fuel rod contains significant amounts of highly radioactive fission products and must be treated with great care. After a fuel rod has been removed from a reactor, it is first kept at the reactor site for several months to give the fission products with short half-lives a chance to decay, and then it is taken to a reprocessing plant where the unused uranium is removed. Some of the gaseous fission products are generally dispersed into the atmosphere, and small quantities of other fission products find their way into local streams. Releases from reprocessing plants do not now constitute a major problem, but this question should be carefully watched as the number of power plants and therefore the amount of fuel to be treated increases.

Finally, the large quantities of radioactive wastes remaining after reprocessing must somehow be stored. (Remember that one reactor produces about one ton of fission products per year. See Question 14.16 for a calculation of how much storage space might be needed.) The wastes produce a significant amount of heat, and this must be taken into account in any proposed storage method. Most significantly, some of the fission products have such long half-lives that the wastes must be stored for hundreds or even thousands of years. At present, most of our long-lived wastes are legacies from the production of nuclear weapons and are stored in steel tanks, but this is not a permanent solution. A number of storage places have been proposed, among them abandoned salt mines and the ice shelves of Antarctica. The engineering problems in waste disposal are many, and the time scale raises political and ethical questions completely unlike those arising from other kinds of technology. The only *permanent* disposal method that has been suggested is to send the wastes off in spaceships, but the possible consequences of a mishap—should one occur during the launching of a rocket carrying such a cargo—seem too horrifying to contemplate.

These are serious problems, but we must remember that other ways of generating energy also have an impact on the environment and on the health of the population. To name but a few of the more obvious advantages of nuclear power, there are no ashes (at least not of the ordinary sort) and no smoke; nor is there any risk of an oil spill or a coal-mine explosion, and no strip mining is necessary.

§14.6 BIOLOGICAL EFFECTS OF RADIATION

Nuclear power is dangerous. No one denies this, least of all those who know most about it. It is in trying to proceed beyond this correct but uninformative statement that the difficulties begin. There are so many kinds of dangers (α-particles, β-particles, γ-rays and neutrons of various energies), so many different ways in which people may be exposed (by standing near a radioactive sample, by eating contaminated food, etc.), so many different kinds of damage that can be done. However, we have much more information about the biological effects of radiation (radiation from radioactive materials and from medical X-rays) than we do about many of the chemicals we eat and breathe. It is a sad fact that much of our best information has been obtained from studies of the effects of the nuclear bombs dropped on Japan in 1945, of the experiences of workers in uranium mines, and of the medical histories of the women whose job it was to paint radium dials of watches in a New Jersey factory during the 1920s.

A "dose" of radiation can be approximately quantified in various ways and given in units called the roentgen, the rad, and the rem. The roentgen is defined in terms of the amount of ionization the radiation produces in air. The rad (radiation absorbed dose) is closer to what we are interested in; 1 rad is the amount of radiation that delivers an energy of 10^{-5} joules per gram of absorbing material. The rem (roentgen equivalent man) is very similar to the rad; in its definition allowance is made for the fact that various kinds of radiation differ in their effectiveness in causing biological damage. No serious error is made if the rad and the rem are used interchangeably, and we will use the rem and the millirem (mrem).*

An energetic α- or β-particle, a γ-ray, a neutron, or an X-ray photon can alter or disrupt molecules within the body. A massive dose leads to such a severe disruption of biological functions that acute radiation sickness or death may follow. A dose of about 500 rem delivered within a brief period of time is followed by death in about 50% of all cases. (The same dose is much less serious, however, if it is spread out over a period of weeks or months; this is one of a number of factors that makes it difficult to quantify the effects of radiation.) A dose of 100 rem or more often results in severe radiation sickness. Doses smaller than about 50 to 100 rem may not be immediately noticed but may cause the beginning of a cancer that may not be apparent for many years. This kind of damage, damage to the body of the individual who receives the dose, is called "somatic" damage. Whether there is a *safe* dose, a "threshold dose" below which no somatic damage is done, is not certain, but it is probable that even a very small dose causes a slight increase in the chance that the individual will develop a cancer. If so, if a large number of people all receive a small dose, it is likely that a few individuals will subsequently die as a result but that most will suffer no harm. At any rate, in the absence of definitive information, health standards are based on the prudent assumption that there is no threshold for somatic damage, but this does not mean that all radiation exposures are forbidden. Establishing standards is a delicate business, calling for much more than just scientific knowledge. Account must be taken not only of the dangers of radiation but also—at least implicitly—of the other risks we run in our lives, of the effects of other pollutants, and of the risks we are willing to live with in order to obtain energy.

A second and very different kind of damage can also be caused by radiation: *genetic* damage. Exposure of the reproductive cells can cause mutations whose effects will not become apparent until the exposed person becomes a parent and may not even show up until many generations later. As far as we know, any dose of radiation, no matter how small, may cause mutations, the probability of genetic damage being approximately proportional to the size of the dose.

---

*The roentgen, the rad, and the rem describe the effects of radiation. Another unit, the *curie*, often appears in discussions of the dangers of radioactivity, but the curie is a very different kind of unit, one that is used to describe the strength of a radioactive source. A 1-curie source is one in which there are 3.7×10^{10} nuclear disintegrations per second. The number of curies does not directly tell us very much about the hazards. A large number of curies behind a lead wall may be totally harmless, whereas a tiny fraction of a curie in the wrong place (e.g., in the lungs) can be very dangerous.

It is important to realize that all of us are constantly exposed to radiation from natural sources and that many of us receive radiation doses in the course of medical and dental diagnosis and treatment. Some of the relevant data are given in Appendix R. We receive an annual dose of about 100 mrem from natural sources. Of this about 44 mrem comes from cosmic rays (a larger than average value being received by those who live in high-altitude areas), a nearly equal amount from radioactive nuclei in the soil and in building materials, and about 18 mrem from naturally occurring radioactive nuclei within the body. On the average, each of us receives about 70 mrem/yr from our doctors and dentists; anyone receiving an extensive series of diagnostic X-rays or X-ray treatments gets much more. Most of us receive no "occupational" dose at all, but some radiation workers may receive several thousand mrem/yr. Another group receiving an occupational dose is that of airline flight personnel; the dose rate from cosmic rays increases with altitude as shown in Appendix R. (Still higher dose rates may be received for brief periods of time by astronauts traversing the earth's radiation belts.)

A truly catastrophic accident, one in which a large fraction of the fission products is released to the environment under the worst possible meteorological conditions, might cause fatalities as far away as 75 miles or so if the population were not promptly evacuated, and might make land within several hundred miles of the plant unfit for habitation or agriculture for several years. This is the kind of accident that one likes to think will never happen (one whose probability of occurrence is almost certainly extremely small), but that *could* happen if everything went wrong. On the other hand, the routine operation of power plants and fuel reprocessing facilities results in very little radiation exposure of the public. Experience to date indicates that the annual dose received even by those living in the immediate vicinity amounts to no more than 5 mrem, and that therefore the average dose received by all members of the population is considerably less than 1 mrem/yr.

Any unnecessary exposure should be avoided, but in comparison with the 180 mrem/yr or so that we receive from other sources, the dose contributed by routinely operating nuclear facilities scarcely seems worth worrying about. If our goal is to reduce the radiation exposure of the population at large, much more attention should be devoted to guarding against the possibility of accidents at nuclear plants, to the safe storage of concentrated radioactive wastes, to avoiding unnecessary medical X-rays, and, above all, to ensuring that nuclear war never occurs.

Two crucial kinds of data have not been given. One is the probability of a serious nuclear accident. Is it one in a thousand per reactor per year? If so, such an accident might well occur in the course of the next decade. Or is it one in a billion, in which case we can afford to be much more sanguine about nuclear power? Unfortunately, no one really knows. The other missing information is that of the risks imposed by the alternatives to nuclear power. How much biological damage results from combustion of coal in electrical power plants? Again, no one knows for sure, although it is certain that some damage is done. Certainly there have been air pollution crises in which coal smoke was important, such as the great London smog of 1952, and it is quite probable that continuous low-level air pollution has an effect on health, but quantitative information is scarce. Another "alternative" to nuclear power is simply that of using less energy. The immediate effect would be to reduce pollution and to improve the public health, but the economic and social consequences and the indirect effects are difficult to calculate. Complete calculations of all the costs and benefits of all the possibilities are virtually impossible; in the absence of these data, completely rational decisions about nuclear power and other energy questions are not possible.

§14.7 NUCLEAR BREEDER REACTORS, THE INEXHAUSTIBLE SOURCE OF ENERGY

One obvious question about nuclear power has not yet been discussed. How much fuel is available? As for all questions about the amounts of various resources, it is impossible to give a simple and satisfactory answer. It is nonetheless important to use what information we have to make an educated guess, since we seem to be at least on the verge of running short of oil, possibly of coal in a century or two. How long will our uranium last? According to a 1973 estimate, there are about 10^6 tons of uranium available in the United States at prices currently considered reasonable. The translation of this figure into more meaningful terms is the subject of the following problem.

QUESTION

14.7 Assume that present reactors use only the U^{235} (0.7% of the uranium) and that nuclear energy can be converted into electrical energy only at an efficiency of about 30%. For how long a time would 10^6 tons of uranium last if we were to continue to use electrical energy at the 1973 rate and if we were to generate *all* this electrical energy with nuclear power plants?

This calculation suggests that our best uranium ores would last not more than a few decades if all our electrical energy were derived from fission of U^{235}. Uranium ores of poorer grade can always be mined, but only at the expense of more money (and energy). Significant increases in the costs of fuel might price nuclear power out of the market, at least if current costs of electricity generated from coal are used for comparison.* Of course we do not now generate a very large proportion of our

*Any discussion of comparative costs could change overnight, in either direction. The government might decide that coal smoke is so unhealthy and coal mining so dangerous or environmentally damaging that a prohibitively large tax should be placed on the use of coal. One can easily imagine many other possibilities.

electricity from nuclear energy (only 4.2% in 1973), and new coal-fueled plants will surely be built. Rough as they are, these calculations indicate that U^{235} fission reactors, safe or unsafe, may represent a technological dead end and that if we are not careful, we may squander our precious U^{235}, the only fissile nucleus found in nature, even more rapidly than we have been burning up our fossil fuels. U^{235} is very definitely *not* an "inexhaustible source of energy."

Reactors of the type built during the 1960s and 1970s may be very important in supplying significant amounts of energy for the rest of this century or a bit longer. Nuclear fission could not possibly be regarded as a truly long-range source of energy were it not for the possibility of greatly extending our supply of nuclear fuel by means of the ingenious idea of "breeding." Although U^{235} is the only naturally occurring species of nucleus that can be used as the fuel in a fission power plant, others that will work equally well can be artificially produced. Of these, the two most important are Pu^{239} and U^{233}. The first is produced as the result of the absorption of a neutron by a U^{238} nucleus, an event that leads to the nucleus U^{239}:

$$_{92}U^{238} + n \longrightarrow {}_{92}U^{239}.$$

U^{239} is unstable and decays by β-emission into a nucleus of the element neptunium (Np, $Z = 93$), the first element beyond uranium in the periodic table:

$$_{92}U^{239} \xrightarrow{\text{24 minutes}} {}_{93}Np^{239} + \beta^- + \text{neutrino}.$$

That is, just as expected by the nuclear physicists of the 1930s, neutron bombardment of uranium does produce transuranium elements. Np^{239} is also unstable and decays into a nucleus of plutonium (Pu, $Z = 94$):

$$_{93}Np^{239} \xrightarrow{\text{2.4 days}} {}_{94}Pu^{239} + \beta^- + \text{neutrino}.$$

Pu^{239} is also unstable; it decays by α-emission, but only with a half-life of 24,000 years. Most significantly, Pu^{239} is a *fissile* nucleus* like U^{235}.

A similar process begins with Th^{232}, the only naturally occurring isotope of the element thorium, an element that is even more abundant than uranium in the earth's crust as a whole. This process leads to the fissile nucleus U^{233}. These two processes ($U^{238} \longrightarrow Pu^{239}$ and $Th^{232} \longrightarrow U^{233}$) are called "uranium-plutonium breeding" and "thorium-uranium breeding." The isotopes U^{238} and Th^{232} are called *fertile* because, although they are not fissile, they can be readily converted into fissile nuclei.

*Because all reactors fueled with U^{235} also contain large amounts of U^{238}, some Pu^{239} is produced in all such reactors, and some of this Pu^{239} undergoes fission and contributes to the amount of energy released. Thus the energy generated by such a plant is somewhat greater than that produced simply by fission of U^{235}. Some of the reactors built during World War II were designed specifically for the production of large amounts of Pu^{239}, material that was used as the explosive in the first nuclear weapon ever set off (in a test in New Mexico in July, 1945) and also in the bomb dropped at Nagasaki the following month, whereas the explosive in the bomb dropped on Hiroshima was U^{235}.

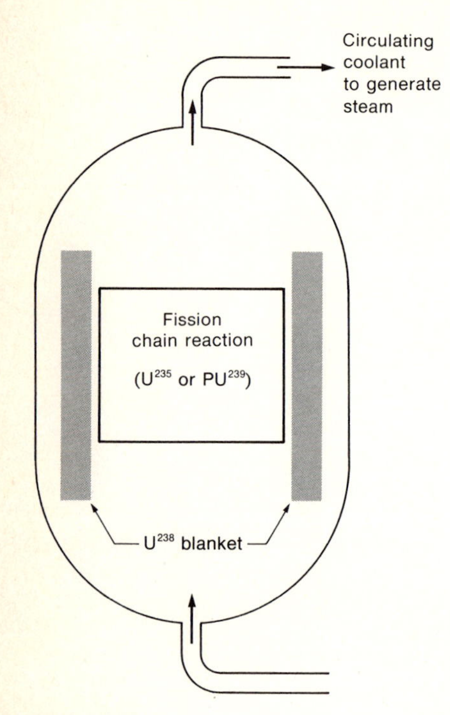

FIGURE 14.12
Central core of a breeder reactor.

The basic plan of a U^{238}–Pu^{239} breeder is shown in Figure 14.12. A self-sustaining chain reaction occurs in a central core. The fuel for this reaction could be U^{235}, but after breeders have been in use for some time, this fuel is more likely to be some previously manufactured Pu^{239}. Around the core is a blanket of uranium. As the chain reaction progresses, neutrons are produced. To keep the reaction going, on the average one neutron from each fission must itself cause another fission. If in addition a *second* neutron is absorbed by one of the U^{238} nuclei in the blanket to produce a Pu^{239} nucleus, then for every fissile nucleus used up, a new fissile nucleus (Pu^{239}) is produced. The reactor can deliver energy with *no* decrease in the total number of fissile nuclei. We can describe the essentials of this process as follows (the β-particles and neutrinos emitted as U^{239} decays into Pu^{239} have been omitted):

$$\mathrm{Pu}^{239} + n \longrightarrow (\mathrm{Pu}^{240}) \longrightarrow \overbrace{\mathrm{X} + \mathrm{Y}}^{\text{fission fragments}} + 2n + 200\ \mathrm{MeV}; \tag{14.2}$$

$$\mathrm{U}^{238} + n \longrightarrow (\mathrm{U}^{239}) \longrightarrow \mathrm{Pu}^{239}. \tag{14.3}$$

Adding these two reactions shows that the net effect is simply this:

$$\mathrm{U}^{238} \longrightarrow \mathrm{X} + \mathrm{Y} + 200\ \mathrm{MeV}.$$

The supply of Pu^{239} nuclei remains unchanged, new ones are created as fast as old ones are used up, but the supply of U^{238} diminishes.

If each fission event results in an average number of neutrons greater than two, then we may even end up with more nuclear fuel, more fissile nuclei, than we began with. Such a reactor is a true breeder; it breeds fissile nuclei from nonfissile (but fertile) nuclei. After a breeder reactor has operated for some length of time, it may have produced enough new fuel so that the excess can be removed and used to start a completely new reactor. Suppose, for simplicity, that each fission event produces *three* neutrons, one of which sustains the chain reaction, the other two being absorbed by U^{238} nuclei. Then every time that Reaction 14.2 occurs, the second reaction occurs *twice*:

$$\begin{aligned} &\mathrm{Pu}^{239} + n \longrightarrow \mathrm{X} + \mathrm{Y} + 3n + 200\ \mathrm{MeV} \\ &\mathrm{U}^{238} + n \longrightarrow \mathrm{Pu}^{239} \\ &\mathrm{U}^{238} + n \longrightarrow \mathrm{Pu}^{239}. \end{aligned}$$

The net effect is simply:

$$2\mathrm{U}^{238} \longrightarrow \mathrm{X} + \mathrm{Y} + 200\ \mathrm{MeV} + \mathrm{Pu}^{239}. \tag{14.4}$$

We generate energy and in addition produce new fuel, new supplies of fissile Pu^{239} nuclei. This plan depends for its success on the number of neutrons available. Each fission must produce at least *two* neutrons, which succeed either in causing fission of another Pu^{239} nucleus or in being absorbed by a U^{238} nucleus. This is a much more stringent requirement than that for a simple chain reaction. If the average number of useful neutrons from each fission event is barely equal to two, we can just hold our own. If this number is greater than two, we have a true breeder. In writing Equation 14.4, we assumed the availability of three neutrons for simplicity, but this is unrealistic. The average number of neutrons produced in a fission event is about 2.5. Some of these are lost in various ways, and a figure of about 2.2 for the average number of useful neutrons is about the best we can hope for. Even this is not easy to achieve. The "doubling time" of a breeder, the time it takes for a plant to double its inventory of fuel, depends on the amount by which the average number of useful neutrons exceeds the critical value of two. If the breeder concept is to be used to supply fuel for an increasing number of power plants, the value of the doubling time is important. A doubling time of 50 years would not be very helpful; doubling times of a few years would permit the construction of new plants at a rate to match growth in consumption of electrical power or permit the rapid substitution of nuclear power plants for coal-burning plants.

It may seem that a power plant that delivers useful energy and at the same time generates more fuel than it uses is a kind of perpetual motion machine. This is not so, however. Breeding simply represents an ingenious way of "unlocking" the nuclear energy of U^{238} nuclei. The real source of energy is U^{238}, and U^{238} *is* consumed in a breeder plant. Although breeder reactors do not violate the law of conservation of energy, they may for all practical purposes be nearly as useful as perpetual motion machines, if very large supplies of otherwise useless U^{238} are available.

There are a number of possible designs for a breeder reactor. Since the mid-1960s, attention in the United States has been centered on one particular concept, the LMFBR (Liquid-Metal Fast Breeder Reactor) (Figure 14.13), using the U^{238}–Pu^{239} breeding cycle. Fast neutrons will be used, because only for high-energy neutrons is the average number of useful neutrons per fission sufficiently greater than two for successful breeding. The chance that any one neutron will induce fission is smaller for fast neutrons; the fuel will therefore have to be more tightly packed, the power generated per unit volume will be larger, and thus the safety problems are different from those in earlier reactors. Because fast neutrons are desired, a moderator is not wanted. Therefore molten sodium will be used as a coolant instead of water. As in the PWR (Figure 14.11), a separate steam-generating loop is added to provide steam to operate the turbine.

Much of the technology developed for U^{235} power plants can be used in breeder reactors, but many new and quite different problems must be faced. Although a good deal of plutonium is inevitably produced in ordinary reactors by the neutrons that strike U^{238} atoms, the widespread use of breeder reactors will result in the production and transportation

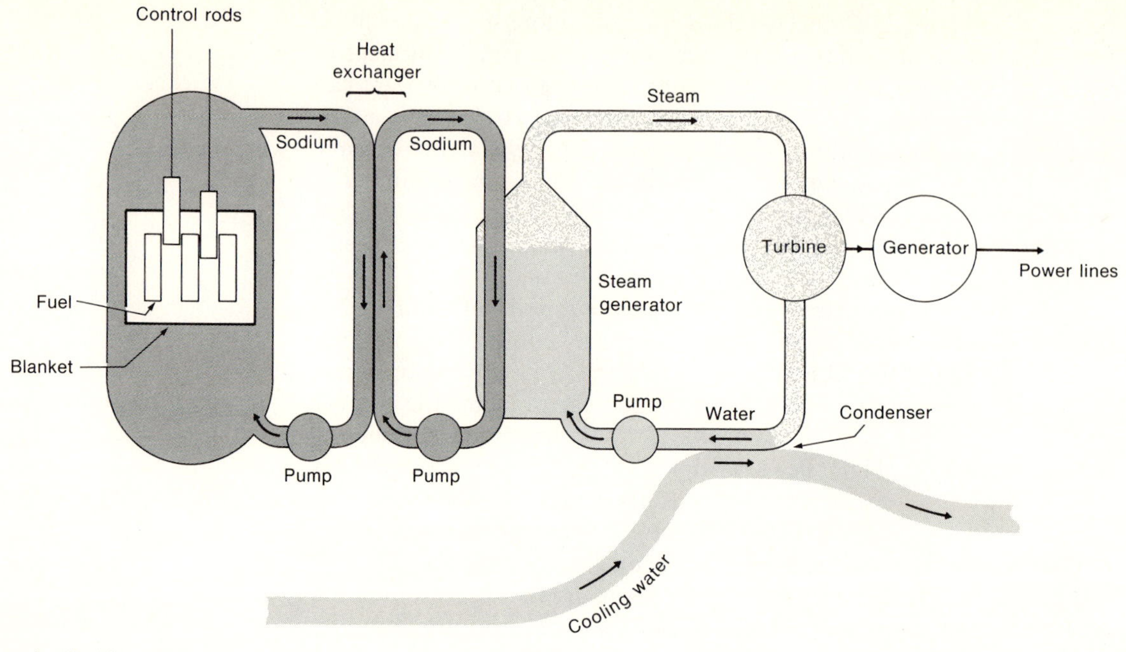

FIGURE 14.13
Schematic diagram of a liquid-metal fast breeder reactor (LMFBR).

of large quantities of plutonium. Plutonium is a highly dangerous material; very small quantities in the lung can be fatal. The uranium fuel used in power plants cannot be used to build bombs without an expensive and difficult process of further isotopic enrichment to make more nearly pure U^{235}; it is unfortunately much easier to build plutonium bombs, although opinions differ as to how difficult it would be for a small group of people to put together a bomb in a basement workshop. It will become increasingly difficult to monitor plutonium shipments; the fear that small nations and even terrorist groups may acquire plutonium to build nuclear bombs is a real one. The fundamental dangers of U^{235} fission remain; nuclear fission of plutonium produces dangerous fission products (Equation 14.2) in almost exactly the same quantities as does fission of U^{235}. There is no doubt that the concept of breeding is technically sound; breeder reactors do work. Whether safe, reliable, and economical breeders can be built is a matter of debate. At any rate, the United States has adopted the LMFBR as a key element in its energy strategy for at least the rest of the 20th century.

If breeders are used, the nuclear fuel picture will be drastically changed. In an economy based on U^{238}–Pu^{239} breeding, nuclear energy can be obtained from all uranium nuclei at the rate of 200 MeV/nucleus, not just from the U^{235} nuclei. This in itself would multiply the effectiveness of our uranium resources by a factor of 140. The impact of breeding will be even greater, however, because it will become economical to mine much poorer grades of uranium ore.

QUESTION

14.8 According to the data in Table J.1, an energy of 1000 Q would be generated by the fission of all the nuclei in the uranium available in the United States at prices as high as $100/lb. Calculate also the energy that could be obtained from fission of all the uranium nuclei in the earth's crust as a whole (see Table E.2) if mined to a depth of 1 km, and discuss the long-range energy problem of the United States and that of the entire world in light of these figures.

Thorium is even more abundant than uranium. Since thorium can be bred into the fissile nucleus U^{233}, it seems important to continue development on both of the two possible types of breeders, at least until it has been demonstrated that one or the other is clearly superior. If safe, reliable breeders allow us to tap even a fraction of the energy available in the thorium and uranium in the earth's crust, then we will have an almost inexhaustible source of energy; if abundant energy is what we need to gain the time to solve our other social problems, then nuclear breeders will do the job. If breeders do not work safely and reliably, a number of unpleasant possibilities can be imagined. The age of nuclear power may gradually end within a few decades as we exhaust our best supplies of U^{235}, leaving us as reminders a few hundred inoperative but radioactive power plants and huge quantities of accumulated radioactive waste that will have to be carefully guarded for centuries. More realistically, having as a nation placed a very large bet on the success of breeders, we may be persuaded by various easily imaginable pressures to go ahead with breeders even in the presence of serious unresolved questions about their safety.

Some of the questions about nuclear power have answers, and many of the answers are reassuring, but many questions are unresolved. On balance, routinely operating nuclear plants are less of a hazard to health than are coal-fueled plants; the most disturbing facts about nuclear power are that catastrophic accidents *could* occur, and that, even if no serious accidents do take place, radioactive wastes will be produced that must be stored almost indefinitely. Even if quantitative answers to all the technical questions were known, who would have the wisdom to weigh the undeniable benefits of nuclear power against the risks and to predict whether or not succeeding generations would guard our radioactive wastes with sufficient care? Many believe that a society that has sent astronauts safely to the moon can certainly deal with nuclear power. Others (and the author, with some ambivalence, is one) feel that a "bargain" in which one party is for the most part unaware that a bargain is being made and in which neither party has access to all the terms of the agreement is not a true bargain at all. Safe breeders may be feasible, present nuclear power plants are probably quite safe, and the hazards of using coal have in the past been consistently overlooked. Nuclear power may provide an easy way to get through the next few decades. But would it not be preferable to look for safer alternatives to both fossil and nuclear fuels, to develop cleaner ways of using coal if we must use coal, and to make concerted efforts to moderate the growth in the consumption of energy? Do we have the right to make a bargain in which *we* get the benefits of nuclear energy and in which our great-great-grandchildren inherit the responsibility of caring for our radioactive wastes?

FURTHER QUESTIONS

14.9 Explain clearly in your own words why the binding energy *per nucleon* (rather than simply the total binding energy of a nucleus) is especially important in considering the possibilities of obtaining energy from fission or fusion.

14.10 The fission of a uranium nucleus results in an energy release of about 200 MeV. If a uranium nucleus could be completely broken up into separated neutrons and protons, would this result in an even greater release of energy? If so, how much?

14.11 Only a small fraction of the mass of a uranium atom is lost during ordinary fission. If *all* the mass were to disappear, a much greater amount of energy would be released. About how much energy would be released per atom of uranium used up if we could carry out the "complete conversion of mass into energy," by having uranium atoms encounter "anti-uranium atoms" and annihilate each other? (It is safe to say that this calculation is only of academic interest.)

14.12 Make a rough check on the figures given in §14.4 for the number and dimensions of the fuel rods in the Vernon power plant by calculating the average density of the fuel if 68 tons of uranium are packed into these fuel rods. (The density of pure uranium metal is 19 g/cm^3; the actual fuel, a chemical compound of uranium, has a somewhat lower density.)

14.13 When a neutron strikes one of the protons in a water molecule, it is sometimes absorbed instead of undergoing an elastic collision. What is the new nucleus that is formed when this occurs?

14.14 Find out as much as you can about the nuclear power plant nearest to you: the type of plant, its electrical output, its capital cost, the amount of fuel it contains, how frequently it is shut down for refueling or repairs, the disposition of its waste heat, and so on.

14.15 In Question 5.18, we estimated the amount of energy that would be needed for disposal of our garbage and trash in outer space. Now that we know the rate at which fission product wastes are generated in nuclear power plants, that calculation can be adapted to find the *minimum* amount of energy needed for disposing of our nuclear wastes in the same way. Discuss this question. In particular, would the minimum amount of energy needed to dispose of one year's worth of fission products from a nuclear plant be much less than, comparable to, or much greater than the amount of energy the plant would generate during the year? If the answer is "much greater," then we can dismiss the idea once and for all; if the answer is "much less," other important questions must be answered before going ahead.

14.16 Make a rough estimate of the amount of storage space that would be needed to store radioactive wastes if *all* our electrical energy were generated with nuclear power plants. How many cubic feet per year? We know the mass of the fission products produced, and if we assume the average density is equal to that of water, we can make a rough estimate of the space required. Is it reasonable to devote this much space to the storage of nuclear wastes?

14.17 As an illustration of some of the analysis needed in dealing with fission product wastes, consider Kr^{85}, a radioactive nucleus with a half-life of 10.8 yr. One of the two fragments resulting from a fission event might be a Kr^{85} nucleus, or—more probably—fission might produce a nucleus with a different proton number that quickly decays into Kr^{85} by one or more β-emissions. The half-life of Kr^{85} is long enough that not many of the Kr^{85} decays will occur inside the power plant, but rather after the fuel rods have been removed and reprocessed. It is thus a reasonable approximation to say that every fission event that produces a fission fragment with $A = 85$ gives rise to one atom of Kr^{85}, which must be treated as part of the wastes that may contaminate the environment.

(a) How many atoms of Kr^{85} will be produced by a single 1000-MW(e) power plant operating for one year? Since each fission event yields 200 MeV, one can calculate the total number of fissions per year, and from Figure 14.7 one can estimate how many of these events produce a fragment with $A = 85$.

(b) Krypton normally exists as a gas, one of the "rare gases," and the present practice is simply to release it into the air during reprocessing of the fuel. If we wish instead to capture the krypton and store it for a few decades until it is harmless, we need to know in practical terms what this number of krypton atoms amounts to. How large a *volume* of gas is this if it is to be stored at a pressure of one atmosphere? Use the ideal-gas law, $pV = NkT$ (Equation 6.10), to estimate this volume, with $T = 300°K$ and $p = 1 \text{ atm} \simeq 10^5 \text{ N/m}^2$. This calculation is important because if it turns out to be 1 cm^3 or so, the problem of storing it is going to be much easier than if the result is 1 km^3! The result of this calculation should be a volume in the neighborhood of 1 m^3, about 30 ft^3, not a tiny volume but not impossibly large. By storing it under increased pressure, we can reduce the volume even more.

(c) Another problem that arises from the storage of all radioactive wastes, including krypton, is due to the heat generated by the radioactive decay. Estimate the rate at which thermal energy will be generated by this much krypton. To do so, we need to know the rate of decay and the amount of energy released in each decay. By examining the masses of Kr^{85} and its daughter nucleus, show that the energy released in each decay is about 0.7 MeV. Of this energy, more than half is carried off by neutrinos and need not concern us, and only about 0.24 MeV on the average is given to the β-particle and converted into thermal energy in the container. Because the half-life is 10.8 yr, half of the Kr^{85} nuclei will decay during the first 10.8 years, and so from this information we can calculate the average rate of production of thermal energy during this initial period; the rate will be slightly higher at the beginning of the period, but this average rate will enable us to make an estimate. Find this average rate in watts. How many light bulbs would produce heat at the same rate? (This comparison may give us an approximate idea of whether it is feasible to keep the krypton cool enough that it does not melt its container.)

14.18 About how many mrem are delivered to a person making a single plane trip from New York to San Francisco? (See Appendix R.) What would be the annual occupational dose of a jet pilot?

14.19 Other things being equal, residents of Denver receive a greater radiation dose because of their altitude than do residents of New York City. About how many more mrem per year?

14.20 Sodium nuclei (Na^{23}) are less effective in slowing down neutrons than are protons. What fraction of a neutron's kinetic energy would be lost in a head-on elastic collision with a Na^{23} nucleus?

14.21 The nucleus Pu^{239} can decay by α-emission. What is the daughter nucleus?

14.22 In Th^{232}—U^{233} breeding, absorption of a neutron by a Th^{232} nucleus eventually results in the production of a U^{233} nucleus. What is the probable sequence of transformations?

14.23 (a) With the successful development of U^{238}–Pu^{239} breeding, an energy of 200 MeV can be obtained from every uranium atom, U^{238} as well as U^{235}. About how many kilowatt-hours of electrical energy could be generated with one pound of uranium?

(b) How much could this electrical energy be sold for?

(c) Assume that uranium costs $10/lb. What do these results suggest about the contribution of the cost of the uranium to the overall cost of the electrical energy generated?

(d) If the cost of uranium rose to $100/lb, by how much would the price of 1 kWh of electrical energy have to be increased to allow for the increased cost of the fuel?

15
Nuclear Fusion Power

High Altitude Observatory, National Center for Atmospheric Research

15
Nuclear Fusion Power

§15.1 INTRODUCTION

§15.1.A Energy from Nuclear Fusion: D-T and D-D Reactions

It was pointed out in §14.1 that nuclei of intermediate mass are more stable than either very light or very heavy nuclei. Either the fission of a heavy nucleus or the fusion of two very light nuclei to make a heavier one leads to a decrease in nuclear energy and the release of a corresponding amount of kinetic energy. As a matter of fact, when the possibilities of obtaining useful energy from fission and fusion were first realized in the 1930s, it was fusion that seemed to be the most promising as a practical source of energy. No one would have guessed that fission would be successfully exploited first; the fact that a slow neutron can induce fission and set off a chain reaction came as a complete surprise. Having discussed fission in Chapter 14, we shall now consider the possibilities of fusion.

Four of the most important fusion reactions are listed in Table 15.1, together with the amounts of energy released. Consider first Equation 15.1, the deuterium-tritium (D-T) reaction, in which a deuteron (d, the nucleus of deuterium, D^2 or H^2) and a triton (t, the nucleus of tritium, T or H^3) interact to form an α-particle and a neutron. Because a neutron is produced as well as an α-particle, this reaction does not quite fit the initial description of fusion as the combination of two light nuclei to form a single heavier nucleus. It does release energy, however, as do the

TABLE 15.1 Four important fusion reactions.

NOTE: d, t, α and p represent the nuclei of the respective atoms; there is no corresponding symbol for the nucleus of a He^3 atom.

$$d + t \rightarrow \alpha + n + 17.6 \text{ MeV} \quad (15.1)$$

$$d + d \rightarrow t + p + 4.0 \text{ MeV} \quad (15.2)$$

$$d + d \rightarrow He^3 + n + 3.3 \text{ MeV} \quad (15.3)$$

$$d + He^3 \rightarrow p + \alpha + 18.3 \text{ MeV} \quad (15.4)$$

others listed in Table 15.1, and the energies listed have been calculated from the relevant mass data.

Remember that in a chemical reaction such as the burning of carbon, the energy released per atom or molecule is a few electron-volts, about a million times smaller than the energy released in a fusion reaction. In a fission reaction, the amount of energy released by fission of one uranium or plutonium nucleus is about 200 MeV. The energy released in one fusion reaction is smaller than this, but because the nuclei in fusion are lighter, the energy per gram of fuel used is of the same order of magnitude for fusion and fission.

The use of the D-T reaction to generate energy has two obvious drawbacks. First, tritium is not a naturally occurring isotope. If the D-T reaction is to be a continuing source of energy, some method must be found to maintain the supply of tritium. Second, tritium is a biological hazard because it is radioactive; it disintegrates by β-emission, with a half-life of 12 years. Deuterium is readily available and (unlike tritium) is not radioactive. It is therefore worth considering the reactions that may result from the interaction of two deuterons, and we will consider these D-D reactions in more detail in §15.1.B.

Other fusion reactions are possible, for instance between two protons. Because protons, the nuclei of normal hydrogen, are 6500 times more abundant than deuterons, it would be very convenient if a proton-proton reaction could be used. Unfortunately, the probability that a proton-proton collision will result in fusion is quite small, and these reactions are therefore not useful for generating energy here on the earth. However, as we will see in §15.5, these reactions are responsible for much of the energy produced in the sun and other stars.

QUESTION

15.1 Consider the fusion reactions listed in Table 15.1. Is the total number of nucleons the same before and after each reaction? Is electrical charge properly accounted for? For any one of the listed reactions, check the value given for the energy released. (Recall the trick used in Chapter 13 to simplify such calculations; by adding electrons to the nuclei on each side of the reaction, we can make use of the tabulated masses of complete neutral atoms, together with the mass of the neutron.)

§15.1.B D-D Fusion and Deuterium Resources

Let us consider more carefully the D-D reactions. There are two possible reactions (Equations 15.2 and 15.3), which occur with roughly equal probabilities:

$$d + d \longrightarrow t + p + 4.0 \text{ MeV};$$
$$d + d \longrightarrow \text{He}^3 + n + 3.3 \text{ MeV}.$$

The triton and the He^3 nucleus then react with deuterons according to Equations 15.1 and 15.4:

$$d + t \longrightarrow \alpha + n + 17.6 \text{ MeV};$$
$$d + \text{He}^3 \longrightarrow p + \alpha + 18.3 \text{ MeV}.$$

By adding these equations, we see that the four reactions together are equivalent to a single reaction:

$$6d \longrightarrow 2p + 2n + 2\alpha + 43.2 \text{ MeV}.$$

If and when D-D fusion power becomes a reality, an energy of 43.2 MeV will be released every time six deuterons undergo fusion, or about 7 MeV per deuteron. Let us consider the implications of this fact for the energy problem. It is the results of the calculations outlined in the following questions that have led to the intense interest in fusion research.

QUESTIONS

15.2 Ordinary water contains vast numbers of deuterons. Show that an energy of about 4×10^{10} J would be released by the fusion of the deuterons in a gallon of water. (Remember that the chemical formula for water is H_2O and so about one-ninth of the mass of the water is hydrogen; of the hydrogen nuclei, about one out of every 6500 is a deuteron.)

15.3 How many gallons of gasoline would have to be burned in order to supply an energy equal to that which would be produced by the fusion of the deuterons in a gallon of water?

15.4 If you are an American, you are responsible, directly or indirectly, for using energy at the average rate of about 10^4 J/sec, day in and day out. That is, you have an average power consumption of about 10 kW. About how long could your own energy needs be supplied by the fusion of the deuterons in a gallon of water?

In order to see how truly inexhaustible an energy supply we would have in the deuterons of the oceans, let us consider what the worldwide situation would be on the basis of some rather extreme assumptions about population and energy use. The present population of the world is about 4×10^9, and most of the people use far less power than the 10 kW used by the average American. Let us make what we may hope is a conservative assumption, that the world population will increase to 30 billion (30×10^9), and that all the people will be using energy at the rate of 10 kW.

QUESTIONS

15.5 Show that, on the basis of these assumptions, the amount of energy used per year in the whole world would be about 10^{22} J, or about 10 Q. (In considering the very long-range importance of various proposals for producing energy on a large scale, and in considerations of the adequacy of various energy resources, a rate of energy use of 10 Q/yr is a useful number to remember.)

15.6 (a) Show that the number of deuterons in the oceans is about 1.4×10^{43}.

(b) Show that if we can obtain an energy of 7 MeV per deuteron, the deuterons in the oceans could provide an energy of about 1.5×10^{10} Q.

(c) Show that on the basis of the preceding assumptions about population increase and energy use, the deuterons in the oceans could supply all the world's energy needs for about 1.5×10^{9} years.

It is almost unnecessary to say that successful tapping of an energy resource of this order of magnitude would have enormous social implications. By comparison, the useful fossil-fuel resources of the world amount to only a few hundred Q. Nuclear fusion could provide energy in sufficient quantity to meet all our imaginable needs for a period that is, for all practical purposes, indefinitely long.

§15.1.C The Energy Barrier and the Need for High Temperatures

However, we do not yet know whether the preceding calculations have any relevance to the world's energy problem. *Can* we get this energy? We might turn the question around, in the following way. If deuterons *can* fuse, why don't they? Why are there any deuterons left in the oceans, why didn't they all fuse long ago? The reason was discussed in §14.1: there is an energy barrier. For two deuterons to fuse, they must get close enough to each other that the strong attractive nuclear force can take effect. At very short distances (about 2×10^{-15} m or less), the attractive force is very strong, but at greater distances, it nearly vanishes. This distance represents the approximate size of the deuteron: $r_d \simeq 2 \times 10^{-15}$ m. At distances greater than r_d, the dominant force between two deuterons is one of electrical repulsion. The repulsive electrical force decreases as the separation between the deuterons increases, but at all distances significantly greater than r_d, it is strong enough to overwhelm the attractive nuclear force. The potential energy of a pair of deuterons resulting from these two forces is shown in Figure 15.1.

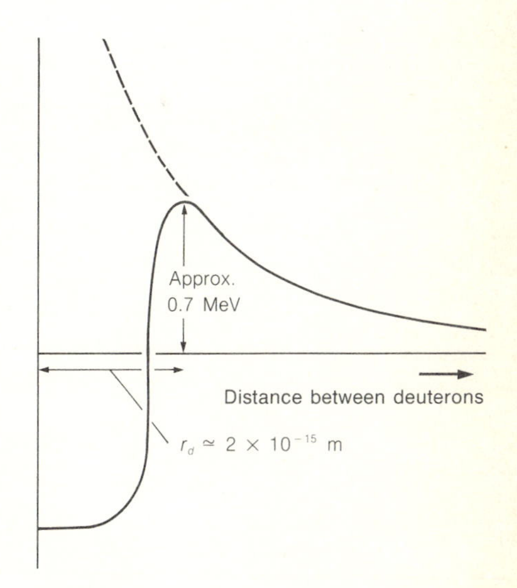

FIGURE 15.1
Interaction energy of a pair of deuterons. (The dashed line shows how the potential energy would continue to increase at shorter distances if the attractive nuclear force were absent.)

An energy barrier also inhibits the fission of heavy nuclei, but for a few nuclei (U^{233}, U^{235}, Pu^{239}) fission can be induced by absorption of a neutron. There is no corresponding trick which will produce the fusion of two deuterons. Instead, in order for two deuterons to get close enough together that the fusion reaction can occur, they must somehow be fired at each other with enough kinetic energy to get over the top of the hill. From very simple energy arguments, we can estimate the necessary amount of kinetic energy.

Suppose for simplicity that one of the deuterons is fixed and that the second deuteron is fired directly at it from a great distance away. If the incoming deuteron does make it to the top of the hill, it will then have an electrical potential energy whose value is represented by the arrow in Figure 15.1 and that is approximately equal to $k_e q_1 q_2 / r_d$, where r_d is about 2×10^{-15} m, q_1 and q_2 the electrical charges of the deuterons (each 1.6×10^{-19} coulombs), and k_e is, in MKS units, 9×10^9. This is an energy of

$$\frac{9 \times 10^9 \times (1.6 \times 10^{-19})^2}{2 \times 10^{-15}} \simeq 1.15 \times 10^{-13}\,\mathrm{J} \simeq 0.7\ \mathrm{MeV}.$$

If the incoming deuteron has this much or more kinetic energy, it can reach the top of the hill, losing kinetic energy as it gains electrical potential energy, and a fusion reaction may then occur. If it has less than this critical amount of kinetic energy, it will slow down, stop, and "roll back down the hill" instead of crossing the barrier. Precisely the same kind of argument was used in Chapter 2. In order for a rock to reach the top of a tree of height, h, its gravitational potential energy must increase by mgh, and therefore its initial kinetic energy must be at least equal to mgh. (This straightforward argument is based on the assumption that the ordinary principles of mechanics, "classical mechanics," can be applied to the behavior of deuterons. We will see later that the conclusions are modified somewhat by the application of quantum mechanics; for the time being, we will pursue the classical argument.)

To produce D-D fusion reactions, we need deuterons with kinetic energies of about 0.7 MeV. One approach is to accelerate deuterons with an electric field, create a beam of high-energy deuterons, and let them hit a deuterium target. Such experiments have been done on a small scale, and have provided us with important information about the fusion reactions. Even when the energy is high enough, however, only a few of the D-D collisions actually lead to fusion with a resulting release of energy. Most of the deuterons are scattered, like the α-particles shown in Figure 13.6, and the energy used to accelerate them is wasted. The amount of energy released in the few fusion reactions that do occur is smaller than the amount of energy required to accelerate the beam of deuterons in the first place.

The most thoroughly studied method of producing large numbers of collisions among high energy deuterons or other nuclei is to heat them up; the temperature must be raised enough that the randomly directed velocities of the particles will be large enough that an occasional head-on encounter will permit the energy barrier to be crossed. Most of the collisions will still lead merely to scattering, not to fusion. If the nuclei can be contained, they will collide again and again, sooner or later undergoing a fusion reaction.

This kind of fusion reaction is often called a "thermonuclear reaction." More precisely, we hope to achieve a *controlled* thermonuclear reaction, as opposed to the *uncontrolled* thermonuclear reactions that take place in the explosion of hydrogen bombs. The temperature required for a thermonuclear reaction is so high that any substance would be vaporized and would act as a gas. In fact, the random thermal energies will be so large that the orbital electrons will be knocked free of the nuclei, and we will have a gas of very high energy nuclei and electrons. Such a gas is called a plasma, and a whole new field of physics, plasma physics, has been developed in recent decades, largely because an understanding of the behavior of plasmas is necessary for achieving successful power production from nuclear fusion. The properties of plasmas are very different from those of common gases; plasmas are excellent conductors of electricity because the nuclei and electrons of which they are composed are free to move rather than being bound into electrically neutral atoms and molecules. The differences between plasmas and the familiar kinds of matter (solids, liquids, and gases) are so great that plasmas are sometimes referred to as the "fourth state of matter." Unusual though plasmas are on earth, in the universe as a whole they are quite common; a star is simply a very large plasma.

How hot must a deuterium plasma be in order for the fusion reaction to occur? We can make an estimate of the required temperature from the kinetic theory discussed in Chapter 6. For a collection of deuterons at temperature T, the average kinetic energy of each deuteron is $\frac{3}{2}kT$, where k is Boltzmann's constant (1.38×10^{-23} J/°K). If we require that this kinetic energy be about 0.7 MeV (10^{-13} J) in order for deuterons to fuse, we can determine the necessary temperature:

$$\frac{3}{2}kT \simeq 10^{-13}\ \text{J},$$

from which it follows that

$$T \simeq 5 \times 10^{9}\ {}^{\circ}\text{K}.$$

QUESTION

15.7 Check the calculation leading to the temperature just stated.

To say that such a temperature is extremely high is hardly an exaggeration. Fortunately, as an estimate of the needed temperature, it is overly pessimistic for several reasons. First, although $\frac{3}{2}\,kT$ is the *average* kinetic energy of the deuterons, some have much higher energies (see Figure

6.17). Even at a temperature of 10^8 °K or so, a few deuterons will have kinetic energies larger than 10^{-13} J. This is a very important correction, because it means that not every collision between deuterons, or even most such collisions, need lead to fusion. To calculate the required percentage of successful collisions is complicated; but it should be clear that, if the plasma can be contained so that the nuclei have many opportunities to collide, the fact that a few deuterons have kinetic energies much higher than average will significantly increase the number of successful collisions.

There is a second reason for considering our first estimate of the required temperature as being too pessimistic: a peculiar quantum-mechanical effect called "barrier penetration" or "tunneling." A deuteron coming in from the right (Figure 15.1) with a kinetic energy less than the critical value "should" slow down, stop, and roll back down the hill. Usually this is exactly what it does, but once in a while, according to quantum-mechanical principles, it can *tunnel* under the hill and appear on the left side of the barrier. The more kinetic energy the deuteron has, the more likely it is to penetrate the barrier, but it has some chance of getting through even though its kinetic energy is too small for it to go over the top. Tunneling is a quantum-mechanical phenomenon, one that is impossible to understand on the basis of ordinary mechanics. When these various factors are taken into account, it is found that our estimate of 5×10^9 °K as the required temperature is quite a bit too high. The temperature needed to make the D-D reaction work is "only" about 4×10^8 °K (400 million degrees!). This too is a very high temperature, higher than that in the interior of the sun.

The prospects are more favorable for D-T fusion. It happens that the chance of a fusion reaction occurring when a deuteron and a triton collide is greater than it is when two deuterons collide. As a result, the minimum temperature needed for D-T fusion is about 4×10^7 °K, ten times smaller than that needed for D-D fusion. It is therefore nearly certain that the first successful fusion reactors will use the D-T reaction. In the immediate future (which means, in this context, the next 50 to 100 years) fusion power, if it works at all, will very likely depend on the D-T reaction in spite of the problems associated with making and handling tritium.

§15.2 BASIC SCIENTIFIC, TECHNOLOGICAL, AND RESOURCE PROBLEMS

In the preceding section, we have seen that energy should be released in a fusion reaction, that a vast amount of fusion energy is potentially available from the deuterium in the oceans, and we have seen what the basic obstacle is to obtaining this energy for useful purposes. Although a practical fusion reactor has not yet been built, and may never be built, we do have three solid pieces of evidence for believing that the fundamental idea is correct, that fusion does release energy. First, as mentioned before, small numbers of fusion reactions have been produced in the laboratory using beams of high-energy deuterons. Second, fusion reactions provide the energy released in the explosion of hydrogen bombs. Third, there are excellent reasons for believing that it is the energy from fusion reactions that is responsible for the light produced by the sun and other stars. With confidence that the essential idea of releasing energy by means of fusion is correct, we can list the basic technical and scientific problems that must be solved before fusion power becomes a reality.

(1) Whatever the fuel is, it must be fed into the reactor and heated to a high enough temperature that a significant number of fusion reactions take place. A certain critical temperature, the "ignition temperature," must be reached for a fusion reactor to be self-sustaining. As noted in the previous section, the ignition temperature for D-T fusion is about 4×10^7 °K, whereas that for D-D fusion is much higher, about 4×10^8 °K.

(2) The plasma density must be high enough that a useful amount of power can be extracted from an apparatus of manageable size.

(3) The nuclei in the hot plasma will have very high speeds. Unless the plasma is somehow contained, it will be rapidly dispersed. If any plasma leaks out, energy is lost, for the lost plasma must be replaced by new fuel, and energy must be expended to heat up the newly added fuel. The energy released in fusion must be large enough to supply this energy, and at least somewhat greater so that useful energy can be extracted. The containment problem has turned out to be an extremely difficult one to solve.

(4) Some way of getting the energy out of the reactor in a useful form must be devised. Current plans call for using the kinetic energy of some of the particles to heat up some substance, and then using the heat to generate electrical energy, just as in a fission or fossil-fuel power plant. It would be highly desirable if such a thermal stage could be avoided, for once the fusion energy has been converted into heat, the second law of thermodynamics prevents us from generating electrical energy at 100% efficiency.

(5) Once we have resolved these problems, a great deal of careful engineering will be needed if we are to build reliable large-scale fusion reactors that will produce power at an acceptable price. What price is "acceptable" will clearly depend on social and economic factors as much as on technological and scientific ones. By the time fusion power becomes a reality, the situation may be very different from what it is now. For example, the realization that coal resources are limited and that the mining and burning of coal are hazardous to human health may be

reflected in a different tax structure favoring a better method of power production. If so, a fusion power plant that would be economically impractical if introduced today might well be the choice of the most hard-headed utility executive thirty years hence.

Once a successful reactor has been developed, new environmental problems must be considered.

(6) Can anything go wrong with a fusion reactor that might result in a catastrophic explosion? A fusion reactor has much in common with an H-bomb. Will it be possible for a fusion power plant to explode in a similar way, or, even if that is inconceivable, is there a possibility of smaller but still disastrous explosions?

(7) Fusion reactors will use and generate radioactive nuclei, and some of the construction materials will become radioactive. The magnitude of the environmental radioactivity problems must be carefully considered.

(8) What will be the situation with regard to resources? As we have seen earlier, there is plenty of deuterium in the oceans, but it will be easier to achieve a successful fusion process using the D-T reaction. It appears, as will be seen in §15.3.F, that lithium will be the key substance for maintaining the necessary supply of tritium, and the availability of lithium may turn out to be a limiting factor.

(9) If a thermal stage is used in the generation of useful energy, a large amount of waste heat must be put into some cooling medium such as an ocean or a river. Fusion power plants will probably be large ones, and this local thermal pollution problem may therefore be at least as serious as it is for fission plants.

(10) As we saw in Chapter 7, virtually all the energy we use is sooner or later dissipated as heat. Although this problem is not peculiar to fusion power, it is important to remember that no matter how much deuterium or lithium may be available, we cannot afford to use energy in unlimited amounts without running the risk of raising the temperature of the earth and producing significant, possibly catastrophic, changes in the climate.

At present, we do not know for certain whether fusion power can be successfully exploited. Much encouraging progress has been made in recent years, but numerous problems remain. Even if success is ultimately achieved, fusion power will not be completely free of environmental problems. In §15.3 we will describe briefly some of the promising approaches to the design problems and indicate how a practical fusion reactor might be built, and the environmental impact of fusion power will be discussed in §15.4.

§15.3 STATUS OF THE SCIENTIFIC AND TECHNOLOGICAL PROBLEMS

In this section we will examine in more detail the technical and scientific problems of fusion power, look briefly at some of the possible solutions, and speculate about the design of practical large-scale fusion power plants for generating electrical energy.

§15.3.A Ignition Temperature and Fuel Injection

As we have noted before, the temperature of the plasma must be at least as great as a critical "ignition temperature" for a fusion reactor to operate. The reason is the following. As the temperature of the plasma increases, more and more of the nuclei will have high enough kinetic energies to penetrate one another's energy barriers and undergo a fusion reaction. Unfortunately, the particles of the hot plasma lose energy in a nonproductive way by electromagnetic radiation, and the amount of energy lost in this way also gets larger as the temperature increases. Luckily, the rate at which energy is released by fusion increases more rapidly with increasing temperature than does the rate at which energy is lost. Figure 15.2 exhibits the trade-off between these two effects. For example, at temperatures below about 4×10^7 °K, a D-T plasma would lose energy faster by radiation than it would release energy by fusion, so

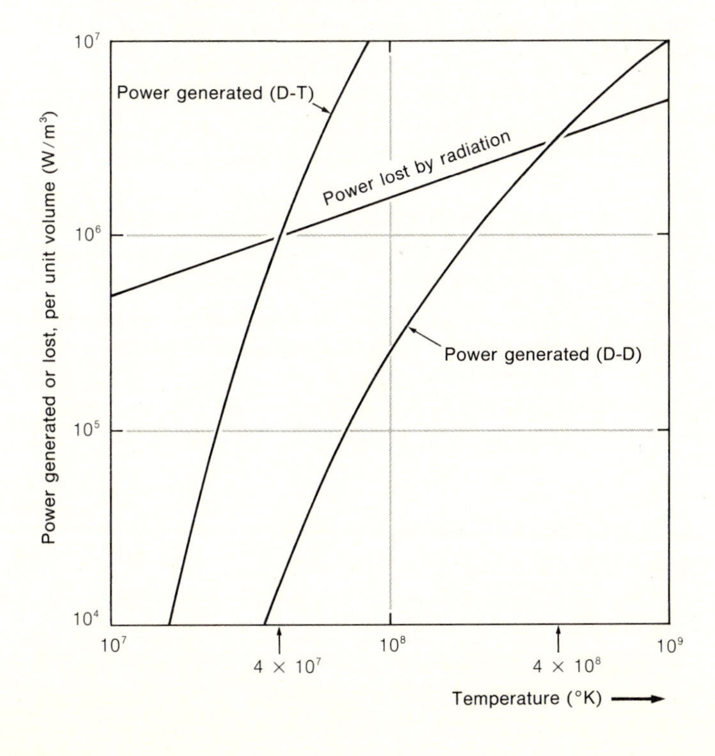

FIGURE 15.2
Power lost by radiation and power generated by D-T and D-D fusion. (The scales on both horizontal and vertical axes are logarithmic.) The numbers on the vertical axis apply specifically to plasmas with 10^{15} nuclei per cubic centimeter, but the temperatures required for achieving D-T and D-D fusion are approximately the same for other densities.

that 4×10^7 °K is the ignition temperature that must be reached or exceeded if a D-T fusion reaction is to be self-sustaining.

Once a fusion reaction has begun, the temperature of the plasma will be maintained by the energy liberated in the fusion reaction itself. There are two obvious problems, however. First, how can such a temperature be achieved in the first place? Second, once the reaction has begun, new fuel will have to be injected and heated, both to replace nuclei that have undergone fusion and to replace those nuclei which have escaped.

Various methods have been suggested for solving these problems. In most plans the initial heating is accomplished by passing an electrical discharge through the gas to ionize the atoms. Once a few atoms have been ionized, the electrical charges of the nuclei and electrons (no longer bound together in electrically neutral atoms or molecules) provide us with a "handle"; by applying electric and magnetic fields we can increase the kinetic energies of the charged particles. Other atoms of the gas will be ionized by collisions to form a plasma, and the temperature of the whole plasma can then be raised. Temperatures in excess of the D-T ignition temperature have been produced in several experimental devices. Unfortunately, although the ignition temperature has been reached, other necessary conditions for a successful fusion reaction (fairly high density and adequate containment) have not yet been achieved in the same devices.

The injection of fresh fuel into the space in which the fusion reaction is going on poses a problem. One cannot, for example, simply use a pipe that leads from external tanks of deuterium and tritium gas and empties into the plasma. The temperatures to which such a pipe would be exposed are much higher than the temperatures at which any known materials can exist without being vaporized. Furthermore, such a pipe would present an easy path by which heat could flow out of the plasma and prevent the ignition temperature from being reached.

It is perhaps premature to speculate now on how fuel will be injected, because the feasibility of any particular method of injection will depend on the size and shape of the plasma, the method by which the plasma is contained, and so on. The fuel might be injected by a continuous beam, either of electrically neutral atoms or of ions, or it might be injected in occasional large bursts; the fuel might even be injected in the form of small pellets of frozen deuterium or tritium, a technique that is sometimes aptly described as the "snowball in hell" method.

§15.3.B Containment of the Plasma

The major difficulty has been that of containing a plasma with a temperature of 5×10^7 °K or so. Of course any substance out of which we might build a "bottle" would melt at such a temperature, but even more serious is the fact that if the plasma were to come in contact with the walls, it would be cooled off so rapidly that the fusion reaction would stop. In other words, the plasma is very hot but has a very low heat capacity; it can be very easily cooled. The main line of attack on the containment problem has been to take advantage of the fact that the plasma is completely ionized, consisting of positively charged nuclei and negatively charged electrons, rather than electrically neutral atoms. If a magnetic field is imposed on charged particles, the particles do not move in straight-line paths but instead in circular or spiral trajectories. This is the basic idea of the "magnetic bottle." Rather than a wall of some physical substance, it is the magnetic field that contains the charged particles. Of course, no magnetic bottle is perfectly tight. A great deal of ingenious work, which we will not describe here, has gone into "plugging the leaks," into designing magnets that will do a sufficiently good job of containing the plasma. A hot dense plasma seems almost to have a mind of its own, finding unanticipated ways of getting out of magnetic bottles which look promising on paper. Plasmas have a distressing tendency to become unstable. That is, when a plasma bulges slightly at one point, the bulge itself may weaken the forces that are supposed to contain it, turning a small leak into a gaping hole. A great deal of theoretical work has been devoted to predicting and correcting for such instabilities, and a wide variety of devices to test proposals for containment have been built.

One intriguing aspect of this work is worth mentioning here. The necessary strong magnetic fields must be produced by an electromagnet, that is, by electrical currents flowing in wires at some distance from the plasma. Normally, whenever currents flow through wires (in a toaster or in an electromagnet), electrical energy is dissipated in heating up the wires. Thus in a conventional electromagnet, a large amount of power must be used just to maintain the magnetic field at a steady value. To avoid this waste of energy, the electromagnets used to contain the plasma may be made of superconducting materials. As discussed in Chapter 8, some materials become superconductors at low temperatures; that is, their electrical resistance vanishes. Thus no energy need be wasted in heating the wires, but refrigeration must be supplied, and some energy must be used for this purpose. In all probability, the first fusion reactors will employ some of the lowest temperatures as well as the highest temperatures ever artificially produced. A hot plasma, at a temperature of about 10^8 °K, will be contained by a magnetic field produced by wires only a short distance away that must be kept at a temperature of only ten degrees or so above absolute zero.

QUESTION

15.8 To see why containment is needed, consider the rate at which a plasma would be lost if it were not contained. Estimate the rms speed of deuterons in a plasma whose temperature is about 10^8 °K. About how long would such a plasma remain within a space whose linear dimensions are 10 cm or so if no steps are taken to prevent it from dispersing?

---

§15.3.C Fuel Density and the Relationship Between Density and Containment Time

Intimately related to the problem of containment is that of achieving a sufficiently high density of the fuel. The higher the plasma density, the more difficult it is to keep it confined with a magnetic field. However, if the density is high, we can afford to be less stringent about containment, for more energy will be generated in the plasma, and we will have more to spare for heating the fresh fuel which must be added. What the actual requirements on density and containment will be depends on the fuel used (and to a lesser extent on the details of the construction of the power plant).

The effectiveness of a containment system can be described by the value of the "containment time," the average length of time a nucleus stays in the plasma before leaking out. It turns out that not only must we reach the ignition temperature, but also we must guarantee that the *product* of the density and the containment time be at least equal to some minimum value. The numerical calculations yield the following approximate value for D-D fusion:

$$n\tau \gtrsim 10^{16}\ \text{sec/cm}^3; \tag{15.5}$$

whereas, for D-T fusion,

$$n\tau \gtrsim 10^{14}\ \text{sec/cm}^3. \tag{15.6}$$

Here τ denotes the containment time discussed earlier, measured in seconds. The plasma density is represented by n, the number of nuclei per cubic centimeter. In ordinary usage, density refers to the *mass* per unit volume, expressed, for instance, in grams per cubic centimeter. It is conventional in discussing fusion reactors to specify instead the *particle* density, the number of particles per unit volume (the number of nuclei per unit volume in this case). Since the number of nuclei is simply a number, independent of what system of units is used, we give the units of the particle density as simply cm^{-3}, and thus the units of $n\tau$ as sec/cm^3. These inequalities do not give sharply defined values of $n\tau$. As noted above, the values needed depend not only on the type of fuel used but also on the details of the reactor. For this reason, the symbol $\gtrsim$ was used, to indicate the approximate value of $n\tau$ that must be achieved, instead of the usual symbol for an inequality.

We can see why the product of density and containment time is the important quantity by the following argument. The issue is the balance between the power generated by nuclear fusion, and the power that must be used to heat up new plasma because of leakage. Let us denote the nuclear fusion power generated by P_g, and the power needed to heat new fuel by P_h. It is surely necessary that P_g be greater than or equal to P_h if fusion power is to be practical:

$$P_g \geq P_h. \tag{15.7}$$

In what way will P_g and P_h vary as the density varies? Consider first the effect of a density variation on P_g, the rate at which energy is generated by fusion. If, for example, the density is doubled, interactions in which nuclei collide and fuse will be four times as frequent, for there will be twice as many nuclei and each nucleus will have twice as many other nuclei to interact with. Other things being equal, P_g will be proportional to the square of the density. That is, we can write an equation:

$$P_g = An^2, \tag{15.8}$$

where A is a factor that depends on quantities other than the density. This factor surely depends on the temperature, for P_g increases with increasing temperature. It also depends on which fusion reactions we are considering, since P_g is greater for the D-T reaction than it is for the D-D reactions. Our concern at the moment is the way in which P_g varies with density, expressed by the factor n^2 in Equation 15.8.

How much power has to be wasted in heating new plasma to replace plasma that has leaked out? The amount of energy stored in the kinetic energy of the plasma particles is proportional to the number of particles, and therefore to the density, n. Let us write:

$$E_{\text{stored}} = Bn,$$

where B is a factor that, like A in Equation 15.8, depends on the temperature but not on the density. (In fact, since the kinetic energy of each nucleus is, on the average, $\frac{3}{2}kT$, B is directly proportional to the temperature.) On the average, every nucleus stays in the plasma for a length of time equal to the containment time, τ. Therefore, in effect, every τ seconds the fuel must be completely replaced, and therefore every τ seconds an amount of energy equal to E_{stored} must be put into newly added fuel to bring it up to the operating temperature. That is, P_h, the rate at which energy must be used for this purpose, is:

$$P_h = \frac{E_{\text{stored}}}{\tau} = \frac{Bn}{\tau}. \tag{15.9}$$

From Equations 15.8 and 15.9, Equation 15.7 can be written as

$$An^2 \geq Bn/\tau$$

or

$$n\tau \geq B/A,$$

where B/A, simply a combination of factors introduced earlier, is itself a quantity that depends on the temperature, the type of fuel, and so on.

From the preceding arguments, we can see why it is that the product of density and containment time is important. In order to find out how large this product must be, for a particular fuel and for a particular kind of reactor, more detailed calculations are necessary. Such calculations lead to the criteria expressed in Equations 15.5 and 15.6. In these equations we can see the trade-off between density and containment time.

If the density is high, we can afford to have a short containment time (and vice versa), while still satisfying the criteria expressed by these equations. It is clear that success will be more easily achieved with deuterium and tritium as a fuel rather than with deuterium alone. Not only is the ignition temperature lower for the D-T reaction but also the density and containment problems are less severe. The difficulties of using the D-D reactions arise primarily from the fact that the kinetic energies of the nuclei must be higher in order that there be appreciable chance of their undergoing fusion reactions (see Figure 15.2). Thus the operating temperature must be higher. With a higher operating temperature, energy is lost more rapidly. In addition, the energy stored in the plasma is larger, and therefore if any plasma leaks out a more serious energy loss is incurred.

§15.3.D Summary of the Critical Problems of Fusion Power

The critical problems, all closely interrelated, are those of reaching a temperature at least equal to the ignition temperature and simultaneously making a plasma for which the density and containment time satisfy the criteria expressed by Equations 15.5 or 15.6. Serious work for the purpose of achieving fusion power began in the early 1950s. Progress has often been frustratingly slow, and after an initial period of optimism, workers in this field have sometimes been extremely gloomy about the prospects for ever satisfying all the necessary criteria for a successful reactor. Many different devices have been built to test various containment schemes and to study the properties of plasmas under various conditions of temperature and density. The long years of frustration are suggested by the names of some of these devices, all of them expensive, that have been built: Scylla, Astron, Stellarator, Perhapsatron, Tokamak, Baseball, Spherator, Levitron, Albedo, Pyrotron, Felix, Ixion, and others. Such a list may give the false impression that workers in this field have merely performed a long series of unsuccessful experiments in vain attempts to make a fusion reactor. Although such an impression may not be altogether inaccurate, most of these machines were built to obtain answers to specific questions about the behavior of plasmas, rather than in the hope that the newest gadget would miraculously turn out to be a successfully operating fusion reactor.

Table 15.2 shows some of the results achieved in several of the best devices. Although these data may give some indication of the status of the field at a given time, it must be realized that specific numbers such as these will, in all probability, very quickly become obsolete. As the table clearly demonstrates, a device that looks promising in terms of one criterion may be quite poor in the light of another criterion. Scylla IV, for example, has been operated at a fairly high temperature and with a high density, but is unable to hold the plasma very long. A high temperature and a fairly long containment time have been achieved in Baseball I,

TABLE 15.2 Results achieved in several fusion devices.

	n (particle density) (particles/cm³)	τ (containment time) (sec)	$n\tau$ (sec/cm³)	Temperature (°K)
Tokamak	5×10^{13}	3×10^{-2}	1.5×10^{12}	5×10^{6}
Scylla IV	5×10^{16}	10^{-5}	5×10^{11}	5×10^{7}
Baseball I	10^{10}	4×10^{-1}	4×10^{9}	2×10^{8}
Minimum requirements for D-T fusion	-	-	10^{14}	4×10^{7}
Minimum requirements for D-D fusion	-	-	10^{16}	4×10^{8}

FIGURE 15.3
A Tokamak at Princeton University. [Plasma Physics Laboratory, Princeton University.]

but only with a very low plasma density. The Tokamak listed in Table 15.2 is a Russian machine that employs a particular type of magnetic containment. Results obtained with the first Tokamak were so promising that a number of Tokamaks based on the same principles have been built in the United States.

Comparison of the results in Table 15.2 with the minimum requirements for D-T fusion indicates that we are still a long way from meeting all the requirements in the same machine, and of course even further away from the necessary conditions for a D-D reactor. Nevertheless, dangerous though it may be to carry out technological and scientific extrapolations, it is hard not to feel optimistic about achieving successful D-T fusion within the next decade or so. Although this may prove to be a false hope, no fundamental principles analogous to the laws of thermodynamics are known that could prevent this achievement.

§15.3.E Extraction of Energy

How will we be able to extract useful energy from a plasma in which a fusion reaction is occurring? Most of the energy released will appear in the form of the kinetic energies of the particles produced. If the magnetic bottle is working, most of the charged particles will be contained within the plasma, but the neutrons, carrying no electrical charge, will not be affected by the magnetic field, and thus energetic neutrons will emerge from the plasma in all directions. Consider, for example, the D-D reaction described by Equation 15.3, and suppose for simplicity that the two deuterons collide head-on, with equal and opposite momenta: the total kinetic energy after the reaction will then be shared by the He^3 nucleus and the neutron, in inverse proportion to their masses. The He^3 nucleus will receive about 25% of the energy, and the neutron about 75%, so that we need only devise a way of making use of the large amount of kinetic energy carried off by the neutrons.

QUESTION

15.9 Why is the kinetic energy shared in inverse proportion to the masses of the outgoing particles?

We already know how to use the kinetic energy of the neutrons because of our experience with fission reactors. The neutrons can be used to heat water, so that steam is generated to drive a turbine and to produce electrical energy just as in a fission plant or a coal-burning plant. We will see in §15.3.F that it may be desirable to have an intermediate step, in which the neutrons are used to heat a stream of molten lithium whose heat is then transferred to water to make steam.

§15.3.F Tritium Breeding

Difficult though it will be to achieve a successful D-T reactor, it will be much easier, as we have seen, to satisfy the criteria for D-T fusion than for D-D fusion. Unfortunately, the D-T reaction requires tritium, an isotope not found in nature, as part of its fuel. In the absence of an abundant supply of tritium, a D-T fusion reactor would simply be a curiosity, a monument to technological and scientific ingenuity but not a useful source of energy. Luckily, nature comes to the rescue here, for the neutrons produced in the D-T reactions can be used to replenish the supply of tritium, and even to "breed" tritium, that is, to make more tritium than is used up. Breeding is essential, for although we could start up a reactor with tritium already made by other methods, we need to do more than simply replace the tritium used up in the fusion reactions, so that we will be able to make up for the inevitable losses of tritium and also produce surplus tritium to start up other fusion reactors.

The fact that makes this possible is that some naturally occurring nuclei produce tritium when bombarded by neutrons. The easiest way to make tritium is to use the reaction that occurs between neutrons and Li^6 (the light isotope of lithium):

$$Li^6 + n \longrightarrow t + \alpha + 4.8\ \text{MeV}. \tag{15.10}$$

Indeed, this process is the way in which most of our present supply of tritium, and the tritium used in H-bombs, has been made, with neutrons from fission reactors.

QUESTION

15.10 Check the value given for the energy released in Equation 15.10.

If every neutron produced in a D-T fusion reaction subsequently interacted in this way with a Li^6 nucleus, the overall sequence of reactions in a D-T reactor would be:

$$\begin{array}{rl} & d + t \longrightarrow \alpha + n + 17.6 \text{ MeV (Equation 15.1)} \\ + & Li^6 + n \longrightarrow t + \alpha + 4.8 \text{ MeV (Equation 15.10)} \\ \hline & d + Li^6 \longrightarrow 2\alpha + 22.4 \text{ MeV} \end{array} \quad (15.11)$$

This is good, but not good enough. This sequence of reactions would just replace the tritium used up, but it would not increase our supply of tritium or leave any margin of error to compensate for tritium losses. Furthermore, neutrons interact with all sorts of nuclei, and any interaction in which a neutron is used up without making a tritium nucleus will be harmful. We can be certain that no matter how closely we surround the fusion reactor with lithium, some of the neutrons will not produce tritium. Fortunately other nuclear reactions, in which one neutron comes in and two neutrons come out, can be used to increase the number of neutrons—for example, the reaction with Li^7 (the heavier, more abundant isotope of lithium):

$$Li^7 + n \longrightarrow Li^6 + 2n. \quad (15.12)$$

The idea, then, is to surround the fusion reactor with a "blanket" of lithium. Because of the excess neutrons from reaction 15.12, the tritium-producing reaction 15.10 can be made to occur somewhat more often than the original D-T fusion reaction (15.1), and thus we can not only maintain the supply of tritium but actually breed extra tritium, just as it is now planned to breed new supplies of fissile material in fission breeders.

§15.3.G A Possible Fusion Power Plant

When one considers the problems still to be solved, it is clear that it would be the height of folly to try to draw up "blueprints" for a fusion power plant at this time. Although we cannot be certain that fusion power plants will *ever* be built (being at the same time quite confident that, if they are, they will incorporate some ideas not yet thought of), it is nonetheless worthwhile to try to synthesize some of our present ideas, to imagine at least in a general way how a fusion power plant might operate.

One possibility is shown in Figure 15.4. Deuterium extracted from water, and tritium (initially from a supply on hand, later from the tritium produced in the plant itself), are somehow injected into the central region where the fusion reaction is occurring. The plasma is contained within this central evacuated region by magnetic fields produced by electrical currents in the surrounding coils. The coils will be made of superconducting materials, and therefore must be kept at very low temperatures. As D-T reactions occur, neutrons are produced that escape from the plasma and are absorbed in the surrounding lithium

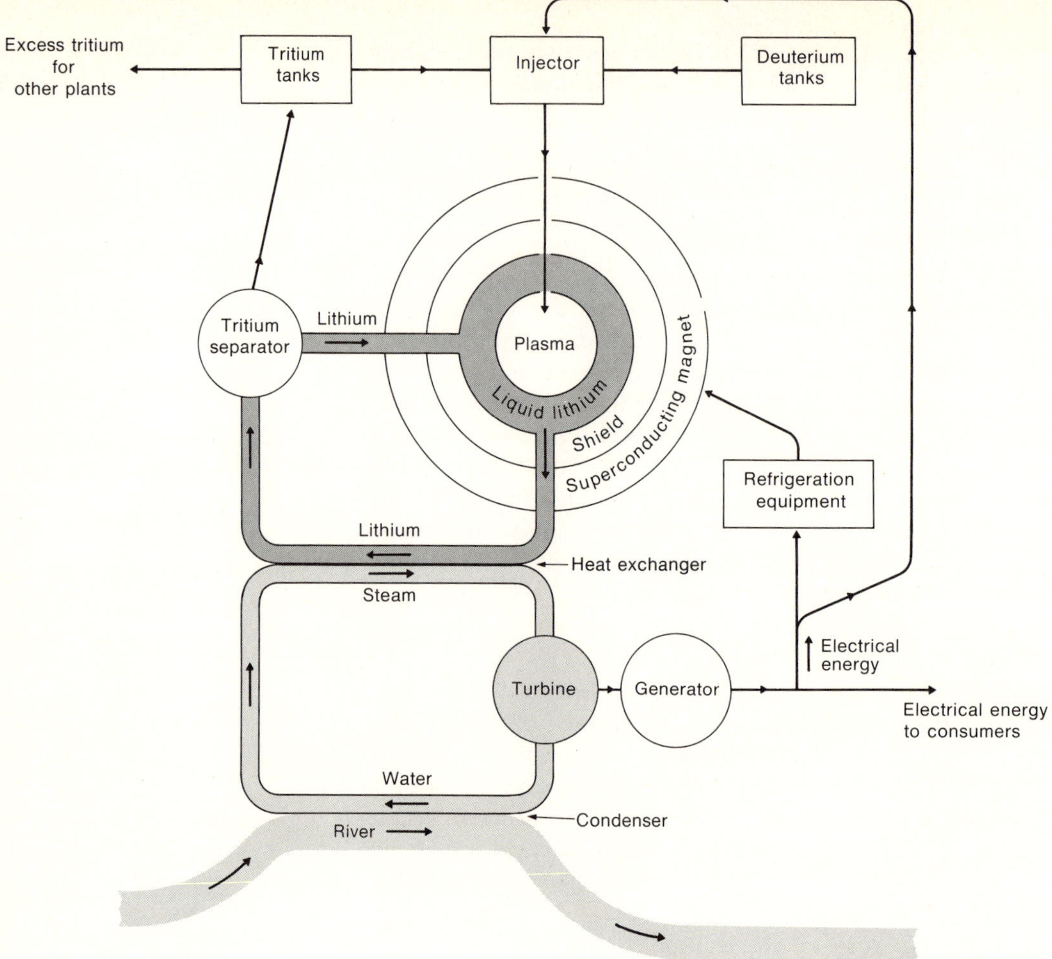

FIGURE 15.4
A schematic diagram of a possible fusion power plant.

blanket. Notice the dual role of the lithium. It is heated up by the neutrons and is thus the means by which we extract the energy liberated by fusion; it also replenishes and even increases the supply of tritium.

One potentially very serious engineering problem results from the fact that the structural materials of the wall between the molten lithium and the plasma will be exposed to intense beams of highly energetic neutrons. Bombardment by neutrons weakens metals, because collisions between the neutrons and the atoms in the metal may knock metal atoms out of their normal positions. Although much is known about "radiation damage" produced by neutrons in fission reactors, we have very little information about how metals would withstand the exposure they would receive in a fusion reactor. The structural materials of the reactor will have to be chosen with care; if the wall has to be replaced every few months, a fusion power plant will not be an attractive investment.

The liquid lithium (kept at a high enough temperature to remain in liquid form by the energy absorbed from the neutrons) is pumped around in a closed loop: first through a heat exchanger where heat flows from the lithium into water to make steam, then through a tritium separator where tritium formed by the processes discussed in §15.3.F is extracted, then back to the reactor to be heated once more. Tritium extracted from the flowing lithium can be reinjected, or if more tritium is extracted than is being used, the excess tritium can be stored to serve as the initial fuel for another plant. (The tritium will certainly not be deliberately released into the atmosphere because of the biological hazard.) Steam will be used to generate electrical energy just as in fission or in fossil-fuel power plants. As indicated in Figure 15.4, some energy will be needed for various parts of the plant itself. These auxiliary energy flows will have to be quantitatively considered in any actual design, lest we end up with an elaborate plant that can produce barely enough electrical energy to keep itself going, with no energy left to be delivered to the customers.

It is indicative of current optimism that serious engineering attention is now being paid not only to the central problem of the fusion reactor but also to peripheral problems such as extraction of the tritium, and also to the question of costs. The question of cost is a highly complicated one, obviously subject to change and dependent on the details of the particular power plant. The attempts that have been made to estimate capital investment and operating costs suggest that electrical energy could be supplied by a fusion power plant at a price not very different from what is currently considered reasonable. One thing that seems clear is that fusion power plants will be large, with electrical outputs of at least 1 GW; family-sized or village-sized units seem to be out of the question.

§15.3.H Lasers—A Different Approach to Fusion Power

In this chapter, we have described the main lines of research which began about 1950 and will undoubtedly continue through the 1970s. The goal of fusion power may be attained, though, by some method that will seem quite strange to those who have struggled for years with the problems of containing hot plasmas in magnetic bottles. We conclude this section by describing one novel approach that began to look quite promising about 1970, the use of high-powered lasers.

A laser produces an intense, narrow beam of light, either continuously or in the form of a series of short, extremely intense, pulses. When it was first suggested that lasers might be useful in producing nuclear fusion, they were seen as having a rather modest function, that of providing the initial heating to bring a sample of fuel up to its ignition temperature so that the fusion reaction would begin. The idea was to drop a frozen pellet of deuterium and tritium into the reactor and then hit the pellet with an intense pulse of light from the laser. The energy of the light pulse would heat the fuel, quickly turning it into a hot plasma, hot enough for the fusion reaction to begin. At that point, a surrounding magnetic bottle would contain the plasma while the fusion reaction proceeded. After a short time, when a significant number of the deuterium and tritium nuclei had undergone fusion, a new pellet of fuel would be inserted and the cycle repeated. (The energy would be delivered in the form of pulses, rather than continuously. This would cause little difficulty, for by the time the energy had been delivered from the neutrons to the circulating lithium and thence to the steam, these pulsations would no longer be noticeable.)

Later it was suggested, and this is the really exciting idea, that if the energy of the light pulse could be delivered quickly enough, it might be possible to dispense with magnetic containment altogether. Even if a plasma is not confined in any way, it takes *time*, admittedly a very short time, for it to be dispersed. In one proposed method, a pellet of deuterium and tritium will be hit simultaneously from many directions with synchronized laser pulses. The pellet will "implode"; that is, the atoms will be driven in toward the center by the laser pulses, acquiring high velocities. In this highly compressed state, before the nuclei have time to turn around and fly apart, fusion reactions will occur.

To anyone familiar with the problem of building a good magnetic bottle, it is obvious that any method in which containment is not needed is one deserving of serious consideration. But like the conventional methods, the laser method poses serious problems that must be solved, such as the construction of higher-powered lasers. Whether successful fusion power will ultimately be attained by pulsed lasers, improved methods of magnetic containment, or a completely different approach is unknown at the present time.

§15.4 ENVIRONMENTAL PROBLEMS AND AVAILABILITY OF RESOURCES

§15.4.A Environmental Radioactivity

The environmental radioactivity problems arising from the use of fusion power will be the result of the neutrons produced in fusion reactions and the tritium that will be used and generated. Neutrons are dangerous to living organisms, and therefore careful shielding must be installed. When the reactor is shut down, the production of neutrons immediately stops, but some of their effects linger on. As we have seen earlier, many nuclei become activated by absorption of a neutron, turning into unstable nuclei that decay at some later time by beta-decay. How intense the resulting radioactivity will be depends critically on the materials used in the construction of whatever structures are exposed to neutrons. During routine operation of the plant, this radioactivity should not be a problem, but it could present a hazard in the event of a catastrophic accident and will certainly make maintenance during a shutdown difficult.

Tritium is a potentially more serious problem. A reactor using deuterium and tritium as fuel will therefore be more dangerous than one using only deuterium. Since D-T reactors will surely be built first, careful attention to the tritium problem is necessary.* Tritium is radioactive, with a half-life of twelve years. As long as it remains outside the body, it is virtually harmless, for in its decay it emits only a β-particle, with no accompanying γ-ray, and the β-particle has such a low energy that even in air it can travel only a few millimeters before being absorbed. Unfortunately, since tritium is a form of hydrogen, it can readily combine with other atoms, for example to form radioactive water molecules, and thus it can become chemically incorporated into living organisms where the energy of its β-particle can do considerable harm.

It is essential to the operation of D-T reactors that the tritium bred from lithium be captured and recycled. But although tritium will not be intentionally released into the atmosphere, small amounts will surely escape. The exact quantity will depend on how much care is taken in the design of the plant. According to one estimate, it is perfectly feasible to keep the leakage rate low enough that, if the whole world were using fusion-produced electricity at the rate at which Americans now use electricity, the resulting dose of radiation would amount to about 0.1 mrem per person per year, in comparison with the 100 mrem per year or so that each of us receives from cosmic rays and other natural sources.

It must be pointed out that in a fusion power plant, the truly serious problems of fission power plants simply do not exist. Neither plutonium nor long-lived fission products are produced, and there is no waste-disposal problem (except for the problem of safely disposing of neutron-activated components that have been replaced). Fusion power is unquestionably cleaner than fission in that the amount of environmental radioactivity is much lower. It has been seriously suggested, however, that one might contemplate a power system based on a combination of fusion and fission. The idea is to use some of the neutrons from a fusion plant to bombard uranium or thorium, thus breeding fissile Pu^{239} or U^{233} as in a fission breeder, for use in other fission power plants. Given all the problems and dangers of fission plants and fission wastes, the wisdom of continuing to use fission if fusion power is available seems questionable, to say the least.

§15.4.B Accidents and Explosions

A *fission* power plant cannot function unless it contains a very large amount of fissile material, an amount much greater than the critical mass necessary to sustain a chain reaction. If some malfunction led to the accumulation at one place of an amount of material greater than the critical mass, or if something should prevent the insertion of

*D-D reactions also produce tritium (Equation 15.2), and even though the tritium thus produced will react with deuterons by the D-T reaction, an operating D-D fusion power plant will contain a substantial amount of tritium. Therefore, even if at some future time our fusion power is generated in D-D reactors, we will still have to be concerned about the effects of tritium.

the control rods, it is conceivable that the fission chain reaction would simply continue in spite of our attempts to stop it. Fusion reactions, however, are quite different. It is difficult to start the fusion reaction and to keep it going; if anything goes wrong, the plasma will immediately be lost and the reaction will stop.

In fission power plants, as we saw in Chapter 14, even after the chain reaction has been stopped, heat continues to be generated, and coolant must be circulated through the core for a considerable period of time. If the coolant should be lost and if the emergency core-cooling system should fail, the decay-heat could lead to a disastrous accident and a large release of radioactivity to the environment. Fusion power plants, too, have a decay-heat problem, one arising from the radioactive nuclei produced by neutron activation. However, given a fission plant and a fusion plant of similar size, the rate at which energy continues to be generated immediately after a shutdown is about ten times smaller for the fusion plant, and the chance of a serious accident therefore much smaller.

Accidents are possible, of course, and it seems prudent to consider the worst imaginable one. Suppose that a series of mishaps or an earthquake were to destroy the plant and disperse the contents into the environment. A large amount of dangerous material, tritium in particular, would be released. Calculations of the resulting hazard to the population are difficult. Much would depend on wind conditions and other such factors, but a number of studies indicate that the amount of radiation delivered to the surrounding population would be smaller by a factor of 100 or more than it would be for a fission plant of comparable size.

§15.4.C Thermal Pollution

There are two general aspects to the thermal pollution problem, as we have seen earlier. First, the generation of electrical energy cannot be carried out with 100% efficiency if a thermal stage is used. Some waste heat is inevitably delivered to the air or to a nearby body of water. In this respect, fusion reactors will not be fundamentally different from coal-fueled power plants or fission power plants, although the efficiencies and therefore the amounts of waste heat may be different. It seems likely that as far as the first fusion plants are concerned, there will not be compelling reasons to opt for or against fusion power on this basis alone.

There are reasons to hope that in the distant future it may be possible to build fusion power plants in which the thermal stage is omitted, thus by-passing (but not violating) the second law of thermodynamics. Normally, a thermal stage is required in a fusion or fission power plant because most of the energy is carried off by neutrons, and there is no practical way of converting the kinetic energies of neutrons into electrical energy without first slowing down the neutrons in some substance (water, sodium, or lithium, for example) which is thereby heated. In some possible sequences of fusion reactions, nearly all the energy released is delivered to *electrically charged* particles; we might be able to take advantage of this by subjecting the particles to electric fields that would slow them down, causing a direct conversion of kinetic energy into electrical potential energy without heating anything up. It may someday be possible to employ these fusion reactions and thus alleviate the thermal pollution problem.

The second aspect to the thermal pollution problem is that virtually all the energy we use sooner or later ends up as heat. The possible global implications of this fact may become serious with continued growth in energy consumption. Whether we obtain energy from coal, oil, fission or fusion, this global thermal pollution is unavoidable. The only possible way of avoiding this effect is to use solar energy, a subject to which we shall return in Chapter 16.

QUESTION

15.11 We saw in §15.1.B that if the D-D fusion reaction could be controlled, we would have an essentially inexhaustible supply of energy, even with a world population of 30 billion, with every person using energy at the rate of 10 kilowatts. How would the world's rate of energy usage under these conditions compare with the rate at which solar energy is incident on the earth?

---

§15.4.D Energy Resources

If D-D fusion can be made to work, we will have what is for all practical purposes an infinite supply of energy in the deuterium of the oceans. However, it is much easier to produce D-T fusion, and it is almost certain that the first fusion reactors will use the D-T reaction. As discussed in §15.3.F, the required tritium will probably be bred from lithium. In this case, the supply of lithium will be the limiting factor. According to Equation 15.11, an energy of 22.4 MeV is released for each Li^6 nucleus used. Only 7.4% of all lithium atoms are of this type, but reactions involving the more abundant isotope, Li^7, will also occur and will contribute to the energy released. Many different reactions are possible, and the value of a given quantity of lithium will depend on which reactions predominate. It is reasonable to expect that on the average, the energy released for every atom of lithium used (either Li^6 or Li^7) will be about 7 MeV.

Estimating the amount of lithium available is difficult. In the past, lithium has not been of interest as a source of energy, and much less effort has been devoted to finding lithium deposits than to exploring for coal, oil and uranium. Even so, it is estimated that in the United States alone, ores are available that would yield about 10^7 tons of lithium metal.

QUESTION

15.12 Show that if an energy of 7 MeV can be obtained from each lithium atom, 10^7 tons of lithium would supply an energy of about 800 Q.

The amount of energy available from D-T fusion using lithium is not nearly as large as the amount we could obtain from D-D fusion, but 800 Q is nevertheless a great deal of energy, since we now use in the United States less than 0.1 Q of energy per year. Successful D-T fusion would provide energy for an extremely long time, and if we succeed in making D-T fusion work, it seems virtually certain that we will eventually be successful in using D-D fusion as well.

§15.4.E The Outlook for Fusion Power

Fusion power may never become a reality. Though there are many reasons for optimism, formidable obstacles remain to be overcome. Not even the most confirmed optimists foresee any significant use of fusion power much before the end of this century; some "hard-headed realists" view research in this area as an interesting field of pure science but deem it folly even to hope that it will lead to useful results.

If we do succeed in using fusion as a source of energy, energy resources will no longer be a problem. Fusion power has sometimes been enthusiastically described as if it were completely clean, as if it produced no environmental and health problems at all. This is not true, but in comparison with the use of coal and fission, the advantages of fusion are impressive. It may be foolish to make our plans on the assumption that fusion power will be a success. It would be truly foolish not to consider fusion power as a world and national goal of the highest priority.

§15.5 FUSION: THE SOURCE OF THE SUN'S ENERGY

The question of the origin of the sun's energy was raised on several earlier occasions (see §5.7, Question 6.31, and Question 12.18). Several possible mechanisms were suggested, all of which failed when subjected to a *quantitative* examination. The problem is that the sun is now emitting energy at the staggering rate of about 4×10^{26} J/sec, and it has apparently been radiating energy at about the same rate for about 10 billion years. This means that the sun has lost altogether about 10^{44} J of energy, and will presumably radiate a good deal more energy in the future.

Previous suggestions have made it possible to understand how the sun might radiate *some* energy, but none of the suggested mechanisms could possibly provide anything close to 10^{44} J. (Fission cannot be the source of the sun's energy, because the sun contains only tiny amounts of the heavy elements that can undergo fission.) Fusion seems to provide a possible answer. A fundamental requirement for producing energy from fusion is a supply of light nuclei, and the sun is composed largely of hydrogen. The crucial question, though, is a quantitative one: could fusion reactions account for the 10^{44} J of energy that the sun has radiated?

Since the sun is composed primarily of ordinary hydrogen (H^1), it is quite certain that if fusion reactions are responsible for the sun's energy, some of the reactions must be ones in which protons interact to form heavier nuclei. In order to decide which reactions might occur in the sun, knowledge of some of the details of nuclear physics is required. Two possible sequences of nuclear reactions could take place in the sun or in other stars. The first is called the proton-proton chain. In the first step of the chain, two protons combine to form a deuteron, a positron, and a neutrino:

$$p + p \longrightarrow d + \beta^{+} + \text{neutrino}.$$

Since plenty of electrons are available, the positron will encounter an electron (β^{-}), and the two will annihilate one another. Thus the net effect is:

$$p + p + \beta^{-} \longrightarrow d + \text{neutrino}. \tag{15.13}$$

Energy is released in this reaction, some being delivered to the deuteron, some to the neutrino, and some appearing in the form of γ-rays produced as a result of the annihilation of the positron and electron. In the next step, the deuteron reacts with another proton:

$$d + p \longrightarrow \text{He}^{3}. \tag{15.14}$$

The resulting He^{3} nucleus can interact with another He^{3} nucleus (produced by the same set of reactions) to make an α-particle and two protons:

$$\text{He}^{3} + \text{He}^{3} \longrightarrow \alpha + 2p. \tag{15.15}$$

Altogether, then, we have reactions 15.13 and 15.14 occurring twice, followed by one occurrence of reaction 15.15:

$$\begin{aligned} p + p + \beta^{-} &\longrightarrow d + \text{neutrino} \\ p + p + \beta^{-} &\longrightarrow d + \text{neutrino} \\ d + p &\longrightarrow \text{He}^{3} \\ d + p &\longrightarrow \text{He}^{3} \\ \text{He}^{3} + \text{He}^{3} &\longrightarrow \alpha + 2p. \end{aligned}$$

By adding these reactions, we see that the result is:

$$4p + 2\beta^{-} \longrightarrow \alpha + 2 \text{ neutrinos} + \text{energy}. \tag{15.16}$$

There is a second set of reactions called the "carbon cycle" (see Question 15.25). This process depends on the presence of a small number of carbon nuclei. The carbon cycle begins with a reaction between a proton and a C^{12} nucleus, but at a later stage, a new C^{12} nucleus is produced so that there is no net consumption of C^{12} nuclei. Although the intermediate steps are different, the overall result of the carbon cycle is precisely the same as that of the proton-proton chain (Equation 15.16). It is believed that both the proton-proton chain and the carbon cycle operate in most stars; which is more important depends on the temperature of the particular star.

Energy is released by either set of reactions, for the sum of the masses of four protons and two electrons is greater than the mass of an α-particle. (The neutrinos do carry off some of the energy, but they have no mass.)

Equation 15.16 describes, in essence, the conversion of protons (hydrogen nuclei) into α-particles (helium nuclei). The process is often referred to as the "burning" of hydrogen to form helium, the helium in this sense playing the role of the "ashes." In order to calculate the amount of energy released every time four protons are "burned," all we need to do is to calculate the decrease in mass that takes place. The result is that the burning of four protons in this way releases about 26.7 MeV of energy. About 5% of this energy is carried off by the neutrinos, almost all of which escape from the sun. The rest of the energy, however, heats the electrons and nuclei in the sun and eventually shows up as electromagnetic radiation from the sun, that is, visible light as well as infrared and ultraviolet radiation.

QUESTIONS

15.13 Check the value 26.7 MeV, which was given as the amount of energy released by reaction 15.16.

15.14 Assume that the sun is composed mostly of ordinary hydrogen. From the known mass of the sun, show that the sun contains about 10^{57} protons.

15.15 Show therefore that fusion of all the protons in the sun would release an energy of about 10^{45} J.

This amount of energy, 10^{45} J, is substantially larger than the estimate of the total amount of energy the sun has given off (10^{44} J). The energy available from the sun's hydrogen is indeed enough that fusion may well be the source of its energy. Of course this calculation does not constitute a proof; there might be some quite different source of energy that we have not considered. But unlike the previously suggested mechanisms that could not possibly supply anything like 10^{44} J, fusion passes this test, and there are other reasons for believing that fusion *is* the source of the sun's energy, and of the energy of other stars as well.

If the sun operates by converting protons into α-particles, one might well ask why we do not build fusion reactors on the earth that use the same set of reactions. Why bother to extract the deuterons from sea water if we could use the protons instead? The reason, perhaps a surprising one, is that the fusion reactions occurring in the sun take place with relatively low probability. The amount of power generated per unit volume is not very high; it is only because the sun is as large as it is that it can produce so much energy. As a matter of fact, the temperature at the center of the sun is estimated to be only about 1.5×10^7 °K, somewhat lower than the temperatures contemplated for a fusion power plant. When one considers the tremendous differences in scale between the sun and one of our power plants, it may not seem surprising after all that the specific fusion reactions that are significant for each are quite different, even though the basic principles are the same.

Virtually all the energy used on the earth comes directly or indirectly from the solar radiation we receive. Not only does the sun keep us warm, but it is the sun's energy that maintains the photosynthetic reactions that enable plants to grow; thus animals, both herbivorous and carnivorous, rely on the sun. Photosynthesis that took place hundreds of millions of years ago provided us with our supplies of fossil fuels. Since the sun's energy is responsible for the winds in our atmosphere, and for evaporation of water from the oceans that leads to rain, it is the source of our hydroelectric power. In short, almost all the energy we now use was originally released in nuclear fusion reactions similar to those we are now attempting to produce and control here on earth.

Recent experimental results have raised some new questions about the sun. The fusion reactions produce neutrinos, virtually all of which escape from the sun and emerge in all directions. Some strike the earth, and almost all of these pass right on through as if the earth were transparent. It is difficult, but (as we saw in §13.6) not completely impossible, to detect a few of these neutrinos. What is perplexing is that the number of solar neutrinos is not nearly as large as it should be. The "missing neutrino" question is an important and unresolved problem. Perhaps the sun's fusion reactions stopped some time ago and the sun is now simply cooling off. Perhaps the sun is a variable star; perhaps energy is generated by fusion in bursts a few million years long, separated by periods in which the sun cools off slightly, and we just happen to live during one of the less active periods. Perhaps neutrinos are unstable and disintegrate (into what?) during the few minutes they need to travel from the sun to the earth. Almost everyone concerned with this problem believes that fusion *is* the original source of the sun's energy, even though it is not at all certain that fusion reactions are now occurring at the expected rate. If we are truly open-minded, we should also entertain the possibility that the sun's energy does not come from fusion at all but from some other process we have not imagined, or even the possibility that the law of conservation of energy fails in the sun, that the sun is a perpetual motion machine. Neither of these possibilities will be taken very seriously unless all of the less drastic explanations have been definitively ruled out.

FURTHER QUESTIONS

15.16 The two important D-D reactions were described in §15.1. There is another conceivable D-D reaction: $d + d \longrightarrow \alpha +$ energy. Unfortunately, this reaction occurs with extremely low probability (if at all) in comparison with the other two D-D reactions. Show that if this reaction did occur, the energy released would be about 24 MeV.

15.17 (a) Another nuclear reaction that has been suggested as a source of energy is the absorption of a proton by a boron nucleus (B^{11}), followed by fission into three α-particles: $B^{11} + p \longrightarrow 3\alpha$. Show that each occurrence of this reaction would release an energy of about 8.7 MeV. This is an attractive possibility, because large quantities of boron are available; its overall abundance in the earth's crust is about equal to that of thorium.

(b) The proposed reaction is quite similar to the fission that follows absorption of a neutron by U^{235}. However, in order for this reaction to occur, the protons and the B^{11} nuclei must have very large kinetic energies; these kinetic energies might be achieved if a fusion "trigger" is used to produce extremely high temperatures. What is the fundamental difference between these two processes that requires the particles to have very large kinetic energies in one process but not in the other?

15.18 The highest particle density listed in Table 15.2 is 5×10^{16} particles/cm³. How does this "dense" plasma compare with an ordinary gas at room temperature at a pressure of one atmosphere?

15.19 Estimate the rate at which water flows from an ordinary faucet. If all the deuterons in this water could be used to release energy by D-D fusion, how many faucets would be needed to supply all the energy used in the United States?

15.20 The deuterons in the oceans would apparently provide an inexhaustible source of energy. According to the results of Question 15.6, the energy released would be 1.5×10^{10} Q, enough to last 1.5×10^{9} yr even with an annual worldwide energy consumption of 10 Q/yr. However, energy consumption, roughly 0.24 Q/yr in 1973, has been increasing since 1950 with a doubling time of about 15 years.

(a) Show that if this rate of growth were to continue, worldwide energy consumption would reach a value of 10^{10} Q per *year* in about 500 years. (Lest there be any misunderstanding, the purpose of this calculation is *not* to show that we might soon run out of deuterium but to dramatize the way in which exponential growth—if unchecked—can rapidly exhaust what seems to be a nearly infinite resource.)

(b) If the earth were a perfect black body, how hot would it have to be to radiate energy at the rate of 10^{10} Q/yr?

15.21 Containment of the plasma is the most serious obstacle to building a fusion reactor on earth, yet this problem has apparently been solved in the sun, our great fusion power plant in the sky. What is a possible mechanism that can hold the sun together but is quite ineffective for containing a plasma in the laboratory?

15.22 In the sun, the "burning" of four protons releases 26.7 MeV of energy. Per gram of fuel used up, how much more productive is fusion than the normal burning of a fuel such as coal?

15.23 About how many solar protons are "burned" every second? How many tons of hydrogen are thus used up every second?

15.24 (a) Consider the sun as a whole, and calculate how much power is generated *per unit volume*.

(b) What would be the approximate physical dimensions of a fairly large power plant (say, 1000 MW) if it generated power at the same rate per unit volume?

(c) A human being who consumes 2000 kcal of food energy per day generates heat at an average rate of about 100 watts. Estimate the volume of a human body, and compare the amounts of energy per unit volume released by the human body and by the sun.

15.25 The probable reactions in the carbon cycle, one of the two sets of fusion reactions that are believed to take place in stars, are listed below. Identify the nuclei that have been omitted. Assuming that the two positrons annihilate with electrons, show that the overall result of this set of reactions is identical to that of the proton-proton chain (Equation 15.16).

$$\begin{aligned} p + C^{12} &\longrightarrow N^{13} \\ N^{13} &\longrightarrow ? + \beta^{+} + \text{neutrino} \\ p + ? &\longrightarrow N^{14} \\ ? + N^{14} &\longrightarrow O^{15} \\ O^{15} &\longrightarrow ? + \beta^{+} + \text{neutrino} \\ p + ? &\longrightarrow C^{12} + \alpha. \end{aligned}$$

15.26 One possible explanation of the missing neutrino problem (§15.5) is that the sun is no longer producing energy by fusion but is simply cooling off. Is there reason to be alarmed? In Question 6.31, it was shown that if the sun is cooling off, its average temperature is decreasing at the rate of about 2×10^{-8} °K/sec. If so, by how much will the temperature of the sun drop during the next 1000 years? Would this decrease be significant in comparison with the sun's present average temperature of 1.5×10^{7} °K?

16
Solar Energy

Nancy Richardot

16 Solar Energy

A major factor contributing to our present energy crisis is that the necessary research and development efforts which could have provided us with the technological options and capabilities we now need so desperately were not undertaken in the past.

—Senator Henry M. Jackson, 1973.

On Monday when the sun is hot, I wonder to myself a lot.

—A. A. Milne, *Winnie-The-Pooh*, 1926

§16.1 SOLAR ENERGY—THE TRULY VAST, INEXHAUSTIBLE, AND CLEAN SOURCE OF ENERGY

Solar energy is abundant, free, and clean. Solar energy is and always has been the energy source that is essential to the existence of life on earth. The amounts of solar energy striking the earth and the significance of solar energy in determining the earth's radiation balance were discussed in Chapter 11. In this section, we will review the overall data, consider the rate at which the land area of the United States receives solar energy, and compare some of these figures to society's energy needs. In the remainder of the chapter, we will discuss ways in which this energy can be used to do some of the jobs now done for us by hydropower, fossil fuels, and nuclear energy.

The sun's total output of power is approximately 4×10^{26} W, energy in the form of electromagnetic radiation, largely in or near the visible part of the spectrum. Of this radiation, all but the tiny fraction aimed directly at the earth (1 part in 2×10^9 of the total) is irretrievably lost; the solar power incident on the top of the earth's atmosphere (P_i in the notation used in Chapter 11) is approximately 1.73×10^{17} W.

QUESTION

16.1 Compare the rate at which solar energy arrives at the earth (P_i) with the rate at which the world as a whole consumes energy in the form of fuels, hydropower, and other sources (about 240 mQ/yr in 1973).

Energy arrives from the sun at a rate greatly exceeding 240 mQ/yr. Although much more must be known before any definitive conclusions can be drawn, this quantitative comparison is important for two reasons. First, it suggests that human consumption of energy at the present rate is probably not an important factor in the overall energy balance of the earth and is probably not producing serious climatic effects on a global scale. There is no reason to be complacent, however, for our climate may be more sensitive to local concentrations of heat than we realize, and consumption of energy will surely increase.

QUESTION

16.2 From 1950 to 1973, worldwide energy consumption increased by about 5% per year. If this exponential growth were to continue, by about what date would our rate of energy consumption equal the rate at which solar energy strikes the earth?

The second implication of the rate at which solar energy reaches the earth is that even very modest successes in using solar energy might go a long way toward providing us with a substantial fraction of our energy needs and reducing the energy problem to manageable size. This is the subject of this chapter.

The total amount of solar power is important, but it is also important to know how this power is distributed, to calculate the amount of power *per unit area*. Consider what the earth looks like from the sun. On an imaginary disk, 4000 miles in radius (located as in Figure 16.1), the total solar power incident on the earth is uniformly distributed; the power per unit area striking this surface is called the solar constant, S_0, and its numerical value is 1.353 kW/m^2. A good approximate value to remember is $S_0 \simeq 1.4$ kW/m^2. The significance of the solar constant is this. If you are at the top of the atmosphere with a surface of some sort whose area is 1 m^2, and if you turn this surface so that it is facing *directly* at the sun (that is, so that the surface is perpendicular to the sun's rays), the solar power striking it is about 1.4 kW. (If the earth had no atmosphere, the same result could be achieved at ground level.) As long as you are somewhere on the illuminated half of the earth (and above the atmosphere), the solar power per unit area on a surface *facing* the sun is the same, whether you are at the equator or somewhere else, whether it is noon or 5:00 P.M. It is important to see why this is so. Consider two sections of the imaginary disk in front of the earth, each of area 1 m^2, as shown in

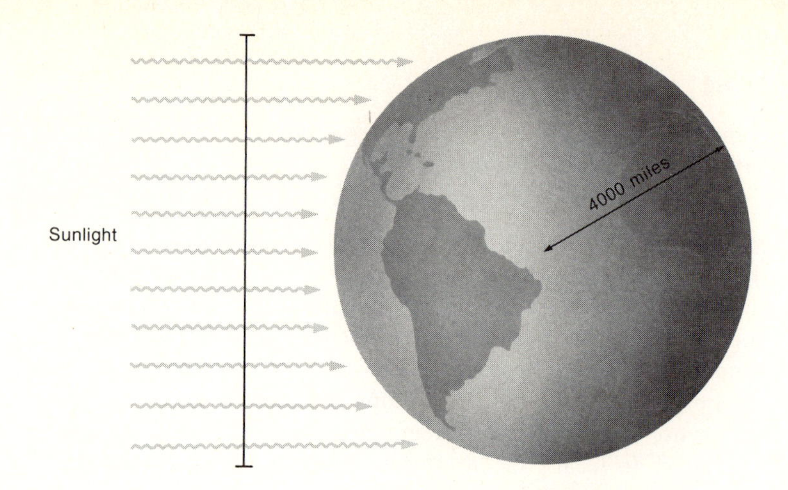

FIGURE 16.1
On an imaginary disk in front of the earth, the solar power would be uniformly distributed.

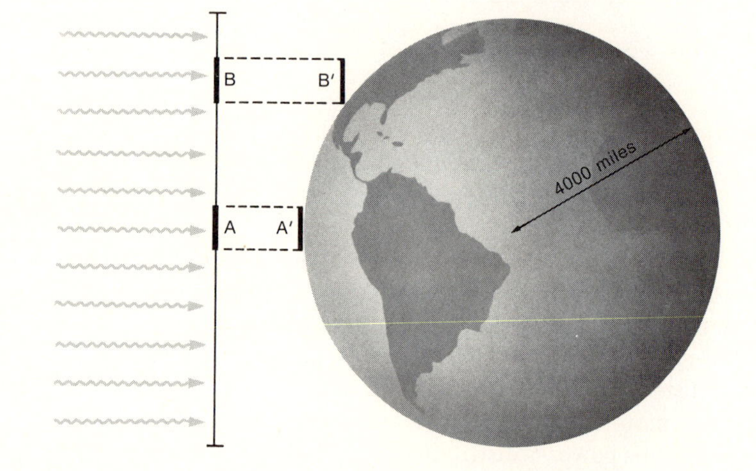

FIGURE 16.2
If A and B both have an area of 1 m² (not to scale!), solar energy crosses both sections at the same rate (1.4 kW). Therefore A′ and B′ each receive 1.4 kW of solar power.

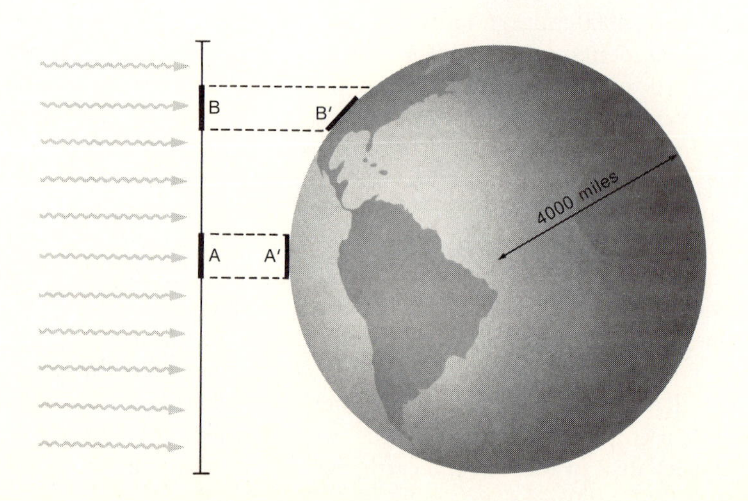

FIGURE 16.3
If B′ is horizontal (not perpendicular to the sun's rays), it receives less energy than does A′, even though A′ and B′ have the same area.

Figure 16.2. Solar energy crosses each of these two sections at a rate of 1.4 kW and continues straight on. Surface A′, of area 1 m^2, will intercept all the power crossing section A, and surface B′, also of area 1 m^2, will intercept all the power crossing section B, 1.4 kW for each surface. The difference between the two surfaces is simply that A′ is "horizontal" (at the equator at noon on March 21, for example), whereas B′ is *tilted* so that it faces the sun. If B′ were horizontal (as in Figure 16.3), or at any other angle than the one shown in Figure 16.2, some of the power coming through section B would miss B′, and the power hitting it would be substantially less than 1.4 kW. This fact must be remembered, since horizontal surfaces on the ground usually do *not* face directly at the sun.

We know how much solar power arrives at the top of the atmosphere, but unless we can get to the top of the atmosphere to use it (and one scheme for doing so is mentioned in §16.6), the amount of solar power reaching the ground is much more relevant. The solar constant ($S_0 \simeq$ 1.4 kW/m^2) is the essential piece of information with which we begin, but a number of important corrections must be made. Some of the solar energy is reflected from the clouds, and some is absorbed as it passes through the atmosphere. On a clear day, at a time not too close to sunrise or sunset, the sunlight striking a surface *facing the sun* delivers energy at a rate of about 1 kW/m^2, 70% of the solar constant. Even for a surface directly facing the sun, the solar power is considerably less in the early morning or late afternoon when the light must travel through a much greater thickness of air, and of course the solar power is greatly reduced on cloudy days. On the average, about 50% of the energy incident at the top of the atmosphere actually reaches the ground. That is, a surface directly facing the sun receives as much as 1 kW/m^2 on a clear day, and about 0.7 kW/m^2 on the average (but only during the 12 hours or so that the sun is up). These figures may be important in assessing the value of an individual solar collector whose orientation is continually varied so that it always faces the sun, but to evaluate the solar energy potential of a large piece of land or of a horizontal collecting surface (or of a collecting surface at any *fixed* angle), we must take into account the angle at which the sun's rays strike the surface, an angle that varies with the time of day and with the season.

Consider the sunlight falling on a horizontal surface on a clear day. If the sun is directly overhead (Figure 16.4a), energy is delivered to the surface at the rate of about 1 kW/m^2. If the sun is not directly overhead (Figure 16.4b), an incoming power of 1 kW is spread out over an area significantly larger than 1 m^2; that is, the power per unit area is significantly reduced, as shown also in Figure 16.3. If you have ever noticed how much more rapidly the snow melts on a hillside facing toward the south than on a level field, you are familiar with the fact that the rate at which solar energy is delivered *per unit area* is larger for a surface directly facing the sun than it is for one with a different orientation.

This correction for the angle the sun's rays make with the ground is an important one. At noon in mid-June (Figure 16.5a), the sun is almost

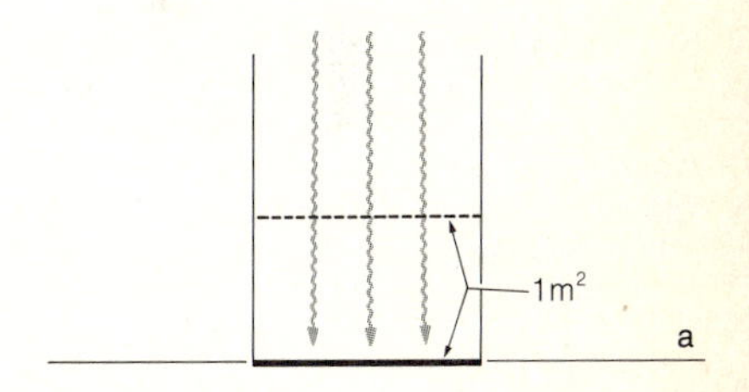

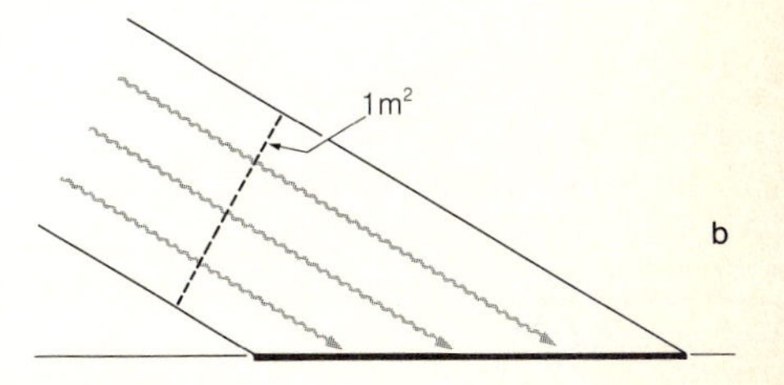

FIGURE 16.4
The solar power per unit area is considerably reduced (in this example by about a factor of two) if the surface is not facing directly at the sun.

directly overhead in Miami (about 88° above the horizon); in Minneapolis at the same time it is 68° above the horizon. In mid-December (Figure 16.5b), the noonday sun is only 41° above the horizon even in Miami, and a mere 21° above the horizon in Minneapolis. The reduction in solar power is not quite as drastic as this geometrical argument would suggest, because some of the radiation striking the surface is "diffuse" radiation, light that has been scattered in the atmosphere and that arrives at the ground from various directions rather than in a direct beam from the sun. Even on a clear day, this diffuse radiation accounts for about 15% of the energy reaching the ground.

In assessing the potentialities of solar energy, it is useful to consider the average rate at which solar energy is incident upon a horizontal surface at ground level, the rate averaged over all hours of the day and night. Consider Miami as an example. At about noon on a clear day in June, solar energy strikes the ground at a rate of about 1 kW/m^2. Because the sun is up for only about 12 hr/day, the average value for all hours of day and night cannot possibly be more than about half this value, 500 W/m^2. During the morning and afternoon, the sun is lower in the sky, and this fact, together with occasional periods of cloudiness, reduces the overall average value for the month of June to about 260 W/m^2. At other times of the year, the situation is less favorable; the average value for Miami in December, for instance, is about 150 W/m^2. For the year as a whole, the average value for Miami is 219 W/m^2.

Detailed solar energy data for a number of cities are given in Table F.2; Figure F.1 shows annual averages for the United States. The overall average for the 48 contiguous states, the rate at which solar energy strikes a horizontal surface at ground level, a rate averaged over the whole year and over all hours of day and night, is approximately 200 W/m^2. This is probably the most useful single number to remember in thinking about the uses of solar energy, although in any specific application the variation from place to place, from month to month, and from hour to hour must be taken into consideration. Using this figure of 200 W/m^2, we can estimate the amount of land area needed to collect solar energy at any particular rate.

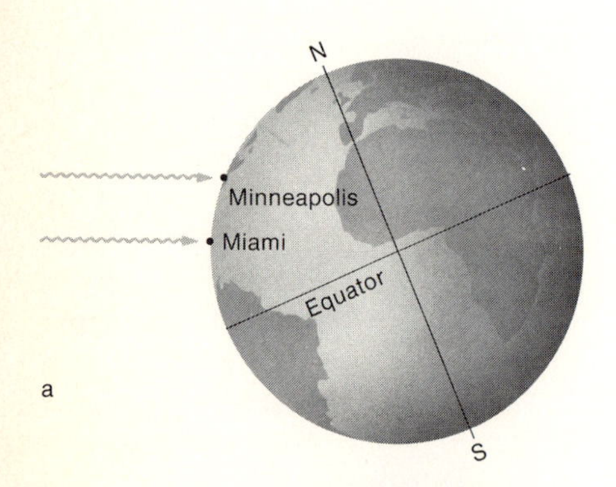

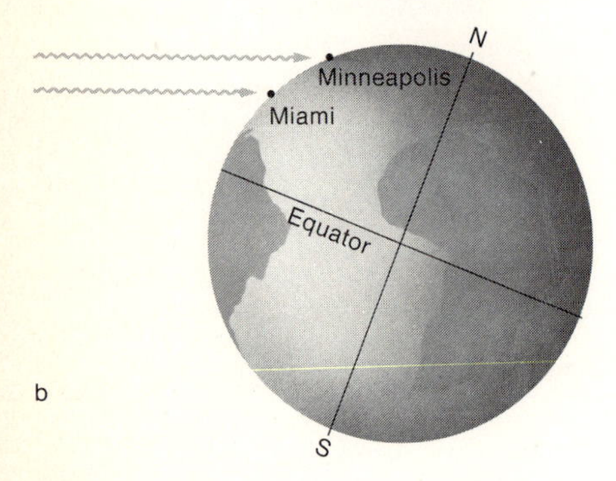

FIGURE 16.5
The noonday sun as seen from Miami and Minneapolis: (a) about June 21; (b) about December 21.

QUESTION

16.3 Using the average value of 200 W/m^2, show that the rate at which solar energy strikes an area the size of Connecticut (an area 1/25 the size of New Mexico) is approximately equal to the total United States rate of energy consumption (in 1973), and that solar energy is incident on the whole country at a rate about 700 times larger.

Such calculations do not show what is practical or even possible, only what is *conceivable* if 100% of the solar energy could be converted to useful forms, but we can also calculate the area required for any assumed value of conversion efficiency. The results of these calculations are shown in Figure F.2. We can see from this figure that even, for instance, at an efficiency of 5%, our total projected energy needs of 178 mQ/yr in the year 2000 could be provided by the solar energy available on an area of about 250,000 square miles, an area about equal to that of Arizona and New Mexico combined.

QUESTION

16.4 In order to obtain substantial amounts of energy, it may be necessary to devote rather large areas to the collection of solar energy. It may seem rather extravagant to use an area equal to that of Connecticut, let alone Arizona and New Mexico, merely for generating energy. A hard-headed energy czar might observe that Arizona and New Mexico are mostly desert anyhow, but it is probably more helpful to realize that solar-collecting areas can be distributed about the country and to remember that we already "waste" large areas of the United States—for power plants, highways and parking lots, for example.

(a) Behind Hoover Dam is Lake Mead (area about 265 square miles), whose major function is to store water for hydroelectric generation. Compare the rate at which solar energy strikes this area with the rate at which electrical energy is generated at the Hoover Dam power plant.

(b) Try to estimate roughly the total area in the United States devoted to purposes such as the following: roads and highways, farming, strip mining of coal, lumbering, or any of our other "needs" that require large amounts of land.

(c) Buildings take up a good deal of space in this country. The rooftops of buildings constitute an area ideally suited for solar energy collection, a subject to be discussed in detail in §16.8. Make a rough estimate of the total amount of roof area in the United States, and estimate the solar power striking this area.

It is sometimes informative to rephrase results, such as those given in kilowatts per square meter, in more picturesque terms. Consider the solar energy striking a football field at Miami during the course of one hour at about noon during the middle of June, and let us convert this amount of energy into an equivalent amount of coal. Solar energy strikes the ground at a rate of about 1 kW/m^2. The area of a football field is approximately 100 yd $\times$ 35 yd $= 3500\ yd^2 \simeq 3000\ m^2$; thus the incident solar power is about $1\ kW/m^2 \times 3000\ m^2 = 3000\ kW = 3 \times 10^6$ J/sec. The number of joules delivered in one hour is therefore $3600 \times 3 \times 10^6 \simeq 10^{10}$ J. The energy content of one ton of coal is about 25×10^6 BTU, or since 1 BTU $\simeq$ 1000 J, about 2.5×10^{10} J. Thus we can convert the solar energy of 10^{10} J into an equivalent amount of coal:

$$10^{10}\ \mathrm{J} = 10^{10}\ \mathrm{J} \times \frac{1\ \text{ton of coal}}{2.5 \times 10^{10}\ \mathrm{J}} = 0.4\ \text{ton of coal.}$$

Thus solar energy strikes this football field at a rate equivalent to about 0.4 ton of coal per hour.

The incoming sunlight certainly carries impressively large amounts of energy. It is essential to realize that these calculations only indicate what *might* be possible. No technological magic is going to cover the state of Connecticut with 100%-efficient solar energy collectors. Proposals for solar energy, no matter how carefully presented, are often regarded with skepticism, and some of the blame for this can be placed on those who get so carried away with the results of calculations like those in this section that they ignore the remaining technological, economic, and social problems. Some solar energy advocates are fond of making comparisons like the one in the preceding paragraph—that at noon during the month of June, solar energy strikes a Miami football field at a rate equivalent to 0.4 tons of coal per hour. Calculations of this sort are fun to do and they can be instructive, but certainly one's knowledge of the subject should not be limited to such exercises. To those who are sufficiently naive and enthusiastic, it sometimes does not seem to matter very much whether the calculations are correct, as long as they sound impressive. The calculation referred to above *is* correct; the following question contains a number of statements of the same general sort, every one of which is *incorrect*.

QUESTION

16.5 All of the following statements are wrong, by at least a factor of 100. Some of them overestimate the amount of solar energy, others are underestimates. Try to guess the category to which each statement belongs, and then make the necessary approximate calculations to correct each one.

(a) The amount of energy emitted by the sun (in all directions) during a time of 10^{-7} sec is equal to all the energy in the world's supplies of fossil fuels.

(b) If all the solar energy striking the state of Texas in the course of a year could be converted to electrical energy and sold at 3¢/kWh, it would be worth 30 billion dollars.

(c) The solar energy incident on the whole earth in one hour is equal to all the energy used by the residents of New York City in the course of a year.

(d) The solar energy arriving at the earth in the course of one year is equivalent to a layer of oil covering the whole earth to a depth of 1.5 miles.

(e) On the average, solar energy strikes the roof of a car in the United States at a rate sufficient to keep 100 cars traveling indefinitely at 40 miles/hr.

(f) The solar energy incident on the whole earth is equal to that which could be carried by a continuous train of coal cars traveling at a speed of 1000 miles/hr.

---

§16.2 POSSIBLE WAYS OF USING SOLAR ENERGY

The amounts of solar energy are very large, but how can we make this energy available for our own purposes? We do of course already use solar energy in very important ways. It keeps the earth warm and without it we could grow no food, but how can we use it to help solve the "energy problem"? Can we convert solar energy into other forms and use it to generate electricity, run our cars, or heat and cool our buildings? Solar energy has its own special characteristics, which must be considered in thinking about each of the many possible ways in which it might be used. Some of the characteristics of solar energy are helpful and others are not; some features that may cause difficulties in some applications may actually be advantageous in others.

Solar energy is virtually inexhaustible, and the amount available next year will not be affected by the amount we use this year. It is free. (At any rate, the "fuel" is free; the equipment needed for converting solar energy to other forms is not free.) It is environmentally "clean," certainly by comparison with other energy sources. It is widely distributed—a disadvantage if we dream of a centralized national power system, but an advantage if we remember that the users are also widely distributed. Solar energy is available in different amounts at different places. Some locations are therefore more favorable than others. In fact, more is available in almost every underdeveloped nation than in the industrialized nations; this must surely be counted as a long-term advantage. The rate at which it arrives at any one place varies with the weather, time of year, and—obviously—time of day. It is impossible to count this variability as an advantage. Since light cannot be stored in the form of light, we can only hope to realize the full value of solar energy if we can convert solar energy into other forms of energy which can be stored, and therefore the general question of energy storage will become increasingly important.

We can divide the possible ways in which solar energy may be used into several broad categories. There are, first of all, indirect methods, in which we take advantage of the conversion of solar energy into other forms that occur all the time in nature. It is solar energy that causes the evaporation of water from the oceans, and so solar energy provides our rainfall and our hydropower. Solar energy is also the cause of the winds and of the fact that the oceans are not all at a uniform temperature; both of these natural phenomena can be exploited as sources of significant amounts of energy.

Human beings can intervene to produce additional conversions of solar energy in three major ways. First, we can do as nature does, and use the chemical reactions of photosynthesis to grow plants; of course we already do this, but we can grow *more* plants, plants to be used as fuels instead of food. Second, with "solar cells" we can convert solar energy directly into electrical energy. Third, we can convert solar energy into thermal energy. Once solar energy has been converted into another form, a wide range of options becomes available. If, for example, we convert solar energy into thermal energy, we can convert some of this thermal energy into electrical energy just as we do in coal or fission power plants, or we can use the thermal energy directly, to heat our homes. We shall examine a number of these possibilities, taking the most detailed look at the plan that will probably be the most important in the near future, that of converting solar energy to thermal energy and using it to heat buildings.

§16.3 WINDMILLS

The windmill is an ancient device for indirectly harnessing the sun's energy on a small scale. Windmills are still used in various places in the United States, usually for pumping water or for generating small amounts of electrical energy, but, where power lines are available, windmills have usually been replaced. The major difficulty with windmills is that the energy we are trying to use is the kinetic energy of air, the density of air is rather low, and rather large windmills are needed to get energy in the quantities we need. Suppose that the wind speed is v, and think of a windmill whose blades cover an area, A, facing directly into the wind. The volume of air that can pass through the blades of the windmill during a time interval Δt is $Av\,\Delta t$ (Figure 16.6). The mass of this air is equal to this volume multiplied by the density of air (ρ_{air}), and so the kinetic energy is

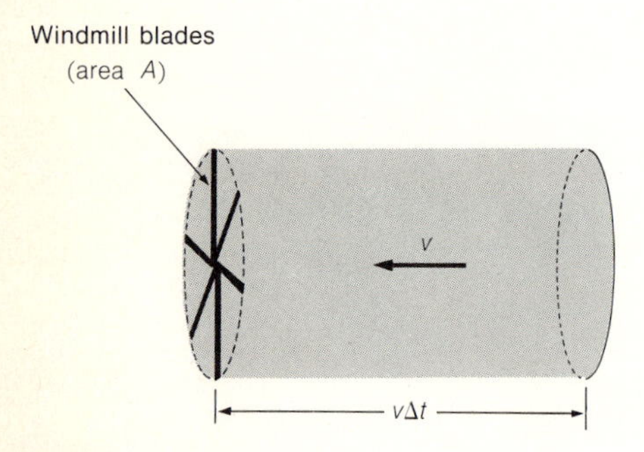

FIGURE 16.6
During a time Δt, the volume of air passing through the blades of the windmill is equal to $Av\,\Delta t$.

$$KE = \frac{1}{2}mv^2 = \frac{1}{2}\rho_{\text{air}}\,Av^3\,\Delta t.$$

If all this kinetic energy can be captured and converted into useful energy, the rate at which energy is converted, the output power of the windmill, is:

$$P = \frac{KE}{\Delta t} = \frac{1}{2}\rho_{\text{air}}\,Av^3. \qquad (16.1)$$

QUESTIONS

16.6 Test Equation 16.1 for dimensional consistency.

16.7 Show that if $v = 10$ miles/hr, the maximum possible output power of a 10-ft diameter windmill is about 400 W.

In this calculation, the most favorable assumption imaginable has been made, that all the wind's kinetic energy could be extracted. We surely cannot do this well, for one thing because the air passing through the windmill must retain some kinetic energy so that it can keep moving and make way for more air. However, if we could count on steady 10 mile/hr winds, a windmill of this size could be expected to deliver a few hundred watts of power, enough to operate a few light bulbs or a small appliance. In order to supply all the electrical energy used, even by a single family, a larger windmill would be needed, and some energy storage system, perhaps storage batteries, would be needed to provide energy on calm days. In order to generate power for an entire city, many very large windmills would be needed.

One of the most ambitious attempts to produce large amounts of energy from the winds was made in Vermont during the early 1940s, with a windmill 110 ft tall, which had two blades, each 90 ft long and weighing 8 tons, and which could generate a maximum electrical power of 1.25 MW. For a brief period of time in 1945, this windmill was part of the regular power distribution system. Then one of the blades broke, and the financial prospects began to look discouraging; the windmill was never repaired, and the experiment was abandoned. In the 1970s, concern about the energy problem has revived interest in this old-fashioned idea; better knowledge of aerodynamics and the availability of stronger materials make large windmills more feasible. By the end of the century, according to one visionary plan, 50,000 or more 800-ft high windmills might be distributed through the Great Plains from Texas to North Dakota, along the Aleutian Islands, and on floating platforms in the Great Lakes and along the Atlantic and Gulf coasts, with a total average electrical output of 2×10^{11} W.

QUESTION

16.8 What fraction of our 1973 electrical energy could be supplied by such an array of windmills?

Wind energy is a renewable resource. If left to themselves, the winds would quickly die out, their kinetic energy being converted to thermal energy by means of friction, but the energy of the winds is constantly replenished from the heat delivered by the sun. It is estimated that (for the world as a whole) solar energy is converted into wind energy at the rate of about 2×10^{15} W.

Large windmills in suitable locations would enable us to use a small fraction of this total and could supply a significant part of our energy needs. Energy storage will be important, because of the variability of the winds. One plan (sketched in Figure 16.8) is to generate electrical energy from the wind and use it to produce hydrogen gas by electrolysis of water. The hydrogen can simply be burned as a fuel or used in fuel cells to generate electrical energy; it can be used nearby, or shipped to a distant location by pipeline.

The costs of such a plan are extremely difficult to estimate. The floating platforms for offshore windmills and the windmills themselves are part of the problem. Another complicating factor is the contemplated production of hydrogen gas and its subsequent use to generate electrical energy in fuel cells; the future of fuel cells is quite uncertain at the present time. Large windmills may present problems of aesthetics, safety and environmental impact as well. They will not be the quaint windmills shown in travel brochures but gigantic structures, perhaps acceptable floating off the coast but not on our scenic mountaintops. The accidental

FIGURE 16.7
The 1.25-MW windmill at Grandpa's Knob, Vermont. [Central Vermont Public Service Corporation.]

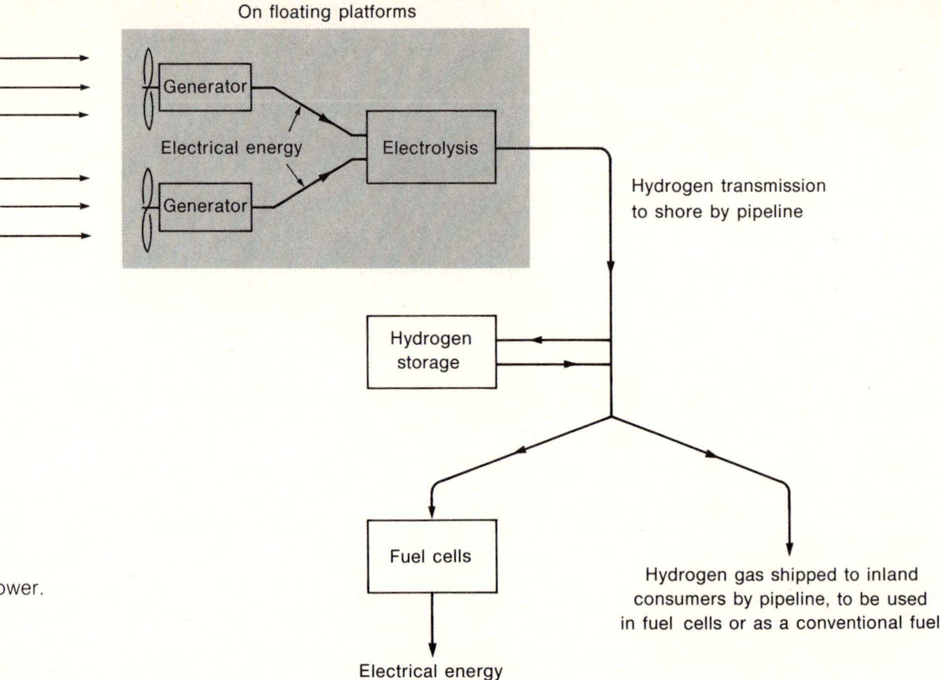

FIGURE 16.8
A proposed system for using wind power.

breaking of a blade could endanger the lives not only of operators but also of farm workers nearby. Furthermore, since the winds have an obvious effect on our climate, both locally and globally, any possible consequences of our tampering should be studied in advance as carefully as possible. In spite of all these questions, the possibility of obtaining a substantial amount of energy from the winds is one that should be fully explored.

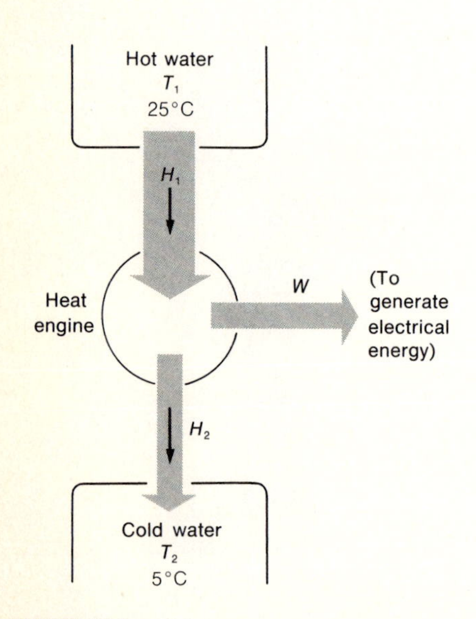

FIGURE 16.9
A heat engine whose operation is based on the temperature differences in the oceans.

§16.4 TEMPERATURE DIFFERENCES IN THE OCEANS

The oceans contain a great deal of thermal energy, but if they were all at a single uniform temperature, the second law of thermodynamics would prevent us from simply extracting this energy and converting it into more useful forms. Fortunately, however, the oceans do not have a uniform temperature. The inevitable tendency of the oceans to achieve a uniform temperature by thermal conduction and mixing is offset by a continuous input of solar energy, which drives the ocean currents and the circulation of the atmosphere. Although it is out of the question to run a heat engine with the tropic seas as the hot reservoir and the Arctic Ocean as the cold reservoir, in many areas significant temperature differences exist between parts of the ocean quite near each other. In the tropics, the surface temperature of the ocean is about 25°C (77°F), and at depths of only about 1000 m the temperature has a nearly constant value of 5°C (41°F). Closer to the United States, the Gulf Stream brings a large supply of warm water, also about 20°C warmer than the surrounding cooler water, through the Straits of Florida and up the east coast.

Two bodies of water at different temperatures can be used as the hot and cold reservoirs for a heat engine (Figure 16.9), and electrical energy can be generated according to the same principles used in steam-electric power plants. Although the principles are the same, the small difference in temperature between the two reservoirs will require power plants quite different in design from conventional steam-electric power plants.

QUESTION

16.9 If T_1 (Figure 16.9) is 25°C and T_2 is 5°C, show that the maximum possible efficiency with which heat removed from the warm water can be converted into electrical energy is about 7%.

This efficiency, the maximum conceivable efficiency, is discouragingly low. However, the basic idea need not be discarded, because a great deal of water is available. This low efficiency *does* mean, though, that for a useful output, W, of, say, 100 MW, the rate of flow of heat, H_1, must be about 15 times larger than W, and H_2 nearly as big, as shown in Figure 16.10. In order that these large heat flows not destroy the temperature difference, fresh supplies of both hot and cold water must be pumped through the plant. Furthermore, in order that large amounts of heat be transferred from the warm water to the heat engine and from the heat engine to the cold water, the heat exchangers must have very large areas in thermal contact with the water on one side and with the working fluid of the heat engine on the other. Also, the heat exchangers must be of very thin material so that not too great an obstacle to the flow of heat is presented.

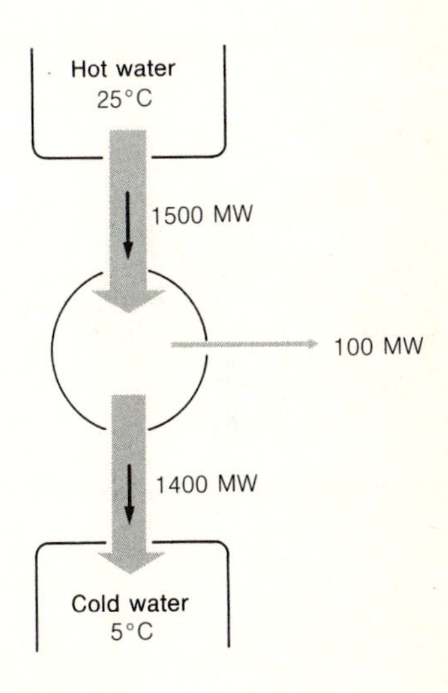

FIGURE 16.10
Energy flows under the most favorable conditions.

A more realistic diagram is shown in Figure 16.11. Warm water flows in at 25°C and emerges after having been cooled to about 23°C. The temperature of the working fluid in the hottest part of its cycle must be somewhat less than that of the warm water; otherwise the heat, H_1, would not flow into the working fluid. A similar problem exists where the waste heat, H_2, is rejected to the cool water. If the various temperatures are those shown in Figure 16.11, then as far as the working fluid is concerned, the maximum and minimum temperatures are 20°C and 10°C, rather than 25°C and 5°C. In an ordinary steam-electric plant, a reduction of the temperature difference by 10°C would not be important, but when the temperature difference is small to begin with, every degree counts.

20°C

3000 MW

100 MW

2900 MW

10°C

FIGURE 16.11
In practice, the full temperature difference is not available to drive the heat engine.

FIGURE 16.12
Numerical values of energy flows, modified to allow for the lower temperature difference.

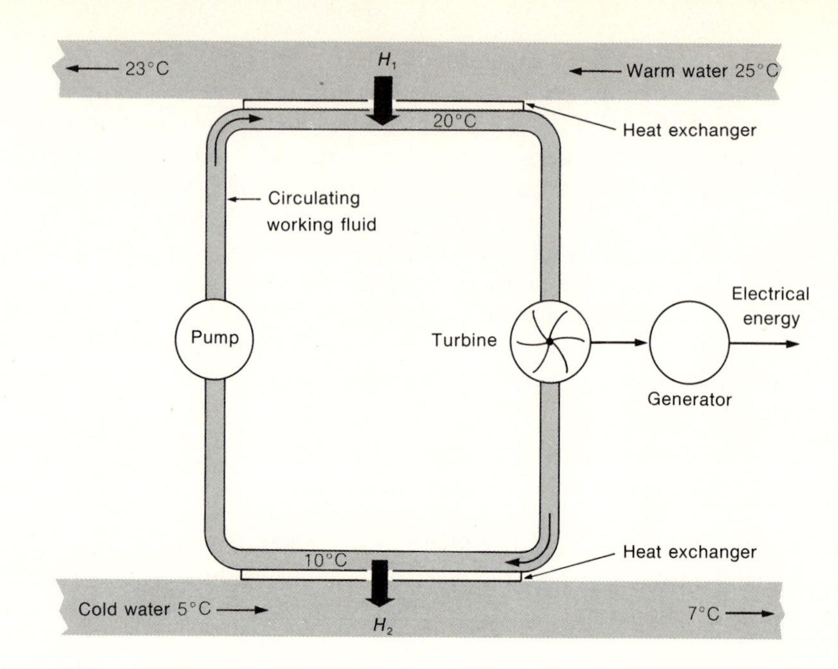

QUESTIONS

16.10 Assuming that the high and low temperatures are 20°C and 10°C, show that the maximum possible efficiency is about 3.5%. Thus the numbers in Figure 16.10 must be revised as shown in Figure 16.12, in order to achieve an output of 100 MW.

16.11 Show that if heat is removed from the warm water flowing through the plant at a rate of 3000 MW and if this results in cooling the water by 2°C, the rate of flow of the water must be about 360 m³/sec.

16.12 Compare this flow of water with the rate at which water flows through a 100-MW hydroelectric plant with a head of 100 ft.

Another problem is the heat engine itself. In a steam-electric plant, water itself is used as the working fluid, and it is steam that drives the turbines. In the oceans, the temperatures are so low that water does not boil, and other substances that do turn into vapors at these low temperatures would be better working fluids for driving the turbines. Ammonia, propane, and freon (commonly used in refrigerators) have all been investigated.

In spite of the low efficiencies, which will in practice be lower than the theoretically possible ones we have calculated above, and in spite of other problems of anchoring the power plant in the ocean and fighting off the corrosive effects of sea water, this method *can* work. Some power was generated by such a plant as long ago as 1926. Whether reliable large-scale plants can be built at a reasonable cost is not clear. The possible climatic effects that might result from manipulation of the oceans and their currents are also difficult to estimate. Nevertheless, it would be foolish not to explore the practical possibilities of using the existing temperature differences in the oceans, because a sizable amount of power might be obtainable.

QUESTION

16.13 The rate of flow of the Gulf Stream is about 2200 km^3/day. How many 100 MW(e) power plants could be driven by this warm water, if we make the assumption (admittedly an extreme one) that all the warm water of the Gulf Stream could be similarly used? How significant a contribution would this many power plants make in supplying energy for the United States?

§16.5 NATURE'S METHOD: PHOTOSYNTHESIS

In using hydroelectric power, wind energy, or the temperature differences in the oceans, we are taking advantage of the conversions of sunlight to other forms of energy that go on all the time. Let us now consider some ways in which we can intervene to convert into useful forms of energy some of the solar energy that would otherwise be wasted. Chemical reactions induced by light partially convert solar energy into stored chemical energy; one of the best reactions is that used in nature, photosynthesis. We could, then, capture some additional solar energy simply by growing more plants and using them as fuels. To get an approximate idea of what might be possible, consider the simple plan of tree farming: grow trees and use the wood as fuel rather than as lumber. (In 1870, when the magnitude of the energy problem was much smaller, wood was the major source of energy in the United States, as shown in Figure K.7. Wood and other plant materials, such as sugar cane stalks, are still important fuels in some parts of the world.)

QUESTIONS

16.14 On the average, an acre of good land can produce about two tons of wood per year. Using the fuel value of wood given in Appendix H, show that the rate (per unit time and per unit area) at which solar energy is converted into chemical energy is about 0.2 W/m^2.

16.15 Do you have any direct knowledge you could use to make an order of magnitude estimate of the rate at which an acre of land can produce wood? Can you think of a lot with young trees on it that you remember as seedlings ten years ago?

16.16 Solar energy strikes the surface of the United States at an average rate of about 200 W/m^2. What is the efficiency of tree farming as a method of collecting solar energy? How much of the area of the United States would have to be devoted to tree farming in order to meet all our current energy needs in this way?

Although it is not completely unrealistic to think of getting useful amounts of energy from wood, simple tree farming does not look like the answer. Can we get more energy per acre by growing plants other than trees? Yes, perhaps about ten times as much, possibly somewhat more, but it should be remembered that the world has a food problem as well as an energy problem. Successful agriculture requires good soil and water as well as sunlight, and the food that can be grown is valuable for many other reasons besides its mere energy content. Food is scarce enough now that the world cannot afford to use much arable land merely for "growing energy," and, if possible, large parts of our forests should be left as wilderness areas. However, wastepaper could be a useful fuel, and materials such as garbage, sewage, and agricultural wastes also contain significant amounts of stored solar energy even though they may be of no further use as food. Small amounts of power are being generated by burning waste products, sometimes directly and sometimes after preliminary treatment to produce a clean fuel such as methane gas. Although we cannot satisfy more than a small fraction of our energy needs from waste products, this kind of recycling is worthwhile if not prohibitively expensive.

FIGURE 16.13
Installation of an array of solar cells to power a rural telephone line. [Bell Laboratories.]

§16.6 THE DIRECT GENERATION OF ELECTRICAL ENERGY: SOLAR CELLS

Light striking a metallic surface can eject electrons and produce an electrical current in an external circuit. This is one of the central features of the photoelectric effect discussed in §11.6. In a similar way, light striking various solid materials can be used to generate electrical currents. In other words, a direct conversion of light energy to electrical energy can be achieved. Devices for this purpose are called photocells, photovoltaic cells (because a voltage can be produced), solar cells, or solar "batteries." The efficiency is not 100%, but because the light energy is converted directly into electrical energy, rather than first being converted into thermal energy, the second law of thermodynamics does not limit the obtainable efficiencies. Modern solar cells made of silicon can convert about 13% of the incident solar energy into electrical energy; efficiencies as high as about 20% may eventually be achieved with silicon or with other materials.

QUESTION

16.17 At an efficiency of 13%, how much land in the United States would have to be devoted to the collection of solar energy for the following purposes:

(a) to provide our presently used amount of electrical energy?

(b) to supply all our estimated energy needs in the year 2000? Besides expressing these results in square miles or square meters, suggest also a geographical area of the required magnitude: for example, Rhode Island? Arizona? Arizona and New Mexico?

Solar cells appear very promising. Indeed, solar cells already have an important function in providing power for spaceships. They also have other small-scale applications (see, for example, Figure 16.13). Silicon is the most abundant element in the earth's crust, and thus the supply of raw material is not a problem. The reason that no large-scale generation of electrical energy from solar cells is currently taking place is simply the overwhelming cost. The "fuel" is free, but exceptionally pure silicon is needed, and the costs of silicon solar cells, as of 1974, are estimated to be about $10,000 per square meter. From these data, we can make a rough estimate of the cost of generating 1 kWh of electrical energy.

QUESTION

16.18 (a) Assume that 1 m² of collecting surface requires a capital investment of $10,000, that its lifetime will be 30 years, and that money must be borrowed at an interest rate of 10%. Show that the annual cost of 1 m² of collector is about $1060.

(b) Because of the high cost, a solar cell power plant would be best in the southwestern United States, and it would be worthwhile to keep the collecting surfaces aimed directly at the sun all day. We might therefore expect an incident power of about 1 kW/m² for about 10 hours per day. Show that at an efficiency of 13%, 1 m² of collector could deliver about 500 kWh of electrical energy per year.

(c) Show that each kilowatt-hour would therefore cost about $2.

In the calculation suggested in the preceding question, an extremely optimistic approach was taken; the cost of the equipment to support the solar cells and to keep them properly oriented, the cost of other auxiliary equipment and the costs of distributing the energy to the consumers were omitted. If solar cells cost $10,000 per square meter, then the true cost of a kilowatt-hour of electrical energy will be at *least* $2, and actually substantially more. If the costs of two sources of energy are equal to within a factor of two or three (like the costs of natural gas and electricity for home heating), other factors such as cleanliness may outweigh the strictly financial considerations. Unfortunately, electrical energy generated with solar cells is in fact 100 or more times as expensive as electrical energy produced in other ways.

Although at present prices, solar cells are unfeasible for generating energy on a large scale, it is important to realize that costs will in all probability decrease. Mass-production methods can lower the costs of solar cells. More significantly, future scientific and technological developments may enable us to use less pure silicon or to make silicon of the necessary purity much more cheaply. No one can say with certainty that this will occur, but the history of other technological innovations is worth remembering. The invention of the transistor, for example, resulted from developments in solid-state physics during the 1930s and 1940s. Over the years, the costs of transistors and similar devices have

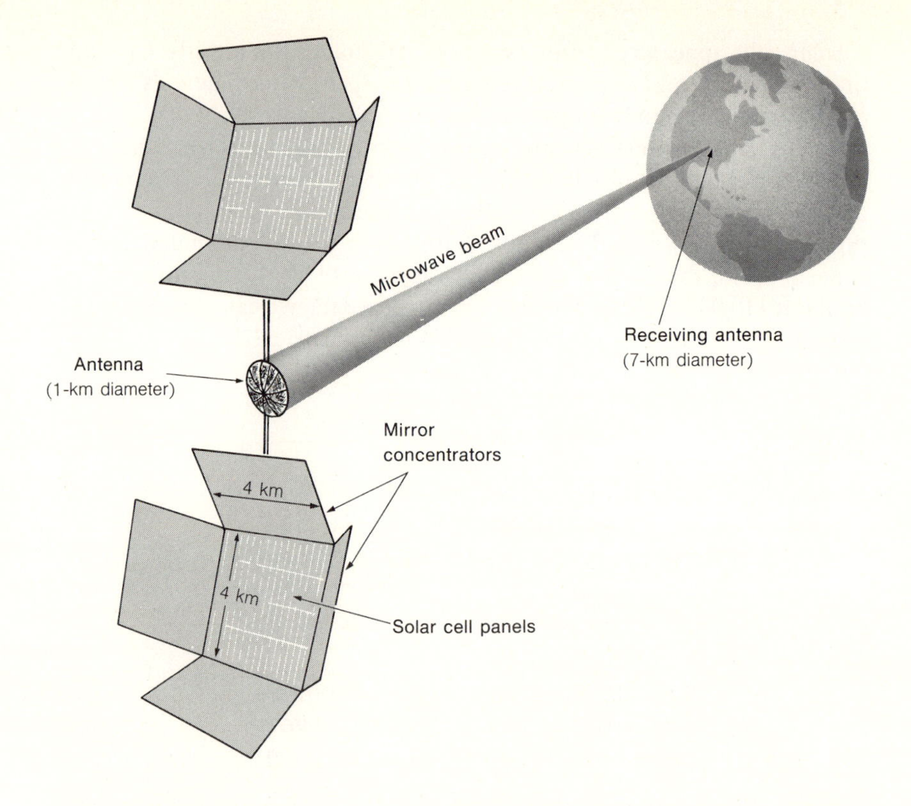

FIGURE 16.14
A proposed method for using solar cells: installation on an orbiting earth satellite. ["Solar Energy as a National Resource," University of Maryland, 1972.]

steadily declined, making possible many applications that would otherwise be prohibitively expensive. If the costs of solar cells could be reduced by a factor of 100 or so, then rooftop solar cells for individual homes as well as for large central power stations would be very attractive.

One imaginative plan for using solar cells is shown in Figure 16.14, that of installing solar cells on an artificial satellite and converting the electrical energy produced into a beam of microwaves that will transmit the energy to the earth. The satellite will be in a "synchronous" orbit; that is, like some communications satellites, it will be above the equator at an altitude such that it takes exactly 24 hours to complete one orbit, thus maintaining a fixed position relative to a receiving station on the ground.

QUESTION

16.19 What must be the radius of the orbit of such a synchronous satellite?

In this orbit, the satellite will be exposed to the sun almost continuously, the solar energy will not be diluted by transmission through the atmosphere, and the cells can always be kept perpendicular to the sun.

Thus the incident energy will strike the solar cells with the full value of the solar constant, 1.4 kW/m^2. Moreover, the mirrors shown in Figure 16.14 will reflect additional sunlight onto the collector, providing altogether about 4 kW on each square meter of collecting surface.

QUESTIONS

16.20 The proposed system shown in Figure 16.14 has a total of 32 km^2 of collectors, exposed to an average solar power of 4 kW/m^2. What will be the electrical output if the solar cells have an efficiency of 13%?

16.21 If, as has been estimated, approximately 75% of this power can be converted to microwaves, if 95% of the microwave power reaches the earth, and if this power can be converted back into useful electrical power at an efficiency of 75%, what will be the net useful amount of electrical power generated? About how many such space stations would be needed to provide the projected energy needs of the United States in the year 2000?

This idea is undeniably intriguing as well as grandiose, but some obvious drawbacks exist. If economical solar cells are available, why go to all the trouble and expense of putting them on satellites when they could be located on the ground? It is true that a satellite above the atmosphere would receive power at the steady rate of 1.4 kW/m^2, seven times as large as the average value of 200 W/m^2 available at the ground. Whether this is a sufficiently large difference to compensate for the additional cost is not clear. Furthermore, use of orbiting collectors would make us more dependent than ever on a small number of large central generating stations, whereas with solar cells on the ground our electrical energy could be generated in many small installations, even on the rooftops of individual houses. At the present time, the feasibility and costs of orbiting solar collectors are the subject of much investigation, and questions about the possible biological effects of the microwave beam also remain unresolved.

§16.7 THERMAL GENERATION OF ELECTRICAL ENERGY

Although the direct conversion of solar energy into chemical energy or electrical energy is difficult, the conversion of solar energy into *thermal* energy is relatively easy. This is simply one of the general consequences of the second law of thermodynamics. If the thermal energy itself is of interest, for instance for heating buildings, no further energy conversion is required, and this important application will be discussed in the following section. Here we shall explore the possibility of using solar energy to generate electrical energy in steam-electric power plants similar to conventional nuclear or fossil-fuel plants.

The moment we consider conversion of thermal energy into electrical energy, the *temperature* at which we can collect the energy becomes important (the higher the better), because this sets the limit to the efficiency with which we can generate electrical energy. In generating electrical energy from the temperature differences in the oceans, the small temperature differences and the resulting very low efficiencies may not be disastrous, because the collection of solar energy and its conversion into thermal energy are done for us at no cost. But if we have to do the initial collecting ourselves (and pay for it), higher efficiencies are desirable, and it is important to collect the solar energy at the highest possible temperature. We will discuss here one proposal for collecting solar energy at temperatures as high as 500°C, with the hope of achieving an overall conversion of 30 to 40% of the solar energy into electrical energy in large-scale power plants.

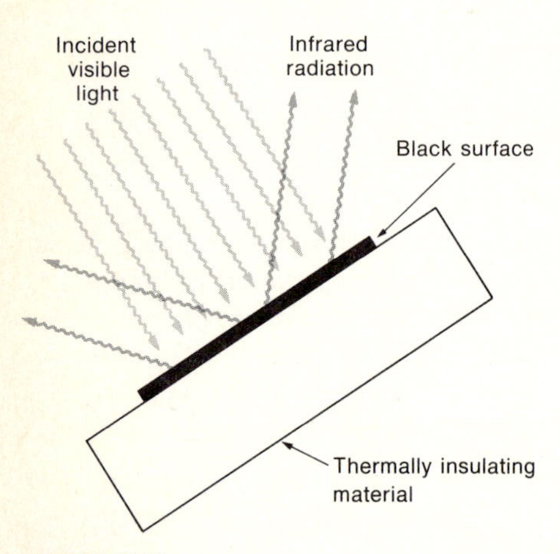

FIGURE 16.15
The radiation balance of a black surface exposed to sunlight.

A simple "black" surface gets hot when exposed to sunlight, as everyone knows who has walked barefoot across a highway on a summer day. Unfortunately, even under the best of circumstances, it does not get hot enough to be useful for producing steam. As the temperature rises, it emits more and more radiant energy of its own, until a steady temperature is reached at which energy is radiated as fast as it is absorbed (Figure 16.15). As shown in the following question, the highest temperature we can hope to achieve in this way is not more than about 90°C.

QUESTION

16.22 Consider an ideally black surface that is aimed directly at the sun on a clear day and that therefore absorbs solar energy at the rate of about 1 kW/m². Assume that the surface loses energy only by emitting its own black-body radiation. (That is, ignore energy losses due to conduction and convection.) Show that in equilibrium, when energy is being absorbed and emitted at equal rates, the temperature is about 364°K or 91°C.

A temperature of 91°C is warm but not warm enough; methods are needed that are more sophisticated than just putting a black surface in the sunlight. There are a number of ways in which we can take advantage of the fact that solar energy consists primarily of radiation in a particular part of the spectrum. As described in Chapter 11, a surface that is "black" in one part of the spectrum may not be black in other parts of the spectrum. The incoming solar radiation is concentrated in or near the visible part of the spectrum, but the radiation emitted, even by an object at temperatures as high as 500°C or so, is mostly in the infrared. Figure 16.16 shows the spectrum of the solar radiation, and also the spectrum of the radiation that would be emitted by a black body at a temperature of 500°C. The two curves barely overlap, a fact that we can utilize if we can make a "selective surface."

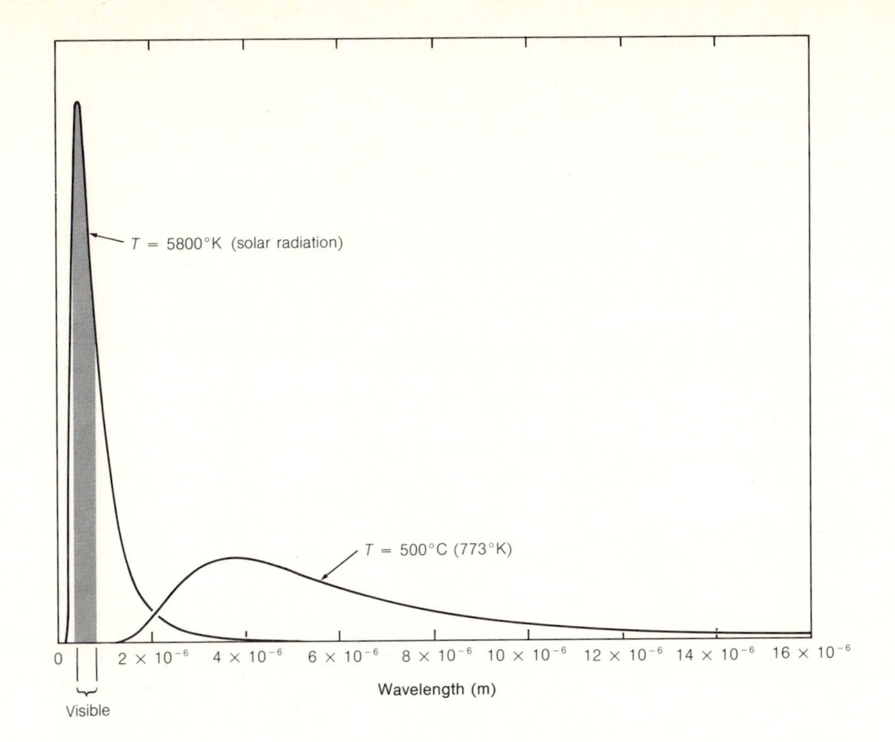

FIGURE 16.16
The spectrum of solar radiation compared with the spectrum of the black-body radiation of a surface at 500°C.

Consider a surface that is a good absorber of visible radiation and a poor absorber of infrared radiation. (That is, it is "black" to visible radiation but not to infrared.) According to the principles discussed in Chapter 11, such a surface would be a poor emitter of infrared radiation and (if the temperature were sufficiently high) a good emitter of visible radiation. No surface can emit more radiation at any wavelength than would a black surface at the same temperature; since even a black surface at a temperature of 500°C emits little visible radiation, the fact that such a surface, if hot enough, *could* lose a great deal of energy by radiating visible light is of no concern. Thus a selective surface could absorb sunlight just as well as an ideal black surface, but would emit infrared radiation at a lower rate and thus reach a higher equilibrium temperature.

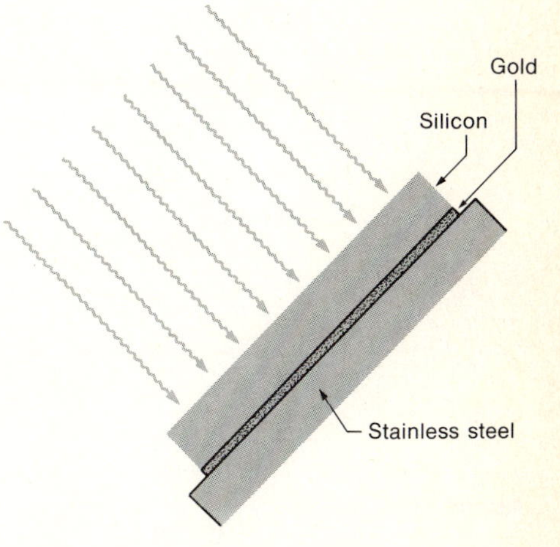

FIGURE 16.17
A selective surface for achieving higher temperatures.

QUESTION

16.23 Consider a surface that is black to the incoming solar energy but that *emits* radiation at only 10% of the rate at which a black body would emit at the same temperature. Show that such a surface absorbing solar energy at the rate of 1 kW/m^2 would reach an equilibrium temperature of about 650°K (377°C).

Composite surfaces with these properties have been made in the way shown in Figure 16.17. A stainless steel sheet is covered with a thin layer of gold, which in turn is covered with a thin layer of silicon. Silicon is a very good absorber of visible light, so this structure is *black* in the visible range. To infrared radiation, however, silicon is almost transparent, so as far as the infrared is concerned, the silicon might as well not be there at all. Gold is an excellent reflector of infrared radiation; that is, it is a poor absorber and hence a poor emitter. Thus this composite structure has the properties desired for a selective surface.

Selective surfaces help in achieving higher temperatures. To obtain still higher temperatures, we can take advantage of another property of visible light, namely that it can be bent by glass or other transparent materials; we can use lenses to concentrate sunlight, to focus the light from a large area onto a smaller area. (However, only light coming almost directly from the sun can be focused, and most of the *diffuse* radiation, which carries about 15% of the energy even on a clear day, is lost.)

A device for putting together these various ideas is shown in Figure 16.18. Incoming sunlight is focused onto a transparent window extending the length of a cylindrical tube. Inside this tube is a smaller tube, coated with a selective surface. The inner surface of the larger tube is coated with a reflecting material (except at the entrance slit). This arrangement permits incoming visible radiation to bounce several times if necessary before striking the inner tube; it also tends to trap whatever infrared radiation is emitted from the inner tube. To reduce heat losses from the inner tube further, the space between the two tubes is evacuated. Energy can be extracted from the inner tube by a stream of nitrogen gas, and delivered to a thermal storage reservoir. Heat can then be extracted from the reservoir to heat water and produce steam for driving turbines, as in other steam-electric plants. A design of a power plant for the southwestern United States with an average electrical output of 1 GW (comparable to that of a large modern coal or fission plant) is shown in Figure 16.19. A large number of collectors, tilted to face the noonday sun, would be spread over an area of about 15 square miles. Because of the spaces left between the collectors to keep them from casting shadows on one another and because of the space required for the other equipment, only about one-third of the solar energy arriving on the total area would be collected, and this would then be converted into electrical energy at an efficiency of about 30%. Energy could be delivered by conventional power lines; alternatively, hydrogen gas could be generated and shipped by pipeline and either burned directly as a fuel or used for generating electrical energy again in a fuel cell.

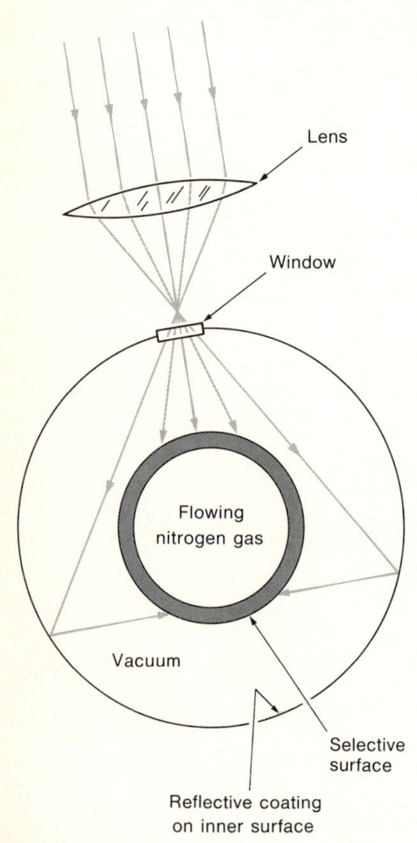

FIGURE 16.18
Plan for achieving high temperatures for the generation of electrical energy (end view).

QUESTIONS

16.24 The overall average rate at which solar energy is incident in the Southwest is about 250 W/m^2. Given this fact, along with the preceding estimate of the fraction of solar energy collected (one-third) and the efficiency with which the energy collected is converted into electrical energy (30%), show that 15 square miles is approximately the correct area for a power plant with an average output of 1 GW.

16.25 (a) How many such plants would be needed to equal our 1973 electrical generating capacity?

(b) How many plants would be needed in the year 2000 if our generating capacity continued to double every 10 years?

(c) It has been proposed that large areas of the Southwest, for instance the regions shown in Figure 16.20, be set aside for solar energy "farming." Estimate the total area of these two regions. Approximately how many power plants of the type described could be accommodated?

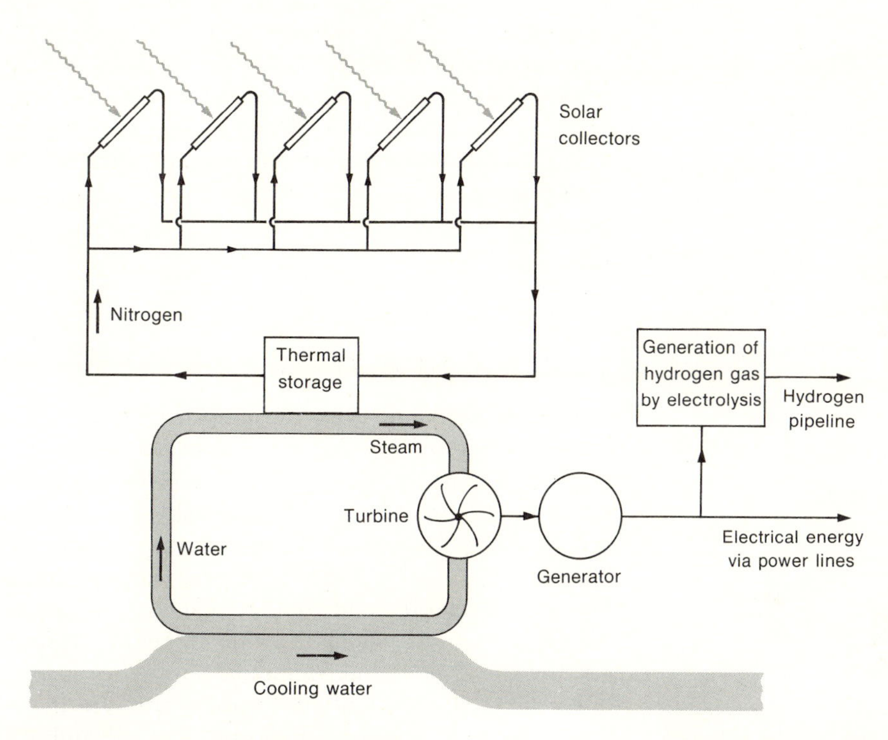

FIGURE 16.19
Diagram of a solar energy power plant.

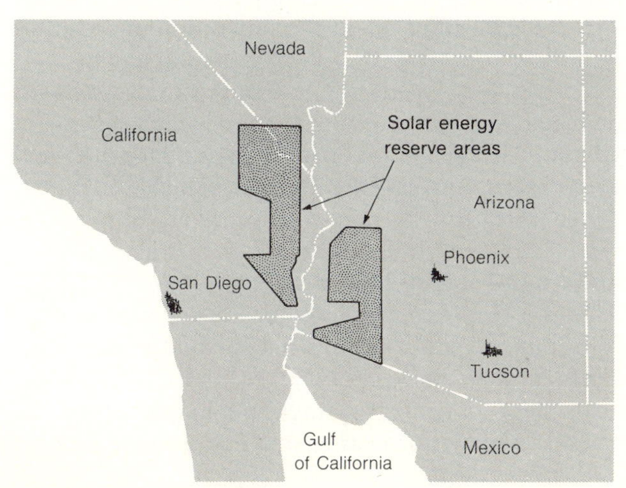

FIGURE 16.20
Proposed areas for solar energy farming. [Adapted from A. B. Meinel and M. P. Meinel, *Bulletin of the Atomic Scientists*, 1971. By permission of Science and Public Affairs. Copyright © 1971 by the Educational Foundation for Nuclear Science.]

Will it work? If so, what will be the cost? Responsible critics believe that the contemplated temperatures may be impossible to achieve, that the efficiency will therefore be much lower and the amount of land required much greater, that reliable selective surfaces will never be cheap enough—in short that a host of technical and economic problems have been treated much too optimistically. Transporting all this energy to the users, many of whom are thousands of miles away, also presents difficulties. Whether energy is transmitted by conventional overhead power lines, by underground superconducting power lines, or in pipelines as hydrogen gas, the surrounding areas of the Southwest, already contaminated with many power lines, will be further subjected to the impact of technology. Advocates of solar energy farming do not have ready answers to all these questions, but they argue that the potential benefits of providing large amounts of clean energy from an inexhaustible source would be so great that the concept should be investigated as thoroughly as possible.

§16.8 SOLAR HEATING

In spite of the vast quantities of solar energy, none of the previously described methods of using this energy is as promising as we might have hoped. For some, the technological problems are unresolved; for others, the costs at present seem astronomically high or at best extremely uncertain. All of these methods, and others, should of course be explored thoroughly, no matter how expensive, impractical, or uncertain they may now appear. In one area, though—the use of solar energy to heat buildings—the technology is quite simple and thoroughly understood, and the costs even today seem competitive (or nearly so) with conventional methods. Most of the changes that are likely to occur in the future—such as continuing shortages of oil and natural gas, further restrictions on the use of high-sulfur oil, or the imposition of pollution taxes—would make the use of solar energy even more attractive.

FIGURE 16.21
A solar cooker.

If the amounts of energy used in heating buildings were very small, there would be little incentive to put much effort into using solar energy for this purpose. One can use a solar cooker as a substitute for the backyard charcoal grill, and solar cookers may be very valuable in India, where all types of fuel are in short supply, but this would do little to relieve the energy problem in the United States. A very substantial fraction of the energy used in this country, about 20% of the total, is devoted to space heating. This is a large enough amount to make solar heating very interesting. Even if *all* our space heating needs could be met with solar energy (and this is impossible), we would still have an energy problem, but the use of solar energy for heating could have a substantial impact. The special characteristics of solar energy that make it difficult

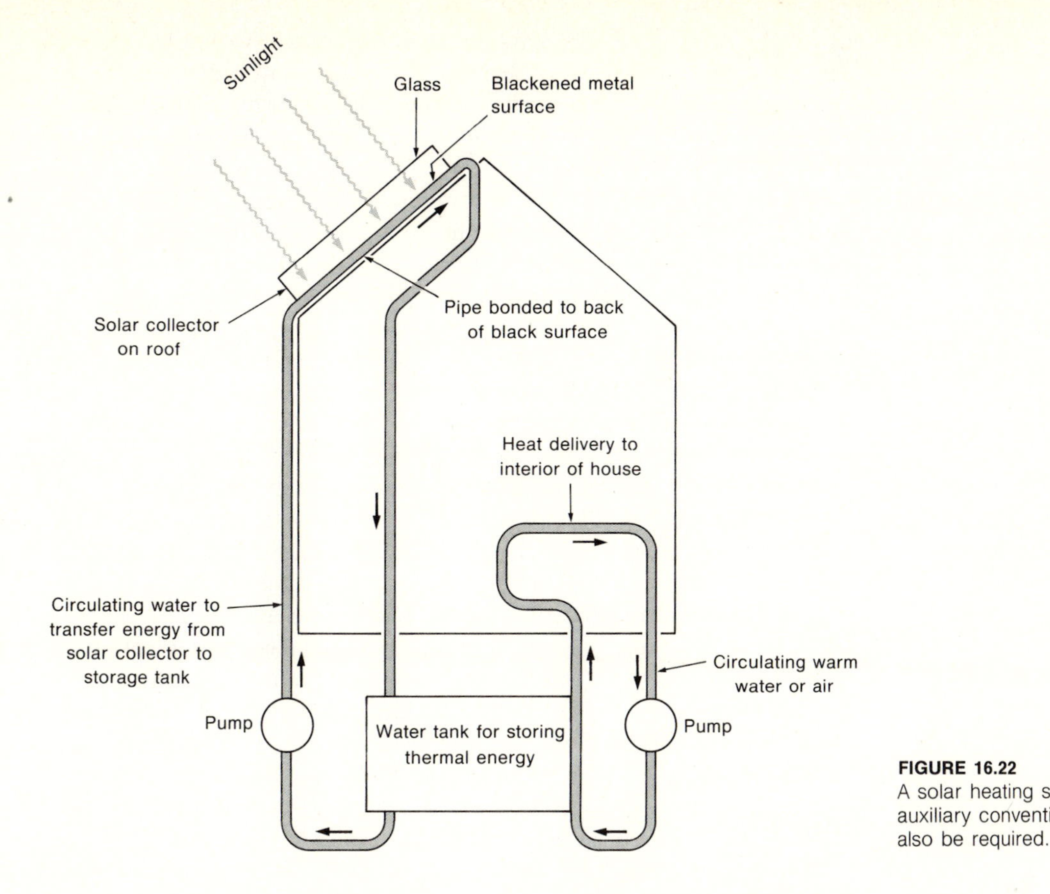

FIGURE 16.22
A solar heating system. In practice, an auxiliary conventional heating system will also be required.

to use for some purposes may be less disadvantageous for solar heating. Indeed, one of the so-called drawbacks—the fact that our solar energy is dispersed across the whole country—is actually quite useful, since the buildings we need to heat are also widely distributed. However, the southern parts of the country, where heating needs are least, receive more sunlight than do northern areas, and of course less sunlight is available during the winter and none at all at nighttime.

It is difficult to use solar energy to get water hot enough to produce steam to drive a turbine. For heating a house, however, we do not need steam; warm water will suffice, and it is quite easy to get warm water with solar energy. The essential elements of one simple method for solar heating are shown in Figure 16.22. Incoming solar energy strikes a roof-top energy collector, which consists of a blackened sheet of metal covered by a sheet of glass to reduce heat losses by conduction and convection. Water pipes are firmly bonded to this sheet, water is pumped through the collector during the day, and warm water is accumulated in a storage tank. (*Some* sort of energy storage is essential for nighttime heating.) Water or air can be passed through the storage tank as needed to deliver heat to various parts of the house. Many variations on this basic plan have been used. For example, it might be air rather than water that is pumped through the tubes on the collector, or the thermal energy could be stored in hot rocks rather than water, but we will not try to study all the possibilities.

What must be done is to see how much solar energy is available and how much of it can be used, and then what it would cost and how that cost would compare with the cost of heating by oil, natural gas, or electrical heat. Instead of analyzing all possible buildings in all possible locations, let us take a fairly careful look at a representative home in a typical location. Consider a house with a thermal conductance of 20,000 BTU/DD in Columbia, Missouri, a city where both the severity of climate and the amount of solar energy available are close to the national average. In the course of an entire year, the average total of degree-days at this location is 5046 DD. Let us first calculate the total annual amount of heat needed and the total amount of incoming solar energy. The total amount of heat needed by this house during a year is $5046 \times 20{,}000 = 101 \times 10^6$ BTU = 101 MBTU (1 MBTU = one million BTU). On the average (an average calculated for the whole year and for all 24 hours of the day), solar energy is incident on a horizontal surface at Columbia at the rate of 184 W/m^2. Suppose for now that the house has an area of about 1300 ft^2 and a *flat* roof. (It is obviously ridiculous to build a solar house with a flat roof, but let us begin with this assumption for simplicity.)

QUESTIONS

16.26 On the basis of these assumptions, show that the total amount of solar energy striking the roof of this house in the course of a year is 665 MBTU. This result is more than six times as large as the total amount of heat needed, sufficiently greater to make solar heating look promising. This calculation must be refined in various ways, but this simple comparison provides ground for optimism.

16.27 What might you conclude if the total amount of incoming solar energy during a year were instead: (a) 100 MBTU? (b) 10^5 MBTU?

If we could store thermal energy accumulated during the summer for use during the winter, solar heating would be easy. Unfortunately, it is simply not feasible to store enough energy. Let us see why not. How much water would be needed to store a winter's worth of thermal energy? First, assume that water can be gotten almost to the boiling point (say, 200°F) during the summer; then let us make the obviously unrealistic assumption that the water can be stored without cooling off at all until we need it, and suppose that we can extract energy by letting the water cool to 70°F.

QUESTION

16.28 (a) Show that the amount of water that can deliver 101 MBTU of heat while cooling from 200°F to 70°F is about 775,000 pounds, or 390 tons.

(b) Show that this amount of water would occupy a volume of about 93,000 gallons, and that a cubical storage tank would have to be 23 ft on a side.

Storage of a 6-month supply of energy seems out of the question, but can we store enough for one day? If we cannot store at least this amount, we will have a problem, since we certainly want to be able to store enough energy during the day to last through the night, and it would be desirable if we could accumulate several days' worth in order to keep warm on cloudy days.

QUESTION

16.29 A winter day with an average temperature of 15°F has 50 degree-days, and the 20,000 BTU/DD house therefore needs $50 \times 20{,}000 = 10^6$ BTU or just 1 MBTU for one day. On the basis of the same assumptions as before, show that the amount of water needed to store heat for one winter day would be about 920 gallons, would weigh about 4 tons, and could be contained in a cubical tank about 5 ft on a side.

A 920-gallon storage tank is much more feasible than a 93,000-gallon tank. We might well use a 2000-gallon tank in order to have enough heat to last through two cloudy days in succession. However, if we insist on getting all of our heat from solar energy, a problem arises. A tank large enough to provide energy for those rare occasions when we have a week or ten days of cloudy cold weather would be very large and expensive. Most of our investment in the storage tank would be wasted. It seems much better to build a tank with a storage capacity of one or two days, thereby meeting a large fraction (but not 100%) of our heating needs with solar energy, and to install an auxiliary gas, oil or electrical heating system that can be used when needed. Every careful analysis of solar heating leads to the same conclusion: namely that, unless the cost is of no concern, it is foolish to try to construct a solar heating system with a capacity sufficient to supply 100% of the heating needs of a building. This conclusion has an important bearing on the cost of solar heating. Even with solar heating, a full-sized auxiliary heating system must be available. Thus the money we invest in solar heating is an *additional* investment, not a substitute for the capital investment in a conventional system. The financial comparison will then be between the annual cost of the capital investment in the solar system and the cost of that amount of fuel (or electricity) we would otherwise have to use during the winter. Since we cannot store thermal energy for long periods of time, it is necessary, in order to carry the analysis further, to consider how heating needs and incident solar energy vary during the year.

QUESTION

16.30 A rough estimate (but one that is more accurate than our initial one) can be made in the following way. Most of the annual heating requirement comes in the five months from November through March. Let us suppose that *all* the heating needs come in this five-month period, and, in an attempt to be conservative, let us assume that during this whole period, the average rate at which solar energy strikes a horizontal surface is equal to the value for the month of December, 82 W/m^2. Show that, even on this assumption, enough solar energy strikes the 1300-ft^2 flat roof of the house to meet the heating needs if all the solar energy can be used.

To make an accurate calculation, it would be best to make a day-by-day (or even hour-by-hour) analysis, but quite accurate estimates can be made by treating each month as a whole. The basic data are shown in columns 1 and 2 of Table 16.1, and the results in columns 3 and 4. All of the numbers in column 4 are larger than the corresponding numbers in column 3. In other words, even in December and January there is enough solar energy striking the roof to keep the house warm *if* we can obtain all of it. Unfortunately, we cannot turn all of the incident solar energy into thermal energy in the collector. Some energy is reflected and some is lost by convection, conduction, and radiation, and a collection efficiency of 50% or so is about the best we can hope for. In addition, although we may get considerably more than we need during a week of sunny weather, some of this energy will have to be thrown away unless we have an extremely large storage tank. The amount of solar energy we can actually collect and use is more likely to be about 30% of the incident energy; if the figures in column 4 of Table 16.1 are multiplied by 0.3, we can immediately see that some auxiliary source of heat will be needed from November through February, as shown in Table 16.2. For the year as a whole, we could heat this house with 65 MBTU of "free" solar energy together with 36 MBTU of auxiliary heat from electricity or gas or oil.

The solar energy is not free, however. The "fuel" is free, and operating and maintenance costs should be quite low, but the capital investment is sizable. Since solar collectors are not now mass-produced, it is not easy to estimate how cheaply they might be produced by the time significant numbers of solar homes are being built. Collectors can be made out of conventional materials (glass, copper, and so on), though, so that cost estimates can be made with more confidence than they can for solar cells or offshore windmills. Estimates by a number of authorities

TABLE 16.1 Solar energy and heating data for a 20,000 BTU/DD house, horizontal 1300-ft^2 collector (Columbia, Missouri).

	(1) Degree-days per month	(2) Solar power per unit area* (W/m^2)	(3) Monthly heating need (MBTU)	(4) Solar energy available† (MBTU)
July	0	278	0	85.4
August	0	255	0	78.2
September	54	217	1.1	64.5
October	251	157	5.0	48.2
November	651	107	13.0	31.8
December	967	82	19.3	25.2
January	1076	87	21.5	26.7
February	874	121	17.5	33.6
March	716	166	14.3	51.0
April	324	210	6.5	62.4
May	121	257	2.4	78.9
June	12	276	0.2	82.0
	Total: 5046	Average: 184	Total: 101	Total: 665

*Incident solar power on a horizontal surface, averaged over 24 hours per day and over the entire month.

†Monthly solar energy incident on a 1300-ft^2 horizontal surface.

TABLE 16.2 Revised solar energy and heating data* for a 20,000 BTU/DD house, horizontal 1300-ft^2 collector (Columbia, Missouri).

	Solar energy available (MBTU)	Monthly heating need (MBTU)	Solar energy used (MBTU)	Auxiliary energy needed (MBTU)
July	25.6	0	0	0
August	23.5	0	0	0
September	19.4	1.1	1.1	0
October	14.5	5.0	5.0	0
November	9.5	13.0	9.5	3.5
December	7.6	19.3	7.6	11.7
January	8.0	21.5	8.0	13.5
February	10.1	17.5	10.1	7.4
March	15.3	14.3	14.3	0
April	18.7	6.5	6.5	0
May	23.7	2.4	2.4	0
June	24.6	0.2	0.2	0
Totals:	200	101	65	36
Fraction of heating supplied by solar energy: 65/101 = 64%				

*Data are taken from Table 16.1, and have been modified in accord with the assumption that at most 30% of the incident solar energy can be collected and used.

suggest that collectors can be manufactured at a cost of about $4 per square foot and that the necessary plumbing, pumps, water storage tank, and other parts for the system described above would cost about $800. The cost data are important, and some persons believe that the true costs will be somewhat higher; let us nevertheless use these numbers to make an estimate of the cost of solar heating. The capital investment would be:

$$(\text{Area} \times \$4) + \$800 = \text{capital investment}, \tag{16.2}$$

or, $(1300 \times \$4) + \$800 = \$6000$. In order to convert this figure into an equivalent annual cost, we can use the same ideas discussed in §4.4.C to estimate the value of a storm window and in §16.6 to estimate the cost of generating electrical energy from solar cells.

QUESTION

16.31 Assume that $6000 is borrowed and is to be paid back over a period of 30 years at an annual interest rate of 10%. Show that the annual cost will be $636/yr.

An annual heating bill of $636 seems rather high for Missouri, and it is. Furthermore, we also have to pay for the 36 MBTU of auxiliary heat. Another way of looking at these data is this. We get 65 MBTU of solar energy per year at a cost of $636, that is, at a price of $636/65 = $9.80/MBTU.

QUESTION

16.32 What is the cost per MBTU of obtaining heat from oil, gas, or electricity? Use current prices if available; otherwise use 1974 prices from Appendix P. Do not forget that oil or gas heating is less than 100% efficient.

When our figure for solar energy, $9.80/MBTU, is compared with gas and oil at 1974 prices, solar energy appears to be a poor choice, and it is even slightly more expensive than electrical heating. But what is extremely significant is that solar heating is not very *much* more expensive than other methods. What can we do to improve the situation? If our goal were simply to collect a great deal of energy, we could build an even larger collector, but the effect would be to increase the cost per unit of energy. One very easy improvement is to use a smaller collector but to tilt it, and this can help a great deal. In the middle of December, just when we need heat, the noonday sun is only 28° above the horizon at Columbia. By merely tilting the collector at an angle of 62° (Figure

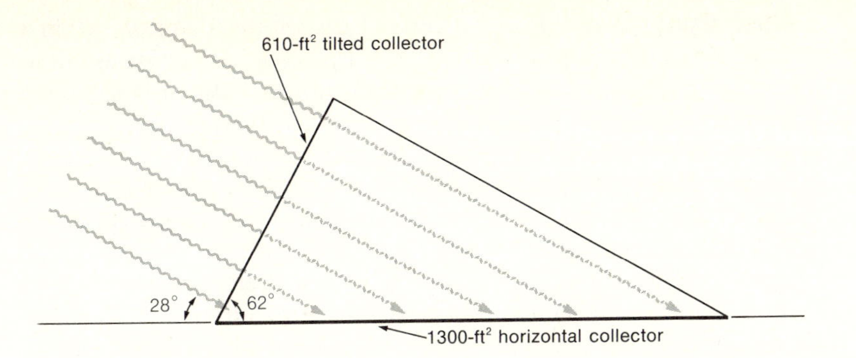

FIGURE 16.23
A small collector, if tilted, can collect as much energy as a larger one.

16.23), we can collect all the energy that would otherwise hit the 1300-ft^2 roof with only 610 ft^2 of collector, and so this obviously makes a great deal of sense.

QUESTION

16.33 It is approximately correct to say that a house with a properly tilted 610-ft^2 roof in Columbia can collect as much useful solar energy (65 MBTU) as a house with a 1300-ft^2 flat roof. Show that reduction of the area in Equation 16.2 from 1300 ft^2 to 610 ft^2 reduces the annual cost of the solar heating installation to \$343/yr, and the cost of the solar energy to \$5.30/MBTU, significantly lower than the previously calculated value of \$9.80.

An angle of 62° is best for December, but a smaller tilt is better during other months. It is not difficult to extend the analysis and decide on the best angle of tilt and the optimum area, but since our purpose has been only to explore the feasibility of solar heating and to arrive at a rough estimate of the costs, we will not pursue these questions. The important point is that it appears quite feasible to use solar energy to provide a substantial fraction of the heating needs of the house and to do so at a cost per MBTU that is not much different from the cost of conventional methods of heating.

QUESTION

16.34 Discuss critically your own understanding of the technical feasibility and costs of solar heating. Which aspects of the discussion in this section must you accept completely on faith? Which aspects can you at least test for consistency with other topics discussed in this book? If you were an advocate of solar energy and wanted to make the result look more favorable, which numbers would you feel justified in altering and why? What changes might you make if you were opposed to the use of solar energy, as you might be if your income came from the profits of a coal company? This analysis of solar heating was written in 1974; have significant changes taken place since that time?

What about solar heating in other parts of the country? At first thought, one might suppose that solar heating would be most useful in the sunniest southernmost regions. This is not so, however, for the following important reason. In Miami, for instance, there is plenty of sunlight but little need for heat. Almost the entire cost of solar heating lies in the capital investment; if you install a solar heating system in Miami, you will be paying interest to the bank every month for an expensive installation that is rarely needed. It makes more sense to buy the cheapest possible oil or gas furnace or to install electric heating, and then pay for the small amount of heat you need. Surprisingly, perhaps, because of this "Miami effect," the prospects for solar heating may be best in the central and northern parts of the United States. The relative importance of capital investment and operating cost is almost exactly the reverse of what it is for conventional heating systems, and this fact complicates the problem of making meaningful comparisons. (However, one factor may modify the Miami effect. Solar energy can be used to provide hot water as well as space heating and, with some increase in complexity, it can be used to provide air conditioning too.* Since residents of Miami need as much hot water as everyone else and more air conditioning than most, we may find that solar energy is advantageous in Miami too. At one time quite a few solar water heaters were being used in Florida, but they seem to have gone out of fashion with the increased availability of low-cost natural gas.)

The three most important points to remember about the use of solar energy for space heating are these. (1) Space heating requires a substantial amount of energy. It is therefore a problem worthy of attention. (2) No technological or scientific advances are required. In contrast to nuclear fusion—which will *probably* work but may in fact not, no matter how much we are willing to pay—solar heating does work. Solar heating is not as exotic as fusion or the installation of solar cells on earth satellites, but that is really the main advantage. Such commonplace materials as copper pipes, glass plates, and so on are all we need. (3) Cost estimates show that although solar heating (as of 1974) may be somewhat more expensive than other methods, it is not ridiculously so. Should the price of gas or oil double a few times, or an industry be developed to mass-produce solar collectors, solar heating may suddenly become the cheapest possible way to heat a house.

---

*It may seem paradoxical that solar energy can be used for cooling as well as for heating. We can see, though, that solar cooling is not out of the question by realizing that we *could* use solar energy to generate electrical energy as discussed in §16.6 and §16.7, and then use this electrical energy to run an ordinary air conditioner. This is not an economical way of providing solar cooling, but it is a possibility. The fundamental fact making solar cooling possible (and, for that matter, making possible the thermal generation of electrical energy with solar energy) is that heat from the sun is high-grade heat, from a very high-temperature source (the sun); it is heat that has little entropy. A practical solar air conditioner might be quite similar to an old-fashioned gas refrigerator in which the necessary high-grade energy is obtained from a high-temperature gas flame; the details of the operation of such devices are too complicated to be described here.

FIGURE 16.24
A solar-heated house in Washington, D.C., built by H. E. Thomason in 1963. [*Popular Science*, © 1974, Times Mirror Magazines, Inc.]

The nation cannot be converted overnight to solar heating. Solar heating will probably never be practical in high-rise apartment buildings in the middle of cities. Rather, the suburbs and small towns seem to be the obvious places. It may not be reasonable to try to convert existing houses, so even if we begin now, some time will elapse before the impact is significant; all the more reason, therefore, to get started as soon as possible. (Of the houses that will be in use in the year 2000, probably half will have been built before 1970.) It is a pity that we did not begin to build solar-heated houses on a large scale a long time ago. As early as 1952, it was suggested in the report of a presidential commission that by 1975 the number of solar-heated houses in the United States might be 13 million; unfortunately the actual figure is closer to 13. Whatever legislative or administrative actions are needed to encourage the use of solar heating should be taken immediately.

It is clear that the underlying reason for the scarcity of solar-heated houses is an economic one. Fuel has been cheap, and little thought has been given to the possibility that this situation might not continue forever. Still, decisions are not always made by choosing the very cheapest option. For example, many persons live in electrically heated homes, even though electric heating is definitely more expensive than oil or gas. Why is this so? The reasons may suggest what steps need to be taken by the nation to implement courses of action that may appear from a narrow point of view to be somewhat more expensive. First, home buyers often have no choice. If the only houses for sale have electric heating, they buy what is available. As a nation, we may find ourselves obliged to opt for solar heating, as a result of the scarcity of oil and gas or because of pollution problems. Second, the home buyer may be ignorant of the true costs. Because the capital cost of electric heating is somewhat lower than that of oil or gas heating, electric heating looks attractive at the time of purchase. This is why it is often chosen by the developer; it permits a lower selling price. If buyers knew enough about how to compare capital and operating costs, they would discover that electric heating is not such a good idea after all. As a nation, we may be in the similar position of not knowing what the future costs of our present practices will be. We do not know how much damage to the environment is being caused by our pollutants, and we will not know how very valuable oil may be for special purposes, such as the manufacture of petrochemicals, until it is too late, after we have burned most of it up. Third, electrical heat is "cleaner," or at any rate it appears cleaner to the home owner. It seems indisputable that solar energy really is a cleaner source of energy, which can be used with little if any harm to the environment.

Most energy questions turn out to be very difficult when we try to balance all the environmental, social, and economic questions, in an effort to determine the best course of action. The use of solar energy to heat buildings does not seem to be such a question. It is hard to imagine a more innocuous source of energy than the sunlight on one's roof. However, if solar energy is more expensive to the user, even if it is only slightly more expensive, it is unrealistic to expect it to be chosen voluntarily by a significant number of builders or buyers. Whether the introduction of solar heating is accomplished by legislative mandate, by tax incentives, or other measures, it seems clear that it is in the interest of the whole nation and the world to take whatever steps are needed. It will be less expensive in the long run to make full use of this clean and inexhaustible source of energy.

§16.9 A SUMMARY OF THE ENTIRE ENERGY PROBLEM

Many of the remarks made about solar heating at the end of the preceding section can be applied to the whole energy problem. We need to take a long-range view, at least 100 years into the future, and consider more carefully the true impact of our present actions. The energy problem of the United States has been emphasized in this book, in part because more information is readily available, but the energy problem of the whole world must be considered. Particularly in this country, energy conservation is important, but few would argue that the underdeveloped nations should be asked not to increase their use of energy, and there are numerous groups in the United States for whom the status quo is not acceptable.

There seem to be only three sources of energy on which we can rely to provide large amounts of energy for a very long period of time: nuclear breeder reactors (if they are safe), nuclear fusion (if it works), and solar energy. Oil and natural gas are very definitely not the fuels of the future. Coal is readily available, at least for the next century or two, but the drawbacks to the mining and burning of coal are obvious. One branch of technology whose development should be emphasized, one that has not been discussed in this book, is that of making clean liquid and gaseous fuels from coal. With breeder reactors, nuclear power could provide abundant energy; in this author's opinion, the risks appear too great at the present time. Research on reactor safety and on the safe storage of radioactive wastes should be pursued at public expense, but we would be well advised to make as little use of nuclear power as possible until it becomes clear that there are no alternatives. It would be more prudent to get by during the rest of this century largely with coal (with due attention to the environmental effects) and our dwindling but still large supplies of oil, while nuclear fusion and solar energy are developed with maximum effort. If fusion can be made to work, abundant energy with minor environmental side effects will be available. Solar energy, however, seems to provide the best long-range solution to our energy problems. Economically, solar heating is already nearly competitive with conventional methods. Advances in solar cell technology during the next few decades, coupled perhaps with the development of giant windmills, may enable us to generate large quantities of electrical energy at a reasonable cost. Should these developments occur, we could gradually come to rely on solar energy for almost all our energy needs. In the meantime, energy conservation is extremely important. There is no single course of action that will itself "solve" the energy problem; we should concentrate on such large-scale areas of energy consumption as space heating and transportation, and hope that a large number of individually small savings will add up to make a significant reduction in the total amount of energy used.

If we can use nuclear fusion or solar energy, will there then be any physical limits to growth? There will be no energy resource problem to speak of, but growth in consumption must nevertheless level off eventually, if only because the generation of thermal energy in large amounts may significantly alter the earth's climate, possibly with disastrous consequences. Whenever we use the energy of nuclear fusion (or of fission or fossil fuels), we are producing heat the earth must somehow get rid of. As we saw in Chapter 11 our contribution to the overall radiation balance of the earth is not now very large, but we cannot afford to take the risk of allowing energy consumption to double another ten times. Even now, the local climate is affected in those areas in which our uses of energy are most highly concentrated. Consumption of solar energy is somewhat different in that we are for the most part using energy that the earth would have absorbed anyhow, although we will absorb slightly more by having large areas of deliberately blackened collectors. We will be redistributing this heat, however, if we collect solar energy in Arizona, convert it into electrical energy, and then dissipate it in New York City. Redistribution of this sort may have a significant effect on the global climate if done on a sufficiently large scale.

What will the world be like when energy consumption and population level off? Will we be able to devise social institutions that can make possible a humane steady-state no-growth society? Historically, inequities in the distribution of material goods among nations and among groups of people within a single nation may conceivably have been tolerable, because as long as the total amount continued to increase, so could the amount available to every individual. Standards of living have been gradually rising even without any significant improvement in the way in which wealth is distributed. In a truly steady-state society, this way of evading the fundamental problems of fairness will be impossible; supplying more goods or energy to one group will require decreasing the amount available to another. This is a problem we have never had to face before.

The implications of growth have been discussed sporadically since 1798 when Malthus wrote his famous essay. It would appear that a careful discussion of these problems should now be given as high a priority as research and development in nuclear fusion and solar power. Making the transition to a condition of no growth will not be easy; it is the author's thesis that the result is much more likely to be a happy one if we have 300 years or so in which to accomplish it than if we have only 50 to 100 years, and an abundant supply of clean energy will be essential in providing the extra time. Our problems are urgent, and if we are to find solutions, we need people with many different kinds of expertise: in solar cell technology, in nuclear physics, in the biological effects of pollutants, in economics and demography, to mention but a few. But scientific and technical knowledge is not enough: the world's problems are too important to be delegated to the experts; we also need people who can stand back and look at the whole picture, and for them as well as for the scientists, a quantitative understanding of energy and its role in society is essential.

FURTHER QUESTIONS

16.35 If a flat pan with a layer of water 1 mm deep were set out in the noonday sun at Miami during the middle of June, approximately how long would it take to raise the temperature from 80°F to the boiling point and then to boil it away? Assume that all the sunlight is absorbed and that no heat is lost by the water.

16.36 Small windmills to generate electricity for individual use are once again becoming popular. One such machine, which can generate on the average 100 W in a typical location, costs about $700 installed, plus $200 for storage batteries. Once installed, the electrical energy is "free." Assume that the storage batteries last 10 years, that the rest of the installation lasts 30 years, and that the interest rate is 10%. On this basis, show that the cost of 1 kWh is about 12¢.

16.37 If the average American generates about 5 lb of garbage and trash per day with an average energy content of about 10^7 BTU/ton, how much energy could be derived from this source per year? Compare this energy with the annual amount used in the United States.

16.38 How does the necessary rate of flow of water through a 100-MW power plant of the type described in §16.4 compare with the average rate of flow of the Colorado River?

16.39 Consider a 20,000 BTU/DD house in the northern part of the United States. If it can be heated by burning wood (with an efficiency of 50%), approximately how large a wood lot would be needed to provide all the heat for this house on a continuing basis?

16.40 The figures in columns 3 and 4 of Table 16.1 are derived from those in columns 1 and 2, respectively. Check the calculation for the month of December.

16.41 Suppose that the Columbia, Missouri house described in Table 16.2 were located somewhere else. Make a revised version of Table 16.2, and find what fraction of the house's heating needs could be supplied by solar energy. If the annual cost of the solar energy installation is the same as for Columbia ($636), what would be the cost per MBTU of the solar energy? Make this analysis for one or more of the following locations: (a) Miami; (b) Barrow, Alaska; (c) a city near you for which data are available in Appendixes F and G.

16.42 Suppose that the United States *were* completely covered with power plants (see Question 10.16). How would the electrical output of these plants compare with the rate at which solar energy strikes the United States?

16.43 What is the approximate amount spent by the federal government per year in developing new sources of energy or reducing the environmental impact of sources we now use? Would a 1% tax on all electric bills provide a significant additional amount of money for research and development?

Appendixes

Appendixes

Let us not underrate the value of a fact;
it will one day flower in a truth.

—Henry David Thoreau, 1842

Note on Accuracy

In the following appendixes, values of most of the physical constants and conversion factors are correct to four significant figures. Much of the other information is known less accurately. Where uncertainty exists or where various authorities differ, an effort has been made to use the most reliable data. In a few places (for example, the amount of water in ice-caps and glaciers), the uncertainty is such that the figure cited may be too large or too small by a factor of two.

Sources of Information

Space does not permit complete documentation of all the figures cited. Among the most important sources of information were the following. (Those marked with an asterisk (*) are issued annually.)

***Statistical Abstract of the United States* (U.S. Government Printing Office, Washington, D.C.)

**Statistical Yearbook* (United Nations, New York).

**Minerals Yearbook* (U.S. Department of the Interior, U.S. Government Printing Office, Washington, D.C.)

**Statistical Yearbook of the Electric Utility Industry* (Edison Electric Institute, New York).

Historical Statistics of the United States (U.S. Government Printing Office, Washington, D.C.)

M.K. Hubbert, "Energy Resources." [National Academy of Sciences—National Research Council, *Resources and Man* (W. H. Freeman and Company, San Francisco, 1969).]

S. H. Schurr and B. C. Netschert, *Energy in the American Economy, 1850–1975* (The Johns Hopkins Press, Baltimore, 1960).

J. Darmstadter, *Energy in the World Economy* (The Johns Hopkins Press, Baltimore, 1971).

Patterns of Energy Consumption in the United States (Executive Office of the President, 1972, U.S. Government Printing Office, Washington, D.C.)

The Potential for Energy Conservation (Executive Office of the President, 1972, U.S. Government Printing Office, Washington, D.C.)

Appendix A
Units and Conversion Factors

List of Conversion Tables

TABLE A.1 Area.

1 square centimeter (cm^2) =	1 square inch ($in.^2$) =	1 square foot (ft^2) =	1 square meter (m^2) =
1 cm^2	6.452 cm^2	929 cm^2	10^4 cm^2
0.155 $in.^2$	1 $in.^2$	144 $in.^2$	1550 $in.^2$
1.076×10^{-3} ft^2	6.944×10^{-3} ft^2	1 ft^2	10.76 ft^2
10^{-4} m^2	6.452×10^{-4} m^2	0.0929 m^2	1 m^2
2.471×10^{-8} acre	1.594×10^{-7} acre	2.296×10^{-5} acre	2.471×10^{-4} acre
10^{-8} hectare	6.452×10^{-8} hectare	9.29×10^{-6} hectare	10^{-4} hectare
10^{-10} km^2	6.452×10^{-10} km^2	9.29×10^{-8} km^2	10^{-6} km^2
3.861×10^{-11} $mile^2$	2.491×10^{-10} $mile^2$	3.587×10^{-8} $mile^2$	3.861×10^{-7} $mile^2$

1 acre =	1 hectare (ha) =	1 square kilometer (km^2) =	1 square mile ($mile^2$) =
4.047×10^7 cm^2	10^8 cm^2	10^{10} cm^2	2.59×10^{10} cm^2
6.273×10^6 $in.^2$	1.55×10^7 $in.^2$	1.55×10^9 $in.^2$	4.014×10^9 $in.^2$
4.356×10^4 ft^2	1.076×10^5 ft^2	1.076×10^7 ft^2	2.788×10^7 ft^2
4047 m^2	10^4 m^2	10^6 m^2	2.59×10^6 m^2
1 acre	2.471 acres	247.1 acres	640 acres
0.4047 hectare	1 hectare	100 hectares	259 hectares
4.047×10^{-3} km^2	0.01 km^2	1 km^2	2.59 km^2
1.562×10^{-3} $mile^2$	3.861×10^{-3} $mile^2$	0.3861 $mile^2$	1 $mile^2$

TABLE A.2 Density.

1 kilogram per cubic meter (kg/m^3) =	1 pound per cubic foot (lb/ft^3) =
1 kg/m^3	16.02 kg/m^3
6.243×10^{-2} lb/ft^3	1 lb/ft^3
8.345×10^{-3} lb/gal	0.1337 lb/gal
0.001 g/cm^3	1.602×10^{-2} g/cm^3
3.613×10^{-5} $lb/in.^3$	5.787×10^{-4} $lb/in.^3$

1 pound per gallon (lb/gal) =	1 gram per cubic centimeter* (g/cm^3) =	1 pound per cubic inch ($lb/in.^3$) =
119.8 kg/m^3	1000 kg/m^3	2.768×10^4 kg/m^3
7.481 lb/ft^3	62.43 lb/ft^3	1728 lb/ft^3
1 lb/gal	8.345 lb/gal	231 lb/gal
0.1198 g/cm^3	1 g/cm^3	27.68 g/cm^3
4.329×10^{-3} $lb/in.^3$	3.613×10^{-2} $lb/in.^3$	1 $lb/in.^3$

*Note that since the density of water is almost precisely 1 g/cm^3, these data give the density of water as 1000 kg/m^3, 62.43 lb/ft^3, etc.

TABLE A.3 Energy.

1 electron-volt (eV) =	1 million electron-volts (MeV) =	1 joule (J) =	1 calorie (cal) =
1 eV	10^{6} eV	6.241×10^{18} eV	2.611×10^{19} eV
10^{-6} MeV	1 MeV	6.241×10^{12} MeV	2.611×10^{13} MeV
1.602×10^{-19} J	1.602×10^{-13} J	1 J	4.184 J
3.829×10^{-20} cal	3.829×10^{-14} cal	0.239 cal	1 cal
1.52×10^{-22} BTU	1.52×10^{-16} BTU	9.485×10^{-4} BTU	3.968×10^{-3} BTU
3.829×10^{-23} kcal	3.829×10^{-17} kcal	2.39×10^{-4} kcal	0.001 kcal
4.451×10^{-26} kWh	4.451×10^{-20} kWh	2.778×10^{-7} kWh	1.162×10^{-6} kWh
1.52×10^{-28} MBTU	1.52×10^{-22} MBTU	9.485×10^{-10} MBTU	3.968×10^{-9} MBTU
1.854×10^{-30} MW-day	1.854×10^{-24} MW-day	1.157×10^{-11} MW-day	4.843×10^{-11} MW-day
5.077×10^{-33} MW-yr	5.077×10^{-27} MW-yr	3.169×10^{-14} MW-yr	1.326×10^{-13} MW-yr
1.52×10^{-37} mQ	1.52×10^{-31} mQ	9.485×10^{-19} mQ	3.968×10^{-18} mQ
1.52×10^{-40} Q	1.52×10^{-34} Q	9.485×10^{-22} Q	3.968×10^{-21} Q

1 British Thermal Unit (BTU) =	1 kilocalorie (kcal or Cal) =	1 kilowatt-hour (kWh) =	1 million BTU (MBTU) =
6.581×10^{21} eV	2.611×10^{22} eV	2.247×10^{25} eV	6.581×10^{27} eV
6.581×10^{15} MeV	2.611×10^{16} MeV	2.247×10^{19} MeV	6.581×10^{21} MeV
1054 J	4184 J	3.6×10^{6} J	1.054×10^{9} J
252 cal	1000 cal	8.604×10^{5} cal	2.52×10^{8} cal
1 BTU	3.968 BTU	3413 BTU	10^{6} BTU
0.252 kcal	1 kcal	860.4 kcal	2.52×10^{5} kcal
2.929×10^{-4} kWh	1.162×10^{-3} kWh	1 kWh	292.9 kWh
10^{-6} MBTU	3.968×10^{-6} MBTU	3.413×10^{-3} MBTU	1 MBTU
1.22×10^{-8} MW-day	4.843×10^{-8} MW-day	4.167×10^{-5} MW-day	0.0122 MW-day
3.341×10^{-11} MW-yr	1.326×10^{-10} MW-yr	1.141×10^{-7} MW-yr	3.341×10^{-5} MW-yr
10^{-15} mQ	3.968×10^{-15} mQ	3.413×10^{-12} mQ	10^{-9} mQ
10^{-18} Q	3.968×10^{-18} Q	3.413×10^{-15} Q	10^{-12} Q

1 megawatt-day (MW-day) =	1 megawatt-year (MW-yr) =	1 milli-Q (mQ) =	1 Q =
5.393×10^{29} eV	1.97×10^{32} eV	6.581×10^{36} eV	6.581×10^{39} eV
5.393×10^{23} MeV	1.97×10^{26} MeV	6.581×10^{30} MeV	6.581×10^{33} MeV
8.64×10^{10} J	3.156×10^{13} J	1.054×10^{18} J	1.054×10^{21} J
2.065×10^{10} cal	7.542×10^{12} cal	2.52×10^{17} cal	2.52×10^{20} cal
8.195×10^{7} BTU	2.993×10^{10} BTU	10^{15} BTU	10^{18} BTU
2.065×10^{7} kcal	7.542×10^{9} kcal	2.52×10^{14} kcal	2.52×10^{17} kcal
2.4×10^{4} kWh	8.766×10^{6} kWh	2.929×10^{11} kWh	2.929×10^{14} kWh
81.95 MBTU	2.993×10^{4} MBTU	10^{9} MBTU	10^{12} MBTU
1 MW-day	365.2 MW-day	1.22×10^{7} MW-day	1.22×10^{10} MW-day
2.738×10^{-3} MW-yr	1 MW-yr	3.341×10^{4} MW-yr	3.341×10^{7} MW-yr
8.195×10^{-8} mQ	2.993×10^{-5} mQ	1 mQ	1000 mQ
8.195×10^{-11} Q	2.993×10^{-8} Q	0.001 Q	1 Q

Other units of energy:

1 erg = 10^{-7} J

1 foot-pound (ft-lb) = 1.356 J

1 therm = 10^{5} BTU, often used in reporting sales of natural gas. (1 therm is almost exactly the energy content of 100 cubic feet of natural gas.)

1 kiloton = 10^{12} cal = 4.184×10^{12} J, approximately the energy released in the explosion of 1000 tons of TNT. The kiloton (and the megaton) are frequently used in referring to A-bomb and H-bomb explosions.

1 horsepower-hour (hp-hr) = 0.746 kWh = 2.686×10^{6} J, the energy delivered by 1 hp acting for 1 hour.

TABLE A.4 Fluid flow rate.

1 gallon per day =	1 acre-foot per year =	1 cubic foot per minute (ft^3/min or cfm) =
1 gal/day	892.2 gal/day	1.077×10^4 gal/day
1.121×10^{-3} acre-ft/yr	1 acre-ft/yr	12.07 acre-ft/yr
9.283×10^{-5} ft^3/min	8.282×10^{-2} ft^3/min	1 ft^3/min
1.547×10^{-6} ft^3/sec	1.38×10^{-3} ft^3/sec	1.667×10^{-2} ft^3/sec
4.381×10^{-8} m^3/sec	3.909×10^{-5} m^3/sec	4.719×10^{-4} m^3/sec
10^{-9} bgd	8.922×10^{-7} bgd	1.077×10^{-5} bgd
1 cubic foot per second (ft^3/sec or cfs) =	**1 cubic meter per second (m^3/sec) =**	**1 billion gallons per day (bgd) =**
6.463×10^5 gal/day	2.282×10^7 gal/day	10^9 gal/day
724.4 acre-ft/yr	2.558×10^4 acre-ft/yr	1.121×10^6 acre-ft/yr
60 ft^3/min	2119 ft^3/min	9.283×10^4 ft^3/min
1 ft^3/sec	35.31 ft^3/sec	1547 ft^3/sec
2.832×10^{-2} m^3/sec	1 m^3/sec	43.81 m^3/sec
6.463×10^{-4} bgd	2.282×10^{-2} bgd	1 bgd

TABLE A.5 Force.

1 gram (g)=	1 newton (N) =	1 pound (lb) =	1 kilogram (kg) =	1 ton =
1 g	102 g	453.6 g	1000 g	9.072×10^5 g
9.807×10^{-3} N	1 N	4.448 N	9.807 N	8896 N
2.205×10^{-3} lb	0.2248 lb	1 lb	2.205 lb	2000 lb
0.001 kg	0.102 kg	0.4536 kg	1 kg	907.2 kg
1.102×10^{-6} ton	1.124×10^{-4} ton	0.0005 ton	1.102×10^{-3} ton	1 ton

Other units of force: 1 dyne = 10^{-5} N; 1 ounce = 1/16 lb = 0.278 N.

Note. The kilogram and the gram are, strictly speaking, units of mass, not of force. A "force of 1 kg" is the gravitational force exerted by the earth on a mass of 1 kg, where the acceleration due to gravity has its standard value.

TABLE A.6 Length.

1 angstrom (Å) =	1 centimeter (cm) =	1 inch (in.) =	1 foot (ft) =
1 Å	10^{8} Å	2.54×10^{8} Å	3.048×10^{9} Å
10^{-8} cm	1 cm	2.54 cm	30.48 cm
3.937×10^{-9} in.	0.3937 in.	1 in.	12 in.
3.281×10^{-10} ft	3.281×10^{-2} ft	8.333×10^{-2} ft	1 ft
10^{-10} m	0.01 m	0.0254 m	0.3048 m
10^{-13} km	10^{-5} km	2.54×10^{-5} km	3.048×10^{-4} km
6.214×10^{-14} mile	6.214×10^{-6} mile	1.578×10^{-5} mile	1.894×10^{-4} mile

1 meter (m) =	1 kilometer (km) =	1 mile =
10^{10} Å	10^{13} Å	1.609×10^{13} Å
100 cm	10^{5} cm	1.609×10^{5} cm
39.37 in.	3.937×10^{4} in.	6.336×10^{4} in.
3.281 ft	3281 ft	5280 ft
1 m	1000 m	1609 m
0.001 km	1 km	1.609 km
6.214×10^{-4} mile	0.6214 mile	1 mile

Other units of length:

1 fermi = 10^{-15} m, often used in referring to sizes of nuclei.
1 micron (μ) = 10^{-6} m
1 millimeter (mm) = 10^{-3} m
1 nautical mile = 1.852 km = 1.151 statute miles
1 light-year = 9.461×10^{15} m, the distance light travels in one year.

Note. The *mile* referred to in this table and everywhere in this book is the common "statute mile" (5280 ft).

TABLE A.7 Mass.

1 atomic mass unit (amu) =	1 gram (g) =	1 pound (lb)* =
1 amu	6.022×10^{23} amu	2.732×10^{26} amu
1.661×10^{-24} g	1 g	453.6 g
3.661×10^{-27} lb	2.205×10^{-3} lb	1 lb
1.661×10^{-27} kg	0.001 kg	0.4536 kg
1.83×10^{-30} ton	1.102×10^{-6} ton	5×10^{-4} ton
1.661×10^{-30} metric ton	10^{-6} metric ton	4.536×10^{-4} metric ton

1 kilogram (kg) =	1 ton*† =	1 metric ton (t or tonne)† =
6.022×10^{26} amu	5.463×10^{29} amu	6.022×10^{29} amu
1000 g	9.072×10^{5} g	10^{6} g
2.205 lb	2000 lb	2205 lb
1 kg	907.2 kg	1000 kg
1.102×10^{-3} ton	1 ton	1.102 tons
0.001 metric ton	0.9072 metric ton	1 metric ton

Other units of mass: 1 ounce = 1/16 lb = 2.835×10^{-2} kg; 1 slug = 32.17 lb.

*The pound and ton are used as units of both force and mass. A force of 1 lb is the force exerted by gravity on an object whose mass is 1 lb at a location where the acceleration due to gravity has its standard value. Other units of mass such as the kilogram are also occasionally used as units of force in this fashion.

†The *ton* in this table is the commonly used "short ton" (2000 lb), as opposed to the "long ton" (2240 lb). The *metric* ton (1000 kg), 10% greater than the short ton, is often used in discussions of energy resources.

TABLE A.8 Mass-energy equivalence.

NOTE: This table can be used to calculate the energy released in a process in which the mass decreases by a known amount. This table cannot be used directly to calculate, for example, the energy available from a given quantity of U^{235}, because in the fission of 1 kg of U^{235}, the decrease in mass is much less than 1 kg.

1 electron mass	1 atomic mass unit (amu)	1 gram (g)	1 kilogram (kg)
0.511 MeV	931.5 MeV	5.61×10^{26} MeV	5.61×10^{29} MeV
8.187×10^{-14} J	1.492×10^{-10} J	8.988×10^{13} J	8.988×10^{16} J
7.765×10^{-17} BTU	1.415×10^{-13} BTU	8.524×10^{10} BTU	8.524×10^{13} BTU
2.274×10^{-20} kWh	4.146×10^{-17} kWh	2.497×10^{7} kWh	2.497×10^{10} kWh
7.765×10^{-35} Q	1.415×10^{-31} Q	8.524×10^{-8} Q	8.524×10^{-5} Q

1 ton	1 MeV	1 joule (J)	1 BTU
5.089×10^{32} MeV	1.957 electron masses	1.221×10^{13} electron masses	1.288×10^{16} electron masses
8.153×10^{19} J	1.074×10^{-3} amu	6.701×10^{9} amu	7.065×10^{12} amu
7.733×10^{16} BTU	1.783×10^{-27} g	1.113×10^{-14} g	1.173×10^{-11} g
2.265×10^{13} kWh	1.783×10^{-30} kg	1.113×10^{-17} kg	1.173×10^{-14} kg
7.733×10^{-2} Q	1.965×10^{-33} ton	1.226×10^{-20} ton	1.293×10^{-17} ton

1 kilowatt-hour (kWh)	1 Q
4.397×10^{19} electron masses	1.288×10^{34} electron masses
2.412×10^{16} amu	7.065×10^{30} amu
4.006×10^{-8} g	1.173×10^{7} g
4.006×10^{-11} kg	1.173×10^{4} kg
4.415×10^{-14} ton	12.93 tons

TABLE A.9 Power.

1 BTU per day =	1 kilowatt-hour per year (kWh/yr) =	1 watt (W) =	1 kilowatt (kW) =
1 BTU/day	9.348 BTU/day	81.95 BTU/day	8.195×10^{4} BTU/day
0.107 kWh/yr	1 kWh/yr	8.766 kWh/yr	8766 kWh/yr
0.0122 W	0.1141 W	1 W	1000 W
1.22×10^{-5} kW	1.141×10^{-4} kW	0.001 kW	1 kW
1.22×10^{-8} MW	1.141×10^{-7} MW	10^{-6} MW	0.001 MW
1.22×10^{-11} GW	1.141×10^{-10} GW	10^{-9} GW	10^{-6} GW
3.652×10^{-13} mQ/yr	3.413×10^{-12} mQ/yr	2.993×10^{-11} mQ/yr	2.993×10^{-8} mQ/yr
3.652×10^{-16} Q/yr	3.413×10^{-15} Q/yr	2.993×10^{-14} Q/yr	2.993×10^{-11} Q/yr

1 megawatt (MW) =	1 gigawatt (GW) =	1 milli-Q per year (mQ/yr) =	1 Q per year (Q/yr) =
8.195×10^{7} BTU/day	8.195×10^{10} BTU/day	2.738×10^{12} BTU/day	2.738×10^{15} BTU/day
8.766×10^{6} kWh/yr	8.766×10^{9} kWh/yr	2.929×10^{11} kWh/yr	2.929×10^{14} kWh/yr
10^{6} W	10^{9} W	3.341×10^{10} W	3.341×10^{13} W
1000 kW	10^{6} kW	3.341×10^{7} kW	3.341×10^{10} kW
1 MW	1000 MW	3.341×10^{4} MW	3.341×10^{7} MW
0.001 GW	1 GW	33.41 GW	3.341×10^{4} GW
2.993×10^{-5} mQ/yr	2.993×10^{-2} mQ/yr	1 mQ/yr	1000 mQ/yr
2.993×10^{-8} Q/yr	2.993×10^{-5} Q/yr	0.001 Q/yr	1 Q/yr

Other units of power: 1 horsepower (hp) = 746 W.

TABLE A.10 Pressure.

1 newton per square meter (N/m^2) =	1 pound per square foot (lb/ft^2) =	1 pound per square inch ($lb/in.^2$ or psi) =	1 atmosphere (atm) =
1 N/m^2	47.88 N/m^2	6895 N/m^2	1.013×10^5 N/m^2
2.089×10^{-2} lb/ft^2	1 lb/ft^2	144 lb/ft^2	2116 lb/ft^2
1.45×10^{-4} $lb/in.^2$	6.944×10^{-3} $lb/in.^2$	1 $lb/in.^2$	14.7 $lb/in.^2$
9.869×10^{-6} atm	4.725×10^{-4} atm	6.805×10^{-2} atm	1 atm

Other units of pressure:

1 bar = 10^5 N/m^2 = 0.9869 atm.

Pressure is often measured by giving the height of a column of water or mercury that exerts such a pressure at its base: 1 atm = 76 cm of mercury = 33.9 ft of water.

TABLE A.11 Temperature.

Temperature intervals

1°C (Celsius or centigrade) = 1°K (Kelvin or absolute) = 1.8°F (Fahrenheit)

$$1°F = \tfrac{5}{9}\,°K = \tfrac{5}{9}\,°C.$$

Correspondence between temperature scales

On the various scales, the normal freezing point of water is: 32°F = 0°C = 273.15°K.

Absolute or Kelvin temperatures are often written without the ° symbol and read as "kelvins." Thus the normal freezing point of water is 273.15 K or 273.15 kelvins.

Conversions between temperature scales

From the relationships between the temperature intervals and the given values of the freezing point of water, it follows that:

$$T\,(°F) = 32 + 1.8 \times T(°C)$$
$$T\,(°C) = \tfrac{5}{9} \times [T(°F) - 32]$$
$$T\,(°K) = 273.15 + T(°C)$$

	°C	°F	°K
Absolute zero	−273.15	−459.67	0
Normal freezing point of water	0	32	273.15
Normal boiling point of water	100	212	373.15

TABLE A.12 Time.

1 second (sec or s) =	1 minute (min) =	1 hour (hr or h) =	1 day (d) =	1 year (yr or y) =
1 sec	60 sec	3600 sec	8.64×10^4 sec	3.156×10^7 sec
1.667×10^{-2} min	1 min	60 min	1440 min	5.259×10^5 min
2.778×10^{-4} hr	1.667×10^{-2} hr	1 hr	24 hr	8766 hr
1.157×10^{-5} day	6.944×10^{-4} day	4.167×10^{-2} day	1 day	365.2 days
3.169×10^{-8} yr	1.901×10^{-6} yr	1.141×10^{-4} yr	2.738×10^{-3} yr	1 yr

TABLE A.13 Velocity.

1 centimeter per second (cm/sec) =	1 kilometer per hour (km/hr) =	1 foot per second (ft/sec) =	1 mile per hour (mile/hr or mph) =
1 cm/sec	27.78 cm/sec	30.48 cm/sec	44.7 cm/sec
0.036 km/hr	1 km/hr	1.097 km/hr	1.609 km/hr
3.281×10^{-2} ft/sec	0.9113 ft/sec	1 ft/sec	1.467 ft/sec
2.237×10^{-2} mile/hr	0.6214 mile/hr	0.6818 mile/hr	1 mile/hr
0.01 m/sec	0.2778 m/sec	0.3048 m/sec	0.447 m/sec
10^{-5} km/sec	2.778×10^{-4} km/sec	3.048×10^{-4} km/sec	4.47×10^{-4} km/sec
6.214×10^{-6} mile/sec	1.726×10^{-4} mile/sec	1.894×10^{-4} mile/sec	2.778×10^{-4} mile/sec

1 meter per second (m/sec) =	1 kilometer per second (km/sec) =	1 mile per second (mile/sec) =
100 cm/sec	10^5 cm/sec	1.609×10^5 cm/sec
3.6 km/hr	3600 km/hr	5794 km/hr
3.281 ft/sec	3281 ft/sec	5280 ft/sec
2.237 miles/hr	2237 miles/hr	3600 miles/hr
1 m/sec	1000 m/sec	1609 m/sec
0.001 km/sec	1 km/sec	1.609 km/sec
6.214×10^{-4} mile/sec	0.6214 mile/sec	1 mile/sec

TABLE A.14 Volume.

1 cubic centimeter (cm^3 or cc) =	1 cubic inch ($in.^3$) =	1 liter (l) =
1 cm^3	16.39 cm^3	1000 cm^3
6.102×10^{-2} $in.^3$	1 $in.^3$	61.02 $in.^3$
0.001 liter	1.639×10^{-2} liter	1 liter
2.642×10^{-4} gal	4.329×10^{-3} gal	0.2642 gal
3.531×10^{-5} ft^3	5.787×10^{-4} ft^3	3.531×10^{-2} ft^3
6.29×10^{-6} barrel	1.031×10^{-4} barrel	6.29×10^{-3} barrel
10^{-6} m^3	1.639×10^{-5} m^3	0.001 m^3
10^{-15} km^3	1.639×10^{-14} km^3	10^{-12} km^3
2.399×10^{-16} $mile^3$	3.931×10^{-15} $mile^3$	2.399×10^{-13} $mile^3$

1 gallon (gal) =	1 cubic foot (ft^3) =	1 barrel (bbl) =
3785 cm^3	2.832×10^{4} cm^3	1.59×10^{5} cm^3
231 $in.^3$	1728 $in.^3$	9702 $in.^3$
3.785 liters	28.32 liters	159 liters
1 gal	7.481 gal	42 gal
0.1337 ft^3	1 ft^3	5.615 ft^3
2.381×10^{-2} barrel	0.1781 barrel	1 barrel
3.785×10^{-3} m^3	2.832×10^{-2} m^3	0.159 m^3
3.785×10^{-12} km^3	2.832×10^{-11} km^3	1.59×10^{-10} km^3
9.082×10^{-13} $mile^3$	6.794×10^{-12} $mile^3$	3.814×10^{-11} $mile^3$

1 cubic meter (m^3) =	1 cubic kilometer (km^3) =	1 cubic mile ($mile^3$) =
10^{6} cm^3	10^{15} cm^3	4.168×10^{15} cm^3
6.102×10^{4} $in.^3$	6.102×10^{13} $in.^3$	2.544×10^{14} $in.^3$
1000 liters	10^{12} liters	4.168×10^{12} liters
264.2 gal	2.642×10^{11} gal	1.101×10^{12} gal
35.31 ft^3	3.531×10^{10} ft^3	1.472×10^{11} ft^3
6.29 barrels	6.29×10^{9} barrels	2.622×10^{10} barrels
1 m^3	10^{9} m^3	4.168×10^{9} m^3
10^{-9} km^3	1 km^3	4.168 km^3
2.399×10^{-10} $mile^3$	0.2399 $mile^3$	1 $mile^3$

Other units of volume:

1 British gallon = 1.2 U.S. gal

1 cord (of wood) = 128 ft^3

1 acre-foot = 1233 m^3, often used in measuring quantities of water. (1 acre-foot is the volume of water that covers an area of 1 acre to a depth of 1 ft.)

1 quart = 1/4 gal = 0.9464 liter; 1 pint = 1/2 qt = 0.4732 liter. The pint, quart and gallon are U.S. units of *liquid* measurement; different "dry" measures are occasionally used.

The barrel used here is the 42-gallon barrel used in measuring petroleum; many other sizes of "barrels" are in use for other special purposes.

Appendix B

Abbreviations and Symbols, Decimal Multiples, and Geometrical Formulas

TABLE B.1 Abbreviations and symbols.

A	mass number or nucleon number
A	ampere
Å	angstrom unit
AC	alternating current
A h	ampere-hour
amu	atomic mass unit
atm	atmosphere
B	magnetic field strength
bbl	barrel
BTU	British thermal unit
BWR	boiling water reactor
c	speed of light
C	heat capacity
C	coulomb
°C	degree-Celsius or degree-centigrade
cal	calorie
Cal	kilocalorie
CE	chemical energy
cm	centimeter
CP	coefficient of performance
d	deuteron
D	deuterium
DC	direct current
DD	degree-day
e	electron or electronic charge
e^-	electron
e^+	positron
E.E.R.	energy efficiency ratio
E	energy
$\mathcal{E}$	electric field strength
emf	electromotive force
EPE	electrical potential energy
eV	electron-volt
F	force
°F	degree-Fahrenheit
ft	foot
g	gram
g	acceleration due to gravity
G	universal gravitational constant
G	giga
gal	gallon
GNP	gross national product
GPE	gravitational potential energy
GW	gigawatt
H	heat
h	Planck's constant
hr	hour
Hz	hertz (cycles per second)
I	current
in.	inch
J	joule
k	kilo
k	Boltzmann's constant or thermal conductance
k_e	Coulomb's law constant
K	K-capture
°K	degree-Kelvin
kcal	kilocalorie
KE	kinetic energy
kg	kilogram
km	kilometer
kV	kilovolt
kW	kilowatt
kWh	kilowatt-hour
l	liter
lb	pound
LMFBR	liquid-metal fast breeder reactor
m	meter
m	mass
M	mega
MBTU	millions of BTU's
MeV	million electron-volts
min	minute
MKS	meter-kilogram-second
mm	millimeter
mph	miles per hour
mQ	milli-Q
mrem	millirem
MW	megawatt
MW(e)	megawatt-electrical
MW(t)	megawatt-thermal
n	neutron
N	neutron number
N	newton
N_A	Avogadro's number
NGL	natural gas liquids
oz	ounce
p	proton
P	power
psi	pounds per square inch
PWR	pressurized water reactor
q	charge
Q	an energy unit, 10^{18} BTU
R	gas constant
R	resistance
rad	radiation absorbed dose
rem	roentgen equivalent man
rms	root-mean-square
S_o	solar constant
sec	second
t	triton
T	tritium
T	temperature
TE	thermal energy
v	velocity or speed
V	volt
V	voltage
W	watt
W	work
yr	year
Z	atomic number or proton number
α	alpha-particle or earth's albedo
β	beta-particle
β^-	electron
β^+	positron
γ	gamma-ray
Δ	calculate the *change* in the following quantity
ϵ	efficiency
λ	wavelength
ν	frequency
ρ	resistivity
σ	Stefan-Boltzmann constant
τ	containment time

TABLE B.2 Prefixes and abbreviations for decimal multiples and submultiples.

Multiple	Prefix	Abbreviation	Example
10^{-12}	pico	p	1 picosecond = 10^{-12} sec
10^{-9}	nano	n	1 nanovolt = 10^{-9} V
10^{-6}	micro	μ	1 μsec = 10^{-6} sec
10^{-3}	milli	m	1 mm = 1 millimeter = 10^{-3} m
10^{-2}	centi	c	1 cm = 1 centimeter = 0.01 m
10^{3}	kilo	k	1 kg = 1 kilogram = 1000 g
10^{6}	mega	M	1 MW = 1 megawatt = 10^{6} W
10^{9}	giga	G	1 GW = 1 gigawatt = 10^{9} W*
10^{12}	tera	T	1 TBTU = 1 teraBTU = 10^{12} BTU*

*In common American usage, 10^9 W = 1 *billion* watts and 10^{12} BTU = 1 *trillion* BTU. The terms thousand and million for 10^3 and 10^6 are universally understood. In the United States, 1 billion = 10^9 (1000 million), 1 trillion = 10^{12} (1000 billion), etc., and whenever these terms are used in this book, it is with these meanings. In some countries, however, 1 billion means 10^{12} (1 million million), 1 trillion = 10^{18}, etc., so that caution should be exercised in comparing American and foreign sources in which these terms are used.

TABLE B.3 Geometrical formulas.

	Values	
Circle of radius r	Circumference = $2\pi r$	Area = πr^2
Sphere of radius r	Surface area = $4\pi r^2$	Volume = $\frac{4}{3}\pi r^3$
Numerical value of π	$\pi \simeq 3.1416$	

Appendix C
Physical and Chemical Data

TABLE C.1 Physical constants.

	Values
Universal gravitational constant	$G = 6.673 \times 10^{-11}$ m³/kg-sec²
Acceleration due to gravity	$g = 9.80665$ m/sec² ("standard" value) $g = 9.8$ m/sec² $= 980$ cm/sec² $= 32$ ft/sec² (for most purposes)
Avogadro's number	$N_A = 6.022 \times 10^{23}$ molecules/mole
Gas constant	$R = kN_A = 8.314$ J/mole-°K
Boltzmann's constant	$k = R/N_A = 1.381 \times 10^{-23}$ J/°K
Black-body radiation	
Stefan-Boltzmann constant (The power radiated by an ideal black surface of area A (in square meters) at an absolute temperature T, is given by $P = \sigma AT^4$.)	$\sigma = 5.670 \times 10^{-8}$ W/m²-°K⁴
Peak wavelength (The equation gives the relationship between the wavelength at the peak of the spectral distribution of black-body radiation, in meters, and the temperature, in°K.)	$\lambda_{max} = 0.0029/T$
Planck's constant	$h = 6.626 \times 10^{-34}$ J-sec
Electronic charge (elementary unit of charge)	$e = 1.602 \times 10^{-19}$ coulombs (C)
Coulomb's law constant (The electrical force between two point charges separated by distance r, with q_1 and q_2 in coulombs, r in meters, and F in newtons, is: $F = k_e q_1 q_2/r^2$.)	$k_e = 9 \times 10^9$ N-m²/C² (approximate value) $k_e = 8.988 \times 10^9$ N-m²/C² (more precise value)
Speed of light in vacuum (The speed of light in air is the same to within 0.1%.)	$c = 3 \times 10^8$ m/sec $= 3 \times 10^{10}$ cm/sec $= 186{,}000$ miles/sec (approximate values) $c = 2.998 \times 10^8$ m/sec (more precise value)
Speed of sound in dry air at standard temperature and pressure ($T = 0$°C, $p = 1$ atm)	331 m/sec = 1087 ft/sec = 741 miles/hr
Density of dry air at standard temperature and pressure	1.29 kg/m³ $= 1.29 \times 10^{-3}$ g/cm³
Atomic mass unit	1 amu $= 1.661 \times 10^{-27}$ kg $= 1.661 \times 10^{-24}$ g

TABLE C.1 (continued)

Masses for particles:

Particle	Mass (amu)	Mass (kg)	Mass (g)
electron	5.48593×10^{-4}	9.110×10^{-31}	9.110×10^{-28}
proton	1.0072766	1.6726×10^{-27}	1.6726×10^{-24}
neutron	1.0086652	1.6749×10^{-27}	1.6749×10^{-24}
hydrogen atom (H^1)	1.0078252	1.6735×10^{-27}	1.6735×10^{-24}

Sizes of atoms and nuclei

Atomic radii: Typically in the range of 1–2 Å for all atoms. (The "radius" of an atom is not a quantity that can be given a precise definition.)

Nuclear radii: Approximate values of nuclear radii are given by the expression:

$$r \simeq r_o A^{\frac{1}{3}},$$

with $r_o = 1.4 \times 10^{-15}$ m and A the atomic mass number (or nucleon number). Thus from the lightest nuclei ($A = 1$) to the heaviest ($A \simeq 240$), r varies from about 1.4×10^{-15} m to 8.7×10^{-15} m.

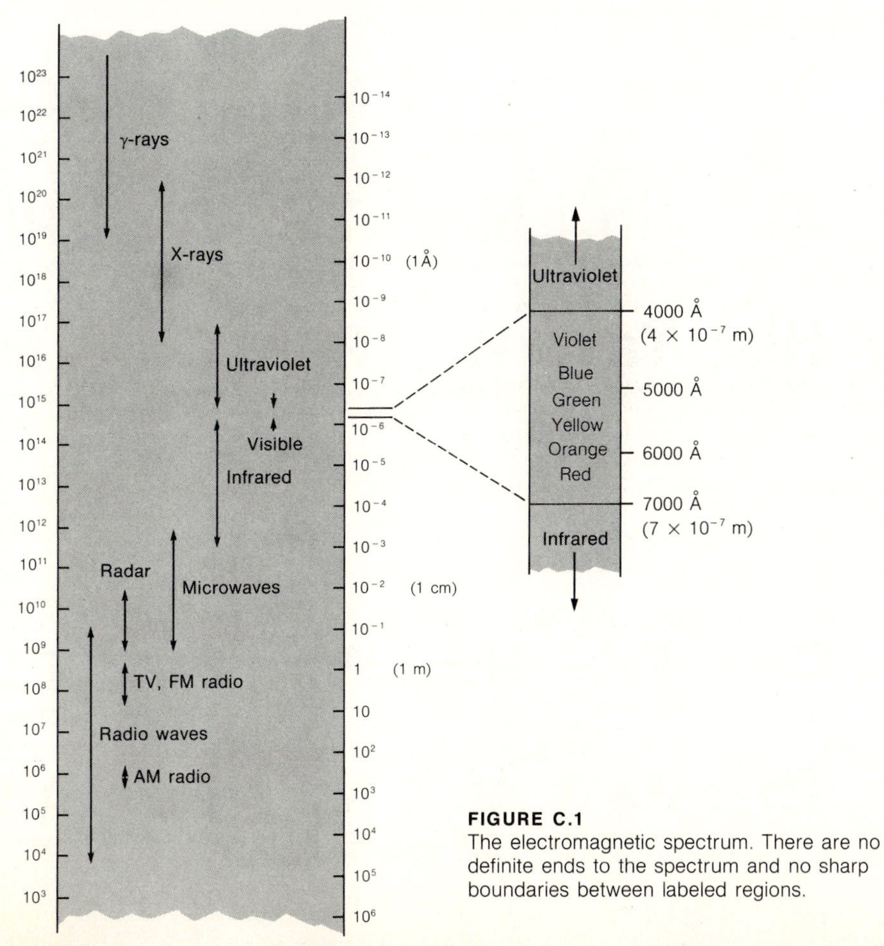

FIGURE C.1
The electromagnetic spectrum. There are no definite ends to the spectrum and no sharp boundaries between labeled regions.

TABLE C.2 Periodic table of the elements.

For each element, the number in the upper left-hand corner is Z (the atomic number or proton number). The number in the upper right-hand corner is the average mass in atomic mass units of the atoms in a sample containing the various isotopes in their usual natural abundance; for those elements not found in nature or for those such as polonium, which are only found as decay products of long-lived elements, the mass number or nucleon number (A) of the isotope with the longest known half-life is given in parentheses.

1 1.008 **H** Hydrogen								
3 6.939 **Li** Lithium	4 9.012 **Be** Beryllium							
11 22.99 **Na** Sodium	12 24.31 **Mg** Magnesium							
19 39.102 **K** Potassium	20 40.08 **Ca** Calcium	21 44.96 **Sc** Scandium	22 47.90 **Ti** Titanium	23 50.94 **V** Vanadium	24 52.00 **Cr** Chromium	25 54.94 **Mn** Manganese	26 55.85 **Fe** Iron	27 58.93 **Co** Cobalt
37 85.47 **Rb** Rubidium	38 87.62 **Sr** Strontium	39 88.91 **Y** Yttrium	40 91.22 **Zr** Zirconium	41 92.91 **Nb** Niobium	42 95.94 **Mo** Molybdenum	43 (97) **Tc** Technetium	44 101.1 **Ru** Ruthenium	45 102.91 **Rh** Rhodium
55 132.91 **Cs** Cesium	56 137.34 **Ba** Barium	57–71 Lanthanide series*	72 178.5 **Hf** Hafnium	73 180.95 **Ta** Tantalum	74 183.85 **W** Tungsten	75 186.2 **Re** Rhenium	76 190.2 **Os** Osmium	77 192.2 **Ir** Iridium
87 (223) **Fr** Francium	88 (226) **Ra** Radium	89–103 Actinide series†	104 (257)	105 (260)	106			

***Lanthanide Series**	57 138.9 **La** Lanthanum	58 140.1 **Ce** Cerium	59 140.9 **Pr** Praseodymium	60 144.2 **Nd** Neodymium	61 (145) **Pm** Promethium	62 150.4 **Sm** Samarium	63 152.0 **Eu** Europium
†Actinide Series	89 (227) **Ac** Actinium	90 232.0 **Th** Thorium	91 (231) **Pa** Protactinium	92 238.0 **U** Uranium	93 (237) **Np** Neptunium	94 (244) **Pu** Plutonium	95 (243) **Am** Americium

								2 4.003 **He** Helium
			5 10.81 **B** Boron	6 12.01 **C** Carbon	7 14.01 **N** Nitrogen	8 15.999 **O** Oxygen	9 19.00 **F** Fluorine	10 20.18 **Ne** Neon
			13 26.98 **Al** Aluminum	14 28.09 **Si** Silicon	15 30.97 **P** Phosphorus	16 32.06 **S** Sulfur	17 35.45 **Cl** Chlorine	18 39.95 **Ar** Argon
28 58.71 **Ni** Nickel	29 63.54 **Cu** Copper	30 65.37 **Zn** Zinc	31 69.72 **Ga** Gallium	32 72.59 **Ge** Germanium	33 74.92 **As** Arsenic	34 78.96 **Se** Selenium	35 79.91 **Br** Bromine	36 83.80 **Kr** Krypton
46 106.4 **Pd** Palladium	47 107.87 **Ag** Silver	48 112.4 **Cd** Cadmium	49 114.8 **In** Indium	50 118.7 **Sn** Tin	51 121.8 **Sb** Antimony	52 127.6 **Te** Tellurium	53 126.9 **I** Iodine	54 131.3 **Xe** Xenon
78 195.1 **Pt** Platinum	79 196.97 **Au** Gold	80 200.6 **Hg** Mercury	81 204.4 **Tl** Thallium	82 207.2 **Pb** Lead	83 209.0 **Bi** Bismuth	84 (209) **Po** Polonium	85 (210) **At** Astatine	86 (222) **Rn** Radon

64 157.3 **Gd** Gadolinium	65 158.9 **Tb** Terbium	66 162.5 **Dy** Dysprosium	67 164.9 **Ho** Holmium	68 167.3 **Er** Erbium	69 168.9 **Tm** Thulium	70 173.0 **Yb** Ytterbium	71 175.0 **Lu** Lutetium
96 (247) **Cm** Curium	97 (247) **Bk** Berkelium	98 (251) **Cf** Californium	99 (254) **Es** Einsteinium	100 (257) **Fm** Fermium	101 (258) **Md** Mendelevium	102 (255) **No** Nobelium	103 (256) **Lw** Lawrencium

TABLE C.3 Masses of atoms and their constituents. Except for the electron, proton, and neutron, the listed masses (in atomic mass units) are those of complete neutral atoms (the nucleus together with the appropriate number of electrons). Many more nuclear species are known than those listed here. For the elements that occur naturally, the natural abundance of each listed isotope is given, the percentage of atoms in a typical sample that have the particular mass number. (Abundances are not given for elements such as polonium that occur naturally but only as the decay products of nuclear species with longer half-lives.) For unstable nuclei, the half-life (in seconds, minutes, hours, days, or years) is given, together with the modes of decay; α, β^-, β^+ and K denote α-decay, ordinary β-decay, positron emission, and K-capture. In most decays, γ-rays are also emitted.

Element or particle	Symbol	Z (atomic number or proton number)	A (mass number or nucleon number)	Mass (atomic mass units)	Percentage abundance	Half-life	Common modes of decay
Electron	e, β^-			5.48593×10^{-4}			
Proton	p			1.0072766			
Neutron	n			1.0086652		11 min	β^-
Hydrogen	H	1	1	1.0078252	99.985		
			2	2.0141022	0.015		
			3	3.0160497		12.3 yr	β^-
Helium	He	2	3	3.0160297	0.00013		
			4	4.002603	99.9999		
Lithium	Li	3	6	6.015123	7.4		
			7	7.016004	92.6		
Boron	B	5	11	11.009305	80.2		
Carbon	C	6	12	12.000	98.9		
			13	13.003354	1.1		
			14	14.003242		5730 yr	β^-
Nitrogen	N	7	13	13.005738		10 min	β^+
			14	14.0030744	99.63		
			15	15.000108	0.37		
Oxygen	O	8	15	15.00307		124 sec	β^+
			16	15.99492	99.759		
			17	16.99913	0.037		
			18	17.99916	0.204		
Sodium	Na	11	23	22.989771	100		
Argon	Ar	18	40	39.962384	99.6		
Potassium	K	19	40	39.964	0.01	1.3×10^9 yr	K, β^-, β^+
			41	40.961827	6.88		
Calcium	Ca	20	40	39.962592	97		
			41	40.962279		7.7×10^4 yr	K
Scandium	Sc	21	41	40.96925		0.6 sec	β^+
Iron	Fe	26	56	55.934934	91.7		
Nickel	Ni	28	58	57.935336	67.9		
			64	63.92796	1.1		
Copper	Cu	29	58	57.94454		3.2 sec	β^+
			63	62.92959	69.1		
			64	63.929757		12.8 hr	K, β^-, β^+
			65	64.92779	30.9		

TABLE C.3 (continued)

Element or particle	Symbol	Z (atomic number or proton number)	A (mass number or nucleon number)	Mass (atomic mass units)	Percentage abundance	Half-life	Common modes of decay
Zinc	Zn	30	64	63.92914	48.9		
			65	64.92923		244 days	K, β^+
Gallium	Ga	31	65	64.93273		15 min	K, β^+
Germanium	Ge	32	65	64.94		1.5 min	β^+
Krypton	Kr	36	85	84.912537		10.8 yr	β^-
Rubidium	Rb	37	85	84.9118	72.2		
			90	89.9148		2.6 min	β^-
Strontium	Sr	38	90	89.90775		28.1 yr	β^-
Yttrium	Y	39	90	89.90716		64.2 hr	β^-
			97	96.918		1.1 sec	β^-
Zirconium	Zr	40	90	89.9047	51.5		
			97	96.91097		17 hr	β^-
Niobium	Nb	41	97	96.9081		72 min	β^-
Molybdenum	Mo	42	97	96.906023	9.5		
Silver	Ag	47	107	106.90509	51.8		
			108	107.90595		2.4 min	β^-
Cadmium	Cd	48	108	107.90419	0.9		
Neodymium	Nd	60	142	141.90777	27.1		
Samarium	Sm	62	146	145.9131		10^8 yr	α
Gold	Au	79	197	196.96655	100		
Thallium	Tl	81	207	206.97745		4.8 min	β^-
Lead	Pb	82	206	205.97447	23.6		
Bismuth	Bi	83	210	209.98413		5 days	β^-, α
			211	210.9873		2.1 min	β^-, α
Polonium	Po	84	210	209.98288		138 days	α
Radium	Ra	88	226	226.02544		1600 yr	α
Thorium	Th	90	232	232.03808	100	1.4×10^{10} yr	α
			233	233.0416		22.2 min	β^-
Protactinium	Pa	91	233	233.04027		27 days	β^-
Uranium	U	92	233	233.03965		1.6×10^5 yr	α
			235	235.04394	0.72	7×10^8 yr	α
			238	238.05082	99.27	4.5×10^9 yr	α
			239	239.05433		23.5 min	β^-
Neptunium	Np	93	239	239.05295		2.35 days	β^-
Plutonium	Pu	94	239	239.05218		2.44×10^4 yr	α

Appendix D
The Solar System

TABLE D.1 The sun, moon, and planets.

	Average radius of orbit			Mass		Radius	
	km	Relative to earth-sun distance	Orbital period (yr)	kg	Relative to earth's mass	km	Relative to earth's radius
Sun	—	—	—	1.99×10^{30}	3.33×10^{5}	6.96×10^{5}	109.2
Mercury	5.79×10^{7}	0.3871	0.2409	3.35×10^{23}	0.056	2.485×10^{3}	0.39
Venus	1.081×10^{8}	0.7233	0.6152	4.89×10^{24}	0.817	6.18×10^{3}	0.97
Earth	1.495×10^{8} (92.9×10^{6} miles)	1.0	1.0	5.98×10^{24}	1.0	6.371×10^{3} (3959 miles)	1.0
Mars	2.278×10^{8}	1.524	1.881	6.46×10^{23}	0.108	3.377×10^{3}	0.53
Jupiter	7.778×10^{8}	5.203	11.862	1.90×10^{27}	318	7.13×10^{4}	11.19
Saturn	1.426×10^{9}	9.539	29.458	5.69×10^{26}	95.2	6.03×10^{4}	9.47
Uranus	2.868×10^{9}	19.182	84.013	8.73×10^{25}	14.6	2.35×10^{4}	3.69
Neptune	4.494×10^{9}	30.058	164.79	1.03×10^{26}	17.3	2.23×10^{4}	3.50
Pluto	5.896×10^{9}	39.439	247.69	5.4×10^{24}	0.9	7×10^{3}	1.1
Moon*	3.844×10^{5} (2.389×10^{5} miles)	0.00257	0.0747 (27.3 days)	7.35×10^{22}	0.012	1.72×10^{3}	0.27

*Data for the moon's orbit refer to its orbit around the earth

Appendix E

The Earth—Its Atmosphere, Continental Crusts, Oceans, Water Resources, and Large-Scale Flows of Energy

TABLE E.1 The earth.

	Values
Average radius:	6371 km = 3959 miles
Mass:	5.98×10^{24} kg = 6.59×10^{21} tons
Volume:	1.083×10^{27} cm^3 = 1.083×10^{12} km^3 = 2.6×10^{11} cubic miles
Average density:	5.52 g/cm^3 = 345 lb/ft^3
Surface area:	5.1×10^8 km^2 = 1.97×10^8 square miles
Land (29%):	1.49×10^8 km^2 = 5.75×10^7 square miles
Oceans (71%):	3.61×10^8 km^2 = 1.39×10^8 square miles
Average distance to sun: (The actual distance varies by about 1.7% above and below this value during the year. The earth is closest to the sun on about January 1 and farthest on about July 1.)	1.495×10^8 km = 92.9×10^6 miles
Orbital velocity of earth around sun:	29.8 km/sec = 18.5 miles/sec
Velocity of a point on the equator due to rotation of earth about its axis:	0.465 km/sec = 0.289 miles/sec
Acceleration due to gravity, g (The "standard" value of g is 9.80665 m/sec^2. The value of g varies with elevation and location; the values listed, correct to two significant figures, are suitable for most purposes.)	g = 9.8 m/sec^2 = 980 cm/sec^2 = 32 ft/sec^2
Average albedo (fraction of incident sunlight reflected)	0.34

TABLE E.2 The earth's continental crusts.

Depth: approximately 35 km
Average density: 2.7 g/cm^3 = 170 lb/ft^3

Element*	Average abundance† (in parts per million by weight)
Oxygen	464,000
Silicon	282,000
Aluminum	82,000
Iron	56,000
Calcium	41,000
Sodium	24,000
Magnesium	23,000
Potassium	21,000
Titanium	5,700
Hydrogen	1,400
Lithium	20
Thorium	9.6
Uranium	2.7

*The elements listed are the ten most abundant plus three others (lithium, thorium, and uranium), which are less common but which may be of special interest as energy sources.

†In practice, a particular element can be mined only from deposits of higher than average concentration.

TABLE E.3 The earth's atmosphere.

	Values
Normal atmospheric pressure at sea-level:	14.7 lb/in.^{2} = 1.013×10^5 N/m^2
Average height (altitude at which pressure is half of its value at sea-level):	5.5 km = 3.4 miles
Volume contained between earth's surface and altitude of 5.5 km:	2.8×10^9 km^3 = 6.7×10^8 cubic miles
Total mass:	5.14×10^{18} kg = 5.66×10^{15} tons
Total water content:	1.3×10^{16} kg = 1.43×10^{13} tons (equivalent to 1.3×10^4 km^3 of liquid), 0.001% of earth's total amount of water
Average density of dry air at sea level at standard temperature and pressure (T = 0°C, p = 1 atmosphere):	1.29×10^{-3} g/cm^3 = 1.29 kg/m^3
Velocity of sound in dry air at standard temperature and pressure:	331 m/sec = 1087 ft/sec = 741 miles/hr

Composition of the earth's atmosphere* (normal composition of clean dry air near sea level)

Substance	Percentage by number of molecules	Percentage by weight
Nitrogen (N_2)	78.084	75.52
Oxygen (O_2)	20.948	23.14
Argon (Ar)	0.934	1.29
Carbon dioxide (CO_2)	0.0314	0.0477
Neon (Ne)	0.0018	0.0013
Helium (He)	0.00052	0.00007
Methane (CH_4)	0.0002	0.0001
Krypton (Kr)	0.00011	0.0003
Hydrogen (H_2)	0.00005	0.000003
Nitrous oxide (N_2O)	0.00005	0.00008
Xenon (Xe)	0.000009	0.00004

*The atmosphere also contains water vapor as an important constituent with variable concentration. Concentrations of carbon dioxide and methane vary significantly from time to time and place to place. Small amounts of sulfur dioxide (SO_2), nitrogen dioxide (NO_2), ammonia (NH_3), carbon monoxide (CO), and iodine (I_2) are also found, and (especially in the upper atmosphere) so is ozone (O_3).

TABLE E.4 The earth's oceans.

	Values
Area (71% of the earth's surface area)	3.61×10^8 km^2 = 1.39×10^8 square miles
Volume (97.2% of all the earth's water)	1.32×10^9 km^3 = 3.17×10^8 cubic miles
Mass	1.36×10^{21} kg = 1.5×10^{18} tons
Average depth	3.7 km = 2.3 miles
Composition* (concentrations of the elements are given in g/m^3)†	
Oxygen	857,000
Hydrogen	108,000
Chlorine	19,000
Sodium	10,500
Magnesium	1,350
Sulfur	885
Calcium	400
Potassium	380
Bromine	65
Carbon	28
Lithium	0.17
Uranium	0.003
Thorium	less than 0.0000005

*Listed, in order of abundance are the ten most abundant elements, including hydrogen and oxygen, the elements that combine to form water (H_2O), as well as lithium, uranium and thorium, which are of interest as possible sources of energy. At least trace amounts of nearly all elements have been found in sea water.

†Since the mass of 1 m^3 of water is very nearly 10^6 g, these figures can also be interpreted as the concentrations in parts per million by weight.

TABLE E.5 The earth's supply of water.

Location	Water volume (km^3)	Percentage of total water	Equivalent depth*
Oceans	1.32×10^9	97.2	2.6 km
Icecaps and glaciers	2.9×10^7	2.14	57 m
Ground water and soil moisture	8.4×10^6	0.62	16.5 m
Fresh-water lakes	1.2×10^5	0.009	24 cm
Saline lakes and inland seas	10^5	0.007	20 cm
Atmosphere	1.3×10^4	0.001	2.5 cm
Rivers (amount flowing in rivers at any moment)	1.2×10^3	0.0001	0.24 cm
Total	1.36×10^9	100.0	2.7 km

*The depth if this amount of water were spread uniformly over the whole surface of the earth.

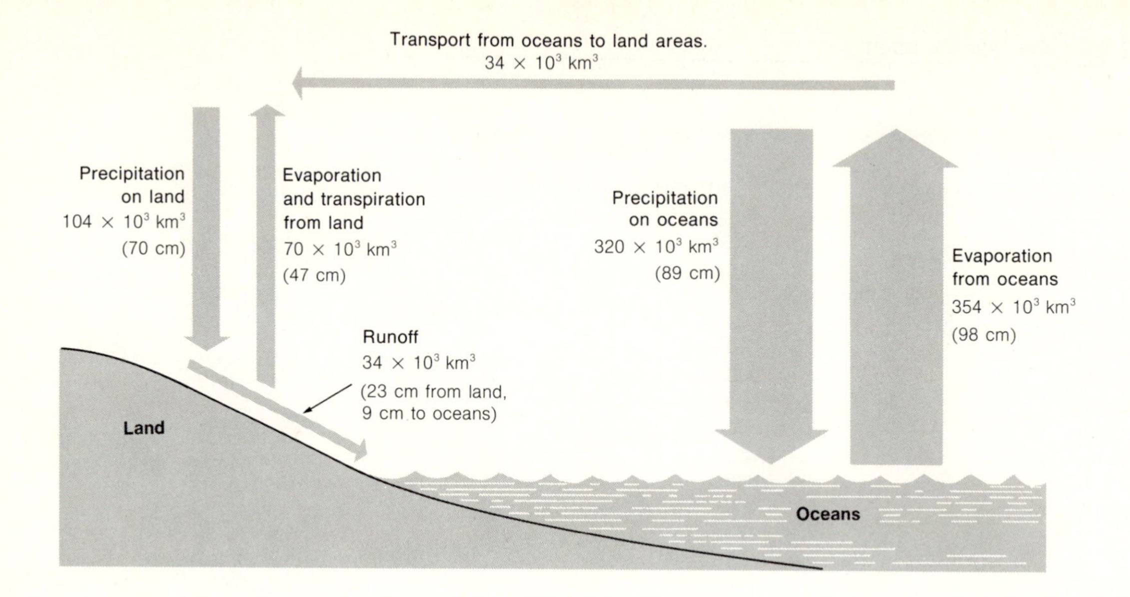

FIGURE E.1
The earth's water cycle. Annual amounts of water transported by precipitation, evaporation, transpiration of plants and animals, and river run-off. Values in parentheses are the depths of the layers of water that would result if the various amounts of water were spread uniformly over the total land area (1.49×10^8 km²) or the total ocean area (3.61×10^8 km²). For example, average annual precipitation is 70 cm on land areas, and 89 cm on the oceans.

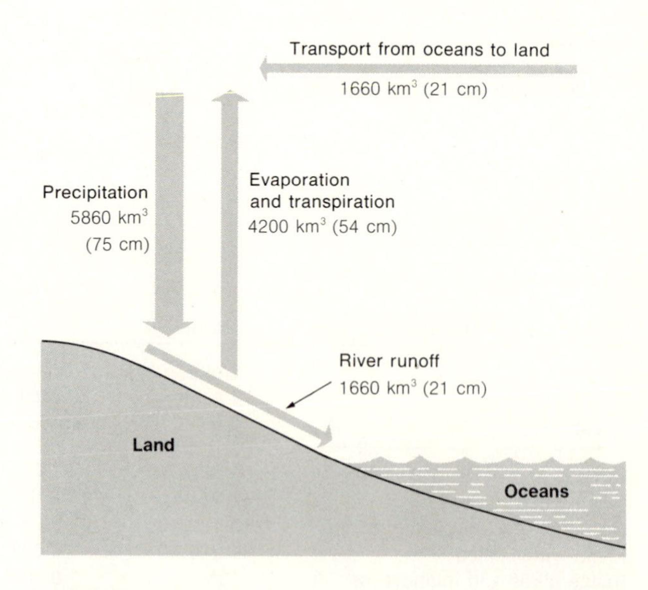

FIGURE E.2
Water flow in the United States (48 contiguous states). Annual amounts of water transported are given in cubic kilometers. Numbers in parentheses are the depths of the layers of water that would result if the various amounts of water were spread uniformly over the total area of the 48 states.

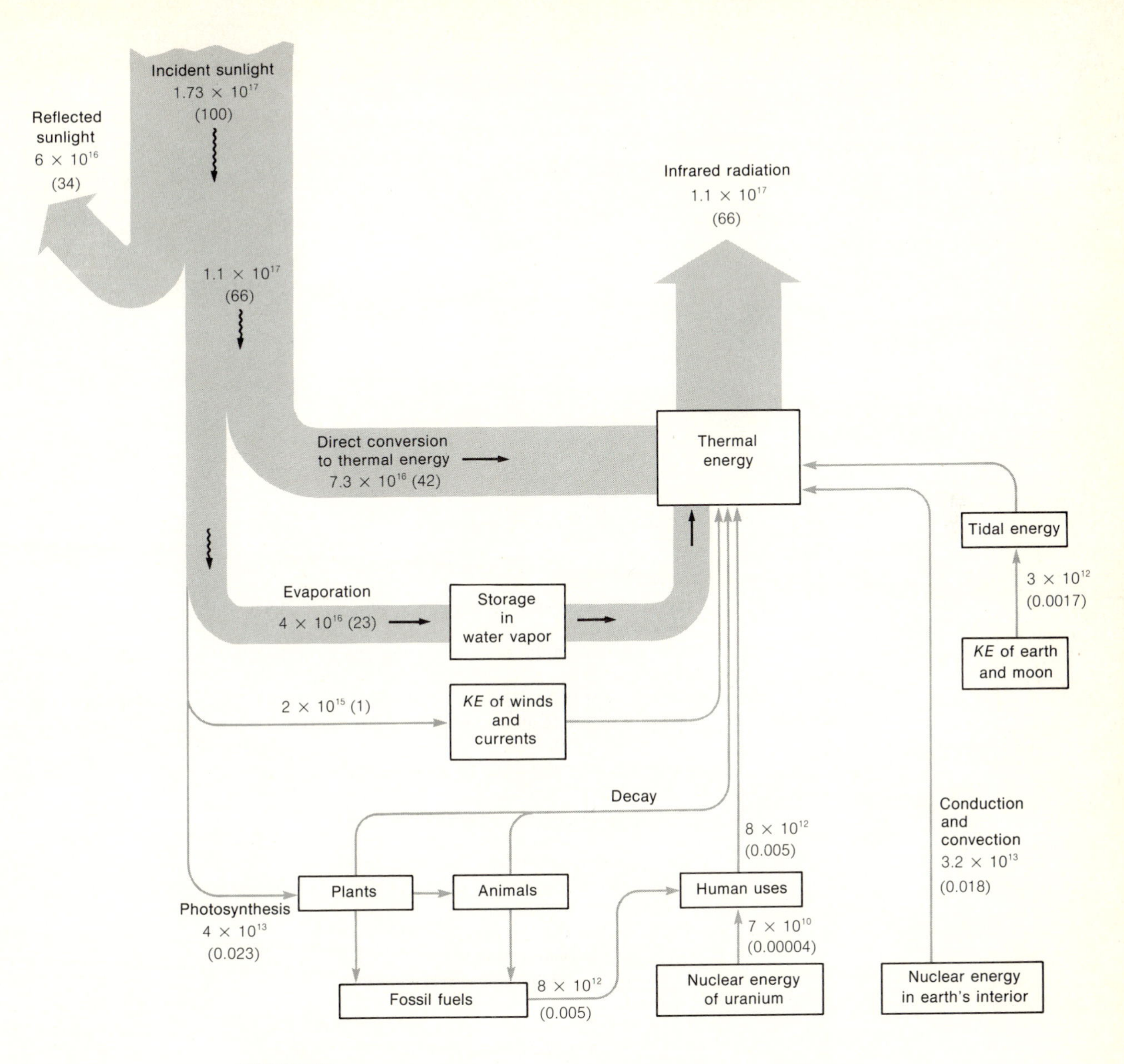

FIGURE E.3
The earth's energy flows, with the details of the exchange of radiation between the surface and the atmosphere omitted. Estimated values are given both in watts and (in parentheses) as percentages of the total power incident from the sun at the top of the atmosphere (1.73×10^{17} W).

Appendix F
Solar Energy

TABLE F.1 Solar energy data.

	Values
Total power radiated from the sun in all directions	3.8×10^{26} W
Power incident on the earth at the top of the atmosphere*	1.73×10^{17} W $= 5.45 \times 10^{24}$ J/yr $= 5174$ Q/yr
Solar constant (S_o)—power per unit area at the top of the earth's atmosphere, for a surface directly facing the sun*	$S_o = 1.353$ kW/m^2 (approximate rounded value: 1.4 kW/m^2)
Amount incident at ground level per unit area, for a surface directly facing the sun (this amount varies with weather conditions and with the amount of atmosphere in the path; the value given here is typical for a time near noon on a clear and cloudless day)	1 kW/m^2
Energy delivered to a horizontal surface (approximate average rate for the 48 contiguous states, averaged over all hours of the day and night and averaged over a full year)†	200 W/m^2

NOTE: The langley. Solar energy data are frequently reported in *langleys* per day, where 1 langley is defined as 1 cal/cm^2. The langley is a unit of energy per unit area, and therefore the number of langleys per unit time is the average power per unit area: 1 langley/day = 0.484 W/m^2. El Paso, for instance, receives during June an average of 730 langleys/day, an average power per unit area of $730 \times 0.484 = 353$ W/m^2.

*These are average values for a whole year. During the year, the rate at which energy is received varies by 3.4% above and below these values, the highest value occurring near January 1 when the earth is closest to the sun, and the lowest value on about July 1 when the earth is farthest from the sun.

†See Table F.2 and Figure F.1 for more detailed data.

TABLE F.2 Average solar radiation for selected cities (incident radiation on a horizontal surface, averaged over all hours of day and night, W/m^2).

	July	Aug.	Sept.	Oct.	Nov.	Dec.	Jan.	Feb.	Mar.	Apr.	May	June	Annual average
Atlanta, Ga.	257	246	201	166	130	102	106	140	184	236	258	272	192
Barrow, Alaska	208	123	56	20	0	0	0	18	87	184	248	256	100
Bismarck, N. Dak.	296	251	185	132	78	60	76	121	170	217	267	284	178
Boise, Idaho	324	275	221	152	88	60	69	113	164	235	284	309	191
Boston, Mass.	240	206	165	115	70	58	67	96	142	176	228	242	150
Caribou, Maine	246	218	161	102	53	51	66	111	178	194	229	232	153
Cleveland, Ohio	267	239	182	127	68	56	60	87	151	182	253	271	162
Columbia, Mo.	278	255	217	157	107	82	87	121	166	210	257	276	184
Dodge City, Kans.	311	287	239	184	138	113	123	153	202	256	275	315	216
El Paso, Tex.	324	309	278	224	178	151	160	209	266	317	346	353	260
Fresno, Calif.	323	293	243	182	117	77	90	143	212	264	308	337	216
Greensboro, N.C.	263	235	197	156	118	95	97	134	171	227	257	273	185
Honolulu, Hawaii	305	293	271	245	208	176	175	200	234	262	300	297	247
Indianapolis, Ind.	262	237	196	142	86	64	70	103	153	192	236	263	167
Little Rock, Ark.	270	250	214	167	118	91	96	127	173	220	256	272	188
Miami, Fla.	260	246	216	188	171	154	166	201	238	263	267	257	219
New York, N.Y.	251	238	175	127	77	62	71	102	151	183	220	255	159
Oklahoma City, Okla.	295	285	234	183	137	115	123	153	197	241	261	302	210
Omaha, Nebr.	275	252	192	142	96	80	99	134	172	224	248	272	182
Rapid City, S. Dak.	288	262	208	152	99	76	90	135	193	235	259	287	190
St. Cloud, Minn.	269	238	174	117	71	60	82	121	177	205	242	262	168
Salt Lake City, Utah	308	274	220	162	103	75	85	127	177	240	288	334	199
San Antonio, Tex.	302	282	237	191	141	122	134	168	203	218	261	292	213
Santa Maria, Calif.	329	297	254	203	151	122	127	167	233	267	307	336	233
Schenectady, N.Y.	215	193	145	106	62	50	63	97	133	165	200	217	137
Seattle, Wash.	242	209	150	84	44	29	34	60	118	174	216	228	132
Tucson, Ariz.	304	286	281	216	172	144	151	195	264	322	358	343	253
Washington, D.C.	267	190	196	145	75	64	101	124	153	182	215	247	163

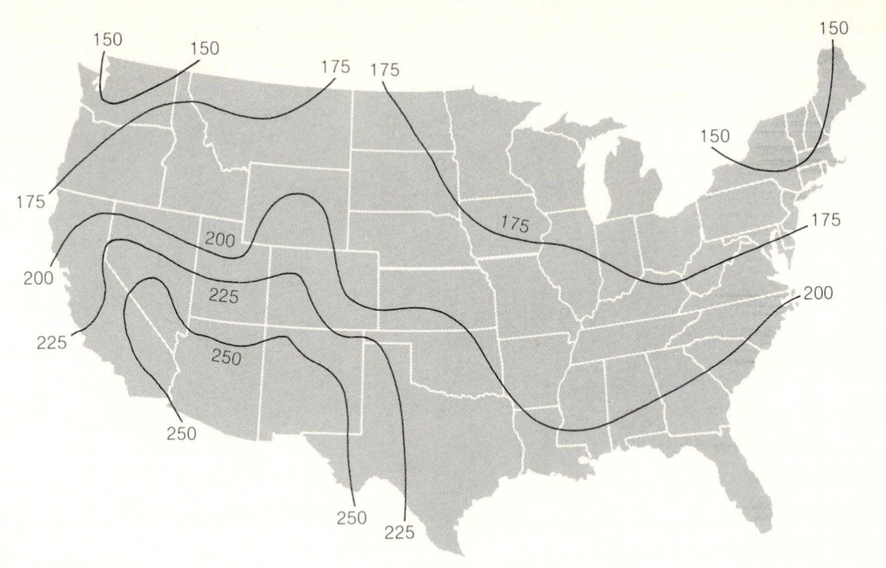

FIGURE F.1
Solar energy zones in the United States. The numbers give the annual average radiation on a horizontal surface, in watts per square meter (averaged over a full year and over all hours of the day and night). *Note*: solar radiation may vary significantly between nearby locations because of local variations in cloudiness. Some of these variations are not shown here, and caution should be used if this map is used to estimate available solar energy for a particular location.

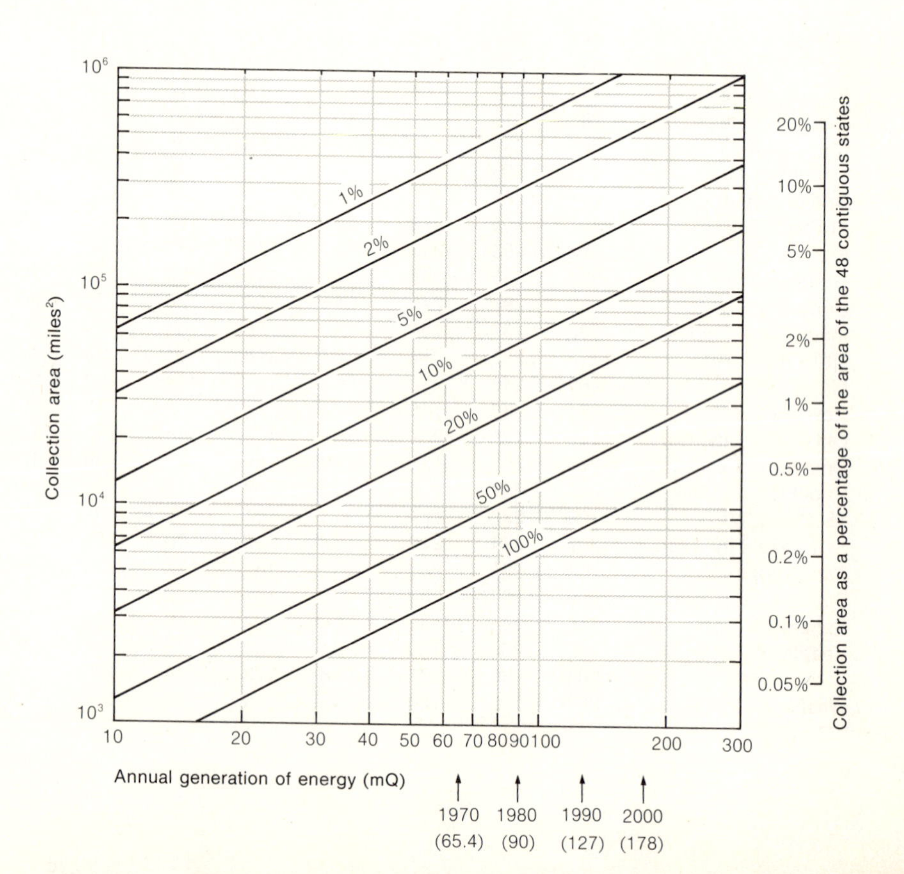

FIGURE F.2
Potentially available solar energy in the United States. The graph shows the required collection area versus the annual amount of energy derived from solar energy for various possible conversion efficiencies. (For example, at an efficiency of 2%, an area of 10^5 square miles could supply 30 mQ/yr.) This figure is based on an average value of 200 W/m² on a horizontal surface. The arrows show *total* annual consumption of energy in the United States for 1970 and projected values for later years.

Appendix G
Degree-Days

FIGURE G.1
Degree-day zones in the United States (annual totals). *Note:* cities that are close together sometimes have significantly different degree-day totals. Some of these local variations are not shown here, and caution should be used if this map is used to estimate the degree-day total for a particular location.

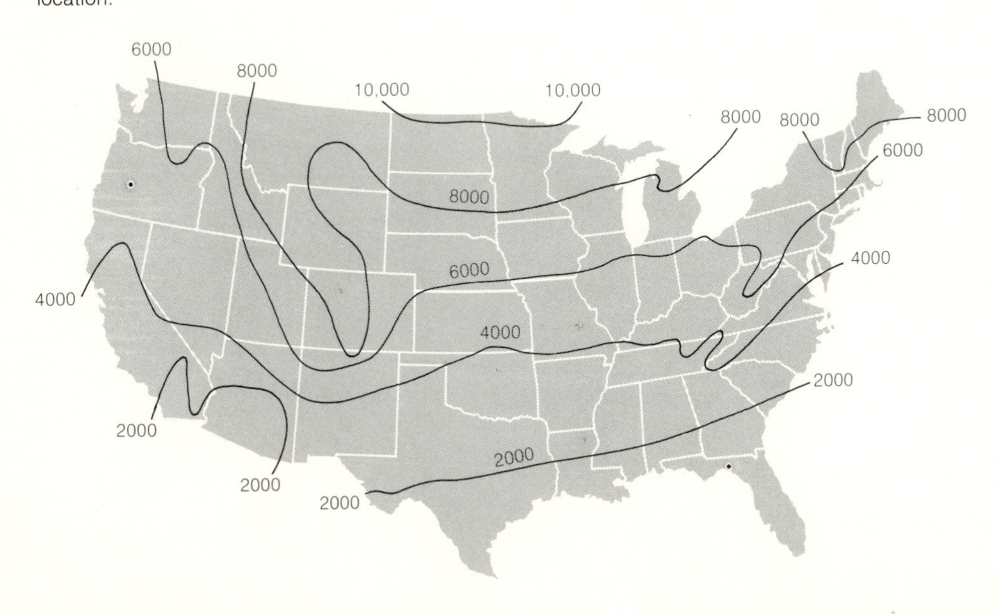

TABLE G.1 Degree-days of selected cities for a typical year.

	July	Aug.	Sept.	Oct.	Nov.	Dec.	Jan.	Feb.	Mar.	Apr.	May	June	Annual total
Atlanta, Ga.	0	0	18	127	414	626	639	529	437	168	25	0	2983
Barrow, Alaska	803	840	1035	1500	1971	2362	2517	2332	2468	1944	1445	957	20174
Bismarck, N. Dak.	34	28	222	577	1083	1463	1708	1442	1203	645	329	117	8851
Boise, Idaho	0	0	132	415	792	1017	1113	854	722	438	245	81	5809
Boston, Mass.	0	9	60	316	603	983	1088	972	846	513	208	36	5634
Caribou, Maine	78	115	336	682	1044	1535	1690	1470	1308	858	468	183	9767
Cleveland, Ohio	9	25	105	384	738	1088	1159	1047	918	552	260	66	6351
Columbia, Mo.	0	0	54	251	651	967	1076	874	716	324	121	12	5046
Dodge City, Kans.	0	0	33	251	666	939	1051	840	719	354	124	9	4986
El Paso, Tex.	0	0	0	84	414	648	685	445	319	105	0	0	2700
Fresno, Calif.	0	0	0	78	339	558	586	406	319	150	56	0	2492
Greensboro, N.C.	0	0	33	192	513	778	784	672	552	234	47	0	3805
Honolulu, Hawaii	0	0	0	0	0	0	0	0	0	0	0	0	0
Indianapolis, Ind.	0	0	90	316	723	1051	1113	949	809	432	177	39	5699
Little Rock, Ark.	0	0	9	127	465	716	756	577	434	126	9	0	3219
Miami, Fla.	0	0	0	0	0	40	56	36	9	0	0	0	141
New York, N.Y.	0	0	30	233	540	902	986	885	760	408	118	9	4871
Oklahoma City, Okla.	0	0	15	164	498	766	868	664	527	189	34	0	3725
Omaha, Nebr.	0	12	105	357	828	1175	1355	1126	939	465	208	42	6612
Rapid City, S. Dak.	22	12	165	481	897	1172	1333	1145	1051	615	326	126	7345
St. Cloud, Minn.	28	47	225	549	1065	1500	1702	1445	1221	666	326	105	8879
Salt Lake City, Utah	0	0	81	419	849	1082	1172	910	763	459	233	84	6052
San Antonio, Tex.	0	0	0	31	207	363	428	286	195	39	0	0	1549
Santa Maria, Calif.	99	93	96	146	270	391	459	370	363	282	233	165	2967
Schenectady, N.Y.	0	22	123	422	756	1159	1283	1131	970	543	211	30	6650
Seattle, Wash.	50	47	129	329	543	657	738	599	577	396	242	117	4424
Tucson, Ariz.	0	0	0	25	231	406	471	344	242	75	6	0	1800
Washington, D.C.	0	0	33	217	519	834	871	762	626	288	74	0	4224

Appendix H
Energy Content of Fuels

TABLE H.1 Energy content.[a]

Fuel		Values	
	(Commonly used units)	(BTU/ton)	(J/kg)
Coal (bituminous and anthracite)		25×10^6	29×10^6
Lignite		10×10^6	12×10^6
Peat		3.5×10^6	4×10^6
Crude oil	5.6×10^6 BTU/barrel	37×10^6	43×10^6
Gasoline	5.2×10^6 BTU/barrel	38×10^6	44×10^6
NGL's (Natural gas liquids)	4.2×10^6 BTU/barrel	37×10^6	43×10^6
Natural gas[b]	1030 BTU/ft^3	47×10^6	55×10^6
Hydrogen gas[b]	333 BTU/ft^3	107×10^6	124×10^6
Methanol (methyl alcohol)	6×10^4 BTU/gal	17×10^6	20×10^6
Charcoal		24×10^6	28×10^6
Wood	20×10^6 BTU/cord	12×10^6	14×10^6
Miscellaneous farm wastes		12×10^6	14×10^6
Dung		15×10^6	17×10^6
Assorted garbage and trash		10×10^6	12×10^6
Bread	1100 kcal/lb	9×10^6	10×10^6
Butter	3600 kcal/lb	29×10^6	33×10^6
Fission	200 MeV/fission	7×10^{13} [c]	8×10^{13} [c]
		5×10^{11} [d]	5.8×10^{11} [d]
D-D Fusion (deuterium)	7 MeV/deuteron	2.9×10^{14} [e]	3.3×10^{14} [e]
		8.6×10^{10} [f]	10^{11} [f]
D-T Fusion (lithium)[g]	7 MeV/Li nucleus	8.4×10^{13}	9.7×10^{13}
Complete "mass-energy conversion"[h]	931 MeV/amu	7.7×10^{16}	9×10^{16}

[a]These data are only intended for use in making estimates of available energy. Various types of wood, for example, have energy values covering a rather wide range; different samples of coal, oil, and other fuels also have varying energy values. Various fuels obtained from the processing of oil (for example residual oil, kerosene, various types of gasoline) have energy values per unit mass within about 20% of those listed for crude oil and gasoline.

[b]Quantities of natural gas are usually reported in cubic feet, the volume of gas at a pressure of 1 atmosphere and a temperature of 60°F, or in thousands of cubic feet (often abbreviated Mcf).

[c]per ton or kilogram of nuclei undergoing fission.

[d]per ton or kilogram of uranium metal, when only the U^{235} (abundance 0.72%) is used.

[e]per ton or kilogram of pure deuterium.

[f]per ton or kilogram of hydrogen, containing 0.015% deuterium.

[g]The data for D-T fusion are based on the assumption that deuterium is available in unlimited quantities, that tritium is produced as discussed in Chapter 15, and that energy production is limited by the availability of lithium.

[h]The data given for complete mass-energy conversion are given for purposes of comparison; no practical "fuel" is known that would yield this much energy.

Appendix I
Fossil Fuels—Resources and Production

TABLE I.1 Fossil fuel resources: estimates of eventual total production.
NOTE: Such estimates are subject to considerable uncertainty and should only be considered as giving the order of magnitude of the eventual total production. Estimates of production of energy from sources already in use are much more reliable than they are for tar sands and oil shales. The amount of oil that *exists* in the world's oil shales is estimated to be about 2×10^{15} barrels, 10,000 times larger than the figure given here; most of this oil is probably unobtainable, but the situation could be changed either by new technological developments or by changing economic conditions.

	United States		World (United States included)	
Fuel	Physical units	Approximate energy content (Q)	Physical units	Approximate energy content (Q)
Coal and lignite	1.6×10^{12} tons	37	8.4×10^{12} tons	170
Crude oil	200×10^{9} barrels	1.1	2100×10^{9} barrels	12
Natural gas liquids (NGL's)	40×10^{9} barrels	0.17	400×10^{9} barrels	1.7
Natural gas	1.1×10^{15} ft^3	1.1	12×10^{15} ft^3	12
Canadian tar sands	–	–	300×10^{9} barrels	1.7
Oil shales	80×10^{9} barrels	0.45	190×10^{9} barrels	1.1
Total	–	40	–	200

TABLE I.2 Major fossil fuels: world and United States consumption rates in relation to resources.

World (United States included)			
	Coal and lignite	Petroleum liquids (crude oil and NGL's)	Natural gas
Resources*	8.4×10^{12} tons	2500×10^{9} barrels	12×10^{15} ft^3
Cumulative production (through 1973)	0.16×10^{12} tons (1.9% of resources)	312×10^{9} barrels (12% of resources)	0.6×10^{15} ft^3 (5% of resources)
Production rate (1973)	3270×10^{6} tons/yr	21.2×10^{9} barrels/yr	44×10^{12} ft^3/yr
Time remaining until resources would be exhausted, if production were to continue at the 1973 rate.†	2500 yr	100 yr	260 yr

United States			
	Coal and lignite	Petroleum liquids (crude oil and NGL's)	Natural gas
Resources*	1.6×10^{12} tons	240×10^{9} barrels	1.1×10^{15} ft^3
Cumulative production through 1973)	0.04×10^{12} tons (2.5% of resources)	117×10^{9} barrels (49% of resources)	0.41×10^{15} ft^3 (37% of resources)
Production rate (1973)	600×10^{6} tons/yr	4.0×10^{9} barrels/yr	21.7×10^{12} ft^3/yr
Time remaining until resources would be exhausted, if production were to continue at the 1973 rate.	2600 yr	30 yr	32 yr

*See Table I.1.
†See also Figures I.7 and I.8.

TABLE I.3 Annual fossil fuel production rates.

	Coal and lignite (10^6 tons)	Crude oil (10^6 barrels)	NGL's* (10^6 barrels)	Natural gas* (10^9 ft^3)
	World (United States included)			
1850	110			
1860	152	0.5		
1870	238	5.7		
1880	371	30		
1890	566	77		249
1900	851	149		251
1905	1041	215		374
1910	1284	328		540
1915	1315	432	1.5	665
1920	1488	689	9.3	835
1925	1512	1069	28	1240
1930	1559	1412	57	2002
1935	1452	1655	45	1973
1940	1854	2150	77	2792
1945	1516	2595	120	4263
1950	2000	3803	190	6683
1955	2345	5626	294	10240
1960	2870	7674	376	16054
1962	2800	8882	415	18933
1964	3026	10310	486	22542
1966	3068	12016	544	26206
1968	3029	14104	672	30380
1970	3219	16690	755	36635
1972	3252	18584	832	41270
1973	3270	20357	883	44160

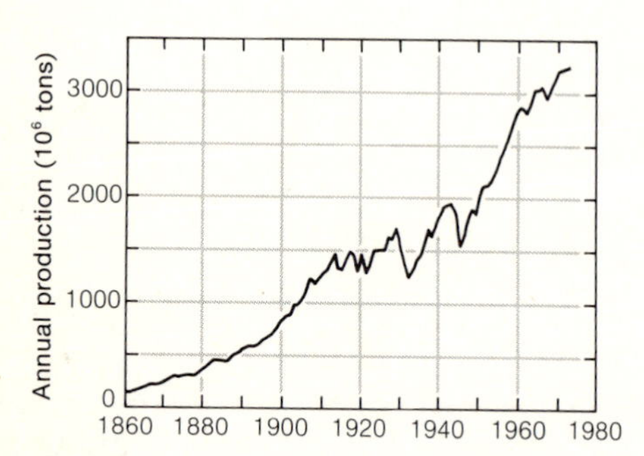

FIGURE I.1
World production of coal and lignite. Cumulative production through 1973: 16×10^{10} tons.

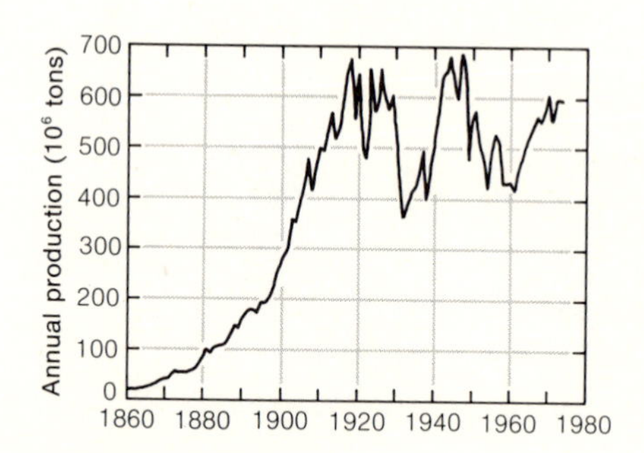

FIGURE I.2
United States production of coal and lignite. Cumulative production through 1973: 4×10^{10} tons.

TABLE I.3 (continued)

	Coal and lignite (10^6 tons)	Crude oil (10^6 barrels)	NGL's* (10^6 barrels)	Natural gas* (10^9 ft^3)
		United States		
1850	8			
1860	20	0.5		
1870	40	5.2		
1880	79	26		
1890	158	46		239
1900	270	63		137
1905	393	135		351
1910	502	210		509
1915	532	281	1.5	627
1920	658	443	9.1	801
1925	582	764	27	1176
1930	537	898	53	1913
1935	425	997	39	1920
1940	512	1353	70	2646
1945	633	1714	112	3902
1950	560	1974	182	6054
1955	491	2484	281	9053
1960	434	2575	340	12333
1962	439	2676	373	13369
1964	504	2787	420	14859
1966	547	3028	465	16511
1968	557	3329	549	18500
1970	613	3517	606	21015
1972	602	3455	638	21580
1973	600	3361	634	21700

*Figures for natural gas production are corrected for "take-off" for production of NGL's. During most of this period of time, much of the natural gas found in foreign oil fields has been wasted by venting or burning; production of natural gas and NGL's outside the United States has therefore been low.

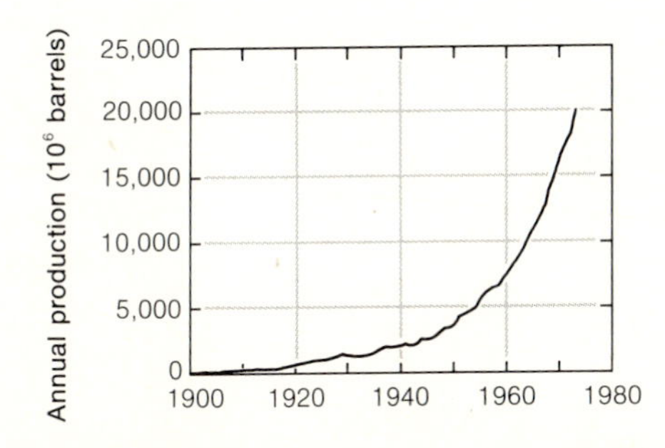

FIGURE I.3
World production of crude oil. Cumulative production through 1973: 300 $\times$ 10^9 barrels.

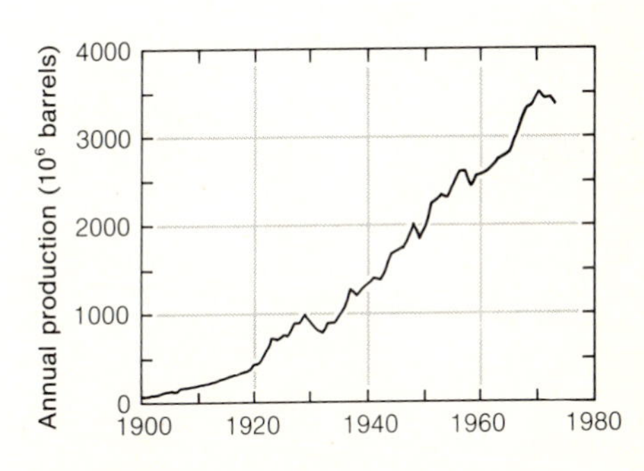

FIGURE I.4
United States production of crude oil. Cumulative production through 1973: 106 $\times$ 10^9 barrels.

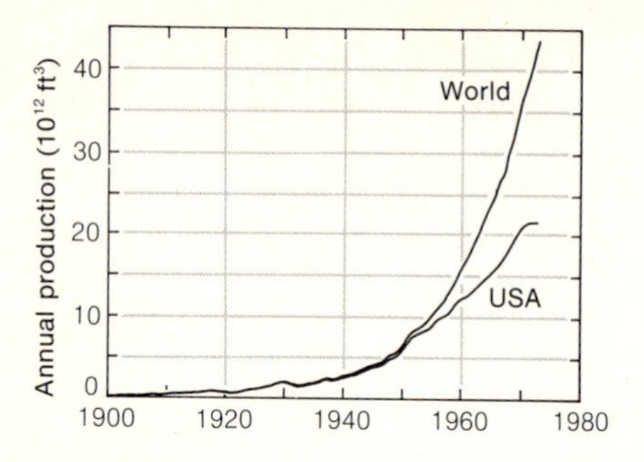

FIGURE I.5
World and United States production of natural gas. Cumulative production through 1973: World, 600 × 10^{12} ft^3; United States, 410 × 10^{12} ft^3.

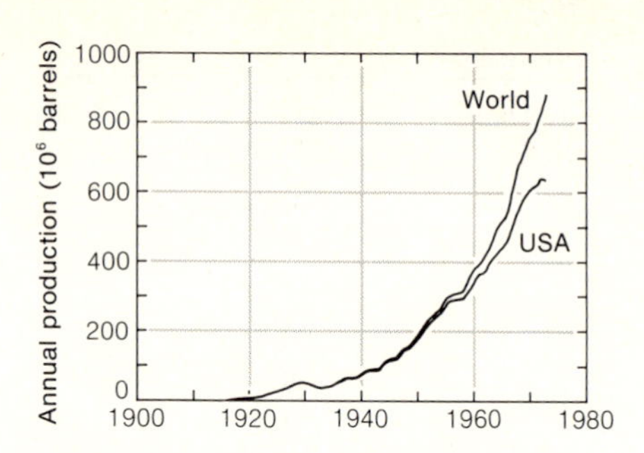

FIGURE I.6
World and United States production of NGL's (natural gas liquids). Cumulative production through 1973: World, 13 × 10^9 barrels; United States, 11 × 10^9 barrels.

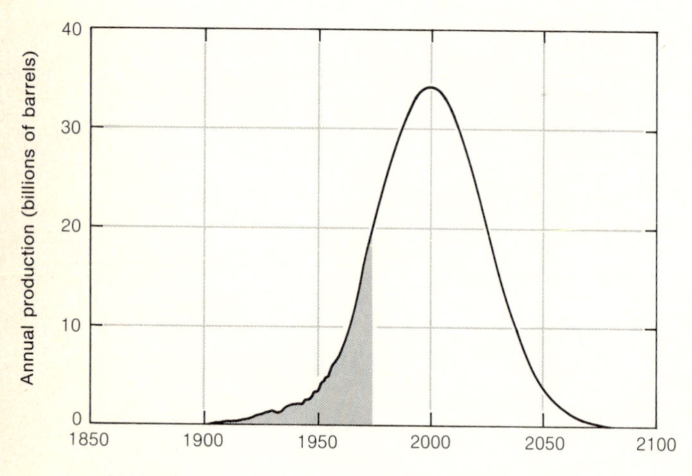

FIGURE I.7
Projected cycle of world oil production (total production = 2100 × 10^9 barrels). For the United States, peak production probably occurred during the early 1970s.

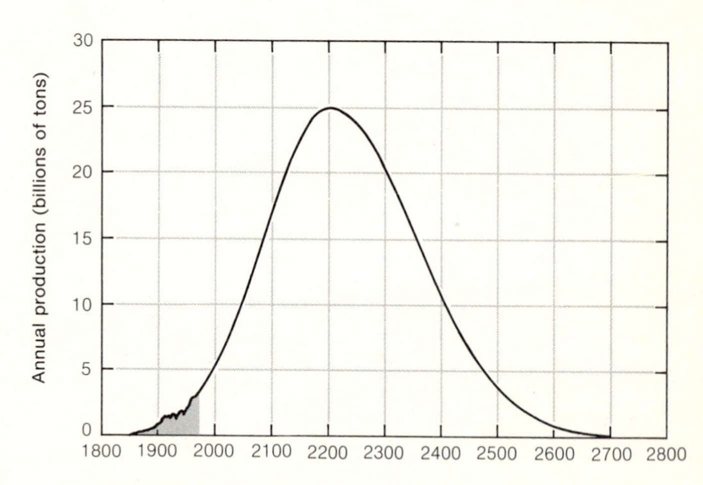

FIGURE I.8
Projected cycle of world coal production (total production = 8.4 × 10^{12} tons). The projected cycle for the United States would have a similar appearance; the size would be reduced because of the lower total production.

Appendix J
Nuclear Fission and Nuclear Fusion—Energy Resources

Uranium and thorium are the important elements for obtaining energy by means of nuclear fission. Present nuclear reactors utilize primarily the isotope U^{235}; U^{238}–Pu^{239} breeders would make available the energy of U^{238} as well. Thorium may become an important source of fission energy if Th^{232}–U^{233} breeders are developed. Lithium will be an important energy resource if the D-T fusion process can be successfully developed. (Deuterium is also required for both D-T and D-D fusion, but the amount of deuterium in the oceans is so large that with D-D fusion essentially no constraints will be imposed by availability of resources, whereas with D-T fusion, production of energy will be limited by the amount of lithium available and not by the supply of deuterium.) Estimates of available resources of uranium, thorium, and lithium are subject to great uncertainties, in part because until recently there has been little interest in these elements as energy resources, and in part because the amount of energy available per unit mass is so large that it may become profitable to extract these elements from very low-grade deposits. Development of the U^{238}–Pu^{239} breeder will have a particularly great impact in determining the value of uranium deposits.

Uranium

The average abundance of uranium in the earth's crust is 2.7 parts per million (Table E.2). More important is the availability of uranium in concentrated deposits. Table J.1 shows an estimate of United States uranium resources, in which uranium deposits are classified in terms of the estimated cost of mining, made by the Atomic Energy Commission in 1973. Uranium resources in the remainder of the world are less well known but are probably at least five times as large as those in the United States.

TABLE J.1 United States uranium resources.

Price ($/lb of ore)	Uranium available at this price or less (10^6 tons)	Fission energy available (Q)	
		From U^{235} only	From U^{235} and U^{238}
10	0.95	0.48	66
15	1.4	0.7	100
30	2.0	1.0	140
50	7	3.5	500
100	15	7.5	1000

Thorium

The average abundance of thorium in the earth's crust is 9.6 parts per million (Table E.2). Rocks containing low-grade thorium deposits are found near the surface in many parts of the world. As one example, the Conway granites in New Hampshire, covering an area of about 750 km^2 and probably extending to a depth of several kilometers, contain 150 g of thorium per cubic meter. The Conway granites down to a depth of 1 km contain approximately 1.2×10^8 tons of thorium, equivalent to a fission energy of approximately 8500 Q.

Lithium

The average abundance of lithium in the earth's crust is 20 parts per million (Table E.2). The extent of high-grade lithium deposits is less well known. According to one estimate, the presently known lithium resources of the United States amount to about 10^7 tons of lithium metal, equivalent to a fusion energy of about 800 Q.

Appendix K
The History of Energy Production and Consumption in the World and the United States

TABLE K.1 Population of the world and the United States.

	World (United States included) (millions)	United States (millions)
1900	1590	76
1905	1636	84
1910	1686	92
1915	1740	101
1920	1811	106
1925	1910	116
1930	2015	123
1935	2129	127
1940	2249	133
1945	2357	140
1950	2486	152
1955	2713	166
1960	2982	181
1962	3101	187
1964	3225	192
1966	3355	197
1968	3490	201
1970	3632	205
1972	3782	209
1973	3860	210
1974	–	212

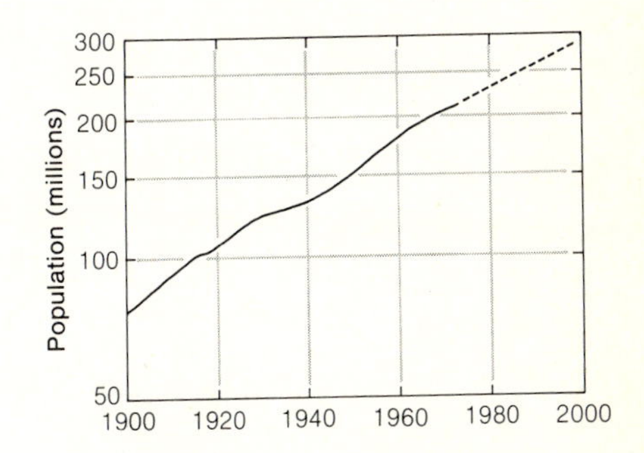

FIGURE K.1
Population of the United States, with extrapolation to 2000.

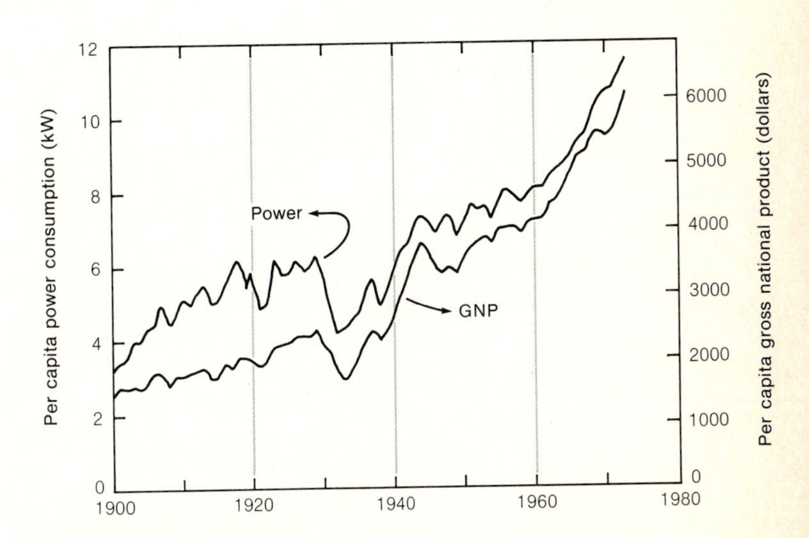

FIGURE K.2
Per capita average power consumption and per capita gross national product in the United States. (Gross national product figures are given in 1973 dollars; dollar figures for previous years have been increased to take account of inflation.)

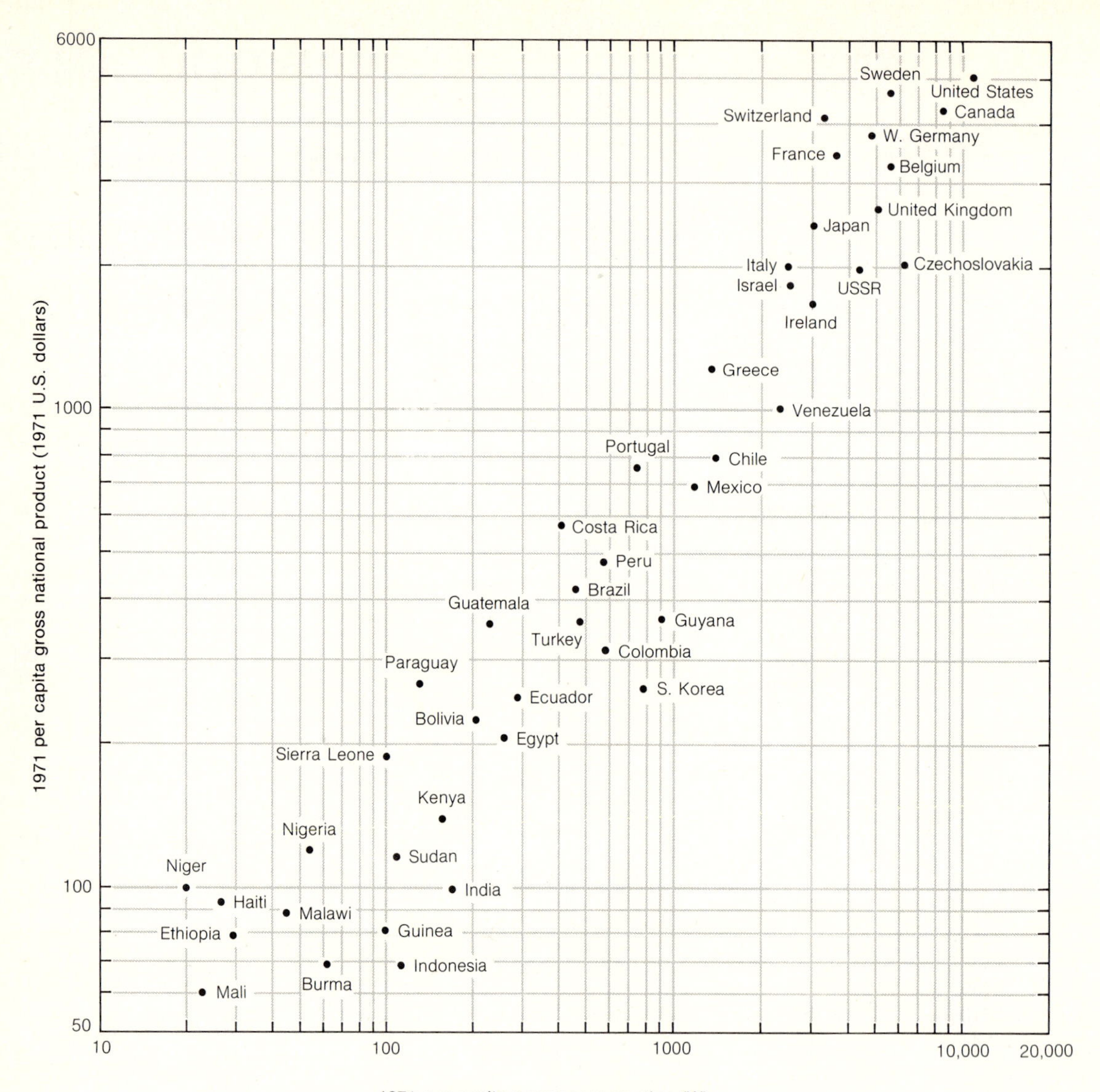

FIGURE K.3
Per capita GNP and per capita average power consumption from fossil fuels, hydropower, and nuclear power for various nations (1971).

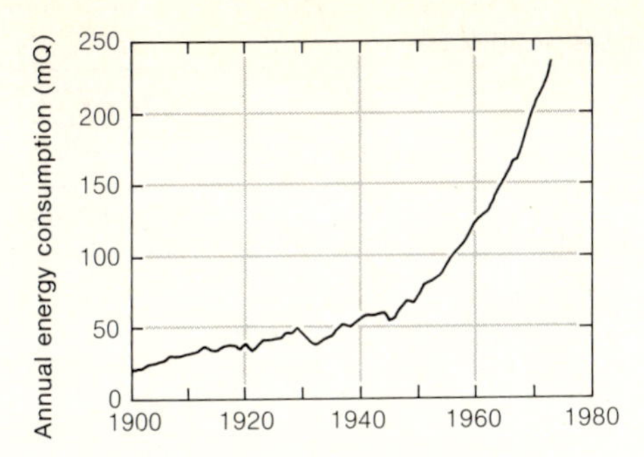

FIGURE K.4
Annual world consumption of energy from fossil fuels, hydropower, and nuclear power.

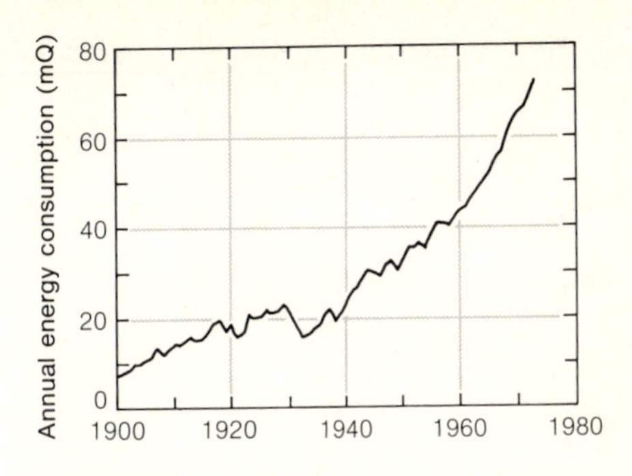

FIGURE K.5
Annual United States consumption of energy from fossil fuels, hydropower, and nuclear power.

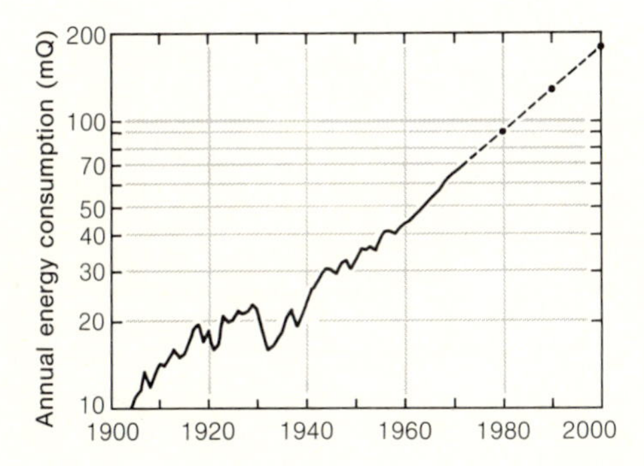

FIGURE K.6
Annual United States consumption of energy from fossil fuels, hydropower, and nuclear power (semilogarithmic graph). Extrapolated values shown: 1980, 90 mQ; 1990, 127 mQ; 2000, 178 mQ.

TABLE K.2 Annual energy consumption by world and United States, 1900–1973.

	Coal and lignite (mQ)	Oil (mQ)	Natural gas (mQ)	Hydropower* (mQ)	Nuclear energy† (mQ)	Totals: Total energy (mQ)	Totals: Average rate (10^{12} W)	Totals: Average per capita rate (W)
				World (United States included)				
1900	19.9	0.84	0.26	0.01		21.0	0.70	442
1905	24.4	1.20	0.39	0.03		26.0	0.87	530
1910	30.0	1.84	0.56	0.06		32.5	1.09	644
1915	30.5	2.42	0.69	0.12		33.8	1.13	649
1920	34.3	3.90	0.86	0.18		39.2	1.31	723
1925	34.4	6.10	1.28	0.27		42.0	1.40	735
1930	35.4	8.15	2.06	0.43		46.0	1.54	763
1935	32.6	9.46	2.03	0.49		44.6	1.49	699
1940	40.6	12.4	2.88	0.68		56.6	1.89	840
1945	34.5	15.0	4.39	0.91		54.8	1.83	777
1950	43.6	22.1	6.89	1.16		73.7	2.46	991
1955	49.2	32.7	10.6	1.62		94.1	3.14	1159
1960	60.6	44.6	16.5	2.35	0.03	124	4.14	1390
1962	58.0	51.5	19.5	2.59	0.07	132	4.40	1419
1964	62.6	59.8	23.2	2.81	0.17	149	4.96	1539
1966	63.8	69.6	27.0	3.37	0.39	164	5.48	1635
1968	62.8	81.8	31.3	3.59	0.60	180	6.02	1724
1970	66.6	96.6	37.8	4.01	0.88	206	6.88	1894
1972	67.2	108	42.5	4.43	1.59	223	7.46	1973
1973	67.6	118	45.5	4.65	2.05	238	7.94	2056

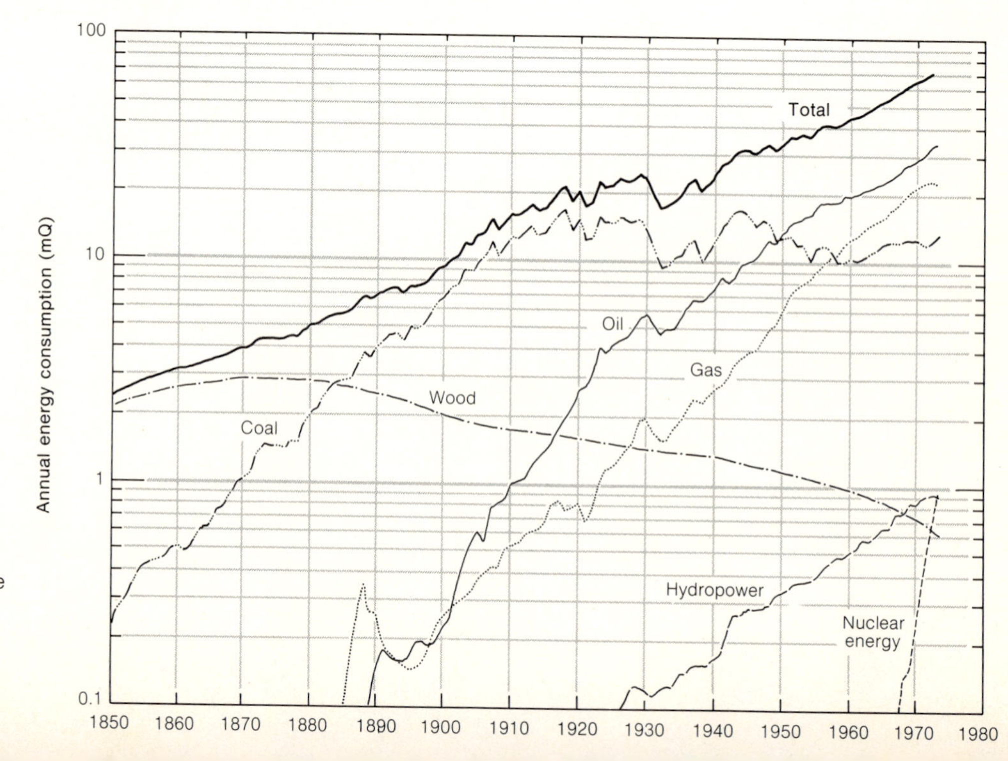

FIGURE K.7
Annual United States consumption of energy from various sources since 1850. Energy from fuel wood is included here, both as a separate item and as part of the total; wood is *not* included in other graphs and tables.

TABLE K.2 (continued)

	Coal and lignite (mQ)	Oil (mQ)	Natural gas (mQ)	Hydropower* (mQ)	Nuclear energy† (mQ)	Totals: Total energy (mQ)	Totals: Average rate (10^{12} W)	Totals: Average per capita rate (W)
United States								
1900	6.84	0.23	0.25	0.01		7.33	0.24	3219
1905	10.0	0.61	0.37	0.02		11.0	0.37	4385
1910	12.7	1.01	0.54	0.03		14.3	0.48	5167
1915	13.3	1.42	0.67	0.05		15.4	0.52	5127
1920	15.5	2.68	0.83	0.06		19.1	0.64	5985
1925	14.7	4.28	1.21	0.09		20.3	0.68	5851
1930	13.6	5.90	1.97	0.12		21.6	0.72	5871
1935	10.6	5.68	1.97	0.15		18.4	0.62	4840
1940	12.5	7.78	2.73	0.17		23.2	0.78	5849
1945	16.0	10.1	3.97	0.29		30.3	1.01	7217
1950	12.9	13.5	6.15	0.34		32.9	1.10	7218
1955	11.7	17.5	9.23	0.40		38.9	1.30	7824
1960	10.4	20.0	12.7	0.51	0.005	43.7	1.46	8081
1962	10.5	21.3	14.1	0.59	0.023	46.5	1.55	8332
1964	11.7	22.4	15.6	0.61	0.034	50.3	1.68	8749
1966	12.5	24.4	17.4	0.66	0.056	55.0	1.84	9349
1968	12.7	27.1	19.6	0.76	0.14	60.2	2.01	10026
1970	12.7	29.6	22.0	0.84	0.25	65.4	2.18	10663
1972	12.4	33.0	23.1	0.93	0.61	70.0	2.34	11200
1973	13.3	34.7	22.9	0.93	0.95	72.8	2.43	11565

*Energy from hydropower is the actual amount of electrical energy generated (3413 BTU per kilowatt-hour generated).

†Nuclear energy used is calculated from the amount of electrical energy generated in nuclear power plants at an average efficiency of 30% (11377 BTU per kilowatt-hour generated).

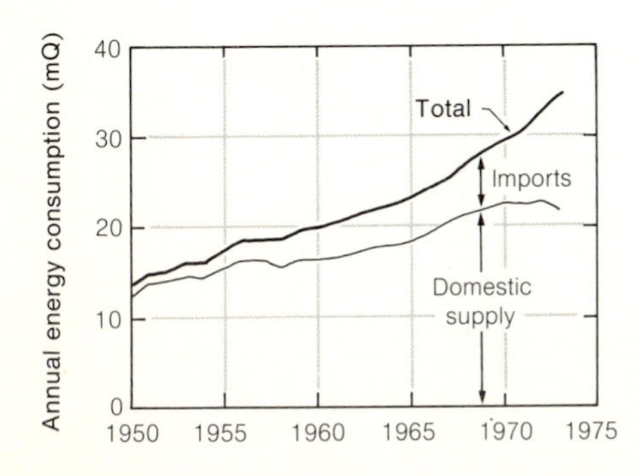

FIGURE K.8
Foreign and domestic supplies of oil. Annual United States consumption of energy from oil (natural gas liquids included).

TABLE K.3 Installed electrical generating capacity, world and United States, 1900–1973.

	Hydroelectric (GW)	Fossil fuel (GW)	Nuclear (GW)	Total (GW)	Per capita total (W)
		World (United States included)			
1900	1.3	1.6		2.9	1.8
1905	2.5	4.3		6.8	4.2
1910	5.1	11.2		16.3	9.7
1915	9.7	21.0		30.7	17.7
1920	14.4	30.4		44.8	24.8
1925	23.2	48.7		71.9	37.6
1930	30.4	84.8		115	57.2
1935	36.9	98.5		135	63.6
1940	44.8	103		148	65.6
1945	54.3	117		171	72.6
1950	76.5	141		218	87.6
1955	112	236		348	128
1960	160	381	0.86	542	182
1962	184	441	2.1	627	202
1964	208	521	3.3	732	227
1966	229	603	8.5	841	251
1968	258	704	12.2	974	279
1970	285	811	18.9	1115	307
1972	313	921	36.7	1271	336
1973	327	989	46.7	1363	353

FIGURE K.9
Installed electrical generating capacity, worldwide.

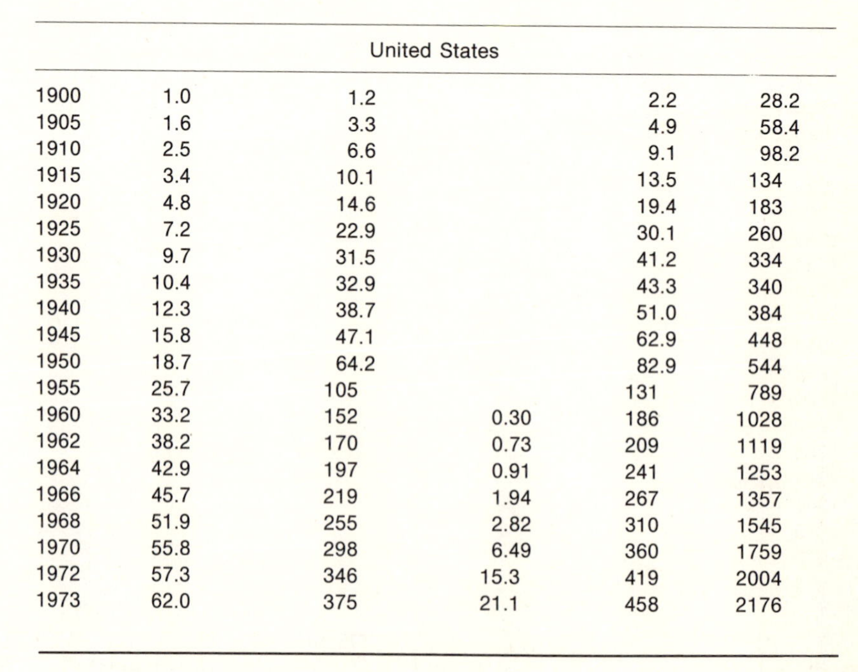

	Hydroelectric (GW)	Fossil fuel (GW)	Nuclear (GW)	Total (GW)	Per capita total (W)
		United States			
1900	1.0	1.2		2.2	28.2
1905	1.6	3.3		4.9	58.4
1910	2.5	6.6		9.1	98.2
1915	3.4	10.1		13.5	134
1920	4.8	14.6		19.4	183
1925	7.2	22.9		30.1	260
1930	9.7	31.5		41.2	334
1935	10.4	32.9		43.3	340
1940	12.3	38.7		51.0	384
1945	15.8	47.1		62.9	448
1950	18.7	64.2		82.9	544
1955	25.7	105		131	789
1960	33.2	152	0.30	186	1028
1962	38.2	170	0.73	209	1119
1964	42.9	197	0.91	241	1253
1966	45.7	219	1.94	267	1357
1968	51.9	255	2.82	310	1545
1970	55.8	298	6.49	360	1759
1972	57.3	346	15.3	419	2004
1973	62.0	375	21.1	458	2176

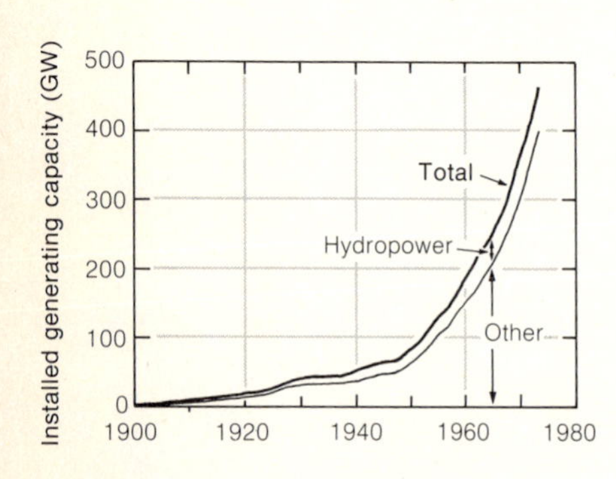

FIGURE K.10
Installed electrical generating capacity of the United States.

TABLE K.4 World and United States electrical generation, annual values (1900–1973).

	Hydroelectric power (10^9 kWh)	Fossil fuel (10^9 kWh)	Nuclear power (10^9 kWh)	Total (10^9 kWh)	Per capita electrical energy per year (kWh)	Per capita average rate of use (W)
			World (United States included)			
1900	3.65	1.35		5.0	3.14	0.36
1905	7.54	5.92		13.5	8.22	0.94
1910	16.9	16.2		33.1	19.6	2.24
1915	39.1	34.6		73.7	42.3	4.83
1920	53.7	68.3		122	67.4	7.69
1925	78.7	110		189	98.9	11.3
1930	127	173		300	149	17.0
1935	145	195		340	160	18.2
1940	199	301		500	222	25.4
1945	265	437		702	298	34.0
1950	339	607		946	380	43.4
1955	475	1065		1540	568	64.8
1960	688	1614	2.73	2304	773	88.2
1962	759	1870	6.50	2636	850	97.0
1964	824	2296	15.0	3135	972	111
1966	988	2614	34.6	3637	1084	124
1968	1053	3100	52.7	4206	1205	137
1970	1176	3654	77.7	4908	1351	154
1972	1298	4209	140	5647	1493	170
1973	1361	4512	180	6053	1568	179
			United States			
1900	2.78	1.17		3.96	52.0	5.93
1905	5.05	5.25		10.3	123	14.0
1910	8.62	11.1		19.7	214	24.4
1915	13.2	21.4		34.6	344	39.3
1920	18.8	37.8		56.6	531	60.6
1925	25.5	59.2		84.7	731	83.4
1930	35.9	78.8		115	931	106
1935	42.7	76.2		119	935	107
1940	50.1	130		180	1357	155
1945	84.7	187		271	1931	220
1950	101	288		389	2553	291
1955	116	513		629	3791	432
1960	149	694	0.52	844	4673	533
1962	172	772	2.0	946	5074	579
1964	178	902	3.0	1084	5648	644
1966	194	1050	5.0	1249	6357	725
1968	222	1200	12.5	1435	7151	816
1970	247	1370	21.8	1639	8000	913
1972	273	1525	54.0	1852	8866	1011
1973	271	1592	83.3	1946	9248	1055

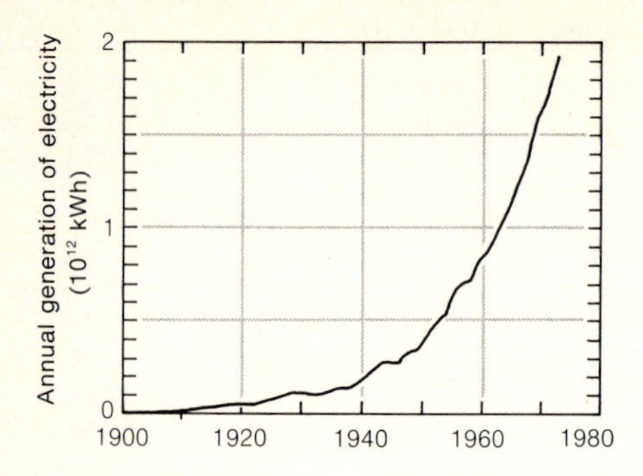

FIGURE K.11
Generation of electrical energy in the United States.

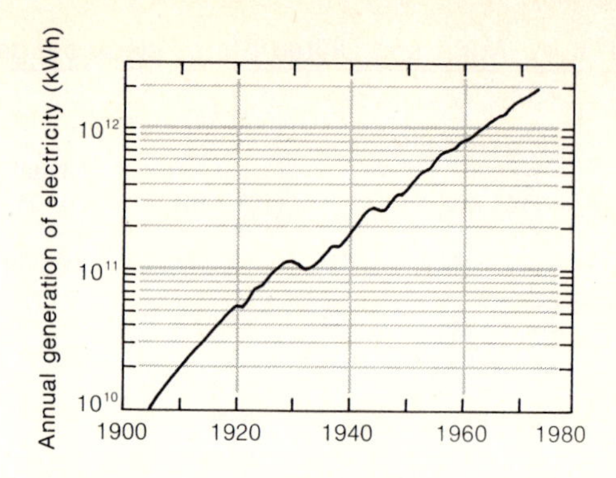

FIGURE K.12
Generation of electrical energy in the United States (semilogarithmic graph).

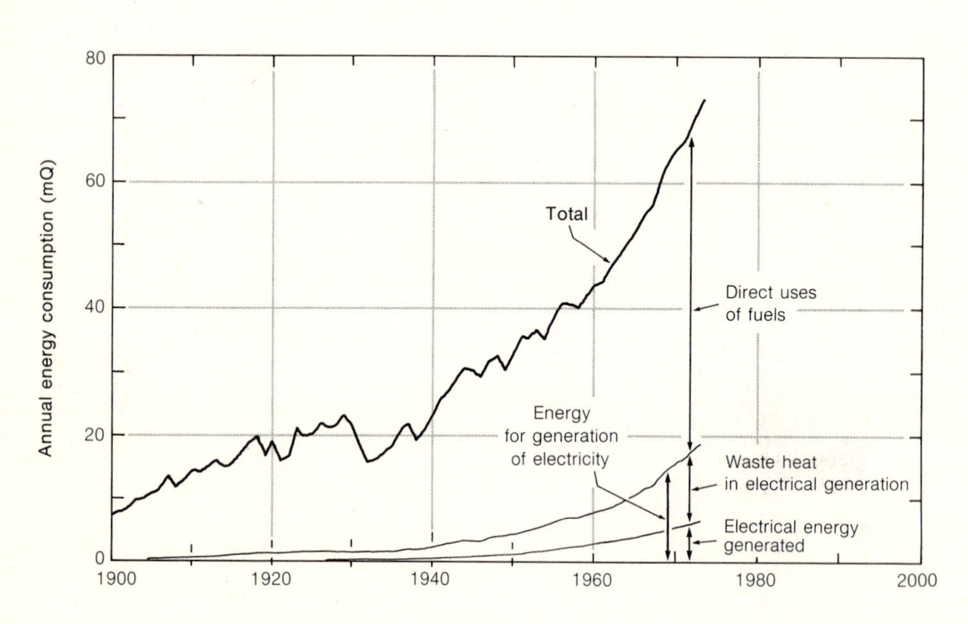

FIGURE K.13
Electrical energy in the United States in relation to overall energy consumption.

TABLE K.5 World and United States energy consumption—direct uses, use for generation of electricity, and total per capita rate of energy consumption—annual values, 1900–1973.

	Electrical energy generated		Energy used for generation of electricity (waste heat included)	Direct uses of energy	Total energy use	Average per capita rate of energy use
	(10^9 kWh)	(mQ)	(mQ)	(mQ)	(mQ)	(W)
			World (United States included)			
1900	5.0	0.017	0.134	20.9	21.0	442
1905	13.5	0.046	0.48	25.5	26.0	530
1910	33.1	0.113	1.07	31.4	32.5	644
1915	73.7	0.25	2.03	31.7	33.8	648
1920	122	0.42	2.87	36.3	39.2	723
1925	189	0.64	3.28	38.7	42.0	735
1930	300	1.02	4.35	41.6	46.0	763
1935	340	1.16	4.39	40.2	44.6	699
1940	500	1.71	6.24	50.3	56.6	840
1945	702	2.40	8.20	46.6	54.8	777
1950	946	3.23	10.3	63.4	73.7	991
1955	1540	5.26	15.3	78.8	94.1	1159
1960	2304	7.86	20.9	103	124	1390
1962	2636	9.00	23.1	109	132	1419
1964	3135	10.7	27.4	121	149	1539
1966	3637	12.4	30.9	133	164	1635
1968	4206	14.4	36.8	143	180	1724
1970	4908	16.8	43.9	162	206	1894
1972	5647	19.3	50.7	172	223	1973
1973	6053	20.7	54.3	184	238	2056
			United States			
1900	3.96	0.013	0.109	7.22	7.33	3219
1905	10.3	0.035	0.40	10.6	11.0	4385
1910	19.7	0.067	0.69	13.6	14.3	5167
1915	34.6	0.12	1.06	14.4	15.4	5127
1920	56.6	0.19	1.47	17.6	19.1	5985
1925	84.7	0.29	1.58	18.7	20.3	5851
1930	115	0.39	1.68	19.9	21.6	5871
1935	119	0.41	1.51	16.9	18.4	4840
1940	180	0.61	2.30	20.9	23.2	5849
1945	271	0.93	3.24	27.1	30.3	7217
1950	389	1.33	4.38	28.5	32.9	7218
1955	629	2.15	6.40	32.5	38.9	7824
1960	844	2.88	7.94	35.8	43.7	8081
1962	946	3.23	8.71	37.8	46.5	8332
1964	1084	3.70	10.0	40.2	50.3	8749
1966	1249	4.26	11.6	43.4	55.0	9349
1968	1435	4.90	13.5	46.7	60.2	10026
1970	1639	5.59	15.7	49.7	65.4	10663
1972	1852	6.32	17.7	52.3	70.0	11200
1973	1946	6.64	18.7	54.1	72.8	11565

Appendix L

Sources and Uses of Energy in the United States, 1973

FIGURE L.1
Energy flow in the United States (1973).

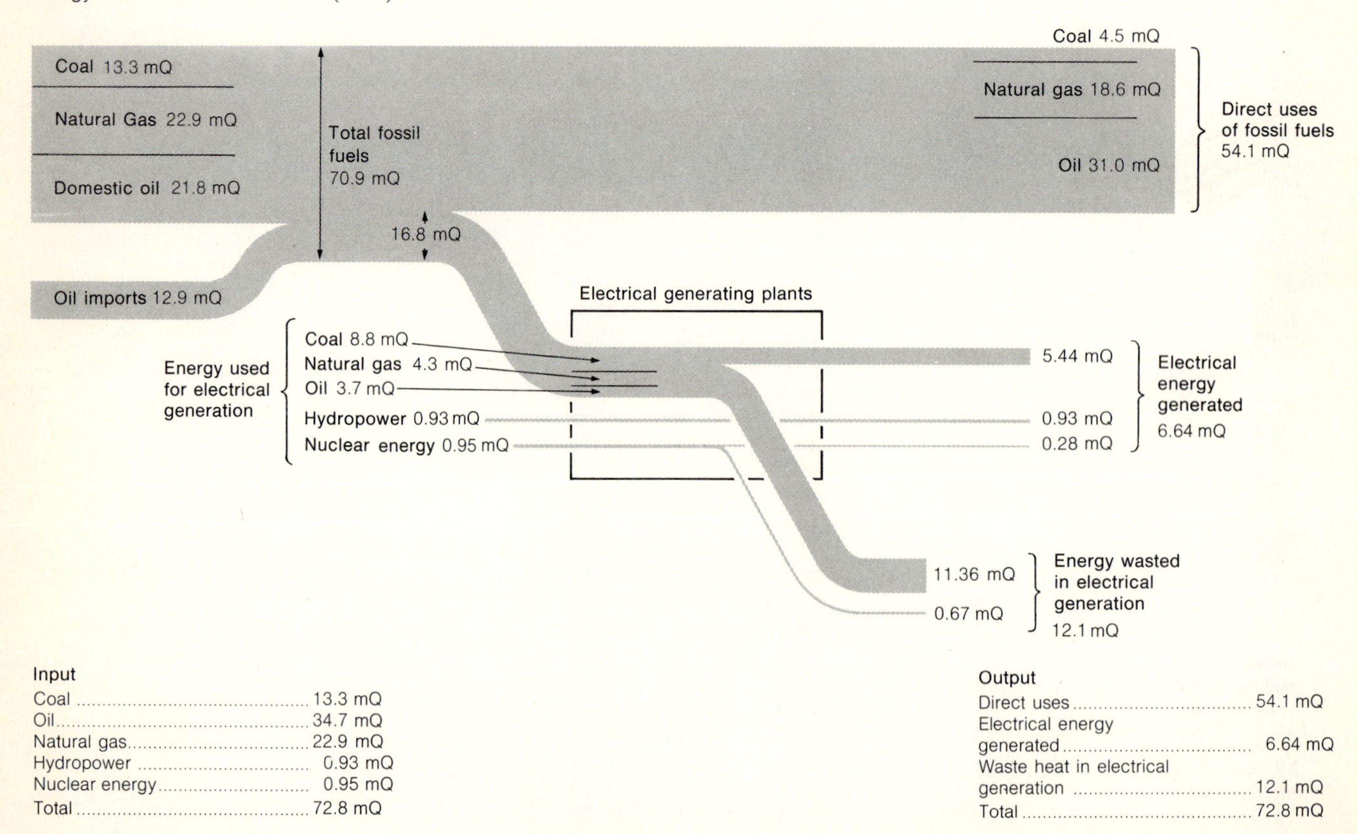

Input	
Coal	13.3 mQ
Oil	34.7 mQ
Natural gas	22.9 mQ
Hydropower	0.93 mQ
Nuclear energy	0.95 mQ
Total	72.8 mQ

Output	
Direct uses	54.1 mQ
Electrical energy generated	6.64 mQ
Waste heat in electrical generation	12.1 mQ
Total	72.8 mQ

FIGURE L.2
Energy consumption in the United States (1973).

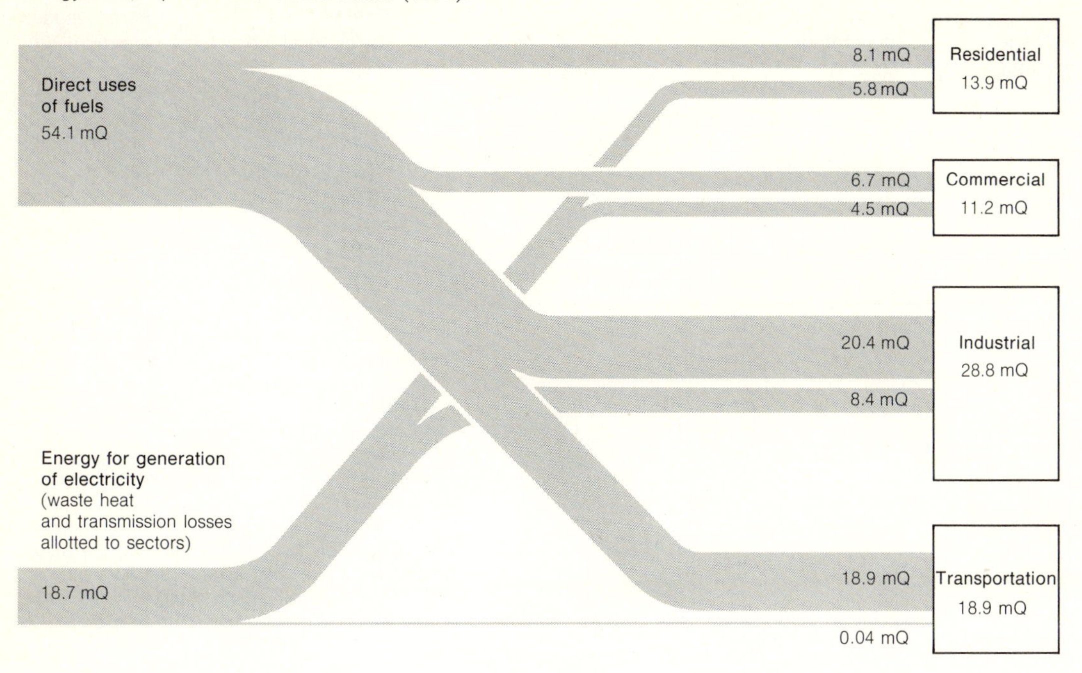

TABLE L.1 Sources of energy, 1973.

Source	Energy* (mQ)	Percentage of national total
Coal	13.3 (+1.7%)	18.3%
Oil†	34.7 (+5.5%)	47.7%
Natural gas	22.9 (+1.2%)	31.4%
Total fossil fuels	70.9 (+3.3%)	97.4%
Hydroelectric power	0.93 (+3.1%)	1.3%
Nuclear power	0.95 (+56%)	1.3%
Total	72.8 (+3.7%)	100%

*Figures in parentheses are average annual percentage rates of change during the period 1970–1973.

†Analysis of the oil supply.

	Energy (mQ)	Percentage of total oil supply
Domestic crude oil	19.3 (−1.4%)	55.6%
Domestic natural gas liquids	2.5 (+0.9%)	7.2%
Total domestic supply	21.8 (−1.1%)	62.8%
Imported oil (crude oil and refined products) (For trends in oil imports, see Figure K.8.)	12.9 (+22.8%)	37.2%
Total	34.7 (+5.5%)	100%

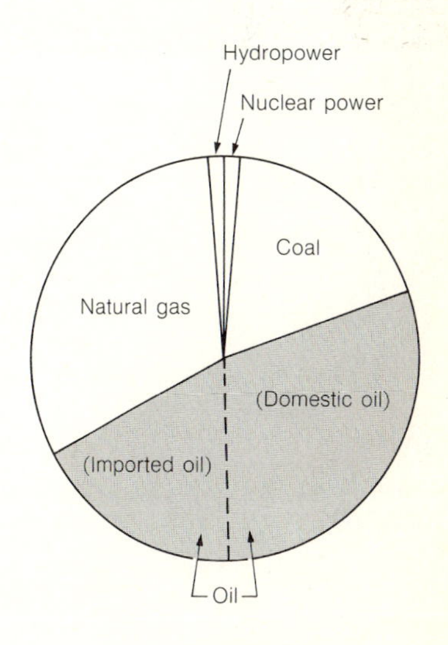

Chart for TABLE L.1

TABLE L.2 Electrical Generation, 1973.

Source of energy	Average efficiency	Electrical energy generated*,† (mQ)	Percentage of electrical energy generated	Energy used for generation* (mQ)	Percentage of energy used for generation
Coal	33%	2.95 (+5.9%)	44.4%	8.82 (+5.6%)	47.2%
Oil	31%	1.16 (+18%)	17.5%	3.71 (+16%)	19.8%
Natural gas	31%	1.33 (−3.7%)	20.0%	4.28 (−3.3%)	22.9%
Hydroelectric power**	100%	0.93 (+3.1%)	14.0%	0.93 (+3.1%)	5.0%
Nuclear power	30%	0.28 (+56%)	4.2%	0.95 (+56%)	5.1%
Total	35.6%††	6.64 (+5.9%)	100%	18.7 (+6.0%)	100%

*Figures in parentheses are average annual percentage rates of change during the period 1970–1973.

†Total amount of electrical energy generated: 6.64 mQ = 1.95×10^{12} kWh. Approximately 95% of this energy is generated by public and private utilities; the remainder is generated by industries for internal use.

**Generation of electrical energy with hydropower is taken to be 100% efficient; the actual efficiency is somewhat smaller, but allowance for this would make only slight changes in the totals.

††Average efficiency *not* including hydropower: 32.2%.

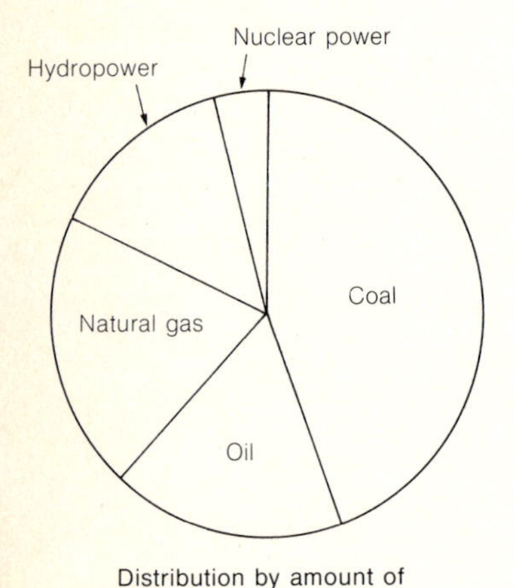

Distribution by amount of electrical energy generated

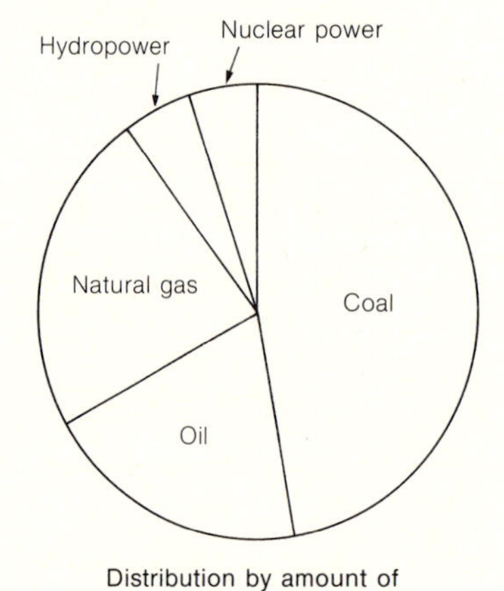

Distribution by amount of energy consumed in generation

Charts for TABLE L.2

TABLE L.3 Sources of energy (1973), divided according to direct uses and use for generation of electrical energy.

Source of energy	Direct uses* (mQ)	Use for generation of electrical energy* (mQ)	Total* (mQ)
Coal	4.54 (−4.5%)	8.82 (+5.6%)	13.3 (+1.7%)
Oil	31.0 (+4.5%)	3.71 (+16%)	34.7 (+5.5%)
Natural gas	18.6 (+2.4%)	4.28 (−3.3%)	22.9 (+1.2%)
Hydroelectric power	0	0.93 (+3.1%)	0.93 (+3.1%)
Nuclear power	0	0.95 (+56%)	0.95 (+56%)
Total	54.1 (+2.9%)	18.7 (+6.0%)	72.8 (+3.7%)

*Figures in parentheses are average annual percentage rates of change during the period 1970–1973.

TABLE L.4 Direct uses of fuels, and use of energy for generation of electricity, 1973.

Uses	Energy* (mQ)	Percentage of national total
Direct uses of fossil fuels		
Coal	4.54 (−4.5%)	6.2%
Oil	31.0 (+4.5%)	42.6%
Natural gas	18.6 (+2.4%)	25.5%
Total direct use	54.1 (+2.9%)	74.3%
Electrical generation		
Electrical energy generated	6.64 (+5.9%)	9.1%
Energy wasted in electrical generation	12.06 (+6.1%)	16.5%
Total for electrical generation	18.7 (+6.0%)	25.7%
Total energy consumption	72.8 (+3.7%)	100%

*Figures in parentheses are average annual percentage rates of change during the period 1970–1973.

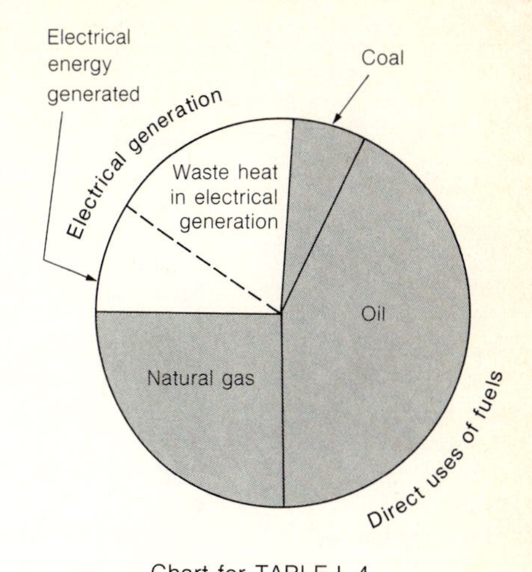

Chart for TABLE L.4

TABLE L.5 Energy consumption by sector, 1973.

Sector	Direct uses of fuels* (mQ)	Electricity (including waste heat from generation)* (mQ)	Total energy consumption* (mQ)	Percentage of national total
Residential	8.1 (+0.8%)	5.8 (+7.5%)	13.9 (+3.3%)	19.1%
Commercial	6.7 (+3.0%)	4.5 (+7.0%)	11.2 (+4.6%)	15.4%
Industrial	20.4 (+2.3%)	8.4 (+4.6%)	28.8 (+3.0%)	39.6%
Transportation	18.9 (+4.4%)	0.04 (0%)	18.9 (+4.4%)	26.0%
Total	54.1 (+2.9%)	18.7 (+6.0%)	72.8 (+3.7%)	100%

*Figures in parentheses are average annual percentage rates of change during the period 1970–1973.

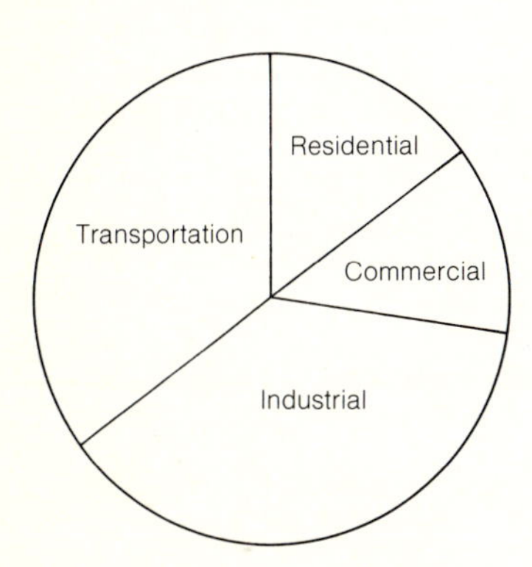

Distribution of energy from direct uses of fossil fuels

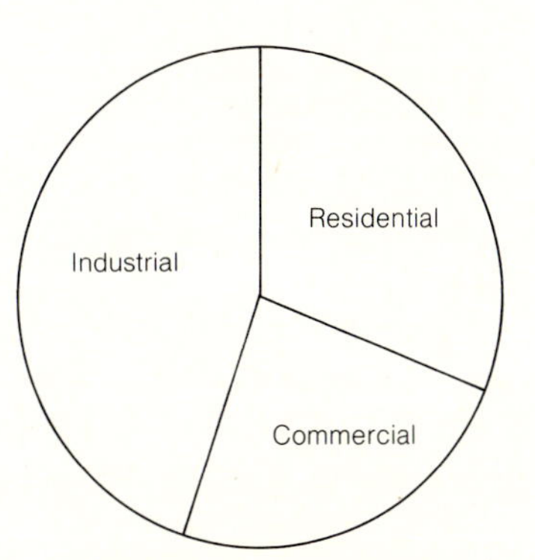

Distribution of energy used for generation of electricity (transportation sector omitted)

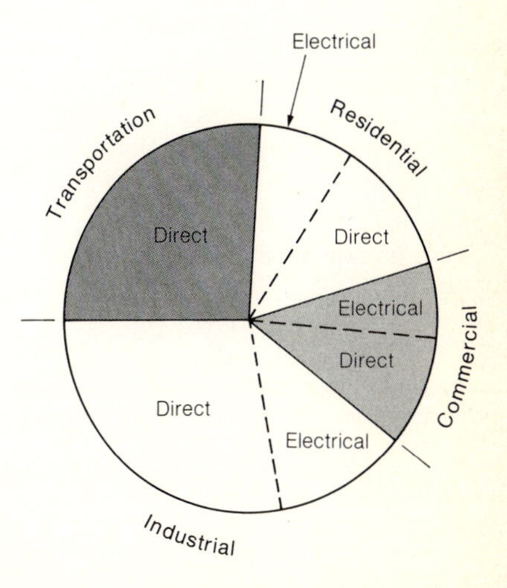

Distribution of total energy

Charts for TABLE L.5

TABLE L.6 Energy consumption in the residential sector, 1973.

Use	Direct uses of fuels* (mQ)	Electricity (including waste heat from generation)* (mQ)	Total energy consumption* (mQ)	Percentage of residential sector	Percentage of national total
Space heating	6.4 (+0.7%)	1.26 (+24.7%)	7.7 (+3.5%)	55.4%	10.6%
Hot water	1.2 (+1.3%)	0.62 (+2.5%)	1.82 (+1.6%)	13.1%	2.5%
Lights	0	1.09 (+0.3%)	1.09 (+0.3%)	7.8%	1.5%
Refrigeration	0	0.76 (+4.8%)	0.76 (+4.8%)	5.5%	1.0%
Cooking	0.36 (0%)	0.28 (+3.6%)	0.64 (+1.6%)	4.6%	0.9%
Air conditioning	0	0.56 (+8.3%)	0.56 (+8.3%)	4.0%	0.8%
Television	0	0.42 (+5.9%)	0.42 (+5.9%)	3.0%	0.6%
Clothes drying	0.08 (+2.6%)	0.19 (+8.8%)	0.27 (+6.8%)	1.9%	0.4%
Freezers	0	0.29 (+8.3%)	0.29 (+8.3%)	2.1%	0.4%
Misc. appliances	0	0.22 (+2.5%)	0.22 (+2.5%)	1.6%	0.3%
Dish washers	0	0.06 (+10.6%)	0.06 (+10.6%)	0.4%	0.08%
Washing machines	0	0.04 (0%)	0.04 (0%)	0.3%	0.05%
Total	8.1 (+0.8%)	5.8 (+7.5%)	13.9 (+3.3%)	100%	19.1%

*Figures in parentheses are average annual percentage rates of change during the period 1970–1973.

TABLE L.7 Energy consumption in the commercial sector, 1973.

Use	Direct uses of fuels* (mQ)	Electricity (including waste heat from generation)* (mQ)	Total energy consumption* (mQ)	Percentage of commercial sector	Percentage of national total
Space heating	4.4 (+1.2%)	1.0 (+30%)	5.4 (+4.7%)	48.2%	7.4%
Non-energy uses†	1.6 (+9.2%)	0	1.6 (+9.2%)	14.3%	2.2%
Air conditioning	0.15 (+8.1%)	1.47 (+7.3%)	1.6 (+7.3%)	14.3%	2.2%
Refrigeration	0	0.68 (0%)	0.68 (0%)	6.1%	0.9%
Hot water	0.43 (+1.0%)	0.22 (+1.0%)	0.65 (+1.0%)	5.8%	0.9%
Lights (except street and highway)	0	0.47 (+0.4%)	0.47 (+0.4%)	4.2%	0.6%
Cooking	0.13 (+1.4%)	0.08 (+4.0%)	0.21 (+2.5%)	1.9%	0.3%
Street and highway lights	0	0.10 (+0.3%)	0.10 (+0.3%)	0.9%	0.14%
Other	0	0.48 (+0.4%)	0.48 (+0.4%)	4.3%	0.7%
Total	6.7 (+3.0%)	4.5 (+7.0%)	11.2 (+4.6%)	100%	15.4%

*Figures in parentheses are average annual percentage rates of change during the period 1970–1973.
†Largely use of petroleum products for highway paving, etc.

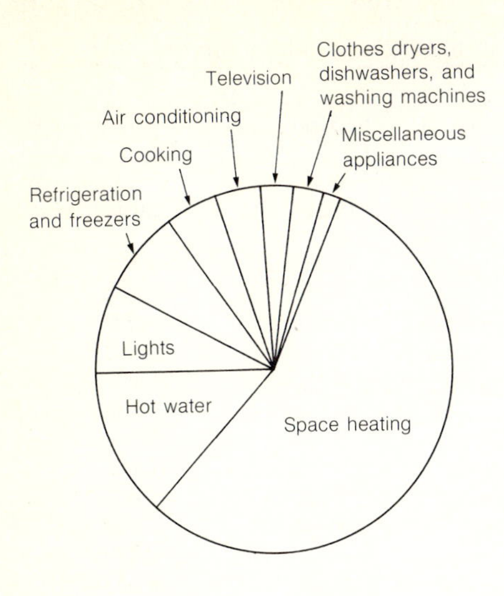

Chart for TABLE L.6

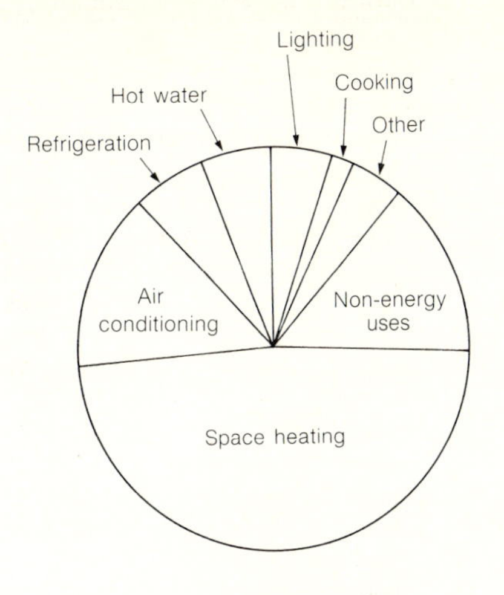

Chart for TABLE L.7

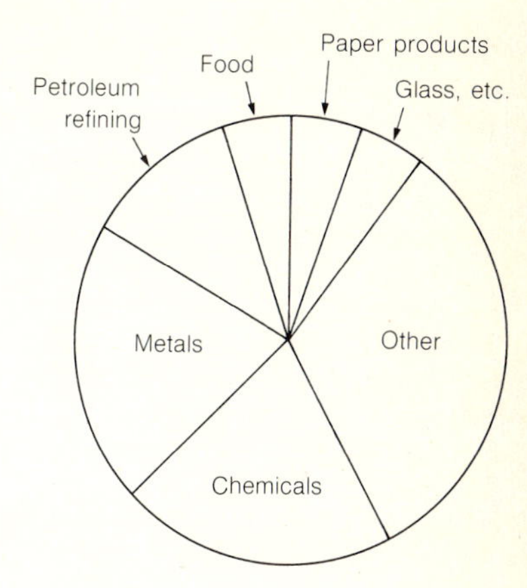

Chart for TABLE L.8

TABLE L.8 Energy consumption in the industrial sector, 1973.

Use	Direct uses of fuels* (mQ)	Electricity (including waste heat from generation)* (mQ)	Total energy consumption* (mQ)	Percentage of industrial sector	Percentage of national total
Chemicals (production of basic chemicals, synthetic fibers, drugs, fertilizers, etc.)	3.6 (+2.7%)	2.4 (+5.0%)	6.0 (+3.6%)	20.8%	8.2%
Metals (smelting and refining, manufacture of basic metal products)	3.9 (+0.5%)	2.0 (+4.7%)	5.9 (+1.9%)	20.5%	8.1%
Petroleum refining, manufacture of paving and roofing materials, lubricants, etc.	3.0 (+4.0%)	0.3 (+5.1%)	3.3 (+4.1%)	11.5%	4.5%
Food manufacturing and processing	1.0 (+2.4%)	0.5 (+4.0%)	1.5 (+2.9%)	5.2%	2.1%
Paper products	1.0 (+1.4%)	0.5 (+4.8%)	1.5 (+2.5%)	5.2%	2.1%
Glass, concrete, asbestos, etc.	1.0 (+1.7%)	0.4 (+4.1%)	1.4 (+2.4%)	4.9%	1.9%
Other industries	6.9 (+2.6%)	2.3 (+4.2%)	9.2 (+3.0%)	31.9%	12.6%
Total	20.4 (+2.3%)	8.4 (+4.6%)	28.8 (+3.0%)	100%	39.6%

*Figures in parentheses are average annual percentage rates of change during the period 1970–1973.

TABLE L.9 Energy consumption in the transportation sector, 1973.

	Energy consumption*,† (mQ)	Percentage of transportation sector	Percentage of national total
Automobiles (intercity travel)	3.61 (+2.3%)	19.1%	5.0%
Automobiles (local travel)	6.80 (+5.9%)	36.0%	9.3%
Buses (intercity travel)	0.04 (0%)	0.21%	0.05%
Buses (local travel)	0.06 (+4.2%)	0.32%	0.08%
Buses (school buses)	0.03 (−5.0%)	0.16%	0.04%
Trucks (intercity)	1.18 (+0.4%)	6.2%	1.6%
Trucks (local)	3.20 (+5.0%)	16.9%	4.4%
Subways	0.03 (−1.0%)	0.16%	0.04%
Railroads (passenger)	0.02 (−9%)	0.11%	0.03%
Railroads (freight)	0.52 (+1.0%)	2.8%	0.7%
Planes (commercial passenger)	0.77 (0%)	4.1%	1.1%
Planes (private)	0.10 (−1%)	0.5%	0.14%
Planes (air freight)	0.18 (+6%)	1.0%	0.25%
Planes (military)	0.52 (−6%)	2.8%	0.7%
Waterways	0.47 (+3.5%)	2.5%	0.6%
Pipelines	0.22 (+4.2%)	1.2%	0.3%
Other transportation	1.04 (+4%)	5.5%	1.4%
Non-energy uses of fossil fuels (motor oils, greases, etc.)	0.16 (+1.8%)	0.8%	0.2%
Total	18.9 (+4.4%)	100%	26.0%

*Figures in parentheses are average annual percentage rates of change during the period 1970–1973.

†Electrical energy uses (approximately 0.2% of energy used for transportation) are not shown separately.

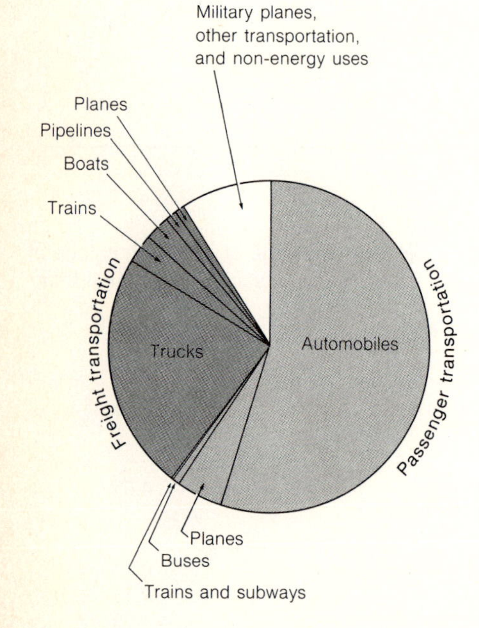

Chart for TABLE L.9

TABLE L.10 Use of fossil fuels, 1973.

Use	Coal* (mQ)	Oil* (mQ)	Natural gas* (mQ)	Total* (mQ)
Direct uses				
Residential sector	0	3.26 (+1.4%)	4.81 (+0.4%)	8.1 (+0.8%)
Commercial sector	0.30 (−11%)	3.86 (+4.6%)	2.56 (+2.8%)	6.7 (+3.0%)
Industrial sector	4.24 (−4%)	5.74 (+5.9%)	10.47 (+3.4%)	20.4 (+2.3%)
Transportation sector	0.003 (−20%)	18.16 (+4.6%)	0.75 (+0.2%)	18.9 (+4.4%)
Total direct uses	4.54 (−4.5%)	31.0 (+4.5%)	18.6 (+2.4%)	54.1 (+2.9%)
Use for generating electrical energy	8.82 (+5.6%)	3.71 (+16%)	4.28 (−3.3%)	16.8 (+4.9%)
Total	13.3 (+1.7%)	34.7 (+5.5%)	22.9 (+1.2%)	70.9 (+3.3%)

*Figures in parentheses are average annual percentage rates of change during the period 1970–1973.

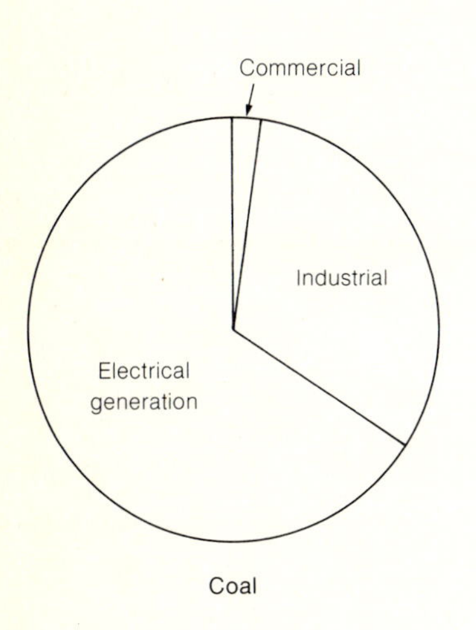

Coal

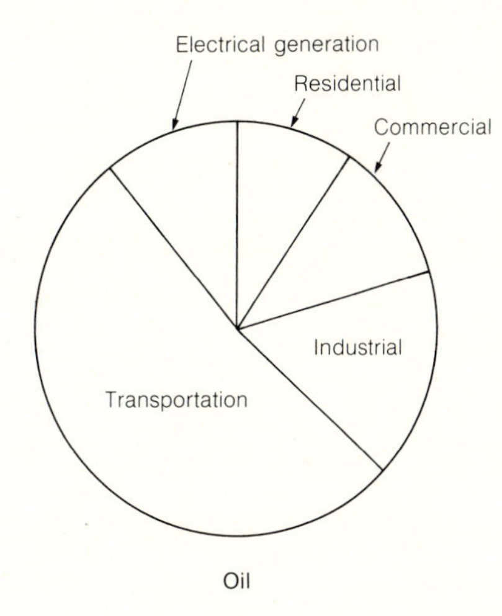

Oil

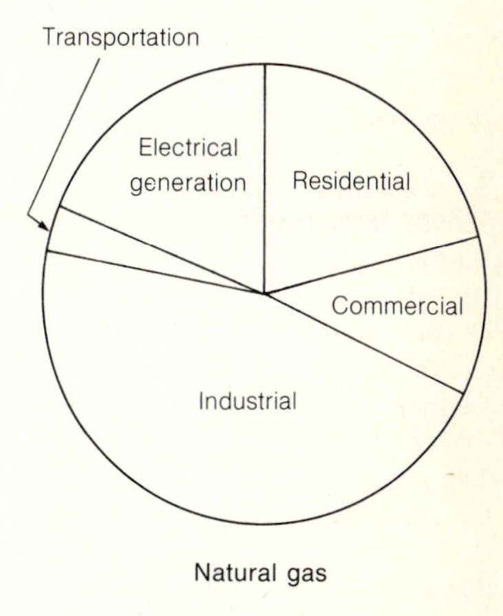

Natural gas

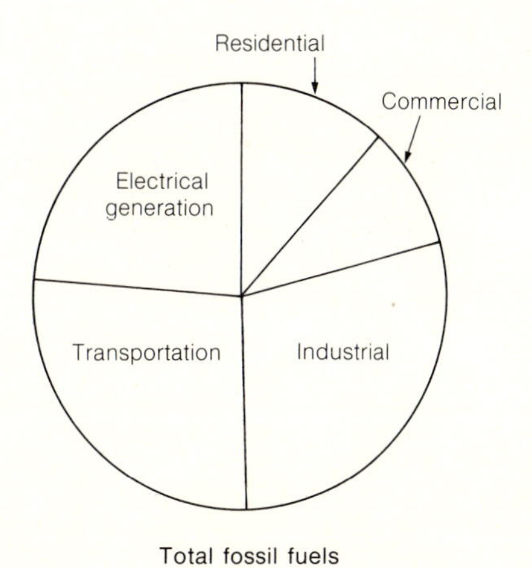

Total fossil fuels

Charts for TABLE L.10

Appendix M

Energy Requirements for Electrical Appliances

TABLE M.1 Average wattage and estimated energy consumption of various appliances.

Appliance	Average power required (W)	Estimated electrical energy used per year (kWh)
Air conditioner (window, 5000 BTU/hr)	1565	1390
Blanket	177	147
Broiler	1436	100
Carving knife	92	8
Clock	2	17
Clothes dryer	4855	990
Coffee maker	895	105
Deep-fat fryer	1450	83
Dishwasher	1200	363
Egg cooker	516	14
Fan (attic)	370	290
Fan (circulating)	88	43
Fan (window)	200	170
Floor polisher	305	15
Food blender	385	15
Food freezer (15 ft^3)	340	1200
Food freezer (15 ft^3, frostless)	440	1760
Food mixer	127	13
Food waste disposer	445	30
Frying pan	1200	185
Grill (sandwich)	1160	33
Hair dryer	380	14
Heat lamp	250	13
Heater (portable)	1320	175
Heating pad	65	10
Hot plate	1260	90
Humidifier	175	163
Iron	1000	144
Microwave oven	1450	190
Radio	71	86
Radio-phonograph	110	110
Razor	14	2
Refrigerator (12 ft^3)	240	730
Refrigerator (12 ft^3, frostless)	320	1215
Refrigerator-freezer (14 ft^3)	325	1135
Refrigerator-freezer (14 ft^3, frostless)	615	1830
Roaster	1333	205
Sewing machine	75	11
Stove	12,200	1175
Sun lamp	280	16
Television		
Black & white, vacuum tubes	160	350
Black & white, solid state	55	120
Color, vacuum tubes	300	660
Color, solid state	200	440
Toaster	1145	40
Tooth brush	7	0.5
Trash compactor	400	50
Vacuum cleaner	630	46
Vibrator	40	2
Waffle iron	1115	22
Washing machine (automatic)	512	103

Source: Electrical Energy Association, New York, N.Y.

Appendix N

Energy Requirements for Passenger and Freight Transportation

TABLE N.1 Energy requirements for passenger transportation.

Mode of transport	Maximum capacity (no. of passengers)*	Vehicle mileage (miles/gal)†	Passenger mileage (passenger-miles/gal)†	Energy consumption (BTU/passenger-mile)
Bicycle	1	1560	1560	80
Walking	1	470	470	260
Intercity bus	45	5	225	550
Commuter train (10 cars)	800	0.2	160	775
Subway train (10 cars)	1000	0.15	150	825
Volkswagen sedan**	4	30	120	1030
Local bus	35	3	105	1180
Intercity train (4 coaches)	200	0.4	80	1550
Motorcycle	1	60	60	2060
Automobile††	4	12	48	2580
747 jet plane	360	0.1	36	3440
727 jet plane	90	0.4	36	3440
707 jet plane	125	0.25	31	3960
United States SST (proposed)	250	0.1	25	4950
Light plane (2 seat)	2	12	24	5160
Executive jet plane	8	2	16	7740
Concorde SST	110	0.12	13	9400
Snowmobile	1	12	12	10,300
Ocean liner	2000	0.005	10	12,400

*The relative effectiveness of various modes of transportation can be drastically altered if a smaller number of passengers is carried.
†Miles per gallon of gasoline or the equivalent in food or in other fuel; all values must be regarded as approximate.
**Long-distance intercity travel.
††Typical American automobile, used partly for local travel and partly for long-distance driving.

TABLE N.2 Energy requirements for freight transportation.

Mode of transport	Mileage (ton-miles/gal)	Energy consumption (BTU/ton-mile)
Oil pipelines	275	450
Railroads	185	670
Waterways	182	680
Truck	44	2800
Airplane	3	42000

Appendix O
Exponential Growth

TABLE O.1 Doubling times for various rates of exponential growth.

Multiplication factor in each unit of time	Percentage increase per unit time	Doubling time
1.0	0	Infinite
1.01	1	69.7
1.02	2	35.0
1.03	3	23.4
1.04	4	17.7
1.05	5	14.2
1.06	6	11.9
1.07	7	10.2
1.08	8	9.0
1.09	9	8.0
1.10	10	7.3
1.12	12	6.1
1.14	14	5.3
1.16	16	4.7
1.18	18	4.2
1.20	20	3.8
1.25	25	3.1
1.30	30	2.6
1.40	40	2.1
1.50	50	1.7
1.75	75	1.2
2.00	100	1.0

Appendix P
Consumer Prices of Common Sources of Energy

The following table lists the average residential prices of electricity, natural gas, and fuel oil, and the retail price of regular grade gasoline (taxes included). The overall consumer price index is given for comparison. Consumer prices vary widely from one part of the United States to another, and prices paid by industrial and commercial users are often much lower than those paid by individuals.

TABLE P.1 Prices of energy sources, 1960–1974.

	Electricity (1 kWh)	Natural gas (1000 ft³)	Fuel oil (1 gal)	Gasoline (1 gal)	Consumer price index (1967=100)
1960	2.47¢	$0.97	15.0¢	31.1¢	88.7
1961	2.45¢	$0.98	15.6¢	30.8¢	89.6
1962	2.41¢	$0.98	15.6¢	30.6¢	90.6
1963	2.37¢	$0.98	16.0¢	30.4¢	91.7
1964	2.31¢	$0.98	15.7¢	30.4¢	92.9
1965	2.25¢	$0.98	16.0¢	31.2¢	94.5
1966	2.20¢	$0.99	16.4¢	32.1¢	97.2
1967	2.17¢	$0.99	17.0¢	33.6¢	100.0
1968	2.12¢	$1.00	17.5¢	33.7¢	104.2
1969	2.09¢	$1.02	17.9¢	34.8¢	109.8
1970	2.10¢	$1.07	18.5¢	35.7¢	116.3
1971	2.19¢	$1.15	19.7¢	36.4¢	121.3
1972	2.29¢	$1.21	19.8¢	37.2¢	125.3
1973	2.38¢	$1.27	23.0¢	39.0¢	133.1
1974	3.20¢	$1.50	36.5¢	53.0¢	147.7

Appendix Q

Comparing Capital and Operating Costs. The Cost of Borrowing Money

In assessing the cost of any proposal requiring a significant capital investment it is important to be able to convert the size of a capital investment into an equivalent *annual* cost. This procedure is essential if one is trying to compare the cost of one method of producing energy, in which the capital investment is large and the fuel costs low, with a second method for which the reverse is true. If money could be borrowed with no interest charges, then, for instance, a capital investment of $1000 in equipment with a useful lifetime of ten years could be financed by paying the lender in ten equal installments of $100 each year. Because interest must be paid on the outstanding balance, the annual cost will be more than expected from this simple calculation, the excess amount depending on the annual interest rate and the length of time over which the loan is repaid.

For the situation in which the borrower agrees to pay the lender an equal sum of money each year (part of each payment as interest on the outstanding balance and part to reduce the size of the balance), Table Q.1 shows, for an initial loan of $1, (1) the amount of each annual payment, and (2) the total amount that must be repaid (the amount of each annual payment multiplied by the number of years). Figures are given for a number of possible rates of interest and time periods. For loans of other sizes, the amounts shown in this table should be increased in proportion to the size of the loan.

With an interest rate of 7% and a time period of 10 years, for example, an initial loan of $1.00 can be paid back in 10 equal installments of 14.2¢ each, a total payment of $1.42. With these figures, a capital investment of $1 is equivalent to an annual cost of 14.2¢/yr, a capital investment of $1000 is equivalent to an annual cost of $142/yr, and so on. Observe that the total amount paid to the lender is 1.42 times as large as the amount borrowed. For a 50-year loan at 15% interest, the total amount paid to the lender is 7.51 times as large as the amount borrowed.

TABLE Q.1 Annual and total cost of a $1 loan.

	Number of years over which loan is repaid									
	10		20		30		40		50	
	Payments		Payments		Payments		Payments		Payments	
Annual interest rate	Annual ($)	Total ($)	Annual ($)	Total ($)	Annual ($)	Total ($)	Annual ($)	Total ($)	Annual ($)	Total ($)
3%	0.117	1.17	0.0672	1.34	0.051	1.53	0.0433	1.73	0.0389	1.94
5%	0.13	1.30	0.0802	1.60	0.0651	1.95	0.0583	2.33	0.0548	2.74
7%	0.142	1.42	0.0944	1.89	0.0806	2.42	0.075	3.00	0.0725	3.62
10%	0.163	1.63	0.117	2.34	0.106	3.18	0.102	4.09	0.101	5.04
15%	0.199	1.99	0.16	3.20	0.152	4.57	0.151	6.02	0.150	7.51

Appendix R
Radiation Exposure in the United States

TABLE R.1 Estimated exposure to ionizing radiation in the United States (1970).

Radiation source	Average dose rate per person (mrem/yr)
Natural sources	
Cosmic rays at ground level*	44
Rocks, soil and building materials†	40
Sources within the body (largely K^{40})	18
Subtotal	102
Artificial sources	
Fallout from nuclear weapons testing	4
Medical and dental diagnosis and treatment	73
Nuclear power installations**	0.003
Occupational exposure	0.8
Miscellaneous††	2
Subtotal	80
Total	182

*Range within the United States: 38–75. See Table R.2 for variation with altitude.

†Range 15–140 depending on type of soil and building material.

**Nuclear power reactors and fuel reprocessing plants; estimated to increase to 0.4 mrem/yr by the year 2000.

††Television sets, airplane travel, etc.

TABLE R.2 Variation of cosmic ray dose rate with altitude above sea level.

Altitude (ft)	Dose rate* (mrem/yr)
0	40
2500	52
5000	68
7500	100
10,000	190
15,000	460
20,000	1100
30,000	4000
45,000	9000
55,000–100,000	13,000

*Values are approximate and vary with latitude and (at high altitudes) vary during the 11-year cycle of solar sun-spot activity. During occasional solar flares, dose rates at jet airplane altitudes of 10^6 mrem/yr may be observed for periods of a few hours. Dose rates in the Van Allen radiation belts (altitude 1000 miles and higher) may be as large as 10^7 mrem/yr.

Index